FIRST-PASSAGE PHENOMENA

AND THEIR APPLICATIONS

FIRST-PASSAGE PHENOMENA
AND THEIR APPLICATIONS

Editors

Ralf Metzler
University of Potsdam, Germany
Tampere University of Technology, Finland

Gleb Oshanin
Université Pierre et Marie Curie, France

Sidney Redner
Boston University, USA

World Scientific

NEW JERSEY • LONDON • SINGAPORE • BEIJING • SHANGHAI • HONG KONG • TAIPEI • CHENNAI

Published by

World Scientific Publishing Co. Pte. Ltd.

5 Toh Tuck Link, Singapore 596224

USA office: 27 Warren Street, Suite 401-402, Hackensack, NJ 07601

UK office: 57 Shelton Street, Covent Garden, London WC2H 9HE

Library of Congress Cataloging-in-Publication Data

First-passage phenomena and their applications / edited by Sidney Redner (Boston University, USA), Ralf Metzler (University of Potsdam, Germany), & Gleb Oshanin (Université Pierre et Marie Curie, France).

pages cm

Includes bibliographical references.

ISBN 978-9814590280 (hardcover : alk. paper)

1. Statistical mechanics--Study and teaching (Graduate) 2. Stochastic processes--Study and teaching (Graduate) 3. Random walks (Mathematics)--Study and teaching (Graduate) I. Redner, Sidney, 1951– editor of compilation. II. Metzler, Ralf, editor of compilation.
III. Oshanin, G. (Gleb) editor of compilation.

QC174.8.F57 2014

530.13--dc23

2013048239

British Library Cataloguing-in-Publication Data

A catalogue record for this book is available from the British Library.

Printed in Singapore by World Scientific Printers.

Preface

In the microscopic world, which lies between the domain of our everyday lives and the submolecular world of elementary particle physics, motion is driven primarily by thermal fluctuations. The stochasticity of these fluctuations has led naturally to fundamental and widely appealing models, including Brownian motion and random walks. In more complex systems such as biological cells, amorphous semiconductors, or groundwater aquifers, more elaborate types of transport processes arise, such as continuous time random walks, Lévy flights and fractional Brownian motion. Since the earliest studies of stochastic motion more than a century ago, much has now been understood about the trajectories of particles that undergo these types of stochastic transport processes. Because the particle motion is inherently non-deterministic, the instantaneous location of a particle does not provide a useful way to characterize the model. Instead, the spatial probability distribution of a particle and multipoint correlation functions are the fundamental quantities of interest. One of the major advances in understanding stochastic transport processes was the determination of when the central limit theory is valid and when it breaks down, leading to more general transport laws.

A complementary—and very fundamental—aspect of stochastic motion is the distribution of times for a particle to reach a fixed spatial position. Of particular interest is the *first-passage probability*, the probability that a diffusing particle reaches a specified spatial position for the *first time*. The first-passage probability is but one aspect of the general topic of first-passage phenomena, which are generally concerned with both the time required for a stochastically driven particle to first reach a given location and the probability that this location is ever reached.

A basic aspect of first-passage phenomena that has the same fundamental character as the central limit theorem is the distinction between recurrence and transience. In the context of discrete random walks, recurrence means that a random walker necessarily returns to its starting point, and indeed reaches any lattice point, when the spatial dimension d is two or less, while the probability to reach a given point is not certain for $d > 2$. This distinction has many important implications for situations as diverse as chemical kinetics and stochastic search processes, where the rate at which an elemental event occurs is controlled by the first passage of a reactant or a searcher to a target. A particularly beautiful feature of first-passage phenomena is its deep connection to harmonic analysis (electrostatics), in which the probability for diffusing particle in a certain geometry to reach a given point is related to the electric field to this target point in the same geometry.

Because of the inherent beauty of first-passage phenomena and the wealth of applications, this topic has seen an upsurge of interest over the past 15 years. In addition to useful information about first passage that is contained in classic monographs, such as Feller's book on probability theory and the books by Van Kampen and by Gardiner on stochastic processes, the monograph "A Guide to First-Passage Processes" (Cambridge Univ. Press, 2001) provides a first introduction to the topic. However, the rapid recent pace of developments and the diversity of applications has made it difficult for an uninitiated researcher to quickly gain a contemporary perspective on the field. The collection of articles in this volume is intended to partially remedy this shortcoming by gathering results that are widely scattered in the literature within a single volume.

The first 11 chapters focus on fundamental theoretical issues associated with first-passage phenomena that go beyond pure diffusive transport and/or ideal Euclidean geometries. Examples of non-diffusive transport include Lévy flights, random acceleration, multidimensional fractional Brownian motion, intermittent random walks, and walks with finite lifetimes. The new types of geometries considered include heterogeneous and scale-free networks, bounded and semi-bounded, e.g., conical, domains. There are also new theoretical questions raised about large fluctuations, as well as order statistics and record phenomena that are induced by first-passage questions.

The following five chapters are concerned with the application of first-passage theory to search processes and biological phenomena. These chapters cover a diverse range of topics associated with biological jump processes, spatial search and foraging, Lévy-based search strategies, as well as exit strategies in visual search. The remaining seven chapters treat applications of first-passage ideas to evolutionary trajectories on fitness landscapes, finance and socio-economic systems, properties of cold atoms in optical traps, the excursion set theory in cosmology, self-organized escape of linear chains in non-linear potentials, and, finally, efficient Monte-Carlo methods for simulating diffusion-reaction processes in complex systems.

We are especially grateful to the contributors to this volume—leading experts in topics related to first-passage phenomena—who devoted the effort required to produce their individual book chapters. Without their dedicated efforts, this book simply could not exist. We hope that this book will be a useful resource for physicists, chemists, mathematicians, engineers, and other quantitative scientists who will profit from a compilation of the state-of-the-art in the field and will also find new sources of inspiration.

Potsdam, Paris, Boston, January 2014
Ralf Metzler, Gleb Oshanin, Sidney Redner

Contents

Chapter 1

Arrival Statistics and Exploration Properties of Mortal Walkers

Santos B. Yuste

Departamento de Física, Universidad de Extremadura, E-06071 Badajoz, Spain

santos@unex.es

Enrique Abad

Departamento de Física Aplicada, Centro Universitario de Mérida, Universidad de Extremadura, E-06800 Mérida, Spain

eabad@unex.es

Katja Lindenberg

Department of Chemistry and Biochemistry and BioCircuits Institute, University of California San Diego, La Jolla, CA 92093-0340, USA

klindenberg@ucsd.edu

We study some of the salient features of the arrival statistics and exploration properties of mortal random walkers, that is, walkers that may die as they move, or as they wait to move. Such evanescence or death events have profound consequences for quantities such as the number of distinct sites visited which are relevant for the computation of encounter-controlled rates in chemical kinetics. We exploit the observation that well-known methods developed decades ago for immortal walkers are widely applicable to mortal walkers. The particular cases of exponential and power-law evanescence are considered in detail. Finally, we discuss the relevance of our results to the target problem with mortal traps and a particular application thereof, namely, the defect diffusion model. Evanescence of defects is postulated as a possible complementary contribution or perhaps even an alternative to anomalous diffusion to explain observed stretched exponential relaxation behavior.

1. Introduction

Random walks play a central role in statistical physics as elementary models describing stochastic transport in a large variety of experimental systems [1]. While there is a large body of literature dealing with the first-passage and exploration behavior of walkers that do not undergo reactions or transformations, considerably less attention has been devoted to the corresponding properties of reactive walkers. In particular, work involving exploration properties of so-called mortal or evanescent walkers, that is, walkers that die in the course of their motion, is surprisingly scarce despite the ubiquitous occurrence of such death processes in many physical, chemical, and biological scenarios.

Ad-hoc models for particular experimental situations have previously been implemented, e.g., to deal with photon scattering and absorption in a tissue [2]. And yet, with very few exceptions (see e.g. [2, 3] or subsection 3.2.4 in [4]) the observation that the effect of elementary death processes on the exploration properties of random walkers can be investigated using arguments similar to those used for immortal walkers seems until recently to have gone largely unnoticed [5]. The aim of the present work is to introduce in a pedagogical way an overarching theoretical framework allowing one to obtain the most interesting properties of mortal walkers. In view of the completely new physics introduced by death processes in the transport behavior and its potential relevance for experiments such as for instance the defect relaxation system discussed in Sec. 4, this endeavor seems very timely.

In what follows we present in a more general framework some recent results first given in [5] in a succinct fashion. More specifically, we shall focus on properties characterizing the arrival statistics of random walkers in discrete time, such as the average number of distinct sites visited after a certain number of steps on perfect lattices of all dimensions and other related quantities. In the continuum time limit one can establish a relationship between these quantities and the number of distinct sites visited up to a given time (rather than step number), a key quantity for the computation of reaction rates in chemical kinetics.

Mortal random walkers have a probability of dying as they walk. This means that their density ρ is not a constant as it is in an ordinary random walk but is instead a decaying function of time, $\rho(t)$, or of step number, $\rho(n)$. One can think of this decay as arising from a process that removes the walkers or transforms them into an inert species This might be the result of a spontaneous random process, e.g., a radioactive decay, or of an interaction with other particles, as is the case, for example, in scavenging reactions. We will later specify particular decay functions $\rho(n)$ or $\rho(t)$.

The present chapter is organized as follows. In Sec. 2 we show how to obtain some properties related to the arrival statistics of mortal walkers at given lattice sites in the framework of the generating function formalism originally developed by Montroll and Weiss [6] and extended in subsequent work (see for example chapter 6 in [4] and references therein). Next, in Sec. 3, we discuss the behavior of some of these quantities for particular forms of the decay law of the walker density, namely, exponential decay and power-law decay. We conclude that the behavior in the case of mortal walkers is very different from that observed for immortal walkers. For instance, as time goes to infinity the number of distinct sites visited tends to infinity in the latter case, while in the former case it goes to a finite value provided that the evanescence is sufficiently fast. In Sec. 4 we discuss the implications of these results for the target problem with mortal traps, in particular, the possibility that the target survives forever if the traps die sufficiently rapidly. On the basis of our results, we suggest that the stretched exponential behavior observed in some relaxational systems driven by defect diffusion can be ascribed to evanescence without the need to invoke anomalous diffusion. Finally, in Sec. 5 we briefly summarize our main conclusions and discuss possible extensions of our work. The appendices are devoted to some technicalities related to the correspondence between the connection of the master equation for mortal walkers to the continuum reaction-diffusion equation with a linear reaction term, as well as to the on-lattice kinetics of mortal walkers and continuum space exploration.

2. Generating function formalism for mortal walkers

2.1. *Master equation for mortal walkers*

Our starting point is a mortal walker which performs a Pólya walk, i.e., a symmetric nearest-neighbor walk on a perfect lattice of dimension d. We introduce the probability $P^*_{m,n-m}(s|s')$ of finding a mortal walker at lattice site s after having taken $n-m$ steps if the walk started at site s' at step m. An asterisk denotes quantities for *mortal* walkers and the same quantities without asterisks are for immortal walkers. The probability P^* for a mortal walker is related to its counterpart P for an immortal walker as follows:

$$P^*_{m,n-m}(s|s') = \frac{\rho(n)}{\rho(m)} P_{m,n-m}(s|s') = \frac{\rho(n)}{\rho(m)} P_{n-m}(s|s'), \tag{1}$$

where the second equality follows from the fact that in the absence of the death process the probability does not depend on when the walk started but only on the number of steps taken by the walker. When $m = 0$ we can more simply write

$$P^*_n(s|s_0) = \frac{\rho(n)}{\rho(0)} P_n(s|s_0), \tag{2}$$

where we have simplified the notation, $P^*_{0,n}(s|s_0) \equiv P^*_n(s|s_0)$. The walker density ratio $\rho(n)/\rho(0)$ can be interpreted as the survival probability of the walker up to step n, or as the fraction of surviving walkers if one considers a statistical ensemble of walkers all with the same initial condition. In what follows, we set $\rho(0) = 1$, and consequently $P^*_n(s|s_0) = \rho(n)\, P_n(s|s_0)$. For simplicity, we also restrict ourselves to the deterministic initial condition

$$P^*_0(s|s_0) = \delta_{ss_0}. \tag{3}$$

From the well-known master equation for the sojourn probabilities for an immortal walker and the definition (2) one gets the corresponding master equation for mortal walkers. For instance, in $d = 1$ one has

$$P_{n+1}(s|s_0) = \frac{1}{2}\left[P_n(s-1|s_0) + P_n(s+1|s_0)\right] \tag{4}$$

and, as a result of Eq. (2),

$$P^*_{n+1}(s|s_0) = \frac{1}{2}\frac{\rho(n+1)}{\rho(n)}\left[P^*_n(s-1|s_0) + P^*_n(s+1|s_0)\right]. \tag{5}$$

This equation simply states that the probability that the particle is found at site s after $n+1$ time steps is equal to the probability that it was either on the left neighbor site $s-1$ or the right neighbor site $s+1$ at the previous time step, then chose to jump to s and did not vanish while doing so (the probability for the particle to survive up to the $n+1$-st step given that it had survived up to the nth step is simply $\rho(n+1)/\rho(n)$). As shown in Appendix A, taking the diffusive limit of (5) yields a continuum reaction-diffusion equation with a linear reaction term. While the reaction-diffusion equation is amenable to exact solution, we shall not pursue this route in what follows, as we are interested in retaining the discrete spatial structure of the lattice for the sake of studying some aspects of the arrival statistics of a mortal walker at particular lattice sites.

2.2. *First-passage probabilities and statistics of site occupation*

Next, we introduce the probability $F_n^*(s|s_0)$ that the evanescent walker arrives at site s *for the first time* at step n if she started at site s_0 at step $n = 0$. The first arrival probabilities in the presence and absence of evanescence are also simply related to one another,

$$F_n^*(s|s_0) = \rho(n)\, F_n(s|s_0), \tag{6}$$

and the initial condition is as for immortal walkers,

$$F_0^*(s|s_0) = 0. \tag{7}$$

The probabilities F^* and P^* are related to one another in the same way as in a non-evanescent walk,

$$P_n^*(s|s_0) = \sum_{j=1}^{n} F_j^*(s|s_0) P_{j,n-j}^*(s|s), \quad n \geq 1. \tag{8}$$

In other words, it still holds that the probability that the evanescent walker is at site s after n steps can be decomposed into the n mutually exclusive events that the walker first arrived at site s after j steps and returned to site s in $n - j$ steps.

Later we will also find it convenient to use the generating functions for these two probabilities,

$$P^*(s|s_0;\xi) = \sum_{n=0}^{\infty} P_n^*(s|s_0)\xi^n = \sum_{n=0}^{\infty} P_n(s|s_0)\rho(n)\xi^n, \tag{9}$$

$$F^*(s|s_0;\xi) = \sum_{n=0}^{\infty} F_n^*(s|s_0)\xi^n = \sum_{n=0}^{\infty} F_n(s|s_0)\rho(n)\xi^n. \tag{10}$$

2.2.1. *Number of distinct sites visited by a mortal walker*

We now introduce two additional quantities, namely, the number of distinct sites visited in an n-step evanescent walk, S_n^*, and the number of new sites Δ_n (sites never visited before) visited in the nth step of an evanescent walk. We follow the convention that the start of the walk counts as a visit to the starting site s_0, i.e., $\Delta_0^* = 1$. The following relations are fairly obvious. The first relates walks with and without evanescence,

$$\langle \Delta_n^* \rangle = \rho(n)\, \langle \Delta_n \rangle, \tag{11}$$

where the brackets denote averages over realizations of the random walk. Clearly, $\langle \Delta_n \rangle$ is a number between zero and one that can be interpreted as the probability to visit a new site at the n-th step. We next relate evanescent walk quantities to one another as follows,

$$S_n^* = \sum_{j=0}^{n} \Delta_j^* \quad \leftrightarrow \quad \langle S_n^* \rangle = \sum_{j=0}^{n} \langle \Delta_j^* \rangle, \tag{12}$$

$$\langle \Delta_n^* \rangle = \langle S_n^* \rangle - \langle S_{n-1}^* \rangle. \tag{13}$$

Let us further introduce the two generating functions associated with these two quantities, namely,

$$S^*(\xi) = \sum_{n=0}^{\infty} \langle S_n^* \rangle \xi^n \tag{14}$$

and

$$\Delta^*(\xi) = \sum_{n=0}^{\infty} \langle \Delta_n^* \rangle \xi^n. \tag{15}$$

It is useful to establish relations between some of the generating functions. First, multiplying the left equation in Eq. (12) by ξ^n, summing over n, and changing the order of summations leads to

$$S^*(\xi) = \frac{\Delta^*(\xi)}{1-\xi}. \tag{16}$$

Next, note that

$$\langle \Delta_j^* \rangle = \Pr(\Delta_j^* = 1) = \sum_{s \neq s_0} F_j^*(s|s_0), \quad j \neq 0. \tag{17}$$

Again multiplying by ξ^j, summing over j, and reversing the order of summation leads to

$$\Delta^*(\xi) = 1 + \sum_{s \neq s_0} F^*(s|s_0;\xi). \tag{18}$$

We now follow with the important observation that Eq. (12) with (11) allows one to calculate the average number of distinct sites visited by a mortal walker up to step n in terms of non-evanescent walk properties,

$$\langle S_n^* \rangle = \sum_{j=0}^{n} \rho(j) \langle \Delta_j \rangle. \tag{19}$$

On the other hand, it is well-known from the theory for immortal walkers that the $\langle \Delta_j \rangle$'s fulfil the recursion relation (see e.g. [4], pp. 324-25)

$$\langle \Delta_j \rangle = 1 - \sum_{k=1}^{j} P_k(0) \langle \Delta_{j-k} \rangle \,, \quad \text{with} \quad j \geq 1, \quad \langle \Delta_0 \rangle = 1, \tag{20}$$

where the sojourn probabilities $P_k(0) \equiv P_k(s_0|s_0)$ are known for many lattices in arbitrary dimension. The calculation of $\langle S_n^* \rangle$ as given in Eq. (19) can now be carried out for any prescribed form of $\rho(n)$. For large values of n, this procedure becomes cumbersome and can be bypassed by the direct use of the generating function $S^*(\xi)$. Indeed, the behavior of this quantity in the limit $\xi \to 1^-$ yields the large n behavior of $\langle S_n^* \rangle$ via a discrete Tauberian theorem (see p. 118 in [4]).

The remainder of the present section deals with additional quantities of interest related to the arrival statistics of a mortal walker which can be straightforwardly computed in the framework of the generating function formalism.

2.2.2. *Expected number of sites visited at least a given number of times*

Let us define $\langle S_n^{*(r)} \rangle$ as the average number of sites visited at least r times in an n-step evanescent walk. Obviously, it is not possible to visit any given site $r \geq 2$ times if no steps have been taken. We may therefore write $\langle S_0^{*(r)} \rangle = \delta_{1r}$.

Following the lines of the theory for immortal walkers we arrive at the relation

$$\langle S_n^{*(r)} \rangle = \sum_{j=0}^{n} F_j^{*(r-1)}(s_0|s_0) + \sum_{s \neq s_0} \sum_{j=0}^{n} F_j^{*(r)}(s|s_0), \tag{21}$$

where $F_n^{*(r)}(s|s_0)$ is the probability that a mortal walker visits site s for the r-th time when performing her n-th step given that she started the walk at s_0. We adopt the convention followed by Hughes[a] [see text below and formula (6.229) in [4]], according to which *the start of the walk is regarded as the zeroth visit to site* s_0, so that $F_0^{*(r)}(s_0|s_0) = \delta_{0r}$ and $F_j^{*(0)}(s_0|s_0) = 0$

[a]This convention differs from the one followed by Montroll and Weiss [6] in that in the latter the start of the walk at s_0 is counted as the first visit to site s_0.

for $j \neq 0$. The generating function of $\langle S_n^{*(r)} \rangle$ is defined in the usual way:

$$S^{*(r)}(\xi) = \sum_{n=0}^{\infty} \langle S_n^{*(r)} \rangle \xi^n = \delta_{1r} + \sum_{n=1}^{\infty} \langle S_n^{*(r)} \rangle \xi^n. \tag{22}$$

Introducing the corresponding generating function

$$F^{*(r)}(s|s_0;\xi) = \sum_{j=0}^{\infty} F_j^{*(r)}(s|s_0)\xi^j, \tag{23}$$

one easily gets from (21) and (22) the following relation,

$$S^{*(r)}(\xi) = \frac{1}{1-\xi} \left\{ F^{*(r-1)}(s_0|s_0;\xi) + \sum_{s \neq s_0} F^{*(r)}(s|s_0;\xi) \right\}. \tag{24}$$

Knowledge of this quantity also allows one to infer the behavior of the number of sites $V_n^{*(r)}$ visited *exactly* a given number of times r in an n-step evanescent walk. Clearly,

$$\langle V_n^{*(r)} \rangle = \langle S_n^{*(r)} \rangle - \langle S_n^{*(r+1)} \rangle, \tag{25}$$

implying

$$V^{*(r)}(\xi) = \sum_{n=0}^{\infty} \langle V_n^{*(r)} \rangle \xi^n = S^{*(r)}(\xi) - S^{*(r+1)}(\xi). \tag{26}$$

2.2.3. *Repeated visits to a given lattice site*

Let $\beta_n^{*(r)}(s|s_0)$ be the probability that site s is visited *exactly* r times in the first n steps of the walk. Since $\sum_{j=1}^{n} F_j^{*(r)}(s|s_0)$ is the probability that site s has been visited *at least* r times, one clearly has

$$\beta_n^{*(r)}(s|s_0) = \sum_{j=1}^{n} \left[F_j^{*(r)}(s|s_0) - F_j^{*(r+1)}(s|s_0) \right], \tag{27}$$

and the corresponding generating function reads

$$\beta^{*(r)}(s|s_0;\xi) = \sum_{n=1}^{\infty} \beta_n^{*(r)}(s|s_0)\xi^n = \frac{1}{1-\xi} \left[F^{*(r)}(s|s_0;\xi) - F^{*(r+1)}(s|s_0;\xi) \right]. \tag{28}$$

An associated quantity is the mean number $\langle \mu_n^*(s) \rangle$ of visits to site s during the first n steps of the walk,

$$\langle \mu_n^*(s) \rangle = \sum_{r=1}^{\infty} r\, \beta_n^{*(r)}(s|s_0), \tag{29}$$

and its generating function

$$\mu^*(s;\xi) = \sum_{n=1}^{\infty} \langle \mu_n^*(s)\rangle \xi^n = \sum_{n=1}^{\infty} \xi^n \sum_{r=1}^{\infty} r\, \beta_n^{*(r)}(s|s_0) = \sum_{r=1}^{\infty} r\beta^{*(r)}(s|s_0;\xi). \tag{30}$$

3. Results for specific decay laws

3.1. *Exponential evanescence*

We now proceed to implement our results for the particular case of exponentially evanescent walkers, a case that allows explicit analytic calculation via the standard generating function formalism. Exponential evanescence is described by the exponentially decaying density $\rho(n) \propto \exp(-\lambda n)$, where λ is the decay rate constant. It is immediately evident that only for this step number dependence is the ratio $\rho(n)/\rho(m)$ a function of the difference $n - m$, $\rho(n)/\rho(m) = \rho(n-m) \propto \exp[-\lambda(n-m)]$. From Eq. (1) it then follows that P^* does not depend on the starting step m, that is, $P^*_{m,n-m}(s|s') = P^*_{n-m}(s|s')$. This greatly facilitates the subsequent calculations. We note also that the model is equivalent to considering an immortal walker stepping on a lattice where *each* site is an imperfect trap whose (constant) absorption probability is $1 - e^{-\lambda}$. A related model for a finite lattice has received some attention in Sec. 5 of [7], where the length of a walk of a random walker moving on an N-site lattice with $N-1$ imperfect traps and a *single* perfectly absorbing trap was calculated.

3.1.1. *Number of distinct sites visited*

We start with Eq. (16) together with Eq. (18),

$$S^*(\xi) = \frac{1}{1-\xi}\left[1 + \sum_{s \neq s_0} F^*(s|s_0;\xi)\right]. \tag{31}$$

Next we focus on Eq. (8), where we now take advantage of the independence of P^* of the starting moment of the walk, so that

$$P_n^*(s|s_0) = \sum_{j=1}^{n} F_j^*(s|s_0) P_{n-j}^*(s|s), \quad n \geq 1. \tag{32}$$

This relation is now identical in form to that of a walk on a perfect lattice with non-evanescent walkers. This means that we can follow the standard steps such as presented in chapter 3 of [4] to "turn the relation around" and arrive at

$$F^*(s|s_0;\xi) = \frac{P^*(s|s_0;\xi) - \delta_{ss_0}}{P^*(s_0|s_0;\xi)}. \tag{33}$$

This expression in Eq. (31) and recognition of the easily proved consequence of normalization,

$$\sum_s P^*(s|s_0;\xi) = \sum_{n=0}^{\infty} \rho(n)\,\xi^n \equiv \rho(\xi), \tag{34}$$

leads to the result

$$S^*(\xi) = \frac{1}{1-\xi}\frac{\rho(\xi)}{P^*(s_0|s_0;\xi)}. \tag{35}$$

In Eq. (34) we have introduced the generating function $\rho(\xi)$ for the decay function $\rho(n)$.

We can finally simplify our expression further by noting that for exponential evanescence the generating function is just

$$\rho(\xi) = \sum_{n=0}^{\infty} e^{-\lambda n}\xi^n = \sum_{n=0}^{\infty} \hat{\xi}^n = \frac{1}{1-\hat{\xi}}, \tag{36}$$

where $\hat{\xi} = e^{-\lambda}\xi$. Furthermore, for exponential evanescence we also have a simple relation between the generating functions for return to the starting site with and without evanescence,

$$\begin{aligned} P^*(s|s_0;\xi) &= \sum_{n=0}^{\infty} P_n^*(s|s_0)\xi^n = \sum_{n=0}^{\infty} P_n(s|s_0)\rho(n)\xi^n \\ &= P(s|s_0;e^{-\lambda}\xi) \equiv P(s|s_0;\hat{\xi}). \end{aligned} \tag{37}$$

With this notation we can finally write

$$S^*(\xi) = \frac{1}{1-\xi}\,\frac{1}{1-\hat{\xi}}\,\frac{1}{P(0;\hat{\xi})}, \tag{38}$$

and all we need to know in order to calculate this generating function is the generating function for returning to the site of origin in the absence of evanescence. We have explicitly recognized independence of the specific starting site by writing

$$P^*(s_0|s_0;\xi) \equiv P^*(0;\xi) = P(0;\hat{\xi}). \tag{39}$$

Equation (38) is the highlight of this subsection. Lattice Green functions $P(0;\xi)$ are well known for the most common d-dimensional lattices (see for instance [4, 8]). The corresponding Green function in Eq. (38) allows one to obtain a large number of results for S_n^*. For example, identifying the n-th coefficient of the power series expansion of $S^*(\xi)$ one gets $\langle S_n^*\rangle$. Besides, from Eq. (38) one immediately realizes that, in contrast to the classical case of immortal walkers, the average number of sites visited by a mortal Pólya walker with exponential evanescence as n goes to infinity is a *finite* quantity given by

$$\langle S_\infty^*\rangle = \frac{1}{1-e^{-\lambda}}\,\frac{1}{P(0;e^{-\lambda})}. \tag{40}$$

For an infinite one-dimensional lattice one has $P(0;\xi) = (1-\xi^2)^{-1/2}$. Employing this in Eq. (40) yields

$$\langle S_\infty^*\rangle = [(1+e^{-\lambda})/(1-e^{-\lambda})]^{1/2}. \tag{41}$$

Expanding Eq. (35) in Taylor series, first around $\hat{\xi} = 1$ and next around $\xi = 0$, one finds

$$\langle S_\infty^*\rangle - \langle S_n^*\rangle = \sum_{k=0}^{\infty} 2^{1/2-k}\binom{k-3/2}{k}\sum_{m=n+1}^{\infty} e^{-\lambda m}\binom{m-k-1/2}{m}. \tag{42}$$

Keeping the first term ($k = 0$), the large n behavior can be shown to be given by the formula

$$\langle S_n^*\rangle \approx \langle S_\infty^*\rangle - \sqrt{2}\,\frac{I_{e^{-\lambda}}(n+1,1/2)}{(1-e^{-\lambda})^{1/2}} \tag{43}$$

where $I_x(a,b)$ stands for the regularized beta function. As seen in Fig. 1, this formula is in excellent agreement with numerical simulations.

For two-dimensional regular lattices the generic form of the generating function is $P(0;\xi) \sim A/\pi \ln[B/(1-\xi)]$ for $\xi \to 1^-$ where A and B are constants which depend on the details of the lattice geometry [4]. From Eq. (40) we find

$$\langle S_\infty^*\rangle \sim \frac{\pi}{\lambda A \log(B/\lambda)}, \qquad \lambda \to 0. \tag{44}$$

In particular, one has e.g. $A = 1$ and $B = 8$ for a square lattice and $A = \sqrt{3}/2$ and $B = 12$ for a triangular lattice.

It is well-known that the probability $\mathcal{R}$ of return to the origin for an immortal walker in $d \geq 3$ is intimately related to the generating function $P(0;\xi)$ evaluated at $\xi = 1$, i.e., $\mathcal{R} = 1 - 1/P(0;1)$. Equation (40) then yields $\langle S_\infty^*\rangle \sim (1-\mathcal{R})\lambda^{-1}$ as $\lambda \to 0$. It is interesting to note that for $d \geq 2$ the value of $\langle S_\infty^*\rangle$ as $\lambda \to 0$ can also be obtained from the main asymptotic term

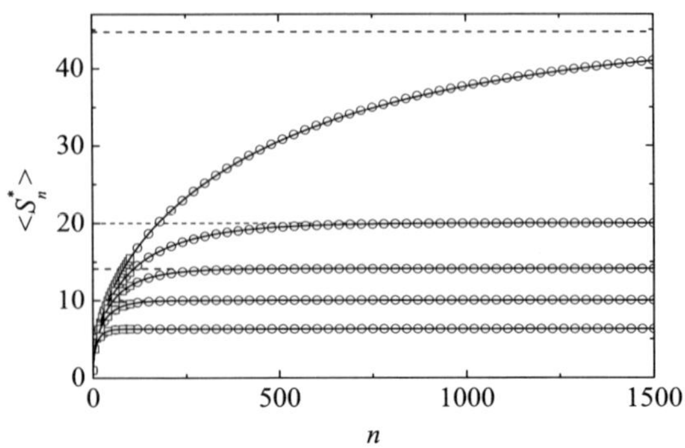

Fig. 1. $\langle S_n^* \rangle$ vs. n for a 1d lattice and, from top to bottom, $\lambda = 0.001, 0.005, 0.01, 0.02$ and 0.05. Solid lines, Eq. (43); broken lines give the values obtained from Eq. (41); circles, simulation values; squares, exact values obtained by identifying the first 100 coefficients in the ξ power expansion of $S^*(\xi)$.

of $\langle S_n \rangle$ by replacing the number of steps n by the average number of steps taken by the walker before it disappears, i.e, by $1/\lambda$. This hand-waving approximation does not work for $d = 1$ as, in this case, $\langle S_\infty^* \rangle \sim \sqrt{2/\lambda}$ whereas $\langle S_{1/\lambda} \rangle \sim \sqrt{8/(\pi\lambda)}$.

3.1.2. *Repeated visits to a given lattice site*

Taking Eq. (24) as a starting point and following the procedure of Montroll and Weiss [6], one easily finds that the generating function for the average number of sites visited at least r times, $\langle S_n^{*(r)} \rangle$, is

$$S^{*(r)}(\xi) = \left[1 - \frac{1}{P(0;\hat{\xi})}\right]^{r-1} S^*(\xi). \tag{45}$$

From here one finds formulas [6] for $\langle S_n^{*(r)} \rangle$ in terms of $\langle S_n^* \rangle$. For instance, in dimension $d = 1$ one obtains $\langle S_n^{*(2)} \rangle = \langle S_n^* \rangle - 1 - e^{-\lambda}$, $\langle S_n^{*(3)} \rangle = 2\langle S_n^* \rangle - e^{-2\lambda}\langle S_n^* \rangle - 2 - 2e^{-\lambda}$, etc. For the average number of visits to site s after n steps, $\langle \mu_n^*(s) \rangle$, one finds the generating function $\mu^*(s;\xi) = (1-\xi)^{-1}P(s|s_0;\hat{\xi})$ for $s \neq s_0$, whereas for the average number of *returns* to the origin after n steps, $\langle \mu_n^*(s_0) \rangle$, one finds $\mu^*(s_0;\xi) = (1-\xi)^{-1}[P(0;\hat{\xi}) - 1]$. (We follow the convention of *not* counting the initial occupancy of the origin by the walker at step zero as the first return [4]). We find that the asymptotic average number of revisits to the origin in any dimension is given by $\langle \mu_\infty^*(s_0) \rangle = [(1-e^{-\lambda})\langle S_\infty^* \rangle]^{-1} - 1$, while the average number of visits to a site s other than the origin is given by $\langle \mu_\infty^*(s)\rangle = P(s|s_0; e^{-\lambda})$, one of the few previously known results for exponentially evanescent walkers (see subsection 3.2.4 in [4]). The average number of sites visited r times before the walker dies is in any dimension given by

$$\langle S_\infty^{*(r)}\rangle = [(1-e^{-\lambda})\langle \mu_\infty^*(s_0)\rangle]^{r-1} \left(\langle S_\infty^*\rangle\right)^r. \tag{46}$$

This result has been tested with numerical simulations for different values of r and of λ, and extremely good agreement between simulations and theory has been found (cf. Fig. 2).

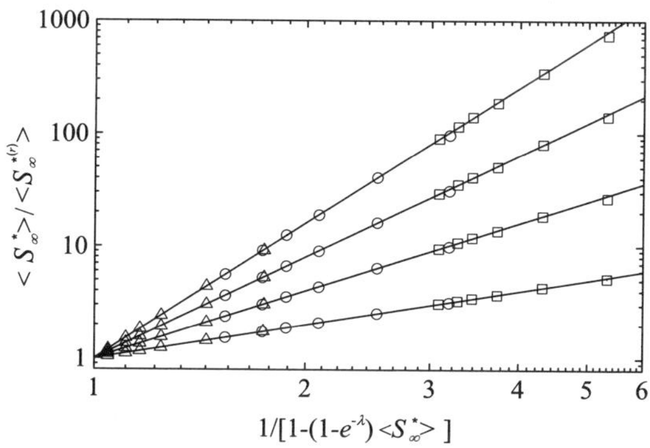

Fig. 2. $\langle S_\infty^*\rangle/\langle S_\infty^{*(r)}\rangle$ vs $1/[1-(1-e^{-\lambda})\langle S_\infty^*\rangle]$ for several values of r and λ and three different lattices. Symbols: numerical simulations for $d=1$ (triangles), $d=2$ (square lattice, circles), and $d=3$ (cubic lattice, squares) for 10^5 runs. The values of $\langle S_\infty^{*(r)}\rangle$ are obtained from the simulation of $\langle S_n^{*(r)}\rangle$ with n sufficiently large to observe no change in at least three significant figures. From left to right: $\lambda = 0.1, 0.05, 0.03, 0.01, 0.05, 0.001$, with from top to bottom $r = 2, 3, 4, 5$. The straight lines of slope $(r-1)$ through the origin are the theoretical predictions.

3.2. *Power-law evanescence*

In what follows we consider power-law evanescence, i.e., a decay law of the form $\rho(n) = (1+\lambda n)^{-\beta}$ with $\lambda > 0$ and $\beta > 0$. We directly use $\langle S_n^*\rangle = \sum_{m=0}^{n}\rho(n)\langle \Delta_n\rangle$ and rely on our knowledge of $\langle \Delta_n\rangle$ for large and small n for the most common lattices [4, 6]. For example, $\langle \Delta_n\rangle \sim (1-\mathcal{R})\,\lambda^{-\beta} n^{-\beta}\left(1 + C n^{-1/2} + \cdots\right)$ for three-dimensional lattices and large n, where C is a constant. Because $\rho(n) \sim (\lambda n)^{-\beta}$ for large n, one sees immediately that $\langle S_\infty^*\rangle$ is finite for $\beta > 1$. For the case of slow evanescence ($\lambda \to 0$) it is not difficult to find that this finite value is $\langle S_\infty^* \rangle \sim (1-\mathcal{R})/[(\beta-1)\lambda]$. However, for $\beta < 1$ the result for slow evanescence is quite different:

$$\langle S_n^* \rangle \sim \frac{1-\mathcal{R}}{1-\beta}\,\lambda^{-\beta}\,n^{1-\beta} \tag{47}$$

for large n. For $\beta = 0$ (no evanescence) one recovers the classical result [4]. For the marginal case $\beta = 1$ one obtains

$$\langle S_n^* \rangle \sim (1-\mathcal{R})\lambda^{-1}\log n. \tag{48}$$

This way we discover that the average number of distinct sites visited by a mortal walker before it dies is infinite for $\beta \geq 1$, whereas this quantity is finite for $\beta < 1$. This is also true for two-dimensional lattices, and for d-dimensional lattices with $d \geq 4$. However, for the one-dimensional lattice the critical value is $\beta = 1/2$.

4. The target problem with evanescent traps

4.1. *Discrete-time problem*

The quantities we have presented so far deal with mortal walkers in a perfect lattice, but there are many situations in which the lattice includes an imperfection of some sort. In particular, one can consider a collection of initially randomly distributed, independently moving walkers ("the traps") distributed on a lattice with a target placed at a given site (the imperfection). The target is instantaneously killed as soon as any of the traps steps on the target site, and one may wish to know the survival probability $Q_T^*(n)$ of the target up to a given step n. This is the discrete version of the so- called target problem, but it differs from the typical setting in that our traps are now mortal. The target problem is clearly a first-passage problem, as $Q_T^*(n)$ is the probability that none of the traps has visited the target site up to time step n. Since we are not interested in the fate of the trap that kills the target after it has "accomplished its goal", for practical purposes we can consider the target site as an absorbing site that terminates the trap's trajectory.

Now, it is well known that in the case of immortal traps the survival probability of the target, $Q_T(n)$, is related to $\langle S_n \rangle$, the average number of distinct sites visited by a *single* trap in the corresponding perfect lattice, i.e., a lattice where the absorbing site is replaced by a regular site. One then has the equality

$$Q_T(n) = \exp\left[-c_0(\langle S_n \rangle - 1)\right]. \qquad (49)$$

Here c_0 is the (constant) number density of traps, that is, the fraction of lattice sites initially occupied by traps. This relation is very useful, since knowledge of $\langle S_n \rangle$, a quantity calculated for a perfect lattice, leads to the determination of $Q_T(n)$ for a defect lattice and the corresponding reaction rate [9]. In this case, the target eventually dies with certainty, since $\langle S_n \rangle \to \infty$ as $n \to \infty$ in all dimensions. Since the link between $Q_T(n)$ and $\langle S_n \rangle$ is purely geometric, one can thus conjecture that a similar relation also holds in the case of mortal traps,

$$Q_T^*(n) = \exp\left[-c_0(\langle S_n^* \rangle - 1)\right], \qquad (50)$$

where c_0 is the *initial value* of the number density of traps; this number now decreases in time. This relation can indeed be rigorously proven. According to the results displayed in the previous section, for exponential evanescence $Q_T^*(n)$ goes to a finite value given by Eq. (40) as $n \to \infty$. Hence, the target has a non-zero chance of survival in all dimensions, as opposed to the classic case with immortal traps. In contrast, when one has power-law evanescence, the target may or may not survive depending on the value of the decay exponent β (see subsection 3.2).

4.2. *Continuous space and time*

The target problem can also be formulated in continuous rather than discrete space and time. For this purpose, it is convenient to consider the relation between $\langle S_n \rangle$ and the mean volume explored by an immortal trap in continuous space up to a given time, $\langle v \rangle_t$ (we are here adopting standard notation). The latter is defined in terms of the Wiener sausage generated by a spherical walker of radius R in time t. When there is no evanescence, this relation is well known and in the long time limit it reads as follows: [10]

$$\langle v \rangle_t \sim B\ell^d \langle S_n \rangle, \qquad (51)$$

where ℓ is the lattice constant in the discrete lattice and d is the dimensionality. The constant prefactor B depends on R, ℓ, dimensionality and the specific lattice structure. The relation is based on the correspondence $n = 2dDt/\ell^2$ when $n \gg 1$ and $R \gg \ell$ (see appendix B). For mortal walkers, the relation between the number of distinct sites visited and the explored volume does not change, i.e., $\langle S_n \rangle$ and $\langle v \rangle_t$ are respectively replaced by $\langle S_n^* \rangle$ and $\langle v^* \rangle_t$ in Eq. (51). Explicit results in continuous space for the survival probability $Q_T^*(t)$ can be found in [9] for both diffusive and subdiffusive walkers.

We have recently suggested [5] that trap evanescence may provide an alternative explanation for the behavior of certain defect-mediated relaxation processes, as in the celebrated model of dielectric relaxation first introduced by Glarum [11] in one dimension and later refined by Bordewijk to account for many-body effects both in one and three dimensions [12]. In such a setting, the relaxation function of a given configuration of a target molecule displays stretched exponential behavior (so-called Kohlrausch-Williams-Watts behavior) when the molecule is subject to a flux of diffusing defects. The identification of these defects (i.e., carriers of free volume or occurrences of elementary relaxation events) as *the traps* is the basis of the defect diffusion model to explain the observed behavior of the survival probability $Q_T(t) \sim \exp(-c_0\langle v\rangle_t)$ of the molecular configuration, where c_0 is the defect concentration. The observed experimental behavior is $\ln Q_T(t) \sim t^\theta$, where θ may take a wide range of values. However, only the values $\theta = 1/2$ or $\theta = 1$ are possible for normal diffusive defects because $\langle S_n\rangle \propto \langle v\rangle_t \propto t^{1/2}$ for $d = 1$ and (and then $\theta = 1/2$) and $\langle S_n\rangle \propto \langle v\rangle_t \propto t$ for $d \geq 2$ (and then $\theta = 1$). The Glarum-Bordewijk model was extended by Shlesinger and Montroll [13] by assuming that the movement of the defects may be described by a CTRW model with a power-law waiting time distribution $\sim t^{-1-\gamma}$, $0 < \gamma < 1$, which leads to $\theta = \gamma/2$ for $d = 1$ and $\theta = \gamma$ for $d \neq 1$ [14]. Thus, the stretched exponential relaxation with $\theta \neq 1/2$ is explained by assuming the diffusion of the defects to be anomalous with anomalous diffusion exponent γ.

Our results suggest that it may not be necessary to invoke anomalous diffusion to explain stretched exponential behavior, as evanescence of defects may lead to a similar qualitative behavior even when defect diffusion is normal. Such evanescence events have indeed been observed in experiments [15, 16]. As we have shown, different types of evanescence lead to different kinds of relaxation; for example, from Eq. (47) we see that $\langle v^*\rangle_t \propto t^{1-\beta}$ for $\beta < 1$ for $d \geq 3$, so that one can get stretched exponential relaxation with exponent $\theta = 1-\beta$ when the concentration of defects decays as a power law. Moreover, when $\beta = 1$, implying that the concentration c of defects decays as $\rho(t) \sim 1/t$ for large t (a decay found in some bimolecular reactions in condensed media), one gets from (48) that $\langle v^*\rangle_t \propto \ln t$, which in turn leads to *algebraic* relaxation [17–19]. This leads one to hypothesize that a situation where the interaction with the target is delayed because of the disappearance of the traps might become experimentally indistinguishable from a scenario where such a delay is caused by the subdiffusive motion of the traps.

Finally, it is worth recalling that the survival probability of the target in the target problem can be considered as a first approximation (Rosenstock approximation) to the survival probability of the target in the so-called trapping problem, in which the target diffuses and the mortal walkers become immobile traps. Our results can also be applied to this fundamental problem when the concentration of traps decreases with time. For the continuous time version of the trapping problem, Den Hollander and Shuler calculated the survival probability of the target in the regime of validity of the Rosenstock approximation [20]. It is worth pointing out that their result for the survival probability is a consequence of the behavior of $\langle v^* \rangle_t$ in the long time regime.

5. Conclusions and outlook

We have reviewed some recent results for the arrival statistics of mortal walkers, a problem which has received very little treatment in the literature. *A posteriori*, the approach used seems straightforward, yet the underlying physics is very different from the case of immortal walkers, thus making this problem a very interesting one. Remarkably enough, the profound modifications introduced by the evanescence reaction in the behavior of evanescent walkers can be studied using the generating function formalism originally developed for immortal walkers. As we have seen, the evanescence reaction dramatically changes the behavior of a number of characteristic quantities, e.g., the distinct number of sites visited as a function of time, or the survival probability of a target surrounded by randomly moving traps. For instance, the average number of distinct sites visited by an immortal walker goes to infinity as $n \to \infty$, while (depending on the speed of evanescence) it may be finite for an evanescent walker. Consequently, the survival probability of an immobile target in the presence of mobile traps is zero if the traps are immortal; however, if the traps evanesce at a sufficiently rapid rate, the target has a chance of survival.

Our results concerning the target problem also provide an alternative explanation to the stretched exponential relaxation observed in the defect diffusion model. In previous work, the possibility of anomalous diffusion induced by long-tailed waiting time distributions was invoked to explain the observed behavior, but we have seen that the same behavior is obtained with normally diffusive defects provided that these disappear at a sufficiently slow rate. Indeed, there is evidence for a decrease in the concentration of defects in time due to different processes such as, for instance, defect coalescence or annihilation at sinks. However, since subdiffusion and evanescence do not exclude each other, both of them could play a role in conjunction and should therefore both be taken into account for quantitative studies.

Finally, the aspects studied here by no means exhaust the family of problems related to explored territory (or distinct sites visited) by mortal walkers. Studies for the case of biased walks, continuous-time random walks, walks in confined spaces, and Lévy flights and walks are underway and will be presented elsewhere.

Acknowledgments

This work was partially funded by the Ministerio de Ciencia y Tecnología (Spain) through Grant No. FIS2010-16587 (partially financed by FEDER funds), by the Junta de Extremadura through Grant. No. GRU10158, and by the US National Science Foundation under Grant No. PHY-0855471.

Appendix A. Derivation of a continuum reaction-diffusion equation from the master equation for mortal walkers

We start from the fundamental equation (5) and introduce the probability density function or "concentration" c via the definition $P_n^*(s|s_0) = \Delta x\, c(s\Delta x, n\Delta t|s_0\Delta x)$, where Δx is the lattice constant. Employing this definition in (5) we get

$$c(x,t+\Delta t|x_0) = \frac{1}{2}\left[c(x-\Delta x, n\Delta t|x_0) + c(x+\Delta x, t|x_0)\right]\frac{\rho(t+\Delta t)}{\rho(t)}, \quad \text{(A.1)}$$

where the notation $x = s\Delta x, x_0 = s_0\Delta x$ and $t = n\Delta t$ has been used. We now perform the usual Taylor expansion in terms of Δx and Δt as follows:

$$c(x,t+\Delta t|x_0) \approx c(x,t|x_0) + \Delta t\, \partial_t c(x,t), \quad \text{(A.2a)}$$

$$c(x+\Delta x,t|x_0) \approx c(x,t|x_0) + \Delta x\, \partial_t p(x,t) + \frac{(\Delta x)^2}{2}\, \partial_x^2 c(x,t), \quad \text{(A.2b)}$$

$$\rho(t+\Delta t) \approx \rho(t) + \Delta t\, \dot{\rho}(t). \quad \text{(A.2c)}$$

Inserting these expressions into (A.1), dividing the resulting equation by Δt and letting Δt and Δx simultaneously go to zero while keeping the ratio

$(\Delta x)^2/\Delta t$ fixed, we get the following continuum reaction-diffusion equation:

$$\partial_t c(x,t|x_0) = D\,\partial_x^2 c(x,t|x_0) - \hat{\lambda}(t)\,c(x,t|x_0), \tag{A.3}$$

where the diffusion coefficient is taken to be $D \equiv \lim_{\Delta x,\Delta t\to 0}(\Delta x)^2/2\Delta t$ and the negative logarithmic derivative $\hat{\lambda}(t) = -\dot{\rho}/\rho$ is a time-dependent rate constant. The latter is assumed to be small in absolute value so that it makes sense to take the diffusion limit (this limit implies that, on average, a sufficient number of time steps $n \gg 1$ must have been taken before reaction takes place). The particular case of exponential evanescence $\rho(t) = \rho_0 \exp(-\hat{\lambda}\,t)$ leads to a time-independent rate constant $\hat{\lambda}$. The above derivation can be straightforwardly generalized for non-nearest neighbor walks or walks in higher dimensions.

Appendix B. Time-discretization of the walker density

The mortal walker performs steps on the d-dimensional lattice at regular time intervals Δt, hence the stepping times are $t_n = n\Delta t$. In the case of an exponential evanescence process, its density decay is given by

$$\rho(t) = e^{-\hat{\lambda}t} \to \rho(t_n) \equiv \rho(n) = \rho(0)e^{-\hat{\lambda}t_n}, \tag{B.1}$$

i.e.,

$$\rho(n) = \rho(0)\,e^{-\lambda n}, \tag{B.2}$$

where we have introduced $\lambda \equiv \hat{\lambda}\Delta t$.

One can proceed in a similar way with a power law decay. One has

$$\rho(t) = \frac{\rho_0}{(1+\hat{\lambda}t)^\beta} \to \rho(n) = \frac{\rho(0)}{(1+\lambda n)^\beta} \tag{B.3}$$

$$\rho(n) = \frac{\rho(0)}{(1+n\hat{\lambda}\Delta t)^\beta}. \tag{B.4}$$

Alternatively, the relation between λ and $\hat{\lambda}$ can be expressed in terms of the diffusion coefficient

$$D = \frac{1}{2d}\frac{\ell^2}{\Delta t}, \tag{B.5}$$

where ℓ^2 is the variance of the step length distribution. Using the resulting expression for Δt, one gets

$$\lambda = \hat{\lambda}\Delta t = \hat{\lambda}\,\frac{\ell^2}{2dD}. \tag{B.6}$$

Conversely, to obtain the continuum limit one must perform the following replacements

$$t_n \to t \qquad \text{and} \qquad n = \frac{t_n}{\Delta t} \to t\,\frac{2dD}{\ell^2}. \tag{B.7}$$

We make use of this correspondence in Sec. 4.2.

References

1. G.H. Weiss, *Aspects and applications of the random walk* (Elsevier, Amsterdam, 1994).
2. R.F. Bonner, R. Nossal, S. Havlin and G. H. Weiss, *J. Opt. Soc. Am. A* **4**, 423 (1987).
3. I. Lohmar and J. Krug, *J. Stat. Phys.* **134**, 307 (2009).
4. B. H. Hughes, *Random Walks and Random Environments, Volume 1: Random Walks* (Clarendon, Oxford, 1995).
5. S. B. Yuste, E. Abad and K. Lindenberg, *Phys. Rev. Lett.* **110**, 220603 (2013).
6. E. W. Montroll and G. H. Weiss, *J. Math. Phys.* **6**, 167 (1965).
7. C. A. Walsh and J. J. Kozak, *Phys. Rev. B* **26**, 4166 (1982).
8. A. J. Guttmann, *J. Phys. A* **43**, 305205 (2010).
9. E. Abad, S. B. Yuste and K. Lindenberg, *Math. Model. Nat. Phenom.* **8**, 100 (2013).
10. A. M. Berezhkovskii, Y. A. Makhnovskii and R. A. Suris, *J. Stat. Phys.* **57**, 333 (1989).
11. S. H. Glarum, *J. Chem. Phys.* **33**, 639 (1960).
12. P. Bordewijk, *Chem. Phys. Lett.* **32**, 592 (1975).
13. M. F. Shlesinger and E. W. Montroll, *PNAS* **81**, 1280 (1984).
14. H. Scher, M. F. Shlesinger and J. T. Bendtler, *Phys. Today* **44**, 26 (1991).
15. M. Heggen, F. Spaepen and M. Feuerbacher, *J. Appl. Phys.* **97**, 033506 (2005).
16. S. Chakravarty, H. Schmidt, U. Tietze, D. Lott, N. P. Lalla and A. Gupta, *Phys. Rev. B* **80**, 014111 (2009).
17. A. Blumen, J. Klafter and G. Zumofen, *Reaction dynamics in glasses* in *Optical Spectroscopy of Glasses*, ed. I. Zschokke (Reidel, Dordrecht, 1986).
18. M. F. Shlesinger, *J. Chem. Phys.* **70**, 4813 (1979).
19. M. F. Shlesinger, *J. Stat. Phys.* **36**, 639 (1984).
20. F. D. Hollander and K. E. Shuler, *J. Stat. Phys.* **67**, 13 (1992).

Chapter 2

First Passage of a Randomly Accelerated Particle

Theodore W. Burkhardt

Department of Physics, Temple University,
Philadelphia, PA 19122, USA
tburk@temple.edu

In the random acceleration process, a point particle is accelerated according to $\ddot{x} = \eta(t)$, where the right hand side represents Gaussian white noise with zero mean. We begin with the case of a particle with initial position x_0 and initial velocity v_0 and review the statistics of its first arrival at the origin and its first return to the origin. Multiple returns to the origin, motion with a constant force in addition to a random force, and persistence properties for several boundary conditions at the origin are also considered. Next we review first-exit properties of a randomly accelerated particle from the finite interval $0 < x < 1$. Then the close connection between the extreme value statistics of a randomly accelerated particle and its first-passage properties is discussed. Finally some applications where first-passage statistics of the random acceleration process play a role are considered.

1. Introduction

For a particle moving in one dimension subject to a random force in the form of Gaussian white noise, the Newtonian equation of motion has the form

$$\ddot{x} = \eta(t), \quad \langle \eta(t) \rangle = 0, \quad \langle \eta(t)\eta(t') \rangle = \Lambda^{-1}\delta(t - t'). \tag{1}$$

This is the random acceleration process considered in this chapter. For a given $\eta(t)$, the position x and velocity v of the particle evolve from the initial values x_0 and v_0 at $t = 0$ according to

$$v(t) = v_0 + \int_0^t \eta(t')\,dt', \tag{2}$$

$$x(t) = x_0 + \int_0^t v(t')\,dt' = x_0 + v_0 t + \int_0^t (t - t')\eta(t')\,dt'. \tag{3}$$

Thus, $v(t)$ corresponds to a Brownian curve or random walk, and $x(t)$ to the integral of a Brownian curve.

The propagator or probability density $P(x, v; x_0, v_0; t)$ for propagation from the initial values x_0, v_0 to x, v in a time t will play a central role in our discussion of first-passage properties. It has the path integral representation

$$P(x, v; x_0, v_0; t) = \int Dx \exp\left[-\frac{\Lambda}{2}\int_0^t \left(\frac{d^2x}{dt^2}\right)^2 dt\right] \tag{4}$$

and satisfies the Fokker-Planck equation [1]

$$\left(\frac{\partial}{\partial t} + v\frac{\partial}{\partial x} - \frac{1}{2\Lambda}\frac{\partial^2}{\partial v^2}\right) P(x, v; x_0, v_0; t) = 0 \tag{5}$$

with initial condition

$$P(x, v; x_0, v_0; 0) = \delta(x - x_0)\delta(v - v_0). \tag{6}$$

The factor $(2\Lambda)^{-1}$ in Eq. (5) can be eliminated by rescaling t, and except in Sec. 5, where applications to semiflexible polymers are considered and Λ corresponds to the persistence length, we set $(2\Lambda)^{-1} = 1$ throughout this chapter.

In the unbounded geometry $-\infty < x < \infty$, Eqs. (5) and (6) have the solution [1]

$$\begin{aligned} P_0(x, v; x_0, v_0; t) &= 3^{1/2}(2\pi)^{-1}t^{-2} \\ &\times \exp\left\{-3t^{-3}\left[(x - x_0 - vt)(x - x_0 - v_0 t) + \frac{1}{3}(v - v_0)^2 t^2\right]\right\}, \end{aligned} \tag{7}$$

which is normalized so that

$$\int_{-\infty}^{\infty} dx \int_{-\infty}^{\infty} dv\; P_0(x, v; x_0, v_0; t) = 1. \tag{8}$$

In this chapter we review theoretical work on the first-passage properties of random acceleration and also consider some applications where these properties play a central role. For a related review of the random acceleration process in bounded geometries, but with less focus on first-passage properties, see [2]. The random acceleration process is also considered in the reviews [3,4] of first-passage properties of a wide variety of processes.

In Sec. 2 we consider the first arrival at the origin of a randomly accelerated particle which is initially released at point x_0 with velocity v_0, the first return to the origin, and multiple returns to the origin. The pioneering contributions of McKean [5] and Marshall and Watson [6] are reviewed. The results are generalized to include a constant force, such as gravity, in addition to the random force. First-passage properties follow from the solution of the Fokker-Planck equation on the half line $x > 0$ with an absorbing boundary condition at the origin. Some results for partially absorbing and inelastic boundaries are also briefly discussed.

In Sec. 3 the initial position x_0 of the randomly accelerated particle is assumed to lie in the finite interval $0 < x < 1$. Theoretical results for the mean exit time of a randomly accelerated particle from the interval and the probabilities that the first exit takes place at $x = 0$ and at $x = 1$ are reviewed.

Section 4 considers the extreme value statistics of the random acceleration process and its close relation to first-passage statistics.

In Sec. 5 several mathematical and physical applications which can be analyzed in terms of first-passage properties of the random acceleration process are discussed or referenced. For example, the partition function of a long, semi-flexible polymer chain, fluctuating in equilibrium in a narrow cylindrical channel, is closely related to the first-exit probability of a randomly accelerated particle from a finite domain.

2. Random acceleration on the half line $x > 0$

2.1. *First arrival at the origin*

We begin by considering the probability density or half-line propagator $P(x, v; x_0, v_0; t)$ for propagation of the particle from x_0, v_0 to x, v in a time t *without leaving the positive x axis.* This quantity satisfies the Fokker-Planck equation (5) with initial condition (6) and boundary condition

$$P(0, v; x_0, v_0; t) = 0, \quad v > 0. \tag{9}$$

This boundary condition rightfully excludes all trajectories with $x = 0$, $v > 0$ at time t from the propagator, since the corresponding particle has not remained on the positive real axis for a time t but is returning after leaving it an earlier time. Following common usage, we refer to Eq. (9)

as the "absorbing" boundary condition, since it obviously applies if the particle sticks permanently at the origin on arrival. In contrast to Eq. (9), $P(0, v; x_0, v_0; t)$ does not vanish for $v < 0$ and is directly related, as indicated in the next paragraph, to the statistics of arrival at the origin. For $v < 0$, $P(0, v; x_0, v_0; t)$ is not known at the outset, but is determined by solving the Fokker-Planck equation (5) with initial condition (6) and boundary condition (9), as constitutes a "well-posed" problem [6].

The "survival probability" or probability $Q(x_0, v_0; t)$ that the randomly accelerated particle has not yet left the positive x axis in a time t is related to the half-line propagator by

$$Q(x_0, v_0; t) = \int_{-\infty}^{\infty} dv \int_0^{\infty} dx\, P(x, v; x_0, v_0; t) \tag{10}$$

and, according to Eq. (5) and the time reversal property $P(x, v; x_0, v_0; t) = P(x_0, -v_0; x, -v; t)$, satisfies the backward Fokker-Planck equation

$$\left(\frac{\partial}{\partial t} - v_0 \frac{\partial}{\partial x_0} - \frac{\partial^2}{\partial v_0^2}\right) Q(x_0, v_0; t) = 0, \tag{11}$$

with the initial condition and absorbing boundary condition

$$Q(x_0, v_0; 0) = 1, \quad x_0 > 0, \tag{12}$$

$$Q(0, v_0; t) = 0, \quad v_0 < 0, \tag{13}$$

respectively. Equations (5), (9), and (10) imply

$$\frac{\partial}{\partial t} Q(x_0, v_0; t) = -\int_0^{\infty} dv\, vP(0, -v; x_0, v_0; t). \tag{14}$$

Thus, the probability that the randomly accelerated particle leaves the the positive x axis for the first time between t and $t+dt$ and with speed between v and $v + dv$, is given by $vP(0, -v; x_0, v_0; t)\, dv\, dt$, where $v > 0$.

In the half space (x, v) with $x > 0$, the propagator $P(x, v; x_0, v_0; t)$ vanishes identically on the positive v axis, according to Eq. (9), but otherwise is positive. Solving the Fokker-Planck equation (5) with this boundary condition is a formidable task. The equation has separable solutions of the form $e^{st-Fx}\mathrm{Ai}(-F^{1/3}v + F^{-2/3}s)$, where Ai(z) is the standard Airy function [7], but these solutions do not satisfy Eq. (9). The problem of superposing the separable solutions in order to satisfy the boundary condition was solved by Marshall and Watson [6] in 1985, as discussed in Sec. 2.4.

2.2. *First return to the origin*

Two decades before the work of Marshall and Watson, the first-return distribution $vP_0(0,-v;0,v_0;t)$ of a particle that begins at the origin with $v_0 > 0$ was already known from the ground-breaking work of McKean [5]. In 1963 he calculated this quantity, but not by solving the Fokker-Planck equation. A randomly accelerated particle which begins at the origin at $t = 0$ with a positive velocity v_0 and crosses the origin at a later time t with a positive velocity v must have left the positive x axis for the the first time at some earlier time t_1. This obvious statement is quantified in the integral equation

$$\begin{aligned} &P_0(0,v;0,v_0;t) \\ &\quad = \int_0^t dt_1 \int_0^\infty dv_1\, P_0(0,v;0,-v_1;t-t_1)\, v_1 P(0,-v_1;0,v_0;t_1) \end{aligned} \tag{15}$$

for positive v and v_0, which relates the free propagator P_0 given in Eq. (7) and the half-line propagator P for the absorbing boundary condition (9).

Performing Laplace and Bessel transformations $t \to s$ and $v \to \gamma$, respectively, McKean converted integral equation (15) to the form

$$\begin{aligned} &[\,2v_0 \cosh(\pi\gamma/3)\,]^{-1} K_{i\gamma}\left(\sqrt{4s}\, v_0\right) \\ &\quad = \int_0^\infty dv_1\, K_{i\gamma}\left(\sqrt{4s}\, v_1\right) \tilde{P}(0,-v_1;0,v_0;s) \end{aligned} \tag{16}$$

for arbitrary positive γ, where $K_\nu(u)$ is the standard modified Bessel function [7] and $\tilde{P}(x,v;x_0,v_0;s) = \int_0^\infty dt\, e^{-st} P(x,v;x_0,v_0;t)$. Solving Eq. (19) with the help of the Kontorovitch-Lebedev transformation [10],

$$f(u) = \frac{2}{\pi^2} \int_0^\infty d\gamma\, \gamma \sinh(\pi\gamma) g(\gamma) K_{i\gamma}(u), \tag{17}$$

$$g(\gamma) = \int_0^\infty du\, u^{-1} f(u) K_{i\gamma}(u), \tag{18}$$

and performing the inverse Laplace transformation $s \to t$, he obtained

$$\begin{aligned} &\tilde{P}(0,-v;0,v_0;s) \\ &\quad = \frac{2}{\pi^2 v v_0} \int_0^\infty d\gamma\, \frac{\gamma \sinh(\pi\gamma)}{2\cosh(\pi\gamma/3)} K_{i\gamma}\left(\sqrt{4s}\, v\right) K_{i\gamma}\left(\sqrt{4s}\, v_0\right) \end{aligned} \tag{19}$$

and

$$\begin{aligned} &vP(0,-v;0,v_0;t) \\ &\quad = \frac{\sqrt{3}}{2\pi} \frac{v}{t^2} \exp\left[-\left(v^2 - v v_0 + v_0^2\right)/t\right] \operatorname{erf}\left[(3 v v_0/t)^{1/2}\right] \end{aligned} \tag{20}$$

for the first-return distribution. Here v and v_0 are both positive, and $\mathrm{erf}(z)$ is the error function [7].

In working with the integrals in this section and Sec. 2.3 involving integration of Bessel functions $K_{i\gamma}(z)$ over γ, such as Eq. (19), it is useful to keep the asymptotic form

$$K_{i\gamma}(z) \approx \left(\frac{2\pi}{\gamma}\right)^{1/2} e^{-\pi\gamma/2} \cos\left(\gamma \ln \frac{2\gamma}{z} - \gamma - \frac{\pi}{4}\right) \tag{21}$$

for $\gamma \to \infty$ with z fixed in mind. For completeness we note that, in contrast to Eq. (19), $\tilde{P}(0,-v;0,v_0;s)$ may be written without special functions in the form

$$\begin{aligned} &\tilde{P}(0,-v;0,v_0;s) \\ &= \frac{3}{2\pi}\,(vv_0)^{1/2} \int_0^1 dy\, \exp\left[-2s^{1/2}\left(v^2 - vv_0 + v_0^2 + 3vv_0y^2\right)^{1/2}\right] \\ &\times \left[\left(v^2 - vv_0 + v_0^2 + 3vv_0y^2\right)^{-3/2} \right. \\ &\left. +2s^{1/2}\left(v^2 - vv_0 + v_0^2 + 3vv_0y^2\right)^{-1}\right]. \end{aligned} \tag{22}$$

This follows from Laplace tranformation of Eq. (20) after substitution of a standard integral representation of the error function [7].

Substituting Eq. (20) into Eq. (14) and integrating over v yields

$$\begin{aligned} &-\frac{\partial}{\partial t} Q(0,v_0;t) = I_1(v_0,t) \\ &= \frac{3}{2\pi^{3/2}}\, v_0^{-2} e^{-v_0^2/t} \sum_{n=0}^{\infty} \frac{\Gamma(\frac{5}{4}+\frac{n}{2})}{n!}\, {}_2F_1(\tfrac{1}{2},-n;\tfrac{3}{2};3) \left(\frac{v_0^2}{t}\right)^{5/4+n/2} \end{aligned} \tag{23}$$

for the rate of first return to the origin. Here ${}_2F_1(a,b;c;z)$ denotes a hypergeometric function [9], and the infinite series in Eq. (23) converges for arbitrary v_0^2/t. Thus, the probability that a particle which leaves the origin with velocity v_0 has not yet returned after a time t decays as [8]

$$Q(0,v_0;t) \approx \frac{3\Gamma(\frac{1}{4})}{2\pi^{3/2}} \left(\frac{v_0^2}{t}\right)^{1/4}, \quad t \gg v_0^2\,. \tag{24}$$

2.3. *Multiple returns to the origin*

The probability density $vP_n(0,-v;0,v_0;t)$ for the $n^{\rm th}$ return of a randomly accelerated particle to the origin, which is determined by the hierarchy

$$P_n(0,-v;0,v_0;t) = \int_0^t dt_1 \int_0^\infty dv_1 \; P(0,-v;0,v_1;t-t_1) \times v_1 P_{n-1}(0,-v_1;0,v_0;t_1), \tag{25}$$

was analyzed by Lachal [11]. Here, v and v_0 are positive, and to avoid complications with signs, we assume that the particle is instantaneously and elastically reflected back onto the positive real axis each time it arrives the origin and just before the $n^{\rm th}$ arrival has speed v and velocity $-v$. Writing Eq. (25) in terms of Laplace transforms, making use of Eqs. (17)–(19), and performing the inverse Laplace transformation, one obtains

$$\tilde P_n(0,-v;0,v_0;s) = \frac{2}{\pi^2 v v_0}\int_0^\infty d\gamma \, \frac{\gamma\sinh(\pi\gamma)}{[2\cosh(\pi\gamma/3)]^n}\, K_{i\gamma}\left(\sqrt{4s}\, v\right) K_{i\gamma}\left(\sqrt{4s}\, v_0\right) \tag{26}$$

and

$$P_n(0,-v;0,v_0;t) = \frac{1}{\pi^2 v v_0 t}\exp\left[-\left(v^2+v_0^2\right)/t\right] \times \int_0^\infty d\gamma\, \frac{\gamma\sinh(\pi\gamma)}{[2\cosh(\pi\gamma/3)]^n}\, K_{i\gamma}\left(2vv_0/t\right). \tag{27}$$

The rate $I_n(v_0,t) = \int_0^\infty dv\, vP_n(0,-v;x_0,v_0;t)$ of $n^{\rm th}$ return to the origin and its Laplace transform follow from Eqs. (26) and (27) and are given by [11]

$$\tilde I_n\left(v_0,s\right) = \frac{1}{\pi v_0 s^{1/2}}\int_0^\infty d\gamma\, \frac{\gamma\sinh(\pi\gamma/2)}{[2\cosh(\pi\gamma/3)]^n}\, K_{i\gamma}\left(\sqrt{4s}\, v_0\right), \tag{28}$$

$$I_n\left(v_0,t\right) = \frac{e^{-v_0^2/2t}}{2\pi^{3/2} v_0 t^{1/2}}\int_0^\infty d\gamma\, \frac{\gamma\sinh(\pi\gamma/2)}{[2\cosh(\pi\gamma/3)]^n}\, K_{i\gamma/2}\left(v_0^2/2t\right). \tag{29}$$

The first return rate $I_1(v_0,t)$ given by Eq. (29) has the series expansion (23) and decays as $t^{-5/4}$ for $t >> v_0^2$. With the approach of [12], one finds that I_2 decays as $t^{-5/4}\ln t$, slower by a factor $\ln t$, and for larger n there are higher powers of $\ln t$. The first and second return rates are compared in Fig. 1.

The time-integrated $n^{\rm th}$ arrival rate $q_n(v_0,t) = \int_0^t dt_1\, I_n\left(v_0,t_1\right)$ follows from Eqs. (28) and (29) and has the form [11]

$$q_n(v_0,t) = \frac{te^{-v_0^2/2t}}{2\pi v_0^2}\int_0^\infty d\gamma\, \frac{\gamma\sinh(\pi\gamma/2)}{[2\cosh(\pi\gamma/3)]^n}\, W_{-1,i\gamma/2}(v_0^2/t), \tag{30}$$

where $W_{\lambda,\mu}(z)$ Whittaker's function [7,9]. In the limit $t \to \infty$, $q_n(v_0, t) \to 1$.

Since $q_n(v_0, t)$ is the probability that a particle which begins at the origin with velocity v_0 makes n or more returns to the origin in a time t, the probability $p_n(v_0, t)$ that it makes n and only n returns is given by

$$p_n(v_0, t) = q_n(v_0, t) - q_{n+1}(v_0, t)\,. \tag{31}$$

The distribution $p_n(v_0, t)$ has been analyzed in detail by Majumdar and Bray [13] and Schehr and Majumdar [14] within the framework of Gaussian stationary processes and by De Smedt et al. [12] on the basis of Eqs. (27)–(31). These studies also utilize results for random acceleration on the half line with partial absorption at the boundary, considered in Sec. 2.6. For $t \gg v_0^2$, $p_n(v_0, t)$ decays as $t^{-1/4}(\ln t)^n$ and $\langle n \rangle$ increases as $\ln t$.

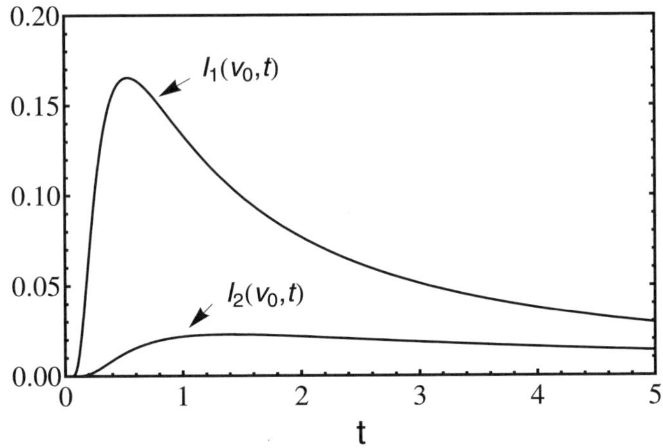

Fig. 1. Time dependence of the first and second return rates to the origin, as given by Eq. (29), for initial velocity $v_0 = 1$. The integrated return rate or area under each curve is 1. As discussed below Eq. (29), I_1 decays as $t^{-5/4}$ and I_2 more slowly by a factor $\ln t$. The two curves cross at $t \approx 50$, with $I_2 > I_1$ after that.

2.4. *Solution of the Fokker-Planck equation*

In 1985 Marshall and Watson [6] derived the Laplace transform $\tilde{P}(x, v; x_0, v_0; s)$ of the propagator for the Langevin equation for Brownian motion, $\ddot{x} + \lambda\dot{x} = \eta(t)$, with an absorbing boundary at the origin. This was accomplished by solving the corresponding Fokker-Planck equation in terms of a special set of basis functions, constructed explicitly in their paper, which vanish at $x = 0$ for positive v but not negative v, so as to fulfill the boundary condition (9).

In the limit in which the damping constant λ approaches zero, the Langevin equation reduces to the random acceleration process (1), and the result of Marshall and Watson simplifies to [15]

$$\begin{aligned}\tilde{P}(x,v;x_0,v_0;s) = \tilde{P}_0(x,v;x_0,v_0;s) - \frac{1}{2\pi}\int_0^\infty dF\, F^{-1/6}\int_0^\infty dG\, G^{-1/6} \\ \times (F+G)^{-1}\exp\left[-Fx - Gx_0 - \frac{2}{3}s^{3/2}\left(F^{-1}+G^{-1}\right)\right] \\ \times \mathrm{Ai}\left(-F^{1/3}v + F^{-2/3}s\right)\mathrm{Ai}\left(G^{1/3}v_0 + G^{-2/3}s\right), \quad (32)\end{aligned}$$

$$\begin{aligned}\tilde{P}_0(x,v;x_0,v_0;s) = \int_0^\infty dF\, F^{-1/3} \\ \times\Big[\theta(x-x_0)e^{-F(x-x_0)}\mathrm{Ai}\left(-F^{1/3}v + F^{-2/3}s\right)\mathrm{Ai}\left(-F^{1/3}v_0 + F^{-2/3}s\right) \\ +\theta(x_0-x)e^{-F(x_0-x)}\mathrm{Ai}\left(F^{1/3}v + F^{-2/3}s\right)\mathrm{Ai}\left(F^{1/3}v_0 + F^{-2/3}s\right)\Big]. \quad (33)\end{aligned}$$

Here $\mathrm{Ai}(z)$ is the Airy function [7], and $\tilde{P}_0(x,v;x_0,v_0;s)$ is the Laplace transform of the free propagator (7) for motion on the unbounded x axis.

Equations (32) and (33) specify the unique superposition of the separable solutions mentioned below Eq. (14) which satisfies the Fokker-Planck equation (5) with the initial condition (6) and the absorbing boundary condition (9). That $\tilde{P}(0,v;x_0,v_0;s)$ vanishes for $v > 0$, in accordance with Eq. (9), is not immediately obvious from Eqs. (32) and (33) but follows from the identity

$$\begin{aligned}\frac{1}{F^{1/6}}\mathrm{Ai}\left(F^{1/3}v + F^{-2/3}s\right) = \frac{1}{2\pi}\int_0^\infty \frac{dG}{F+G}\exp\left[-\tfrac{2}{3}s^{3/2}\left(F^{-1}+G^{-1}\right)\right] \\ \times\frac{1}{G^{1/6}}\mathrm{Ai}\left(-G^{1/3}v + G^{-2/3}s\right). \quad (34)\end{aligned}$$

This relation only holds for $v > 0$ and may be derived by rewriting the right-hand side of Eq. (34) as an integral in the complex G plane with contour surrounding the positive G axis, deforming the contour so it surrounds the pole at $G = -F$, and then invoking Cauchy's residue theorem.

Expressions (32) and (33) provide the extension of McKean's results (19) and (20) for the half space propagator with $x = x_0 = 0$ to arbitrary points x and x_0 on the positive x axis. Inverting the Laplace transform in Eq. (32) analytically appears out of the question, but for short times $P(x,v;x_0,v_0;t)$

approaches the free propagator (7), and for long times $t \gg x^{2/3}$, $x_0^{2/3}$, v^2, and v_0^2,

$$P(x,v;x_0,v_0;t) \approx 3^{5/3}\pi^{-3/2}t^{-5/2}\left(xx_0\right)^{1/6} \times U\left(-\frac{1}{6},\frac{2}{3},-\frac{v^3}{9x}\right)U\left(-\frac{1}{6},\frac{2}{3},\frac{v_0^3}{9x_0}\right), \tag{35}$$

as shown in [15], where $U(a,b,c)$ is Kummer's confluent hypergeometric function [7].

The Laplace transform of the survival probability follows from Eqs. (10), (32), and (33) and is given by [15]

$$\tilde{Q}(x_0,v_0;s) = s^{-1} - \int_0^\infty dF\,F^{-5/3}e^{-Fx_0}\mathrm{Ai}\left(F^{1/3}v + F^{-2/3}s\right) \times\left[1 + \frac{1}{4\pi^{1/2}}\Gamma\left(-\tfrac{1}{2},\tfrac{2}{3}F^{-1}s^{3/2}\right)\right]. \tag{36}$$

The small s behavior of $\tilde{Q}(x_0,v_0;s)$, which diverges as $s^{-3/4}$, implies the long time behavior

$$Q(x_0,v_0;t) \approx \frac{3^{4/3}\Gamma(\frac{1}{4})}{2\pi^{3/2}}\left(\frac{x_0^{2/3}}{t}\right)^{1/4} U\left(-\frac{1}{6},\frac{2}{3},\frac{v_0^3}{9x_0}\right) \tag{37}$$

for $t \gg x_0^{2/3}$, $t \gg v_0^2$. Expression (37) extends the McKean-Goldman result (24) for $x_0 = 0$ to arbitrary x_0.

Due to the slow $t^{-1/4}$ decay in Eq. (37), the average time of first arrival at the origin,

$$T(x_0,v_0) = \int_0^\infty dt\,t\left[-\frac{\partial}{\partial t}Q(x_0,v_0;t)\right] = \int_0^\infty dt\,Q(x_0,v_0;t), \tag{38}$$

is infinite.

The $t^{-1/4}$ dependence of the survival probability in Eqs. (24) and (37) also follows from the argument, see, e.g., [4,16], that Q decays as $N^{-1/2}$ for large N, where N is the number of zero crossings of the velocity, in accordance with the Sparre-Anderson theorem [17], and that N scales as $t^{1/2}$. Sinai [18] has given a rigorous derivation of the $t^{-1/4}$ behavior and other first-passage properties along these lines. For a derivation [20] that leads to the asymptotic form (37) in just a few steps (but does not determine the multiplicative constant) see Sec. 2.6.

2.5. *Particle subject to a constant force and a random force*

For a particle subject to a constant force, such as gravity, in addition to a random force, and governed by the equation of motion $\ddot{x} = g + \eta(t)$, the Fokker-Planck equation (5) is replaced by

$$\left(\frac{\partial}{\partial t} + v\frac{\partial}{\partial x} + g\frac{\partial}{\partial v} - \frac{\partial^2}{\partial v^2}\right) P_g(x, v; x_0, v_0; t) = 0\,. \qquad (39)$$

Here, by rescaling of x and t, we have chosen $g = \pm 1$ and $2\Lambda = 1$ with no loss of generality. The initial condition for $P_g(x, v; x_0, v_0; t)$ is the same as in Eq. (6), and for studying the first exit from the positive real axis, the same absorbing boundary condition (9) applies. Keeping this in mind and comparing Eqs. (5) and (39), we see that the propagators with and without the constant force g are related by

$$P_g(x, v; x_0, v_0; t) = \exp\left[\tfrac{1}{2}g(v - v_0) - \tfrac{1}{4}t\right] P(x, v; x_0, v_0; t). \qquad (40)$$

Making use of Eq. (40) and expressions (20) and (32) for the propagator on the right-hand side, Burkhardt [19] has studied the first exit from the positive x axis of a particle subject to a constant force and a random force. In contrast to the long time behavior discussed in connection with Eqs. (37) and (38), the survival probability $Q_g(x_0, v_0; t)$ for $g = 1$, corresponding to a constant force driving the particle in the $+x$ direction away from the origin, does not vanish for $t \to \infty$, and in the case $g = -1$ of a constant force toward the origin, the mean time $T_g(x_0, v_0)$ of arrival at the origin is finite.

2.6. *Partially absorbing, inelastic boundary*

Consider a randomly accelerated particle on the half line $x > 0$, which, on reaching the origin, is absorbed with probability $1 - p$ and reflected inelastically with probability p and velocity $-rv$, where r is the coefficient of restitution. The probability $Q(x_0, v_0; t)$ that the particle has not yet been absorbed after a time t satisfies the backward Fokker-Planck equation (11) with initial condition (12) and boundary condition

$$Q(0, -v_0; t) = pQ(0, rv_0; t), \quad v_0 > 0. \qquad (41)$$

Burkhardt [20] showed that $Q(x_0, v_0; t)$ decays asymptotically as $t^{-\phi}$ for large t, with persistence exponent $\phi(p)$ given by

$$2\sin\left[\frac{\pi}{6}(1 - 4\phi)\right] = pr^{2\phi}. \qquad (42)$$

Making use of Eqs. (27)–(31), De Smedt et al. [12] confirmed the persistence exponent (42) and derived the functional form of the survival probability $\tilde{Q}(0, v_0; s)$ for the case $r = 1$ of absorption with probability $1 - p$ and elastic reflection with probability p. This quantity has the expansion

$$Q(0, v_0; t) = \sum_{n=0}^{\infty} p^n p_n(v_0, t) \tag{43}$$

in terms of the probability $p_n(v_0, t)$ of n returns in a time t introduced in Eq. (31) and can be used as a generating function in calculating the moments of n. The studies [12–14] of the distribution $p_n(v_0, t)$ cited at the end of Sec. 2.3 are carried out in the context of random acceleration with partial absorption and make use of Eqs. (42) and (43).

A short route to the persistence exponent (42), following [20], is as follows: Scaling invariance restricts $Q(x_0, v_0; t)$ to a function of two scale independent variables v_0^3/x_0 and t^3/x_0^2. This and the assumption of a $t^{-\phi}$ decay imply

$$Q(x_0, v_0; t) \approx \left(t x_0^{-2/3}\right)^{-\phi} \mathcal{F}\left(v_0^3/9x_0\right). \tag{44}$$

Substituting this in the Fokker-Planck equation (11) and dropping all terms which decay faster than $t^{-\phi}$ leads to the confluent hypergeometric [7] differential equation $\left[z\partial_z^2 + \left(\frac{2}{3} - z\right)\partial_z + \frac{2}{3}\phi z\right]\mathcal{F}(z) = 0$. The solution which is finite for $v_0 \to \infty$ is given by $\mathcal{F} = CU\left(-\frac{2}{3}\phi, \frac{2}{3}, v_0^3/9x_0\right)$, where C is a constant and $U(a, b, z)$ is Kummer's function [7]. Imposing the boundary condition (41) leads directly to expression (42) for the persistence exponent ϕ. For $p = 0$ the first-passage probability given in Eq. (37) is reproduced, apart from the multiplicative constant.

A randomly accelerated particle moving between two walls from which it is reflected inelastically is expected to reach a steady-state, in which the kinetic energy gain due to the random force is lost in the inelastic collisions. The question whether there is "inelastic collapse" or localization of the particle at the boundaries has been considered by several authors, see [21-24] and references therein. According to [23], the equilibrium rate of collision with the boundaries is infinite for coefficient of restitution $r < r_c = \exp\left(-\pi/\sqrt{3}\right) = 0.163..$, as first pointed out by Cornell et al. [21], but this does not lead to localization of the particle at the boundary.

Similar conclusions were reached [25] for a particle which is accelerated on the half line $x > 0$ by both a random force and a constant force directed to the origin, where it is reflected inelastically. The collision rate at the origin is infinite for $r < r_c = \exp\left(-\pi/\sqrt{3}\right)$, but this does not lead to localization at the origin. For two values $r = \frac{1}{2}$ and $\frac{1}{3}$, the steady state distribution function $P_{\rm eq}(x, v)$ was calculated explicitly.

2.7. *Windy cliff problem*

Consider a particle in the half plane (x, y) with $x > 0$ which moves diffusively in the y direction and and is transported in the x direction by a flow field $f(y)$ according to

$$\dot{y} = \eta(t), \qquad \dot{x} = f(y). \tag{45}$$

In the case $f(y) = y$ of uniform shear flow, Eq. (45) is equivalent to the random acceleration process $\ddot{x} = \eta(t)$, and the probability that the particle has not yet reached the boundary $x = 0$ after a time t decays asymptotically as $t^{-1/4}$ for long times, as in Eq. (37).

Redner and Krapivsky [26] have shown that for the flow field $f(y) = v_0 \operatorname{sgn}(y)$, the survival probability also decays as $t^{-1/4}$, and Bray and Gonos [27] have extended this result to any $f(y)$ which is an odd function of y and has the same sign as y. Results for an even larger class of flow fields and for partially absorbing boundaries are given in [4,28].

3. First exit from the interval $0 < x < 1$

3.1. *Mean exit time*

Unlike the mean exit time (38) from the positive x axis, the mean exit time $T(x_0, v_0)$ of a randomly accelerated particle from a line segment of finite length is finite. Rescaling of both x and t allows us to set the segment length equal to 1 and $2\Lambda = 1$ in Eqs. (1) and (5), and with no loss of generality we consider the interval $0 < x < 1$. From Eq. (38) and the backward Fokker-Planck equation (11) for the survival probability, we obtain the inhomogeneous partial differential equation

$$\left(v_0 \frac{\partial}{\partial x_0} + \frac{\partial^2}{\partial v_0^2}\right) T(x_0, v_0) = -1, \tag{46}$$

for the mean exit time, to be solved with the boundary conditions

$$T(0, v_0) = 0, \quad v_0 < 0, \tag{47}$$

$$T(1, v_0) = 0, \quad v_0 > 0. \tag{48}$$

This differential equation looks simpler than the Fokker-Planck equation (11), since there is no time derivative, but a complicating feature is that there are two boundaries at which the particle can exit. The system of equations (46)-(48) was solved by Franklin and Rodemich [29] and Masoliver and Porrà [30], and the mean escape time is

$$\begin{aligned}T(x_0, v_0) = T(1-x_0, -v_0) &= \frac{1}{3}(\pi v_0)^{1/2} \int_0^{1-x_0} dy\, \frac{e^{-v_0^3/18y}}{y^{1/2}} I_{-1/6}\left(\frac{v_0^3}{18y}\right) \\ &+ \frac{2^{-7/3}3^{5/6}}{\Gamma(\frac{2}{3})^2} \int_{x_0}^1 dy\, \frac{e^{-v_0^3/9(y-x_0)}}{(y-x_0)^{2/3}} [y(1-y)]^{-1/6} \\ &\times \left[{}_2F_1(1, -\tfrac{2}{3}; \tfrac{5}{6}; 1-y) - {}_2F_1(1, -\tfrac{2}{3}; \tfrac{5}{6}; y)\right], \quad v_0 > 0. \end{aligned} \tag{49}$$

3.2. *Splitting probabilities*

The probabilities $q_0(x_0, v_0)$ and $q_1(x_0, v_0)$ that the particle makes its first exit from the line segment $0 < x < 1$ at $x = 0$ and $x = 1$, respectively, are determined by the sum rule $q_0(x_0, v_0) + q_1(x_0, v_0)$, the reflection symmetric property $q_0(x_0, v_0) = q_1(1 - x_0, -v_0)$, and the differential equation

$$\left(v_0 \frac{\partial}{\partial x_0} + \frac{\partial^2}{\partial v_0^2}\right) q_0(x_0, v_0) = 0, \tag{50}$$

with boundary condition

$$q_0(0, v_0) = 1, \quad v_0 < 0. \tag{51}$$

The solution, obtained by Bicout and Burkhardt [31], is

$$\begin{aligned} q_0(x_0, v_0) &= 1 - q_0(1 - x_0, -v_0) \\ &= \frac{1}{2\pi} \int_{x_0}^1 dy\, \frac{e^{-v_0^3/9(y-x_0)}}{(y-x_0)^{2/3}} [y(1-y)]^{-1/6}, \quad v_0 > 0. \end{aligned} \tag{52}$$

3.3. *Long time behavior*

In the long time limit, the probability density for propagation of the particle from x_0, v_0 to x, v without leaving the line segment $0 < x < 1$ decays as

$$P(x, v; x_0, v_0; t) \approx \psi_0(x, v)\psi_0(x_0, -v_0)e^{-E_0 t}, \tag{53}$$

where E_0 is the eigenvalue with the smallest real part of the time-independent Fokker-Planck equation

$$\left(-E + v\frac{\partial}{\partial x} - \frac{\partial^2}{\partial v^2}\right)\psi(x, v) = 0\,, \tag{54}$$

with the absorbing boundary condition $\psi(0,v) = \psi(1,-v) = 0$ for $v > 0$, analogous to Eq. (9). From Eqs. (10) and (53) it is clear that the survival probability $Q(x_0, v_0; t)$ also decays as $e^{-E_0 t}$. Solving numerically an integral equation for E_0 and ψ_0 that arises in an exact analytic approach, Burkhardt [32] obtained $E_0 = 1.3904$ to 5 significant figures, a result which is in excellent agreement with precision simulations. As discussed below in Sec. 5.1, this number is of interest in connection with the equilibrium statistical mechanics of semi-flexible polymers confined in narrow channels.

4. Extreme statistics of the random acceleration process

Consider the family of possible trajectories of a randomly accelerated particle which lead from x_0, v_0 to x_1, v_1 in a time $t = T$. Each trajectory $x(t)$ reaches a maximum value of x at some time in the interval $0 \le t \le T$. The probability that the maximum value lies between X and $X + dX$ may be expressed as $\mathcal{P}(X, T|x_1, v_1; x_0, v_0)\, dX$. Probability distributions such as $\mathcal{P}$ play a central role in the field of extreme value statistics [33–36]. We will see that the extreme value and first-passage statistics of random acceleration are closely related.

We begin with the relation

$$\mathcal{P} = \frac{\partial \mathcal{F}}{\partial X}\,, \tag{55}$$

between the desired distribution and the cumulative probability $\mathcal{F}(X, T|x_1, v_1; x_0, v_0)$ that the maximum displacement of a trajectory from x_0, v_0 to x_1, v_1 in time T is less than X. From invariance under the translation $x \to x - X$ and the reflection $x \to -x$, $v \to -v$, $\mathcal{F}(X, T|x_1, v_1; x_0, v_0)$ is the same as the probability of propagation from $X - x_0, -v_0$ to $X - x_1, -v_1$, without ever leaving the positive x axis. Thus, in the regime $X > \max(x_1, x_2)$ in which $\mathcal{F}$ does not vanish identically,

$$\mathcal{F}(X, T|x_1, v_1; x_0, v_0) = \frac{P(X - x_1, -v_1; X - x_0, -v_0; T)}{P_0(X - x_1, -v_1; X - x_0, -v_0; T)}. \tag{56}$$

Here $P(x, v; x_0, v_0; t)$ is the half-line propagator with absorbing boundary condition at the origin, discussed at length in Sec. 2, and $P_0(x, v; x_0, v_0; t)$ is the corresponding propagator in the absence of boundaries, given explicitly in Eq. (7).

The asymptotic form of $\mathcal{F}(X, T|x_1, v_1; x_0, v_0)$ for large T follows directly from Eqs. (7), (35), and (56), and for general T, $\mathcal{F}(X, T|x_1, v_1; x_0, v_0)$ can be evaluated by numerical inversion of the Laplace transform (32).

It is simple to derive analogous extreme distributions for other boundary conditions of interest. For example, the probability that the maximum displacement of a trajectory which begins at x_0, v_0 does not exceed X in a time T, with no restrictions on x_1, v_1, is given by

$$\mathcal{F}(X,T|x_0,v_0) = \frac{\int_{-\infty}^{\infty} dv_1 \int_{-\infty}^{X} dx_1\ P(X-x_1,-v_1;X-x_0,-v_0;T)}{\int_{-\infty}^{\infty} dv_1 \int_{-\infty}^{\infty} dx_1\ P_0(X-x_1,-v_1;X-x_0,-v_0;T)} \\ = Q(X-x_0,-v_0;t), \tag{57}$$

where we have used the normalization condition (8), and where Q is the survival probability introduced in Eq. (10), with Laplace transform and asymptotic form for large t given in Eqs. (36) and (37), respectively. Although the distribution $\mathcal{P}(X,T|x_0,v_0)$ defined by Eqs. (36), (55), and (57) has not been evaluated analytically, Reymbaut et al. [38] have derived all its moments $\langle (X-x_0)^n\rangle_T$ for $v_0=0$. The first two are given by

$$\langle X-x_0\rangle_T = \sqrt{\frac{3}{8\pi}}\,T^{3/2}, \quad \langle (X-x_0)^2\rangle_T = \left(\frac{2}{3}-\frac{103}{384}\sqrt{\frac{3}{2}}\right)T^3. \tag{58}$$

The extreme displacement of trajectories which start at the origin, return after a time T, and are periodic, with $x(t)=x(t+T)$ and $x(0)=0$ is analyzed in Ref. [37]. In analogy with Eqs. (56 and (57), the probability such a periodic trajectory has maximum displacement less than X is given by

$$\mathcal{F}_{\rm per}(X,T) = \frac{\int_{-\infty}^{\infty} dv\ P(X,v;X,v;T)}{\int_{-\infty}^{\infty} dv\ P_0(X,v;X,v;T)}. \tag{59}$$

Evaluating the right-hand side of Eq. (59) using Eqs. (7) and (32) and differentiating with respect to X, in accordance with Eq. (55), leads to the extreme value distribution [37]

$$\mathcal{P}_{\rm per}(X,T) = \frac{2^{4/3}3^{5/6}}{\sqrt{\pi}X}\left(\frac{X}{T^{3/2}}\right)^{2/3} e^{-48X^2/T^3}\ U(-\frac{1}{6},\frac{2}{3},\frac{48X^2}{T^3}). \tag{60}$$

Here again, $U(a,b,z)$ is Kummer's confluent hypergeometric function [7]. For more details and a discussion of the extreme statistics of the process $d^n x/dt^n = \eta(t)$ for general integer and non-integer n, not just the case $n=2$ of random acceleration, see [37].

The distribution $\mathcal{P}_g(X,v_0)$ of the maximum height X attained by a particle which is thrown vertically upwards with velocity v_0 and is subject to both a random force and a downward gravitational force $g=-1$ is studied in [19]. As in the examples considered above, the extreme distribution follows from the ratio of the propagators, discussed in Sec. 2.5, for motion on the half line and entire line and is given by [19]

$$\mathcal{P}_g(X, v_0) = \frac{e^{v_0/2}}{\sqrt{2\pi}} \int_0^\infty dF\, F^{-1/6} \times \exp\left(-\frac{1}{12F} - FX\right) \mathrm{Ai}\left(-F^{1/3}v_0 + \tfrac{1}{4}F^{-2/3}\right) \qquad (61)$$

for positive v_0. In the limit of large v_0, the mean height and standard deviation vary according to $\langle X \rangle \approx \frac{1}{2}v_0^2 + v_0$ and as $\sigma \approx \left(\frac{2}{3}v_0^3\right)^{1/2}$, as compared to the familiar maximum height $\frac{1}{2}v_0^2$ in the absence of the random force.

Finally we note that Reymbaut et al. [38] make use of extreme value statistics in their analysis of the convex hulls of trajectories of randomly accelerated particles in two dimensions, which begin at the origin with velocity zero and continue for a time T. For the mean values of the perimeter L and area A of the hull, they obtain

$$\langle L \rangle_T = \sqrt{\frac{3\pi}{2}}\, T^{3/2}, \quad \langle A \rangle_T = \frac{5\pi}{192}\sqrt{\frac{3}{2}}\, T^3. \qquad (62)$$

5. Applications

5.1. *Semi-flexible polymer in a cylindrical channel*

Consider a long, semi-flexible polymer or worm-like chain fluctuating in equilibrium in a narrow cylindrical channel. If the width of the channel is much smaller than the persistence length P, typical polymer configurations hardly deviate from straight lines and correspond, as shown in Fig. 2, to single valued functions $\vec{r}(t)$, where (x, y, t) are Cartesian coordinates, and $\vec{r} = (x, y)$ specifies the transverse displacement of the polymer from the symmetry axis or t axis of the tube. The bending energy $\mathcal{H} = \frac{1}{2}\kappa \int ds\, (d\hat{\tau}/ds)^2$, where s is the arc length and $\hat{\tau}$ is a unit tangent vector, simplifies to to $\mathcal{H} = \frac{1}{2}\kappa \int dt\, \left(d^2\vec{r}/dt^2\right)^2$, and the polymer partition function is given by

$$Z(\vec{r}, \vec{v}; \vec{r}_0, \vec{v}_0; t) = \int D^2 r \exp\left[-\frac{P}{2}\int_0^t \left(\frac{d^2\vec{r}}{dt^2}\right)^2\right], \qquad (63)$$

where $P = \kappa/k_BT$, and satisfies the Fokker-Planck equation

$$\left(\frac{\partial}{\partial t} + \vec{v}\cdot\nabla_{\mathbf{r}} - \frac{1}{2P}\nabla_{\mathbf{v}}^2\right) Z(\vec{r}, \vec{v}; \vec{r}_0, \vec{v}_0; t) = 0. \qquad (64)$$

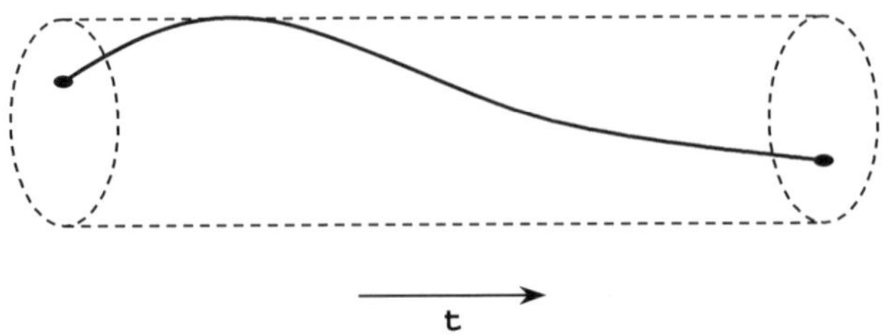

Fig. 2. The curve may be interpreted as the trajectory $\vec{r}(t)$ of a randomly accelerated particle moving in two dimensions, plotted as a function of t, or as the configuration of a tightly confined semi-flexible polymer in a channel.

Since discontinuities in slope of the polymer at the boundary cost an infinite bending energy and are suppressed, $Z(\vec{r},\vec{v};\vec{r}_0,\vec{v}_0;t)$ vanishes as $\vec{r}$ approaches the wall of the tube for $\hat{n}\cdot\vec{v}>0$, where $\hat{n}$ is normal to the wall and directed toward the interior of the tube.

From comparing the path integrals (4) and (63), the Fokker-Planck equations (5) and (64), and the boundary condition (9) with the one in the preceding paragraph, one concludes [32] that the polymer partition function is identical with the probability that a randomly accelerated particle moving in two dimensions propagates from $(\vec{r}_0,\vec{v}_0)$ to $(\vec{r},\vec{v})$ without leaving a two-dimensional domain corresponding to the cross section of the channel. As illustrated in Fig. 2, each of the possible trajectories of the randomly accelerated particle may be interpreted as a polymer configuration.

In the long-polymer limit, the partition function (63) has the asymptotic form $Z\approx\psi_0\left(\vec{r},\vec{v}\right)\psi_0\left(\vec{r}_0,-\vec{v}_0\right)e^{-E_0t}$, where E_0 is the smallest eigenvalue of the t-independent Fokker-Planck equation and $\psi_0\left(\vec{r},\vec{v}\right)$ is the corresponding eigenfunction. Thus, we interpret $f=k_BTE_0$ as the free energy per unit length of confinement of the polymer. According to the correspondence in Fig. 2, the probability that the randomly accelerated particle has not yet left the two-dimensional domain corresponding to the channel cross section in a time t decays as e^{-E_0t}.

For a tightly confined polymer polymer in a channel with a rectangular cross section with edges D_1 and D_2, the partition function (63) factors in the form $Z(x,v_x;x_0,v_{0x};t)Z(x,v_x;x_0,v_{0x};t)$, and the value of E_0 follows from the numerical result for random acceleration in one dimension quoted just below Eq. (54). Going from dimensionless variables back to the original variables, one obtains [32]

$$f \approx A_{\Box} \frac{kT}{P^{1/3}} \left(\frac{1}{D_1^{2/3}} + \frac{1}{D_2^{2/3}} \right) \tag{65}$$

for the free energy of confinement per unit length, with $A_{\Box} = 2^{-1/3}(1.3904) = 1.1036$.

By performing computer simulations of randomly accelerated motion, one can study properties of tightly confined polymers in channels, using the correspondence illustrated in Fig. 2. The estimates $A_{\Box} = 1.1038 \pm 0.0006$ for the amplitude in Eq. (65) and the corresponding amplitude $A_{\circ} = 2.3565 \pm 0.0004$ in the free energy of confinement

$$f \approx A_{\circ} \frac{kT}{P^{1/3} D^{2/3}}, \tag{66}$$

for a channel with a circular cross section of diameter D were obtained this way [39–41], and the mean value and fluctuations of the length of channel occupied by the polymer were also studied.

The dependence on the persistence length and channel dimensions in Eqs. (65) and (66), which follows from the polymer partition function (63), is completely consistent with scaling predictions of Odijk [42]. Finally, we call attention to recent experimental studies of single biopolymers, for example DNA [43] and actin filaments [44–46], in channels with widths smaller than or comparable with the persistence length.

5.2. *Spatial persistence in interface growth*

Majumdar and Bray [4,47] have studied the persistence of fluctuating interfaces which evolve according to the Langevin equation

$$\frac{\partial h}{\partial t} = -\left(-\nabla^2\right)^{z/2} h + \xi, \tag{67}$$

where $h(\vec{r}, t)$ denotes the height of the interface above point $\vec{r}$ in a d-dimensional hyperplane, z is a dynamical exponent, and ξ is Gaussian white noise with zero mean and $\langle \xi(\vec{r}, t)\, \xi(\vec{r}\,', t')\rangle = \Lambda \delta(\vec{r} - \vec{r}\,')\, \delta(t-t')$. The values $z = 2$ and $z = 4$ correspond to the well-known cases of Edward-Wilkinson and Mullins-Herring growth, respectively.

Majumdar and Bray analyzed the correlations of the quantity $g_n(x_1, x_2, x_3, ...) = \partial^n h/\partial x_1^n$ with respect to an arbitrary direction x_1 in the hyperplane. In the steady-state limit $t \to \infty$, the equal time pair correlation function $\langle g_n(x_1, x_2, x_3, ...) g_n(x_1', x_2, x_3, ...)\rangle$ turns out to be proportional to

$\delta(x_1 - x_1')$, corresponding to Gaussian white noise, for $n = \frac{1}{2}(z - d + 1)$. In particular, in the case $z = 4$, $d = 1$, corresponding to Mullins-Herring or Golubović-Bruinsma-Das Sarma-Tamborenea [48,49] growth, this implies $d^2h/dx^2 = \eta(x)$, where $\eta(x)$ is Gaussian white noise with zero mean. Since this is the same as the random acceleration process (1) with variables (h, x) in place of (x, t), the spatial persistence of the interface and the temporal persistence of the randomly accelerated particle are the same. Thus, the probability, in steady state, that an interface $h(x)$ with height h_0 and slope h_0' at x_0 does not return to the same height h_0 within a distance $|x - x_0|$ of x_0 decays as $(h_0'^2/|x - x_0|)^{1/4}$ for large distances, as in Eq. (24).

For more details and an analysis of other growth models and other types of persistence in interface growth, see Refs. [4,47].

5.3. *Other applications*

In closing we reference several other applications where first-passage properties of random acceleration play a role:

Several authors [50–55] have considered the desorption or depinning transition in a rather artificial $1+1$ dimensional model of a semi-flexible polymer fluctuating without overhangs in the half plane $x > 0$, $-\infty < t < \infty$ with partition function

$$Z(x, v; x_0, v_0; t) = \int Dx \exp\left\{-\int_0^t dt \left[\frac{P}{2}\left(\frac{d^2x}{dt^2}\right)^2 + V(x)\right]\right\}, \qquad (68)$$

reminiscent of the propagator (4) for random acceleration and the partition function (63) for a nearly straight polymer. Here, x and t are Cartesian coordinates, and $V(x)$ is a short range interaction favoring absorption of the polymer along the boundary $x = 0$. The desorption transition, as the temperature is raised, appears to be first-order, in contrast to the second-order desorption transition, in both two and three spatial dimensions and in the limit of infinite contour length, of a flexible self-avoiding, semi-flexible polymer, fluctuating with overhangs included [56].

Schwarz and Maimon [16] make use of first-passage properties of random acceleration in analyzing a model for crack propagation in elastic media.

The propagator for random acceleration on the half line with absorbing boundary condition appears in Valageas' analysis [57] of the statistical properties of the Burgers equation with Brownian initial velocity.

Convex hulls have applications [38] in crystallography, computer image processing, ecology, and polymer statistics. Hilhorst et al. [58] have found a connection between the random acceleration process and Sylvester's problem of convex polygons. Reymbaut et al. [38] map the convex hull of a two-dimensional random acceleration process onto a one-dimensional problem and make use of first-passage properties and extreme statistics of random acceleration in their analysis.

Finally we note that Majumdar et al. [59] have analyzed the distribution $p(t_m|T)$ of the time t_m at which a particle reaches its maximum displacement if it begins at the origin with velocity zero and is randomly accelerated for a time T. They obtain simple analytic expressions both for integrals of Brownian bridges (trajectories constrained to return to $v = 0$ in a time T) and integrals of free Brownian motion (no restrictions on the trajectory at T). For ordinary unrestricted Brownian motion beginning at the origin, both distributions $p(t_m|T)$ and $p(t_{\rm occup}|T)$, where $t_{\rm occup}$ is the total time spent on the positive x axis, are given by [60,61] Lévy's arcsine law $p(t|T) = \pi^{-1}\left[t(T-t)\right]^{-1/2}$. For random acceleration, on the other hand, Majumdar et al. conclude, on the basis of computer simulations, that $p(t_{\rm occup}|T)$ does not coincide with either of their results for $p(t_m|T)$. The derivation of the exact analytical form of $p(t_{\rm occup}|T)$ for random acceleration is a challenging, currently unsolved problem.

Acknowledgments

It is a pleasure to thank Dominique Bicout, Alan Bray, Dieter Forster, Claude Godrèche, Jerrold Franklin, Richard Gawronski, Gerhard Gompper, Stanislav Kotsev, Satya Majumdar, Zoltan Rácz, and Yingzi Yang for stimulating discussions and correspondence. I also thank Satya for useful comments on a draft of this chapter and help with references.

References

1. Risken, H. (1989). *The Fokker-Planck Equation: Methods of Solution and Applications*, 2nd edn., Springer.
2. Burkhardt, T. W. (2007). *The random acceleration process in bounded geometries*, J. Stat. Mech. P07004 (16 pp.).
3. Redner, S. (2001). *A Guide to First-Passage Properties*, Cambridge.
4. Bray, A. J., Majumdar, S. N. and Schehr, G. (2013). *Persistence and First-Passage Properties in Non-equilibrium Systems*, Advances in Physics, in press.
5. McKean, H. P. (1963). *A winding problem for a resonator driven by a white noise*, J. Math. Kyoto Univ. **2**, pp. 227-235.

6. Marshall, T. W. and Watson, E. J. (1985). *A drop of ink falls from my pen...It comes to earth, I know not when*, J. Phys. A **18**, pp. 3531-3559.
7. Abramowitz, M. and Stegun, I. A. eds. (1965). *Handbook of Mathematical Functions*, Dover.
8. Goldman, M. (1971). *On the first passage of the integrated Wiener process.* Ann. Math. Stat. **42**, pp. 2150-2155.
9. Gradshteyn, I. S. and Ryzhik, I. M. (1980). *Table of Integrals, Series, and Products*, Academic.
10. Erdélyi. A. (1953). Tables of Integral Transforms, Vols. 1 and 2, Bateman Manuscript Project, McGraw-Hill.
11. Lachal, A. (1997). *Les temps de passage successifs de l'intégrale du mouvement brownien*, Ann. Inst. Henri Poincaré B **33**, pp. 1-36.
12. De Smedt, G., Godrèche, C. and Luck, J. M. (2001) *Partial survival and inelastic collapse for a randomly accelerated particle*, Europhys. Lett. **53**, pp. 438-443.
13. Majumdar, S. N. and Bray A. J. (1998). *Persistence with partial survival,* Phys. Rev. Lett. **81**, pp. 2626-2629.
14. Schehr, G. and Majumdar, S. N. (2007). *Statistics of the number of zero crossings: From random polynomials to the diffusion equation*, Phys. Rev. Lett. **99**, 060603 (4 pp.).
15. Burkhardt, T. W. (1993). *Semiflexible polymer in the half plane and statistics of the integral of a Brownian curve*, J. Phys. A **26**, pp. L1157-L1162.
16. Schwarz, J. M. and Maimon, R. (2001) *First-passage-time exponent for higher-order random walks: Using Lévy flights*, Phys. Rev. E **64** 016120 (pp. 10).
17. Sparre-Andersen, E. (1953). *On the fluctuations of sums of random variables*, Math. Scand. **1** pp. 263-285; Math. Scand. **2** pp. 195-223.
18. Sinai, Y. G. (1992). *Distribution of some functionals of the integral of a random walk*, Theor. Math. Phys. **90**, pp. 219-241. (Teor. Mat. Fiz. **90**, pp. 323-353).
19. Burkhardt, T. W. (2008). *First-passage and extreme-value statistics of a particle subject to a constant force plus a random force*, J. Stat. Phys. **133**, pp. 217-230.
20. Burkhardt, T. W. (2000). *Dynamics of absorption of a randomly accelerated particle*, J. Phys. A **33**, pp. L429-L432.
21. Cornell, S. J., Swift, M. R. and Bray, A. J. (1998). *Inelastic collapse of a randomly forced particle*, Phys. Rev. Lett. **81**, pp. 1142-1145.
22. Burkhardt T. W., Franklin, J. and Gawronski, R. R. (2000). *Statistics of a confined, randomly accelerated particle with inelastic boundary conditions*, Phys. Rev. E. **61**, pp. 2376-2381.
23. Burkhardt, T. W. and Kotsev, S. N. (2004) *Equilibrium of a confined, randomly accelerated, inelastic particle: Is there inelastic collapse?*, Phys. Rev. E **70**, 026105 (6 pp.).
24. Kotsev, S. N. and Burkhardt, T. W. (2005). *Randomly accelerated particle in a box: Mean absorption time for partially absorbing and inelastic boundaries*, Phys. Rev. E **71**, 046115 (7 pp.).

25. Burkhardt, T. W. and Kotsev, S. N. (2006). *Equilibrium statistics of an inelastically bouncing ball, subject to gravity and a random force*, Phys. Rev. E **73**, 046121 (7 pp.).
26. Redner, S. and Krapivsky, P. L. (1996). *Diffusive escape in a nonlinear shear flow: Life and death at the edge of a windy cliff*, J. Stat. Phys. **82**, pp. 999-1014.
27. Bray, A. J. and Gonos, P. (2005). *Survival of a diffusing particle in a transverse flow field*, J. Phys. A. **38**, pp. 5617-5626.
28. Bray, A. J. and Majumdar, S. N. (2006). *Partial survival and crossing statistics for a diffusing particle in a transverse shear flow*, J. Phys. A. **39**, L625-L631.
29. Franklin, J. N. and Rodemich, E. R. (1968). *Numerical analysis of an elliptic-parabolic partial diffeerential equation*, SIAM J. Numer. Anal. **4**, pp. 680-716.
30. Masoliver, J. and Porrà, J. M. (1995). *Exact solution to the mean exit time problem for free inertial processes driven by Gaussian white noise*, Phys. Rev. Lett. **75**, pp. 189-192.
31. Bicout, D. J. and Burkhardt, T. W. (2000). *Absorption of a randomly accelerated particle: gambler's ruin in a different game*, J. Phys. A **33**, pp. 6835-6841.
32. Burkhardt, T. W. (1997). *Free energy of a semiflexible polymer in a tube and statistics of a randomly accelerated particle*, J. Phys. A **30**, pp. L167-L172.
33. Gumbel, E. J. (1958). *Statistics of Extremes*, Columbia University.
34. Galambos, J. (1978). *The Asymptotic Theory of Extreme Order Statistics*, Wiley.
35. Coles, S. (2001). *An Introduction to Statistical Modeling of Extreme Values*, Springer.
36. Castillo. E., Hadi, A. S., Balakrishnan, N. and Sarabia, J. M. (2005). *Extreme Value and Related Models with Applications in Engineering and Science*, Wiley.
37. Burkhardt, T. W., Györgyi, G., Moloney, N. R. and Rácz, Z. (2007) *Extreme statistics for time series: Distribution of the maximum relative to the initial value*, Phys. Rev. E **76**, 041119 (12 pp.).
38. Reymbaut, S., Majumdar, S. N. and Rosso, A. (2011). *The convex hull for a random acceleration process in two dimensions*, J. Phys. A **44**, 415001 (15 pp.).
39. Bicout, D. J. and Burkhardt, T. W. (2001). *Simulation of a semiflexible polymer in a narrow cylindrical pore*, J. Phys. A **34**, pp. 5745-5750.
40. Yang, Y., Burkhardt, T. W. and Gompper, G. (2007). *Free energy and extension of a semiflexible polymer in cylindrical confining geometries*, Phys. Rev. E **76**, 011804 (7 pp.).
41. Burkhardt, T. W., Yang, Y. and Gompper, G. (2010). *Fluctuations of a long, semiflexible polymer in a narrow channel*, Phys. Rev. E **82**, 041801 (9 pp.).
42. Odijk, T. (1983) *On the statics and dynamics of confined or entangled stiff polymers*, Macromolecules **16**, pp. 1340-1344; (1986). *Theory of lyotropic polymer liquid crystals*, Macromolecules **19**, pp. 2313-2329.
43. Reisner, W., Morton, K. J., Riehn, R., Wang, Y. M., Yu, Z., Rosen, M.,

Sturm, J. C., Chou, S. Y., Frey, E. and Austin, R. H. (2005). *Statics and dynamics of single DNA molecules confined in nanochannels*, Phys. Rev. Lett. **94**, 196101 (4 pp.).

44. Köster, S., Steinhauser, D. and Pfohl, T. (2005). *Brownian motion of actin filaments in confining microchannels*, J. Phys. Condens. Matter **17**, S4091-S4104.
45. Köster, s., Kierfeld, J and Pfohl, T. (2008). *Characterization of single semi-flexible filaments under geometric constraints*, Eur. Phys. J. E **25**, pp. 439-449.
46. Köster S. and Pfohl, T. (2009). *An in vitro model system for cytoskeletal confinement*, Cell Motility and the Cytoskeleton **66**, pp. 771-776.
47. Majumdar, S. N. and Bray, A. J. (2001). *Spatial persistence of fluctuating interfaces*, Phys. Rev. Lett. **86**, pp. 3700-3703.
48. Golubović, L. and Bruinsma, R. (1991). *Surface diffusion and fluctuations of growing interfaces*, Phys. Rev. Lett. **66**, pp. 321-324.
49. Das Sarma, S. and Tamborenea P. (1991). *A new universality class for kinetic growth: One-dimensional molecular-beam epitaxy*, Phys. Rev. Lett. **66**, pp. 325-328.
50. Maggs, A. C., Huse, D. A. and Leibler S. (1989) *Unbinding transitions of semi-flexible polymers*, Europhys. Lett. **8**, pp. 615-620.
51. Gompper, G. and Burkhardt, T. W. (1989) *Unbinding transition of semiflexible membranes in* $(1+1)$ *dimensions*, Phys. Rev. A, **40**, 6124-6127.
52. Bundschuh, R., Lässig, M. and Lipowsky, R. (2000). *Semi-flexible polymers with attractive interactions*, Eur. Phys. J. E **3**, pp. 295-306.
53. Bundschuh, R. and Lässig, M. (2002). *Delocalization transitions of semiflexible manifolds*, Phys. Rev. E **65**, 061502 (7 pp.).
54. Benetatos, P. and Frey, E. (2003). *Depinning of semiflexible polymers*, Phys. Rev. E **67**, pp. 051108 (7 pp.).
55. Caravenna, F. and Deuschel, J.-D. (2008). *Pinning and wetting transition for* $(1+1)$*-dimensional fields with Laplacian interaction*, Ann. Probab. **36**, pp. 2388-2433.
56. Eisenriegler, E. (1993) *Polymers near Surfaces*, World Scientific.
57. Valageas, P. (2009). *Statistical properties of the Burgers equation with Brownian initial velocity*, J. Stat. Phys. **134**, pp. 589-640.
58. Hilhorst, H. J., Calka, P. and Schehr, G. (2008). *Sylvester's question and the random acceleration process*, J. Stat. Mech. P10010 (25 pp.).
59. Majumdar, S. N., Rosso, A. and Zoia, A. (2010). *Time at which the maximum of a random acceleration process is reached*, J. Phys. A **43** 115001 (16 pp.).
60. Lévy, P. (1939). *Sur certains processus stochastiques homogénes*, Comp. Math. **7**, pp. 283-339.
61. Feller W. (1968). *An Introduction to Probability Theory and Its Applications*, Wiley.

Chapter 3

First Passage Problems in Anomalous Diffusion

Alberto Rosso
CNRS - Université Paris-Sud, LPTMS, UMR8626, 91405 Orsay, France
alberto.rosso@u-psud.fr

Andrea Zoia
CEA/Saclay, DEN/DM2S/SERMA/LTSD, 91191 Gif-sur-Yvette, France
andrea.zoia@cea.fr

Many physical and biological systems can be described in terms of a particle undergoing random displacements. When the spread of these particles at long times asymptotically scales as $x^2(t) \sim t^{2H}$, with $H \neq 1/2$, the walker diffusion is said to be anomalous, as opposed to regular diffusion, which corresponds to plain Brownian motion. Examples of anomalous diffusion emerge for instance when considering the motion of polymers through nanopores, or the percolation of tracer particles in heterogeneous soils. Determining the first-passage properties of such walkers is particularly difficult, because anomalous particles often display long-range correlations and/or wild distributions of time and space increments. Here we will review some relevant examples of anomalous diffusion and show by which mathematical and numerical tools the corresponding first-passage properties can be assessed.

1. Beyond Brownian motion: anomalous diffusion

Brownian motion plays a key role in modern theoretical physics and it has proved very successful in explaining several key features of physical and biological systems. Actually, it is currently used in various fields of science so as to interpret, for instance, the dynamics of complex molecules within the cell, animal food-searching strategies, or tracer dispersion in soils, only to name a few [1, 2]. In all such cases, the system can be conveniently represented as a fluctuating particle, whose position has a stochastic evolution in time. The signature of Brownian motion is that the mean square displacement of the particle grows linearly as a function of time. However, fluctuations of real-world systems are often observed to grow in a nonlinear fashion, which demands to go beyond the Brownian motion paradigm. This situation is referred to as *anomalous diffusion*, where by opposition *normal diffusion* corresponds to regular Brownian motion; more precisely, *subdiffusion* if the mean square displacement grows slower than linearly in time, and *superdiffusion* if it grows faster [3].

Consider for example the random walker depicted in Fig. 1. The walker starts at the origin at time $t = 0$ and holds its position until a random time τ_1, whereupon it makes a random jump to the position $x_1 = \xi_1$. The walker then waits at x_1 up to a random time $\tau_1 + \tau_2$, whereupon it jumps to the new random position $x_2 = \xi_1 + \xi_2$, and the process is renewed. The times $\tau_1, \tau_2, \ldots \tau_n$ are named *waiting times* and the jumps $\xi_1, \xi_2, \ldots, \xi_n$ are named *increments*. Both waiting times and increments are random variables. For the sake of simplicity, we assume that the waiting times are independent and identically distributed, and that the increments are symmetric (i.e., the set $\xi_1, \xi_2, \ldots, \xi_n$ is generated with the same probability as $-\xi_1, -\xi_2, \ldots, -\xi_n$).

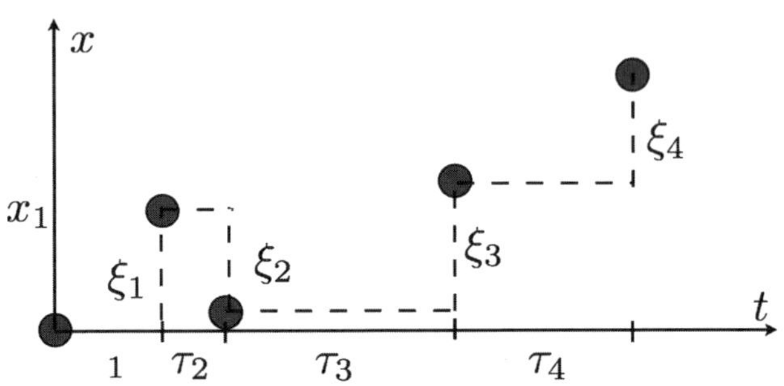

Fig. 1. (Color online) Sketch of a generic random walk in one dimension. The positions of the walker after the jumps are marked with a red dot.

Let $x(t)$ be the position held by the walker at time t. At large times, the fluctuations of $x(t)$ typically become independent of (most of) the microscopic details of the waiting times and increments distributions [1, 2]. In particular, if the increments are independent and identically distributed, and if both $\langle \tau_i \rangle = \tau$ and $\langle \xi_i^2 \rangle = \sigma^2$ are finite [a], the process exhibits a diffusive behavior, i.e., $x(t)$ asymptotically obeys a Gaussian [b] probabil-

[a]Here $\langle \ldots \rangle$ stays for the average over all realizations and $\langle \xi_i \rangle = 0$ because by assumption the process is symmetric.

[b]Because of the Central Limit Theorem.

ity density function as $t \to \infty$, and the memory of the waiting times and increments distributions is kept only in σ^2 and τ:

$$\langle x^2(t) \rangle = 2\frac{\sigma^2}{2\tau}t = 2\,D\,t, \tag{1}$$

where D is the diffusion coefficient [1, 2]. Brownian motion corresponds to the continuum limit [c] of this family of random walks and it is self affine with a characteristic exponent $H = 1/2$. Self-affinity with a characteristic (Hurst) exponent $H > 0$ means that $x(bt)$ and $b^H x(t)$ have the same probability density for $b > 0$.

Anomalous diffusion occurs whenever the process $x(t)$ is self-affine (at least at large times) with $H \neq 1/2$, so that the mean square displacement grows with time as t^{2H} [3]. This is achieved whenever the random walk $x(t)$ falls out of the basin of attraction of the Central Limit Theorem for one of the following reasons:

(1) the increments are not independent and the process is non-Markovian: for instance, the motion of a fluorescent monomer in a polymer is subdiffusive because of the interactions with the other monomers;
(2) the increments can be very large and $\langle \xi_i^2 \rangle = \infty$ (e.g., wild fluctuations in the stock market prices), and/or the waiting times can be very long $\langle \tau_i \rangle = \infty$ (e.g., high retention rates of colloids in porous media);
(3) the increments are not identically distributed: for instance, a small particle in a non-viscous fluid is subject to a random force which induces increments growing with $\sqrt{t}$.

Anomalous diffusion has been reported to occur in a huge number of physical and biological systems [3, 4]. Knowledge of the walker spread $\langle x^2 \rangle$ in unbounded domains typically allows inferring the exponent H. However, real-world systems often involve confined geometries, where the microscopic behavior of the walker (hence the spread) is not directly observable. When the process itself is not accessible, one can try to measure the time taken for the walker to reach some detector, e.g., the outer surface of the domain where particles evolve. Given the stochastic nature of the underlying process, such *first-passage times* are also random variables, whose exact distribution would provide invaluable information on the precise nature of the anomalous dynamics [5]. Despite the fact that characterizing the full

[c] The continuum limit is obtained by taking both τ and σ^2 vanishing small, and keeping $x(t)$ and t fixed. This implies for standard diffusion that the ratio $D = \sigma^2/2\tau$ remains finite.

distribution of first-passage times for an anomalous diffusion process is generally a prohibitively difficult task, some ingenious general results have been recently achieved [6–8]. A unified description of non-Brownian behavior is nonetheless still missing: possibly, one of the reasons is that very different systems can exhibit anomalous diffusion with the same exponent H, which makes the interpretation of the physical origins of such behavior tougher than in the case of (somehow universal) Brownian motion.

In the following, we review the first-passage properties of some physically relevant examples of anomalous walkers, namely, fractional Brownian motion (having non-independent increments, as in case (1)), Lévy flights and Continuous Time Random Walks (having very large increments and waiting times, respectively, as in case (2)), and random acceleration processes (having non identically distributed increments, as in case (3)). Our aim is twofold: on one hand, to illustrate the rather intriguing features of anomalous walkers, and on the other hand to show by which mathematical and numerical tools the properties of the resulting first-passage distributions can be singled out.

2. Polymer translocation and fractional Brownian motion

A prominent goal of biophysics is to set up efficient DNA sequencers. A fundamental prerequisite is the comprehension of the transfer mechanism of a molecular fragment through a nanopore, the so-called *translocation*. The translocation of a single polymer through a pore in a hard wall is the most schematic modelization of this sequencer; yet, the mechanics of this process is still poorly understood [9]. Direct simulations of the dynamics of the entire polymer are cumbersome, because of the large number of degrees of freedom involved [10]. The translocation coordinate $s(t)$, namely the label of the monomer crossing the pore at time t, has been shown to be key in understanding the translocation process [11, 12], which begins when $s = 1$ and ends when $s = N$, i.e., when the first and the last monomer of the chain, respectively, enter the pore (see Fig. 2 left).

Various dynamical regimes of $s(t)$ have been identified: in the absence of driving forces and hydrodynamic effects (free polymer), fluctuations dominate, and $s(t)$ can be regarded as a stochastic process, whose features vary with the polymer length N [11, 12]. Understanding the full dynamics of $s(t)$ represents a challenging problem. A free polymer is characterized by two natural time scales (see Fig. 2 right). First, the intrinsic equilibration time t_{eq} required by the center of mass of the polymer R_{cm} to travel a distance

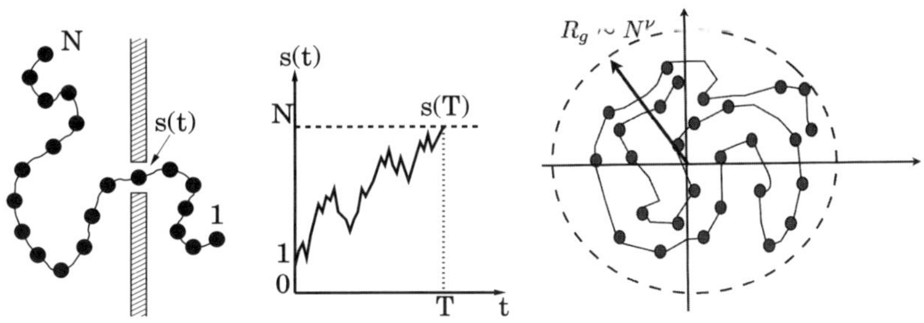

Fig. 2. (Color online) Left: Translocation of a polymer chain through a pore. Middle: The translocation coordinate $s(t)$ denotes the number of the monomer that is crossing the pore at time t. Right: Sketch of a free self-avoiding polymer.

of the order of the typical size of the chain. This size is given by the radius of gyration R_g, which scales as $R_g \sim N^\nu$ in the large N limit. In a good solvent, $\nu = 3/4$ in $2d$ and $\nu \simeq 0.59$ in $3d$ when excluded-volume effects for the monomers are considered; moreover, $\nu = 1/2$ for an ideal ('phantom') polymer. The center of mass diffuses with a diffusion coefficient $\sim 1/N$. Then, $R_{cm} \sim \sqrt{t/N}$. Hence, $\sqrt{t_{eq}/N} \sim R_g$ and $t_{eq} \sim N^{2\nu+1}$ for large N. On the other hand, the translocation time T (much longer than t_{eq}) is the time required by the polymer to go through the pore, so that $s(T) = N$ (Fig. 2 middle). Under the hypothesis that the translocation is a self-affine process, i.e., $s(t) \sim t^H$, with Hurst exponent H, it follows that $T \sim N^{1/H}$.

For short polymers, excluded-volume effects are negligible, $s(t)$ undergoes diffusion, and $T \sim N^2$. However, as N increases, the excluded-volume interactions become relevant and $s(t)$ undergoes subdiffusion, $0 < H < 1/2$ [11]. Numerical simulations (mostly $2d$) support the following conclusions:

i) T and t_{eq} have the same scaling, up to a large prefactor, i.e., $T \sim t_{eq}$. Hence, $H = 1/(1 + 2\nu)$ [11]; (Note that for 'phantom' polymers $s(t)$ always diffuses, even for large N.)

ii) the probability $P(s,t)$ of finding the monomer s in the pore at time t for an infinite chain (i.e., in the absence of boundaries) is Gaussian [10].

Upon gathering these hints from simulations, a natural candidate for $s(t)$ appears to be the fractional Brownian motion (fBm) [13], whose properties are briefly recalled in the following.

2.1. *The fractional Brownian motion*

Fractional Brownian motion (fBm) is a self-affine Gaussian process with $0 < H < 1$ [14, 15]. A Gaussian process is completely defined by its autocorrelation function, which for fBm writes

$$\langle x(t_1)x(t_2)\rangle = D\left(t_1^{2H} + t_2^{2H} - |t_1 - t_2|^{2H}\right) . \tag{2}$$

As a particular case, for $H = 1/2$ the fBm corresponds precisely to regular Brownian motion, namely,

$$\langle x(t_1)x(t_2)\rangle = 2D\min(t_1, t_2) . \tag{3}$$

Note that the process $x(t)$ is Markovian only for $H = 1/2$, whereas for $H \neq 1/2$ the process is non-Markovian. The peculiarity of the fBm is that the process is *homogeneous*, i.e., its increments are identically distributed. To check this property, it is useful to sample a fBm process $x(t)$ at discrete times $t_1 = 1, t_2 = 2, \ldots$, as proposed in Fig. 1 (with $\tau_1 = \tau_2 = \ldots = 1$).

Using Eq. (2), we can compute the autocorrelation function of the Gaussian increments $\xi_t = x(t) - x(t-1)$,

$$\langle \xi_{t_0}\xi_{t_0+t}\rangle = C(t) = D\left[|t-1|^{2H} + (t+1)^{2H} - 2\,t^{2H}\right] , \tag{4}$$

We note that this function is independent of the initial time t_0 and thus the increments are identical Gaussian numbers with variance $2D$, displaying power-law correlations. By taking the limit $t \to \infty$, the power-law decay of these correlations can be easily singled out. For superdiffusive fBm (i.e., $H > 1/2$), $C(t)$ is positively correlated with a decay $t^{-2(1-H)}$. Positive correlations mean that there is a high probability to observe a long sequence of increments with the same sign. For subdiffusive fBm (i.e., $H < 1/2$), $C(t)$ is negatively correlated and decays as $-t^{-2(1-H)}$. Negative correlations mean that there is a high probability to observe a long sequence of increments of alternating signs.

Discrete fBm can be easily implemented numerically by observing that the autocorrelation matrix C is symmetric and has positive eigenvalues; it is thus possible to find a positive and symmetric matrix A such that $C = A^2$. The matrix A is called the square root of C. The paths of fBm are simulated by using the standard procedure for Gaussian correlated processes:

(i) determine A, the square root of C
(ii) a set of increments $\vec{\xi} = \{\xi_1, \xi_2 \ldots\}$ is given by the matrix multiplication $\vec{\xi} = A\vec{\eta}$, where the vector $\vec{\eta} = \{\eta_1, \eta_2, \ldots\}$ is a set of independent Gaussian numbers with variance equal to $2D$ and zero mean.

It is easy to verify that these paths are characterized by the exact autocorrelation function as in Eq. (2) [d]. The outcome is shown in Fig. 3, where we compare an example of fBm generated with the correlated increments $\xi_1, \xi_2, \ldots$ with the Brownian motion generated with the uncorrelated increments $\eta_1, \eta_2, \ldots$. In the subdiffusive case the negative correlations prevent the occurrence of a long sequence of consecutive positive (or negative) increments, whereas in the superdiffusive case the positive correlations enhance the occurrence of such events.

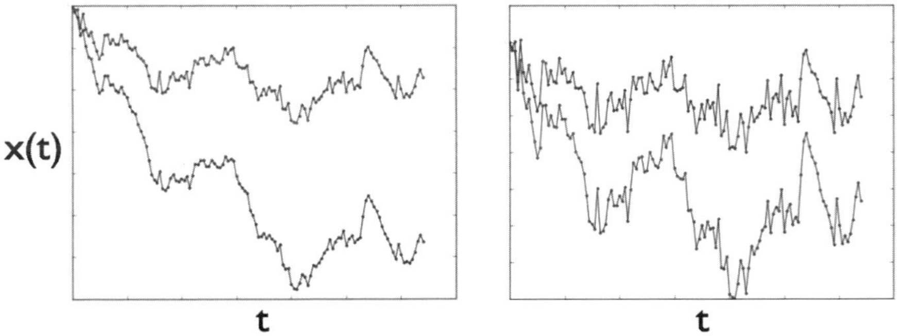

Fig. 3. (Color online) Fractional Brownian motion (blue) generated with the correlated increments $\xi_1, \xi_2, \ldots$ versus Brownian motion (red) generated with the uncorrelated increments $\eta_1, \eta_2, \ldots$. Left: $H = 1/4$, subdiffusive. Right: $H = 3/4$, superdiffusive.

2.2. *First-passage properties: the hitting probability*

On the basis of the considerations exposed above, fBm satisfies conditions *i)* and *ii)* upon taking $H = 1/(1 + 2\nu)$, which relates the Hurst exponent to the scaling parameter ν. Identification of the underlying stochastic process governing the motion of $s(t)$ allows extracting information about translocation without having to deal with the cumbersome study of the full polymer dynamics. In particular, it is possible to answer some relevant questions connected to the first-passage properties of the anomalous diffusing polymer. A natural question is whether a finite polymer chain will ultimately succeed in translocating through a pore, which corresponds to

[d] Unfortunately, this procedure is exact but time-consuming, since for step (i) it demands the full diagonalization of C. Better results are obtained by observing that fBm is homogeneous, so that C is a Toeplitz matrix. Efficient numerical methods for Toeplitz matrices allow avoiding the full diagonalization of C. For instance, one can resort to the Levinson algorithm (for a practical implementation see [16] and [17]).

characterizing the *hitting probability* of the process [18].

Moreover, numerical simulations for a finite chain yet to have completed translocation show that the distribution of $s(t)$ converges to a non-Gaussian form at long times [10]. In particular, this distribution vanishes *nonlinearly* as s^ϕ and $(N-s)^\phi$ at the two boundaries $s=0$ and $s=N$ respectively, with $\phi \simeq 1.44$ in $2d$ [10]. In the next section, we will show that is actually possible to precisely compute the exponent ϕ [18, 19]. Here, we begin our analysis by first addressing the hitting probability.

The translocation coordinate $s(t)$ can be seen as a random walker evolving in a finite box of size L (L being the polymer length), starting from some initial value $X(0)=x$, $0<x<L$, and terminated upon touching either boundary for the first time (Fig. 4 left). We define the hitting probability $Q(x,L)$ as the probability of exiting the domain through the boundary at L, which corresponds to the polymer completing the translocation. If the stochastic process associated with $s(t)$ is self-affine, the only length scale in the problem is L; thus, $Q(x,L)$ is a function only of the scaled variable x/L: $Q(x,L)=Q(x/L=z)$. For a Brownian motion, $Q(z)=z$, a simple linear function [5]. For a generic stochastic process, $Q(z)$ is non trivial (see for example Fig. 4 right) [18].

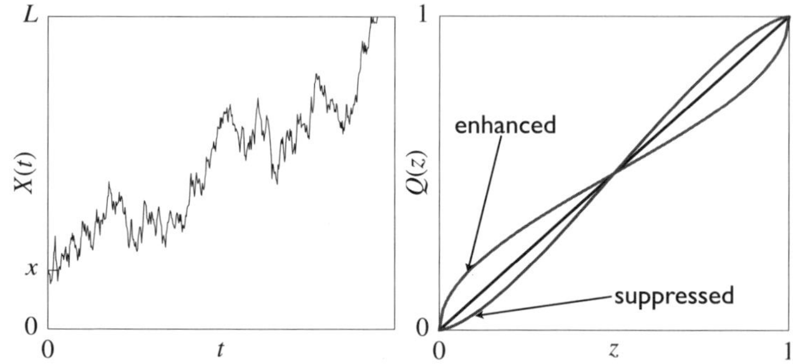

Fig. 4. (Color online) Left. The evolution of a stochastic process initiated at $X(0)=x$ and terminated upon exiting from the box of size L. Right. The hitting probability $Q(z)$ of a self-affine walker. The solid black line corresponds to the solution for standard diffusion. When $z\to 0$, the hitting probability in presence of anomalous diffusion can be enhanced (red curve) or depressed (blue curve).

Observe that the hitting probability is precisely equal to the probability that the global maximum of the process $s(t)$ stays below L until the first time at which it crosses the origin (see Fig. 5). This time is actually the so-called first passage time, during which the process does not change sign, and is therefore related to the persistence of the process. The cumulative distribution of the first-passage time for this problem is the survival probability $S(x,t)$ on the half-axis, i.e., the probability that the walker has not left the positive half-axis up to time t. For self affine processes, $S(t)$ is known to have a power law decay at long times, namely,

$$S(x,t\to\infty)\to\frac{1}{t^{\theta}}, \tag{5}$$

where $\theta>0$ is the so called *persistence exponent* [20]. This non trivial exponent can be computed for a few processes: for instance, the Sparre-Andersen theorem shows that for Brownian motion $\theta=1/2$ [21]. The value of θ for homogeneous and continuous processes, like fBm, is known explicitly, and reads $\theta=1-H$ [15] (precise asymptotic bounds for the first passage distribution of fBm are discussed in [22]). The relation between persistence properties and hitting probability can be then used to infer the properties of the hitting probability close to the origin: it has been shown in [18] that for a self-affine symmetric process the hitting probability vanishes as

$$Q(z\to 0)\sim z^{\theta/H}. \tag{6}$$

In particular, for fBm one gets $Q(z\to 0)\sim z^{(1-H)/H}$, a result which is also supported by direct numerical simulations of fBm paths [18].

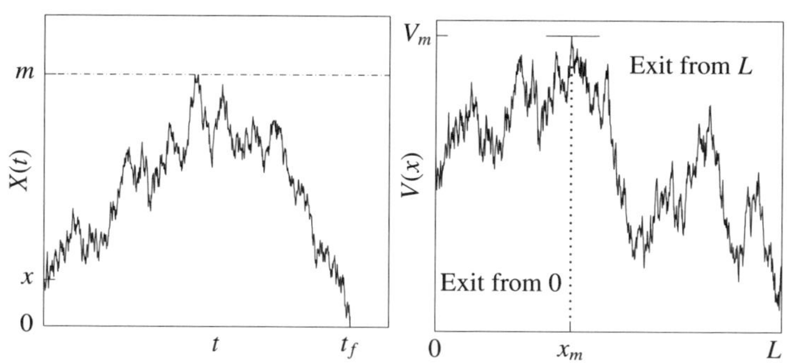

Fig. 5. Hitting probability. Left: A stochastic process starting at x leaves the positive half-axis for the first time at t_f; m denotes its maximum till t_f (see Sec. 2.2). Right: A self-affine disordered potential with maximum at x_m: when L is large, the diffusing particle, starting at $0<x<L$, exits the box through 0 for $x<x_m$ and through L for $x>x_m$ (see Sec. 5).

For long polymers, the excluded volume effects make the translocation process subdiffusive ($H<1/2$), which implies that $Q(z)$ vanishes faster than linearly with an exponent $(1-H)/H$ smaller than 1. This means that the translocation of a real self-avoiding polymer is not only longer (because of subdiffusion), but also less likely with respect to the phantom-polymer ideal case.

2.3. *The effects of the boundaries: the exponent ϕ*

We now address a somewhat complementary first-passage problem: we are interested in computing the probability $P(s,t)$ to find a portion s of the polymer translocated at time t, for a finite chain yet to have completed translocation. Up to a normalization constant, $P(s,t)$ coincides with the *propagator* [e] of the process $s(t)$ in the presence of two absorbing boundaries at $s=0$ and $s=N$. When the method of images holds, the propagator in the presence of a boundary can be expressed as a linear combination of free propagators, i.e., propagators of the same process in the absence of boundaries. For a standard Brownian motion, for instance, the free propagator is the Gaussian distribution, and in the presence of two absorbing boundaries the method of the images gives $P(s,t\to\infty)\propto\sin(\pi s/N)$. By construction, this method predicts that $P(s,t)$ vanishes linearly close to an absorbing boundary.

Numerical simulations of the polymer translocation show in contrast that the distribution of $s(t)$ has a non-Gaussian shape [10]. In particular, $P(s,t)$ vanishes *nonlinearly* as s^ϕ and $(N-s)^\phi$ at the two boundaries $s=0$ and $s=N$ respectively, with $\phi\simeq 1.44$ in $2d$ [10]. This is a rather general feature: actually, it turns out that the method of images systematically fails when the stochastic process displays anomalous diffusion [f]. The simplest geometry for studying the behavior of anomalous walkers close to an absorbing boundary is that of a particle moving on the positive half-axis and killed upon crossing the origin. In this case, the probability density $P_+(s,t)$ of the particle position on the positive half-axis has the scaling form

$$P_+(s,t)\sim t^{-H}R_+(s/t^H), \tag{7}$$

where the function $R_+(y)$ vanishes as $\sim y^\phi$. Based on scaling arguments, it has been shown that for homogeneous processes the exponent ϕ is related

[e] The only difference between the distribution $P(s,t)$ and the propagator is that the latter is not normalized to 1. For a fixed time t, the integral over s of the propagator gives the probability that the process has not been absorbed, i.e., the survival probability in presence of absorbing boundaries.

[f] With the remarkable exception of subdiffusive CTRW, which we will discuss later.

to θ by

$$\phi = \frac{\theta}{H} \tag{8}$$

close to the absorbing boundary $y = 0$ [13]. In particular, for fBm one has $\phi = (1-H)/H$. For $2d$ polymers with excluded-volume effects, using $\nu = 3/4$ one gets $H = 2/5$ and $\phi = 3/2$, which is coherent with the value observed in numerical simulations, namely, $\phi \simeq 1.44$ [10]. In $3d$, using $\nu \simeq 0.59$, one predicts $H \simeq 0.46$ and $\phi \simeq 1.18$. More generally, for all homogeneous processes with Hurst exponent H the exponent ϕ is related to the persistence exponent θ of the process via $\phi = \theta/H$. These results are well supported by direct numerical simulations of fBm paths [13].

2.4. *Perturbation methods*

The exact form of $R_+(y)$ is known for Brownian motion, and reads $R_+(y) = y\exp(-y^2/2)$ (setting the diffusion constant $D = 1$). To gain some analytical intuition of non-Markovian processes and to go beyond the scaling arguments or the numerical simulations, one can develop perturbation schemes around the known Brownian solution [19].

Our starting point will be the propagator $P(x, x_0, t)$ of a standard Brownian motion, defined as the probability to find the particle in the interval $[x, x+dx]$ at time t, knowing that the particle was in x_0 at time 0 and is confined by a region of the space. Using the Gaussianity of Brownian motion we can write the propagator by resorting to the path integral formalism [23], namely,

$$P(x, x_0, t) = \int_{x(0)=x_0}^{x(t)=x} \mathcal{D}[x]\, e^{-\mathcal{S}[x]}\, \Theta[x] \, . \tag{9}$$

Note that we use the field-theoretic notation: $f(x)$ is a function of the variable x, and $\mathcal{S}[x]$ is a functional (the action), depending on the function $x(t')$, with $0 < t' < t$. Here $\Theta[x]$ is an indicator function that is 1 if the path $x(t')$ stays inside the allowed region over the interval $[0, t]$, and 0 otherwise. For regular Brownian motion the action writes as

$$\mathcal{S}[x] = \frac{1}{4D} \int_0^t dt' \left(\partial_{t'} x\right)^2 , \tag{10}$$

and the path integral can be computed.

The first step consists in writing the action for a fractional Brownian motion. To this aim, recall that for any Gaussian process the statistical weight of a path $x(t')$ without any boundary is proportional to $\exp(-\mathcal{S}[x])$, where the action $\mathcal{S}[x]$ is quadratic in x and given by

$$\mathcal{S}[x] = \int_0^t dt_1 \int_0^t dt_2 \, \frac{1}{2} x(t_1) G(t_1, t_2) x(t_2). \tag{11}$$

The kernel $G(t_1, t_2)$ of the action is related to the auto-correlation function of the process via $G^{-1}(t_1, t_2) = \langle x(t_1) x(t_2) \rangle$. Then, for a generic fBm we need to invert Eq. (2). While it is possible to find the exact action for a generic H, it will be very hard to compute the path integral in Eq. (10) in a confined geometry. For this reason, it is preferable to expand the action around the well-known Brownian solution. For $H = 1/2 + \epsilon$ one can write

$$\mathcal{S}[x] = \mathcal{S}^{(0)}[x] + \epsilon \, \mathcal{S}^{(1)}[x] + \ldots, \tag{12}$$

where $\mathcal{S}^{(0)}[x]$ is the action in Eq. (10) and $\mathcal{S}^{(1)}[x]$ reads

$$\mathcal{S}^{(1)}[x] = -\frac{1}{2} \int_0^t dt_1 \int_{t_1}^t dt_2 \, \frac{\partial_{t_1} x(t_1) \partial_{t_2} x(t_2)}{|t_1 - t_2|} - 2\mathcal{S}^{(0)}[x](1 + \log \tau) \,. \tag{13}$$

Note that a regularization for coinciding times $t_1 = t_2 \to \log |t_1 - t_2| = \log \tau$ has been introduced, where $\tau > 0$ is the so-called UV cutoff. By using the expansion $e^{-\mathcal{S}[x]} \sim e^{-\mathcal{S}^{(0)}[x]} \left(1 + \epsilon \, \mathcal{S}^{(1)}[x]\right)$ in Eq. (9), one realizes that the expression of the first correction $P^{(1)}(x, x_0, t)$ is

$$P^{(1)}(x, x_0, t) = \int_{x(0)=x_0}^{x(t)=x} \mathcal{D}[x] \mathcal{S}^{(1)}[x] \, e^{-\mathcal{S}^{(0)}[x]} \, \Theta[x] \,. \tag{14}$$

Here, the path integral is performed over the standard Brownian motion, which is the cornerstone of the proposed perturbation approach.

Setting $H = 1/2 + \epsilon$, $P^{(1)}$ and the scaling function $R_+(y)$ can be computed explicitly to the first order in ϵ [19]. The asymptotic expansions of $R_+(y)$ read

$$\begin{aligned} R_+(y) &\xrightarrow{y \to 0} y \left[1 - 4\epsilon \log y - 2\epsilon(\gamma_{\mathrm{E}} + \log 2) + \ldots\right], \\ R_+(y) &\xrightarrow{y \to \infty} y e^{-y^2/2} \left[1 - 2\epsilon \log y + \epsilon(1 - \log 2 - \gamma_{\mathrm{E}})\right] \end{aligned} \tag{15}$$

and can be recast to

$$\begin{aligned} R_+(y) &\sim y^{\phi}, \quad y \to 0 \\ R_+(y) &\sim y^{\gamma} e^{-\frac{y^2}{2}}, \quad y \to \infty \,, \end{aligned} \tag{16}$$

where the two exponents ϕ and γ are given by

$$\phi = 1 - 4\epsilon + O(\epsilon^2) \,, \quad \gamma = 1 - 2\epsilon + O(\epsilon^2) \tag{17}$$

at the first order in ϵ. Using $H = 1/2 + \epsilon$, from Eq. (8) one expects $\phi = (1-H)/H = 1 - 4\epsilon + O(\epsilon^2)$ for fBm. This is in agreement with the results in Eq. (17), which sets the scaling arguments on a firmer footing.

Remark that the scaling function $R_+(y)$ given in Eq. (15) displays the same leading large-y behavior $\sim e^{-y^2/2}$ as in absence of boundary, at least to $O(\epsilon)$. This behavior can be understood by a simple heuristic argument: the process is not yet aware of the boundary, when being located far from it. This calculation reveals that the process nevertheless knows about the boundary, and $R_+(y)$ correspondingly shows a subleading power-law prefactor y^γ, where γ is a new (independent) exponent, whose expression to the order ϵ is given in Eq. (17).

2.5. *The tagged monomer*

So far, we have considered the first-passage properties of fBm in connection with the problem of polymer translocation. Fractional Brownian motion is actually key to the description of several physical and biological systems [25–28]. Among non-Markovian processes, fBm is rather unique, due to its continuous and homogeneous nature: this entails, in particular, the persistence exponent $\theta = 1 - H$. In order to better understand the relevance of these properties, we conclude by discussing in detail an example where fBm plays an important role: the dynamics of a tagged monomer in a standard Rouse chain. This model describes the conformational dynamics of an ideal polymer. Excluded volume interactions are neglected, and each monomer is subject to a random thermal force and an elastic force induced by the interactions with the neighbours. The Langevin equation for the polymer reads

$$\partial_t U(x,t) = \partial_x^2 U(x,t) + \eta(x,t), \tag{18}$$

where U is the position of monomer x at time t, and η is a Gaussian white noise, namely, $\langle \eta(x,t)\eta(x',t')\rangle = 2T\delta(x-x')\delta(t-t')$; here we will take the temperature $T = 1$. Eq. (18) is linear with respect to U: this means that the process $X(t) = U(x=0,t) - U(x=0,0)$ associated to the tagged monomer is Gaussian and entirely characterized by its autocorrelation function $\langle X(t_1)X(t_2)\rangle$. It is possible to integrate Eq. (18) by resorting to the Fourier transform $U_p(t) = \int dx \exp(-ipx) U(x,t)$. Then, Eq. (18) becomes

$$\partial_t U_p(t) = -p^2 U_p(t) + \eta_p(t), \tag{19}$$

where $\langle \eta_p(t)\eta_{p'}(t')\rangle = 4\pi\,\delta(p+p')\delta(t-t')$, and we get

$$U_p(t) = U_p(0)e^{-p^2 t} + \tilde{U}_p(t) \quad \text{with} \quad \tilde{U}_p(t) = \int_0^t dse^{-p^2(t-s)}\eta_p(s). \tag{20}$$

The autocorrelation function in Fourier space can be computed, and yields

$$\langle \tilde{U}_p(t_1)\tilde{U}_{p'}(t_2)\rangle = 2\pi\,\delta(p+p')\frac{e^{-p^2|t_1-t_2|} - e^{-p^2(t_1+t_2)}}{p^2}. \tag{21}$$

Let us recall that $X(t) = \int dp\left[\tilde{U}_p(t) + (e^{-p^2 t}-1)U_p(0)\right]/(2\pi)$. Now, two relevant cases can be considered: i) an equilibrated initial condition, corresponding to $\langle U_{p'}(0)U_p(0)\rangle = 2\pi\,\delta(p+p')/p^2$, and ii) a flat initial condition, corresponding to $U_p(0)=0$. In the former case,

$$\langle X(t_1)X(t_2)\rangle = \frac{1}{\sqrt{\pi}}\left[(t_1)^{1/2} + (t_2)^{1/2} - |t_1-t_2|^{1/2}\right], \tag{22}$$

and the tagged monomer is precisely a fBm with $H=1/4$, whereas in the second case

$$\langle X(t_1)X(t_2)\rangle = \frac{1}{\sqrt{\pi}}\left[(t_1+t_2)^{1/2} - |t_1-t_2|^{1/2}\right], \tag{23}$$

a Gaussian process with $H=1/4$ that is not a fBm. In case i), the dynamics is equilibrated and the tagged monomer displacements are identically distributed. In case ii), the system evolves from a far-from-equilibrium condition (the flat initial condition) and the jumps are not identically distributed. This provides an example of ageing system: with time, the polymer equilibrates on larger and larger scales. Even in the Gaussian case, very little is known about the first-passage properties of such a process.

3. Continuous time random walks and Lévy flights

The continuous time random walk (CTRW) was initially introduced in [29], and is defined as the process sketched in Fig. 1. In its simplest version, one can assume that increments are independent and identically distributed. While in principle CTRW allows for spatially and/or temporally correlated displacements, a fully decoupled version of the model is often considered, which considerably lightens the treatment. Under such assumption, assuming waiting times to have a finite mean and increments to have a finite variance would asymptotically lead to regular Brownian motion, as recalled above. To mimic the trapping effects often encountered by walkers when traversing complex and/or sticky media, it is common to assume that waiting times have a power-law decay of the kind $\tau^{-1-\alpha}$ with $0 < \alpha < 1$, so that the first moment $\langle\tau\rangle$ is divergent (this implies that very long trapping times become rather frequent). Power-law waiting times coupled with finite-variance spatial increments lead to a subdiffusive behavior, with $H = \alpha/2$. In the limit of a large number of increments, the probability of finding the particle in x at time t converges to

$$P(x,t) = \frac{1}{t^{\alpha/2}} R\left(\frac{x}{t^{\alpha/2}}\right) , \tag{24}$$

where $R(y)$ is the Mittag-Leffler function (thus the process is non-Gaussian). CTRW is a well-known model for anomalous diffusion, for which a large number of exact results are available; for instance, the persistence exponent reads $\theta = 2/\alpha$ [3, 4], and the first-passage distribution for confined geometries has been thoroughly discussed [30].

Conversely, one can conceive a walker with finite-mean waiting times and (independent and identically distributed) increments having a broad distribution $\varphi(\xi)$, with a diverging second moment, i.e.,

$$\varphi(\xi) \sim \frac{c}{|\xi|^{1+\alpha}} , \; |\xi| \gg 1 , \tag{25}$$

with $0 < \alpha < 2$. For the sake of simplicity, we will assume that the process is discrete in time ($\tau_1 = \tau_2 = \ldots = 1$). In this case, the random walk is *Markovian*, *homogeneous* and exhibits a superdiffusive behavior, namely, $x(t) \sim t^{1/\alpha}$, i.e., $H = 1/\alpha$. Power-law distributions such as in Eq. (25) have been initially studied in the early Sixties in economics [31] and in financial theory [32]. Later on, these processes have become very common in Physics, where they have found many applications, encompassing laser-cooling of cold atoms [33], random matrices [34], disordered systems [35], photons in hot atomic vapours [36], and many others. One striking feature of such processes is that their statistical behavior is dominated by a few rare and very large events, whose occurrence is thus governed by the *tail* of the distribution.

Analogously as for Brownian motion, after a large number of steps, we expect the statistical properties of $x(t)$ to become independent of the details of $\varphi(\xi)$, except for the index α and the constant c. In this limit, the process is named Lévy flight. In particular, the probability density of finding the particle in x at a time t converges to

$$P(x,t) = \frac{1}{t^{1/\alpha}} R\left(\frac{x}{t^{1/\alpha}}\right) , \tag{26}$$

where the function $R(y)$ is a Lévy stable distribution (thus the process is non-Gaussian). Although the Fourier transform of $R(y)$ (the characteristic function) has a very simple expression, in general there is no closed-form expression for $R(y)$. An asymptotic expansion shows that

$$R(y) \sim \frac{c}{y^{1+\alpha}} + \mathcal{O}(y^{-1-2\alpha}) \,. \tag{27}$$

The amplitude c plays the same role as σ^2 for the Brownian motion.

3.1. *First-passage properties of Lévy Flights*

By taking the continuous time limit of the hopping process described above, one can derive the so-called Fractional Fokker-Planck equation for a Lévy flight propagator [3], namely,

$$\frac{\partial}{\partial t}P(x,t) = \frac{\pi\, c}{\sin\left(\frac{\alpha\pi}{2}\right)\Gamma(\alpha+1)}\frac{\partial^\alpha}{\partial|x|^\alpha}P(x,t)\;,\quad P(x,t=0)=\delta(x)\;, \tag{28}$$

where $\frac{\partial^\alpha}{\partial|x|^\alpha}$ is the Riesz-Feller derivative of fractional order $\alpha > 0$ [37]. In the Fourier space, $\frac{\partial^\alpha}{\partial|x|^\alpha}$ has the simple form $-|k|^\alpha$, whereas in the real space it takes an integral representation involving a singular kernel of power-law form, which intuitively stems from the long-range increments distribution.

The first-passage properties of Lévy flights, which typically involve eigenvalue problems for the operator $\mathcal{H} = \frac{\pi\, c}{\sin(\frac{\alpha\pi}{2})\Gamma(\alpha+1)}\frac{\partial^\alpha}{\partial|x|^\alpha}$, are particularly difficult to solve, precisely due to the non-local nature of the spatial increments (infinite variance implies that walkers can jump across a boundary without touching it, which is not the case of regular Brownian motion; as a consequence, the method of images does not apply [38]). The solution is known in the absence of boundaries, where the Fourier representation is diagonal and the eigenfunctions are simple plane waves. On the contrary, only a very limited number of results are known for Lévy flights on bounded and semi-bounded domains, even for the very simplest geometries; for a survey, see, e.g., [37–41]. The Sparre-Andersen theorem shows that symmetric Lévy flights share the same persistence exponent as the Brownian motion, namely, $\theta = 1/2$ [21]. As a particular case, for $\alpha = 2$ one recovers the standard Laplacian, for which exact results are often known even in confined geometries.

A possible strategy to gain some insight on the behavior of a Lévy flight in the presence of boundaries is to use a perturbation scheme around the $\alpha = 2$ case [42], not dissimilarly as done for fBm. Our starting point will be again the regular Brownian motion. Using the Markovian property of

Brownian motion, we can write the Fokker-Planck equation for the propagator P in the familiar Schrödinger form

$$\partial_t P(x, x_0, t) = \mathcal{H} P(x, x_0, t) , \tag{29}$$

$$P(x, x_0, t = 0) = \delta(x - x_0) , \tag{30}$$

where the operator $\mathcal{H}$ is the standard Laplacian. In Quantum Mechanics, Eq. (29) corresponds to the Schrödinger equation of the element (x, x_0) of the density matrix P at the temperature $1/t$, whose general solution reads

$$P(x, x_0, t) = \int dq\, \psi_q^*(x_0) \psi_q(x) e^{E(q)t} , \tag{31}$$

where $E(q)$ are the eigenvalues and $\psi_q(x)$ the associated eigenfunctions of the operator $\mathcal{H}$. These are the solutions of the eigenvalue problem $\mathcal{H}\psi_q(x) = E(q)\psi_q(x)$, *with the appropriate boundary conditions*, and satisfy the ortho-normality $\int \psi_q(x)\psi_{q'}^*(x)dx = \delta(q - q')$ and closure $\int \psi_q(x)\psi_q^*(x')dq = \delta(x - x')$ relations. Here, the domain of integration over x depends on the boundary conditions and the integration over q is meant over the entire spectrum. If the spectrum is discrete, the integral is replaced by a discrete sum.

Then, we write the Fokker-Planck equation associated to the Lévy flight propagator. For $\alpha = 2 - \epsilon$ the operator $\mathcal{H}$ and the propagator can be expanded in powers of ϵ, namely,

$$\begin{aligned} P(x, x_0, t) &= P^{(0)}(x, x_0, t) - \epsilon P^{(1)}(x, x_0, t) + \mathcal{O}(\epsilon^2) \\ \mathcal{H} &= \mathcal{H}_0 - \epsilon \mathcal{H}_1 + O(\epsilon^2) , \end{aligned} \tag{32}$$

where $P^{(0)}(x, x_0, t)$ is the propagator associated to $\mathcal{H}_0 = \partial_x^2$, with the prescribed boundary conditions. The expression of the first correction $P^{(1)}(x, x_0, t)$ in Eq. (32) is known from Quantum Mechanics, and reads

$$P^{(1)}(x, x_0, t) = \int_q \int_{q'} \psi_q(x) \psi_{q'}^*(x_0) \frac{e^{E(q)t} - e^{E(q')t}}{E(q') - E(q)} \langle q|\mathcal{H}_1|q'\rangle, \tag{33}$$

where we use the notation $\int_q \equiv \int dq$ and it is understood that $\psi_q(x)$ are the eigenvectors and $E(q)$ the corresponding eigenvalues of $\mathcal{H}_0$. For the matrix elements $\langle q|\mathcal{H}_1|q'\rangle$ we use the bra-ket notation, borrowed from Quantum Mechanics, with $\langle x|q\rangle = \psi_q(x)$. The formula in Eq. (33) is the cornerstone of the subsequent perturbation approach.

Setting $\alpha = 2 - \epsilon$, one can compute $P^{(1)}$ and $R_+(y)$ at the first order in ϵ. Close to the origin, one gets [42]

$$R_+(y) \sim y - \frac{\epsilon}{2} y \log(y) \sim y^{\alpha/2} , \tag{34}$$

in agreement with the results $\phi = \theta/H = \alpha/2$. The (non-normalized) propagator in presence of a single boundary has been studied for generic α [39]: it has been shown that $R_+(y)$ vanishes as $\sim y^{\alpha/2}$ when $y \to 0$. This perturbative approach allows conjecturing the exact behavior of the tail of $R_+(y)$, which governs the statistics of rare events:

$$R_+(y) = \frac{c_+}{y^{1+\alpha}} + \begin{cases} \frac{d_+}{y^{2+\alpha}} + o(y^{-2-\alpha}) \ , \ 2 > \alpha > 1 \ , \\ \\ \frac{d_+}{y^{1+2\alpha}} + o(y^{-1-2\alpha}) \ , \ 1 > \alpha > 0 \ , \end{cases} \tag{35}$$

where d_+ is a constant. The leading term $R_+(y) \propto y^{-1-\alpha}$ is coherent with previous investigations [39]. The perturbative approach allows also the amplitude c_+ to be explicitly computed [42], namely, $c_+ = 2c$, where c is the amplitude of the tail of the jump distribution in Eq. (25). Observe that the first sub-leading corrections in Eq. (35) are also affected by the absorbing wall, as seen by comparison with the free case of Eq. (27). These results are supported by numerical simulations.

4. The random acceleration model

The random acceleration model is a stochastic process defined by

$$\ddot{x}(t) = \eta(t), \tag{36}$$

where $\eta(t)$ is a Gaussian white noise with $\langle \eta(t) \rangle = 0$ and $\langle \eta(t)\eta(t') \rangle = 2D\delta(t-t')$ [43, 44]. The process starts at $x(0) = 0$ with the initial velocity $v(0) = 0$, and it is superdiffusive, with $x^2(t) \sim t^3$, i.e., $H = 3/2$. The random acceleration process in the single variable $x(t)$ is non-Markovian, whereas the position-velocity process is Markovian and does not depend on the past history. The increments $\xi_t = x(t) - x(t-1) = \sum_0^t \eta(t)$ are Gaussian numbers growing as $\sqrt{t}$, so that ξ_t are random numbers not identically distributed. The random acceleration process appears for instance in the continuum description of the equilibrium Boltzmann weight of a semiflexible polymer chain with non-zero bending energy [44]. It can also describe the steady state profile of a (1 + 1)-dimensional Gaussian interface [45] with dynamical exponent $z = 4$, and the continuum version of the Golubovic-Bruinsma-Das Sarma-Tamborenea model [46]. As we will see in the following, in addition to being relevant in many applications the random acceleration model represents a simple, yet nontrivial, non-Markov process where some observables of physical interest can be explicitly computed.

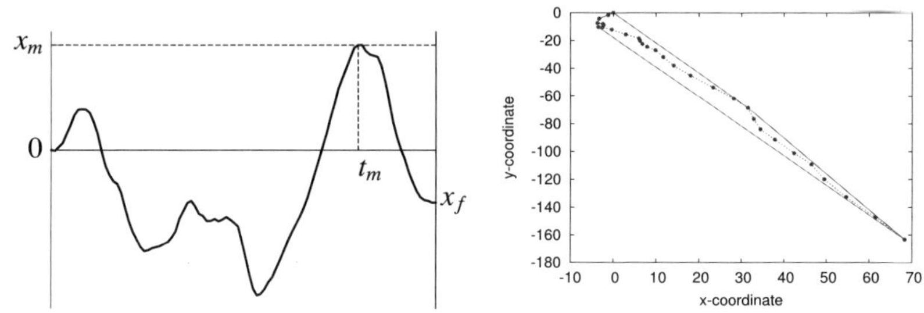

Fig. 6. (Color online) Left: A realization of a random acceleration process reaching its maximum x_m at t_m. Right: Convex hull of 30 points generated using the discrete random acceleration process. The convex hull is constructed using the Graham span algorithm [49].

4.1. *Extreme statistics*

Recently, the exact distribution of the time t_m at which a random acceleration process attains its maximum has been computed [47] (see Fig. 6 left). This result represents for the random acceleration process the equivalent of the arcsine law emerging when considering the first-passage properties of Brownian motion and Lévy flights. Further on, we will show that the fluctuations of t_m appear when computing the hitting probability of a particle moving in a random self-affine potential [18], and also in connection with the properties of the convex hull of a two dimensional random walk [48] (see Fig. 6 right).

Consider a process $x(t)$, starting from $x(0) = 0$ and observed over a fixed time interval $[0, T]$. Let t_m denote the time at which the process achieves its maximum value x_m during the interval $[0, T]$ (see Fig. 6 left). Since the only time scale in the problem is T, the probability density $p(t_m|T)$ must have the scaling form

$$p(t_m|T) = \frac{1}{T} p\left(\frac{t_m}{T}\right), \tag{37}$$

where the scaling function $p(z)$ is defined over $0 \leq z \leq 1$. For a random acceleration process, it is possible to compute $p(z)$ with two different boundary conditions [47]: the integral of a Brownian bridge and the integral of a free Brownian motion.

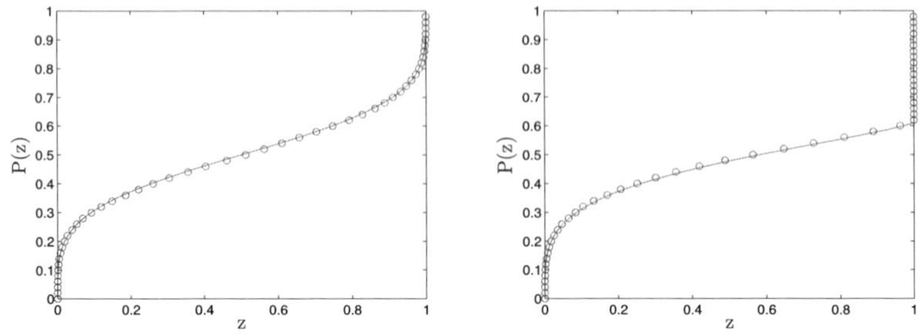

Fig. 7. (Color online) Simulation results for the cumulative distribution $P(z)$ compared to the analytical results. Left: Integral of a Brownian Bridge, the solid line corresponds to Eq. (39). Right: Integral of a free Brownian Motion, the solid line corresponds to Eq. (41).

4.1.1. *Integral of a Brownian bridge*

When the final velocity of the random acceleration process vanishes, $v_f = 0$, the process formally corresponds to the integral of a Brownian bridge. In this case, it has been shown that

$$p(z) = \frac{\Gamma(1/2)}{\Gamma^2(1/4)} \frac{1}{[z(1-z)]^{3/4}}, \tag{38}$$

which is evidently *symmetric* around the mid-point $z = 1/2$ and diverges at the two end-points as $z^{-3/4}$ and $(1-z)^{-3/4}$, respectively [47]. It is also useful to consider the cumulative distribution $P(z) = \int_0^z p(z')\,dz'$, which reads

$$P(z) = \frac{\Gamma(1/2)}{\Gamma^2(1/4)} B_z\left(\frac{1}{4}, \frac{1}{4}\right). \tag{39}$$

Here $B_z(p,q) = \int_0^z x^{p-1}(1-x)^{q-1}\,dx$ is the incomplete Beta function.

4.1.2. *Integral of a free Brownian motion*

When v_f is arbitrary, the random acceleration process formally coincides with the integral of a free Brownian motion. In this case, it is possible to show that

$$p(z) = C\,\delta(z-1) + \frac{(1-C)}{\pi\sqrt{2}} z^{-3/4}(1-z)^{-1/4}, \tag{40}$$

where $C = 1 - \sqrt{\frac{3}{8}}$ [47]. The density is thus *asymmetric* around the mid-point $t_m = T/2$ (i.e., $z = 1/2$), and the maximum may either occur at some time *strictly* shorter than T, namely $t_m < T$ (i.e., $z < 1$), or with a finite non-vanishing probability $C = 0.387628..$ at the end point of the interval $t_m = T$ (or equivalently $z = 1$). In other words, roughly 38.67% of all trajectories, starting at $x(0) = 0$ and $v(0) = 0$, achieve their maximum only at the end of the interval $[0, T]$. The corresponding cumulative distribution $P(z) = \int_0^z p(z')\,dz'$ is given by

$$P(z) = C\,\Theta(z-1) + \frac{(1-C)}{\pi\sqrt{2}}\,B_z\left(\frac{1}{4}, \frac{3}{4}\right), \tag{41}$$

where $\Theta(z-1)$ vanishes for $z < 1$ and is equal to 1 for $z = 1$, i.e, $P(z)$ exhibits a discontinuous jump at $z = 1$ from $1 - C = \sqrt{3/8}$ to 1.

A plot of $P(z)$ for both boundary conditions is provided in Fig. 7, where it is also compared to direct simulation results: an excellent agreement between analytical and numerical results is found.

4.2. *The convex hull of a random acceleration process*

It has been recently shown that the problem of computing the mean perimeter and area of the convex hull of *any two dimensional* stochastic process can be mapped, using Cauchy's formulae [50], to the problem of computing the moments of the maximum and the time at which the maximum occurs for the associated one dimensional component stochastic process [51, 52]. Following this general mapping, the average perimeter and area of the convex hull are given by

$$\langle L(t)\rangle = 2\pi\langle x_m(t)\rangle \tag{42}$$

$$\langle A(t)\rangle = \pi\left[\langle x_m^2(t)\rangle - \langle y^2(t_m)\rangle\right], \tag{43}$$

where $y(t_m)$ is the ordinate of the process at t_m, i.e., when the displacement along the x direction is maximal. Using the results obtained for the extreme statistics of the one dimensional random acceleration process it is possible to show that the mean perimeter and area of the convex hull associated to a two-dimensional random acceleration process of duration T, starting from the initial zero velocity conditions $v_x(0) = v_y(0) = 0$, are given by the following exact expressions

$$\langle L(T)\rangle = \sqrt{\frac{3\pi}{2}}\,T^{3/2} = (2.1708\ldots)\,T^{3/2} \tag{44}$$

$$\langle A(T)\rangle = \frac{5\pi}{192}\sqrt{\frac{3}{2}}\,T^3 = (0.100199\ldots)\,T^3. \tag{45}$$

5. A digression on the hitting probability

We have previously discussed the fundamental role played by the hitting probability in the context of polymer translocation. More generally, the hitting probability $Q(x, L)$ of a particle undergoing anomalous diffusion is key to understanding phenomena as diverse as the classical gambler's ruin problem in finance and risk management [5], the transport of charge carriers in conductors with disordered impurities [35], or reaction kinetics in cells [7]. In this latter case, it is important to discuss the behavior of a particle in a correlated disordered potential $V(x)$. To provide an example, we consider the case of a strongly correlated self-affine disordered potential $V(x) \sim x^{H_V}$ with $H_V > 0$. For $H_V = 1/2$, the potential $V(x)$ is a trajectory of a Brownian motion in space and one recovers the Sinai model. In this class of models, the maximal barrier that the particle has to overcome grows with the size of the system as L^{H_V}. By the Arrhenius law for the activated dynamics, the time required for a particle diffusing in $V(x)$ to overcome an energy barrier scales as $t \sim e^{V(x)}$. We deduce that

$$x(t) \sim [\log(t)]^{1/H_V}, \tag{46}$$

where for $H_V = 1/2$ one recovers the well-known $x(t) \sim \log^2(t)$. Thus, the particle motion is a self-affine process as a function of the variable $\log(t)$, with a Hurst exponent $H = 1/H_V$. It is actually possible to show that for this problem the hitting probability is related to the statistics of the *maximum location* x_m of the process $V(x)$ (see Fig. 5 right) [18, 53]. The location of the extreme value of a process plays an important role in many situations. In the context of finance, for example, a trader is interested in knowing the time at which the stock price is at its highest. For symmetric Lévy flights and Brownian motion, the distribution of such maxima is the celebrated Lévy arcsine law [54], namely,

$$p(x_m) = \frac{1}{\pi} \frac{1}{\sqrt{x_m(L - x_m)}}. \tag{47}$$

Computing this quantity for a generic process is a formidable task (we have actually shown that $p(x_m)$ can be computed for the random acceleration process). Using the relation between the hitting probability and the maximum location it is possible to prove that the survival probability of a particle in a correlated disorder potential $V(x)$ behaves as

$$S(x, t \to \infty) \to \frac{1}{[\log(t)]^{\theta_V/H_V}}, \tag{48}$$

where θ_V is the persistence exponent of the process $V(x)$. For instance, in the Sinai model, using $\theta_V = 1/2$, $H_V = 1/2$ we get $S(t) \sim 1/\log(t)$, in agreement with the exact result [55, 56]. A random acceleration process potential would be self-affine with $H_V = 3/2$ and persistence exponent $\theta_V = 1/4$ [44]. Thus, for this potential the survival probability up to time t would decay as $\sim (\log t)^{-1/6}$, with $\theta = 1/6$.

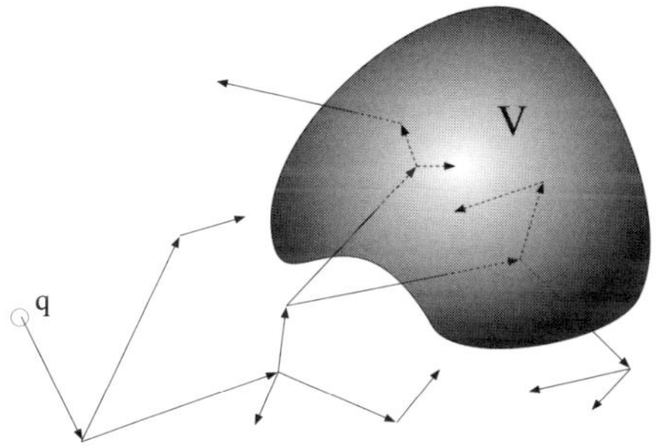

Fig. 8. A random walk with branching and death, starting from a source q and traversing a volume V.

6. Conclusions and perspectives

We have analyzed some of the first-passage properties of a few relevant anomalous diffusion processes, namely, fractional Brownian motion, Lévy flights and Continuous Time Random Walks, and the random acceleration model. To this aim, different and often complementary approaches have been adopted: scaling arguments, numerical simulations, perturbation schemes and mapping to extreme values statistics. Several extensions of the results recalled here are possible: for instance, the analytical and numerical tools described above are invaluable also in the analysis of the first-passage properties of branching random walks with death (see Fig. 8), which emerge in relation with, e.g., the spread of epidemics, the migration of reproducing bacteria and the multiplication of neutrons in nuclear reactors [57, 58]. In this context, some results have been recently obtained concerning the statistical fluctuations of the convex hull of a branching Brownian motion (via the mapping based on Cauchy's formulae) [59], and the residence time spent by branching processes within a given volume (by resorting to a path-integral formulation) [60, 61].

References

1. B. D. Hughes, *Random walks and random environments* Vol. I (Clarendon Press, Oxford, 1995).
2. G. H. Weiss, *Aspects and applications of the random walk* (North Holland Press, Amsterdam, 1994).
3. R. Metzler, J. Klafter, Phys. Rep. **339**, 1 (2000).
4. R. Metzler, J. Klafter, J. Phys. A: Math. Gen. **37**, R161 (2004).
5. S. Redner, *A Guide to First-Passage Processes*, Cambridge University Press, New York, (2001).
6. S. Condamin, O. Bénichou, V. Tejedor, R. Voituriez, J. Klafter, Nature **450**, 77 (2007).
7. S. Condamin, V. Tejedor, R. Voituriez, O. Bénichou, J. Klafter, Proc. Natl. Acad. Sci. U.S.A. **105**, 5675 (2008).
8. O. Bénichou, C. Chevalier, J. Klafter, B. Meyer, R. Voituriez, Nature Chemistry **2**, 472 (2010).
9. J. J. Kasianowicz, E. Brandin, D. Branton, D. W. Deamer, Proc. Natl. Acad. Sci. U.S.A. **93**, 13770 (1996).
10. C. Chatelain, Y. Kantor, M. Kardar, Phys. Rev. E **78**, 021129 (2008).
11. J. Chuang, Y. Kantor, M. Kardar, Phys. Rev. E **65**, 011802 (2001).
12. Y. Kantor, M. Kardar, Phys. Rev. E **69**, 021806 (2004).
13. A. Zoia, A. Rosso, S.N. Majumdar, Phys. Rev. Lett. **102**, 120602 (2009).
14. B. B. Mandelbrot, J. W. van Ness, SIAM Rev. **10**, 422 (1968).
15. J. Krug, H. Kallabis, S. N. Majumdar, S. J. Cornell, A. J. Bray, C. Sire, Phys. Rev. E **56**, 2702 (1997).
16. T. Dieker, 'Simulation of fractional Brownian motion', master thesis, *http://www2.isye.gatech.edu/ adieker3.*
17. R. Garcia-Garcia, A. Rosso, G. Schehr, Phys. Rev. E **81**, 010102(R) (2010).
18. S. N. Majumdar, A. Rosso, A. Zoia, Phys. Rev. Lett. **104**, 020602 (2010).
19. K. J. Wiese, S. N. Majumdar, A. Rosso, Phys. Rev. E **83**, 061141 (2011).
20. S. N. Majumdar, Current Science **77**, 370 (1999).
21. E. Sparre-Andersen, Math. Scand. **1**, 263 (1953); Math. Scand. **2**, 195 (1954).
22. G. M. Molchan, Comm. Math. Phys., **205**, 97 (1999).
23. S. N. Majumdar, Curr. Sci. **89**, 2076 (2005).
24. J.-H. Jeon, A. V. Chechkin, R. Metzler, Europhys. Lett. **94**, 20008 (2011).
25. J. Szymanski, M. Weiss, Phys. Rev. Lett. **103**, 038102 (2009).
26. S. C. Weber, A. J. Spakowitz, J. A. Theriot, Phys. Rev. Lett. **104**, 238102 (2010).
27. J.-H. Jeon, H. Monne, M. Javanainen, R. Metzler, Phys. Rev. Lett. **109**, 188103 (2012).
28. G. Oshanin, A. Rosso, G. Schehr, Phys. Rev. Lett. **110**, 100602 (2013).
29. E. W. Montroll, G. H. Weiss, J. Math. Phys. **6**, 167 (1965).
30. S. Condamin, O. Bénichou, J. Klafter, Phys. Rev. Lett. **98**, 250602 (2007).
31. V. Pareto, *Cours d'économie politique.* Reprinted as a volume of *Oeuvres complètes* (Droz, Geneva, 1896-1965).
32. B. B. Mandelbrot, Journal of Business **36** 394 (1963).

33. M. F. Shlesinger, G. M. Zaslavsky, U. Frisch Eds., *Lévy Flights and Related Topics in Physics* (Proc. Nice, France; Springer Verlag, 1994).
34. G. Biroli, J.-P. Bouchaud, M. Potters, Europhys. Lett. **78**, 10001 (2007).
35. J.-P. Bouchaud, A. Georges, Phys. Rep. **195**, 127 (1990).
36. N. Mercadier, W. Guérin, M. Chevallier, R. Kaiser, Nature Physics **5**, 602 (2009).
37. A. Zoia, A. Rosso, M. Kardar, Phys. Rev. E **76** 021116 (2007).
38. A. V. Chechkin, R. Metzler, V. Y. Gonchar, J. Klafter, L. V. Tanatarov, J. Phys. A **36**, L537 (2003).
39. G. Zumofen, J. Klafter, Phys. Rev. E **51**, 2805 (1995).
40. S. V. Buldyrev, S. Havlin, A. Ya. Kazakov, M.G.E. da Luz, E. P. Raposo, H. E. Stanley, and G. M. Viswanathan, Phys. Rev. E **64**, 041108 (2001).
41. S. N. Majumdar, Physica A **389** (20), 4299 (2010).
42. R. Garcia-Garcia, A. Rosso, G. Schehr, Phys. Rev. E **86**, 011101 (2012).
43. Ya. G. Sinai, Theor. Math. Phys. **90**, 219 (1992).
44. T. W. Burkhardt, J. Phys. A: Math. Gen. **26**, L1157 (1993).
45. S. N. Majumdar, A. J. Bray, Phys. Rev. Lett. **86**, 3700 (2001).
46. L. Golubovic, R., Bruinsma, Phys. Rev. Lett. **66**, 321 (1991); S. Das Sarma, P. Tamborenea, Phys. Rev. Lett. **66** 325 (1991).
47. S. N. Majumdar, A. Rosso, A. Zoia, J. Phys. A: Math. Theor. **43**, 115001 (2010).
48. A. Reymbaut, S. N. Majumdar, A. Rosso, J. Phys. A: Math. Theor. **44**, 415001 (2011).
49. R. Graham, Inf. Process. Lett. **1**, 132-133 (1972).
50. A. Cauchy, Mem. Acad. Sci. Inst. Fr. **22**, 3 (1850); see also the book by L. A. Santaló, *Integral Geometry and Geometric Probability* (addison-Wesley, Reading, MA, 1976).
51. J. Randon-Furling, S. N. Majumdar, A. Comtet, Phys. Rev. Lett. **103**, 140602 (2009).
52. S. N. Majumdar, A. Comtet, J. Randon-Furling, J. Stat. Phys. **138**, 955 (2010).
53. G. Oshanin, S. Redner, Europhys. Lett. **85**, 10008 (2009).
54. A. Comtet, C. Texier, in *Supersymmetry and Integrable Models*, Aratyn *et al.* (eds.), Lecture Notes in Physics **502**, 313, Springer (1998). cond-mat:/9707313.
55. Y. G. Sinai, Theory Probab. Appl. **27**, 256 (1982).
56. A. Comtet, D.S. Dean, J. Phys. A: Math. Gen. **31** 8595 (1998).
57. T. E. Harris, *The Theory of Branching Processes* (Springer, Berlin, 1963).
58. P. Jagers, *Branching Processes with Biological Applications* (Wiley Series in Probability and Mathematical Statistics, London, 1975).
59. E. Dumonteil, S. N. Majumdar, A. Rosso, A. Zoia, Proc. Natl. Acad. Sci. U.S.A. **110**, 4239 (2013).
60. A. Zoia, E. Dumonteil, A. Mazzolo, EuroPhys. Lett. **100** 40002 (2012).
61. A. Zoia, E. Dumonteil, A. Mazzolo, S. Mohamed, J. Phys. A: Math. Theor. **45** 425002 (2012).

Chapter 4

First-Passage Times of Intermittent Random Walks

O. Bénichou[1] and R. Voituriez[1,2]

[1] *Laboratoire de Physique Théorique de la Matière Condensée, CNRS/UPMC, 4 place jussieu 75005 Paris, France*

[2] *Laboratoire Jean Perrin, CNRS/UPMC, 4 place jussieu 75005 Paris, France*

This chapter discusses intermittent target search strategies, which combine phases of slow motion, allowing a searcher to detect a target, and phases of fast motion during which targets cannot be detected. It is first shown that such intermittent random walks are widely observed at various scales, ranging from animals looking for food to the intermittent transport patterns involved in the search for a specific sequence on DNA by a protein. Next, generic stochastic models are introduced, which show that intermittent random walks are efficient strategies that enable the minimization of the mean first-passage time to a target. This suggests that the intrinsic efficiency of intermittent search strategies could justify their frequent observation in nature.

1. Introduction

How long does it take a "searcher" to find a "target" of unknown location? This apparently simple question arises in various contexts, such as the search for victims in avalanches, the search for food by animals, or at the microscopic scale the search for a target sequence on DNA by a regulatory protein. In all these examples, the search time is a limiting quantity that often needs to be optimized.

The question of optimizing search strategies was formalized early by US-Navy mathematicians during world war II : the purpose was then to locate as fast as possible the enemy submarines. The problem was to determine the optimal spatial distribution of searching ships, compatible with a given likelihood map of the target position. More recently, these methods have been generalized in the context of rescue operations by coast guards, and also employed for example to recover the wreck of the submarine USS Scorpio in 1968. However, such methods are useful only when at least a partial information about the target location is available.

In the opposite case where no information on the target location is available so that no specific region should be favored, the searcher can adopt a random trajectory. Such random trajectories are also involved usually at the microscopic scale when the searcher is subject to thermal fluctuations, as in the case of a reactive molecule searching for its reaction partner. These so-called "random search strategies" have been the subject of an increasing number of publications in statistical physics over the recent years, including influential works from the group of G. Viswanathan [1–3].

These authors consider a searcher that is able to detect targets *while it moves*, and whose trajectory is made of a sequence of straight ballistic moves of random direction and random length l. Viswanathan and coworkers studied the case where the probability distribution $p(l)$ is a Lévy law defined by $p(l) \propto l^{-\mu}$, where $\mu > 1$ is a given parameter. In the particular case of so called "revisitable" targets, which immediately reappear at the same position after each visit of the searcher, they showed that the Lévy law that minimizes the search time corresponds to $\mu \approx 2$.

This situation of revisitable targets is however rather artificial since in most of real situations targets disappear once they are discovered : this is for instance the case of a prey captured by a predator. In this last more realistic situation, which we will consider in this chapter, the model of Viswanathan and coworkers leads to an expected result : the search time is minimized by a simple uninterrupted ballistic motion[a]. We present here search strategies that are alternative to the celebrated Lévy strategies, which in fact are optimal only in the particular case of revisitable targets.

More precisely, we address the following question : what happens if the searcher is not able to detect targets *while it moves*? This situation is quite common since in general motion degrades perception abilities. Often, motion and detection are in fact intrinsically incompatible : it is the case of targets hidden in the ground, as would be the case of keys lost in a sandy beach (Fig. 1). Recent ideas, introduced in [5] and reviewed in [6, 7], put

[a]Note that in the case of a single target, Levy strategies are still not optimal, and outperformed by persistent random walks with exponentially distributed mean free path l [4].

forward a new type of random search, exemplified by the searcher for keys of figure 1, called intermittent search strategies : the searcher combines phases of careful inspection to detect the target and phases of rapid motion during which target detection is not possible. As we shall see, intermittent search strategies are widely observed in nature at various scales. One can then suggest that the reason why such strategies are so common in nature is because they are *extremely efficient*.

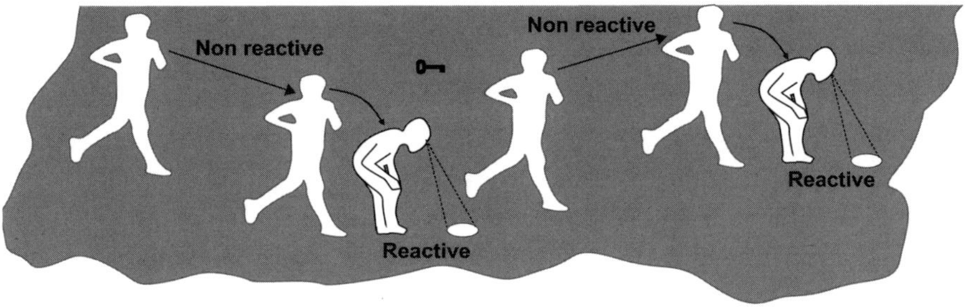

Fig. 1. Intermittent search strategy : the searcher alternates phases of careful inspection to detect the key, and phases of fast motion during which he is unable to detect the target.

2. Animals looking for food

The target search problem is especially relevant in the context of animals looking for food. The efficiency of this search behavior is of vital importance for the animal, and one can suggest that fast behaviors have been selected by evolution. Analysing the search behavior of animals and assessing its efficiency is therefore very important, and behavioral ecologists have embarked in the study of the behaviors of animals looking for food [8–10]. Such studies schematically put forward two extreme types of search strategies. The first one corresponds to predators that search for preys while they move. It is the case for example of pelagic fish like tuna that never stop swimming. The second strategy, called ambush strategy, corresponds to animals such as rattle snakes, which stay immobile for long periods of time, waiting for a prey. These two strategies in fact should be considered as extreme situations.

A few decades ago, O'Brien and colleagues [8] showed that in fact the search behavior of many species couldn't be described by these extreme strategies. They observed, first on the example of planktivorous fish and next on examples of lizards and bird species, that many animals in fact displayed an intermediate behavior, named "saltatory", defined by the following observations : the animal alternates phases of fast motion and "stationary" phases, during which the animal moves much less, but during which targets can be detected. The duration of each of the phases can vary significantly from species to species, but observations reveal the existence of correlations between the times spent in each phase, independently of the size or weight of the animal [4].

To explain such observations, behavioral ecologists have developed models based in general on the idea of maximization of the energy gain [8] : the energy provided by food consumption should compensate at best the energy lost during the search for preys. Most of these models involve parameters that are difficult to access experimentally. In addition, they are specific to a given species. Is it possible to propose a different approach relying on a generic modeling of such saltatory search behaviors?

2.1. *Model of intermittent search*

A model of random search accounting for available experimental observations has been proposed in [5, 11]. This model relies directly on the idea of intermittent search suggested on the example of the key searcher. The searcher alternates between two states (Fig. 2) : in state 1, the searcher hardly moves and carefully scans the environment, while in state 2 it is unable to detect targets but moves fast. A key quantity in the problem of animals searching for food is the *search time* for a target. The search strategy that minimizes the search time can then be called optimal, since it enables the searcher to get food rapidly and outperform its competitors.

Let us consider more carefully the rules defining the model. First, available experimental observations indicate that after each stationary phase the animal in practice only slightly modifies its direction of motion, which suggests a 1-dimensional model. Second, state 2 is described as a ballistic phase of constant velocity V. This in fact corresponds to the most efficient way of exploring space. We also make the assumption that the time spent in this state 2 is a random variable distributed according to an exponential law of mean τ_2. The time spent in state 1 is also taken as exponentially distributed, of mean τ_1, and we hypothesize in addition that this state 1 can be described as a Brownian motion of diffusion coefficient D. Such hypothesis calls for comments. First, the model obviously does not aim at describing precisely the detection phase, which involves very complex biological mechanisms, but only at extracting the main characteristics of

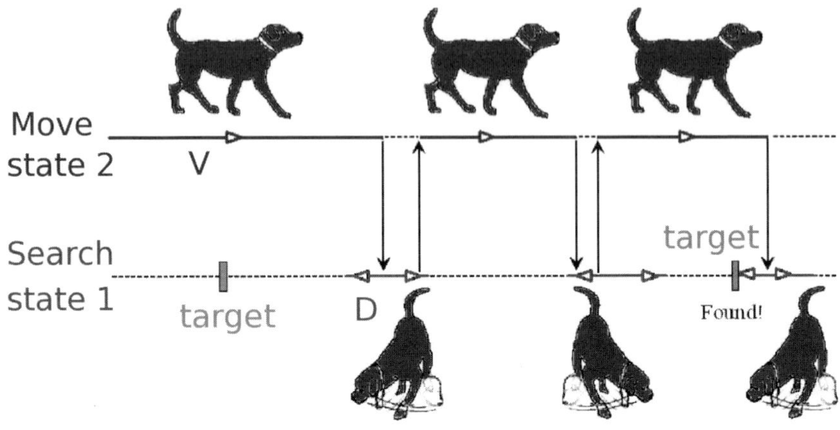

Fig. 2. The two-state model of intermittent search.

these mechanisms in order to estimate the time needed to detect a target. Second, we stress that it is not necessarily the very position of the searcher that performs diffusion, but rather the focus point of the detection system, such as the focus point of the eyes. Next, we underline that this process of visual search is often described in terms of random walks, which makes our hypothesis in agreement with previous studies. Finally, as focusing and processing the information received by sensory organs require a minimum time, the search phase cannot be too short, which implies a lower bound τ_1^{min} for the time τ_1.

2.2. *Minimizing the search time*

Can one define an optimal strategy? In other words, is it possible to adjust the times τ_1 and τ_2 spent in each state to minimize the search time?

The search time can actually be calculated. It is easily found that the mean first-passage time (MFPT) $t_i(x)$ starting from x in state i satisfies

$$D\frac{d^2t_1}{dx^2} + \frac{1}{\tau_1}(t_2(x) - t_1(x)) = -1 \tag{1}$$

$$-V\frac{dt_2}{dx} + \frac{1}{\tau_2}(t_1(x) - t_2(x)) = -1. \tag{2}$$

These differential equations have to be completed by boundary conditions. We rely here on periodic boundary conditions. As the target at $x = 0$ can be found only in state 1, we get $t_1(0) = t_1(L) = 0$, $t_2(0) = t_2(L)$.

The average search time $\langle t\rangle$ is defined as the average of $t_1(x)$ over the initial position x of the searcher, which is uniformly distributed over the segment $[0, L]$, as the searcher initially does not know the target's location. It is found to be given by [11] :

$$\langle t\rangle = (\tau_2+\tau_1)\left(\frac{L}{2}\frac{\left(e^{\alpha+\beta}-1\right)\sqrt{1+4r}+(1+2r)\left(e^{\beta}-e^{\alpha}\right)}{\sqrt{1+4r}\left(e^{\beta}-1\right)\left(e^{\alpha}-1\right)\tau_2 V}-\frac{1}{r}-1\right), \tag{3}$$

with :

$$r = \tau_2^2 V^2/(D\tau_1), \tag{4}$$

$$\alpha = \frac{L}{2}\left(-\frac{1}{\tau_2 V}+\sqrt{\frac{1}{\tau_2{}^2V^2}+4\frac{1}{D\,\tau_1}}\right), \tag{5}$$

$$\beta = -\frac{L}{2}\left(\frac{1}{\tau_2 V}+\sqrt{\frac{1}{\tau_2{}^2V^2}+4\frac{1}{D\,\tau_1}}\right). \tag{6}$$

In the limit of low target density $L \gg V\tau_2, \sqrt{D\tau_1}, D\tau_1/(V\tau_2)$, this expression simplifies :

$$\langle t\rangle \simeq \frac{L\,(\tau_2+\tau_1)\left(D\,\tau_1+2\,\tau_2{}^2V^2\right)}{2\tau_2\,V\,\sqrt{D\,\tau_1}\sqrt{D\,\tau_1+4\,\tau_2{}^2V^2}}. \tag{7}$$

Note that thanks to intermittence, $\langle t\rangle \propto L$, whereas for diffusion alone the mean detection time is $t_{\text{diff}} = L^2/(12D)$. Intermittence is thus favorable (meaning that the gain, defined as $t_{\text{diff}}/\langle t\rangle$ is greater than 1), at least for L large enough.

Intermittence is favorable and the strategy can even be optimized. The mean search time is minimized for $\tau_1^{opt} = \tau_1^{min}$, and τ_2^{opt} satisfying the relation :

$$\tau_1^3 + 6\frac{\tau_1^2\tau_2^2}{\tau} - 8\frac{\tau_2^5}{\tau^2} = 0, \tag{8}$$

where $\tau = D/V^2$ is an extra characteristic time, depending on the searcher's characteristics. This minimum takes a simple form in two different regimes.

- If $\tau_1 \gg \tau$, the minimum of the search time is for $\tau_1 = \tau_1^{min}$, and

$$\tau_2^{opt} = \left(\frac{3\tau\tau_1^2}{4}\right)^{1/3}. \tag{9}$$

  In this regime, one has $\tau_1 > \tau_2$: the searcher spends more time scanning than moving.

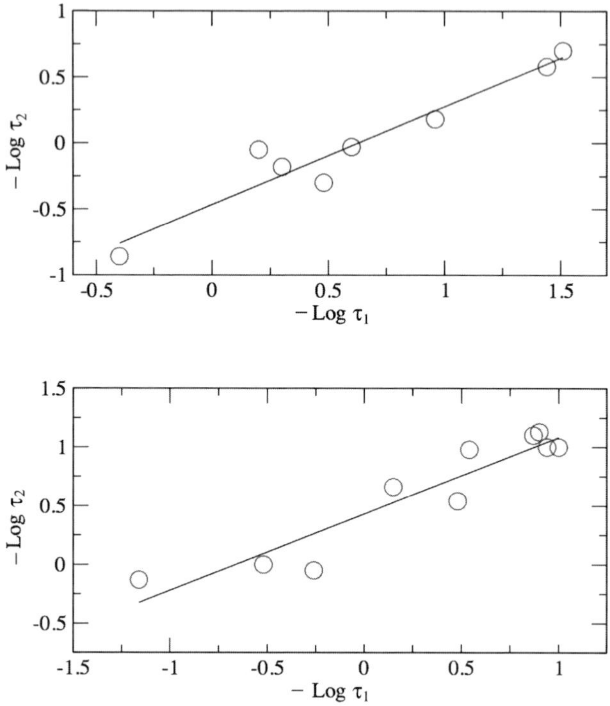

Fig. 3. Plots (in log/log scale) of the mean time τ_2 spent in the fast moving phase as function of the mean time τ_1 spent in the careful scanning phase. The top panel corresponds to animals that spend more time moving than searching, and the bottom panel to animals that spend more time searching than moving. Each point stands for given species ; lines are linear fits and yield the slopes 0.7 (top) and 0.6 (bottom).

- If $\tau_1 \ll \tau$, the minimum of the search time is for $\tau_1 = \tau_1^{min}$, and

$$\tau_2^{opt} = \left(\frac{\tau^2 \tau_1^3}{8}\right)^{1/5} . \tag{10}$$

  In this regime, one has $\tau_1 < \tau_2$, which means that the searcher spends more time moving than scanning.

Available experimental data provide for a set of 18 animal species [9], including lizards, fish and birds the mean times spent in each of the two phases. The comparison of these experimental data with the power laws (9) and (10) predicted by the theoretical analysis yields a satisfactory agreement (cf. Fig. 3). Hence, it appears that animals as various as fish, lizards and birds perform intermittent search strategies to minimize their search time for preys. Note that since then, similar 1-dimensional models of intermittent search have been developed [12–19] and are reviewed in [7].

2.3. *Should animals really perform Lévy strategies?*

The question of optimizing search strategies in the context of behavioral ecology has recently been very controversial. The status of Lévy walks as optimal search strategies has in fact been questioned in 2006 from a theoretical point of view [20], essentially because in most of realistic cases the optimal strategy predicted by the Lévy walk model is trivially the straight ballistic motion. Since then the Lévy walk model has also been questioned experimentally [21]. In fact, it was shown that some of the data that had been used to identify Lévy walks in animal trajectories were misinterpreted, and most likely were consistent with classical exponential random walks. Finally, as opposed to what is commonly stated, animals have no reason to perform Lévy strategies, since they are not optimal. In turn, intermittent strategies, which can really be minimized, constitute an alternative model of animal trajectories.

3. How does a protein find its target sequence on DNA?

Target search problems are also involved at much smaller scales, such as the scale of a single molecule. In fact, a chemical reaction can be seen as the search of a reactive site by a reactant. Living cells provide striking examples of chemical reactions whose kinetics must be optimized in order to efficiently fulfill a biological function. A prototypical example is given by proteins looking for specific target sequences on DNA. It is the case of transcription factors, which bind to specific DNA sequences in order to regulate - either promote or repress - the transcription of a given gene. Another well known example is provided by the restriction enzyme EcoRV, which enables bacteria to fight viral attacks : EcoRV is designed to cut the viral DNA at a specific site as soon as it is introduced in the bacterium, thus preventing its replication. Such reactions involve very short target sequences - typically a few base pairs long. Given that such targets are sometimes unique, from the point of view of the protein the search problem is therefore very hard, and a simple diffusive strategy, as usually observed in chemistry, would yield search times of the order of hours, far longer than what is observed in vivo (of the order of seconds). This paradox has long remained unresolved.

Recently, the development of new experimental techniques such as the micro manipulation of single DNA molecules or the tracking of fluorescent proteins at the single molecule level have shown that two mechanisms can be involved in the process of target search by proteins : a slow diffusive motion along the DNA molecule during which the protein remains non specifically bound, and a much faster free diffusive motion in the surrounding medium. The protein can then switch randomly between these two phases by binding and unbinding DNA, as was early suggested by O. Berg and collaborators [22]. In fact, proteins such as EcoRV have a site that is locally positively charged, which induces an affinity for DNA, which is negatively charged. The binding of such proteins to DNA is therefore energetically favored, and is called non specific since it is largely independent of the sequence. Thermal fluctuations, that are due to random collisions with the molecules of the surrounding medium, are then responsible for the diffusive motion of the protein along the DNA, and sometimes for its unbinding. This type of trajectories is in fact intrinsically intermittent, and can be compared to our first example of key searcher. The protein carefully scans the DNA while it is bound, and moves much faster but without any chance to detect the target when it is not bound to DNA.

3.1. *Definition of the model*

We present here in details a simple model of intermittent reaction paths with minimal ingredients introduced in [23]. We consider a generic protein searching for its target site on a DNA molecule (see Fig. 5). The pathway followed by the protein, considered as a point-like particle, is a succession of 1D diffusions along the DNA strand (sliding phases denoted phases 1) and 3D excursions in the surrounding solution (denoted phases 2, during which the target is not accessible). In this minimal model we assume that the target site is a perfect reactive point of the DNA molecule, which means that reaction occurs as soon as the target is reached by the protein. Note that in this case the protein can find the target site only by diffusing along DNA, and therefore follows the scheme of intermittent search presented in introduction. The key quantity that we evaluate in this section is the search time, or reaction time, defined as the mean first-passage time [24] of the protein at the target, here denoted by $\langle T \rangle$. This quantity gives directly access in a mean field approximation to the first order reaction constant $K = 1/\langle T \rangle$ [25].

We now introduce further ingredients of the model. The time spent by the protein on DNA during each sliding phase is assumed to follow

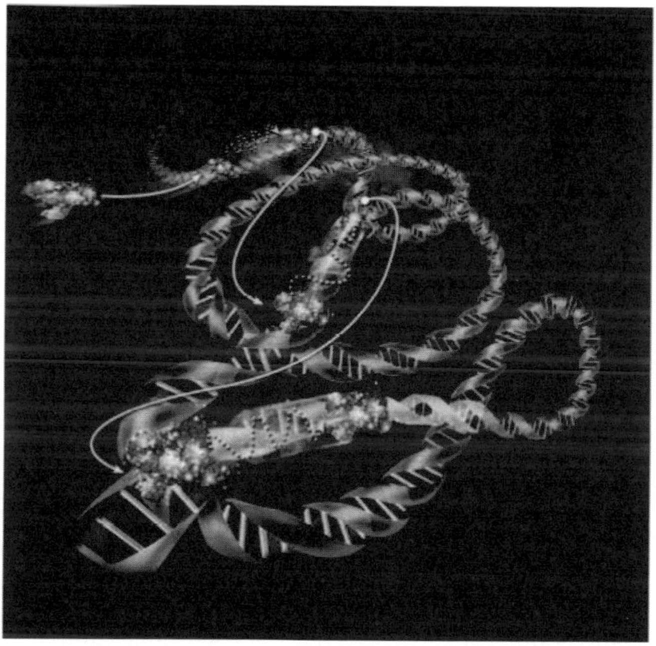

Fig. 4. Artist view of a protein looking for a target site on DNA. Artwork by Virginie Denis/Pour la Science n° 352, February 2007.

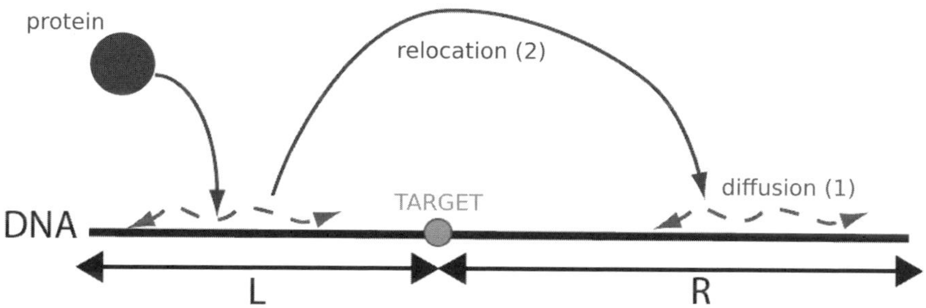

Fig. 5. A model of intermittent transport for DNA/protein interactions.

an exponential law with dissociation frequency λ_1. This law relies on a Markovian description of the chemical bond which is commonly used for bimolecular association. The probability density that the protein leaves the DNA at a random time t is then given by $\lambda_1 \exp(-\lambda_1 t)dt$, and the mean duration of a sliding event reads $\tau_1 = 1/\lambda_1$.

The one–dimensional motion on DNA during sliding phases 1 is modeled by a continuous Brownian motion with diffusion coefficient D. We assume that the ends of the DNA chain act on the protein as reflecting boundaries, but in practice this assumption is unimportant for long DNA molecules. Moreover, the case of circular DNA, such as plasmids, is readily obtained by taking the particular case $L = R$. We next assume that the 3D excursions of phase 2 are uncorrelated in space. This means that after dissociation from DNA, the protein will rebind the DNA at a random position independently of its starting position. This is justified when the DNA is in a random coil conformation, as in this case even short 3D excursions can lead to a long effective translocation of the linear position of the protein on DNA. We further assume that the probability density $P_{3D}(t)$ of the duration t of such 3D excursions is exponentially distributed, and write $P_{3D}(t) = \lambda_2 \exp(-\lambda_2 t)$. This assumption is justified, at least for the tail of the distribution, as soon as the 3D excursions of the protein are confined in a closed volume, for instance an experimental volume in vitro, or in vivo the cell or a cell compartment. The mean time spent in the surrounding solution in phase 2 is denoted $\tau_2 = 1/\lambda_2$.

We next introduce $P_{1D}(t|x)$, which is the conditional probability density that the protein, being on the DNA at position x and at time $t = 0$, will dissociate at time t before any encounter with the target site. We rewrite this quantity as :

$$P_{1D}(t|x) = \lambda_1 \exp(-\lambda_1 t) Q(t|x) \tag{11}$$

where $Q(t|x)$ is the conditional probability density that the protein, starting from the position x, does not meet the target site during a single sliding event. The probability density $j(t|x)$ of the first passage to the target site position at time t without dissociation is then related to $Q(t|x)$ according to $Q(t|x) = 1 - \int_0^t j(t'|x)dt'$.

Last, we introduce $\bar{P}_{1D}(t|x)$, which is the conditional probability density that the protein, being on DNA at position x at time $t = 0$, will find the target site for the first time at time t within a single sliding phase 1, without leaving the DNA:

$$\bar{P}_{1D}(t|x) = \exp(-\lambda_1 t) j(t|x). \tag{12}$$

Given these quantities, we show below that the first-passage density of the protein to the target site, and consequently the reaction constant, can be calculated explicitly.

3.2. First-passage density

By calculating the first-passage density, we obtain the mean reaction time, as well as all associated moments. We assume that the protein starts at $t = 0$ in state 1 (bound to the DNA) at position x. We consider a generic event whose number of 3D excursions is $n-1$, and denote the duration of successive sliding phases $t_1, \ldots, t_n$, and the duration of successive 3D excursions $\theta_1, \ldots, \theta_{n-1}$. The probability density of such an event, for which the protein finds the target site for the first time at time $t = \sum_{i=1}^{n} t_i + \sum_{i=1}^{n-1} \theta_i$ is:

$$P_n(t|x) = \bar{P}_{1D}(t_n)P_{3D}(\theta_{n-1})P_{1D}(t_n)\ldots P_{1D}(t_2)P_{3D}(\theta_1)P_{1D}(t_1|x) \quad (13)$$

where $P_{1D}(t)$ and $\bar{P}_{1D}(t)$ are averaged over the initial position of the protein: $P_{1D}(t) = \langle P_{1D}(t|x)\rangle_x$ and $\bar{P}_{1D}(t) = \langle \bar{P}_{1D}(t|x)\rangle_x$. We denote by L the DNA length on the "left" side of the target site and by R the length on the "right" side of the target site. The average of a function f over the initial position x is given by $\langle f(t|x)\rangle_x \equiv \frac{1}{L+R}\int_{-L}^{R} f(t|x)\mathrm{d}x$.

To obtain the density of first passage to the target site, $F(t|x)$, we sum over all possible numbers of excursions and we integrate over all intervals of time, ensuring that $t = \sum_i^n t_i + \sum_i^{n-1}\theta_i$. The average over the initial position of the protein, $F(t) = \langle F(t|x)\rangle_x$, can be expressed as:

$$F(t) = \sum_{n=1}^{\infty}\int_0^{\infty} dt_1 \ldots dt_n d\theta_1 \ldots d\theta_{n-1}\delta\left(\sum_{i=1}^{n} t_i + \sum_{i=1}^{n-1}\theta_i - t\right) \left[\prod_{i=1}^{n-1} P_{3D}(\theta_i)\right]\left[\prod_{i=1}^{n-1} P_{1D}(t_i)\right]\bar{P}_{1D}(t_n). \quad (14)$$

Taking the Laplace transform of $F(t)$, $\widehat{F}(s) = \int_0^{\infty} dt e^{-st}F(t)$, we obtain:

$$\widehat{F}(s) = \left\langle \widehat{j}(\lambda_1 + s|x)\right\rangle_x \left\{1 - \frac{1 - \left\langle \widehat{j}(\lambda_1 + s|x)\right\rangle_x}{(1+s/\lambda_1)(1+s/\lambda_2)}\right\}^{-1}, \quad (15)$$

where $\widehat{j}(s|x)$ is the Laplace transform of $j(t|x)$. This expression completely solves the problem *for any 1D motion.* We will see next that the main quantities of physical interest can be extracted from this formula.

3.3. *Optimal search strategy*

The relevant quantity to describe the protein/DNA association reaction is the mean time $\langle T\rangle$ necessary for the protein to find the target site (see above). This mean time is obtained from the derivative of the first passage density by the following relation:

$$\langle T\rangle = -\left(\frac{\partial \widehat{F}(s)}{\partial s}\right)_{s=0} \tag{16}$$

which combined with equation (15) gives:

$$\langle T\rangle = \frac{1-\left\langle \widehat{j}(\lambda_1|x)\right\rangle_x}{\left\langle \widehat{j}(\lambda_1|x)\right\rangle_x}\left(\frac{1}{\lambda_1}+\frac{1}{\lambda_2}\right). \tag{17}$$

This expression is general and holds for any $1D$ motion in the slow phase 1. Now, we calculate this quantity in the case where the phase 1 is a free $1D$ diffusion. The one–dimensional Laplace transform of the first passage probability density is well known (see [24]), and leads to, after averaging over the starting position x:

$$\langle T\rangle = \left(\frac{1}{\lambda_1}+\frac{1}{\lambda_2}\right)\left\{\frac{\sqrt{\frac{\lambda_1}{D}}(L+R)}{\tanh\left(\sqrt{\frac{\lambda_1}{D}}L\right)+\tanh\left(\sqrt{\frac{\lambda_1}{D}}R\right)}-1\right\}, \tag{18}$$

where D is the diffusion coefficient. This defines as a by product the association constant of the reaction as $K = 1/\langle T\rangle$. Several comments are in order. (i) First, as soon as the length of the DNA strand is large enough (more precisely as soon as $\sqrt{\frac{\lambda_1}{D}}L \gg 1$ or $\sqrt{\frac{\lambda_1}{D}}R \gg 1$), $\langle T\rangle$ grows linearly with the length of the DNA. This mirrors the efficiency of intermittent reaction paths, as compared to the quadratic growth obtained in the case of pure sliding. In particular, the boundary effects are negligible for this quantity as soon as the overall length is large enough. (ii) Second, this expression is valid for a very large class of 3D motions. More precisely, it holds as soon as the mean first return time τ_2 corresponding to the 3D motion is finite and independent of the departure and arrival points. (iii) Last, these results can be generalized to the case of extended targets and higher dimensions (see [26]). See also [27] for a related search mechanism.

We now come to an important question addressed in [23, 28], which concerns the *optimization* of such intermittent reaction paths. We here assume that the mean search time is a limiting quantity which might have been minimized in the course of evolution. In this context, we consider $\lambda_1 = 1/\tau_1$, which characterizes the protein/DNA affinity, as the adjustable parameter. Indeed, this quantity depends strongly both on the structure of the protein and on physiological conditions such as the ionic strength, and therefore could widely vary from one protein to another. On the contrary $\lambda_2 = 1/\tau_2$ depends mostly on the properties of the environment, such as the DNA conformation, which is itself subject to very stringent constraints and therefore much less likely to be varied. Another adjustable parameter is the 1D diffusion coefficient D. Optimizing the search time with respect to this parameter is trivial : it is found that D should be as large as possible (assuming that D and λ_1 are independent), but obviously one should keep in mind that D is controlled by the hydrodynamic radius of the protein which can not be too small. For these reasons we focus here on λ_1.

It can be seen qualitatively that $\langle T\rangle$ is large for both λ_1 very large (in the limit λ_1 infinite, the protein is never on the DNA), and λ_1 very small (pure sliding limit which gives a quadratic growth with the DNA length), and could therefore be minimized for and intermediate value of λ_1. The sign of the derivative of the mean search time at $\lambda_1 = 0$ shows that it can indeed be minimized provided that

$$\lambda_2 > 15D\,\frac{L^2+R^2-LR}{L^4+R^4+4LR(L^2+R^2)-9R^2L^2}. \tag{19}$$

This condition means that bulk excursions, to be favorable, should be shorter than a fraction of the typical time needed to scan the full DNA molecule by 1D diffusion. In particular, it requires that the DNA length is long enough. If this condition is fulfilled, the search time can indeed be minimized (see Fig. 6). A careful analysis of the implicit equation satisfied by λ_1 at the optimum leads to the following expansion for large $\ell = L + R$

$$\lambda_1 = \lambda_2 - 4\frac{\sqrt{D\lambda_2}}{\ell} - \frac{8D}{\ell^2} - \frac{40D^{3/2}}{\sqrt{\lambda_2}\ell^3} + O\left(\frac{1}{\ell^4}\right). \tag{20}$$

Equations (19) and (20) refine the result of [28], which predicts than the optimal strategy is realized when $\lambda_1 = \lambda_2$. This result actually holds in the large ℓ limit, or more precisely for $\sqrt{\frac{\lambda_1}{D}}\ell \gg 1$. For intermediate values of ℓ boundary effects become important and the minimum can be significantly different.

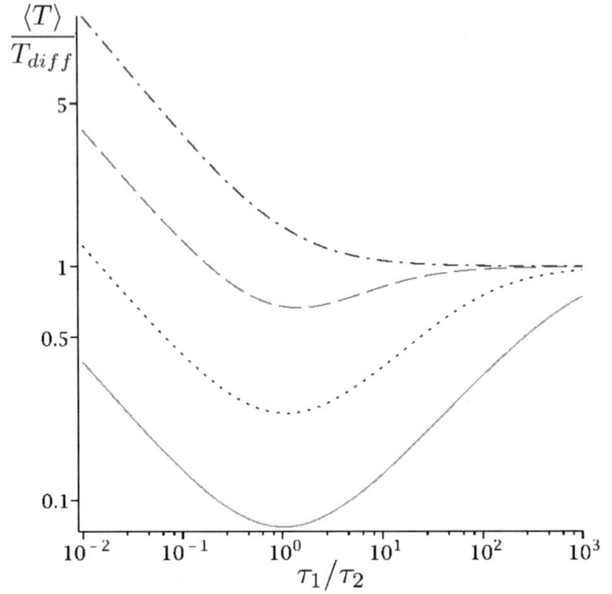

Fig. 6. Mean first passage time of a DNA–binding protein to its target site, renormalized by the mean first passage time by diffusion alone (T_{diff}), as a function of τ_1/τ_2. $T_{diff}/\tau_2 = 1$ (red dashed-dotted line), $T_{diff}/\tau_2 = 10$ (violet dashed line), $T_{diff}/\tau_2 = 100$ (blue dotted line), $T_{diff}/\tau_2 = 1000$ (green solid line). For a small DNA length ℓ (and therefore small T_{diff}) (red dashed-dotted line), the reaction constant depends monotonously on τ_1 and intermittent reaction paths are inefficient. For larger values of l (other curves), the reaction rate can be optimized as a function of τ_1. Here we averaged other the initial position of the target.

The $\langle T\rangle$ value at the minimum is particularly interesting. We compare it to the case of pure sliding where $\langle T_s\rangle = \ell^2/(3D)$:

$$\frac{\langle T\rangle}{\langle T_s\rangle} = \frac{6}{\ell}\sqrt{\frac{D}{\lambda_1}}\,. \tag{21}$$

The efficiency of the 3D mediated strategy is therefore much more important when the DNA chain is long. For example, using standard values for λ_1 (a few 10^{-2} $s.$) and D (typically 10^{-2} $\mu^2/s.$) and for a DNA substrate of length 10^6 bp, the mean reaction time is three orders of magnitude smaller than for a pure sliding strategy. By comparing the theoretical model to experimental data, it can be shown that the typical DNA length that is scanned by the protein during a single binding phase is of the order of

300 base pairs, and therefore much shorter than the total length of a DNA molecule (which is typically larger than 10^6 base pairs). This very short length of diffusive excursions along the DNA has since then been measured directly by analyzing trajectories of single EcoRV molecules on DNA in the team of P. Desbiolles at ENS Paris [29, 30], which confirms that the restriction enzyme EcoRV makes use of the extreme efficiency of intermittent strategies to find its target on DNA. Beyond the importance of such results for understanding the kinetics of gene transcription, this first minimal model shows that intermittent reactive paths are indeed very efficient, and that they can even allow to optimize the reaction kinetics. Further models have since extended these results [31–45].

4. Active intermittent transport in cells

At the microscopic scale, intermittent search strategies are also observed in the context of active transport of vesicles (or other molecular assemblies) in cells. In this section we briefly summarize the results published in [46]. Because of the intrinsically out of equilibrium nature of the cellular medium, the motion of a typical tracer particle in a cell can be very different from the usual thermal diffusion. In fact, in a cell, a tracer particle such as a vesicle can diffuse, but it can also bind to molecular motors, which thanks to ATP hydrolysis (the chemical energy of the cell) are able to perform a ballistic directional motion along the cytoskeleton filaments (Fig. 7). Such motion, called active, is indeed intermittent in the general sense defined above, since it involves a slow phase, here thermal diffusion, and a fast phase of ballistic motion induced by molecular motors. We assume here that the tracer is not reactive when it is bound to motors, as in the examples where the targets are membrane proteins. Such intermittent trajectories have been observed at the single vesicle level in [47], and these experiments indeed indicate that reaction (in that case exocytosis of the vesicle) occurs mostly in the diffusive phase. In a general context, we aim here at discussing the impact of active transport on reaction kinetics in cells. For that purpose, we estimate the rate constant of a simple first order chemical reaction, which is given by the inverse mean first-passage time to the target.

In the case of a tracer moving in 3 dimensions, for example in the volume V of the cytoplasm (Fig. 7), we have shown [46] that intermittent transport is more efficient than a simple diffusion of diffusion coefficient D, and that it makes possible the maximization of the kinetic rate constant as soon as the reaction radius (distance below which reaction can take place) verifies

$a > a_c \simeq 6{,}5D/V$. In usual cellular conditions, D is comprised between 10^{-2} $\mu m^2\,s^{-1}$ for vesicles and 10 $\mu m^2\,s^{-1}$ for small proteins, and the typical velocity of a molecular motor is of the order of $V \sim 1$ $\mu m\,s^{-1}$. The critical reaction radius is then comprised between 10 nm, which is smaller than any cell organelle, and 10 μm, which is comparable to the cell size. For tracers of the size of a vesicle active transport therefore enables an optimal exploration of space, whereas for individual molecules simple diffusion is more efficient. In addition, it should be noted that in usual cellular conditions the optimal strategy is realized in the case of a reaction radius of 0,1 μm when the typical interaction time with molecular motors is of the order of 0,1 s. This value is compatible with experimental observations and suggest that intracellular transport could be close to optimal efficiency. Last, it can be shown that in these conditions the optimal kinetic rate constant can be up to 10 times larger than in the absence of active transport.

In the case of a tracer evolving in a 2-dimensional space, typically along a membrane, or in a 1 dimensional structure such as axons or dendrites (Fig. 7), the kinetic rate constant can also be maximized, and in fact the optimal rate constant can become much larger. In 1 dimension, for a low enough concentration of targets, it can be 100 times larger than in the absence of active transport, since in low space dimension the gain can be shown to increase when the target concentration is decreased.

5. Robustness of the results

The previous examples show that intermittent search strategies are observed at various scales. This suggests that intermittent search is a general mechanism, which enables the optimization of the success of random searches for hidden targets. To support this hypothesis, a minimal model has been proposed, which only relies on the essential ingredients of intermittent strategies introduced up to now on the example of animals looking for food, proteins looking for a target sequence on DNA, or of active intermittent transport in cells, and which avoids the specificities of these examples. We discuss below this model introduced in [20, 48].

Following the above examples, we consider a 2-state searcher, which alternates between phases of slow but reactive motion (which enable target detection) (state 1) and phases of fast but non reactive motion (state 2). The random durations of each phase i are assumed to be exponentially distributed of mean τ_i, which means that the searcher has no memory in time : it does not remember when the last transition between phases

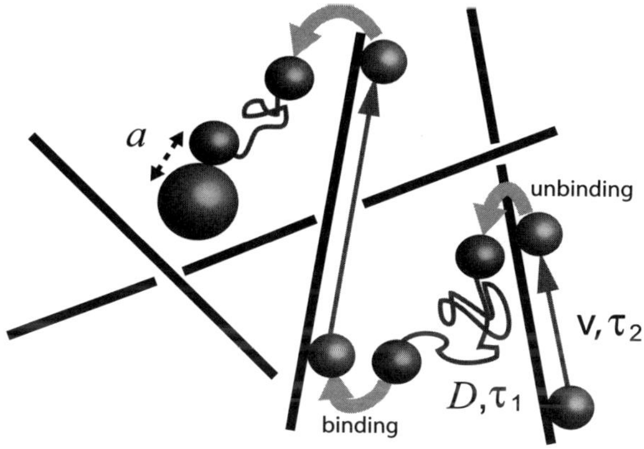

Active transport in the cell bulk

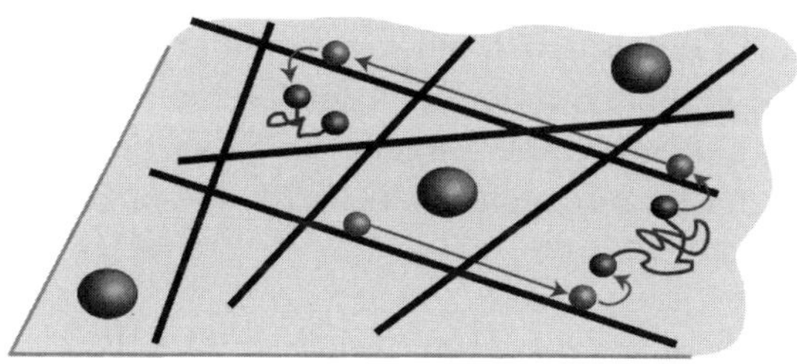

Active transport on 2-dimensional structures such as membranes.

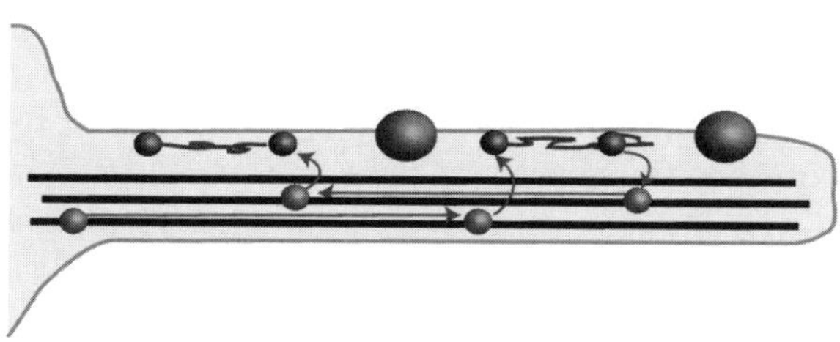

Active transport along 1-dimensional structures such dendrites or axons.

Fig. 7. Active transport in cells : different relevant geometries.

occurred. The fast phase (state 2) is described as a ballistic motion of constant velocity V, performed in a direction drawn at random. Hence, one can consider that the searcher has no memory in space either. In that sense it can be considered as minimal, since it has minimal cognitive abilities.

In order to assess the robustness of this minimal model, we consider two limiting modes of target detection (Fig. 8). The first one, called "dynamic" mode, is a direct generalization of the model introduced to describe animals looking for food : the detection phase is assumed to be *diffusive*, characterized by a diffusion coefficient D. In addition reactivity is assumed to be infinitely efficient, in the following sense : as soon as the target is reached in state 1, it is found. The second one, called "static" mode, is very different. During detection phases, the searcher is assumed *immobile*, but reactivity is now imperfect : the target is detected only with finite probability k per unit time if it is within the reach of the searcher.

5.1. *Basic equations*

We now present the basic equations combining the two search modes introduced above in the case of a spherical target of radius a centered in a spherical domain of radius b with reflexive boundary. Note that this geometry mimics both relevant situations of a single target and of infinitely many regularly spaced non revisitable targets. As in section 2.2, the MFPT to a target satisfies backward equations (see [20, 48] for details), which write here :

$$D\nabla_{\mathbf{r}}^2 t_1 + \frac{1}{2\pi\tau_1}\int_0^{2\pi}(t_2 - t_1)d\theta_{\mathbf{V}} - k\mathrm{I}_a(\mathbf{r})t_1 = -1 \qquad (22)$$

$$\mathbf{V}\cdot\nabla_{\mathbf{r}} t_2 - \frac{1}{\tau_2}(t_2 - t_1) = -1 \qquad (23)$$

where t_1 stands for the mean first-passage time starting from state 1 at position $\mathbf{r}$, and t_2 for the mean first passage time starting from state 2 at position $\mathbf{r}$ with velocity $\mathbf{V}$, of direction characterized by the angle $\theta_{\mathbf{V}}$. Here $\mathrm{I}_a(\mathbf{r}) = 1$ if $|\mathbf{r}| \leq a$ and $\mathrm{I}_a(\mathbf{r}) = 0$ if $|\mathbf{r}| > a$. In the present form, these integro-differential equations do not seem to allow for an exact resolution with standard methods. We thus resort to an approximate decoupling scheme, which relies on the following idea. If the searcher initially starts in phase 2, and if the target is close, its initial direction matters. But as soon as the initial position is far from the target, there are numerous reorientations before finding the target, implying that the initial direction does not matter. Consequently, if $b \gg a$ and once the mean search time has been averaged over the starting position, the effect of the initial direction can be neglected. It allows us to make an approximation and solve the system.

5.2. *Results for the diffusive mode of detection*

For the diffusive mode of detection ($k \to \infty$), an analytical approximation for the search time can be obtained. In the case of low target density ($a \ll b$), which is the most relevant for hidden target search problems, three regimes arise. In the first regime $a \ll b \ll D/V$, the relocating phases are not efficient and intermittence is useless. In the second regime $a \ll D/V \ll b$, it can be shown that the intermittence can significantly speed up the search (typically by a factor 2), but that it does not change the order of magnitude of the search time. On the contrary, in the last regime $D/V \ll a \ll b$, the optimal strategy, obtained for

$$\tau_1^{opt} \sim \frac{D}{2V^2}\frac{\ln^2(b/a)}{2\ln(b/a)-1}, \quad \tau_2^{opt} \sim \frac{a}{V}(\ln(b/a)-1/2)^{1/2}, \tag{24}$$

leads to a search time arbitrarily smaller than the non intermittent search time when $V \to \infty$. Note that this optimal strategy corresponds to a scaling law

$$\frac{\tau_1^{opt}}{\tau_2^{opt}} \sim \frac{D}{a^2}\frac{1}{(2-1/\ln(b/a))^2}, \tag{25}$$

which does not depend on V.

5.3. *Results for the static mode of detection*

We now turn to the static mode ($D \to 0$). In this case, intermittence is trivially necessary to find the target, and the optimization of the search time leads for $b \gg a$ to :

$$\tau_{1,\min} = \left(\frac{a}{Vk}\right)^{1/2}\left(\frac{2\ln(b/a)-1}{8}\right)^{1/4}, \tag{26}$$

$$\tau_{2,\min} = \frac{a}{V}\left(\ln(b/a)-1/2\right)^{1/2}, \tag{27}$$

which corresponds to the scaling law $\tau_{2,\min} = 2k\tau_{1,\min}^2$, which still does not depend on V.

5.4. *Discussion*

This bidimensional two-state model of search processes for non-revisitable targets closely relies on the experimentally observed intermittent strategies adopted by foraging animals. Using a decoupling approximation numerically validated (see [49] for a related approximation, and [50] for similar equations), it can be analytically solved, allowing us to draw several conclusions. (i) The mean search time $\langle t \rangle$ presents a global minimum for finite values of the τ_i, which means that intermittent strategies constitute optimal strategies, as opposed to Lévy walks which are optimal only for revisitable targets. (ii) The optimal τ_1^{opt} obtained for two modes of detection are different and depend explicitly on D and k, leading to different scaling laws which are susceptible to discriminate between the two search modes. (iii) A very striking and non intuitive feature is that both modes of search studied lead to the *same optimal value* of τ_2^{opt}. As this optimal time does not depend on the specific characteristics D and k of the search mode, it seems to constitute a general property of intermittent search strategies. The robustness of these conclusions have been discussed further in [51] in the framework of a more general model.

6. Conclusion

Finally, we have shown that intermittent search strategies are efficient search strategies that enable the minimization of the search time for hidden targets. Hence, even if it may seem counterintuitive, the key searcher should indeed interrupt his phases of careful search to perform fast displacement phases. Even if they may seem to be a loss of time, such "blind" phases of displacement can in fact speed up the search process. In addition we have seen that a model searcher with minimal cognitive abilities can significantly improve its search efficiency by tuning the characteristic times it spends in each of the careful search and displacement phases. Such strategy seems easily implementable since it does not involve potentially costly cognitive skills. It therefore seems favorable from an evolutionary point of view, which could explain the observation of such intermittent strategies in situations as various as animals looking for food or proteins looking for a target sequence on DNA.

Intermittent search strategies could also be used in the context of human search tasks, such as search for explosives or search for victims in avalanches or earthquakes, as suggested by M. Schlesinger from the research department of US navy [52].

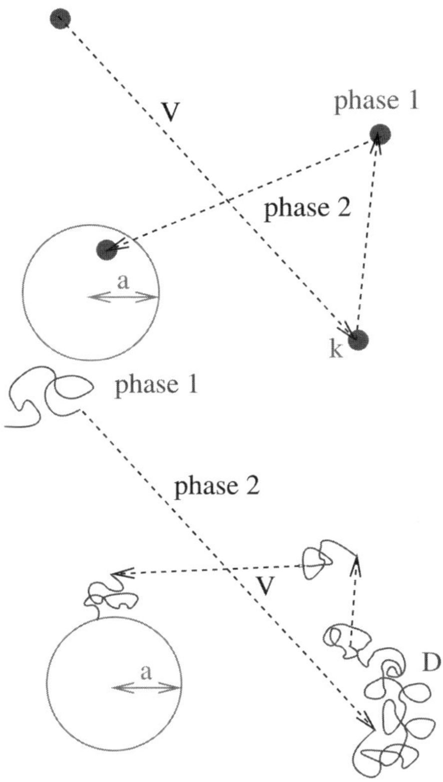

Fig. 8. A minimal search model : the searcher alternates between detection phases and motion phases during which targets cannot be detected. top : static mode of detection. bottom : dynamic mode of detection.

References

1. G. M. Viswanathan, V. Afanasyev, S. V. Buldyrev, E. J. Murphy, P. A. Prince, and H. E. Stanley, Levy flight search patterns of wandering albatrosses, *Nature*. **381**(6581), 413–415 (05, 1996). URL `http://dx.doi.org/10.1038/381413a0`.
2. G. M. Viswanathan, S. V. Buldyrev, S. Havlin, M. G. E. da Luz, E. P. Raposo, and H. E. Stanley, Optimizing the success of random searches, *Nature*. **401** (6756), 911–914 (10, 1999). URL `http://dx.doi.org/10.1038/44831`.
3. G. M. Viswanathan, E. P. Raposo, and M. G. E. da Luz, Lèvy flights and superdiffusion in the context of biological encounters and random searches, *Physics of Life Reviews*. **5**(3), 133–150 (2008). URL `http://www.sciencedirect.com/science/article/B75DC-4S5FJG2-1/2/3482e34ecc00cdb0ce8f1771ea95fca1`.

4. V. Tejedor, R. Voituriez, and O. Bénichou, Optimizing persistent random searches, *Physical Review Letters.* **108**(8), 088103– (02, 2012). URL http://link.aps.org/doi/10.1103/PhysRevLett.108.088103.
5. O. Benichou, M. Coppey, M. Moreau, P.-H. Suet, and R. Voituriez, Optimal search strategies for hidden targets., *Phys Rev Lett.* **94**(19), 198101 (2005). ISSN 0031-9007 (Print).
6. O. Benichou, C. Loverdo, M. Moreau, and R. Voituriez, Optimizing intermittent reaction paths, *Physical Chemistry Chemical Physics.* **10**(47), 7059–7072 (2008). URL http://dx.doi.org/10.1039/b811447c.
7. O. Bénichou, C. Loverdo, M. Moreau, and R. Voituriez, Intermittent search strategies, *Reviews of Modern Physics.* **83**(1), 81–129 (03, 2011). URL http://link.aps.org/doi/10.1103/RevModPhys.83.81.
8. W. J. O'Brien, H. I. Browman, and B. I. Evans, *Am. Sci.* **78**, 152 (1990).
9. L. D. Kramer and R. L. McLaughlin, *Am. Zool.* **41**, 137 (2001).
10. J. W. Bell, *Searching behaviour, the behavioural ecology of finding resources, Animal Behaviour Series.* Chapman and Hall, London (1991).
11. O. Bénichou, M. Coppey, M. Moreau, P. H. Suet, and R. Voituriez, A stochastic model for intermittent search strategies, *Journal of Physics: Condensed Matter.* **17**(49), S4275–S4286 (2005).
12. O. Bénichou, M. Coppey, M. Moreau, and R. Voituriez, Intermittent search strategies: When losing time becomes efficient, *EPL (Europhysics Letters).* **75**(2), 349–354 (2006).
13. G. Oshanin, H. S. Wio, K. Lindenberg, and S. F. Burlatsky, Intermittent random walks for an optimal search strategy: one-dimensional case, *Journal of Physics: Condensed Matter.* **19**(6), 065142 (2007).
14. M. A. Lomholt, K. Tal, R. Metzler, and K. Joseph, Lévy strategies in intermittent search processes are advantageous, *Proceedings of the National Academy of Sciences.* **105**(32), 11055–11059 (2008). URL http://www.pnas.org/content/105/32/11055.abstract.
15. D. Campos, V. Méndez, and F. Bartumeus, Optimal intermittence in search strategies under speed-selective target detection, *Phys Rev Lett.* **108**(2), 028102 (Jan, 2012).
16. P. Bressloff and J. Newby, Directed intermittent search for hidden targets, *New Journal of Physics.* **11**(2) (2009).
17. F. Rojo, C. E. Budde, and H. S. Wio, Optimal intermittent search strategies, *Journal of Physics A: Mathematical and Theoretical.* **42**(12) (2009).
18. G. Oshanin, K. Lindenberg, H. S. Wio, and S. Burlatsky, Efficient search by optimized intermittent random walks, *Journal of Physics A: Mathematical and Theoretical.* **42**(43), 434008 (2009). URL http://stacks.iop.org/1751-8121/42/i=43/a=434008.
19. F. Rojo, J. Revelli, C. E. Budde, H. S. Wio, G. Oshanin, and K. Lindenberg, Intermittent search strategies revisited: effect of the jump length and biased motion, *Journal of Physics A: Mathematical and Theoretical.* **43**(34), 345001 (2010). URL http://stacks.iop.org/1751-8121/43/i=34/a=345001.
20. O. Benichou, C. Loverdo, M. Moreau, and R. Voituriez, Two-dimensional intermittent search processes: An alternative to levy flight strategies., *Phys*

Rev E Stat Nonlin Soft Matter Phys. **74**(2 Pt 1), 020102 (2006). ISSN 1539-3755 (Print).

21. A. M. Edwards, R. A. Phillips, N. W. Watkins, M. P. Freeman, E. J. Murphy, V. Afanasyev, S. V. Buldyrev, M. G. E. da Luz, E. P. Raposo, H. E. Stanley, and G. M. Viswanathan, Revisiting levy flight search patterns of wandering albatrosses, bumblebees and deer, *Nature.* **449**(7165), 1044–1048 (10, 2007). URL `http://dx.doi.org/10.1038/nature06199`.
22. O. G. Berg, R. B. Winter, and P. H. von Hippel, Diffusion-driven mechanisms of protein translocation on nucleic acids. 1. models and theory, *Biochemistry.* **20**, 6929 (1981).
23. M. Coppey, O. Benichou, R. Voituriez, and M. Moreau, Kinetics of target site localization of a protein on dna: A stochastic approach, *Biophys. J.* **87** (3), 1640–1649 (Sept., 2004).
24. S. Redner, *A guide to First- Passage Processes.* Cambridge University Press, Cambridge, England (2001).
25. O. Berg and C. Blomberg, *Biophysical Chemistry.* **4**, 367 (1976).
26. O. Benichou, M. Moreau, P.-H. Suet, and R. Voituriez, Intermittent search process and teleportation., *J Chem Phys.* **126**(23), 234109 (2007). ISSN 0021-9606 (Print). doi: 10.1063/1.2741516.
27. M. R. Evans and S. N. Majumdar, Diffusion with stochastic resetting, *Physical Review Letters.* **106**(16), 160601 (04, 2011). URL `http://link.aps.org/doi/10.1103/PhysRevLett.106.160601`.
28. M. Slutsky and L. A. Mirny, Kinetics of protein-dna interaction: Facilitated target location in sequence-dependent potential, *Biophysical Journal.* **87**(6), 4021–4035 (2004). URL `http://www.biophysj.org/cgi/content/abstract/87/6/4021`.
29. I. Bonnet, A. Biebricher, P.-L. Porte, C. Loverdo, O. Benichou, R. Voituriez, C. Escude, W. Wende, A. Pingoud, and P. Desbiolles, Sliding and jumping of single ecorv restriction enzymes on non-cognate dna, *Nucleic Acids Research.* **36**(12), 4118–4127 (7, 2008). URL `http://nar.oxfordjournals.org/cgi/content/abstract/36/12/4118`.
30. C. Loverdo, O. Benichou, R. Voituriez, A. Biebricher, I. Bonnet, and P. Desbiolles, Quantifying hopping and jumping in facilitated diffusion of dna-binding proteins, *Physical Review Letters.* **102**(18), 188101–4 (05, 2009). URL `http://link.aps.org/abstract/PRL/v102/e188101`.
31. M. A. Lomholt, T. Ambjornsson, and R. Metzler, Optimal target search on a fast-folding polymer chain with volume exchange, *Phys. Rev. Lett.* **95**(26), 260603–4 (Dec., 2005).
32. I. Eliazar, T. Koren, and J. Klafter, Searching circular dna strands, *Journal of Physics: Condensed Matter.* **19**(6) (2007).
33. T. Hu, A. Y. Grosberg, and B. I. Shklovskii, How proteins search for their specific sites on dna: The role of dna conformation, *Biophysical Journal.* **90**(8), 2731–2744 (2006). URL `http://www.biophysj.org/cgi/content/abstract/90/8/2731`.
34. M. Sheinman and Y. Kafri, The effects of intersegmental transfers on target location by proteins, *Physical Biology.* **6**(1) (2009). URL `http://stacks.`

iop.org/1478-3975/6/i=1/a=016003.

35. O. Bénichou, Y. Kafri, M. Sheinman, and R. Voituriez, Searching fast for a target on dna without falling to traps, *Physical Review Letters.* **103** (13), 138102–4 (09, 2009). URL `http://link.aps.org/abstract/PRL/v103/e138102`.
36. B. van den Broek, M. A. Lomholt, S. M. J. Kalisch, R. Metzler, and G. J. L. Wuite, How dna coiling enhances target localization by proteins, *Proceedings of the National Academy of Sciences.* **105**, 15738 (2008). URL `http://www.pnas.org/content/early/2008/10/03/0804248105.abstract`.
37. M. A. Lomholt, B. van den Broek, S.-M. J. Kalisch, G. J. L. Wuite, and R. Metzler, Facilitated diffusion with dna coiling, *Proc Natl Acad Sci U S A.* **106**(20), 8204–8 (May, 2009). doi: 10.1073/pnas.0903293106.
38. A. B. Kolomeisky, Physics of protein-dna interactions: mechanisms of facilitated target search, *Phys Chem Chem Phys.* **13**(6), 2088–95 (Feb, 2011). doi: 10.1039/c0cp01966f.
39. O. Bénichou, C. Chevalier, B. Meyer, and R. Voituriez, Facilitated diffusion of proteins on chromatin, *Physical Review Letters.* **106**, 038102 (2011). URL `http://link.aps.org/doi/10.1103/PhysRevLett.106.038102`.
40. M. Sheinman, O. Bénichou, Y. Kafri, and R. Voituriez, Classes of fast and specific search mechanisms for proteins on dna, *Reports on Progress in Physics.* **75**(2), 026601 (2012). URL `http://stacks.iop.org/0034-4885/75/i=2/a=026601`.
41. A. B. Kolomeisky, Mechanisms of protein binding to dna: statistical interactions are important, *Biophys J.* **104**(5), 966–7 (Mar, 2013). doi: 10.1016/j.bpj.2013.01.039.
42. O. Bénichou, D. Grebenkov, P. Levitz, C. Loverdo, and R. Voituriez, Optimal reaction time for surface-mediated diffusion, *Physical Review Letters.* **105**, 150606 (2010). URL `http://link.aps.org/doi/10.1103/PhysRevLett.105.150606`.
43. O. Bénichou, D. Grebenkov, P. Levitz, C. Loverdo, and R. Voituriez, Mean first-passage time of surface-mediated diffusion in spherical domains, *Journal of Statistical Physics.* **142**(4), 657–685–685 (2011). URL `http://dx.doi.org/10.1007/s10955-011-0138-6`.
44. G. Foffano, D. Marenduzzo, and E. Orlandini, Facilitated diffusion on confined dna, *Phys Rev E Stat Nonlin Soft Matter Phys.* **85**(2 Pt 1), 021919 (Feb, 2012).
45. C. A. Brackley, M. E. Cates, and D. Marenduzzo, Facilitated diffusion on mobile dna: configurational traps and sequence heterogeneity, *Phys Rev Lett.* **109**(16), 168103 (Oct, 2012).
46. C. Loverdo, O. Bénichou, M. Moreau, and R. Voituriez, Enhanced reaction kinetics in biological cells, *Nature Physics.* **4**, 134 – 137 (2008). doi: 10.1038/nphys830. URL `http://www.nature.com/nphys/journal/v4/n2/full/nphys830.html`.
47. S. Huet, E. Karatekin, V. S. Tran, I. Fanget, S. Cribier, and J.-P. Henry, Analysis of Transient Behavior in Complex Trajectories: Application to Secretory Vesicle Dynamics, *Biophys. J.* **91**(9), 3542–3559 (2006).

48. O. Benichou, C. Loverdo, M. Moreau, and R. Voituriez, A minimal model of intermittent search in dimension two, *Journal Of Physics-Condensed Matter.* **19**(6), 065141 (Feb., 2007).
49. P. C. Bressloff and J. M. Newby, Quasi-steady-state analysis of two-dimensional random intermittent search processes, *Physical Review E.* **83**(6) (06, 2011). URL `http://link.aps.org/doi/10.1103/PhysRevE.83.061139`.
50. J. Reingruber and D. Holcman, Gated narrow escape time for molecular signaling, *Physical Review Letters.* **103**(14), 148102–4 (10, 2009). URL `http://link.aps.org/abstract/PRL/v103/e148102`.
51. C. Loverdo, O. Benichou, M. Moreau, and R. Voituriez, Robustness of optimal intermittent search strategies in one, two, and three dimensions, *Physical Review E (Statistical, Nonlinear, and Soft Matter Physics).* **80**(3), 031146–29 (09, 2009). URL `http://link.aps.org/abstract/PRE/v80/e031146`.
52. M. F. Shlesinger, Mathematical physics: Search research, *Nature.* **443**(7109), 281–282 (Sept., 2006).

Chapter 5

First-Passage Phenomena on Finite Inhomogeneous Networks

Elena Agliari* and Davide Cassi†

*Dipartimento di Fisica, Sapienza Università di Roma, P. le Aldo Moro 5, 00185 Roma, Italy

†DiFeST, Università di Parma, Viale Usberti 7/a 43100, Parma, Italy

*elena.agliari@roma1.infn.it

†davide.cassi@fis.unipr.it

A sequence of strongly inhomogeneous networks, with increasing sizes, can lead to very different thermodynamic limits, depending on the site chosen as the origin. This singular behavior reflects in remarkable phenomena even at finite sizes. Indeed, the finite networks belonging to such sequences, even with a small number of sites, presents significant inhomogeneities in first-passage related quantities, depending on the starting site of the random walker. This kind of behavior is of primary importance for the understanding of diffusion phenomena on real inhomogeneous systems, which are finite and cannot be univocally described by an infinite structure in the thermodynamic limit. Here, we explore the properties of the above mentioned quantities on a collection of the most typical and known inhomogeneous finite networks, both analytically and numerically, evidencing their unique features.

1. Introduction

In order to study first passage and trapping phenomena, as well as other non Markovian random walks quantities [1], on large discrete structures, it is usually very useful to introduce the thermodynamic limit, considering the infinite network obtained as the limit of a sequence of similar, increasing, finite structures. On (statistically) homogeneous networks, i.e. networks where all points are statistically equivalent as far as large scale properties are concerned (see Sec. 2 for a formal treatment), the results obtained in the thermodynamic limit are unambiguous, and differ from the exact results for finite structures only for small quantities vanishing with increasing number of sites. On the other hand, when considering strongly inhomogeneous structures, it can happen that two (or more) points, with topologically different neighborhood, become more and more distant from each other when increasing the network size, and their neighborhoods evolve to two deeply different infinite graphs in the thermodynamic limit. In such cases, the thermodynamic limit is not unique and it depends on the specific point chosen as the origin of the intrinsic coordinate system. Of course, when this happens, the choice of the starting point gives rise to completely different time evolution of random walks [2].

Based on these considerations, it appears of primary importance to investigate the random walk problem on finite networks that can be viewed as elements of a graph sequence with non-unique thermodynamic limit, considering the possibility of evidencing very different site-dependent behaviours. Consider, as a first example, the sequence of networks depicted in Fig. 1. The end points A and B of the basis segment AB become more and more distant as the sequence evolves. In the meanwhile, the length of the vertical tooth starting from A is always 1, while the length of the tooth starting from B tends to infinite. The same happens to the distances from A and B and their nearest branching points. Therefore, the thermodynamic limit of the network sequence seen from A and from B are deeply different. For realistic applications, we are obviously interested only in finite, and perhaps small, graphs belonging to such a sequence. Therefore, the distance from B to its nearest branching point will be large but finite. In any case, this can lead to strong differences in the behaviour of random walks starting from A and B, in particular if such a distance exceeds some critical length scale of the problem.

In this chapter, we will focus on deterministic structures for simplicity's sake. In particular, we will concentrate on the so-called extended Cayley tree [3], which was introduced to describe dendrimers which have a completely ordered structure but, different from (compact) Cayley tree, they display only an approximate self-similarity.

In the following, we will first review some basic definitions concerning the thermodynamic limit of graphs (Sec. 2) and we will describe some paradigmatic examples of structures for which the thermodynamic limit is non-unique (Sec. 3). Then, we will describe some analytical and numerical results for mean first-passage times on the extended dendrimer showing that they provide a clear signature for the non-uniqueness of the thermodynamic limit on these structures (Sec. 4). Finally, we discuss results and possible perspectives (Sec. 5); the most technical parts are left to appendices.

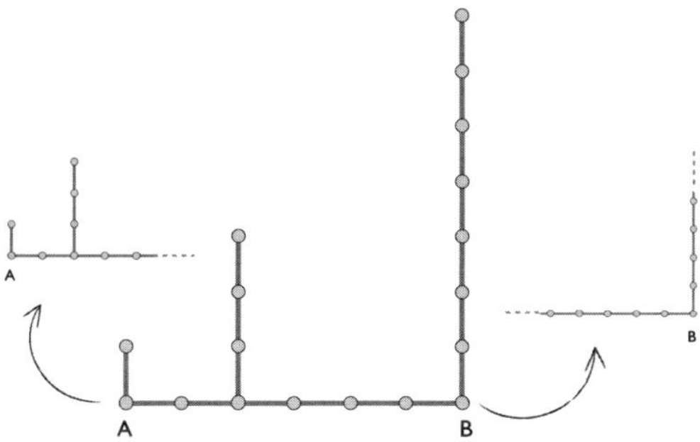

Fig. 1. As this structure is enlarged, the distance between A and B progressively grows and teeth of progressively larger length are inserted. As a result, the neighborhoods of A and B correspond to a finite branched structure and to an infinite chain, respectively.

2. Infinite graphs

Let us define a graph $\mathbb{G}$ as a set $\mathcal{V}$ of sites connected pairwise by a set $\mathcal{E}$ of unoriented links. A couple of sites $i, j \in \mathcal{V}$ is said to be connected whenever $(i, j) = (j, i) \in \mathcal{E}$. The overall number of sites making up the graph is called the volume V of the graph and it is given by the cardinality of the set $\mathcal{V}$, i.e. $V \equiv |\mathcal{V}|$. When V is finite (infinite), $\mathbb{G}$ is called a finite (infinite) graph.

A graph can be algebraically described through the *adjacency matrix* $\mathbf{A}$ of size $V \times V$ and whose entry A_{ij} equals k iff i and j are connected through k edges. It follows that $z_i = \sum_{j=1}^{V} A_{ij}$ is the number of links stemming from i, namely the *degree* (also called coordination number) of i. A graph is said *locally finite* if $z_i < \infty, \forall i \in \mathcal{V}$.

Another fundamental matrix associated with a locally finite graph $\mathbb{G}$ is the *transition matrix* $\mathbf{P}$, whose entries are defined as

$$P_{ij} = \frac{A_{ij}}{z_i}. \tag{1}$$

This is a stochastic matrix and, as such, it generates a Markov chain with state space $\mathcal{V}$. This process is just a simple random walk on $\mathbb{G}$: the sequence of states $X_t \in \mathcal{V}$, $t = 0, 1, ...$, fulfilling the Markov property $\Pr[X_t = i | X_k = i_k, k = 0, 1, ..., t-1] = \Pr[X_t = i | X_{t-1} = j] = P_{ij}, \forall i_0, i_1, ..., i_{t-1} = i, j \in \mathcal{V}$, corresponds to the sequence of positions, reached at time $t = 0, 1, ...$, by a random walk moving along the edges according to probabilities $P_{ij}, \forall i, j \in \mathcal{V}$ (see also [2]).

When dealing with macroscopic systems, composed of a very large number V of sites, one usually takes the thermodynamic limit $V \to \infty$. This procedure has the advantage of providing an infinite structure which effectively describes a number of large systems displaying similar geometrical features, i.e. in infinite graphs local details get negligible. Also, as for the random walk problem, the asymptotic behavior expected on infinite graphs follows the same laws characterizing the behavior of a random walk embedded in large, real structures when the time scale considered is long enough to explore large portions of the systems, but still too short to experience the finite size effects; such time scale is often the most interesting in physical processes.

Formally, the thermodynamic limit of a graph $\mathbb{G}$ can be defined by means of a sequence of generalized Van Hove spheres $S_{o,r} \subset \mathbb{G}$ of center $o \in \mathcal{V}$ and radius[a] r, that is, $S_{o,r}$ is the subgraph of $\mathbb{G}$ containing all nodes $i \in \mathcal{V}$ whose distance from o is not larger than r. Then, an arbitrary observable can be evaluated by restricting the model under study to the finite spheres and then taking the thermodynamic limit by letting $r \to \infty$.

[a] A connected graph is endowed with an intrinsic metric generated by the chemical distance r_{ij}, which is defined as the number of links in the shortest path connecting vertices i and j.

3. Examples

In this section we review some basic examples of structures where the thermodynamic limit is non-unique. Of course, starting from these examples one may figure out several variations on theme exhibiting analogous ambiguities.

3.1. *Trivial non-uniqueness*

Probably, the easiest example of structure where the thermodynamic limit can not be uniquely defined is the semi-infinite chain: here, nodes belonging to the "bulk" of the structure, will see a long (eventually infinite) chain on both sides, while nodes finitely close to the origin will see a long (eventually infinite) chain on one side but the other side will remain finite and unchanged.

When considering, for instance, the mean time τ to first return to the starting point, when starting from a bulk node, we get τ_{bulk}, which, in the thermodynamic limit recovers the result expected for an infinite chain. On the other hand, when starting from a peripheral node, say the one at distance 1 from the origin, we get $\tau_{periph} \sim \tau_{bulk}/2$.

This example can be easily generalized to the case of semi-infinite Euclidean lattices. The T-fractal (compare the central node with any node farthest from the center) and the Sierpinski gasket (compare the apex with the node in the middle of any of the three main sides) also work like that.

In general, such structures display a "soft" inhomogeneity, which, once the thermodynamic limit is taken, may affect quantitatively the main physical quantities, although their asymptotic laws are preserved.

3.2. *Non-trivial non-uniqueness*

When inhomogeneity is strong, the difference between the asymptotic behaviors can not be absorbed into a simple factor. In the following we outline some examples.

3.2.1. *Hierarchical comb*

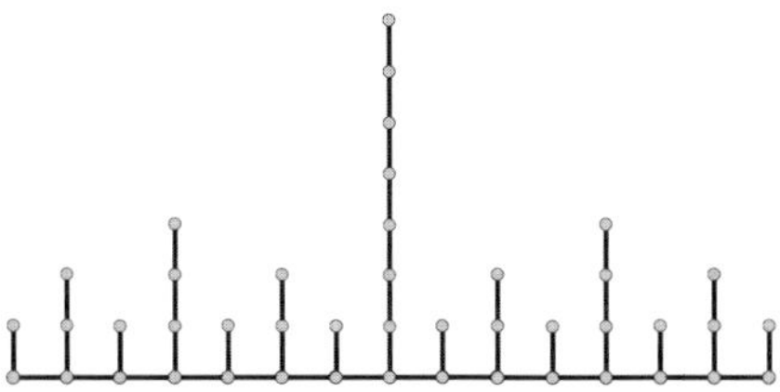

Fig. 2. Hierarchical combs can be built by starting from a chain of length $2^{g+1}-1$, which plays as backbone. Then, we attach to its middle node a chain of length $L_g^{HC} = R^g$; in both half-sides of the backbone, we can outline a middle point, and we attach a chain of length R^{g-1} to both. Now, the backbone is divided into four "bare" segments, we detect the related middle points and we attach a chain of length R^{g-2} to each of the them. We proceed recursively, up to the exhaustion of all bare nodes of the backbone. In the last step there will be 2^g nodes and these will be connected to a chain of length R^0, i.e. to a single node. In this figure we show an example of hierarchical comb with $R = 2$ and $g = 3$.

Combs displaying side branches with a hierarchical distribution of lengths (see Fig. 2) have attracted much attention, due to their ability to mimic the organization of percolation clusters and ultrametricity in spin glasses and related systems, see [1]. Interestingly, the moments of the first passage time of random walks on such structures scale with the size of the system with an infinite hierarchy of exponents characterizing a multifractal behavior [4].

The thermodynamic limit of this structure is non-unique: according to the node where we sit, we may see a semi-infinite chain, or an infinite chain, or a comb with teeth of different (possibly infinite) length.

3.2.2. *Cayley tree*

The Cayley tree (CT) can be built recursively starting from a single node, called root, and connecting it to z nodes, hence realizing a star with z arms of unitary length. Then, at the next step, we attach to each end-node (or "rim") $z-1$ nodes. We proceed iteratively in this way up to the desired generation g.

This structure displays a radial symmetry with respect to the root: nodes added at the g-th step have chemical distance g from the root and can be arranged in shells around the central one. Thus, given a random walk on the CT, if we are simply interested in its position in terms of the distance from the root, we can simplify the problem by collapsing the arms of the tree into a chain of length $L_g^{CT} = g$, where the probability of moving outwards and inwards is $(z-1)/z$ and $1/z$, respectively.

The overall number of nodes V_g^{CT} at generation g is

$$V_g^{CT} = 1 + \sum_{k=0}^{g} z(z-1)^{k-1} = \frac{z(z-1)^g - 2}{z-2}, \tag{2}$$

which is comparable with the number of rims $z(z-1)^{g-1}$. The large number of peripheral nodes may have dramatic consequences in the thermodynamic limit $g \to \infty$ [5].

The non-uniqueness of the thermodynamic limit is highlighted by Fig. 3.

3.2.3. NT_D

NT_D are infinite fractal trees (see Fig. 4) which have been shown to display remarkable properties such as dynamical dimension splitting [6]. Indeed,

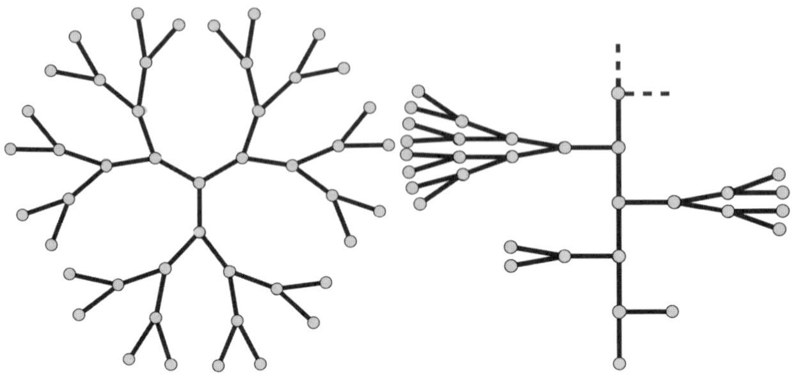

Fig. 3. Cayley tree of generation $g = 4$ and $z = 3$. The topological perspective from the central node (left panel) and from a rim (right panel) are compared.

NT_D are not exactly decimable and therefore the simple decimation techniques cannot be applied, however, NT_D are invariant under a more complex transformation consisting of the product of a cutting transform and a decimation [6]. Along the same line of NT_D one can define the n-polygon NT_D, where each chain of length g is replaced by a polygon of n vertices, each of length g.

Notice that, in the thermodynamic limit, if we start from the root, we will see a branched structure, while if we start from a peripheral node we will see a (semi)infinite chain.

3.2.4. *Extended Cayley tree*

The extended Cayley tree (often refereed to as extended dendrimer, ED) can be built recursively (see Fig. 5) starting from a root and attaching to it z chains of length g. Then, we attach to each end-node of this star $z-1$ chains of length $g-1$. In the following step we attach to each end-node $z-1$ chains of length $g-2$ and so on up to the g-th step where $z(z-1)^{g-2}$

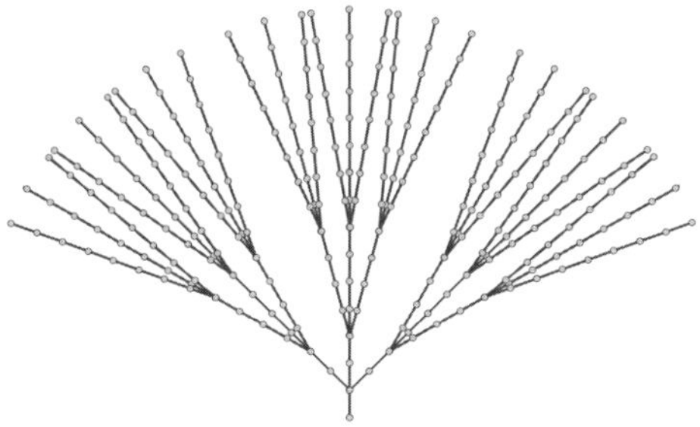

Fig. 4. NT_D fractal trees can be defined recursively as follows: an origin point is connected to a point by a link of unitary length; to the latter point $z-1$ chains of length 2 are attached; the ends of these branches split into $z-1$ branches of length 4 and so on; each endpoint of a branch of length 2^g splits into $z-1$ branches of length 2^{g+1}. In this figure we show an example of NT_D of generation $g=4$ and $z=4$.

end-nodes are connected to $z-1$ chains of length 1, i.e. to single nodes. The overall number of nodes is

$$V_g^{ED} = 1+\sum_{k=0}^{g} z(z-1)^k(g-k) = 1+\frac{z[(z-1)^{g+1}-(g+1)(z-1)+g]}{(z-2)^2}. \quad (3)$$

Given a graph of generation g, we distinguish different classes of nodes: i. rims, ii. branching points (and we refer to this set as $\mathcal{B}$), iii. inner nodes (and we refer to this set as $\mathcal{C} = \cup_{k=1}^{g-1}\mathcal{C}_k$, being $\mathcal{C}_k$ the chain of length $g-k$).

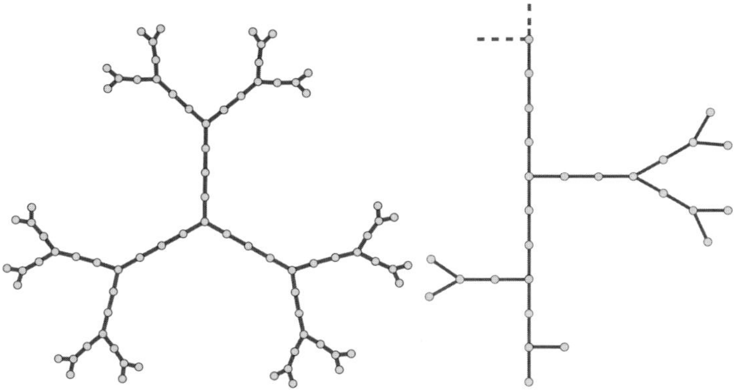

Fig. 5. Extended dendrimer of generation $g=4$ and $z=3$. The topological perspective from the central node (left panel) and from a rim (right panel) are compared.

Extended dendrimers have a completely ordered structure just like CT, but only an approximate self-similarity. Moreover, analogously to the CT, the ED is endowed with a radial symmetry so that the motion of a random walk on it can be looked at as the motion of a random walk on a one-dimensional structure of length $L_g^{ED} = 1 + g(g+1)/2$ and endowed with a bias. However, while for the CT the bias is homogeneous, applying equivalently to any node, but the boundaries, in the ED the bias is effective only on nodes $\in \mathcal{B}$ corresponding to a fraction $\sim g^{-1}$ of the whole one-dimensional structure (see Fig. 6). Of course, one may think of an extended Cayley tree where the length of branches, instead of growing according to an arithmetic progression, grows algebraically. In the latter case, the resulting one-dimensional analogue would be the same as the one obtained by collapsing the arms of the NT_D, but with bias working in the opposite way, namely towards the short-branch side.

In the following we will focus on the extended Cayley tree with arithmetic growth of branches due to their interest from an experimental point of view. In fact, such structures are easily engineered by repeating units in a self-similar way. Dendrimers are the best-known synthetic realization of (compact or extended) CTs and their special structures have been widely shown to present a wide range of potential applications [7–10]. However, probably the most exploited applications concern the context of energy storage and transfer where dendrimers are designed and exploited as efficient light harvesting complexes and artificial antennae [3]. Indeed, while the number of absorbing units grows exponentially with the generation, their distance from the center, where a fluorescent trap, a reaction center, or a chemical sensor, can be located, remains relatively short. Dendrimers are therefore expected to harvest light and transfer it to a trap at the center with high efficiency.

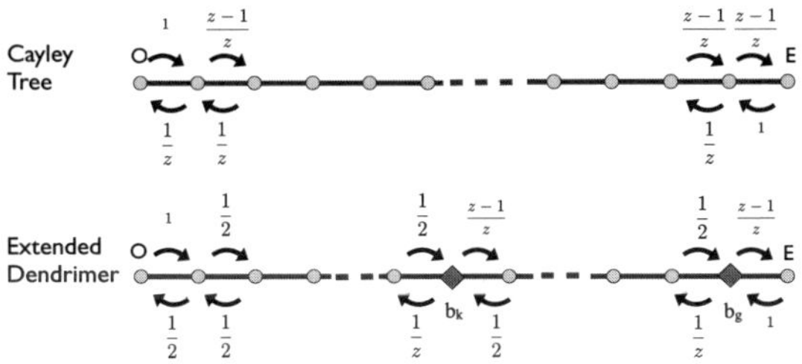

Fig. 6. One-dimensional analogues for CT and ED. In this representation each node is labelled by $i = 1, ..., L_g$, in such a way that the root O corresponds to $i = 1$ and any end-node E to $i = L_g$. In the ED the k-th branching point is located at $i = b_k = k\,g-(k+1)(k-2)/2$; we also fix $b_0 \equiv 1$. Hence, nodes corresponding to $i \in [b_{k-1}+1, b_k-1]$ belong to $\mathcal{C}_k$.

Experiments concerning the electronic absorption on extended dendrimers revealed that, although the slight difference between extended and compact structures, the effects can be significantly different. In particular, the absorption spectra of the latter suggest the existence of an energetic funnel that may enhance absorption of energy from the periphery to the core [11]. This is corroborated by analytical investigations [8] showing that the probability of the excitation being trapped on the extended dendrimer has a decay characterized by a longer tail than is the case for compact dendrimers.

4. Results

We now present the results obtained for the ED and, in order to highlight the peculiarity of this structure, we will often compare with the CT; for all the details of calculations we refer to appendices.

4.1. *Stationary probability*

The random walk diffusion on a locally finite graph $\mathbb{G}$ can be described by means of the probability $p_i(t)$ that the random walk is on site i at time t. Being $\mathbf{P}$ the transition matrix, we get the following vectorial equation

$$p(t+1) = \mathbf{P}\, p(t) = \mathbf{P}^2\, p(t-1) = \cdots = \mathbf{P}^{t+1}\, p_0, \tag{4}$$

where p_0 represents the initial state at $t = 0$.

As mentioned in Sec. 2, a random walk on a graph is a Markov process and we can therefore apply the fundamental limit theorem which states, for regular, finite chains, the existence of a stationary distribution π corresponding to the unique (normalized such that the entries sum to 1) left eigenvector of $\mathbf{P}$ with eigenvalue 1. Moreover, $\lim_{t\to\infty} p(t) = \pi$ and one expects [12]

$$\pi_i = \frac{z_i}{\sum_{i\in\mathcal{V}} z_i} = \frac{z_i}{2|\mathcal{E}|}; \tag{5}$$

when $\mathbb{G}$ is a tree $|\mathcal{E}| = V - 1$.

Such stationary probabilities are directly related to the mean return times by the Kac formula [13]: being τ_i the mean return time to i one has

$$\tau_i = \frac{1}{\pi_i}. \tag{6}$$

In Appendix A, exploiting the radial symmetry of the ED and of the CT, we consider the related one-dimensional structures (see Fig. 6) and get their stationary probabilities η and μ, respectively; the stationary probability for the true CT and ED, i.e. $\tilde{\eta}$ and $\tilde{\mu}$, respectively, will then be recovered by a proper normalization.

In particular, for the CT, in order to get $\tilde{\eta}$ we rescale each entry η_i (see Eqs. (A.6)-(A.8)) by the number of nodes sharing the same distance $i - 1$

from the root, i.e. $z(z-1)^{i-2}$, that is,

$$\tilde{\eta}_O \equiv \eta_1 = \frac{z-2}{2[(z-1)^g - 1]}, \tag{7}$$

$$\tilde{\eta}_i = \frac{\eta_i}{z(z-1)^{i-2}} = \frac{z-2}{2[(z-1)^g - 1]}, \quad \forall i : 1 < r_{i,O} < g \tag{8}$$

$$\tilde{\eta}_E = \frac{\eta_g}{z(z-1)^{g-2}} = \frac{z-2}{2z[(z-1)^g - 1]}. \tag{9}$$

As expected, the asymptotic probability of finding the random walk on any node but the rims is the same.

Recalling Eq. (6) and Eq. (2), we have that the mean time to return to node i on the CT scales as $\tau_i^{\mathrm{CT}} \sim (z-1)^g \sim V_g^{CT}$.

As for the ED, in order to get $\tilde{\mu}$, we rescale each entry μ_i (see Eqs. (A.11)-(A.14)) by the number of nodes sharing the same distance from the root, i.e. $z(z-1)^{k-1}$ for $i \in \{\mathcal{C}_k, b_k\}$, that is,

$$\tilde{\mu}_O \equiv \mu_1 \sim \frac{1}{(z-1)^g}, \tag{10}$$

$$\tilde{\mu}_{b_k} = \frac{\mu_{b_k}}{z(z-1)^{k-1}} \sim \frac{1}{(z-1)^g}, 1 \leq k < g \tag{11}$$

$$\tilde{\mu}_i = \frac{\mu_i}{z(z-1)^{k-1}} \sim \frac{1}{z(z-1)^g}, \quad i \in \mathcal{C}_k \tag{12}$$

$$\tilde{\mu}_E = \frac{\mu_{b_g}}{z(z-1)^{g-1}} \sim \frac{1}{z(z-1)^g}. \tag{13}$$

Again, nodes with analogous degree display the same stationary probability and this is of course mirrored by the mean return times (see Eq. (6)). In the limit of large g we can derive $\tau_i^{ED} \sim (z-1)^g \sim V_g^{ED}$, for any node i (see Eq. (3)), similarly to the CT case.

4.2. *Mean first-passage times*

In the previous subsection we analyzed the mean return time $\tau_i \equiv t(i \to i)$. Another interesting quantity is the mean time to reach the root (O) starting from any end-node (E), as it provides information about the expected time to cross the whole structure and it is often used as a measure for the absorption efficiency of the structure. Again, this time can be calculated exploiting the underlying radial symmetry and, as shown in Appendix B, for the ED it turns out to be

$$t^{ED}(E \to O) \sim g(z-1)^g. \tag{14}$$

Analogous calculations for the CT lead to

$$t^{CT}(E \to O) \sim (z-1)^g. \tag{15}$$

In both cases the crossing time depends exponentially on the number of branching points in between O and E, which corresponds to the generation of the graphs. However, in the two structures the chemical distance $r_{O,E} \equiv L_g$ between O and E scales differently (i.e., quadratically and linearly, respectively) with g. Thus,

$$t^{ED}(E \to O) \sim L_g^{ED}(z-1)^{\sqrt{L_g^{ED}}}, \tag{16}$$

$$t^{CT}(E \to O) \sim (z-1)^{L_g^{CT}}. \tag{17}$$

From this point of view the ED is more efficient.

4.3. *Splitting probability*

Let us consider a random walk starting from a node i on a dendrimer of generation g. We are interested in the so-called *splitting probability* $P_{O|E}(i)$ of ever reaching the root O before visiting any rim (of course, since the structure is finite, its complementary is the probability $P_{E|O}(i)$ of ever reaching any rim before visiting the root, i.e., $P_{E|O}(i) = 1 - P_{O|E}(i)$, for any i). Assuming that the root and all rims are occupied by reaction centers, if the kind of reaction is different according to whether it occurs on either the root or a rim, $P_{O|E}(i)$ provides information on what kind of reaction is expected given the starting point. As all rims are assumed to be equally active we can restrict calculations to the case of a biased chain.

For the CT the solution is straightforward and given by [b]

$$P^{CT}_{E|O}(i) = \frac{1-(\frac{1}{z-1})^i}{1-(\frac{1}{z-1})^g}. \tag{18}$$

This means that the probability of being trapped at any rim grows exponentially with i; from the minimum value $P^{CT}_{E|O}(1) \approx (z-2)/(z-1)$ it saturates to 1.

Now, let us consider the ED case. As explained in Appendix C, it is convenient to treat separately the case $i \in \mathcal{C}$ and $i \in \mathcal{B}$. More precisely,

$$P^{ED}_{E|O}(b_k) \approx \frac{1-(\frac{1}{z-1})^k}{1-(\frac{1}{z-1})^g}, \quad 1 \leq k \leq g, \tag{19}$$

$$P^{ED}_{E|O}(i) \approx P^{ED}_{E|O}(b_{k-1}) + \frac{i}{(g-k)(z-1)^k}, \quad i \in \mathcal{C}_k. \tag{20}$$

Therefore, $P^{ED}_{E|O}(b_k) \approx P^{CT}_{E|O}(k)$, yet, differently from the CT, here the splitting probability to the origin can even assume values larger than $(z-2)/(z-1)$, due the long chain $\mathcal{C}_1$. Indeed, for a large (actually infinite, in the thermodynamic limit) set of starting nodes the probability to first reach the root is larger than the probability to first visit any end node, notwithstanding the bias or, when referring to the true extended dendrimer, the macroscopic amount of traps. This difference between CT and ED ultimately stems from the non-uniqueness of the thermodynamic limit of the latter: for nodes $\in \mathcal{C}_1$ the splitting probability $P^{ED}_{E|O}(i)$ scales linearly with i as it happens on a chain, while for the remaining nodes the splitting

[b]The problem can be seen as the gambler's problem with unfair game.

probability scales exponentially with i as it happens on the CT.

In Fig. 7 (left panel), we show numerical results for $P_{E|O}(i)$, also comparing different sizes. Notice that, at $i = b_1$, $P_{E|O}(i)$ exhibits a flex and it gets steeper and steeper as g grows. This suggests a singular behavior in the thermodynamic limit: walks starting from nodes belonging to the Van Hove sphere centered in O and with radius g are unlikely to be absorbed by rims, while for the remaining nodes the probability to reach rims without ever seeing the root approaches 1.

We also show the mean time to absorption distinguishing between absorption on the root (Fig. 7, middle panel) and on the rim (Fig. 7, right panel). Again, we notice qualitatively different behaviors for nodes $\in \mathcal{C}_1$ and all the remaining nodes.

Finally, we outline a possible application for these features. Let us consider a system where a species A undergoes the following reactions: perfect absorption $A \to \emptyset$ at rims; autocatalytic reaction $A \to \alpha A$ with growth factor $\alpha > 1$ at the root.

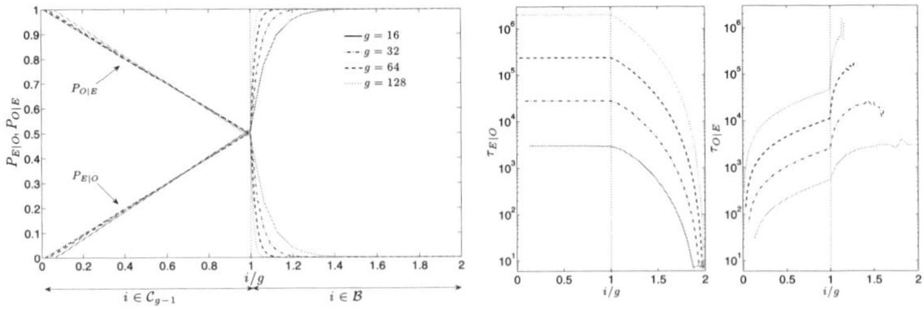

Fig. 7. Splitting probability (left panel), mean time to be absorbed in O (middle panel) and in E (right panel) as a function of the distance (properly normalized) between the starting point and O. Different line styles denote different sizes, as explained by the legend. Notice that the x-axis is not homogeneous: in the first half (referring to nodes $i \in \mathcal{C}_1$) we plot versus the index i divided by g; in the second half (referring to nodes $i \in \mathcal{B}$) we plot versus the branching index divided by g. Thus, the singularity at $i/L_g = 1$ corresponds to the first branching node b_1. This normalization allows to effectively compare several generations.

For the CT, being i the starting point, the amount of particles M, as a function of the number k of passages to the root, is expected to vary according to $M(k) = P_{O|E}(i)\alpha M(k-1) = P_{O|E}(i)M(0)\alpha[\alpha P_{O|E}(1)]^k$. Recalling Eq. (18), we need to choose $\alpha > z-1$ for the particle to be expected to survive. As for the ED, analogous calculations hold and, in this case, for the particle to survive it suffices that $\alpha > 1 + 1/g$, namely, asymptotically, that α is not smaller than one.

5. Discussion

From our analysis, it comes out as a primary result that topological inhomogeneities, usually studied on infinite networks, can play a crucial role in random walks behaviour even on finite graphs. This is of particular relevance if we consider that real systems, described by such mathematical models, are normally not only finite, but even small. Future investigations, exploring the effects of inhomogeneity on random structures, could unveil unexpected synergies, due to its interplay with disorder, and leading to a possibly new phenomenology.

Acknowledgments

Financial support from the FIRB project RBFR08EKEV is acknowledged.

Appendix A. Stationary probability

Given the transition matrices, $\mathbf{P}^{\mathbf{CT}}$ and $\mathbf{P}^{\mathbf{ED}}$, for the one-dimensional structures related to the CT and to the ED (see Fig. 6), we can calculate the stationary probabilities η and μ, which satisfy $\eta\mathbf{P}^{\mathbf{CT}} = \eta$ and $\mu\mathbf{P}^{\mathbf{ED}} = \mu$, respectively. Otherwise stated, η and μ are the Frobenius-Perron vectors for the transition matrices $\mathbf{P^{CT}}$ and $\mathbf{P^{ED}}$, respectively.

Here, we calculate η and μ from which we derive the stationary probabilities $\tilde{\eta}$ and $\tilde{\mu}$ for the original CT and ED, hence recovering the expected result of Eq. (5); this exercise allows to get familiar with the algebraic properties of $\mathbf{P^{CT}}$ and of $\mathbf{P^{ED}}$, which shall be exploited also in the following.

Now, for the Cayley tree of generation g the non-null elements of the transition matrix $\mathbf{P^{CT}}$ are: $P_{1,2}^{CT} = P_{g,g-1}^{CT} = 1$, $P_{i,i+1}^{CT} = \frac{z-1}{z}$, for $2 \leq i \leq g-1$, $P_{i,i-1}^{CT} = \frac{1}{z}$, for $2 \leq i \leq g-1$, namely,

$$\mathbf{P^{CT}} = \begin{pmatrix} 0 & 1 & 0 & 0 & \cdots & 0 \\ \frac{1}{z} & 0 & \frac{z-1}{z} & 0 & \cdots & 0 \\ 0 & \frac{1}{z} & 0 & \frac{z-1}{z} & \cdots & 0 \\ \vdots & \vdots & \ddots & \vdots & 0 & 0 \\ 0 & 0 & \cdots & \frac{1}{z} & 0 & \frac{z-1}{z} \\ 0 & 0 & \cdots & 0 & 1 & 0 \end{pmatrix}. \tag{A.1}$$

Therefore, we have

$$\eta_2 = z\eta_1, \tag{A.2}$$

$$\eta_3 = (z-1)\eta_2 = z(z-1)\eta_1, \tag{A.3}$$

$$\vdots$$

$$\eta_{g-1} = (z-1)\eta_{g-2} = z(z-1)^{g-3}\eta_1, \tag{A.4}$$

$$\eta_g = \frac{z-1}{z}\eta_{g-1} = (z-1)^{g-2}\eta_1. \tag{A.5}$$

Imposing the normalization one finds

$$\eta_1 = \frac{z-2}{2[(z-1)^g - 1]}, \tag{A.6}$$

$$\eta_j = \frac{z(z-1)^{i-2}(z-2)}{2[(z-1)^g - 1]}, \quad 2 \le i \le g-1, \tag{A.7}$$

$$\eta_g = \frac{(z-1)^{g-2}(z-2)}{2[(z-1)^g - 1]}. \tag{A.8}$$

The latter expressions have been used to get Eqs. (7)–(9).

For the extended dendrimer of total length $L = 1 + g(g+1)/2$, the transition probability $\mathbf{P^{ED}}$ displays the following non-null elements: $P^{ED}_{1,2} = P^{ED}_{L,L-1} = 1$, $P^{ED}_{i,i+1} = P^{ED}_{i,i-1} = \frac{1}{2}$, for $i \in \mathcal{C}$, $P^{ED}_{i,i+1} = \frac{z-1}{z}$, for $i \in \mathcal{B}$, $P^{ED}_{i,i-1} = \frac{1}{z}$, for $i \in \mathcal{B}$, namely,

$$\mathbf{P^{ED}} = \begin{pmatrix} 0 & 1 & 0 & 0 \cdots & & 0 \\ \frac{1}{2} & 0 & \frac{1}{2} & 0 \cdots & & 0 \\ 0 & \frac{1}{2} & 0 & \frac{1}{2} \cdots & & 0 \\ \vdots & \vdots & \ddots & \vdots & 0 & 0 \\ 0 & 0 & \cdots & \frac{1}{z} & 0 & \frac{z-1}{z} \\ 0 & 0 & \cdots & 0 & 1 & 0 \end{pmatrix}. \tag{A.9}$$

We can detect, within $\mathbf{P^{ED}}$, the tridiagonal blocks $\mathbf{C}_k$, $k = 1, ..., g-1$ of size $(g-k) \times (g-k)$, corresponding to the transition rates of an unbiased random walks embedded in chains of length $g - k$ and open boundary conditions, respectively; each block $\mathbf{C}_k$ corresponds to the branch $\mathcal{C}_k$.

Now, in each branch $\mathcal{C}_k$, we can find the stationary probability $\mu^{(k)}$ ($\mu^{(k)} = \{\mu^{(k)}_1, ..., \mu^{(k)}_{g-k}\}$) as a function of the boundary probabilities, $\mu^{(k)}_-$ and $\mu^{(k)}_+$, namely the probabilities for the adjacent branching points b_{k-1} and b_k ($\mu^{(k)}_- \equiv \mu_{b_{k-1}}, \mu^{(k)}_- \equiv \mu_{b_k}$, $b_0 \equiv 1$). Notice that from $\mu^{(k)}$ we can recover the vector μ via the mapping $\mu^{(k)}_i = \mu_{b_k + i}$. For the arbitrary k-th branch we can write $\mu^{(k)} = \mathbf{C}_k \mu^{(k)} + \theta^{(k)}$, where $\theta^{(k)}$ is a column vector of length $g-k$ given by $\theta^{(k)} = \left(\mu_-^{(k)}(z-1)/z, 0, ..., \mu_+^{(k)}/z\right)^T$. Therefore, for any branch, we have the vectorial equation

$$\mu^{(k)} = (\mathbf{I} - \mathbf{C}_k)^{-1}\theta^{(k)}, \tag{A.10}$$

where $\mathbf{I}$ is the identity matrix [here it has size $(g-k)$]. The boundary conditions are provided by $\mu_-^{(k)} = \frac{1}{2}[\mu_{g-k+1}^{(k-1)} + \mu_1^{(k)}]$ and $\mu_+^{(k)} = \frac{1}{2}[\mu_{g-k}^{(k)} + \mu_1^{(k+1)}]$. Now, the inverse of $(\mathbf{I} - \mathbf{C}_k)$, referred to as $\Phi^{(k)}$, can be easily calculated because it is a symmetric tridiagonal matrix. Exploiting [14], we find that $\Phi^{(k)}$ has diagonal elements: $\Phi_{1,1}^{(k)} = \Phi_{g-k,g-k}^{(k)} = \frac{2k}{k+1}$, $\Phi_{i,i}^{(k)} = \frac{2i(k-i+1)}{k+1}$, and its off-diagonal elements are $\Phi_{ij}^{(k)} = \Phi_{ji}^{(k)} = \frac{2i(k-i+1)}{k+1}$. Imposing the boundary conditions we find

$$\mu_{b_0} \equiv \mu_1 = \frac{(z-2)^2}{2[(z-1)^{g+1} - (g+1)(z-1) + g]}, \tag{A.11}$$

$$\mu_{b_k} = \mu_1 z(z-1)^{k-1}, \ 1 \leq k < g \tag{A.12}$$

$$\mu_i^{(k)} = 2\mu_1(z-1)^{k-1}, \ i \in \mathcal{C}_k \tag{A.13}$$

$$\mu_{b_g} \equiv \mu_L = \mu_1(z-1)^{g-1}. \tag{A.14}$$

The latter expressions have been used to get Eqs. (10)–(13).

Appendix B. Crossing time

A key time scale characterizing the trapping processes in dendrimers is the mean time to reach the root given that the walker was initially placed at a boundary site. In order to calculate this quantity, different techniques can be adopted.

For instance, a very powerful matricial approach is based on the recurrent equations $\tau_i^{(j)} = \sum_k P_{i,k}^{(j)} \tau_k^{(j)} + 1, \quad i \neq j$, being j the trap location and

$\tau_i^{(j)}$ the mean time to absorption from initial site $i \neq j$, moreover, $\mathbf{P}^{(j)}$ is obtained from $\mathbf{P}$ by deleting the j-th row and the j-th column (this equals to say that j is a trap). In matrix notation, $\tau^{(j)} = \mathbf{P}^{(j)}\tau^{(j)} + \mathbf{e}$, where $\tau^{(j)} = (\tau_1^{(j)}, ..., \tau_{j-1}^{(j)}, \tau_{j+1}^{(j)}, ..., \tau_V^{(j)})^T$ is a $(V-1)-$dimensional vector and $\mathbf{e}$ is the $(V-1)-$dimensional "one" vector $\mathbf{e} = (1, 1, ..., 1)^T$. Therefore,

$$\tau^{(j)} = (\mathbf{I} - \mathbf{P}^{(j)})^{-1}\mathbf{e} \equiv \mathbf{Q}\mathbf{e}, \tag{B.1}$$

where $\mathbf{Q}$ is often referred to as *fundamental matrix* (see e.g. [12]).

For the special case of the ED with trap on the root, one can consider the matrix $(\mathbf{I} - \mathbf{P}^{\mathbf{CT}^{(1)}})$ and proceed with the evaluation of its inverse, exploiting the fact that the matrix to be inverted is tridiagonal [14].

From another perspective, one can notice that the mean absorbing times $\tau_i^{(1)}$ fulfill analogous recurrence equations: focusing on the $(k+1)$-th branch bounded by b_k and b_{k+1},

$$\tau_{i+2}^{(1)} = 2\tau_{i+1}^{(1)} - \tau_i^{(1)} - 2, \quad b_k + 1 < i \leq b_{k+1}, \tag{B.2}$$

$$\tau_i^{(1)} = \frac{z\tau_{i-1}^{(1)} - \tau_{i-2}^{(1)} - z}{z-1}, \quad i = b_k + 1. \tag{B.3}$$

From such equations one can derive that the mean time to be absorbed saturates to the maximum value $\tau_E^{(1)}$ exponentially with k, while it grows algebraically with respect to i for a given k.

Finally, we sketch an approach to get the crossing time, which is based on a renormalization procedure which maps the branching points b_k of the graph at generation g into the branching points b_{k+1} of the graph at generation $g+1$. We introduce the notation $t_g(i \to j)$, where i and j refer to the g-th generation of the graph. For the crossing time we have

$$t_g(E \to O) = t_g(E \to b_1) + t_g(b_1 \to O), \tag{B.4}$$

where, $t_g(E \to b_1) \equiv t_{g-1}(E \to O)$ while

$$t_g(b_1 \to O) = 1 + \frac{1}{z} t_g(b_1 - 1 \to O) + \frac{z-1}{z}[t_g(b_1 + 1 \to b_1) + t_g(b_1 \to O)].$$

Now, $t_g(b_1 + 1 \to b_1) \equiv t_{g-1}(2 \to O)$ is simply the mean return time to 1 at generation $g-1$, i.e. $\tau_{1,g-1}$, minus one, moreover, by recurrence, one finds $t_g(i \to O) = (i-1) + [(i-1)/i]t_g(i+1 \to O)$, for $1 \leq i < g$, from which $t_g(b_1 - 1 \to O) = (g-1) + [(g-1)/g]t_g(b_1 \to O)$. Exploiting such results, with some algebra, we get

$$t_g(E \to O) = t_{g-1}(E \to O) + g[1 + (z-1)\tau_{1,g-1}], \tag{B.5}$$

which, recalling results for mean return time (see Eq. (5) and Appendix A), becomes

$$\begin{aligned} t_g(E \to O) = \frac{1}{6(z-2)^4}\{&6g(z-2)(z-1)^{2+g} - 6(z-1)^g \\ &- 2g^3(z-2)^3(z-1) + g(z-2)^2[3g(2z-3) - 2z(z-6) - 13] \\ &+ 6(z-1)^2\} \sim \frac{g}{(z-2)^3}(z-1)^{g+2}. \end{aligned}$$

Analogous calculations for the CT lead to

$$\begin{aligned} t_g^{CT}(E \to O) &= \frac{1}{(z-2)^2}\{2(z-1)^g + z[z-4-g(z-2)] + 2\} \\ &\sim \frac{2}{(z-2)^2}(z-1)^g. \end{aligned}$$

Appendix C. Splitting probability

To calculate the splitting probabilities $P_{O|E}$ and $P_{E|O}$ we can exploit the properties of absorbing Markov chains (see e.g. [12]). In particular, for an arbitrary chain displaying a set $\mathcal{A} \in \mathcal{V}$, with $|\mathcal{A}| = k < V$, of absorbing states, and the remaining transient states, we properly reshuffle the columns and the rows of $\mathbf{P}$, in such a way that the first $V-k$ states are transient and the last k states are absorbing. In this way we get the so-called canonical form

$$\mathbf{P} = \left(\begin{array}{c|c} \mathbf{Q} & \mathbf{R} \\ \hline \mathbf{0} & \mathbf{I} \end{array}\right), \tag{C.1}$$

where $\mathbf{I}$ is a $k \times k$ identity matrix, $\mathbf{0}$ is a $k \times (V-k)$ zero matrix, $\mathbf{R}$ is a non-zero $(V-k) \times k$ matrix and $\mathbf{Q}$ is a $(V-k) \times (V-k)$ matrix.

Now, when building $\mathbf{P}^n$, the entries of $\mathbf{Q}^n$ give the probabilities of being in each of the transient states after n steps for each possible transient starting state. Then, the entry N_{ij} of the fundamental matrix $\mathbf{N} \equiv (\mathbf{I} - \mathbf{Q})^{-1} = I + Q^2 + Q^3 + ...$, is the expected number of times that the chain is in state j, given that it starts in state i. We can therefore derive that $\tau^{(\mathcal{A})} = \mathbf{Ne}$ is the expected number of steps before the chain is absorbed. More precisely, $\tau_i^{(\mathcal{A})}$ is the expected time when the chain starts in i. Finally, we can also write $\mathbf{B} = \mathbf{NR}$, where B_{ij} is the probability that an absorbing chain will be absorbed in the state j if it stars in the transient state i.

For the particular case of the ED, where the root and all end-nodes are absorbing, one properly reshuffles $\mathbf{P^{ED}}$ (see Eq. (A.9)) to obtain the canonical form and sees that to get the fundamental matrix it is necessary to invert a matrix in the form $\left(\begin{array}{c|c} 1 & -\mathbf{a} \\ \hline -\mathbf{b} & \mathbf{I} - \mathbf{U} \end{array}\right)$, where $\mathbf{a} = (0, 0, ..., 1/z)$, $\mathbf{b} = (0, 0, ..., 1/z, 0, (z-1)/z)^T$ and $\mathbf{U}$ corresponds to the bulk of $\mathbf{P^{ED}}$ (namely the square block of size $V-2$ unaltered by reshuffling). Thus, by calling $\mathbf{N} = (\mathbf{I} - \mathbf{U})^{-1}$, the fundamental matrix gets $\left(\begin{array}{c|c} \mathbf{I} - \mathbf{aNb} & \mathbf{Ia}(\mathbf{N}^{-1} - \mathbf{bIa})^{-1} \\ \hline \mathbf{Nb}(\mathbf{I} - \mathbf{aNb}) & (\mathbf{U} - \mathbf{bIa})^{-1} \end{array}\right)$. After inversion we need to multiply the fundamental matrix by $z^{-1}\begin{pmatrix} (z-1)\,0\,0\,...\,0 \\ 0 \quad 1\,0\,...\,0 \end{pmatrix}^T$, to get the splitting probability. We therefore need only the first two columns of the fundamental matrix, which can be calculated exactly exploiting the tridiagonal form of $\mathbf{U}$ [14]. Calculations are somehow analogous to those implemented for the crossing time in Appendix B: $P_{E|O}(i)$ grows linearly as i is varied within any $\mathcal{C}_k$, but it grows exponentially when branching points are considered.

References

1. S. Redner, *A Guide to First-passage Processes.* Cambridge University Press (2007).
2. W. Woess, *Random walks on infinite graphs and groups.* Cambridge Tracts in Mathematics, Cambridge University Press (2000).
3. R. Kopelman, M. Shortreed, Z.-Y. Shi, W. Tan, Z. Xu, and J. S. Moore, Spectroscopic evidence for excitonic localization in fractal antenna supermolecules, *Phys. Rev. Lett.* **78**, 1239–1242 (1997).
4. S. Havlin and O. Matan, Multifractal nature of diffusion on hierarchical structures, *J. Phys. A: Math. Gen.* **21**, L307 (1988).
5. R. J. Baxter, *Exactly solved models in statistical mechanics.* London: Academic Press Inc. (1982).
6. R. Burioni and D. Cassi, Random walks on graphs: ideas, techniques and results, *J. Phys. A: Math. Gen.* **38**, R45–R78 (2005).
7. V. Balzani, P. Ceroni, M. Maestri, and V. Vicinelli, Light-harvesting dendrimers, *Current Opinion in Chemical Biology.* **7**, 657–665 (2003).
8. J. L. Bentz, F. N. Hosseini, and J. J. Kozak, Influence of geometry on light harvesting in dendrimeric systems, *Chem. Phys. Lett.* **370**, 319–326 (2003).

9. D. Katsoulisa, P. Argyrakis, A. Pimenovb, and A. Vitukhnovsky, Diffusion and trapping in dendrimer structures, *Chem. Phys.* **275**, 261–269 (2002).
10. A. A. Gurtovenko, D. A. Markelov, Y. Y. Gotlib, and A. Blumen, Dynamics of dendrimer-based polymer networks, *J. Chem. Phys.* **119**, 7579 (2003).
11. D. Rana and G. Gangopadhyay, Steady-state spectral properties of dendrimer supermolecule as a light harvesting system, *Chem. Phys. Lett.* **334**, 314–324 (2001).
12. J. G. Kemeny and J. L. Snell, *Finite Markov Chains*, Springer-Verlag (1981).
13. S. Condamin, O. Bénichou, and M. Moreau, Random walks and brownian motion: a method of computation for first-passage times and related quantities in confined geometries., *Phys. Rev. E: Stat. Nonlin. Soft Matter Phys.* **75**, 021111 (2007).
14. Y. Huang and W. F. McColl, Analytical inversion of general tridiagonal matrices, *J. Phys. A: Math. Gen.* **30**, 7919–7933 (1997).

Chapter 6

Effective Spectral Dimension in Scale-Free Networks

S. Hwang

Department of Physics and Astronomy, Seoul National University, Seoul 151-747, Korea

D.-S. Lee

Department of Natural Medical Sciences and Department of Physics, Inha University, Incheon 402-751, Korea

B. Kahng

Department of Physics and Astronomy, Seoul National University, Seoul 151-747, Korea

1. Introduction

During the last decade, it was revealed that many of social, technological, and biological networks exhibit scale-free nature in their connectivity patterns [1]. Such heterogeneous topology significantly influences the dynamical processes. To understand the fundamental features of dynamics on scale-free networks, we consider random walks in this chapter as the simplest model dynamics.

The return to origin (RTO) probability and the first passage time (FPT) distribution are two major quantities of interest in the study of random walks. The RTO probability is defined as the probability that the random walker returns to the starting node after t time steps, and the FPT distribution is the probability that the random walker reaches the target node for the first time at time t. Those two quantities are closely related to each other. In this work, we show that they exhibit crossover behaviors as functions of time in scale-free networks, which are absent in regular or fractal lattices. We first review the basic properties of random walks on regular lattices in Sec. 2. Then we introduce prototypical models of complex networks and their properties in Sec. 3. In Sec. 4, we derive the crossover behavior of the RTO probability by introducing the effective degree of the starting node. An exactly solvable case is presented as well. In turn, we use the obtained behaviors of the RTO probability to derive the global FPT (GFPT) and the GFPT distribution in Sec. 5, which behave differently depending on the spectral dimension and the degree exponent.

2. Random walks on regular and disordered lattice in Euclidean space

2.1. *Spectral dimension*

Let us consider simple random walks on regular and disordered lattice structures in d dimensional Euclidean space. Initially, a random walker (RWer) is positioned at one of the lattice points, which we denote as s. At each time, the RWer hops to one of the nearest neighbor nodes with equal probability. The occupations probability $P_{s\to i}(t)$, which is the probability that the random walker locates at the lattice point i after time step t, evolves with time as

$$P_{s\to i}(t+1) = \sum_{j\in \mathrm{n.n}(i)} \frac{1}{k_j} P_{s\to j}(t), \tag{1}$$

where n.n(i) denotes the set of the nearest neighbor lattice points of i and k_i is the number of n.n(i). In disordered lattices, k_i may be different depending on node.

Analogous to the Laplacian operator, ∇^2, in the continuum Euclidean space, one can define the normalized Laplacian matrix L, of which the element is $L_{ij} = \delta_{ij} - A_{ij}/k_j$, and A_{ij} is 1 if the lattice points i and j are connected to each other and 0 otherwise [2]. Using the normalized Laplacian matrix, we rewrite Eq. (1) as

$$P_{s\to i}(t+1) - P_{s\to i}(t) = -\sum_{j=1}^{N} L_{ij} P_{s\to j}(t), \tag{2}$$

with the initial condition $P_{s\to j}(0) = \delta_{sj}$.

The formal solution of Eq. (2) can be represented as

$$P_{s\to i}(t) = \sum_{j=1}^{N} (I-L)^t_{ij} P_{s\to j}(0) = (I-L)^t_{is}. \tag{3}$$

Therefore, the occupation probability $P_{s\to i}(t)$ is obtained in terms of the eigenvalues and the eigenvectors of the matrix L. In particular, for large t, the contribution from the large eigenvalues becomes increasingly small, and the distribution of small eigenvalues essentially determines the behavior of $P_{s\to i}(t)$. In a wide class of disordered lattices, the spectral density distribution $\rho(\mu) \equiv \frac{1}{N}\sum_{i=1}^{N}\delta(\mu-\lambda_i)$ of the eigenvalues of L behaves as

$$\rho(\mu) \sim \mu^{d_s/2-1} \tag{4}$$

for $\mu \ll 1$, where d_s is called the spectral dimension, which coincides with the dimension d in regular lattices [3, 4]. We will show that the spectral dimension plays an important role in classifying the behavior of random walks in both regular and disordered lattices.

2.2. *Mean-square-root distance vs time*

In random walks with finite step lengths, the occupation probability after time step t can be approximated as a Gaussian distribution with the standard deviation $\sigma \sim \sqrt{t}$ by the central limit theorem. The mean square displacement $\langle r^2\rangle_c \equiv \langle r^2\rangle - \langle r\rangle^2$ is proportional to the variance and scales linearly with time steps as

$$\langle r^2\rangle_c \sim t, \tag{5}$$

in any d dimensional regular lattices.

However, for random walks in fractals such as Sierpinski gasket or percolation clusters [3], Eq. (5) does not hold generally, but is written in a general form as

$$\langle r^2\rangle_c \sim t^{2/d_w}, \tag{6}$$

where we define the random walk dimension d_w, which is not always two. Comparing both Eqs. (5) and (6), one finds that the random walk dimension d_w is two in regular lattices. If $d_w > 2$, the process is regarded as subdiffusive; if $d_w < 2$ it is as superdiffusive. Subdiffusion is typically encountered in disordered media such as porous systems and proteins in the nucleus. Superdiffusion appears in turbulent system, surfactant diffusion in living polymers, and so on. For a review of the processes displaying anomalous diffusion, see, e.g., [3, 5].

2.3. Probability to return to the origin

One of the important quantity characterizing the random walks in a given substrate is the return-to-origin (RTO) probability $R_s(t) \equiv P_{s\to s}(t)$ [6]. The inverse of RTO probability may be interpreted as the effective number of lattice points that random walker can cover until time step t.

In d dimensional regular lattice with the periodic boundary condition, since all lattice points are topologically identical, the occupation probability in the steady state is given by $P_{si} = 1/N$ for all i.[a] The RTO probability $R_s(t)$ decays from 1 to $1/N$ and thus the inverse of $R_s(t)$ quantifies the effective number of lattice points. From Eq. (3), we find that $R_s(t) = (I-L)^t_{ss}$. Since all the nodes are topologically identical in regular lattices, the diagonal element $(I-L)^t_{ss}$ can be replaced by $\frac{1}{N}\operatorname{Tr}(I-L)^t$. In the limit $N \to \infty$, we have,

$$R_s(t) = \frac{1}{N}\operatorname{Tr}(I-L)^t = \int \mathrm{d}\mu\rho(\mu)(1-\mu)^t$$
$$\simeq \int \mathrm{d}\mu\rho(\mu)\exp(-\mu t), \tag{7}$$

where the last approximation holds for large t so that the contributions for μ not close to zero become negligible. Eq. (7) shows that the long time behavior of $R_s(t)$ and the small-μ behavior of $\rho(\mu)$ are related via the Laplace transformation.

In the limit of large system size, the normalized Laplacian matrix L in regular lattices has Fourier components as its eigenvalues and thus one obtains $\rho(\mu)\,\mathrm{d}\mu \sim k^{d-1}dk$, where $\mu = k^2$ and $k = |\vec{k}|$ with $\vec{k}$ being the wave vector. Then we obtain

$$\rho(\mu) \sim \mu^{\frac{d}{2}-1}. \tag{8}$$

Inserting Eq. (8) into Eq. (7), we have

$$R_s(t) \sim t^{-d/2}. \tag{9}$$

Since the number of lattice points within the radius r scales as r^d, the total number of lattices points covered by random walker can be obtained

[a] Depending on the lattice structure, the event that the RWer returns to the origin can only occur if t is an even integer. This means that spectral density function is symmetric about $\mu = 1$. In this case, there are two sets of lattices points where the RWer can reach the lattice points in the first set at even t while it can in the second set at odd t. Even though this periodicity can be found in many lattice structures, we are only interested in the envelope of the RTO probability. Therefore, we ignore the contribution from $\rho(\mu)$ where μ is near two.

by $(t^{1/2})^d$ from Eq. (5), which gives the same result as in Eq. (9).

In the disordered media, we consider the node-averaged RTO probability $R(t) \equiv \frac{1}{N}\sum_{i=1}^{N} R_i(t)$. Following Eqs. (4) and (7),

$$\begin{aligned} R(t) = \mathrm{Tr}(I-L)^t &\simeq \int \mathrm{d}\mu\rho(\mu)\exp(-\mu t) \\ &\sim \int \mathrm{d}\mu\mu^{d_s/2-1}\exp(-\mu t) \sim t^{-d_s/2}. \end{aligned} \tag{10}$$

In [4, 6], it was shown that the node-specific RTO, $R_s(t)$ also exhibits the same scaling as in Eq. (10) in a large class of fractal lattices embedded in Euclidean space,

$$R_s(t) \sim t^{-d_s/2}. \tag{11}$$

The spectral dimension d_s is related to classifying recurrent and transient random walks. The probability to return to the origin, independent of the return time, is evaluated by $1 - \frac{1}{\int_0^\infty R(t)\,\mathrm{d}t}$. If the integral $\int_0^\infty R(t)\,\mathrm{d}t$ diverges, then the random walk is called recurrent, otherwise, it is called transient. From Eq. (11), we can conclude that $d_s = 2$ is the critical dimension so that the RWs are recurrent if $d_s < 2$ and are transient otherwise.

In disordered lattices, the nodes are not topologically identical, the occupation probability in the steady state is given by $P_{s\to i}(\infty) = k_i/2L$ with k_i the degree of node i [7]. Then the probability π_{ij} that the random walker crosses a link from node i to j in the steady state can be written as

$$\pi_{ij} = P_{s\to i}(\infty)\frac{1}{k_i} = \frac{1}{2L}. \tag{12}$$

This means that π_{ij} is uniform for all links (ij). Thus, the time evolution of the occupation probability of random walks can be considered as the relaxation of a state having such equivalent link-crossing probability. This suggests that we can define the effective total number of links $\hat{L}(t)$ passed through by the RWer, which satisfies,

$$R_s(t) \simeq \frac{k_s}{2\hat{L}(t)}, \tag{13}$$

and, from Eq. (10),

$$\hat{L}(t) \sim t^{d_s/2}. \tag{14}$$

2.4. *First passage time probability*

The first-passage properties (FPT) have been of great interest in the study of random walks. Especially, the FPT probability $F_{s\to m}(t)$, the probability that the random walker, initially placed at s, reaches the target lattice point m for the first time has drawn much attention [8]. The FPT probability and the occupation probability are closely related,

$$P_{s\to m}(t) = \delta_{sm}\delta_{t0} + \sum_{t'=0}^{t} F_{s\to m}(t')P_{m\to m}(t-t'). \tag{15}$$

Then the generating function $\mathcal{F}_{s\to m}(z) \equiv \sum_t z^t F_{s\to m}(t)$ is related to the generating function of the occupation probability $\mathcal{P}_{s\to m}(z) \equiv \sum_t z^t P_{s\to m}(t)$ as

$$\mathcal{F}_{s\to m}(z) = \begin{cases} 1 - 1/\mathcal{R}_m(z) & \text{for} \quad m = s, \\ \mathcal{P}_{s\to m}(z)/\mathcal{R}_m(z) & \text{for} \quad m \neq s. \end{cases} \tag{16}$$

Here, $\mathcal{R}_m(z)$ is the generating function of the RTO probability: $\mathcal{R}_m(z) = \sum_t z^t R_m(t)$. If we define the global first passage time (GFPT) distribution as

$$F_m(t) \equiv \sum_{s=1}^{N} \frac{k_s}{2L} F_{s\to m}(t), \tag{17}$$

then it can represent the accessibility to a node m for the first time. Note that this formula is valid for finite N and L. The GFPT distribution is the average of the FPT distribution over all sites s with the weight of $\frac{k_s}{2L}$, which is the probability of finding the RWer at node i in the steady state [7]. Then, the generating function of the GFPT distribution $\mathcal{F}_m(z) \equiv \sum_{t=0}^{\infty} F_m(t)z^t$ can be written in a closed form:

$$\begin{aligned} \mathcal{F}_m(z) &= \frac{k_m}{2L}\left(1 - \frac{1}{\mathcal{R}_m(z)}\right) + \sum_{i\neq m} \frac{k_i}{2L}\frac{\mathcal{P}_{i\to m}(z)}{\mathcal{R}_m(z)} \\ &= \frac{k_m z}{2L(1-z)}\frac{1}{\mathcal{R}_m(z)}, \end{aligned} \tag{18}$$

where we used the relations $k_i\mathcal{P}_{m\to i}(z) = k_m\mathcal{P}_{i\to m}(\text{z})$ [7] and $\sum_{i\neq m} \mathcal{P}_{i\to m}(z) = (1-z)^{-1} - \mathcal{R}_m(z)$. It should be noted that $\mathcal{F}_m(z)$ depends only on $\mathcal{R}_m(z)$.

Plugging Eq. (9) into Eq. (18), we obtain that,

$$F_m(t) \sim \begin{cases} t^{-1+\frac{d_s}{2}} & \text{for} \quad d_s < 2, \\ \dfrac{1}{\ln t} & \text{for} \quad d_s = 2, \\ \text{const.} & \text{for} \quad d_s > 2. \end{cases} \tag{19}$$

This expression is valid for $1 \ll t \ll \tau$ with some characteristic time τ which can be obtained [9].

3. Complex networks

The constituents and their interaction in a given complex system can be represented by nodes and links, respectively, offering the change to view the global organization of the system. The concepts and methods of the graph theory turn out to be very useful to analyze the structure of those networks, and have become widespread to various disciplines including physics, molecular biology, computer science, sociology and many other fields [10, 11]. The topologies of the underlying networks are crucial for the dynamical processes occurring on them. Random walks are the simplest dynamics influenced by network topologies; but they still capture the essence of diverse dynamical properties.

3.1. *Erdős-Rényi random graphs and scale-free networks*

In investigating network structures, it is useful to consider an ensemble of graphs with some fixed macroscopic variables. One natural choice is to fix the total number of nodes N and links L. The corresponding graph ensemble was originally proposed by Erdős and Rényi [12], hence we call the ensemble $G_{N,L}$ and a graph an Erdős-Rényi(ER) graph. An important variation is $G_{N,p}$, which is the graph ensemble with a fixed number of nodes N and connection probability p [13]. In the limit $N \to \infty$, $G_{N,L}$ and $G_{N,p}$ are statistically equivalent [14].

For a graph in $G_{N,p}$, every node can have $N-1$ possible links which are connected with probability p. Therefore, the probability for a node to have k links or the degree distribution in $G_{N,p}$ is given by the binomial distribution

$$P_d(k) = \binom{N-1}{k} p^k (1-p)^{N-1-k}. \tag{20}$$

Here, the index d is used to identify $P_d(k)$ as the degree distribution. In the limit $N \to \infty$ with fixed $Np = \langle k \rangle$, one gets

$$P_d(k) = \exp(-\langle k \rangle)\frac{\langle k \rangle^k}{k!}, \tag{21}$$

which is the Poisson distribution. The degree fluctuation is small compared to that of the power-law distribution, which we will consider.

One of the striking feature of the ER graphs is that the mean distance between nodes increases logarithmically with system size N. Since ER graphs are locally of tree structure in the limit $N \to \infty$ with mean degree $\langle k \rangle$ finite, the number of nodes within the hopping distance ℓ from one randomly chosen node is given by $\langle k \rangle^\ell$ [15]. If D is the mean distance, it holds that $\langle k \rangle^D \sim N$, which gives

$$D \sim \frac{\ln N}{\ln \langle k \rangle}. \tag{22}$$

Therefore ER random graphs are small-world networks [16, 17].

In contrast to Eq. (21), the degree distribution of the WWW turned out to take a power-law of the form $P_d(k) \sim k^{-\gamma}$ with γ called the degree exponent and there have been a plenty of papers suggested that such power-law degree distribution is ubiquitous, giving birth to a novel class of networks called scale-free (SF) networks. Various models have been proposed for scale-free networks [1, 18]. The Barabasi-Albert model showed that the preferential attachment rule can lead to a power-law degree distribution in growing networks. For static networks, the configuration model [19] generates the ensemble of networks with a preassigned degree distribution. Also a generalization of $G_{N,p}$ was introduced which assigns unequal selection probabilities to nodes so that the resulting degree distribution follows a power law [18].

For such random SF networks, the mean distance of a randomly chosen pair follows Eq. (22) for $\gamma > 3$ [20]; however

$$d \sim \ln \ln N \tag{23}$$

for $2 < \gamma < 3$, implying that the corresponding networks are ultra-small world networks.

3.2. *Deterministic networks: flower network*

Aside from the networks constructed stochastically, there are deterministic networks models for SF networks as well [21–25]. Due to the deterministic nature, many properties of/on the networks can be obtained analytically.

Table 1. Basic properties of the flower network at generation n.

Number of links	$L^{(n)} = (u+v)^n$
Number of nodes	$N^{(n)} = \frac{u+v-2}{u+v-1}(u+v)^n + \frac{u+v}{u+v-1}$
Degree of the oldest node	$k^{(n)} = 2^n$
Degree distribution	$p(k) \sim k^{-\gamma}$ with $\gamma = 1 + \frac{\log(u+v)}{\log 2}$
Spectral dimension	$d_s = 2\frac{\log(u+v)}{\log(uv)}$

We consider a deterministic network - the *flower* network. Its construction method, also given in detail in [25–28], is as follows: Initially, at generation $n = 0$, two nodes are connected by a link. For $n = 1, 2, \ldots$, the network at the nth generation is constructed iteratively by replacing each link by a *leaf* as shown in Fig. 1. Each leaf consists of two old nodes at both ends, which existed in the $(n-1)$th generation, and two linear chains of length u and v, respectively. Here u and v are the model parameters. The first four generations of a flower network with $u = 2$ and $v = 3$ are drawn in Fig. 1. Basic properties of the flower networks are studied in [25–27] and also listed in Table 1. They are scale-free networks with the degree exponent varying with u and v. Also, as noted in Table 1, the spectral dimension d_s may be larger or smaller than 2 depending on u and v. Note that $d_s = 2$ is the critical dimension of random walks distinguishing the recurrent and transient random walks. Therefore, the flower network is a useful for studying how the spectral dimension and the degree exponent influence the random walks.

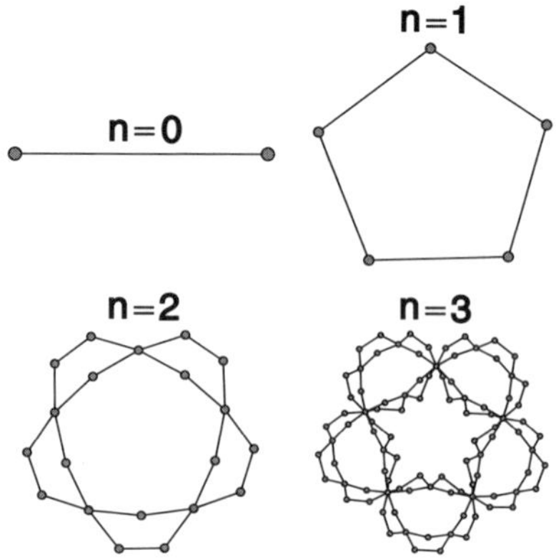

Fig. 1. The evolution of (2, 3) flower model network.

3.3. *Spectral dimensions of complex networks*

As shown in Sec. 2, the spectral dimension d_s plays an important role in random walks. Unlike the ubiquitous power-law behavior in Eq. (11), the RTO probability follows a stretched exponential function in ER graphs and random scale-free networks constructed in the static model [18], i.e.,

$$R_s(t) \sim t^{\eta} \exp(-at^{1/3}), \tag{24}$$

where a and η are constants [29, 30]. In Ref [31], it was shown that the spectral density forms a spectral gap in the mean-degree limit, which again means that the spectral dimension is infinite in random networks.

In contrast, the RTO probability on many real world networks decays in a power-law manner. The main reason is the locality in real-systems [32, 33] due to the presence of modular structures.

Unlike random network ensembles, deterministic scale-free networks often have finite spectral dimensions [28, 34–36]. In [28], we obtained that $d_s = \frac{2\ln(u+v)}{\ln(uv)}$ for the flower network models which covers a large range of spectral dimension depending on the parameters. Therefore, we will study the random walk problems on the flower model networks to see how the spectral dimension affects the RTO probability and the GFPT.

4. Probability to return to the origin of random walks in scale-free networks

4.1. *Heuristic derivation*

We consider a scale-free network of N nodes and L links with the degree distribution $P_d(k) \sim k^{-\gamma}$. For random walks in such inhomogeneous networks, one can also represent the RTO probability at node s in terms of the degree of s and the effective total number of links. While one can expect $R_s(t)$ to be proportional to degree k_s of s sufficiently long time later, it can happen that the RWer does not pass all links incident on the starting point and thereby goes around the starting node even for $t \gg 1$. In scale-free networks, such trapping effect may happen more likely around hub nodes. This motivates us to define the effective degree $\hat{k}_s(t)$. Using it, we can express $R_s(t)$ in a similar fashion to Eq. (13) as

$$R_s(t) = \frac{\hat{k}_s(t)}{2\hat{L}(t)}, \tag{25}$$

and

$$\hat{k}_i(t) = \sum_{j \in \text{n.n}(i)} W_{ij}(t), \tag{26}$$

where we considered the accessibility W_{ij} for a link between i and j (jump from j to i)

$$W_{ij}(t) \propto \frac{1}{k_j} P_{s\to j}(t-1). \tag{27}$$

The proportional coefficient of Eq. (27) will be introduced below to complete the definition of $W_{ij(t)}$. For consistency, the effective degree should be summed up to give the effective number of links as

$$\hat{L}(t) = \frac{1}{2}\sum_i \hat{k}_i(t). \tag{28}$$

We remark that W_{ij} may differ from W_{ji} even in undirected networks.

To introduce the proportional coefficient of Eq. (27), we consider the following conditions. The effective degree $\hat{k}_s(t)$ increases from 1 to k_s, indicating the increase of the number of explored links of node s with time and its saturation at k_s. From this consideration we choose to define $W_{ij}(t)$ as

$$W_{ij}(t) \equiv \frac{\frac{1}{k_j} P_{s\to j}(t-1)}{\left\langle \frac{1}{k_\ell} P_{\ell\to\ell}(t-2) \right\rangle_{\ell\in \text{n.n}(s)}} \quad \text{for} \quad j \in \text{n.n}(i). \tag{29}$$

The denominator in Eq. (29) corresponds to the average conditional link-crossing probability for the first-crossed link, the link crossed at the first time step. The time evolution is depicted in Fig. 2.

Initially, the link accessibilities are small or zero, indicating that only a few links are accessed by the RWer to a different extent. Note that

$$W_{ij}(t) = \frac{\frac{1}{k_j k_s}}{\langle \frac{1}{k_\ell} \rangle_{\ell\in \text{n.n}(s)}}$$

for $j \in \text{n.n}(s)$ and 0, otherwise at $t = 2$. In the steady state, on the other hand, $W_{ij}(t \to \infty) = 1$ for all links (i, j). In this sense, we call $W_{ij}(t)$ the link accessibility. With $W_{ij}(t)$ in Eq. (29), one can see that $\hat{k}_s(t)$ increases from 1 to k_s.

In the early-time regime, we show in [37] that $W_{s\ell}$ behaves as

$$W_{s\ell}(t) \simeq \frac{\kappa_\ell(t)}{k_s} + o\left(\frac{1}{k_s}\right), \tag{30}$$

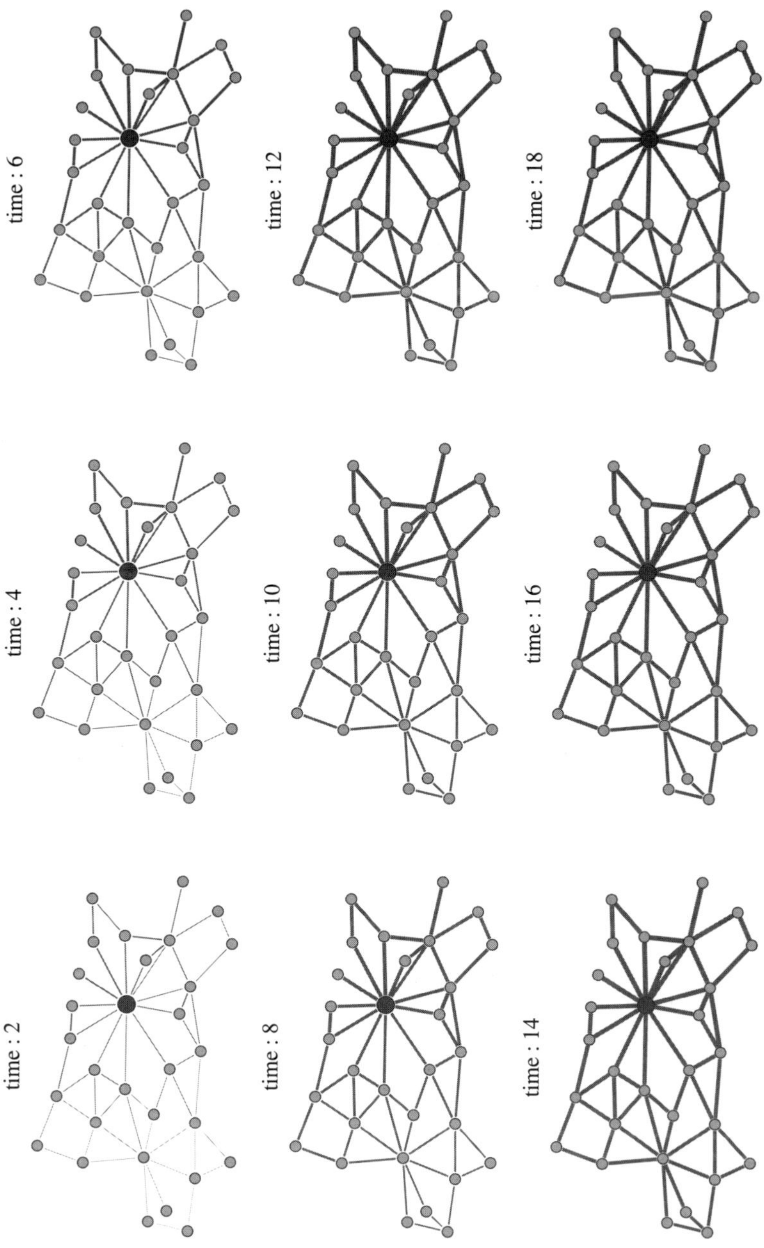

Fig. 2. (Color online) The time evolution of $W_{ij}(t)$. The node which RWer is initially placed is marked as larger pink circle.

where ℓ is a neighbor node of the starting node s and $\kappa_\ell(t)$ is a function independent of the starting node s. This condition can be understood by considering two terms contributing to $W_{s\ell}(t)$. One comes from the contribution by the RWers who move to node ℓ at time step 1 and then returns to ℓ at time step $t-1$ and finally arrive at s. Since the probability to arrive ℓ at time $t=1$ is $1/k_s$, this contribution is proportional to $1/k_s$. The other contribution is made by the RWers who choose to move to a neighbor node $m \neq \ell$ at time step 1 but arrives at ℓ at $t-1$. Since the networks are locally tree-like, the probability that a RWer finds a path connecting ℓ and m but not passing s is negligible in the early-time regime. Therefore, a RWer should reach ℓ via the starting node s, which implies that the probability to walk along such a path is proportional to $1/k_s^2$. Since there are $k_s - 1$ neighbors, except ℓ, this contribution is also proportional to $1/k_s$. Therefore both contributions are proportional to $1/k_s$. Apart from the overall factor $1/k_s$, we argue in [37] that the average of $\kappa_\ell(t)$ is independent of k_s.

Thus, the time-dependent behavior of the effective degree of s is independent of the raw degree of s in the early time regime as

$$\hat{k}_s(t) \simeq \frac{1}{k_s} \sum_{\ell \in \mathrm{n.n}(s)} \kappa_\ell(t) = \langle \kappa_\ell \rangle_\ell \equiv \kappa(t). \tag{31}$$

We summarize the behavior of the effective degree $\hat{k}_s(t)$ for different time regimes as (i) $\hat{k}_s(t) \simeq \kappa(t)$ in early-time regime, and (ii) $\hat{k}_s(t) = k_s$ in long-time regime.

To obtain $\kappa(t)$ and eventually $R_s(t)$, we consider a particular case that a RWer starts from the hub node, i.e., $s = h$. In networks without strong degree-degree correlation, we showed in [37] that, the maximum effective degree or the cutoff of effective degree distribution can be determined by the natural cutoff condition, yielding the scaling of the effective degree of the hub as $\hat{k}_h \sim \hat{L}^{\frac{1}{\gamma-1}}$. We assume that the total number of effective links is insensitive to the starting node and is given as the inverse of the node-averaged RTO probability

$$\hat{L}(t) \sim \begin{cases} t^{d_s/2} & \text{for} \quad t \ll t_x, \\ L & \text{for} \quad t \gg t_x, \end{cases} \tag{32}$$

where $t_x \sim L^{2/d_s}$.

Then, one can obtain that

$$\hat{k}_h(t) \sim \hat{L}(t)^{1/(\gamma-1)} \sim \begin{cases} t^{d_s/2(\gamma-1)} & \text{for} \quad t \ll t_x, \\ L^{1/(\gamma-1)} & \text{for} \quad t \gg t_x. \end{cases} \tag{33}$$

Both $\hat{k}_h(t)$ and $\hat{L}(t)$ increase with time and saturate to their values k_h and L at $t \simeq t_x$, respectively. The RTO probability for the hub behaves as

$$R_h(t) = \frac{\hat{k}_h(t)}{2\hat{L}(t)} \sim \begin{cases} t^{-d_s^{(\text{hub})}/2} & \text{for} \quad t \ll t_x, \\ \frac{k_h}{2L} & \text{for} \quad t \gg t_x, \end{cases} \tag{34}$$

where

$$d_s^{(\text{hub})} = d_s \frac{\gamma - 2}{\gamma - 1}. \tag{35}$$

We call $d_s^{(\text{hub})}$ the local spectral dimension of the hub. When d_s and γ are finite, $d_s^{(\text{hub})} < d_s$, and thus $R_h(t)$ decays more slowly than $R(t)$. That is, the RWer wanders around the hub for a long time. Particularly, when $\gamma \to 2$, $R_h(t) \sim$ const., implying that the RWer is effectively trapped at the hub.

Let us return to the general case of arbitrary starting node. The effective degree of the starting node s evolves with time as $\hat{k}_s(t) \sim \kappa(t)$, independent of s, until it reaches the value k_s. Therefore the behavior of $\hat{k}_h(t)$ for $t \ll t_x$ in Eq. (33) appears in $\hat{k}_s(t)$ in the early-time regime for arbitrary s as

$$\hat{k}_s(t) \sim \begin{cases} t^{d_s/2(\gamma-1)} & \text{for} \quad t \ll t_c(s), \\ k_s & \text{for} \quad t \gg t_c(s), \end{cases} \tag{36}$$

where $t_c(s)$ is the crossover time between the early- and the intermediate-time regime, which depends on k_s as

$$t_c(s) \sim k_s^{2(\gamma-1)/d_s}. \tag{37}$$

The time regime $t \gg t_x$ for which $\hat{L}(t) \simeq L$ is called the late-time regime. From Eqs. (25), (32) and (36), we obtain the RTO probability $R_s(t)$ for an arbitrary starting node s to be given by

$$R_s(t) \sim \begin{cases} t^{-d_s^{(\text{hub})}/2} & \text{for} \quad t \ll t_c(s), \\ k_s t^{-d_s/2} & \text{for} \quad t_c(s) \ll t \ll t_x, \\ \frac{k_s}{2L} & \text{for} \quad t \gg t_x. \end{cases} \tag{38}$$

The intermediate time regime, $t_c(s) \ll t \ll t_x$, disappears when the starting node is the hub; the crossover time t_c reduces to t_x when $k_s = k_h \sim L^{1/(\gamma-1)}$. The numerical simulation for various networks can be found in Fig. 3.

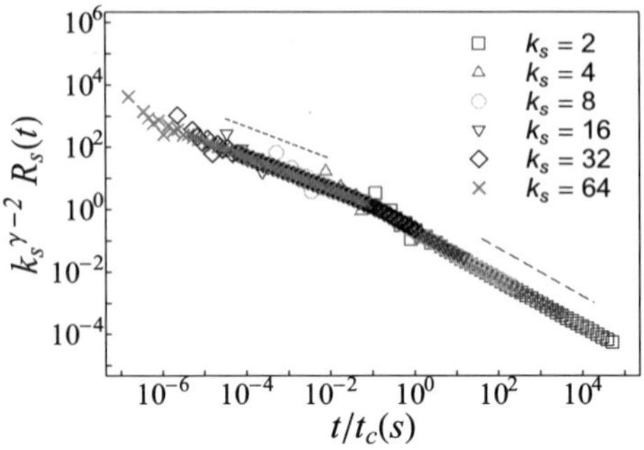

Fig. 3. The scaling behavior of RTO probability in (3, 5) flower model network. The two dashed lines correspond to the theoretical exponents of spectral dimension and hub spectral dimension.

4.2. *Exactly solvable case*

In this subsection, we obtain the RTO probability of one of the two oldest nodes in the (u, v) flower network using the probability-flux method [38]. The two oldest nodes are indexed by 0 and 1, respectively, which are hubs having the largest degree 2^n at the nth generation. Here, the flux is defined in terms of occupation probability,

$$Q_i \equiv \sum_{j \in \text{n.n}(i)} \frac{P_{0\to i}(t)}{k_i} - \frac{P_{0\to j}(t-1)}{k_j}, \tag{39}$$

which is often used in the probability-flux method. If there is no source or sink node in a system, the sum of the occupation probabilities are conserved and the flux of every node is always zero. From the definition, one finds that the generating functions $\mathcal{Q}_i(z) \equiv \sum_{t=0}^{\infty} Q_i(t) z^t$ and $\mathcal{P}_{0\to i}(z)$ are related to each other by

$$\mathcal{Q}_i(z) = \sum_{j \in \mathcal{V}^{(n)}} (k_j^{(n)} \delta_{ij} - z A_{ij}^{(n)}) \frac{\mathcal{P}_{0\to j}(z)}{k_j^{(n)}}, \tag{40}$$

where we set $P_{0\to i}(-1) = 0$ for all i. We introduced $j \in \mathcal{V}^{(n)}$ to denote the indices of the nodes present in the nth generation of the flower network. We will use the superscript (n) to denote different generations like e.g., $A_{ij}^{(n)}$ and $k_j^{(n)}$.

Below we solve Eq. (40) by performing decimation from the youngest to the oldest nodes to finally obtain $\mathcal{R}_0(z) = \mathcal{P}_{0\to 0}(z)$ of the oldest (hub) node 0 in the (u, v) flower network. For the purpose, we introduce the decimation-step-dependent variables $a(s)$ and $\phi_j(s)$'s at each decimation step s with

$$a(0) = \frac{1}{z} \text{ and } \phi_j(0) = \frac{\mathcal{P}_{0\to i}(z)}{k_i^{(n)} a(0)} \tag{41}$$

to rewrite Eq. (40) as

$$\mathcal{Q}_i = \sum_{j\in\mathcal{V}^{(n)}} \left(a(0)k_j^{(n)}\delta_{ij} - A_{ij}^{(n)}\right)\phi_j(0). \tag{42}$$

Here we do not present explicitly the z-dependence of $\phi(s)$ and $a(s)$ to focus on their dependencies on the decimation step s. We first eliminate $\phi_j(0)$ for the youngest nodes, i.e., those with index j in $\mathcal{V}^{(n)}\backslash\mathcal{V}^{(n-1)}$ and rescale the remaining variables such that the equations take the same form as in Eq. (42). The decimation process is described in [39]. After decimating the youngest nodes s times, the equations for the remaining ϕ's read as

$$q_i = \sum_{j\in\mathcal{V}^{(n-s)}} \left(a(s)k_j^{(n-s)}\delta_{ij} - A_{ij}^{(n-s)}\right)\phi_j(s), \tag{43}$$

where $\phi(s)$ and $a(s)$ follow the recursive relations

$$a(s+1) = \frac{\sinh\left((u+v)\cosh^{-1}(a(s))\right)}{\sinh\left(v\cosh^{-1}(a(s))\right) + \sinh\left(u\cosh^{-1}(a(s))\right)} \tag{44}$$

and

$$\phi_j(s+1) = \xi(a(s))\phi_j(s), \tag{45}$$

where

$$\xi(x) = \sqrt{x^2-1}\left[\operatorname{csch}\left(u\cosh^{-1}x\right) + \operatorname{csch}\left(v\cosh^{-1}x\right)\right]. \tag{46}$$

We repeat the decimation process until we have the equation only for the two initial nodes. The remaining equation is

$$\begin{pmatrix}1\\0\end{pmatrix} = \begin{pmatrix}a(n) & -1\\ -1 & a(n)\end{pmatrix}\begin{pmatrix}\phi_0(n)\\ \phi_1(n)\end{pmatrix}, \tag{47}$$

and its solution gives $\phi_0(n) = \frac{a(n)}{a(n)^2-1}$. Using $\phi_0(n)$ and Eq. (45), we find that $\phi_0(0) = \phi_0(n)\prod_{s=0}^{n-1}[\xi(a(s))]^{-1}$ and

$$\mathcal{R}_0^{(n)}(z) = \frac{2^n a(n)a(0)}{a(n)^2-1}\prod_{s=0}^{n-1}\frac{1}{\xi(a(s))}, \tag{48}$$

where $a(0) \equiv 1/z$ and $a(s)$'s for $s \geq 1$ are obtained by Eq. (44). Similarly, we have

$$\mathcal{R}_0^{(n-1)}(z') = \frac{2^{n-1}a(n)a(1)}{a(n)^2 - 1} \prod_{s=1}^{n-1} \frac{1}{\xi(a(s))}$$
$$= \mathcal{R}_0^{(n)}(z) \frac{a(1)}{2a(0)} \xi(a(0)). \tag{49}$$

With $\delta \equiv -\ln z$ being small around $z = 1$, z' is expanded as $z' \simeq 1 - uv\delta + \mathcal{O}(\delta^2) \simeq \exp(-uv\delta)$. Taking the limit $n \to \infty$ with Eq. (49), we find that

$$\mathcal{R}_0^{(\infty)}(e^{-\delta}) \sim \delta^{\frac{\ln(u+v)-\ln 2}{\ln(uv)} - 1}. \tag{50}$$

The singularity of $\mathcal{R}^{(\infty)}(z)$ around $z = 1$ in Eq. (50) gives the RTO probability $R_0(t)$ via the relation $(d^t/dz^t)(1-z)^\zeta \sim t!t^{-\zeta-1}$ for large t as

$$R_0^{(\infty)}(t) \simeq c\, t^{-\frac{\ln(u+v)-\ln 2}{\ln(uv)}} = c\, t^{-\frac{d_s^{(\text{hub})}}{2}}, \tag{51}$$

which reduces to Eq. (35).

5. First passage time probability of random walks in scale-free networks

5.1. *Mean global first passage time and classification of networks in terms of spectral dimension and the degree exponent*

In Sec. 2.3, we showed that the GFPT distribution $F_m(t)$ at a node m satisfies the following relation with the RTO probability,

$$\mathcal{F}_m(z) = \frac{k_m z}{2L(1-z)} \frac{1}{\mathcal{R}_m(z)}. \tag{52}$$

The mean GFPT $T_m = \sum_t tF_m(t)$, corresponds to the first derivative of $\mathcal{F}_m(z)$ at $z = 1$. Considering that $R_m(\infty) = k_m/(2L)$, we introduce $\mathcal{R}_m^*(z) \equiv \sum_t z^t[R_m(t) - R_m(\infty)] = \mathcal{R}_m(z) - k_m/[2L(1-z)]$. Then Eq. (52) can be rewritten as $\mathcal{F}_m(z) = k_m z/[k_m + 2L(1-z)\mathcal{R}_m^*(z)]$. The mean GFPT is in turn obtained as

$$T_m = \frac{\partial}{\partial z}\mathcal{F}_m(z)\bigg|_{z=1} \approx \frac{2L}{k_m}\mathcal{R}_m^*(1) + 1$$
$$= \frac{2L}{k_m} \sum_{t=0}^{\infty} (R_m(t) - R_m(\infty)) + 1. \tag{53}$$

This relation is well known in [7, 40], and corresponds to the inverse of the RW centrality defined in Ref. [7].

Plugging $R_m(t)$ as in Eq. (38) into Eq. (53), we find that the behavior of the mean GFPT may be classified into three cases, depending on the spectral dimension d_s:

$$T_m \approx \frac{2L}{k_m} \int_1^{t_x} [R_m(t) - R_m(\infty)]\, \mathrm{d}t \\ \sim \begin{cases} N^{2/d_s} & \text{(I)} \quad d_s < 2, \\ N k_m^{-\alpha} & \text{(II)} \; 2 < d_s < d_c, \\ N k_m^{-1} & \text{(III)} \quad d_s > d_c, \end{cases} \tag{54}$$

where $d_c = 2(\gamma - 1)/(\gamma - 2)$ and $\alpha = (1 - 2/d_s)(\gamma - 1)$. To derive this result, we used $t_x \sim L^{2/d_s}$ and $L \sim N$. It is noteworthy that the mean GFPT does not depend on the degree of the target for case (I) (compact exploration); however, the mean GFPT does depend on the degree of the target for cases (II) and (III) for which $d_s > 2$. Applying this to real-world networks, we can discover the following. The spectral dimension of the World-Wide Web is $d_s \approx 1.8 < 2$ [9], and thus, the time taken to access a node is independent of their degree. Therefore, the hub would not be a better location for posting information than other nodes. However, for the Internet in the AS system for which $d_s > 2$, the hub can be very quickly accessible. For artificial networks such as the Barabási-Albert (BA) model or the static model, $d_s = \infty$, and thus, the mean GFPT is the minimum of the nodes.

The plot of $\log_N T_m$ vs. $\log_N k_m$ for various networks, displays different slopes depending on the phase each belongs to. These slopes are in agreement with the theoretical values. Finally, we note that logarithmic corrections to the mean GFPT can appear at the boundaries between the regions, (I), (II), and (III).

If the target is the hub, then using the relation $k_h \sim N^{1/(\gamma-1)}$, we obtain that

$$T_h \sim \begin{cases} N^{2/d_s} & \text{(I and II)}, \\ N^{(\gamma-2)/(\gamma-1)} & \text{(III)}. \end{cases} \tag{55}$$

Thus, the mean travel time is sub-linear for (III), that is, a RWer can reach the hub without visiting all nodes.

Table 2. The GFPT distribution for the corresponding cases. $\tau_{(\mathrm{I})} \sim N^{d_s/2}$, $\tau_{(\mathrm{II})} \sim N k_m^{-\alpha}$ and $\tau_{(\mathrm{III})} \sim N k_m^{-1}$.

$F_m(t)$	$1 \ll t \ll t_c(k_m)$	$t_c(k_m) \ll t \ll t_x$	$t \gg t_x$
(I)	$\frac{k_m}{2N} t^{-(1-d_s^{(\mathrm{hub})}/2)}$	$\frac{1}{2N} t^{-(1-d_s/2)}$	$\tau_{(\mathrm{I})}^{-1} \exp(-\frac{t}{\tau_{(\mathrm{I})}})$
(II)	$\frac{k_m}{2N} t^{-(1-d_s^{(\mathrm{hub})}/2)}$	$\tau_{(\mathrm{II})}^{-1} \exp(-\frac{t}{\tau_{(\mathrm{II})}})$	
(III)	$\tau_{(\mathrm{III})}^{-1} \exp(-\frac{t}{\tau_{(\mathrm{III})}})$		

5.2. *The global first passage time distribution*

Next, we solve the long time behavior of the GFPT distribution $F_m(t)$ using Eqs. (38) and (52), which can be determined by $\mathcal{R}_m(z)$ for small $\epsilon = 1 - z$. To implement this behavior, we use the approximation $\mathcal{R}_m(z = 1 - \epsilon) \simeq R_m(\infty)/\epsilon + \int_1^{t_x} (R_m(t) - R_m(\infty)) e^{-\epsilon t}\, dt$, and determine the leading behavior of $\mathcal{R}_m(z)$ by comparing different time scales ϵ^{-1}, t_c, and t_x. $\mathcal{R}_m(z)$, which depends on the magnitude of ϵ, is determined as follows:

$$\mathcal{R}_m(z = 1 - \epsilon) - \frac{k_m}{2N\epsilon} \tag{56}$$
$$\sim \begin{cases} \max(1, \epsilon^{d_s^{(\mathrm{hub})}/2-1}) & \text{for} \quad \epsilon_c \ll \epsilon \ll 1, \\ \max(1, k_m \epsilon_c^{d_s/2-1}, k_m \epsilon^{d_s/2-1}) & \text{for} \quad \epsilon_x \ll \epsilon \ll \epsilon_c, \\ \max(1, k_m \epsilon_c^{d_s/2-1}, k_m \epsilon_x^{d_s/2-1}) & \text{for} \quad \epsilon \ll \epsilon_x, \end{cases}$$

where $\epsilon_c = 1/t_c(k_m)$ and $\epsilon_x = 1/t_x$.

Inserting this result for $\mathcal{R}_m(z = 1 - \epsilon)$ into Eq. (52), one finds the leading singularity of $\mathcal{F}_m(z = 1 - \epsilon)$ for small values of ϵ. Next, applying the Tauberian theorem to $\mathcal{F}_m(z)$ for each case, we obtain $F_m(t)$ as listed in Table I. We note that the prefactor of $F_m(t)$ in the early-time regime $t \ll t_c(k_m)$ for cases (I) and (II) or $t \ll \tau_{(\mathrm{III})}$ for case (III) is commonly $k_m/(2L)$, suggesting that $F_m(t)$ is proportional to k_m/N using $N \sim L$ for finite t. These results are from the fact that a RWer far from the target cannot reach it within a finite number of time steps; $F_m(t)$ for finite t is contributed mainly by a RWer who is located close to the target node, which is proportional to the degree of the target node k_m. On the other hand, in the long time regime, for $t \gg \tau_{(\mathrm{I,II,III})}$, the RTO probability converges to the non-zero value $k_m/2L$, which causes the GFPT distribution to decay exponentially in finite networks via the inverse Laplacian transformation. The characteristic times $\tau_{(\mathrm{I,II,III})}$ show the same scaling property as the mean GFPT T_m given in Eq. (54). It is also worth mentioning that for case (I) i.e., $d_s < 2$, the exponent in the intermediate time regime $1 - d_s/2$ is different from $2 - d_s/2$ obtained for the FPT between two given nodes a finite distance apart [41]. This difference implies that a large delay is more likely to occur when receiving information from a source of unknown position.

If the degree k_m of the target node is on the order of one, the crossover time $t_c(k_m)$ is of order one as well, resulting in the behavior in the intermediate time regime $F_m(t) \sim t^{-(1-d_s/2)}$ for $d_s < 2$ and $F_m(t) \sim$ const. for $d_s > 2$ dominating the whole non-stationary time regime, $t \ll \tau_{(\mathrm{I,II,III})}$. In case of (I) and (II), i.e., for $d_s < d_c$, the different functional behaviors of the GFPT distributions between the early- and the late-time regimes can be represented by using a scaling function as

$$F_m^{(\mathrm{I,II})}(t) \sim \frac{k_m^\alpha}{N} H_{(\mathrm{I,II})}\left(\frac{t}{t_c(k_m)}\right), \tag{57}$$

where the scaling function $H_{(\mathrm{I,II})}(x)$ is defined as

$$H_{(\mathrm{I})}(x) = \begin{cases} x^{-1+d_s^{(\mathrm{hub})}/2} & x \ll 1, \\ x^{-1+d_s/2} & x \gg 1, \end{cases} \tag{58}$$

$$H_{(\mathrm{II})}(x) = \begin{cases} x^{-1+d_s^{(\mathrm{hub})}/2} & x \ll 1, \\ e^{-x/(N k_m^{1-\gamma})} & x \gg 1. \end{cases} \tag{59}$$

These cross-over behaviors are confirmed by numerical simulation in Fig. 4.

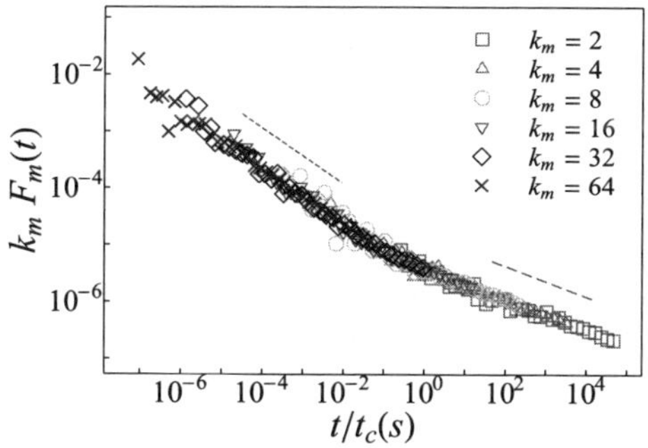

Fig. 4. The scaling behavior of GFPT distribution in (3, 5) flower model network. The two dashed line shows the exponents of scaling function in Sec. 5.2.

6. Conclusions

We have explored the scaling properties of the first passage time in scale-free network. The most striking feature we obtained is that there is notable dependency between degree of target node and first passage time. The degree dependence is clearly seen when $d_s > d_c$, becomes weaker when $2 < d_s < d_c$, and disappears when $d_s < 2$. These behaviors suggest that to understand the dynamics on the underlying network, the spectral dimension d_s as well as the degree exponent γ are important.

Other peculiar feature is that there exists the effective spectral dimension $d_s^{(\mathrm{hub})}$, in the early-time regime which leads to the slow decay for the RTO probability. Since $d_s^{(\mathrm{hub})}$ vanishes as $\gamma \to 2$, the RWer are wandering around the hub so the effective trapping occurs. Since γ is around two for the WWW network, the slow decay explains why Google search algorithm has to introduces random jump with some probability.

We also presented the exact derivation for the RTO probability and the FPT distribution on a deterministic network. Since we can trace the origin of term by term in the derivation, one can know how $d_s^{(\mathrm{hub})}$ forms during the decimation step. In scale-free networks, to fix mean degree $\langle k \rangle$, the degree distribution should be rescaled which leads to the rescaling of degree in the process of decimation. This additional rescaling leads to the additional factor which forms $d_s^{(\mathrm{hub})}$.

Acknowledgments

This work was supported by NRF research grants funded by MEST (Nos. 2010-0015066 (BK, SH), 2011-0003488 (DSL) and 2011-35B-C00014 (SH)). D.-S.L. acknowledges the TJ Park Foundation for support.

References

1. A. Barabási and R. Albert, Emergence of scaling in random networks, *Science.* **296**, 509–512 (1999).
2. B. Bollobás, *Random Graphs (Cambridge Studies in Advanced Mathematics).* Cambridge University Press (2001). ISBN 0521797225.
3. D. ben-Avraham and S. Havlin, *Diffusion and Reactions in Fractals and Disordered Systems.* Cambridge University Press (2005). ISBN 0521617200.
4. R. Burioni and D. Cassi, Random walks on graphs: ideas, techniques and results, *Journal of Physics A: Mathematical and General.* **38**(8), R45 (2005).
5. V. Mendez, S. Fedotov, and W. Horsthemke, *Reaction-Transport Systems: Mesoscopic Foundations, Fronts, and Spatial Instabilities (Springer Series in Synergetics).* Springer (2010). ISBN 3642114423.
6. B. D. Hughes, *Random Walks and Random Environments: Volume 1: Random Walks.* Oxford University Press, USA (1995). ISBN 0198537883.
7. J. D. Noh and H. Rieger, Random walks on complex networks, *Phys. Rev. Lett.* **92**, 118701 (Mar, 2004). doi: 10.1103/PhysRevLett.92.118701.

8. S. Redner, *A Guide to First-Passage Processes*. Cambridge University Press (2007). ISBN 0521036917.
9. S. Hwang, D.-S. Lee, and B. Kahng, First passage time for random walks in heterogeneous networks, *Phys. Rev. Lett.* **109**, 088701 (Aug, 2012). doi: 10.1103/PhysRevLett.109.088701.
10. M. Newman, Power laws, pareto distributions and zipf's law, *Contemporary Physics*. **46**(5), 323–351 (2005). doi: 10.1080/00107510500052444.
11. A. Clauset, C. Shalizi, and M. Newman, Power-law distributions in empirical data, *SIAM Review*. **51**(4), 661–703 (2009). doi: 10.1137/070710111.
12. P. Erdős and A. Rényi, On random graphs, *Publ. Math.* **6**, 290297 (1959).
13. E. Gilbert, Random graphs, *Ann. Math. Stat.* **30**, 1141–1144 (1959).
14. A. Barrat, M. Barthlemy, and A. Vespignani, *Dynamical Processes on Complex Networks*. Cambridge University Press (2012). ISBN 1107626250.
15. S. Bornholdt and H. S. Schuster, *Handbook of Graphs and Networks*. Wiley-VCH (2003). ISBN 3527403361.
16. F. Chung and L. Lu, The diameter of sparse random graphs, *Advances in Applied Mathematics*. **26**(4), 257 – 279 (2001). ISSN 0196-8858. doi: 10.1006/aama.2001.0720.
17. N. Riordan, Oand Wormald, The diameter of sparse random graphs, *Combinatorics, Probability and Computing*. **19**, 835–926 (10, 2010). ISSN 1469-2163. doi: 10.1017/S0963548310000325.
18. K.-I. Goh, B. Kahng, and D. Kim, Universal behavior of load distribution in scale-free networks, *Phys. Rev. Lett.* **87**, 278701 (Dec, 2001). doi: 10.1103/PhysRevLett.87.278701.
19. M. Molloy and B. Reed, A critical point for random graphs with a given degree sequence, *Random Struct. Alg.* **6**(2-3), 161–180 (1995). ISSN 1098-2418.
20. R. Cohen and S. Havlin, Scale-free networks are ultrasmall, *Phys. Rev. Lett.* **90**, 058701 (Feb, 2003). doi: 10.1103/PhysRevLett.90.058701.
21. S. N. Dorogovtsev, A. V. Goltsev, and J. F. F. Mendes, Pseudofractal scale-free web, *Phys. Rev. E.* **65**, 066122 (Jun, 2002). doi: 10.1103/PhysRevE.65.066122.
22. J. D. Noh and H. Rieger, Constrained spin-dynamics description of random walks on hierarchical scale-free networks, *Phys. Rev. E.* **69**, 036111 (Mar, 2004). doi: 10.1103/PhysRevE.69.036111.
23. E. M. Bollt and D. ben-Avraham, What is special about diffusion on scale-free nets?, *New Journal of Physics*. **7**(1), 26 (2005).
24. C. Song, S. Havlin, and H. A. Makse, Origins of fractality in the growth of complex networks, *Nat. Phys.* **2**(4), 275–281 (Apr., 2006). ISSN 1745-2473.
25. H. Rozenfeld, S. Havlin, and D. ben Avraham, Fractal and transfractal recursive scale-free nets, *New J. Phys.* **9**(6), 175 (2007).
26. M. Hinczewski and B. Nihat, Inverted berezinskii-kosterlitz-thouless singularity and high-temperature algebraic order in an ising model on a scale-free hierarchical-lattice small-world network, *Phys. Rev. E.* **73**, 066126 (Jun, 2006). doi: 10.1103/PhysRevE.73.066126.

27. H. D. Rozenfeld and D. ben Avraham, Percolation in hierarchical scale-free nets, *Phys. Rev. E.* **75**, 061102 (Jun, 2007). doi: 10.1103/PhysRevE.75.061102.
28. S. Hwang, C.-K. Yun, D.-S. Lee, B. Kahng, and D. Kim, Spectral dimensions of hierarchical scale-free networks with weighted shortcuts, *Phys. Rev. E.* **82**, 056110 (Nov, 2010). doi: 10.1103/PhysRevE.82.056110.
29. G. J. Rodgers and A. J. Bray, Density of states of a sparse random matrix, *Phys. Rev. B.* **37**, 3557–3562 (Mar, 1988). doi: 10.1103/PhysRevB.37.3557.
30. A. N. Samukhin, S. N. Dorogovtsev, and J. F. F. Mendes, Laplacian spectra of, and random walks on, complex networks: Are scale-free architectures really important?, *Phys. Rev. E.* **77**, 036115 (Mar, 2008). doi: 10.1103/PhysRevE.77.036115.
31. D. Kim and B. Kahng, Spectral densities of scale-free networks, *Chaos: An Interdisciplinary Journal of Nonlinear Science.* 17(2):026115 (2007). doi: 10.1063/1.2735019.
32. M. A. Serrano, D. Krioukov, and M. Boguñá, Self-similarity of complex networks and hidden metric spaces, *Phys. Rev. Lett.* **100**, 078701 (Feb, 2008). doi: 10.1103/PhysRevLett.100.078701.
33. M. Boguná, D. Krioukov, and K. C. Claffy, Navigability of complex networks, *Nat Phys.* **5**(1), 74–80 (Jan., 2009). ISSN 1745-2473.
34. Z. Zhang, Y. Qi, S. Zhou, W. Xie, and J. Guan, Exact solution for mean first-passage time on a pseudofractal scale-free web, *Phys. Rev. E.* **79**, 021127 (Feb, 2009). doi: 10.1103/PhysRevE.79.021127.
35. Z. Zhang, W. Xie, S. Zhou, M. Li, and J. Guan, Distinct scalings for mean first-passage time of random walks on scale-free networks with the same degree sequence, *Phys. Rev. E.* **80**, 061111 (Dec, 2009). doi: 10.1103/PhysRevE.80.061111.
36. Z. Zhang, S. Zhou, W. Xie, L. Chen, Y. Lin, and J. Guan, Standard random walks and trapping on the koch network with scale-free behavior and small-world effect, *Phys. Rev. E.* **79**, 061113 (Jun, 2009). doi: 10.1103/PhysRevE.79.061113.
37. S. Hwang, D.-S. Lee, and B. Kahng, Effective trapping of random walkers in complex networks, *Phys. Rev. E.* **85**, 046110 (Apr, 2012). doi: 10.1103/PhysRevE.85.046110.
38. C. P. Haynes and A. P. Roberts, Global first-passage times of fractal lattices, *Phys. Rev. E.* **78**, 041111 (Oct, 2008). doi: 10.1103/PhysRevE.78.041111.
39. S. Hwang, D.-S. Lee, and B. Kahng, Origin of the hub spectral dimension in scale-free networks, *Phys. Rev. E.* **87**, 022816 (Feb, 2013). doi: 10.1103/PhysRevE.87.022816.
40. V. Tejedor, O. Bénichou, and R. Voituriez, Global mean first-passage times of random walks on complex networks, *Phys. Rev. E.* **80**, 065104 (Dec, 2009). doi: 10.1103/PhysRevE.80.065104.
41. Y. Meroz, I. M. Sokolov, and J. Klafter, Distribution of first-passage times to specific targets on compactly explored fractal structures, *Phys. Rev. E.* **83**, 020104 (Feb, 2011). doi: 10.1103/PhysRevE.83.020104.

Chapter 7

First-Passage Statistics for Random Walks in Bounded Domains

R. Voituriez[1,2] and O. Bénichou[1]

[1] *Laboratoire de Physique Théorique de la Matière Condensée, CNRS/UPMC, 4 place jussieu 75005 Paris, France*

[2] *Laboratoire Jean Perrin, CNRS/UPMC, 4 place jussieu 75005 Paris, France*

We present a general theory which allows one to accurately evaluate the mean FPT (MFPT) for regular random walks in bounded domains, and its extensions to related first-passage observables like splitting probabilities and occupation times. It is showed that this analytical approach provides a universal scaling dependence of the MFPT on both the volume of the confining domain and the source-target distance in the case of general scale invariant processes. This analysis is applicable to a broad range of stochastic processes characterized by scale invariant properties. The full distribution of the FPT can be obtained using similar tools, and displays universal features. This allows to quantify the fluctuations of the FPT in confinement, and to reveal the key role that can be played by the starting position of the random walker.

1. Introduction

How long does it take for a drunkard to go from a given bar to another one? This time is known in the random walk literature as a first passage time (FPT), and it has generated a considerable amount of work for many years [1, 2]. The importance of FPT relies on the fact that many physical properties, including fluorescence quenching [3], neuron dynamics [4] or resonant activation [5] to name a few, are controlled by first passage events. More recently, FPTs have regained interest in the context of random search strategies [6–8]. Unfortunately, explicit determinations of FPT are most of the time limited to simple geometries, such as 1D and spherically symmetric problems [1, 2].

The determination of FPT for random walks in realistic *confined* geometries is not just a theoretical challenge in its own right. It is actually a very general issue involved as soon as molecules diffuse in confined media as, for example, biomolecules diffusing in the cell and undergoing a series of transformations at precise regions of the cell. An estimation of the time needed to go from one point to another is then an essential step in the understanding of the kinetics of the whole process.

Over the last decade, important advances in the analysis of FPT statistics in confinement have been performed. In this chapter, we present first a general theory which allows one to accurately evaluate the mean FPT (MFPT) for regular random walks in bounded domains. We then briefly mention extensions of these results to related first-passage observables like splitting probabilities and occupation times. We next show that this analytical approach provides a universal scaling dependence of the MFPT on both the volume of the confining domain and the source-target distance in the case of general scale invariant processes. This analysis is applicable to a broad range of stochastic processes characterized by scale invariant properties. Last, we show that the full distribution of the FPT can be obtained, and discuss its universal features. This allows finally to quantify the fluctuations of the FPT in confinement, and to reveal the key role that can be played by the starting position of the random walker.

2. Discrete random walks on regular lattices

In this section we consider the case of discrete time random walks on discrete structures, discussed for example in [9]. We introduce both physical observables considered in this chapter (first-passage times, occupation times, splitting probabilities) as well as the tools used to determine them (pseudo-Green functions, exact expressions in the case of rectangular domains, large volume asymptotic expansions in the case of arbitrary domains). The formalism is introduced for general graphs, but explicit results are derived in this section for regular euclidean lattices. The cases of fractal structures and complex networks will be presented in the forthcoming sections. Note that the case of continuous time random-walks (CTRWs) can easily be derived from the formalism introduced in this section. Its analysis can be found in [10, 11].

2.1. *Mean-first passage time with a single target*

We consider a discrete time random walk on a discrete lattice of N sites, and we denote w_{xy} the probability to jump from site y to site x in one step, which satisfies

$$\sum_x w_{xy} = 1, \tag{1}$$

where the sum goes over all the sites x of the lattice. Let $P_n(x|y)$ be the probability to be at time n at site x knowing that initially the walker was at site y, $F_n(x|y)$ the probability to visit site x for the fist time at time n knowing that initially the walker was at site y and $P(x)$ the stationary probability to be at site x. We first derive an exact expression of the mean first-passage time $\langle \mathbf{T}(x|y) \rangle$ from site y to site x in terms of pseudo-Green functions of the problem, which will be used all along this chapter.

2.1.1. *Expression in terms of the pseudo-Green function*

The starting point is the renewal equation [12]

$$P_n(x|y) = \sum_{k=0}^{n} F_k(x|y) P_{n-k}(x|x), \tag{2}$$

where we have used the usual conventions that $F_0(x|y) = 0$ and $P_0(x|y) = \delta_{x,y}$ and assumed that $x \neq y$. This equation is obtained by partionning over the first time k the site x is visited, and writing that the walker has to come back at site x in the $n-k$ remaining steps to be still at site x at time n. Laplace transforming this temporal convolution equation, we get

$$\widehat{F}(x,s|y) = \frac{\widehat{P}(x,s|y)}{\widehat{P}(x,s|x)}, \tag{3}$$

where we have introduced the discrete Laplace transforms

$$\widehat{P}(x,s|y) \equiv \sum_{n=0}^{\infty} P_n(x|y) e^{-sn} \text{ and } \widehat{F}(x,s|y) \equiv \sum_{n=0}^{\infty} F_n(x|y) e^{-sn}. \tag{4}$$

It is useful to consider the small Laplace variable s expansions of these quantities:

$$\widehat{P}(x,s|y) = \sum_{n=0}^{\infty} [P(x) + (P_n(x|y) - P(x))] e^{-sn} = \frac{P(x)}{s} + H_{xy} + o(1), \tag{5}$$

where

$$H_{xy} \equiv \sum_{n=0}^{\infty} (P_n(x|y) - P(x)), \tag{6}$$

and

$$\widehat{F}(x, s|y) = 1 - s\langle \mathbf{T}(x|y)\rangle + o(s). \tag{7}$$

Expanding Eq. (3) to first order in s leads to

$$\langle \mathbf{T}(x|y)\rangle = \frac{1}{P(x)}(H_{xx} - H_{xy}). \tag{8}$$

This equation, which will play a key role in this chapter, has been obtained by Noh and Rieger in the physics literature [13], and also seems to be known in the mathematics literature (see [14] and references therein). Several comments are in order:

(i) A first important property of the H function resulting directly from its definition (6) and the normalization of $P_n(x|y)$ and $P(x)$ is that

$$\sum_x H_{xy} = 0. \tag{9}$$

A second useful property resulting from the detailed balance hypothesis and shown in [11] is that

$$H_{xy}P(y) = H_{yx}P(x). \tag{10}$$

(ii) This H function actually admits a simple physical interpretation. It can be conveniently obtained by first writing a backward equation for P_n

$$P_{n+1}(x|y) = \sum_z w_{zy} P_n(x|z), \tag{11}$$

resulting from a partition of the first step of the walk, and the trivial equation

$$P(x) = \sum_z w_{zy} P(x). \tag{12}$$

Summing over n the difference of equations (11) and (12) gives:

$$-\sum_z w_{zy}(H_{xz} - H_{xy}) = \delta_{xy} - P(x), \tag{13}$$

which indicates that H is actually a pseudo-Green function (see [15]) of the operator involved in the lhs of Eq. (13), which is a discrete Laplacian. Physically, it corresponds to the stationary solution of a discrete diffusion equation in the presence of a uniformly distributed sink term.

(iii) Expression (8) of the MFPT can be shown using Eq. (13) to verify the classical backward equation:

$$-\sum_z w_{zy}(\langle \mathbf{T}(x|z)\rangle - \langle \mathbf{T}(x|y)\rangle) = 1 \tag{14}$$

with boundary condition $\langle \mathbf{T}(x|x)\rangle = 0$.

(iv) The so-called Kac formula [14] is easily derived from this formalism, since the mean first return time to site x can be written

$$\begin{aligned}\langle \mathbf{T}_r(x)\rangle &= 1 + \sum_z w_{zx}\langle \mathbf{T}(x|z)\rangle \\ &= 1 + \sum_z w_{zx}\frac{H_{xx} - H_{xz}}{P(x)} \\ &= \frac{1}{P(x)},\end{aligned} \tag{15}$$

where the first line is a backward equation for the mean return time similar to Eq. (14), and Eqs. (8) and (13) have been used successively.

(v) The global MFPT [16–18], defined as the MFPT averaged over the initial position of the random walk (with the stationary weight) can be written as:

$$\begin{aligned}\overline{\langle \mathbf{T}(x)\rangle} &\equiv \sum_y \langle \mathbf{T}(x|y)\rangle P(y) \\ &= \frac{1}{P(x)}\sum_y (H_{xx} - H_{xy})P(y) \\ &= \frac{H_{xx}}{P(x)} - \frac{1}{P(x)}\sum_y H_{yx}P(x) \\ &= \frac{H_{xx}}{P(x)}\end{aligned} \tag{16}$$

where we have used successively the symmetry relation of the pseudo-Green function (9) and the relation (10).

Finally, we stress that Eq. (8) is exact and general. However, since it involves the a priori unknown H function, it remains somehow formal at this stage. In the following, we show how explicit expressions of the MFPT can be obtained. Otherwise explicitly stated, we will assume that the stationary probability $P(x)$ is uniform

$$P(x) = \frac{1}{N}, \tag{17}$$

which is the case as soon as the probabilities w are symmetrical (i.e. $w_{xy} = w_{yx}$ for all x and y).

2.1.2. *Exact calculation of the MFPT: the case of rectangular domains*

In the particular case of a rectangular domain, exact expressions of the pseudo-Green function H can be obtained, both for periodic and reflecting boundary conditions [19, 20]. In turn, using Eq. (8), this provides exact expressions for the MFPT (and for several other first-passage observables as we will show later). For the sake of self-consistency, we remind in this section how these exact expressions are obtained in 2D (see [19, 20] for the 3D case).

The simplest computation is for periodic boundary conditions. Let us consider a domain, with X sites in the x direction, Y sites in the y direction. In this case, $w_{\mathbf{r},\mathbf{r}'} = 1/4$ if $\mathbf{r}$ and $\mathbf{r}'$ are nearest neighbors and zero elsewhere. We decompose H into a Fourier sum:

$$H_{\mathbf{r},\mathbf{r}'} = \sum_{m=0}^{X-1} \sum_{n=0}^{Y-1} a_{mn}(\mathbf{r}') \exp\left(\frac{2im\pi x}{X} + \frac{2in\pi y}{Y}\right), \tag{18}$$

where the coordinates of $\mathbf{r}$ are (x, y, z) and the coordinates of $\mathbf{r}'$ are (x', y', z'). If we substitute this into Eq. (13), we obtain an explicit expression of $a_{mn}(\mathbf{r}')$, which leads to

$$H_{\mathbf{r},\mathbf{r}'} = \frac{1}{N} \sum_{m=0}^{X-1} \sum_{m=\delta_{n0}}^{Y-1} \frac{\exp\left(\frac{2im\pi(x-x')}{X} + \frac{2in\pi(y-y')}{Y}\right)}{1 - \frac{1}{2}\left(\cos\frac{2m\pi}{X} + \cos\frac{2n\pi}{Y}\right)}. \tag{19}$$

The case of reflecting boundary conditions is slightly more complicated to implement. We use semi-integer coordinates for the points: x (resp. y) varies with unit increments between $1/2$ and X(resp. Y) $-1/2$. In this case, we assume that, if $\mathbf{r}'$ is not on the boundary of the domain, $w_{\mathbf{r},\mathbf{r}'} = 1/4$ if $\mathbf{r}$ and $\mathbf{r}'$ are nearest neighbors and zero elsewhere. When $\mathbf{r}'$ is on the boundary of the domain, it is assumed that $w_{\mathbf{r},\mathbf{r}'} = 1/4$ if $\mathbf{r}$ and $\mathbf{r}'$ are nearest neighbors but also if $\mathbf{r} = \mathbf{r}'$, and zero elsewhere. This choice of transition probabilities ensures in particular that the stationary probability $P(\mathbf{r})$ is uniform. The H function can be conveniently obtained by mapping the problem into a problem with periodic boundary conditions, but with x (resp. y) ranging from $-X$ (resp. Y) $+1/2$ to X (resp. Y) $-1/2$, for which H satisfies the following equation:

$$H_{\mathbf{r},\mathbf{r}'} = \frac{1}{4} \sum_{\mathbf{r}'' \in N(\mathbf{r})} H(\mathbf{r}''|\mathbf{r}) + \sum_{\mathbf{s} \in S(\mathbf{r})} \delta_{\mathbf{s}\mathbf{r}'} - \frac{1}{N} \tag{20}$$

where $N(\mathbf{r})$ is the ensemble of neighbours of $\mathbf{r}$, and $S(\mathbf{r})$ is the ensemble of points symmetrical to $\mathbf{r}$, i.e. the points of coordinates $(\pm x, \pm y)$. Because of the symmetry in the problem, the H restricted to the original domain satisfies the reflective boundary conditions. We decompose H into a Fourier sum:

$$H_{\mathbf{r},\mathbf{r}'} = \sum_{m=-X+1}^{X} \sum_{n=-Y+1}^{Y} a_{mn}(\mathbf{r}') e^{\frac{im\pi x}{X}} e^{\frac{im\pi y}{Y}}. \tag{21}$$

We then obtain an explicit expression for $a_{mn}(\mathbf{r}')$ which yields

$$\begin{aligned} H_{\mathbf{r},\mathbf{r}'} &= \frac{4}{N} \sum_{m=1}^{X-1} \sum_{n=1}^{Y-1} \frac{\cos\frac{m\pi x'}{X} \cos\frac{n\pi y'}{Y} \cos\frac{m\pi x}{X} \cos\frac{n\pi y}{Y}}{1 - \frac{1}{2}\left(\cos\frac{m\pi}{X} + \cos\frac{n\pi}{Y}\right)} \\ &+ \frac{4}{N} \sum_{m=1}^{X-1} \frac{\cos\frac{m\pi x'}{X} \cos\frac{m\pi x}{X}}{1 - \cos\frac{m\pi}{X}} + \frac{4}{N} \sum_{n=1}^{Y-1} \frac{\cos\frac{n\pi y'}{Y} \cos\frac{n\pi y}{Y}}{1 - \cos\frac{n\pi}{Y}}. \end{aligned} \tag{22}$$

2.1.3. *Large volume asymptotics of the MFPT*

In the case of general confining domains, exact expressions of the pseudo-Green function H are not available. However, large volume asymptotics of the MFPT can be obtained. As we proceed to show, they provide accurate expressions even for reasonably small systems and allow us to give first answers to the questions raised above concerning the influence of geometrical parameters on MFPTs.

Actually, even if formal, Eq. (8) is very well suited to large volume asymptotics of the MFPT. In the following, we consider two prescriptions for this large volume limit: the case when the volume goes to infinity with the constraint that both the target and the starting points of the walk remain infinitely distant from any point of the boundary (target and starting points "in the bulk"); the case when the volume goes to infinity with the constraint that both the target and the starting points of the walk remain close to a single planar boundary of the domain.

Target and starting points in the bulk. In this case, (8) leads to

$$\lim_{N\to\infty} \frac{\langle \mathbf{T}(x|y)\rangle}{N} = \sum_{n=0}^{\infty} (P_n^{\infty}(x|x) - P_n^{\infty}(x|y)), \tag{23}$$

where $P_n^{\infty}(x|y)$ stands for the propagator on an infinite lattice. Using standard large distance expressions of these quantities [12], we finally obtain for hypercubic lattices:

$$\lim_{N\to\infty} \frac{\langle \mathbf{T}(x|y)\rangle}{N} = G_0(0) - \frac{\Gamma(d/2)}{2\pi^{d/2}} \frac{1}{|x-y|^{d/2}} + o(\frac{1}{|x-y|^{d/2}}) \tag{24}$$

if $d > 2$, and

$$\lim_{N\to\infty} \frac{\langle \mathbf{T}(x|y)\rangle}{N} = \frac{2}{\pi} \ln|x-y| + \frac{3}{\pi}\ln 2 + \frac{2\gamma}{\pi} + o(1) \tag{25}$$

if $d = 2$. In these expressions,

$$G_0(x) \equiv \sum_{n=0}^{\infty} P_n^{\infty}(x|0) \tag{26}$$

is the Green function of the unconfined random walk and γ is the Euler constant (with $3/\pi \ln 2 + 2\gamma/\pi = 1.029374...$). Note that in $d = 1$, it is found that

$$\lim_{N\to\infty} \frac{\langle \mathbf{T}(x|y)\rangle}{N} = |y-x|, \tag{27}$$

in agreement with well-known results [2].

Several important conclusions can be reached from these expressions.

(i) In any dimension, the MFPT depends linearly on the volume ;

(ii) In dimension $d > 2$, the MFPT saturates at large distance $x - y$. In other words, the MFPT loses the memory of the starting point at large separations between the starting and the target points ;

(iii) The situation is very different in dimensions $d \leq 2$, where the MFPT strongly depends on the initial position of the random walker, even at large distances.

As we will show in section 3, these conclusions originate from the very nature of the random walk in the absence of any confinement, the so-called type problem. In this context, a random walk is called compact (or recurrent) if it comes back with probability one to its starting point and non compact (or transient) in the opposite case. The saturation of the MFPT with distance mentioned above is in fact characteristic of non compact random walks as is the case of 3D regular random walks, while the strong dependence on the starting position observed for 1D and 2D is a hallmark of compact transport. As we will show later, these conclusions will also hold for any Markovian scale-invariant processes (including Levy flights, random walks on fractal structures ...).

Figure 1 displays the accuracy of this large volume approximation even in the case of a small confining cube and illustrates the saturation of the MFPT with the distance mentioned above.

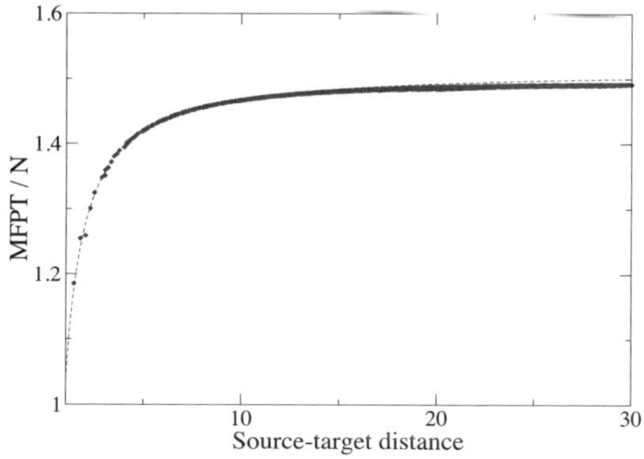

Fig. 1. MFPT in 3D parallelepipedic domains - Influence of the distance between the source and the target. Red crosses: simulations; blue dashed line: large volume approximation (24). The domain is a cube of side 41, the target being in the middle of it. All the simulation points correspond to different positions of the source.

Target and starting points close to a boundary. We now assume that the volume goes to infinity with the constraint that both x and y remain close to an infinite reflecting plane (P). In other words, the confining domain tends towards a semi infinite space. The corresponding propagator, denoted by $P_n^{\infty/2}(x|y)$ is then easily written in terms of the infinite space propagator $P_n^{\infty}(x|y)$ by using the method of images [12]:

$$P_n^{\infty/2}(x|y) = P_n^{\infty}(x|y) + P_n^{\infty}(x|s(y)), \tag{28}$$

where $s(y)$ stands for the symmetrical point with respect to the plane (P) of point y. Finally, for hypercubic lattices,

$$\begin{aligned}\lim_{N\to\infty} \frac{\langle \mathbf{T}(x|y)\rangle}{N} &= G_0(0) - \frac{\Gamma(d/2)}{2\pi^{d/2}} \frac{1}{|x-y|^{d/2}} \\ &+ \frac{\Gamma(d/2)}{2\pi^{d/2}} \left(\frac{1}{|x-s(x)|^{d/2}} - \frac{1}{|x-s(y)|^{d/2}} \right) + \ldots \end{aligned}\tag{29}$$

if $d > 2$, while

$$\lim_{N\to\infty} \frac{\langle \mathbf{T}(x|y)\rangle}{N} = \frac{2}{\pi}\ln|x-y| + \frac{3}{\pi}\ln 2 + \frac{2\gamma}{\pi} + \frac{2}{\pi}\ln\frac{|x-s(y)|}{|x-s(x)|} + o(1) \tag{30}$$

in $d = 2$.

These expressions allow one to discuss several aspects of the influence of the boundaries of the domain.

(i) The qualitative effect of the boundaries is to increase the MFPT when a target is near a boundary, or decrease it when the source is near a boundary.

(ii) The first effect is much more important than the second (because $|x - s(x)|$ is small in the first case, while neither $|x - s(y)|$ nor $|x - s(x)|$ is small in the second one).

(iii) These effects are as much marked than the dimension is small.

2.2. *Higher order moments and the entire first-passage time distribution*

In fact, the approach developed in section 2.1 for the mean of the first-passage time can be extended to determine the generalization of Eq. (8) to higher order moments. In the large volume limit, it is found that [9, 20]:

$$\langle \mathbf{T}^n(x|y) \rangle = n! N^n \{ H_{xx}(H_{xx} - H_{xy})^{n-1} + \mathcal{O}(nN^{-2/3}) \}. \tag{31}$$

The more general case of a scale-invariant processes will be discussed in section 3.

2.3. *Situations with several targets: splitting probabilities*

We consider the case when N targets of positions $x_1, x_2, ..., x_N$ are present in the problem, and focus on the splitting probability $\pi_i^{x_1,\cdots,x_N}(y)$ that a random walker starting from site y reaches the target x_i before the others. These quantities are especially useful to quantify the kinetics of competitive reactions [3], and may be useful in biology to determine to which extent cellular variability may be controlled by diffusion [21, 22]. It is shown in [9, 20] by writing a renewal equation with N targets that the splitting probabilities can be expressed exactly in terms of pseudo-Green functions only. In the particular case of two targets, it is found that

$$\pi_1^{x_1,x_2}(y) = \frac{P(1)(H_{22} - H_{2y}) + P(2)(H_{1y} - H_{12})}{P(1)(H_{22} - H_{21}) + P(2)(H_{11} - H_{12})}, \tag{32}$$

$$\pi_2^{x_1,x_2}(y) = \frac{P(2)(H_{11} - H_{1y}) + P(1)(H_{2y} - H_{21})}{P(1)(H_{22} - H_{21}) + P(2)(H_{11} - H_{12})} \tag{33}$$

where we have used the notations $H_{ij} \equiv H_{x_i x_j}$. As in the case of MFPTs with a single target presented in section 2.1, exact expressions are obtained in the case of rectangular domains, and large volume asymptotics can be derived in the general case. These two targets results are displayed in figure 2, which shows that the accuracy of the large volume asymptotic expressions even for finite domains.

An important qualitative difference between the 2D and 3D cases is revealed by the large volume asymptotics of the pseudo-Green function. In

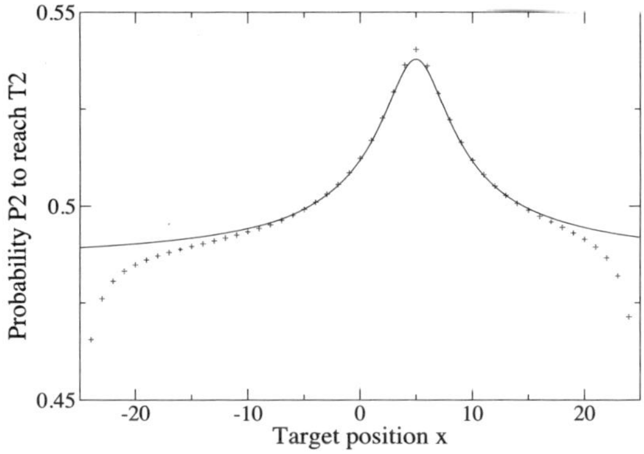

Fig. 2. Splitting probability in 3D domains. Simulations (red crosses) vs. theory with the large volume approximation of the pseudo-Green function $H = G_0$ (solid line). One target is fixed in (-5,0,0); the source is fixed in (5,0,0); the other target is at (x,3,0). The domain is a cube of side 51, the middle is the point (0,0,0).

3D, the furthest target always has a significant probability to be reached first, since the most important terms in the probabilities P_i are H_{11} and H_{22}. In 2D, if a target is much closer from the source than the other, it will almost certainly be reached first, since $H_{\mathbf{r}_i,\mathbf{r}_j}$ scales as $\ln|\mathbf{r}_i - \mathbf{r}_j|$. Actually, the probability for the furthest target to be reached first decreases logarithmically. As will be discussed on more general grounds in section 3, these properties are related to the non compact character of the infinite 3D walk, and the compact character of the 2D walk. Indeed, an infinite 2D walk explores all the sites of the lattice, whereas an infinite 3D walk does not; we may thus consider that the 2D walk will explore most of the sites surrounding the source before going much further, whereas the 3D walk will not, which qualitatively explains the difference of behaviour. This behavior parallels the dependance of the MFPT on the starting distance described above.

2.4. *Occupation times*

We now consider the time $N^{(x)}(z|y)$ spent on a given site z by a random walker starting from site y before reaching a target x. The study of the statistics of this general quantity, known in the random walk literature as the occupation time, has been a subject of interest for long, both for mathematicians [12, 14] and physicists [23–30]. As a matter of fact, the occupation time has proven to be a key quantity in various fields, ranging from astrophysics [31], transport in porous media [32] and diffusion limited reactions [33]. The point is that as soon as some specific sites of a system have different physical or chemical properties, it becomes crucial to know precisely how many times each site is visited by the random walker.

As the MFPT, the mean occupation time can easily been written in terms of the pseudo-Green functions [34]:

$$\langle N^{(x)}(z|y)\rangle = \frac{P(z)}{P(x)}(H_{xx} - H_{xy}) + H_{zx} - H_{zy}. \tag{34}$$

Note that this result also gives the mean occupation time of a subdomain, which is simply the sum of the mean occupation time of all the sites in the subdomain. In particular, using Eq. (9), we can check that the mean occupation time for the whole domain,

$$\sum_z \langle N^{(x)}(z|y)\rangle = \frac{1}{P(x)}(H_{xx} - H_{xy}) \tag{35}$$

gives back the MFPT from y to z provided by Eq. (8). The information provided by the mean occupation is therefore finer than the MFPT.

2.5. *Extension to continuous Brownian motion*

The expressions obtained earlier in the case of discrete time and space random walks can be extended to the case of continuous time and space Brownian motion. We therefore now consider a continuous Brownian motion of diffusion coefficient D in a confining domain with reflecting boundary conditions, and we still want to know how long it takes to reach a given target. A fundamental difference with the discrete case is that, except in the 1D case, the target must have a finite size to be reached, and this size a will be an important parameter of the problem. This question has also been addressed in the mathematical literature (see for instance [35–38]).

In fact, taking the limit of small target size makes it possible to directly transpose essentially all the results on discrete random walks obtained in the previous section [20, 39]. More precisely, in the small target size limit, the expressions of the MFPT and splitting probabilities obtained in the discrete case hold also in this continuous case by making use of

$$H(\mathbf{r}_T|\mathbf{r}_T) \equiv G_0(a) + H^*(\mathbf{r}_T|\mathbf{r}_T), \tag{36}$$

where $H^*(\mathbf{r}|\mathbf{r}_T)$ is defined as the regular part of $H(\mathbf{r}|\mathbf{r}_T)$ as $\mathbf{r} \to \mathbf{r}_T$:

$$H^*(\mathbf{r}|\mathbf{r}_T) = H(\mathbf{r}|\mathbf{r}_T) - G_0(\mathbf{r} - \mathbf{r}_T), \tag{37}$$

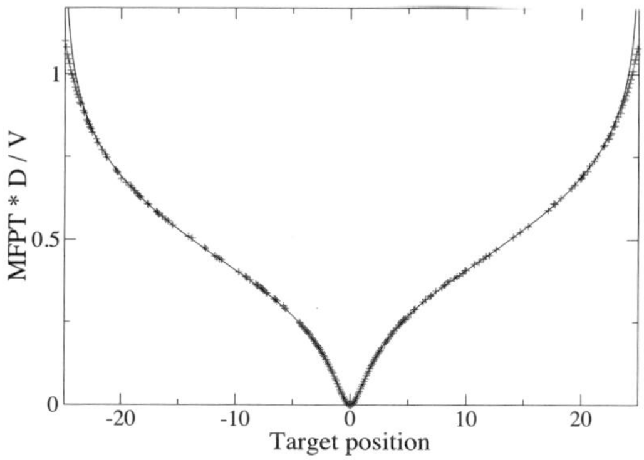

Fig. 3. (Color online) Brownian motion on a 2D disk of radius 25 centered at (0,0); the source is at (0,1) and the target of radius 1 is at $(x, 0)$. Red crosses: simulations; black solid line: theoretical prediction of equation (8) with the substitution of equation (36) where the exact function H given by equation (38).

and $G_0(\mathbf{r} - \mathbf{r}_T)$ is the infinite space Green function.

As in the discrete case, the pseudo-Green function H is known exactly for domains with simple shapes such as spheres. For example the pseudo-Green function for the unit disk of surface area $A = \pi$ reads [15]:

$$H(\mathbf{r}|\mathbf{r}') = \frac{1}{2\pi}\left[-\ln|\mathbf{r} - \mathbf{r}'| - \ln B(\mathbf{r}, \mathbf{r}') + \frac{1}{2}\left(r^2 + r'^2\right) - \frac{3}{4}\right], \tag{38}$$

where B is the modulus of

$$\mathbf{B} = r'\,\mathbf{r} - \frac{\mathbf{r}'}{r'}. \tag{39}$$

Figure 3 shows the accuracy of the method on the example of a 2D disk with an off centered target.

For a domain of arbitrary shape, there is no explicit analytical expression of the pseudo-Green function. However, as in the discrete case, large volume asymptotics of first-passage observables can be derived. Indeed, one has to leading order $H \approx G_0$ where G_0 is the infinite space Green function. This leads to explicit expressions of first-passage observables in terms of G_0 only (see [20, 40]). For example, the MFPT to a single target starting from S writes:

$$\langle \mathbf{T}(\mathbf{r}_S) \rangle = \frac{V}{4\pi D}\left(\frac{1}{a} - \frac{1}{r}\right) \quad \text{(3D)} \tag{40}$$

$$\langle \mathbf{T}(\mathbf{r}_S) \rangle = \frac{A}{2\pi D}\ln\frac{r}{a} \quad \text{(2D)} \tag{41}$$

r being the source-target distance.

3. First-passage times in complex scale-invariant media

So far we have presented general results on first-passage observables valid for regular diffusion processes, either discrete or continuous. We now show that such results can be generalized to the much wider class of scale invariant processes that include Levy flights, random walks on fractals and in disordered media. Note that these results can be straightforwardly generalized to CTRWs on fractals [11]. In this section we present results for the first-passage observables introduced earlier: FPTs, splitting probabilities and occupation times. Technically, we derive general large volume asymptotics, by directly paralleling the approach introduced in the case of regular diffusion processes. These results in fact provide explicitly the dependence of these observables on the geometrical (volume and source to target distance) and dynamical (fractal and walk dimension to be defined below) parameters.

3.1. *MFPT*

As in section 2, we consider a Markovian random walker moving in a bounded domain of N sites, with reflecting walls. Let $P_n(x|y)$ be the propagator at time n, starting from the site y at time 0. For the sake of simplicity we assume that the walker performs symmetric jumps and that the stationary distribution is homogeneous $\lim_{n\to\infty} P_n(x|y) = P(x) = 1/N$. For an analysis of non homogeneous networks see for instance [41]. We start from the exact expression for the MFPT derived earlier (see section 2), which in fact only requires that the processes is Markovian:

$$\langle \mathbf{T}(x|y)\rangle = \frac{1}{P(x)}(H_{xx} - H_{xy}). \tag{42}$$

Here again, H is the pseudo-Green function [42] of the domain:

$$H_{xy} \equiv \sum_{n=0}^{\infty}(P_n(x|y) - P(x)). \tag{43}$$

As in section 2.1.3, we take the large volume limit which yields:

$$\lim_{N\to\infty} \frac{\langle \mathbf{T}(x|y)\rangle}{N} = \sum_{n=0}^{\infty}(P_n^{\infty}(x|x) - P_n^{\infty}(x|y)), \tag{44}$$

where $P_n^{\infty}(x|y)$ stands for the propagator on an infinite lattice.

To go further and obtain explicit results, we follow [43] and assume that the problem is scale-invariant. More precisely, we use for P^∞ the standard scaling [44]:

$$P_n^\infty(x|y) \sim n^{-d_f/d_w}\Pi\left(\frac{|x-y|}{n^{1/d_w}}\right), \tag{45}$$

where the fractal dimension d_f characterizes the accessible volume $V_r \sim r_f^d$ within a sphere of radius r, and the walk dimension d_w characterizes the distance $r \sim n^{1/d_w}$ covered by a random walker in a given time n. The form (45) ensures the normalization of P^∞ by integration over the whole fractal set. Note that the MSD is then given by $\langle \Delta \mathbf{r}^2 \rangle \sim n^\beta$ with $\beta = 2/d_w$. A derivation given in [43] then yields for the MFPT:

$$\langle \mathbf{T} \rangle \sim \begin{cases} N(A - Br^{d_w-d_f}) & \text{for } d_w < d_f \\ N(A + B\ln r) & \text{for } d_w = d_f \\ N(A + Br^{d_w-d_f}) & \text{for } d_w > d_f \end{cases}, \tag{46}$$

where explicit expressions of A and B are given in [43]. This result generalizes the asymptotics derived in the case of regular random walks ($d_w = 2$) on euclidean lattices of dimension $d_f = d$ discussed in section 2.

Several comments are in order, which complement the analysis of section 2. First, we point out that equation (46) gives the large N asymptotics of the MFPT as a function of N and r as *independent* variables. In particular the volume dependence is linear with N *for r fixed* in any case, which can not be inferred from the standard scaling $\langle \mathbf{T} \rangle \propto L^{d_w}$, L being the characteristic length of the domain of order N^{1/d_f}. However, a global rescaling of the problem $r \to \lambda r, L \to \lambda L$, when applied to equation (46), gives the standard form $\langle \mathbf{T} \rangle \sim \lambda^{d_w}$ for $d_w > d_f$ and $\langle \mathbf{T} \rangle \sim \lambda^{d_f}$ for $d_w < d_f$ in accord with [45, 46]. Second, equation (46) shows two regimes, which rely on infinite space properties of the walk: in the case of compact exploration [44] ($d_w \geq d_f$) where each site is eventually visited, the MFPT behaves like $\langle \mathbf{T} \rangle \propto N r^{d_w-d_f}$ ($\langle \mathbf{T} \rangle \propto N \ln r$ for $d_w = d_f$) at large distance, so that the dependence on the starting point always matters; in the opposite case of non-compact exploration $\langle \mathbf{T} \rangle$ tends to a finite value for large r, and the dependence on the starting point is lost.

These analytical results can be confirmed by Monte Carlo simulations and exact enumeration methods applied to various models which cover the three previous cases [43, 47].

- The random barrier model in 2 dimensions [44] is a widespread model of transport in disordered systems whose MFPT properties remain widely unexplored. It is defined by a lattice random walk with nearest neighbors symmetrical transition rates Γ distributed according to some distribution $\rho(\Gamma)$. Even for a power law distribution $\rho(\Gamma)$ the scaling function $\Pi(\xi)$ can be shown to be Gaussian [32] ($d_f = d_w = 2$), which allows us to explicitely compute the constant B and obtain $\langle \mathbf{T} \rangle \sim N\,(A + (1/2\pi D_{\text{eff}})\ln r)$. Here D_{eff} is a diffusion constant depending on $\rho(\Gamma)$ which can be determined by an effective medium approximation [32, 43].

- The Sierpinski gasket [44] of finite order is a representative example of deterministic fractal. In this case $d_f = \ln 3/\ln 2 < \ln 5/\ln 2 = d_w$, so that the theory predicts the scaling $\langle \mathbf{T} \rangle \sim N r^{(\ln 5 - \ln 3)/\ln 2}$.
- The Lévy flight model of anomalous diffusion [12, 48, 49] is based on a fat-tailed distribution of jump lengths $p(l) \propto l^{-d-\beta}$ $(0 < \beta \leq 2)$. The walk dimension is now $d_w = \beta$, while the fractal dimension is the dimension of the Euclidian space d . In dimensions $d \geq 2$, or in 1D when $\beta < 1$, one has $d_f > d_w$ and the theory gives $\langle \mathbf{T} \rangle \sim N\left(A - B r^{\beta - d}\right)$.
- Critical percolation clusters (see figure 1b) constitute a representative example of random fractals [44, 50, 51]. Here we consider the case of bond percolation, where the bonds connecting the sites of a regular lattice of the d–dimensional space are present with probability p. The ensemble of points connected by bonds is called a cluster. If p is above the percolation threshold p_c, an infinite cluster exists. If $p = p_c$, this infinite cluster is a random fractal characterized by its fractal dimension d_f. We consider a nearest neighbor random walk on such critical percolation cluster, with the so–called "blind ant [12]" dynamics : on arrival at a given site $\mathbf{s}$, the walker attempts to move to one of the adjacent sites on the original lattice with equal probability. If the link corresponding to this move does not exist, the walker remains at site $\mathbf{s}$. This walk is characterized by the walk dimension d_w. In the example of the 3–dimensional cubic lattice, one has $d_f = 2.58...$, and $d_w = 3,88...$ [51] and the motion is subdiffusive with $\beta = 2/d_w \simeq 0.51...$

Such models have been frequently used to describe transport processes in real complex media, for example in the case of exciton trapping on percolation systems [52] or in the case of anomalous motions induced by obstruction and binding in crowded environments [53–55] such as biological cells [47, 56–59]. In particular fractal models have regained interest in the

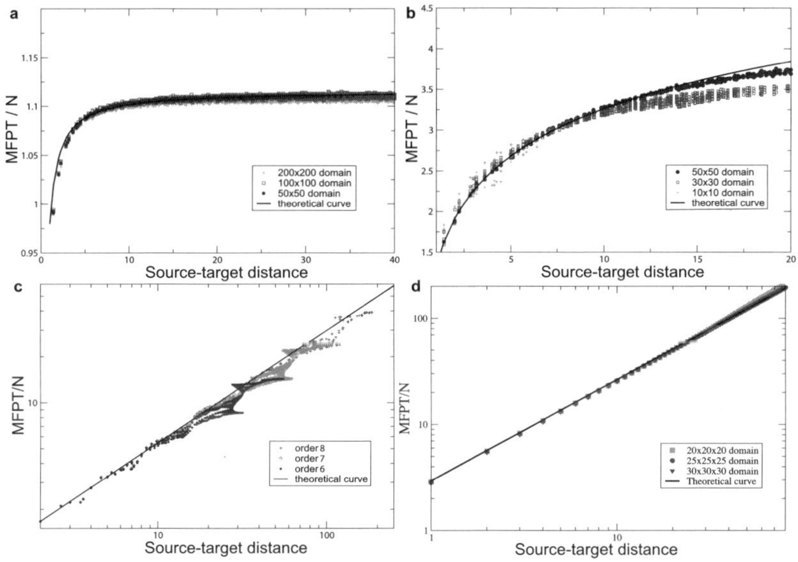

Fig. 4. MFPT in scale invariant media. **a** Lévy flights on a 2D square lattice ($\beta = 1$). The confining domains are a 50×50, 100×100 and 200×200 squares, with the target approximately in the middle. The MFPT is presented as a function of the source-target distance for different source points. Simulation points are fitted with $\langle \mathbf{T} \rangle \sim N \left(A - Br^{\beta - 2} \right)$. **b** Random barrier model with a transition rate distribution $\rho(\Gamma) = (\alpha/\Gamma)(\Gamma/\Gamma_0)^\alpha$, with $\Gamma_0 = 1$ and $\alpha = 0.5$. The confining domain is a $L \times L$ square with the target point in the middle. Numerical simulations of the MFPT rescaled by the volume N, averaged over the disorder, for 3 different domain sizes. The theoretical curve (black line) is given by $\langle \mathbf{T} \rangle / N \sim (A + (1/2\pi D_{\text{eff}}) \ln r)$, where the only fitting parameter is A. **c** Numerical simulations of random walks on a Sierpinski gasket (log/log plot) for 3 different system sizes (order 6,7 and 8). For each set of points, the size of the Sierpinski gasket and the target point are fixed, and the starting point takes various positions on the Sierpinski gasket. The plain line corresponds to the theoretical scaling $r^{d_w - d_f}$. **d** 3-D critical percolation clusters. For each size of the confining domain, the MFPT, normalized by the number of sites N, is averaged both over the different target and starting points separated by the corresponding chemical distance, and over percolation clusters. The black plain curve corresponds to the theoretical prediction of equation (46) with $d_w - d_f \simeq 1$.

context of the nuclear organization of DNA [60]. Indeed, recent experiments based on neutron scattering [61], rheology technics [62] and more recently the Hi-C method [63] revealed independently a fractal structure of the chromatin.

Figure 4 reveals an excellent quantitative agreement between the analytical predictions and the numerical simulations. Both the volume dependence and the source-target distance dependence are unambiguously captured by the theoretical expressions (46), as shown by the data collapse of the numerical simulations. We emphasize that the very different nature of these examples demonstrates that the range of applicability of the approach, which mainly relies on the length scale-invariant property of the infinite space propagator (45), is wide.

These results can be partly reproduced by using other methods such as electrical analogies [64, 65] and elastic vibrations spectrum analysis [66]. It has also been extended in several directions. First, in the case where the scaling form of Eq. (45) holds at all scales it is shown in [67] that in the compact case $A = 0$. Approximations of the constant B, independent of the scaling function Π can also be obtained by using the Kac formula. This yields in the compact case a zero constant formula for the MFPT, which has been shown to be in very good agreement with numerical results in various examples [67]. Note that these results rely on the scale invariance hypothesis; examples of anisotropic media have been studied in [65]. Last, we note that in the case of deterministic fractals, the scaling relation Eq. (45) actually only holds for discrete rescaling of the distance $|x - y|$. This can be shown to induce small amplitude oscillations of the MFPT with the source to target distance, which can be analytically derived by adapting the above formalism [68].

3.2. *Further first-passage observables*

Following the methods developed in section 2, the explicit geometric dependence of further first-passage observables of scale-invariant Markovian processes can be derived. Indeed, the large volume asymptotics of the exact expressions of first-passage observables in terms of pseudo-Green functions derived earlier in section 2 are straightforwardly obtained by making the substitution $H(x|y) \to G_0(x|y) \equiv \sum_{n=1}^{\infty} P_n^{\infty}(x|y)$ and using Eq. (45). Note that only differences of H functions are involved, which always remain finite in the large volume limit.

For example, it is possible to compute splitting probabilities within this framework. We consider here 2 targets located at x_1 and x_2 and a source point y. The probability π_1 of first reaching target 1, starting from y, is then [47]:

$$\pi_1 \sim \begin{cases} \dfrac{A + B(r_{1y}^{d_w-d_f} - r_{2y}^{d_w-d_f} - r_{12}^{d_w-d_f})}{2(A - Br_{12}^{d_w-d_f})} & \text{for } d_w < d_f \\ \dfrac{A + B\ln(r_{2y}r_{12}/r_{1y})}{2(A + B\ln(r_{12}))} & \text{for } d_w = d_f \\ \dfrac{1}{2}\left((r_{2y}/r_{12})^{d_w-d_f} - (r_{1y}/r_{12})^{d_w-d_f} + 1\right) & \text{for } d_w > d_f \end{cases} \tag{47}$$

where $r_{ij} = |\mathbf{r}_i - \mathbf{r}_j|$ is different from 0. The mean occupation time of site i starting form site S before reaching site T can also be calculated [47], and reads:

$$\langle \mathbf{N_i} \rangle \sim \begin{cases} A + B(r_{iS}^{d_w-d_f} - r_{iT}^{d_w-d_f} - r_{ST}^{d_w-d_f}) & \text{for } d_w < d_f \\ A + B\ln(r_{iT}r_{ST}/r_{iS}) & \text{for } d_w = d_f \\ B(r_{iT}^{d_w-d_f} + r_{ST}^{d_w-d_f} - r_{iS}^{d_w-d_f}) & \text{for } d_w > d_f \end{cases} . \tag{48}$$

Note that the entire distribution of $\mathbf{N_i}$, which involves only splitting probabilities, can be derived [47].

3.3. *Higher order moments and the entire first-passage time distribution*

As soon as several time-scales are involved, the MFPT is not sufficient to characterize the effect of confinement on reactivity and the full distribution is needed. Here we present the universal expressions of this distribution. These results can be applied to various models of disordered media [32, 44], such as fractal networks, percolation clusters, random trap and random barrier models, and verified by numerical simulations. As we will show next, fluctuations of the FPT, encoded in the distribution, can play a major role in reaction kinetics in confinement, and even lead to so-called "geometry-controlled kinetics" [69].

3.3.1. *Universality classes of FPT distributions.*

It is showed in [11, 69] that the natural time variable to use is the rescaled variable θ defined by:

$$\theta = \frac{t}{\overline{\langle \mathbf{T} \rangle}_T}, \tag{49}$$

where $\overline{\langle \mathbf{T} \rangle}_T$ is the global MFPT defined in equation (16) as the MFPT to reach the target T averaged over the starting positions. The FPT probability distribution then reads for the two types of exploration, in the large-volume limit:

$$q^*_{TS}(\theta) = (1 - \Pi_{TS})\,\delta(\theta) + \Pi_{TS}\,\psi(\theta). \tag{50}$$

It characterizes universality classes of FPT distributions in confined media.

The dependence on the reduced time θ is fully contained in the functions δ and ψ while the spatial dependence on r, lies entirely in the function Π_{TS}, which we consequently call the geometric factor. Last, the dependence on N enters both the reduced time θ and the geometric factor Π_{TS}. The specific shape of the function ψ and that of the geometric factor Π_{TS} characterize the type of exploration (non-compact or compact).

For the non-compact exploration, which qualitatively corresponds to trajectories leaving many sites unvisited, one has:

$$\psi(\theta) = \exp(-\,\theta) \tag{51a}$$

$$\overline{\langle \mathbf{T}\rangle}_T = \frac{H_{TT}}{W_T^{\mathrm{stat}}} \propto N \tag{51b}$$

$$\Pi_{TS} = \frac{\langle \mathbf{T}_{TS}\rangle}{\overline{\langle \mathbf{T}\rangle}_T} \propto 1 - \left(\frac{a}{r}\right)^{d_f - d_w}, \tag{51c}$$

where a is a typical length of the order of one step size. We have used that H_{TT} remains finite in the large volume limit, and the scaling form of the first-moment of the FPT [17, 43] derived in section 3.1. The simple exponential form of ψ can be verified numerically on various examples (see [11] and figure 5 for examples).

In the case of compact exploration, it can be shown that:

$$\psi(\theta) \sim \zeta(\theta) \quad \text{for } \theta \gg \theta_c \tag{52a}$$

$$\text{with} \quad \zeta(\theta) = \frac{2\,d_f\,d_w}{d_w^2 - d_f^2}\,\frac{\Gamma(\nu)}{\Gamma(2-\nu)} \sum_{k=0}^{\infty} \left(\frac{\alpha_k}{2}\right)^{3-2\nu} \times \frac{J_\nu(\alpha_k)}{J_{1-\nu}(\alpha_k)} \exp\left(-\frac{\alpha_k^2\,d_w\,d_f}{2\,(d_w^2 - d_f^2)}\,\theta\right) \tag{52b}$$

$$\overline{\langle \mathbf{T}\rangle}_T = \frac{H_{TT}}{W_T^{\mathrm{stat}}} \propto N^{d_w/d_f} \tag{52c}$$

$$\Pi_{TS} = \frac{2d_f^2}{d_w(d_f + d_w)}\,\frac{\langle \mathbf{T}_{TS}\rangle}{\overline{\langle \mathbf{T}\rangle}_T} \propto \left(\frac{r}{R}\right)^{d_w - d_f}, \tag{52d}$$

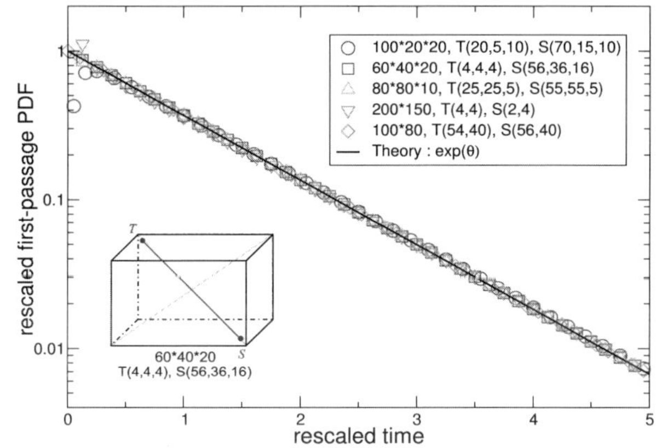

Fig. 5. 3D and 2D Euclidean lattices. Simulated rescaled first-passage PDFs (equation (50)) obtained by exact enumeration (symbols) are plotted against theoretical distribution $\psi(\theta) = e^{-\theta}$ (equation (51a)) (solid line) for several confining domain shapes and source/target positions. Domain dimensions $L_x L_y (L_z)$ and source/target coordinates are indicated in the legend inset. Rescaled time is $\theta = t/\overline{\langle \mathbf{T} \rangle}_T$ (equation (49)). For this type of networks, moments $\langle \mathbf{T}_{TS} \rangle$ and $\overline{\langle \mathbf{T} \rangle}_T$ are calculated exactly [20], so that the theoretical first-passage distribution is known exactly.

and

$$\theta_c = \frac{2(d_w^2 - d_f^2)}{d_w d_f} \left(\frac{r}{R}\right)^{d_w}. \tag{53}$$

In the relations above, $\alpha_0 < \alpha_1 < \ldots$ are the real zeros of $J_{-\nu}$; $\nu = 1 - d_f/d_w$. The form of the function ψ can be verified numerically on various examples (see [11] and figure 6 for examples).

Finally, in the marginal case, *i.e.* when $d_f = d_w$, the same method applies and leads to an exponential scaling function ψ as for the non-compact case, with logarithmic corrections in the volume-dependence of $\overline{\langle \mathbf{T} \rangle}_T$ as well as in the volume and source-target dependences of Π_{TS}. More explicitly one gets

$$\psi(\theta) = e^{-\theta} \tag{54a}$$

$$\overline{\langle \mathbf{T} \rangle}_T = \frac{H_{TT}}{W_T^{\mathrm{stat}}} \propto N \ln N \tag{54b}$$

$$\Pi_{TS} = \frac{\langle \mathbf{T}_{TS} \rangle}{\overline{\langle \mathbf{T} \rangle}_T} \propto \frac{\beta - \ln(r)}{\beta - \ln(R)}, \tag{54c}$$

where β is a lattice dependent constant.

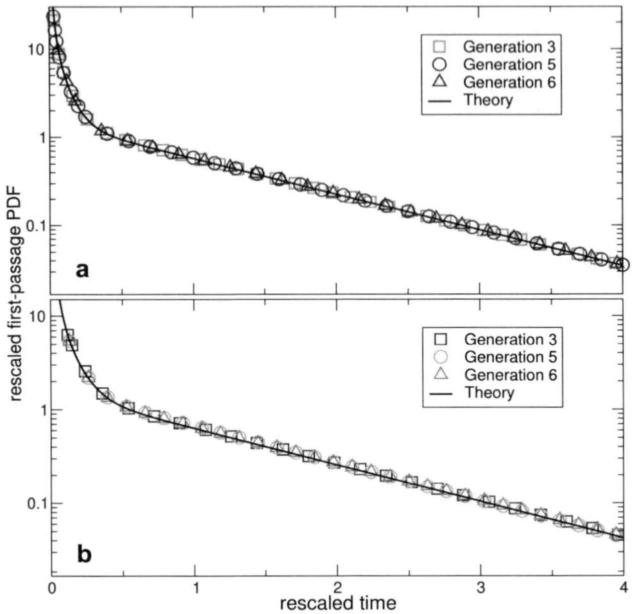

Fig. 6. FPT distribution on deterministic fractals. Rescaled first-passage densities (symbols, see equation (50)) obtained by exact enumeration are drawn versus the theoretical prediction $\psi(\theta)$ (solid lines, see equation (52b)) for different domain sizes (see insets). The target is fixed and the curves are averaged over all the sources of the network to fulfill the isotropy hypothesis of the propagator (in other words, for each generation, the plotted points result from an average over the rescaled PDFs of the walks starting from every site of the network, and ending at the chosen target). The quantity $\overline{\langle \mathbf{T} \rangle}_T$ that is needed to rescale the results is calculated exactly [70, 71]. **a.** Sierpinki gasket, the target is set at the apex. **b.** T-graph, the target is set at the central node.

3.3.2. *Discussion.*

First, a few comments about expression (50). (i) Strikingly, the FPT distribution is entirely determined as soon as the first-moment $\langle \mathbf{T}_{TS} \rangle$ and averaged first-moment $\overline{\langle \mathbf{T} \rangle}_T$ of the FPT distribution are known. (ii) The first term (Dirac δ function) in the right-hand takes into account trajectories reaching the target site within a time $t \ll \overline{\langle \mathbf{T} \rangle}_T$ (corresponding to $\theta \ll 1$), which more precisely scales as $t \propto \mathcal{O}(r^{d_w})$. Such trajectories can be interpreted as trajectories reaching the target site before touching the boundaries and the factor $1 - \Pi_{TS}$ can be seen as the weight of these trajectories. On the other hand, the second term describes trajectories reaching the boundaries before the target, with weight Π_{TS}.

Let us now discuss the differences between the two types of exploration, starting with their temporal dependence contained in $\psi(\theta)$. In the non-compact case, one can see that, for $t > 0$, when the source-target distance r is much larger than the step size a, the FPT distribution reduces to a simple time-decreasing exponential with weight unity. Remarkably, the exponential form for the function ψ holds universally for any values of the dimensions d_f and d_w satisfying $d_w < d_f$. Transport processes as varied as regular diffusion in 3 dimensions (for which $d_w = 2$ and $d_f = 3$), transport on supercritical percolation cluster, as well as transport in some quenched disordered media, fall into that universality class. The single-exponential scaling of ψ with time could be expected since both $\overline{\langle \mathbf{T} \rangle}_T$ and $\langle \mathbf{T}_{TS} \rangle$ [43] scale linearly with N. In the compact case on the other hand, the function ψ is not given by a single exponential, as for the non-compact case, but rather exhibits a countable infinity of time scales, which range between r^{d_w} and R^{d_w}. This complex behaviour, parametrized by d_w and d_f, is induced by the interplay between the characteristic times $\overline{\langle \mathbf{T} \rangle}_T$ and $\langle \mathbf{T}_{TS} \rangle$, which do not share the same dependence on N. Note that most fractal sets (even if embedded in 3D space, like the critical percolation cluster) lead to compact exploration.

Let us now focus on the r-dependence of Π_{TS}. As expected one finds that Π_{TS} increases with r in both cases, *i.e.* the further the source from the target, the higher the probability to reach the boundaries before the target site. In the non-compact case, that dependence of Π_{TS} on r is significant only for r of the order of the step size a, and is lost for $r \gg a$. In the compact case on the other hand, Π_{TS} increases with r, and quite remarkably tends to 0 for r small. Overall the dependence on r of the geometric weight is important for all values of r, in contrast with the non-compact case. This suggests that in the case of compact exploration, the initial distance between the reactants plays a crucial role in the kinetics of transport-limited reactions. The latter comment is reflected by the scaling of the reduced variance $(\langle \mathbf{T}_{TS}^2 \rangle - \langle \mathbf{T}_{TS} \rangle^2) / \langle \mathbf{T}_{TS} \rangle^2$: in the case of non-compact exploration that quantity is of order 1 at any distance, whereas in the compact case it scales like $(R/r)^{d_w - d_f}$, so that very large fluctuations occur when $r \ll R$. This regime was called geometry controlled kinetics in [69] (see figure 7). Note that while the distribution of the FPT fully quantifies the fluctuations of the FPT, other observables such as participation ratios have recently been studied, leading to similar conclusions [72].

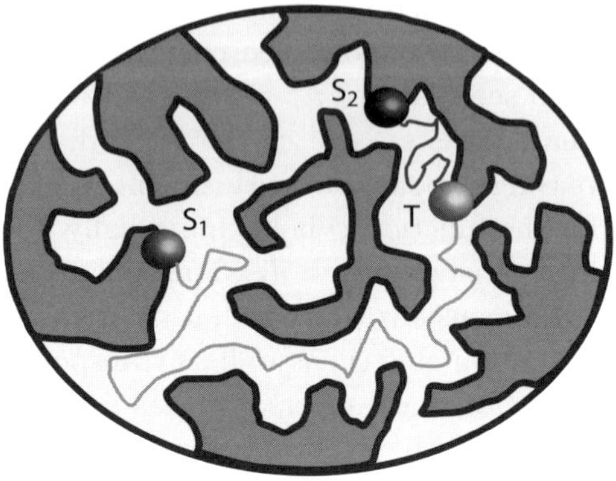

Fig. 7. **First-passage time distribution (FPT) and geometry controlled kinetics.** Is (or is not) the initial position of the particle an important parameter of the kinetics? We show quantitatively that for dilute systems (eg non compact case), the kinetics turns out to be widely independent of the starting point (S_1 or S_2), whereas in the case of crowded environments (eg compact case), the position of the starting point strongly influences the search time of the target, leading to "geometry controlled kinetics". In the context of gene transcription, this result implies that the kinetics of activation of a gene T by a transcription factor (TF) can be orders of magnitudes faster if the TF is released from a site $S \equiv S_2$ which is colocalized with (*i.e.* in the vicinity of) T, as compared to the case where the TF is released from a remote site $S \equiv S_1$.

4. Conclusion

To conclude let us summarize the main content of this chapter. We have presented a general theory which allows one to accurately evaluate the mean FPT (MFPT) for regular random walks in bounded domains. We then briefly mentioned extensions of these results to related first-passage observables like splitting probabilities and occupation times. It was showed that this analytical approach provides a universal scaling dependence of the MFPT on both the volume of the confining domain and the source-target distance in the case of general scale invariant processes. This analysis is applicable to a broad range of stochastic processes characterized by length scale invariant properties. Last, the full distribution of the FPT can be obtained, and its universal features have been discussed. This makes it possible to quantify the fluctuations of the FPT in confinement, and to reveal the key role that can be played by the starting position of the random walker, in a regime called geometry controlled kinetics. Finally, while all the results presented in this chapter concern Markovian random walks, we stress that first extensions to non Markovian processes have recently been obtained in the cases of persistent random walks [73] and Rouse chains [74–76].

Acknowledgments

Support from European Research Council starting Grant FPTOpt-277998 is acknowledged.

References

1. N. V. Kampen, *Stochastic Processes in Physics and Chemistry.* North-Holland (1992).
2. S. Redner, *A guide to First-Passage Processes.* Cambridge University Press, Cambridge, England (2001).
3. S. Rice, *Diffusion-Limited Reactions.* Elsevier, Amsterdam (1985).
4. H. Tuckwell, *Introduction to Theoretical Neurobiology.* Cambridge University Press (1988).
5. C. Doering and J. Gadoua, *Phys. Rev. Lett.* **69**, 2318 (1992).
6. O. Benichou, M. Coppey, M. Moreau, P.-H. Suet, and R. Voituriez, Optimal search strategies for hidden targets., *Phys Rev Lett.* **94**(19), 198101 (2005). ISSN 0031-9007 (Print).
7. O. Benichou, C. Loverdo, M. Moreau, and R. Voituriez, Optimizing intermittent reaction paths, *Physical Chemistry Chemical Physics.* **10**(47), 7059–7072 (2008). URL `http://dx.doi.org/10.1039/b811447c`.
8. O. Bénichou, C. Loverdo, M. Moreau, and R. Voituriez, Intermittent search strategies, *Reviews of Modern Physics.* **83**(1), 81–129 (03, 2011). URL `http://link.aps.org/doi/10.1103/RevModPhys.83.81`.
9. S. Condamin, O. Benichou, and M. Moreau, First-passage times for random walks in bounded domains., *Phys Rev Lett.* **95**(26), 260601 (2005). ISSN 0031-9007 (Print).
10. S. Condamin, O. Benichou, and J. Klafter, First-passage time distributions for subdiffusion in confined geometry., *Phys Rev Lett.* **98**(25), 250602 (2007). ISSN 0031-9007 (Print).
11. B. Meyer, C. Chevalier, R. Voituriez, and O. Bénichou, Universality classes of first-passage-time distribution in confined media, *Physical Review E.* **83** (5), 051116 (2011). URL `http://link.aps.org/doi/10.1103/PhysRevE.83.051116`.
12. B. D. Hughes, *Random Walks and Random Environments.* Oxford Press (1995).
13. J. D. Noh and H. Rieger, Constrained spin-dynamics description of random walks on hierarchical scale-free networks, *Physical Review E.* **69**(3) (03, 2004). URL `http://link.aps.org/abstract/PRE/v69/e036111`.
14. D. Aldous and J. Fill, *Reversible Markov chains and random walks on graphs.* http://www.stat.berkeley.edu/users/aldous/RWG/book.html (1999).

15. G. Barton, *Elements of Green functions and propagation: potentials, diffusion and waves*. Oxford University Press, New-York (1989).
16. C. P. Haynes and A. P. Roberts, Global first-passage times of fractal lattices, *Phys. Rev. E*. **78**(4), 041111 (Oct, 2008). doi: 10.1103/PhysRevE.78.041111.
17. V. Tejedor, O. Bénichou, and R. Voituriez, Global mean first-passage times of random walks on complex networks, *Physical Review E*. **80**(6), 065104 (2009). URL `http://link.aps.org/abstract/PRE/v80/e065104`.
18. B. Meyer, E. Agliari, O. Bénichou, and R. Voituriez, Exact calculations of first-passage quantities on recursive networks, *Physical Review E*. **85**(2), 026113– (02, 2012). URL `http://link.aps.org/doi/10.1103/PhysRevE.85.026113`.
19. S. Condamin and O. Benichou, Exact expressions of mean first-passage times and splitting probabilities for random walks in bounded rectangular domains., *J Chem Phys*. **124**(20), 206103 (2006). ISSN 0021-9606 (Print). doi: 10.1063/1.2192770.
20. S. Condamin, O. Benichou, and M. Moreau, Random walks and brownian motion: a method of computation for first-passage times and related quantities in confined geometries., *Phys Rev E Stat Nonlin Soft Matter Phys*. **75** (2 Pt 1), 021111 (2007). ISSN 1539-3755 (Print).
21. J. S. van Zon, M. J. Morelli, S. Tanase-Nicola, and P. R. ten Wolde, Diffusion of transcription factors can drastically enhance the noise in gene expression, *Biophysical Journal*. **91**(12), 4350–4367 (12, 2006). URL `http://www.sciencedirect.com/science/article/pii/S000634950672149X`.
22. B. Meyer, O. Bénichou, Y. Kafri, and R. Voituriez, Geometry-induced bursting dynamics in gene expression, *Biophysical Journal*. **102**(9), 2186–2191 (5, 2012). URL `http://www.sciencedirect.com/science/article/pii/S0006349512004018`.
23. C. Godrèche and J. Luck, Statistics of the occupation time of renewal processes, *J. Stat. Phys*. **104**, 489 (2001).
24. S. Majumdar and A. Comtet, Local and occupation time of a particle diffusing in a random medium, *Phys. Rev. Lett*. **89**, 60601 (2002).
25. O. Benichou, M. Coppey, J. Klafter, M. Moreau, and G. Oshanin, On the joint residence time of n independent two-dimensional brownian motions, *Journal of Physics A: Mathematical and General*. **36**(26), 7225–7231 (2003).
26. S. Blanco and R. Fournier, An invariance property of diffusive random walks, *EPL (Europhysics Letters)*. **61**(2), 168–173 (2003).
27. O. Bénichou, M. Coppey, M. Moreau, P. H. Suet, and R. Voituriez, Averaged residence times of stochastic motions in bounded domains, *EPL (Europhysics Letters)*. **70**(1), 42–48 (2005).
28. S. Condamin, O. Benichou, and M. Moreau, First-exit times and residence times for discrete random walks on finite lattices., *Phys Rev E Stat Nonlin Soft Matter Phys*. **72**(1 Pt 2), 016127 (2005). ISSN 1539-3755 (Print).
29. S. Blanco and R. Fournier, Short-path statistics and the diffusion approximation, *Physical Review Letters*. **97**(23), 230604–4 (2006). URL `http://link.aps.org/abstract/PRL/v97/e230604`.
30. S. Burov and E. Barkai, Occupation time statistics in the quenched trap

model, *Physical Review Letters*. **98**(25), 250601–4 (2007). URL `http://link.aps.org/abstract/PRL/v98/e250601`.
31. M. Ferraro and L. Zaninetti, *Physica A*. **338**, 307 (2004).
32. J.-P. Bouchaud and A. Georges, Anomalous diffusion in disordered media: Statistical mechanisms, models and physical applications, *Physics Reports*. **195**(4-5), 127–293 (1990). URL `http://www.sciencedirect.com/science/article/B6TVP-46SXPMN-7F/2/691b5d0049eff14ec4a52e098fae92db`.
33. O. Benichou, M. Coppey, M. Moreau, and G. Oshanin, Kinetics of diffusion-limited catalytically activated reactions: an extension of the wilemski-fixman approach., *J Chem Phys*. **123**(19), 194506 (2005). ISSN 0021-9606 (Print). doi: 10.1063/1.2109967.
34. S. Condamin, V. Tejedor, and O. Benichou, Occupation times of random walks in confined geometries: from random trap model to diffusion-limited reactions., *Phys Rev E Stat Nonlin Soft Matter Phys*. **76**(5 Pt 1), 050102 (2007). ISSN 1539-3755 (Print).
35. M. J. Ward and J. B. Keller, Strong localized perturbations of eigenvalue problems, *SIAM Journal on Applied Mathematics*. **53**(3), 770–798 (06, 1993). URL `http://link.aip.org/link/?SMM/53/770/1`.
36. A. F. Cheviakov and M. J. Ward, Optimizing the principal eigenvalue of the laplacian in a sphere with interior traps, *Mathematical and Computer Modeling*. **53**, 1394 (2011).
37. Z. Schuss, A. Singer, and D. Holcman, The narrow escape problem for diffusion in cellular microdomains, *Proc Natl Acad Sci U S A*. **104**(41), 16098–103 (Oct, 2007). doi: 10.1073/pnas.0706599104.
38. Z. Schuss, The narrow escape problem—a short review of recent results. **53** (1), 194–210 (2012). doi: 10.1007/s10915-012-9590-y. URL `http://dx.doi.org/10.1007/s10915-012-9590-y`.
39. C. Chevalier, O. Bénichou, B. Meyer, and R. Voituriez, First-passage quantities of brownian motion in a bounded domain with multiple targets: a unified approach, *Journal of Physics A: Mathematical and Theoretical*. **44**, 025002 (2011). URL `http://stacks.iop.org/1751-8121/44/i=2/a=025002`.
40. O. Benichou and R. Voituriez, Narrow-escape time problem: Time needed for a particle to exit a confining domain through a small window, *Physical Review Letters*. **100**(16), 168105–4 (2008). URL `http://link.aps.org/abstract/PRL/v100/e168105`.
41. V. Tejedor, O. Bénichou, and R. Voituriez, Close or connected: Distance and connectivity effects on transport in networks, *Physical Review E*. **83** (6), 066102 (2011). URL `http://link.aps.org/doi/10.1103/PhysRevE.83.066102`.
42. G. Barton, *Elements of Green's Functions and Propagation*. Oxford Science Publications (1989).
43. S. Condamin, O. Benichou, V. Tejedor, R. Voituriez, and J. Klafter, First-passage times in complex scale-invariant media., *Nature*. **450**(7166), 77–80 (2007). ISSN 1476-4687 (Electronic). doi: 10.1038/nature06201.
44. D. Ben-Avraham and S. Havlin, *Diffusion and reactions in fractals and disordered systems*. Cambridge University Press (2000).

45. E. W. Montroll, Random walks on lattices. iii. calculation of first-passage times with application to exciton trapping on photosynthetic units, *Journal of Mathematical Physics.* **10**(4), 753–765 (1969). URL `http://link.aip.org/link/?JMP/10/753/1`.
46. E. M. Bollt and D. ben Avraham, What is special about diffusion on scale-free nets?, *New Journal of Physics.* **7**, 26–26 (2005). ISSN 1367-2630.
47. S. Condamin, V. Tejedor, R. Voituriez, O. Benichou, and J. Klafter, Probing microscopic origins of confined subdiffusion by first-passage observables, *Proceedings of the National Academy of Sciences.* **105**, 5675 (2008). URL `http://www.pnas.org/cgi/content/abstract/0712158105v1`.
48. R.Metzler and J.Klafter, The random walk's guide to anomalous diffusion: a fractionnal dynamics approach, *Phys. Rep.* **339**(1), 1–77 (2000).
49. R.Metzler and J.Klafter, The restaurant at the end of the random walk: recent developments in the description of anomalous transport by fractional dynamics, *J.Phys.A.* **37**, R161–R208 (2004).
50. S. Havlin and D. ben Avraham, Diffusion in disordered media, *Adv.in Phys.* **36**(6), 695 (1987).
51. A. Bunde and S. Havlin, eds., *Fractals and disordered systems.* Springer-Verlag, Berlin (1991).
52. R. P. Parson and R. Kopelman, Percolative versus homogenous energy transport kinetics: time-resolved donor and acceptor fluoresence of isotopic mixed naphthalene crystals, *Chemical Physics Letters.* **87**(6), 528–532 (4, 1982). URL `http://www.sciencedirect.com/science/article/B6TFN-44P8CMJ-HY/2/0e75636f821b8c378f70b042b875daec`.
53. S. B. Zimmerman and A. P. Minton, Macromolecular crowding: Biochemical, biophysical, and physiological consequences, *Annual Review of Biophysics and Biomolecular Structure.* **22**(1) (06, 1993). URL `http://dx.doi.org/10.1146/annurev.bb.22.060193.000331`.
54. R. Metzler and J. Klafter, The restaurant at the end of the random walk: recent developments in the description of anomalous transport by fractional dynamics, *Journal of Physics A: Mathematical and General.* **37**(31) (2004). URL `http://stacks.iop.org/0305-4470/37/i=31/a=R01`.
55. H.-X. Zhou, G. Rivas, and A. P. Minton, Macromolecular crowding and confinement: Biochemical, biophysical, and potential physiological consequences*, *Annual Review of Biophysics.* **37**(1), 375–397 (05, 2008). URL `http://dx.doi.org/10.1146/annurev.biophys.37.032807.125817`.
56. M. J. Saxton, A biological interpretation of transient anomalous subdiffusion. i. qualitative model, *Biophysical Journal.* **92**(4), 1178–1191 (2007). URL `http://www.biophysj.org/cgi/content/abstract/92/4/1178`.
57. M. J. Saxton, A biological interpretation of transient anomalous subdiffusion. ii. reaction kinetics. **94**(3), 760–771 (02, 2008). URL `http://linkinghub.elsevier.com/retrieve/pii/S0006349508706775`.
58. N. Malchus and M. Weiss, Elucidating anomalous protein diffusion in living cells with fluorescence correlation spectroscopy—facts and pitfalls, *Journal of Fluorescence* (2009). URL `http://dx.doi.org/10.1007/s10895-009-0517-4`.

59. J. Szymanski and M. Weiss, Elucidating the origin of anomalous diffusion in crowded fluids, *Physical Review Letters*. **103**(3), 038102–4 (07, 2009). URL `http://link.aps.org/abstract/PRL/v103/e038102`.
60. O. Bénichou, C. Chevalier, B. Meyer, and R. Voituriez, Facilitated diffusion of proteins on chromatin, *Physical Review Letters*. **106**, 038102 (2011). URL `http://link.aps.org/doi/10.1103/PhysRevLett.106.038102`.
61. D. Lebedev, M. Filatov, A. Kuklin, A. Islamov, J. Stellbrink, R. Pantina, Y. Denisov, B. Toperverg, and V. Isaev-Ivanov, Structural hierarchy of chromatin in chicken erythrocyte nuclei based on small-angle neutron scattering: Fractal nature of the large-scale chromatin organization, *Crystallography Reports*. **53**(1), 110–115 (01, 2008). URL `http://dx.doi.org/10.1134/S1063774508010136`.
62. A. Bancaud, S. Huet, N. Daigle, J. Mozziconacci, J. Beaudouin, and J. Ellenberg, Molecular crowding affects diffusion and binding of nuclear proteins in heterochromatin and reveals the fractal organization of chromatin, *EMBO J.* **28**(24), 3785–3798 (12, 2009). URL `https://acces-distant.upmc.fr:443/http/dx.doi.org/10.1038/emboj.2009.340`.
63. E. Lieberman-Aiden, N. L. van Berkum, L. Williams, M. Imakaev, T. Ragoczy, A. Telling, I. Amit, B. R. Lajoie, P. J. Sabo, M. O. Dorschner, R. Sandstrom, B. Bernstein, M. A. Bender, M. Groudine, A. Gnirke, J. Stamatoyannopoulos, L. A. Mirny, E. S. Lander, and J. Dekker, Comprehensive mapping of long-range interactions reveals folding principles of the human genome, *Science*. **326**(5950), 289–293 (10, 2009). URL `http://www.sciencemag.org/cgi/content/abstract/326/5950/289`.
64. C. P. Haynes and A. P. Roberts, Generalization of the fractal einstein law relating conduction and diffusion on networks, *Physical Review Letters*. **103** (2), 020601–4 (07, 2009). URL `http://link.aps.org/abstract/PRL/v103/e020601`.
65. A. P. Roberts and C. P. Haynes, Electrostatic approximation of source-to-target mean first-passage times on networks, *Physical Review E*. **83**(3) (03, 2011). URL `http://link.aps.org/doi/10.1103/PhysRevE.83.031113`.
66. S. Reuveni, R. Granek, and J. Klafter, Vibrational shortcut to the mean-first-passage-time problem, *Physical Review E*. **81**, 040103 (2010). URL `http://link.aps.org/doi/10.1103/PhysRevE.81.040103`.
67. O. Benichou, B. Meyer, V. Tejedor, and R. Voituriez, Zero constant formula for first-passage observables in bounded domains, *Physical Review Letters*. **101**(13), 130601–4 (2008). URL `http://link.aps.org/abstract/PRL/v101/e130601`.
68. E. Akkermans, O. Benichou, G. V. Dunne, A. Teplyaev, and R. Voituriez, Spatial log-periodic oscillations of first-passage observables in fractals, *Physical Review E*. **86**(6), 061125– (12, 2012). URL `http://link.aps.org/doi/10.1103/PhysRevE.86.061125`.
69. O. Bénichou, C. Chevalier, J. Klafter, B. Meyer, and R. Voituriez, Geometry-controlled kinetics, *Nat Chem*. **2**(6), 472–477 (06, 2010). URL `http://dx.doi.org/10.1038/nchem.622`.
70. C. P. Haynes and A. P. Roberts, Global first-passage times of fractal lattices,

Physical Review E (Statistical, Nonlinear, and Soft Matter Physics). **78**(4), 041111–9 (2008). URL `http://link.aps.org/abstract/PRE/v78/e041111`.
71. E. Agliari, Exact mean first-passage time on the t-graph, *Physical Review E (Statistical, Nonlinear, and Soft Matter Physics).* **77**(1), 011128–6 (2008). URL `http://link.aps.org/abstract/PRE/v77/e011128`.
72. T. G. Mattos, C. Mejía-Monasterio, R. Metzler, and G. Oshanin, First passages in bounded domains: When is the mean first passage time meaningful?, *Physical Review E.* **86**(3), 031143– (09, 2012). URL `http://link.aps.org/doi/10.1103/PhysRevE.86.031143`.
73. V. Tejedor, R. Voituriez, and O. Bénichou, Optimizing persistent random searches, *Physical Review Letters.* **108**(8), 088103– (02, 2012). URL `http://link.aps.org/doi/10.1103/PhysRevLett.108.088103`.
74. T. Guérin, O. Bénichou, and R. Voituriez, Non-markovian polymer reaction kinetics, *Nat Chem.* **4**(7), 568–573 (07, 2012). URL `http://dx.doi.org/10.1038/nchem.1378`.
75. T. Guérin, O. Bénichou, and R. Voituriez, Reactive conformations and non-markovian reaction kinetics of a rouse polymer searching for a target in confinement, *Physical Review E.* **87**(3), 032601– (03, 2013). URL `http://link.aps.org/doi/10.1103/PhysRevE.87.032601`.
76. T. Guerin, O. Benichou, and R. Voituriez, Reactive conformations and non-markovian cyclization kinetics of a rouse polymer, *The Journal of Chemical Physics.* **138**(9), 094908–18 (03, 2013). URL `http://dx.doi.org/10.1063/1.4792437`.

Chapter 8

First Passage Behavior of Multi-Dimensional Fractional Brownian Motion and Application to Reaction Phenomena

Jae-Hyung Jeon[1,*], A. V. Chechkin[2,3,†], and Ralf Metzler[4,1,‡]

[1] *Department of Physics, Tampere University of Technology Tampere 33101, Finland*

[2] *Akhiezer Institute for Theoretical Physics NSC KIPT Akademicheskaya Str.1, 61108 Kharkov, Ukraine*

[3] *Max-Planck Institute for the Physics of Complex Systems, Nöthnitzer Straße 38, 01187 Dresden, Germany*

[4] *Institute for Physics & Astronomy, University of Potsdam 14476 Potsdam-Golm, Germany*

**jae-hyung.jeon@tut.fi*

†achechkin@kipt.kharkov.ua

‡rmetzler@uni-potsdam.de

Fractional Brownian motion is a generalized Gaussian diffusive process that is found to describe numerous stochastic phenomena in physics and biology. Here we introduce a multi-dimensional fractional Brownian motion (FBM) defined as a superposition of conventional FBM for each coordinate in analogy to multi-dimensional Brownian motion, and study its first passage properties. Starting from the well-established first passage time statistics of one-dimensional FBM and the associated approximation schemes, we explore the first passage time behavior of multi-dimensional FBM and compare these results with simulations. The asymptotic kinetic behavior of diffusion-limited reactions of reactant particles performing FBM in a one- and multi-dimensional space is studied based on the corresponding first passage time statistics.

1. Introduction

Theories of anomalous diffusion and active transport of physical particles are becoming increasingly important concepts and quantitative tools in many branches of the physical sciences, especially biology-related fields, as deviations from normal free and directed Brownian motion have been ubiquitously observed in many systems, in the wake of significant advances of experimental techniques such as single particle tracking [1–6]. Anomalous dynamic phenomena are manifested through a departure from the linear time dependence of the mean squared displacement of Brownian motion, $\langle \mathbf{r}^2(t)\rangle \simeq t$ [1–3]. In many cases, data follow the scaling law

$$\langle \mathbf{r}^2(t)\rangle \propto t^{\alpha}, \text{ with } \alpha \neq 1. \tag{1}$$

Here the anomalous diffusion exponent α is in the range $0 < \alpha < 1$ for subdiffusion and $1 < \alpha < 2$ for superdiffusion. $\alpha = 2$ denotes ballistic motion, while $\alpha > 2$ is often referred to as hyperdiffusion. The generalized diffusion law (1) is nonuniversal in the sense that it may be based on multiple physical mechanisms and equally well be described by several, physically different, prominent diffusion models. More specifically, fractional Brownian motion (FBM) gives rise to anomalous diffusion processes (1) in which the displacement autocorrelation function (DAF) has a power-law decay with a positive or negative sign at long times. In one spatial coordinate x, this implies DAF $\sim \alpha(\alpha-1)t^{\alpha-2}$. Meanwhile, continuous time random walks (CTRWs) describe processes in which the diffusion is hindered by multiple trapping events whose duration is distributed according to $\psi(\tau) \propto \tau^{-1-\alpha}$ with $0 < \alpha < 1$, such that the mean waiting time $\langle\tau\rangle$ diverges [3, 7, 8]. Superdiffusive CTRWs are either based on the spatiotemporally coupled Lévy walk scheme with finite moments of all orders, such that the MSD (1) is sub-ballistic with $1 < \alpha < 2$. Alternatively, they are described by Lévy flights corresponding to CTRW processes with a broad jump length distribution of the power-law form $\lambda(x) \propto |x|^{-1-\mu}$ with $0 < \mu < 2$, leading to a diverging variance $\langle x^2\rangle$.[a] Within the model of diffusion on fractals, the subdiffusive motion of the form (1) is induced by the fractal geometry of the space in which the diffusion process takes place [1].

While all of the aforementioned anomalous diffusion models share the scaling law (1), they differ in their stochastic properties and thus exhibit distinct diffusion-controlled dynamics. The first passage behavior is an important dynamic measure that pinpoints stochastic subtleties of the underlying anomalous diffusive process [4, 11]. First passage theory is a powerful stochastic concept that can be applied to attaining a quantitative description of numerous dynamic problems including diffusion-limited reaction kinetics [12–14], polymer cyclization [15–17], target search [18–20], barrier crossing [21, 22], and polymer translocation [23], among many others. In

[a]In this case, the conventional mean squared displacement (1) is replaced by rescaled fractional order moments $\langle|x(t)|^{\delta}\rangle^{2/\delta} \propto t^{2/\mu}$, with $0 < \delta < \mu$ [3, 4, 9, 10].

this chapter we study the first passage properties of FBM-type anomalous diffusions with applications to diffusion-limited reaction kinetics.

First passage statistics are systematically obtained using standard methods for Markovian processes [11]. However, their analytic derivation becomes challenging for anomalous diffusive processes. In some cases, though, full analytical solutions are available. This is true for subdiffusive CTRW processes [20, 24–27], for which the first passage behavior is obtained by standard techniques such as the separation of variables or the images method. Alternatively, it can be directly derived from the corresponding Brownian boundary value problem by subordination [4]. For Lévy flights some results have been derived such as the Sparre Andersen 3/2 scaling of the first passage density and the slower decay of the probability density of first arrival [28, 29]. For diffusion in complex environments, general statements on the mean first passage time are available [19]. In the case of FBM, however, analytic approaches are notoriously limited, the best known analytical result being Molchan's asymptotic scaling result of the first passage behavior of FBM in a semi-infinite, one-dimensional interval with an absorbing boundary [30, 31]. Except for this case, generally the first passage behavior of FBM is quite poorly understood and mostly limited to the one-dimensional case [23, 32–34]. Based on the fact that many dynamic processes in nature, including intracellular diffusive processes (see below), are governed by FBM, its first passage behavior in higher dimensions is of particular interest and relevance. This is especially so as the sample path of FBM has fractal dimension $2/\alpha$: FBM thus explores the space both compactly (recurrently) and non-compactly (transiently) for different anomalous diffusion exponents α when the dimensionality of the embedding space is greater than unity.

In this chapter we collect recent theoretical results for the first passage behavior of FBM in one and higher spatial dimensions and discuss their implications to diffusion-limited reaction phenomena. Our study focuses on the anomalous reaction kinetics resulting from FBM-type anomalous diffusion of reactant particles. Anomalous reaction problems based on other models such as CTRW or diffusion on fractals can be found, for instance, in [35–43]. In Sec. 2, we start from an overview of some basic facts on FBM and the related motion governed by the fractional Langevin equation, present its generalisation to d-dimensional FBM, and comment on the fractal dimension of FBM. In Sec. 3 we define the first passage process with mathematical rigour and then provide exact results and approximations of first passage of one-dimensional FBM, along with a comparison to simulations results. In the subsequent section, we consider the first passage problem of FBM in d-dimensional space, seeking the first passage time statistics based on the results of the one-dimensional case and the approximation theory presented in Sec. 3. Consequences for diffusion-limited reaction kinetics of FBM reactants in one and higher dimensional space are also discussed. Finally, our concluding remarks are presented in Sec. 5.

2. Fractional Brownian motion

One-dimensional FBM $x_\alpha(t)$ with the anomalous diffusion exponent α ($0 < \alpha < 2$), was originally studied by Kolmogorov [44] and Yaglom [45], and later became famous through the work of Mandelbrot and van Ness [46]. FBM may be viewed as a natural extension of normal Brownian motion (BM). Thus, FBM is a Gaussian process satisfying the criteria that $x_\alpha(0) = 0$ and for all t and Δt the increment $x_\alpha(t+\Delta t) - x(t)$ is stationary and given by the normal distribution of zero mean and variance $2K_\alpha(\Delta t)^\alpha$ so that its probability density function satisfies [45]

$$G(x_\alpha(t+\Delta t)|x_\alpha(t)) = \frac{1}{\sqrt{4\pi K_\alpha(\Delta t)^\alpha}} \exp\left(-\frac{[x_\alpha(t+\Delta t) - x_\alpha(t)]^2}{4K_\alpha(\Delta t)^\alpha}\right). \quad (2)$$

Here K_α is the generalized diffusivity of physical dimension $\mathrm{cm}^2/\mathrm{sec}^\alpha$. By this definition FBM satisfies the expected properties of a generalized Brownian motion, i.e.,

$$\langle x_\alpha^2(t)\rangle = 2K_\alpha t^\alpha \quad (3)$$

as well as

$$\langle [x_\alpha(t+\Delta t) - x_\alpha(t)]^2\rangle = 2K_\alpha(\Delta t)^\alpha. \quad (4)$$

It is immediately inferred from these relations that the covariance of FBM is uniquely given by [47, 48]

$$\langle x_\alpha(t_1)x_\alpha(t_2)\rangle = K_\alpha(|t_1|^\alpha + |t_2|^\alpha - |t_1 - t_2|^\alpha). \quad (5)$$

Thus, except for the special case $\alpha = 1$ of ordinary BM, the increments of FBM are correlated. FBM was shown to quantitatively describe numerous anomalous diffusive phenomena found in nature, for instance, annual river discharges [49], diffusion of a tracer particle in a single file [50–53] or in a viscoelastic environment [54, 55], conformational dynamics of proteins [56], diffusion of macromolecules in crowded or intracellular environments [57–61], or the lateral diffusion of phospholipid molecules in lipid bilayers [62, 63].

Mandelbrot and van Ness [46] presented an explicit form of the above FBM process via *fractional integration* of a white and Gaussian noise:

$$x_\alpha(t) = \frac{1}{\Gamma(\frac{\alpha+1}{2})}\left(\int_0^t d\tau (t-\tau)^{(\alpha-1)/2}\xi(\tau) + \int_{-\infty}^0 d\tau\left[(t-\tau)^{(\alpha-1)/2} - (-\tau)^{(\alpha-1)/2}\right]\xi(\tau)\right) \tag{6}$$

where $\Gamma(z)$ is the Gamma function and $\xi(\tau)$ represents white Gaussian noise with zero mean $\langle\xi(t)\rangle = 0$ and δ-correlation $\langle\xi(t)\xi(t')\rangle = \sqrt{2K_1}\delta(t-t')$. Expression (6) leads to FBM with the generalized diffusivity $K_\alpha = K_1[\Gamma((\alpha+1)/2)]^{-2}\left\{\int_{-\infty}^0[(1-\tau)^{(\alpha-1)/2} - (-\tau)^{(\alpha-1)/2}]^2 d\tau + 1/\alpha\right\}$ [46]. In the special case $\alpha = 1$, the integral representation of FBM (6) reduces to the familiar expression of BM, $x_1(t) = \int_0^t d\tau\xi(\tau)$.

Alternatively, FBM can be expressed in terms of its incremental sequence, i.e., fractional Gaussian noise (FGN) $\xi_\alpha(t)$, such that

$$x_\alpha(t) = \int_0^t dt'\xi_\alpha(t'). \tag{7}$$

Here FGN $\xi_\alpha(t)$ with the anomalous diffusion exponent $0 < \alpha < 2$ is a generalized Gaussian process of zero mean $\langle\xi_\alpha(t)\rangle = 0$ and satisfies the autocorrelation of the form[b] [46, 64, 65]

$$\langle\xi_\alpha(t_1)\xi_\alpha(t_2)\rangle = \begin{cases} \alpha K_\alpha(\alpha-1)|t_1-t_2|^{\alpha-2}, & \alpha\neq 1,\ |t_1-t_2|\to\infty \\ 2K_\alpha\delta(t_1-t_2), & \alpha = 1 \end{cases}. \tag{8}$$

Note that FGN with anomalous diffusion exponent $\alpha \neq 1$ has a power-law decay of the correlations with exponent $\alpha - 2$ in the long-time limit. The prefactor tells that FGN is negatively (positively) correlated for $0 < \alpha < 1$ ($1 < \alpha < 2$) and, in turn, gives rise to subdiffusive (superdiffusive) FBM. For $0 < \alpha < 1$ FGN is often called antipersistent, and persistent in the case $1 < \alpha < 2$. According to Eq. (8) the limiting case $\alpha = 2$ of ballistic motion is fully persistent. At $\alpha = 1$ FGN becomes ordinary white and Gaussian noise. Figure 1(Left) illustrates sample paths of FBM $x_\alpha(t)$ obtained from FGN for cases of subdiffusion ($\alpha = 1/2$), normal diffusion ($\alpha = 1$), and superdiffusion ($\alpha = 3/2$). The significantly enhanced exploration of space for growing α is distinct.

[b] An important property of FGN for $0 < \alpha < 1$ is that the integral over the autocorrelation function vanishes [46]

$$\int_0^\infty \langle\xi_\alpha(t)\xi_\alpha(0)\rangle dt = 0.$$

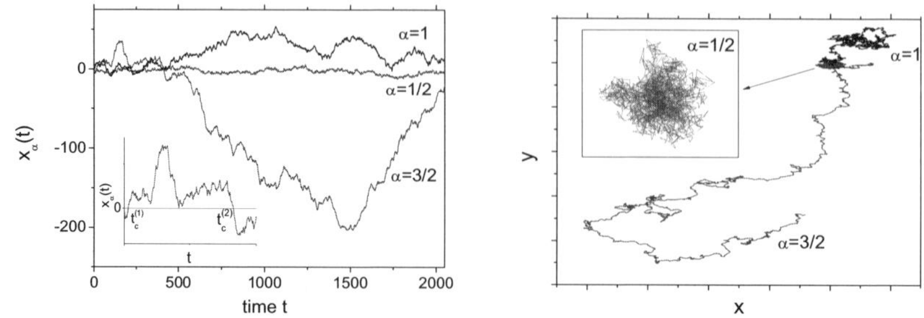

Fig. 1. (Color online) (Left) Sample FBM trajectories $x_\alpha(t)$ for three different cases: subdiffusion ($\alpha = 1/2$), normal diffusion ($\alpha = 1$), and superdiffusion ($\alpha = 3/2$). Inset: Zero-crossing times $t_c^{(i)}(> 0)$ are defined as the intersects of $x_\alpha(t)$ and the horizontal line of $y = 0$ for a given FBM trajectory. (Right) Two-dimensional FBM trajectories $\mathbf{x}_\alpha(t)$ with the anomalous diffusion exponent $\alpha = 1/2$ (red or light grey), 1 (black), 3/2 (blue or dark grey). Inset: Zoom-in motion of the subdiffusive walk of $\alpha = 1/2$.

We note that in our simulations we generate FGN using the Hosking method [66]. Alternative methods are described in [67, 68].

2.1. *Relation to fractional Langevin equation driven by FGN*

FBM is a nonequilibrium stochastic process driven by FGN in the sense that it is not subject to the fluctuation-dissipation theorem [see Eq. (7)]. The thermal motion $y_\alpha(t)$ of a particle of mass m driven by FGN is described by the fractional Langevin equation (FLE) with anomalous diffusion exponent $0 < \alpha < 1$ [54, 55, 64, 69–71]

$$m\frac{d^2y_\alpha(t)}{dt^2} = -\gamma\int_0^t d\tau(t-\tau)^{\alpha'-2}\frac{dy_\alpha}{d\tau} + \left(\frac{k_B\mathcal{T}\gamma}{\alpha'(\alpha'-1)K_{\alpha'}}\right)^{1/2}\xi_{\alpha'}(t). \quad (9)$$

Here $\gamma(> 0)$ is the generalized friction coefficient, k_B the Boltzmann constant, and $\mathcal{T}$ the absolute temperature. Equation (9) is a special form of the generalized Langevin equation with power-law noise correlations. As the integral in Eq. (9) defines the fractional Caputo operator $\partial^{2-\alpha'}/\partial t^{2-\alpha'}$, we can rephrase Eq. (9) in the fractional form

$$m\frac{d^2y_\alpha(t)}{dt^2} = -\gamma\Gamma(\alpha'-1)\frac{\partial^{2-\alpha'}}{\partial t^{2-\alpha'}}y_\alpha(t) + \left(\frac{k_B\mathcal{T}\gamma}{\alpha'(\alpha'-1)K_{\alpha'}}\right)^{1/2}\xi_{\alpha'}(t), \quad (10)$$

hence the name FLE. The above FLE is constructed with the persistent FGN $\xi_{\alpha'}(t)$ of $1 < \alpha' < 2$ and obeys the fluctuation-dissipation theorem

$\eta^2\langle\xi_{\alpha'}(t)\xi_{\alpha'}(t')\rangle = \gamma k_B\mathcal{T}|t-t'|^{\alpha'-2}$ where $\eta \equiv (k_B\mathcal{T}/[\alpha'(\alpha'-1)K_{\alpha'}])^{1/2}$.[c] Using the generalized Mittag-Leffler function $E_{a,b}(z) = \sum_{n=0}^{\infty} z^n/\Gamma(an+b)$, one finds that the variance of the above FLE is [70, 74]

$$\langle y_\alpha^2(t)\rangle = 2\frac{k_B\mathcal{T}}{m}t^2 E_{\alpha',3}\left[-\frac{\gamma\Gamma(\alpha'-1)}{m}t^{\alpha'}\right] \tag{11}$$

with the assumption of thermal initial velocity $\langle\dot{y}_\alpha^2(0)\rangle = k_B\mathcal{T}/m$. Equation (11) states that FLE motion shows a crossover from the short-time ballistic motion $\langle y_\alpha^2(t)\rangle \sim t^2$ to the long-time subdiffusion of the exponent $0 < \alpha < 1$

$$\langle y_\alpha^2(t)\rangle \sim t^\alpha \tag{12}$$

in the overdamped limit ($t \to \infty$). Note that the subdiffusion exponent α of the FLE process relates with the anomalous diffusion exponent α' of the driving FGN in Eq. (8) through

$$\alpha = 2 - \alpha'. \tag{13}$$

It turns out that the overdamped FLE process $y_\alpha(t)$ fulfils all the criteria of FBM, thus behaving as a subdiffusive FBM $x_\alpha(t)$ of the same anomalous diffusion exponent α [75, 76].

FLE motion was simulated via the method discussed in [64]. Thus, the stochastic Volterra integral equation of $dy(t)/dt$ derived from Eq. (9) is numerically solved, while the FGN is created with the Hosking method. Alternative methods to generate FLE motion are found in [54, 55, 77].

2.2. *Multi-dimensional fractional Brownian motion*

One may introduce FBM in d-dimensional space $\mathbf{x}_\alpha(t)$ as a superposition of independent FBM $x_\alpha(t)$ for each Cartesian coordinate [64, 78–80], namely,

$$\mathbf{x}_\alpha(t) = \sum_{i=1}^{d}\int_0^t dt'\xi_\alpha^{(i)}(t')\hat{x}_i \tag{14}$$

where $\hat{x}_i$ is the ith component in the Cartesian coordinate and $\xi_\alpha^{(i)}(t)$ is FGN of zero mean $\langle\xi_\alpha^{(i)}(t)\rangle = 0$ and covariance $\langle\xi_\alpha^{(i)}(t_1)\xi_\alpha^{(j)}(t_2)\rangle = \langle\xi_\alpha(t_1)\xi_\alpha(t_2)\rangle\delta_{ij}$. From this definition, d-dimensional FBM $\mathbf{x}_\alpha(t)$ is shown

[c] FLE motion $y_\alpha(t)$ with anomalous diffusion exponent $1 < \alpha < 2$ leading to superdiffusion in the long-time limit and satisfying the fluctuation-dissipation theorem can be defined with the anti-persistent FGN in a similar way, see [72, 73].

to satisfy the three criteria [64]:

$$\langle \mathbf{x}_\alpha(t) \rangle = 0, \tag{15a}$$

$$\langle \mathbf{x}_\alpha^2(t) \rangle = 2dK_\alpha t^\alpha, \tag{15b}$$

$$\langle \mathbf{x}_\alpha(t_1) \cdot \mathbf{x}_\alpha(t_2) \rangle = dK_\alpha(|t_1|^\alpha + |t_2|^\alpha - |t_1 - t_2|^\alpha). \tag{15c}$$

Note that the absolute magnitude of $\mathbf{x}_\alpha(t)$, $|\mathbf{x}_\alpha(t)|$, is not a one-dimensional FBM. In analogy to this, d-dimensional FLE motion can also be defined via $\mathbf{y}_\alpha(t) = \sum_{i=1}^d y_\alpha^{(i)}(t)\hat{x}_i$ where $y_\alpha^{(i)}(t)$ is the FLE process (9) with FGN $\xi_{\alpha'}^{(i)}(t)$ and $1 < \alpha' < 2$.

2.3. *Fractal properties*

FBM is a self-affine process whose trajectory has a fractal dimension [46, 47]. As seen in the distribution (2), the rescaled FBM process $\gamma^{-\alpha/2}x_\alpha(\gamma t)$ is statistically identical to FBM $x_\alpha(t)$ for any scaling factor $\gamma > 0$, e.g., $\langle x_\alpha^2(\gamma t)\rangle = \gamma^\alpha \langle x_\alpha^2(t)\rangle$. The fractal dimension of an FBM path can be defined by the walk dimension d_w of the path, i.e., the distance $\sqrt{\mathbf{x}^2} \sim t^{1/d_w}$ travelled by a random walker. Relating this formula with the scaling behavior (15b), the walk dimension of FBM $\mathbf{x}_\alpha(t)$ in d-dimensional space is identified as

$$d_w = \frac{2}{\alpha}. \tag{16}$$

When $d_w \geq d$, d-dimensional FBM is a *compact* (or *recurrent*) random process in the sense that it can visit all sites in the space in the long-time limit. In contrast, when $d_w < d$, d-dimensional FBM becomes *non-compact* (or *transient*), and the stochastic trajectory does not fill the whole space even in the infinite time limit. In two-dimensional space, subdiffusive FBM with $0 < \alpha < 1$ is always compact while superdiffusive FBM with $1 < \alpha < 2$ is non-compact. The difference in the spatial exploration for two-dimensional FBM is depicted in Fig. 1.

3. First passage theory and approximation skills

The first passage time is the time t at which a stochastic process crosses a given threshold value for the first time or escapes from a spatial domain through a prescribed boundary. The associated first passage time density $\wp(t)$ is defined via the survival probability $S(t)$ through the relation

$$\wp(t) = -\frac{dS(t)}{dt}. \tag{17}$$

The survival probability $S(t)$ is the probability that the random walker survives up to time t within the domain, not being absorbed at the absorbing boundary. In terms of the density $G(\mathbf{x}, t)$ to find the particle at position $\mathbf{x}$ at time t, the survival probability is given by

$$S(t) = \int_D G(\mathbf{x}, t) d^d x. \tag{18}$$

Because of the absorbing boundary the survival probability decays with time and G is not normalized, such that $\lim_{t\to\infty} S(t) = 0$ with $S(t = 0) = 1$. In principle, once the solution of the density G, satisfying the given boundary conditions, is identified, the survival probability and the first passage time density can be straightforwardly obtained. However, to actually perform this calculation is generally nontrivial for many anomalous diffusion processes [19, 24, 25, 27–29].

3.1. *Exact result for FBM in one-dimensional domain*

Let us consider the first passage problem of an FBM process in a semi-infinite interval with absorbing boundary at $x = 0$. This problem was initially studied by Ding and Yang as a first return problem of an FBM path [30]. In this study, as illustrated in Fig. 1(inset), the first return time t_R was defined through the difference of consecutive zero-crossing times $t_R = t_c^{(i+1)} - t_c^{(i)}$ where $t_c^{(i)}$ represents the sequence of the zero-crossing times satisfying $x_\alpha(t_c^{(i)}) = 0$ of a given FBM path. Using scaling arguments and numerical simulations, it was conjectured that the corresponding first return time density at long times has a power-law decay of the form

$$\wp(t) \simeq t^{\alpha/2-2}. \tag{19}$$

This conjecture was later confirmed by Molchan's work on the maximum value M_t of FBM on the interval $[0, t]$. It was proven that the probability of M_t being less than a threshold value c, say $c = 1$, follows the asymptotics

$$\ln \mathrm{Prob}(M_t < 1) = (1 - \alpha/2) \ln t^{-1} \left[1 + O\Big((\ln t)^{-1/2}\Big)\right] \tag{20}$$

as $t \to \infty$ [31]. In the context of the first passage process, the probability $\mathrm{Prob}(M_t < c)$ is the survival probability of an FBM not being absorbed up to time t at the location $x = c$ of the absorbing boundary. Therefore, according to Eq. (20) the long-time scaling law of the survival probability of FBM in semi-infinite interval is identified as [26, 31]

$$S(t) \simeq t^{\alpha/2-1}, \tag{21}$$

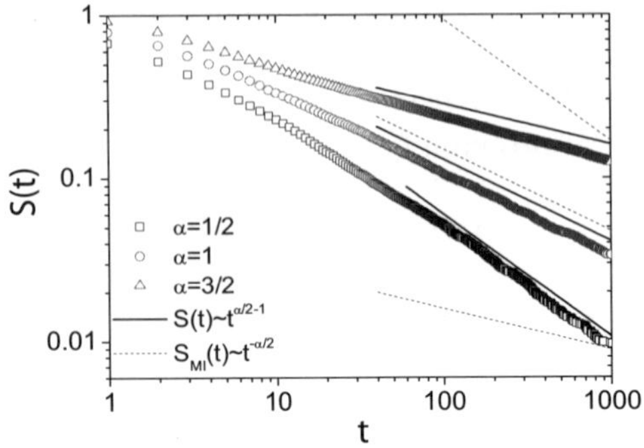

Fig. 2. Numerical results for the survival probabilities $S(t)$ of FBM in a semi-infinite interval with absorbing boundary at $x = 0$ for the cases $\alpha = 1/2,\ 1,\ 3/2$ (from bottom to top). Scaling laws of the survival probability are inserted according to Molchan's result (21) (solid line) and to the incorrect solution via the method of image (dotted line).

and the corresponding first passage time density is given by the scaling law (19). Figure 2 presents the numerically obtained survival probability $S(t)$ of one-dimensional FBM with anomalous diffusion exponents $\alpha = 1/2$ (subdiffusion), 1 (normal diffusion), and 3/2 (superdiffusion). For all cases Molchan's scaling form (21) is in good agreement with the simulations results in the long-time limit.

Let us briefly discuss the first passage behavior of FBM in a semi-infinite interval. At $\alpha = 1$ it reproduces the universal Sparre-Andersen 3/2 scaling for symmetric Markovian processes, a special case of which is the asymptotic behavior of the Lévy-Smirnov result for the first passage density of BM. Somewhat counterintuitively, Molchan's result suggests that subdiffusive FBM particles with stronger anti-correlation have more frequent first passage events, while superdiffusive FBM particles ($1 < \alpha < 2$) tend to survive longer within the confining domain as α gets larger. This is an essential feature of FBM first passage processes, with profound consequences in concrete applications. An analogous case is the Kramers escape problem of an FBM particle, studied for an harmonic potential $U(x) = x^2/2$ with a cutoff (i.e. absorbing boundary) at $x_a \neq 0$ in [22]. There, it was numerically shown that the first passage time density has the exponential decay $\wp(t) \sim \exp(-t/\langle t\rangle)$ consistent with the Kramers' theory, whereas the escape rate $R = \langle t\rangle^{-1}$ is sensitive to the degree of persistence or antipersistence of the FBM processes. In accordance to the first passage behavior in Eq. (19) the escape rate R monotonically increases when FBM is slower (α decreases below unity) while it decreases as the FBM is faster (α increases above unity). We shall see below that such first passage behavior is conserved for FBM processes in multi-dimensional space.

3.2. *Failure of the method of images*

For Markovian particles, the method of images is a useful technique to obtain the first passage time density in cases when the geometry of the problem is relatively simple [11]. Also it is still a valid technique for subordinated BMs such as subdiffusive CTRWs with a diverging mean waiting time [4, 10, 24]. However, it is not applicable to deeply non-Markovian process of FBM type. To show this, we apply the images method to the above first passage problem in a semi-infinite domain. Assuming $G(x, t; x_0)$ is the correct probability density of an FBM particle located at $x_0 > 0$ at $t = 0$ in a domain $x > 0$ with an absorbing boundary at $x = 0$, the method of images yields

$$G(x,t;x_0) = \frac{1}{2\sqrt{4\pi K_\alpha t^\alpha}}\left[\exp\left(-\frac{(x-x_0)^2}{4K_\alpha t^\alpha}\right) - \exp\left(-\frac{(x+x_0)^2}{4K_\alpha t^\alpha}\right)\right]. \tag{22}$$

Using the survival probability $S(t) = \int_0^\infty G(x, t; x_0)$, it is easily shown that the first passage time density corresponding to this process G has the complete functional form [77, 81]

$$\wp_{\mathrm{MI}}(t) = \frac{\alpha x_0}{\sqrt{4K_\alpha \pi} t^{1+\alpha/2}} \exp\left(-\frac{x_0^2}{4K_\alpha t^\alpha}\right). \tag{23}$$

Thus, in the long-time limit $t \to \infty$, the method of images yields that the first passage time density scales as

$$\wp_{\mathrm{MI}}(t) \sim t^{-1-\alpha/2}. \tag{24}$$

Except for BM ($\alpha = 1$), the scaling of the obtained form does not agree with the exact result in Eq. (19). We demonstrate this fact in Fig. 2, in which the survival probability $S_{\mathrm{MI}}(t) \sim t^{-\alpha/2}$ due to the images method exhibits major deviations in the scaling behavior from the simulation results for both subdiffusive and superdiffusive cases. The first passage behavior predicted from $\wp_{\mathrm{MI}}(t)$ is qualitatively opposite to the correct scaling, in the sense that first passage events are more frequent as the diffusion is faster (α is larger). Finally we note that the correct result of Molchan for $\wp(t)$ is related with the asymptotic scaling of $\wp_{\mathrm{MI}}(t)$ through the substitution $\alpha \to 2 - \alpha$.

3.3. *The pseudo-Markovian approximation*

The technique of the pseudo-Markovian approximation[d] was originally introduced to describe the cyclization dynamics of two end monomers of a polymer chain which is known to be a non-Markovian process [12–16]. Within the pseudo-Markovian approximation scheme, the conditional probability $G(\mathbf{x}, t|\mathbf{x}_0, t_0)$ that the particle is found to be at $\mathbf{x}$ at time t given that it was located at $\mathbf{x}_0$ at time t_0 is self-consistently written through the renewal equation [11, 15, 82]

$$G(\mathbf{x}, t|\mathbf{x}_0, 0) = \delta(\mathbf{x} - \mathbf{x}_0)\delta(t) + \int_0^t dt' \mathcal{P}(\mathbf{x}, t'; \mathbf{x}_0) G(\mathbf{x}, t|\mathbf{x}, t'). \qquad (25)$$

Here $\mathcal{P}(\mathbf{x}, t; x_0)$ defines the first passage time density that the diffusing particle embedded in d-dimensional space starts to move at $\mathbf{x}_0$ at $t = 0$ and visits the site $\mathbf{x}$ for the first time at time t'. For Markovian processes such as BM, Eq. (25) is an exact description relating the conditional probability density G to the corresponding first passage probability [11, 82]. For non-Markovian processes, however, it neglects the fact that the returning probability $G(\mathbf{x}, t|\mathbf{x}, t')$ in the above expression does depend on all the visited sites of the particle before t'. In this sense, the first passage density $\mathcal{P}$ is a Markovian approximation to the true first passage solution for non-Markovian processes. This technique has been exploited to relate the first passage problems of FBM and the fractional Langevin equation associated with FGN [22, 77, 83, 84]. It turns out that the pseudo-Markovian approximation often works reasonably well to explain the first passage processes for some problem. It is known that the pseudo-Markovian scheme is the first approximation to the exact perturbative method introduced by Likhtman and Margues [17].

The first passage problem for FBM in a semi-infinite interval ($d = 1$) was studied within the scheme of the pseudo-Markovian approximation [77, 84]. In particular, in [77] the complete functional form of the first passage time density was obtained by this method, finding that it agrees with the simulations over the entire time domain explored. Except for certain values of $\alpha = (n + 1)/(4m + 2)$ with $n, m \in \mathbb{Z}^+$ the pseudo-Markovian solution of the corresponding first passage time density, $\wp_{\mathrm{PM}}(t) \equiv \mathcal{P}(0, t; x_0)$ ($x = 0$ is

[d] It is also often referred to as the Wilemski-Fixman approximation in the literature.

the absorbing boundary), is found to follow the series expansion [77]

$$\wp_{\rm PM}(t) = \frac{\sigma^{-1/\alpha}}{\Gamma(1-\alpha/2)}\left([K(t)]^{2-\alpha}\frac{1}{\alpha}\sum_{n=0}^{\infty}[K(t)]^n\frac{(-1)^n}{n!}\frac{\Gamma([\alpha/2-1-n]/\alpha)}{\Gamma(\alpha/2-1-n)}\right.$$
$$\left. + K(t)\sum_{m=1}^{\infty}[K(t)]^{\alpha m}\frac{(-1)^m}{m!}\frac{\Gamma(1-\alpha/2-\alpha m)}{\Gamma(-\alpha m)}\right) \tag{26}$$

with

$$K(t) = \frac{\sigma^{1/\alpha}}{t}. \tag{27}$$

Here $\sigma = x_0^2/(4K_\alpha)$ and $\Gamma(z)$ is the Gamma function. In the long-time limit where $K(t) \to 0$, remarkably, the above pseudo-Markovian solution (26) yields the correct scaling behavior Eq. (19) of FBM

$$\wp_{\rm PM}(t) \sim t^{\alpha/2-2} \tag{28}$$

in the range $1/6 \le \alpha < 2$ of the anomalous diffusion exponent. However, it gives an inconsistent result

$$\wp_{\rm PM}(t) \sim t^{-\alpha-1} \tag{29}$$

for strong anti-persistent FBM with $0 < \alpha < 1/6$. Numerical tests showed that $\wp_{\rm PM}(t)$ describes remarkably well the simulated first passage behavior of FBM in the entire time window for the cases with anomalous diffusion exponent $1/6 \le \alpha \le 1$. However, such good agreement is not fulfilled for superdiffusive FBM with $1 < \alpha < 2$, where $\wp_{\rm PM}(t)$ only correctly captures the long-time scaling law (19).

3.4. *First passage process of fractional Langevin equation*

As mentioned previously, the overdamped FLE process $y_\alpha(t)$ driven by a persistent FGN $\xi_{\alpha'}(t)$ with $1 < \alpha' < 2$ behaves analogously to a subdiffusive FBM $x_\alpha(t)$ of $0 < \alpha < 1$ provided that $\alpha = 2 - \alpha'$ [75, 76]. Both processes share the same stochastic properties, and thus one can identify the first passage behavior of FLE motion from that of its counterpart FBM process, and vice versa [76]. Using this correspondence, we find that the survival probability and the associated first passage time density of the overdamped FLE process $y_\alpha(t)$ in a semi-infinite interval are respectively governed by

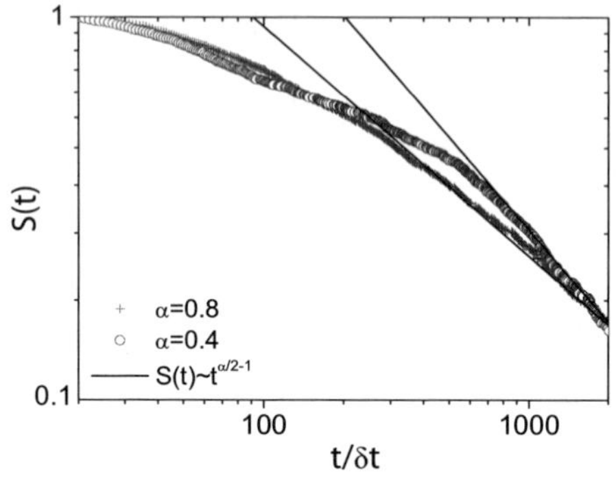

Fig. 3. Numerical results for the survival probabilities $S(t)$ of FLE motion with anomalous diffusion exponent $\alpha = 0.4$ and 0.8 (from top to bottom) on a semi-infinite interval with absorbing boundary at $x = 0$. Here the anomalous diffusion exponent α is associated with the FLE motion $y_\alpha(t)$ satisfying the scaling relation $\langle y_\alpha^2(t)\rangle \sim t^\alpha$ in the overdamped limit, see Sec. 2. The predicted scaling law $S(t) \sim t^{\alpha/2-1} = t^{-\alpha'/2}$ compares favourably with the simulation results. Parameter values in the simulation: integration time step $\delta t = 0.001$, particle mass $m = 0.1$, generalized frictional coefficient $\gamma = 100$, and $k_B\mathcal{T} = 1$.

Molchan's scaling results (21) and (19), i.e., the first passage process of FLE motion is governed by the asymptotic laws

$$S(t) \simeq t^{-\alpha'/2}, \tag{30a}$$

$$\wp(t) \simeq t^{-\alpha'/2-1} \tag{30b}$$

in terms of the anomalous diffusion exponent α' of the persistent FGN. We emphasise here that the above scaling forms (30) should not be confused with Eq. (24) obtained by the method of images, albeit they appear to have the identical exponents. In other words, the scaling form (30) does not imply that the method of images can be successfully applied to FLE motion.

The first passage time behavior Eq. (30) of the FLE process was numerically validated in terms of the single-file diffusion scheme, which is known to be governed by FLE motion $y_\alpha(t)$ of the anomalous diffusion exponent $\alpha = 1/2$ [76, 77]. For further numerical confirmation of the equivalence of the first passage behavior between FBM and FLE motion, in Fig. 3 we simulate the first passage process of FLE motion $y_\alpha(t)$ with $\alpha = 0.4$ and 0.8. It shows that the long-time scaling behavior of FLE processes are in good agreement with Molchan's scaling laws [i.e., Eq. (30)]. It is worthwhile mentioning that on short time scales, when inertia effects are relevant, the

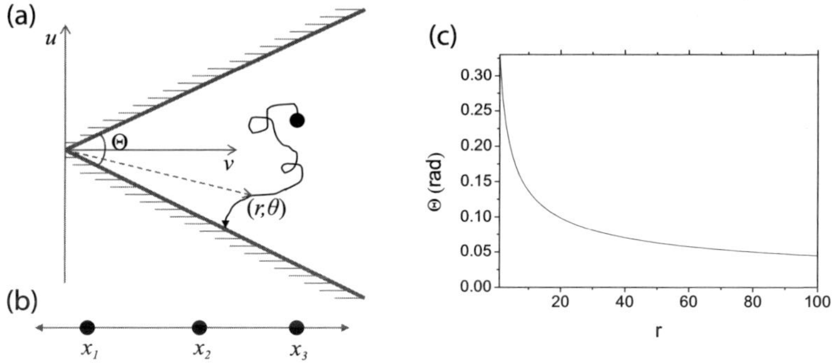

Fig. 4. (a) First passage process of an FBM particle confined to a two-dimensional open wedge of angle Θ with absorbing walls. The trajectory of the particle is described in polar coordinates by its radius r and angle θ. (b) Three-body reaction of FBM particles in a one-dimensional interval. The location of the ith particle is given by x_i. (c) Dependence of the wedge angle Θ as a function of the ratio $r = K_\alpha^{(2)}/K_\alpha^{(1)}$ in (b). See text for details.

first passage behavior of FLE motion naturally deviates significantly from those of FBM, as shown in our simulations.

4. Diffusion-limited reaction of FBM particles

Based on the first passage behavior of an FBM particle studied in the previous section, here we explore the kinetics of diffusion-limited reaction for FBM particles in d-dimensional space. Under the assumption that a reaction spontaneously occurs whenever two diffusing reactants meet, our reaction problem can be regarded as the problem of a stochastic searcher to find one of multiple target sites distributed in space.

4.1. *Three-body reactions in one-dimensional space*

Before we embark for the multi-particle case, let us consider the first passage problem of a single FBM particle that is confined to a two-dimensional open wedge of angle Θ [see Fig. 4(a)]. The particle is assumed to escape out of the wedge when it hits its absorbing wall. It turns out that this wedge problem is mapped onto the diffusion-limited three-body reaction problem in one-dimensional space [11, 81, 85, 86]. In the special case of BM with $\alpha = 1$, the first passage statistics is analytically solvable using the Green's function formalism. For the general case of FBM with $\alpha \neq 1$ the first passage statistics is not analytically derivable from an equation, but it can be conjectured from combination of Molchan's result for the one-dimensional problem and the exact solution at $\alpha = 1$ [81].

4.1.1. *Case I: Exact solution for BM*

The motion of a diffusing particle in an absorbing wedge [Fig. 4(a)] satisfies the diffusion equation in polar coordinates,

$$\frac{\partial}{\partial t}G(r,\theta;t) = K_1\left(\frac{\partial^2}{\partial r^2} + \frac{1}{r}\frac{\partial}{\partial r} + \frac{1}{r^2}\frac{\partial^2}{\partial \theta^2}\right)G(r,\theta;t) \tag{31}$$

with the boundary conditions $G(r,0;t) = G(r,\Theta;t) = 0$ with $0 < \Theta < 2\pi$ and the initial condition $G(r,\theta;0) = \delta(r-r_0)\delta(\theta-\theta_0)/r_0$. It can be shown that at large time t the solution G has the asymptotic form [11]

$$G(r,\theta;t) \simeq \frac{\pi\sin\left(\frac{\pi\theta}{\Theta}\right)}{4K_1\Theta t}\exp\left(\frac{-r^2+r_0^2}{4K_1 t}\right)I_{\pi/\Theta}\left(\frac{rr_0}{2K_1 t}\right), \tag{32}$$

where $I_\nu(\text{z})$ is the modified Bessel function of the first kind. Using the solution (32) we are able to obtain the survival probability of the Brownian particle $S(t) = \int_0^\Theta\int_0^\infty rG(r,\theta;t)drd\theta$ and the corresponding first passage time density $\wp(t) = -\frac{dS(t)}{dt}$. From the approximations $I_\nu(z) \approx (z/2)^\nu/\Gamma(1+\nu)$ and $e^{-r_0^2/[4K_1 t]} \approx 1$ we further obtain the long-time scaling forms of the two quantities such that

$$S_\Theta(t) \simeq \left(\frac{r_0}{\sqrt{K_1}}\right)^{\pi/\Theta} t^{-\pi/(2\Theta)}, \tag{33a}$$

$$\wp_\Theta(t) \simeq \frac{\pi}{2\Theta}\left(\frac{r_0}{\sqrt{K_1}}\right)^{\pi/\Theta} t^{-1-\pi/(2\Theta)}. \tag{33b}$$

The obtained expressions reproduce the Lévy-Smirnov scaling form for BM in a semi-infinite interval when the wedge becomes half-infinite at $\Theta = \pi$, as it should be. It also gives an important general relation of the survival probability for the wedge of a right angle, $S_{\pi/2}(t) = S_\pi^2(t)$, expected from the independence of x and y motion. Importantly, the scaling exponents depend on the wedge angle Θ inverse-proportionally. The first passage behavior in a wedge of angle $\Theta < \pi/2$ is qualitatively different from those in a wedge of angle $\Theta \geq \pi/2$; in the former case the mean first passage time

$$\langle t\rangle = \int_0^\infty t\wp_\Theta(t)dt \tag{34}$$

is finite while it diverges for the latter case.

4.1.2. *Case II: Exact solution for FBM with wedge angles* $\Theta = \pi$ *and* $\Theta = \pi/2$

From Molchan's exact results (19) and (21), the scaling of the first passage of FBM confined to a wedge of angle π due to the independence of spatial directions is necessarily

$$S_\pi(t) \simeq t^{\alpha/2-1}, \tag{35a}$$

$$\wp_\pi(t) \simeq t^{\alpha/2-2} \tag{35b}$$

for any exponents $0 < \alpha < 2$. Making use of the general relation of the survival probability $S_{\pi/2}(t) = S_\pi^2(t)$, we also find the exact scaling forms of the first passage events for a wedge of angle $\Theta = \pi/2$,

$$S_{\pi/2}(t) \simeq t^{\alpha-2}, \tag{36a}$$

$$\wp_{\pi/2}(t) \simeq t^{\alpha-3}. \tag{36b}$$

4.1.3. *Case III: Conjecture for FBM with wedge angle* $0 < \Theta < 2\pi$

Using the exact first passage scaling for the special cases presented in Secs. 4.1.1 & 4.1.2 we now conjecture its complete functional form for FBM confined to a wedge of any angle $0 < \Theta < 2\pi$. To get a clue for its scaling form, we start to derive the survival probability and the first passage time density for the scaled Brownian motion process $\overline{\mathbf{x}}_\alpha(t)$ whose probability density $\overline{G}(\mathbf{x}, t; \mathbf{x}_0)$ satisfies the generalized diffusion equation [70]

$$\frac{\partial}{\partial t}\overline{G}(\mathbf{x}, t; \mathbf{x}_0) = K(t)\nabla^2\overline{G}(\mathbf{x}, t; \mathbf{x}_0). \tag{37}$$

Here the time-dependent diffusion constant is given by $K(t) \sim K_1 t^{\alpha-1}$ so that the mean squared displacement of the process has the scaling behavior $\langle \overline{\mathbf{x}}_\alpha^2(t)\rangle \simeq 4K_1 t^\alpha$. It is known that the scaled Brownian motion process described by Eq. (37) has non-stationary increments [48], thus it is not an FBM. Following the steps introduced in Sec. 4.1.1 we find that the first passage time quantities of this process are

$$\overline{S}_\Theta(t) \propto t^{-\pi\alpha/(2\Theta)}, \tag{38a}$$

$$\overline{\wp}_\Theta(t) \propto t^{-1-\pi\alpha/(2\Theta)}. \tag{38b}$$

The obtained scaling behaviors are identical to those for BM when $\alpha = 1$, but inconsistent with the results of FBM for wedges of $\Theta = \pi$ and $\Theta = \pi/2$. Albeit incorrect, from (38) we infer that the correct scaling for FBM is also expressed in terms of the product of α and Θ^{-1}. Under this assumption, there exists only one unique functional form that is consistent with all of the above special cases, and this is our conjecture for the first passage time quantities of FBM confined to a 2D wedge:

$$S_\Theta(t) \simeq t^{\pi(\alpha-2)/(2\Theta)}, \tag{39a}$$

$$\wp_\Theta(t) \simeq t^{-1+\pi(\alpha-2)/(2\Theta)}. \tag{39b}$$

The validity of this conjecture was numerically tested in [81]. We note that through the correspondence $\alpha \to 2-\alpha$ the conjecture (39) is obtained from Eq. (38), in the same way as between the solution (19) by Molchan and the approximation (24) by the images method for the one-dimensional case. Also consistent to the first passage behavior discussed in the previous section, the conjecture (39) states that first passage events occur more frequently as the FBM has stronger anti-persistent memory (smaller α). The mean first passage time $\langle t\rangle$ defined by Eq. (34) is finite for wedge angles $\Theta < \Theta_c$ where the critical angle

$$\Theta_c = \pi\left(1-\frac{\alpha}{2}\right) = \pi\left(\frac{d_w-1}{d_w}\right) \tag{40}$$

depends on α, or, alternatively on the walk dimension d_w of the FBM.

4.1.4. *Reaction kinetics of three FBM particles*

Consider a coalescence reaction of N FBM particles diffusing in one-dimensional space. For simplicity, it is assumed that a reaction occurs whenever any two of them meet, and after the reaction the coalesced particle (or cluster) is also governed by FBM. As for its reaction dynamics, this problem can be reduced to a three-body problem where the middle FBM particle is surrounded by two coalesced FBM particles. The survival condition of the middle particle is given by two inequalities, $x_1 < x_2$ and $x_2 < x_3$, where x_i represents the position of the ith particle [see Fig. 4(b)]. To solve the problem we assume that $\left(\frac{x_1(t)}{\sqrt{K_\alpha^{(1)}}}, \frac{x_2(t)}{\sqrt{K_\alpha^{(2)}}}, \frac{x_3(t)}{\sqrt{K_\alpha^{(3)}}}\right)$ are the rescaled Cartesian components of a single FBM particle $\mathbf{x}_\alpha(t)$ in a three-dimensional space, where $K_\alpha^{(i)}$ defines the generalized diffusivity of the ith particle in the above one-dimensional space [11, 85]. Then, the original survival problem of the three one-dimensional particles can be understood such that the *single* three-dimensional FBM particle diffuses inside the domain defined by the intersects of two absorbing planes $x_1 = x_2$ and $x_2 = x_3$, in general with different diffusivities in the different spatial directions [11]. Since the particle's motion in the direction perpendicular to both normal vectors of the two planes is independent of the absorbing event, the survival problem in the three-dimensional domain reduces to the two-dimensional wedge problem studied above. Here the wedge angle in the new orthonormal coordinate (u, v) on the plane spanned by the two normal vectors can be unambiguously found to be

$$\Theta = \arccos\left(\frac{K_\alpha^{(2)}}{\sqrt{\left(K_\alpha^{(1)} + K_\alpha^{(2)}\right)\left(K_\alpha^{(2)} + K_\alpha^{(3)}\right)}}\right) \tag{41}$$

from the geometry of the confining domain [11].

Let us discuss the meaning of these results: (1) In the case when all three particles have identical diffusivity $K_\alpha^{(1)} = K_\alpha^{(2)} = K_\alpha^{(3)}$, the wedge angle is given by $\Theta = \pi/3$. Then the survival probability of the middle particle decays as

$$S(t) \simeq t^{-3(2-\alpha)/2}. \tag{42}$$

It reproduces the classical result $S(t) \sim t^{-3/2}$ of the decay of the concentration of the remaining reactant particles for $\alpha = 1$ [85]. The decay law is modified for FBM particles in a seemingly counterintuitive fashion: for the superdiffusive FBM particle the reaction occurs more slowly, resulting in a slower decay of the surviving particles. In contrast, the subdiffusive FBM particles have increased likelihood to meet each other and produce a sharp decay of the survival probability. Indeed, this property is related to the higher recurrence probability of an FBM motion discussed above.

(2) In the case when the middle particle is immobile $K_\alpha^{(2)} = 0$, the wedge angle is $\Theta = \pi/2$. It corresponds to the situation that the two surrounding particles independently search their target and thus the decay law of the survival probability is given by

$$S(t) = S_\pi^2(t) \simeq t^{-(2-\alpha)}. \tag{43}$$

For both subdiffusive and superdiffusive motions the general tendency is that the decay of the survival probability is always slowed down by the immobilization of the target particle. In other words, a moving target always has a higher likelihood to be captured.

(3) The surrounding particles have an identical diffusivity $K_\alpha^{(1)} = K_\alpha^{(3)}$ and their diffusivity is smaller than that of the middle particle, $K_\alpha^{(1)} < K_\alpha^{(2)}$. A physical scenario of this case is that the surrounding particles are already

coalesced FBM particles with a decreased diffusivity due to the larger particle size. The dependence of the wedge angle $\Theta = \cos^{-1}[r/(1+r)]$ on the ratio $r = K_\alpha^{(2)}/K_\alpha^{(1)}$ is plotted in Fig. 4(c). The fact that Θ monotonically decreases from $\pi/3$ to 0 as r increases from 1 to infinity implies that slower movement of the surrounding particles always leads to an increased hitting probability between the middle and surrounding particles. In a naive expectation it may be assumed that the diffusivity of the surrounding particles is decreased inverse-proportionally to the number of the coalesced particles. Then, the profile of Θ in Fig. 4(c) can be understood such that their reaction kinetics is noticeably accelerated only up to a few tens of particle coalescence events.

4.2. *Many-body reactions of FBM particles in multi-dimensional space*

In the previous section we studied the reaction kinetics of FBM particles in one-dimensional using the analogy to the first passage process of an FBM particle that escapes from a confined d-dimensional domain through absorbing walls. For the study of the reaction kinetics of d-dimensional FBM (with $d \geq 2$), we now consider the first arrival problem in free d-dimensional space that an FBM particle with the initial condition $\mathbf{x}_\alpha(0) = 0$ arrives at a specific target site $\mathbf{x} = \mathbf{x}_T$ at time t for the first time. For BM this problem is exactly solvable using the renewal equation (25) [11]. For non-Markovian FBM, to our knowledge, this question represents an open problem. Here, we employ the pseudo-Markovian approximation scheme to find the scaling laws of the corresponding first arrival time quantities for FBM with $d > 1$ and test their validity to simulation results.

To do this, let us first consider the Laplace transform of $G(\mathbf{x} = 0, t) = (4\pi K_\alpha t^\alpha)^{-d/2}$ of a free d-dimensional FBM. Using the relation (16) between the anomalous diffusion exponent α and the walk dimension d_w and $z \equiv \exp(-u)$ we arrive at

$$G(0, z) = \int_{l_c}^{\infty} G(\mathbf{x} = 0, t) e^{-ut} dt = (4\pi K_\alpha)^{-d/2} \int_{l_c}^{\infty} t^{-\frac{d}{d_w}} z^t dt. \qquad (44)$$

Here the lower cutoff l_c of the order of unity is introduced to prevent the physically irrelevant divergence of the integral as $t \to 0$, which does not affect the long-time behavior. This integral behaves differently depending on the ratio d/d_w: (i) For $d < d_w$ (compact exploration), the integral diverges like $G(0, z) \sim A_d (1-z)^{(d-d_w)/d_w}$ with a proportionality factor $A_d = (\frac{d}{d_w - d})(4\pi K_\alpha)^{-d/2}$ as $z \to 1$ (equivalently $u \to 0$). (ii) For

$d = d_w$, Eq. (44) diverges logarithmically $G(0, z) \sim -A_{dw} \log(1 - z)$ where $A_{dw} = (4\pi K_\alpha)^{-d_w/2}$. (iii) For $d > d_w$ (non-compact exploration), $G(0, z)$ converges to $(1 - \mathcal{R})^{-1} + B_d(1 - z)^{(d-d_w)/d_w}$, where $\mathcal{R}$ turns out to be the non-vanishing probability that the random walk eventually returns to the starting site $\mathbf{x} = 0$ and $B_d = d_w/(d - d_w)$. Following the general strategy presented in [11] of finding the long-time scaling behavior of first passage processes from $G(0, z)$, it is straightforward to show that the survival probability behaves as

$$S(t) \sim \begin{cases} A_d^{-1} t^{\frac{d-d_w}{d_w}}, & d < d_w \\ A_{dw}^{-1} (\log t)^{-1}, & d = d_w \\ (1 - \mathcal{R}) + C_d t^{\frac{d_w - d}{d_w}}, & d > d_w \end{cases} \tag{45}$$

C_d ia an irrelevant constant and the corresponding first arrival time density becomes

$$\wp(t) \sim \begin{cases} t^{\frac{d}{d_w}-2}, & d < d_w \\ t^{-1}(\log^2 t)^{-1}, & d = d_w \\ t^{-\frac{d}{d_w}}, & d > d_w \end{cases} . \tag{46}$$

For the case of the compact exploration $d < d_w$ and the borderline case $d = d_w$, the survival probability eventually decays to zero, but the mean first arrival time $\langle t \rangle = \int_{l_c}^{\infty} t p(t) dt$ is always infinite irrespective of the ratio $d/d_w \leq 1$. It reflects the fact that the diffusing particle always finds its target only in the limit of infinite time. Consequently, in the reaction kinetics scenario the compact exploration of a reactant particle facilitates local reaction with the nearby reactant particles while it eventually slows down the total reaction kinetics over the space by preventing the well-mixing of the reactant particles. Consistent results were reported in the numerical study of reaction kinetics of FBM particles in two-dimensional space [87]. Note that the slow-down of the reaction kinetics of the compact-exploring FBM particle is mainly due to the effect of higher dimensionality $d \geq 2$. When $d = 1$, reactions only occur among neighbouring particles and thus the reaction kinetics becomes faster for compact-exploring FBM particles. This is also consistent with the fact that they have a finite mean first passage time.

For non-compact exploring FBM particles with $d > d_w$, the survival probability decays towards a nonzero constant $1 - \mathcal{R}$ as $t \to \infty$. Thus, the non-compact searcher never locates its target with a probability $1 - \mathcal{R}$, even in the limit of $t \to \infty$. For the particles contributing to the reaction with a

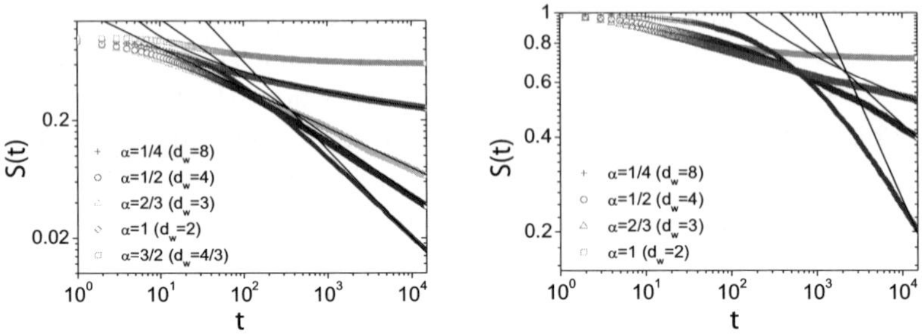

Fig. 5. Survival probability $S(t)$ of a d-dimensional FBM starting at $\mathbf{x} = 0$ and not being absorbed at $\mathbf{x} = \mathbf{x}_T (\neq 0)$ up to time t, (Left) for $d = 2$ and (Right) for $d = 3$. The first arrival processes of FBM with different values of α are simulated to check all cases: $d < d_w$ (compact exploration), $d = d_w$, and $d > d_w$ (non-compact exploration). Left: two-dimensional FBM with $\alpha = 1/4,\ 1/2,\ 2/3,\ 1$, and $3/2$ (bottom to top). Right: three-dimensional FBM with $\alpha = 1/4,\ 1/2,\ 2/3$, and 1 (bottom to top). In each panel, solid lines depict the solution using the pseudo-Markovian approach (45).

probability $\mathcal{R}$, the mean first arrival time $\langle t \rangle$ becomes finite if $d_w < d/2$. In this case, the non-compact exploration enables a reactant particle to find distant counterpart particles.

For BM with walk dimension $d_w = 2$, the above pseudo-Markovian results (45) and (46) are evidently exact in any spatial dimension [11]. In the case of FBM in a one-dimensional space, the pseudo-Markovian approximations are also consistent with Molchan's first passage time results. Are the approximation still sufficiently accurate in explaining at least the long-time scaling laws of the first arrival processes for d-dimensional FBM with $d \geq 2$? We numerically investigate this with simulations of the first arrival processes of FBM in two- and three-dimensional space. Figure 5 presents the comparison of the corresponding survival probabilities between simulations and the pseudo-Markovian scaling law (45). A notable general tendency is that the pseudo-Markovian approximation gets worse as the dimensionality is increased. In $d = 3$ the power-law exponents of $S(t)$ are significantly different from $(d - d_w)/d_w$ for compact exploration and $S(t)$ does not logarithmically decay for $d_w = d(= 3)$. In $d = 2$ the pseudo-Markovian approximation produces reasonable results compared to the simulations. Moreover, the theory tends to be exact as $d_w \to d$, at which $S(t)$ decays logarithmically with time t.

The fractal kinetics of a reaction $A + B \to C$ for subdiffusive FBM reactants in a two-dimensional space was numerically studied in [87]. The reaction kinetics was shown to follow the general rule $[C] = k_0 t^{-h}[A][B]$

with $0 \leq h \leq 1$ and a constant k_0. The parameter h tends to increase as the FBM reactant has a stronger anti-persistency (α is decreased). The simulation also showed that the subdiffusive character of the reactants leads to strong segregation of reactants and, in turn, slower sampling process. As a consequence, the total amount of products, $N_C(t) = N(0) - N_A(t) - N_B(t)$, decreases slower as α is decreased. These results are in line with the reaction kinetics obtained from the approximation above. Clearly, a full understanding of this topic requires further studies.

5. Conclusions

In this chapter we provided an overview of some recent progress in the theoretical understanding of the first passage time behavior of FBM in one- and higher-dimensional space. We also introduced an application to the kinetics of diffusion-limited reaction of FBM reactants. We showed that the first passage processes of FBM (and overdamped FLE motion) qualitatively possess generic features in the sense that the first passage events tend to occur more frequently when FBM attains stronger antipersistence in its displacement correlation, in other words, when the anomalous diffusion exponent α of the motion decreases. Our simulations suggest that the same tendency still holds for multi-dimensional FBM, but in spaces of dimension higher than one the first passage processes turn out to possess three distinct statistical regimes depending on the ratio of the dimensionality of the embedding space and the walk dimension associated with FBM. We found that the first passage behavior is qualitatively explained in terms of the pseudo-Markovian approximation of the first passage processes. However, the scaling law of the first passage time quantities are quantitatively consistent with the pseudo-Markovian approximation only for the one-dimensional case. Intriguingly, the pseudo-Markovian results tend to deviate more from the simulations results as the dimensionality of the space is increased, albeit each component of a multi-dimensional FBM is independent. This discrepancy is worthwhile investigating in detail in the future.

Acknowledgments

This work is supported by the Academy of Finland within the Finland Distinguished Professor (FiDiPro) programme. We gratefully acknowledge fruitful discussion with Otto Pulkkinen and Michael Lomholt.

References

1. S. Havlin and D. Ben-Avraham, Diffusion in disordered media, *Adv. Phys.* **36**, 695-788 (1987).
2. J.-P. Bouchaud and A. Georges, Anomalous diffusion in disordered media: statistical mechanisms, models and physical applications, *Phys. Rep.* **195**, 127-293 (1990).
3. R. Metzler and J. Klafter, The random walk's guide to anomalous diffusion: a fractional dynamics approach, *Phys. Rep.* **339**, 1-77 (2000).
4. R. Metzler and J. Klafter, The restaurant at the end of the random walk: recent developements in the description of anomalous transport by fractional dynamics, *J. Phys. A: Math. Gen.* **37**, R161-R208 (2004).
5. X. S. Xie, P. J. Choi, G.-W. Li, N. K. Lee, and G. Lia, Single-molecule approach to molecular biology in living bacterial cells, *Annu. Rev. Biophys.* **37**, 417-444 (2008).
6. P. C. Bressloff and J. M. Newby, Stochastic models of intracellular transport, *Rev. Mod. Phys.* **85**, 135-196 (2013).
7. E. W. Montroll, Random Walks on Lattices. III. Calculation of first-passage times with application to exciton trapping on photosynthetic units, *J. Math. Phys.* **10**, 753-765 (1969).
8. H. Scher and E. W. Montroll, Anomalous transit-time dispersion in amorphous solids, *Phys. Rev. B* **12**, 2455-2477 (1975).
9. H. C. Fogedby, Lévy flights in random environments, *Phys. Rev. Lett.* **73**, 2517-2520 (1997).
10. R. Metzler and J.-H. Jeon, *Anomalous diffusion and fractional transport equations*, In eds. J. Klafter, S. C. Lim, and R. Metzler, *Fractional dynamics*, World Scientific (2012).
11. S. Redner, *A guide to first-passage processes*, 1st edn. Cambridge University Press (2001).
12. G. Wilemski and M. Fixman, Diffusion-controlled intrachain reactions of polymers. I Theory, *J. Chem. Phys.* **60**, 866-877 (1974).
13. G. Wilemski and M. Fixman, Diffusion-controlled intrachain reactions of polymers. II Results for a pair of terminal reactive groups, *J. Chem. Phys.* **60**, 878-890 (1974).
14. A. Szabo, K. Schulten, and Z. Schulten, First passage time approach to diffusion controlled reactions, *J. Chem. Phys.* **72**, 4350-4357 (1980).
15. I. M. Sokolov, Cyclization of a polymer: first-passage problem for a non-Markovian process, *Phys. Rev. Lett.* **90**, 080601-4 (2003).
16. T. Guérin, O. Bénichou, and R. Voituriez, Non-Markovian polymer reaction kinetics, *Nature Chem.* **4**, 568-573 (2012).
17. A. E. Likthman and C. M. Marques, First-passage problem for the Rouse polymer chain: an exact solution, *Europhys. Lett.*, **75**(6), 971-977 (2006).
18. M. A. Lomholt, T. Koren, R. Metzler, and J. Klafter, Lévy strategies in intermittent search processes are advantageous, *Proc. Nat. Acad. Sci. USA* **105**, 11055-11059 (2008).
19. S. Condamin, O. Bénichou, V. Tejedor, R. Voituriez, J. Klafter, First-passage

times in complex scale invariant media, *Nature* **450**, 77-80 (2007).
20. S. Condamin, V. Tejedor, R. Voituriez, O. Bénichou, and J. Klafter, Probing microscopic origins of confined subdiffusion by first-passage observables, *Proc. Nat. Acad. Sci. USA* **105**, 5675-5680 (2008).
21. I. Goychuk and P. Hänggi, Anomalous escape governed by thermal $1/f$ noise, *Phys. Rev. Lett.* **99**, 200601 (2007).
22. O. Y. Sliusarenko, V. Y. Gonchar, A. V. Chechkin, I. M. Sokolov, and R. Metzler, Kramers-like escape driven by fractional Gaussian noise, *Phys. Rev. E* **81**, 041119 (2010).
23. J. L. A. Dubbeldam, V. G. Rostiashvili, A. Milchev, and T. A. Vilgis, Fractional Brownian motion approach to polymer translocation: the governing equation of motion, *Phys. Rev. E* **83**, 0118028 (2011).
24. R. Metzler and J. Klafter, Boundary value problems for fractional diffusion equations, *Physica A* **278**, 107-125 (2000).
25. G. Rangarajan and M. Ding, Anomalous diffusion and the first passage time problem, *Phys. Rev. E* **62**, 120-133 (2000).
26. G. Rangarajan and M. Ding, First passage time distribution for anomalous diffusion, *Phys. Lett. A* **273**, 322-330 (2000).
27. S. Condamin, O. Bénichou, and J. Klafter, First-passage time distributions for subdiffusion in confined geometry, *Phys. Rev. Lett.* **98**, 250602 (2007).
28. A. V. Chechkin, R. Metzler, V. Y. Gonchar, J. Klafter, and L. V. Tanatarov, First passage and arrival time densities for Lévy flights and the failure of the method of images, *J. Phys. A.:Math. Gen.* **36**, L537-L544 (2003).
29. T. Koren, M. A. Lomholt, A. V. Chechkin, J. Klafter, and R. Metzler, Leapover lengths and first passage time statistics for Lévy flights, *Phys. Rev. Lett.* **99**, 160602 (2007).
30. M. Ding and W. Yang, Distribution of the first return time in fractional Brownian motion and its application to the study of on-off intermittency, *Phys. Rev. E* **52**, 207-213 (1995).
31. G. M. Molchan, Maximum of a fractional Brownian motion: probabilities of small values, *Commun. Math. Phys.* **205**, 97-111 (1999).
32. Y. Kantor and M. Kardar, Anomalous diffusion with absorbing boundary, *Phys. Rev. E* **76**, 061121 (2007).
33. K. J. Wiese, S. N. Majumdar, and A. Rosso, Perturbation theory for fractional Brownian motion in presence of absorbing boundaries, *Phys. Rev. E* **83**, 061141 (2007).
34. G. Oshanin, A. Rosso, and G. Schehr, Anomalous fluctuations of currents in Sinai-type random chains with strongly correlated disorder, *Phys. Rev. Lett.* **110**, 100602 (2013).
35. S. B. Yuste, E. Abad, and K. Lindenberg, *Reactions in subdiffusive media and associated fractional equations*, In eds. J. Klafter, S. C. Lim, and R. Metzler, *Fractional dynamics*, World Scientific (2012).
36. S. Schnell and T. E. Turner, Reaction kinetics in intracellular environments with macromolecular crowding: simulations and rate laws, *Prog. in Biophys. & Mol. Biol.* **85**, 235-260 (2004).
37. D. ben-Avraham and S. Havlin, *Diffusion and reactions in fractals and dis-*

ordered systems, 1st edn. Cambridge University Press (2000).
38. S. B. Yuste and K. Lindenberg, Subdiffusion-limited $A + A$ reactions, *Phys. Rev. Lett.* **87**, 118301 (2001).
39. S. B. Yuste and K. Lindenberg, Subdiffusion-limited reactions, *Chem. Phys.* **284**, 169-180 (2002).
40. K. Seki, M. Wojcik, and M. Tachiya, Fractional reaction-diffusion equation, *J. Chem. Phys.* **119**, 2165-2170 (2003).
41. I. M. Sokolov, M. G. W. Schmidt, and F. Sagués, Reaction-subdiffusion equations, *Phys. Rev. E* **73**, 031102 (2006).
42. D. Froemberg and I. M. Sokolov, Stationary fronts in an $A + B \to 0$ reaction under subdiffusion, *Phys. Rev. Lett.* **100**, 108304 (2008).
43. T. A. Langlands, B. I. Henry, and S. L. Wearne, Anomalous subdiffusion with multispecies linear reaction dynamics, *Phys. Rev. E* **77**, 021111 (2008).
44. A. N. Kolmogorov, Curves in Hilbert space, which are invariant with respect to a one-parameter group of motions, *Dok. Akad. Nauk SSSR* **26**, 6 (1940).
45. A. M. Yaglom, *Correlation theory of stationary and related random functions*, 1st edn. Springer (1987).
46. B. B. Mandelbrot and J. W. van Ness, Fractional Brownian motions, fractional noises and applications, *SIAM Rev.* **10**, 422-437 (1968).
47. K. Falconer, *Fractional geometry: Mathematical Foundations and Applications*, 2nd edn. Wiley (2003).
48. I. M. Sokolov, Models of anomalous diffusion in crowded environments, *Soft Matter* **8**, 9043-9052 (2012).
49. H. E. Hurst, The long-term storage capacity of reservoirs, *Trans. Am. Soc. Civ. Eng.* **116**, 770-799 (1951).
50. T. E. Harris, Diffusion with collisions between particles, *J. Appl. Prob.* **2**, 323-338 (1965).
51. J. Lebowitz and J. Percus, Kinetic equations and density expansions: exactly solvable one-dimensional system, *Phys. Rev.* **155**, 122-138 (1967).
52. Q. H. Wei, C. Bechinger, and P. Leiderer, Single-file diffusion of colloids in one-dimensional channels, *Science* **287**, 625-627 (2000).
53. L. Lizana and T. Ambjörnsson, Single-file diffusion in a box, *Phys. Rev. Lett.* **100**, 200601 (2008).
54. I. Goychuk, Viscoelastic subdiffusion: from anomalous to normal, *Phys. Rev. E* **80**, 046125 (2009).
55. I. Goychuk, Viscoelastic subdiffusion: Generalized Langevin equation approach, *Adv. Chem. Phys.* **150**, 187-249 (2012).
56. S. C. Kou and X. S. Xie, Generalized Langevin equation with fractional gaussian noise: subdiffusion within a single protein molecule, *Phys. Rev. Lett.* **93**, 180603 (2004).
57. V. Tejedor, O. Bénichou, R. Voituriez, R. Jungmann, F. Simmel, C. Selhuber-Unkel, L. B. Oddershede, and R. Metzler, Quantitative analysis of single particle trajectories: Mean maximal excursion method, *Biophys. J.* **98**, 1364-1372 (2010).
58. J.-H. Jeon, V. Tejedor, S. Burov, E. Barkai, C. Selhuber-Unkel, K. Berg-Sørensen, L. Oddershede, and R. Metzler, In vivo anomalous diffusion and

weak ergodicity breaking of lipid granules, *Phys. Rev. Lett.* **106**, 048103 (2011).

59. S. Weber, A. J. Spakowitz, and J. A. Theriot, Bacterial chromosomal loci move subdiffusively through a viscoelastic cytoplasm, *Phys. Rev. Lett.* **104**, 238102 (2010).
60. J. Szymanski and M. Weiss, Elucidating the origin of anomalous diffusion in crowded fluids, *Phys. Rev. Lett.* **103**, 038102 (2009).
61. J.-H. Jeon, N. Leijnse, L. B. Oddershede, and R. Metzler, Anomalous diffusion and power-law relaxation of the time averaged mean squared displacement in worm-like micellar solutions, *New Journal of Physics* **15**, 045011 (2013).
62. G. R. Kneller, K. Baczynski, and M. Pasenkiewicz-Gierula, Consistent picture of lateral subdiffusion in lipid bilayers: Molecular dynamics simulation and exact results, *J. Chem. Phys.* **135**, 141105 (2011).
63. J.-H. Jeon, H. M.-S. Monne, M. Javanainen, and R. Metzler, Anomalous diffusion of phospholipids and cholesterols in a lipid bilayer and its origins, *Phys. Rev. Lett.* **109**, 188103 (2012).
64. J.-H. Jeon and R. Metzler, Fractional Brownian motion and motion governed by the fractional Langevin equation in confined geometries, *Phys. Rev. E* **81**, 021103 (2010).
65. H. Qian, *Process with Long-Range Correlations: Theory and Applications*, In eds. G. Rangarajan and M. Z. Ding, *Lecture Notes in Physics*, vol. 621, Springer (2003).
66. J. R. M. Hosking, Modeling persistence in hydrological time series using fractional differencing, *Water Resour. Res.* **20**, 1898-1908 (1984).
67. C. C. Fritsch and J. Langowski, Kinetic lattice Monte Carlo simulation of viscoelastic subdiffusion, *J. Chem. Phys.* **137**, 064114 (2012).
68. A. V. Chechkin and V. Yu. Gonchar, Approximation to fractional Brownian motion based on fractional integration of a white noise, *Chaos, Solitons & Fractals* **12**, 391-398 (2001).
69. R. Zwanzig, *Nonequilibrium Statistical Mechanics*, 1st edn. Oxford University Press (2001).
70. E. Lutz, Fractional Langevin equation, *Phys. Rev. E* **64**, 051106 (2001).
71. A. Gemant, A method of analyzing experimental results obtained from elasto-viscous bodies, *Phys.* **7**, 311-317 (1936).
72. P. Siegle, I. Goychuk, and P. Hänggi, Origin of hyperdiffusion in generalized Brownian motion, *Phys. Rev. Lett.* **105**, 100602 (2010).
73. P. Siegle, I. Goychuk, and P. Hänggi, Markovian embedding of fractional superdiffusion, *Europhys. Lett.* **93**, 20002 (2011).
74. N. Pottier, Aging properties of an anomalously diffusing particle, *Physica A* **317**, 371-382 (2003).
75. W. Deng and E. Barkai, Ergodic properties of fractional Brownian-Langevin motion, *Phys. Rev. E* **79**, 011112 (2009).
76. A. Taloni, A. Chechkin, and J. Klafter, Generalized elastic model yields a fractional Langevin equation description, *Phys. Rev. Lett.* **104**, 160602 (2010).

77. L. P. Sanders and T. Ambjörnsson, First passage time for a tracer particle in single file diffusion and fractional Brownian motion, *J. Chem. Phys.* **136**, 175103 (2012).
78. A. M. Yaglom, Some classes of random fields in n-dimensional space, related to stationary random processes, *Theor. Prob. & Its Applications* **11**, 273-320 (1957).
79. J. Unterberger, Stochastic calculus for fractional Brownian motion with Hurst exponent $H > 1/4$: A rough path method by analytic extension, *Ann. Probab.* **37**, 565-614 (2009).
80. H. Qian, G. M. Raymond, and J. B. Bassingthwaighte, On two-dimensional fractional Brownian motion and fractional Brownian random field, *J. Phys. A* **31**, L527-L535 (1998).
81. J.-H. Jeon, A. Chechkin, and R. Metzler, First passage behaviour of fractional Brownian motion in two-dimensional wedge domains, *Europhys. Lett.* **94**, 20008 (2011).
82. B. D. Hughes, *Random walks and random environments*, 1st edn. Oxford University Press (1995).
83. D. Chatterjee and B. J. Cherayil, Anomalous reaction-diffusion as a model of nonexponential DNA escape kinetics, *J. Chem. Phys.* **132**, 025103 (2010).
84. M. Bologna, F. Vanni, A. Krokhin, and P. Grigolini, Memory effects in fractional Brownian motion with Hurst exponent $H < 1/3$, *Phys. Rev. E* **82**, 020102(R) (2010).
85. D. ben-Avraham, Computer simulation methods for diffusion controlled reactions, *J. Chem. Phys.* **88**, 941-948 (1988).
86. S. Redner and P. L. Krapivsky, Capture of the lamb: Diffusing predators seeking a diffusing prey, *Am. J. Phys.* **67**, 1277-1283 (1999).
87. M. Hellmann, D. W. Heermann, and M. Weiss, Anomalous reaction kinetics and domain formation on crowded membranes, *Europhys. Lett.* **94**, 18002 (2011).

Chapter 9

Trajectory-to-Trajectory Fluctuations in First-Passage Phenomena in Bounded Domains

Thiago G. Mattos

Max Planck Institute for Intelligent Systems, 70569 Stuttgart, Germany
tgmattos@is.mpg.de

Carlos Mejía-Monasterio

Department of Mathematics and Statistics, University of Helsinki, 00014 Helsinki, Finland & Laboratory of Physical Properties, Technical University of Madrid, 28040 Madrid, Spain
carlos.mejia@helsinki.fi

Ralf Metzler

Institute for Physics & Astronomy, University of Postdam, 14476 Potsdam-Golm, Germany & Physics Department, Tampere University of Technology, 33101 Tampere, Finland
rmetzler@uni-potsdam.de

Gleb Oshanin

CNRS – Université Pierre et Marie Curie, LPTMC UMR7600, 75252 Paris, France
oshanin@lptmc.jussieu.fr

Grégory Schehr

CNRS – Université Paris-Sud, LPTMS, 91405 Orsay, France
gregory.schehr@u-psud.fr

We study the statistics of the first passage of a random walker to absorbing subsets of the boundary of compact domains in different spatial dimensions. We describe a novel diagnostic method to quantify the trajectory-to-trajectory fluctuations of the first passage, based on the distribution of the so-called uniformity index ω, measuring the similarity of the first passage times of two independent walkers starting at the same location. We show that the characteristic shape of $P(\omega)$ exhibits a transition from unimodal to bimodal, depending on the starting point of the trajectories. From the study of different geometries in one, two and three dimensions, we conclude that this transition is a generic property of first passage phenomena in bounded domains. Our results show that, in general, the Mean First Passage Time (MFPT) is a meaningful characteristic measure of the first passage behaviour only when the Brownian walkers start sufficiently far from the absorbing boundary. Strikingly, in the opposite case, the first passage statistics exhibit large trajectory-to-trajectory fluctuations and the MFPT is not representative of the actual behaviour.

1. Introduction

The concept of first passage underlies diverse stochastic phenomena for which the crucial aspect is an event when some random variable of interest reaches a preset value for the first time. A few stray examples across disciplines include chemical reactions [1–6], the firing of a neuron [7, 8], random search of a mobile or an immobile target [9–28], diffusional disease spreading [29], DNA bubble breathing [30, 31], dynamics of molecular motors [32–34], the triggering of a stock option [35], etc. One distinguishes also between continuously varying variables for which the first passage across a given preset value coincides with the first arrival to exactly this value, and discontinuous processes for which this value can be overshoot, as it happens, *e.g.*, for Lévy flights [36, 37]. A variety of first passage time phenomena and different related results have been discussed in [38, 39]. This volume presents an even broader exposition of relevant areas and describes the state of the art in physical and mathematical comprehension of such phenomena.

In this chapter we will be concerned with first passage phenomena for particles executing Brownian motion. The distribution of first passage times for such processes occurring in unbounded domains is typically broad, such that not even the mean first passage time exists [39]. In particular, in one-dimensional, semi-infinite domains the first passage time distribution of a Markovian process is universally dominated by the $t^{-3/2}$ scaling nailed down by the Sparre Andersen theorem [39]. A similar divergence of the mean first passage time occurs in stochastic processes characterised by scale-free distributions of waiting times [40, 41]. This signifies that sample-to-sample or trajectory-to-trajectory fluctuations are of a crucial importance for such processes so that the mean values — the mean first passage times, even if they exist — have little physical meaning.

However, in many practically important situations first passage processes involve Brownian particles moving in bounded domains (see, *e.g.*, [42–47]). In this case the random variable of interest, *e.g.*, the first passage time τ to a boundary, a target chemical group, a binding site on the surface of the domain or elsewhere within the domain, etc., has a distribution $\Psi(\tau)$ which possesses moments of arbitrary, negative and positive order. Such distributions are usually considered as *narrow*, as opposed to *broad* distributions, which do not possess all moments [36–41].

Nonetheless, even for such distributions first passage variables do fluctuate, of course, from sample to sample or from trajectory to trajectory, and an important question is how to quantify such fluctuations. One usually resorts to standard measures of the statistical analysis, such as, *e.g.*, standard deviations, the skewness or the kurtosis of the distribution. But these quantities are often not very instructive since they just produce numbers and it is not always evident how to interpret or compare them. Clearly, quantifying the trajectory-to-trajectory fluctuations in a robust and meaningful way is a challenging problem with many important applications.

In this chapter we focus on this non-trivial problem and discuss a recently proposed procedure which presents a lucid illustration of the effect of the trajectory-to-trajectory fluctuations. It allows to quantify their impact and, moreover, shows that in some cases the distributions considered as *narrow* behave effectively as broad distributions [22, 23]. We employ a novel diagnostic method based on the concept of *simultaneity* of first passage events. Instead of the original first passage problem of quantifying the statistical outcome for a single Brownian walker, in this procedure one simultaneously launches two identical, independent Brownian particles at the same position $\mathbf{r_0}$, or, in other words, considers two different realisations of a single Brownian Motion (BM) starting at $\mathbf{r_0}$. The corresponding outcomes are the first passage times τ_1 and τ_2. One defines then the random variable

$$\omega \equiv \frac{\tau_1}{\tau_1 + \tau_2}, \tag{1}$$

such that ω ranges in the interval $[0, 1]$. The *uniformity index* ω measures the *likelihood* that both walkers arrive to the target simultaneously: when ω is close to $1/2$, the process is uniform and the particles behave as if they were almost performing a Prussian *Gleichschritt*. In contrast, values of ω close to 0 or 1 mean highly non-uniform behaviour, strongly affecting the trajectory-to-trajectory fluctuations. We note parenthetically that similar random variables have been used in the analysis of random probabilities induced by normalization of self-similar Lévy processes [48], of the fractal characterisation of Paretian Poisson processes [49], and of the so-called Matchmaking paradox [50, 51], and more recently, to quantify sample-to-sample fluctuations in mathematical finances [52, 53], chaotic systems [54], the analysis of distributions of the diffusion coefficient of proteins diffusing along DNAs [55] and optimal estimators of the diffusion coefficient of a single Brownian trajectory [56].

To illustrate the concept of simultaneity, consider a generic, *generalized inverse Gaussian* form of $\Psi(\tau)$ (see *e.g.*, the discussion in [22, 23] and references therein)

$$\Psi(\tau) \sim \exp\left(-\frac{a}{\tau}\right) \frac{1}{\tau^{1+\mu}} \exp\left(-\frac{\tau}{b}\right), \tag{2}$$

where a and b are some constants. Note that the first passage time distribution in Eq. (2) is exact in the particular case of Brownian motion on a semi-infinite line in presence of a bias pointing towards the target site, or, equivalently, for the celebrated integrate-and-fire model of neuron firing by Gerstein and Mandelbrot [7]. In general, the detailed form of $\Psi(\tau)$ for diffusion in bounded domains is obviously much more complex than given by Eq. (2), depending on the shape of the domain under consideration and the exact boundary value problem and is typically given in terms of an infinite series.

Nonetheless, on a *qualitative* level Eq. (2) provides a clear picture of the actual behaviour of the first passage time distribution in bounded domains. Namely, $\Psi(\tau)$ consists of three different parts: a singular decay for small values of τ, which mirrors the fact that the first passage to some point starting from a distant position cannot occur instantaneously. This is followed at intermediate times by a generic power-law decay with exponent μ — the so-called persistence exponent [57, 58], depending on the exact type of random motion and spatial dimension. Finally, an exponential decay at long τ cuts off the power-law. A crucial aspect is that the exponential cutoffs at both short and long τ ensure that in bounded domains $\Psi(\tau)$ possesses moments of arbitrary positive or negative order. The parameters a and b depend on the shape of the domain, its typical size, and the starting position of the particle within the domain. When the linear size of the domain (say, the radius R of a circular or a spherical domain) diverges (*i.e.*, $R \to \infty$), the parameter b also diverges such that the long-time asymptotic behaviour of the first passage time distribution is of power-law form without a cutoff. In this case, at least some, if not all, of the moments of $\Psi(\tau)$ diverge.

Now, given the distribution $\Psi(\tau)$, the distribution of the uniformity index ω in Eq. (1) can be readily calculated to give [53, 54]

$$P(\omega) = \frac{1}{(1-\omega)^2} \int_0^\infty \tau \Psi(\tau) \Psi\left(\frac{\omega}{1-\omega}\tau\right) d\tau \,. \tag{3}$$

As we explicitly show in section 2, for the generic distribution in Eq. (2) one finds the following form [23]:

$$P(\omega) = \frac{1}{2K_\mu^2\left(2\sqrt{a/b}\right)} \frac{1}{\omega(1-\omega)} K_{2\mu}\left(2\sqrt{\frac{a}{b\omega(1-\omega)}}\right) , \tag{4}$$

where $K_{2\mu}(\cdot)$ is the modified Bessel function of the second type.

One notices first that the form of the distribution $P(\omega)$ in Eq. (4) is distinctly sensitive to the value of the persistence exponent μ, which characterises the scaling behaviour of the first passage time distribution $\Psi(\tau)$ at intermediate times. Thus, for $\mu > 1$, $P(\omega)$ is always a unimodal, bell-shaped function with a maximum at $\omega = 1/2$, which implies that in such a situation both walkers (or two distinct trajectories of one walker) will most likely arrive to the target simultaneously. For $\mu = 1$, $P(\omega)$ is almost uniform, $P(\omega) \approx 1$, apart from narrow regions at the corners $\omega = 0$ and $\omega = 1$, for $b/a \gg 1$. Curiously, for $\mu < 1$, which corresponds to the most common case, there exists a critical value p_c of the ratio $p = b/a$ such that for $p > p_c$ the distribution $P(\omega)$ has a characteristic M-shaped form with two maxima close to 0 and 1, while at $\omega = 1/2$ one finds a local minimum. Such a transition from a unimodal, bell-shaped to bimodal, M-shaped form mirrors a significant manifestation of sample-to-sample fluctuations.

In what follows we survey known results and further explore this intriguing behaviour of the first passage time distribution and of the corresponding distribution of the uniformity index ω for diffusion in bounded domains, focusing on the effects of the domain shape, dimensionality, location of the target, the type of the boundary conditions, and of the initial position of the walker.

2. First passage times and the uniformity distribution

Consider a BM inside a general d-dimensional domain $\mathcal{S}$, whose boundary $\partial\mathcal{S} \equiv \partial\mathcal{S}_a \cup \partial\mathcal{S}_r$ comprises reflecting, $\partial\mathcal{S}_r$, and absorbing, $\partial\mathcal{S}_a$, parts. At time $t = 0$, the BM initiates at $\mathbf{r}_0 \in \mathcal{S}$ and evolves within the domain until the trajectory hits $\partial\mathcal{S}_a$ for the first time at some random instant τ. Furthermore, let $P(\mathbf{r}, t|\mathbf{r}_0)$ denote the conditional probability distribution for finding the Brownian walker at position $\mathbf{r}$ at time t, provided the initial condition was at $\mathbf{r}_0$ at $t = 0$. The distribution $P(\mathbf{r}, t|\mathbf{r}_0)$ is the solution of the diffusion equation

$$\partial_t P(\mathbf{r}, t|\mathbf{r}_0) = D\ \partial_{\mathbf{r}}^2 P(\mathbf{r}, t|\mathbf{r}_0) \tag{5}$$

on $\mathcal{S}$, subject to the initial condition as well as the boundary conditions at $\partial\mathcal{S}$. Here D is the diffusion coefficient and $\partial_{\mathbf{r}}^2$ is the d-dimensional Laplacian. The solution of this initial boundary value problem is, in the best case, cumbersome, and explicit solutions may be obtained for only few simple geometries, (see *e.g.*, [59]).

If a finite part of the boundary is absorbing, *i.e.*, $\partial\mathcal{S}_a$ is not empty, then the distribution $P(\mathbf{r}, t|\mathbf{r}_0)$ is no longer normalised. The survival probability $\mathscr{S}_{\mathbf{r}_0}(t)$ that the walker has not reached $\partial\mathcal{S}_a$ up to time t, is defined by

$$\mathscr{S}_{\mathbf{r}_0}(t) = \int_{\mathcal{S}} P(\mathbf{r}, t|\mathbf{r}_0) d\mathbf{r}. \tag{6}$$

$\mathscr{S}_{\mathbf{r}_0}(t)$ is a monotonically decreasing function of time, eventually reaching zero value, $\lim_{t\to\infty} \mathscr{S}_{\mathbf{r}_0}(t) = 0$. In terms of the survival probability, the distribution of first passage times to the absorbing boundary becomes

$$\Psi_{\mathbf{r}_0}(\tau) = -\frac{d\mathscr{S}_{\mathbf{r}_0}(\tau)}{d\tau}, \tag{7}$$

and the MFPT associated with the distribution $\Psi(\tau)$ is defined as the first moment

$$\langle\tau\rangle(\mathbf{r}_0) = \int_0^\infty \tau\Psi_{\mathbf{r}_0}(\tau)d\tau = \int_0^\infty \mathscr{S}_{\mathbf{r}_0}(\tau)d\tau. \tag{8}$$

In most of the existing literature, the dependence of the MFPT on the starting position of the walker is either simply neglected, or it is assumed that the starting point is randomly distributed within the domain $\mathcal{S}$. However, as we proceed to show, the $\mathbf{r}_0$-dependence of the first passage time distribution is a crucial aspect which cannot be neglected.

From the First Passage Time Distribution (FPTD) one readily obtains the distribution of the uniformity index ω defined in Eq. (1), for two independent BM, in terms of its moment generating function

$$\Phi(\lambda) = \int_0^1 P(\omega) \exp(-\lambda\omega)\, d\omega, \tag{9}$$

with $\lambda \geq 0$. Since τ_1 and τ_2 are independent, identically distributed random variables, expression (9) can be formally represented as

$$\Phi(\lambda) = \int_0^\infty \int_0^\infty \Psi(\tau_1)\Psi(\tau_2) \exp\left(-\lambda\frac{\tau_1}{\tau_1 + \tau_2}\right) d\tau_1 d\tau_2. \tag{10}$$

Integrating over $d\tau_1$ we change the integration variable, $\tau_1 \to \omega$, so that Eq. (10) is rewritten in the form

$$\Phi(\lambda) = \int_0^1 \exp(-\lambda\omega) \frac{d\omega}{(1-\omega)^2} \int_0^\infty \tau_2 \Psi(\tau_2) \Psi\left(\frac{\omega}{1-\omega}\tau_2\right) d\tau_2. \tag{11}$$

From comparison with Eq. (9), one obtains $P(\omega)$ in Eq. (3).

2.1. *Trajectory-to-trajectory fluctuations and the shape of $P(\omega)$*

To better grasp the relation between the trajectory-to-trajectory fluctuations and the distribution of the uniformity index, consider $P(\omega)$ of Eq. (4) corresponding to the FPTD of Eq. (2). First, one readily notices that $P(\omega)$ vanishes exponentially fast when $\omega \to 0$ or $\omega \to 1$ so that $P(\omega = 0) = P(\omega = 1) = 0$. Second, $P(\omega)$ is symmetric under the replacement $\omega \to 1 - \omega$. Since the distribution $\Psi(\tau)$ possesses moments of arbitrary order, our first guess would be that $P(\omega)$ is always a bell-shaped function with a maximum at $\omega = 1/2$.

To study the shape of the distribution of the uniformity index, we expand $P(\omega)$ in Taylor series around its symmetric point $\omega = 1/2$ up to second order in $\left(\omega - \frac{1}{2}\right)^4$,

$$P(\omega) \approx \frac{2K_{2\mu}(2C)}{K_\mu^2(C)} \left[1 + 4\left(1 - \mu - C\frac{K_{2\mu-1}(2C)}{K_{2\mu}(2C)}\right)\left(\omega - \frac{1}{2}\right)^2\right], \tag{12}$$

where $C \equiv 2\sqrt{a/b}$. Inspecting the sign of the coefficient before the quadratic term, *i.e.*,

$$g = 1 - \mu - C\frac{K_{2\mu-1}(2C)}{K_{2\mu}(2C)}, \tag{13}$$

we notice that

- For $\mu > 1$, g is always negative for any value of b/a so that here the distribution $P(\omega)$ is a bell-shaped function with a maximum at $\omega = 1/2$.
- For $\mu = 1$, g is negative and approaches 0 from below when $b/a \to \infty$. It means that $P(\omega)$ is generally a bell-shaped function with a maximum at $\omega = 1/2$, but it becomes progressively flatter when b/a is increased, so that ultimately $P(\omega) \approx 1$ apart from very narrow regions at the edges for $b/a \gg 1$.

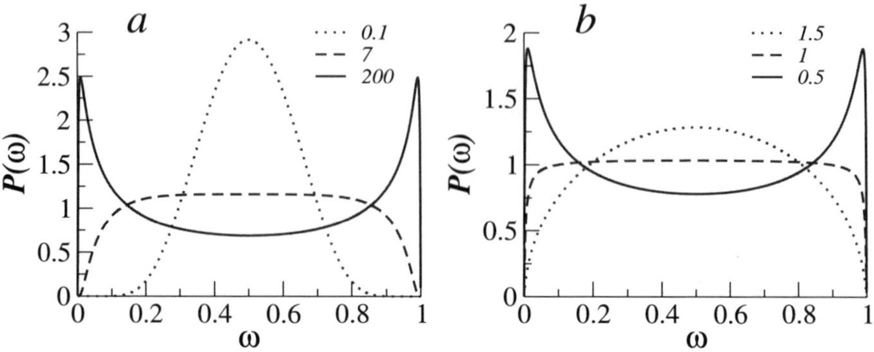

Fig. 1. The distribution $P(\omega)$ of Eq. (4): Panel (a), for $\mu = 0.25$ and different values of b. Panel (b), for $b = 200$ and different values of μ. Parameter a is set equal to 1.

- For $0 \leq \mu < 1$ there always exists a critical value $y_c(\mu)$ of the parameter $y = b/a$ which is defined implicitly as the solution of Eq. (13) for $g = 0$. For $b/a < y_c(\mu)$, the distribution $P(\omega)$ is unimodal with a maximum at $\omega = 1/2$. For $b/a = y_c(\mu)$, the distribution is nearly uniform except for narrow regions in the vicinity of the edges. Finally, which is quite surprising in view of the fact that in this case $\Psi(\tau)$ possesses all moments, for $b/a > y_c(\mu)$ the distribution $P(\omega)$ is bimodal with a characteristic M-shaped form, two maxima close to 0 and 1 and $\omega = 1/2$ being the least probable value.

We depict in Fig. 1 three characteristic forms of $P(\omega)$ for $\mu = 1/2$ and three different values of b/a.

Therefore, for $\mu \geq 1$ two independent BMs will most probably reach the target simultaneously. For $0 \leq \mu < 1$ two distinctly different situations are possible: if b/a is less than some well-defined critical value $y_c(\mu)$, then most likely both BM will arrive to the target for the first time together. If, on the contrary, b/a exceeds this critical value, the event in which both BMs arrive to the location of the target simultaneously will be the least probable one, yielding very large fluctuations in the MFPT obtained from different trajectories.

In the rest of this chapter we present several concrete examples in which the transition between unimodal and bimodal distributions of $P(\omega)$ is observed.

3. One-dimensional Brownian motions

In this section we study the trajectory-to-trajectory fluctuations of the first passage time, in terms of the distribution of the uniformity index ω for two independent identical BM.

3.1. *BM in the semi-infinite interval with a bias*

We first consider a BM in the semi-infinite interval $\mathcal{S} \equiv [0,\infty)$ with $\partial\mathcal{S}_a = \{0\}$ and $\partial\mathcal{S}_r = \{\infty\}$, in the presence of a constant bias pointing towards the target, for which the FPTD in Eq. (2) with $\mu = 1/2$ is exact. In this case one has $a = x_0^2/4D$, where x_0 is the starting point, D the diffusion coefficient, and $b = 4D/v^2$ (v being the drift velocity). Hence, the Péclet number, defined as the ratio between the rate of advection and the rate of diffusion, is $\text{Pe} = 2\sqrt{a/b} = x_0|v|/2D$ (see, *e.g.*, [39]). Consequently, we can make the following statement:

Consider two independent identical BMs on a semi-infinite interval, starting at the same point x_0, having the same diffusion coefficient D and experiencing the same bias F which points towards the origin so that the drift velocity of both BMs is $v < 0$. Then, an event in which both BMs arrive for the first time to the origin simultaneously is

— the *least* probable if $\text{Pe} < \text{Pe}_c$,
— the *most* probable if $\text{Pe} > \text{Pe}_c$,

where $\text{Pe}_c \approx 0.66$ is the solution of the transcendental equation

$$1 = 2\text{Pe}_c \frac{K_0(2\text{Pe}_c)}{K_1(2\text{Pe}_c)} \,. \tag{14}$$

Therefore, the MFPT to the target, which in this case equals $x_0/|v|$, might be an appropriate measure of the search efficiency for sufficiently large Péclet numbers, but definitely is not in case of small Pe. In the latter case the trajectory-to-trajectory fluctuations are significant and the mean value is not representative of the actual behaviour.

3.2. *BM on a finite interval*

We consider now BMs in the finite interval $\mathcal{S} \equiv [0, L]$ with $\partial\mathcal{S}_a = \{0\}$ and $\partial\mathcal{S}_r = \{L\}$, that are initiated at $x = x_0$ at time $t = 0$. In this case the

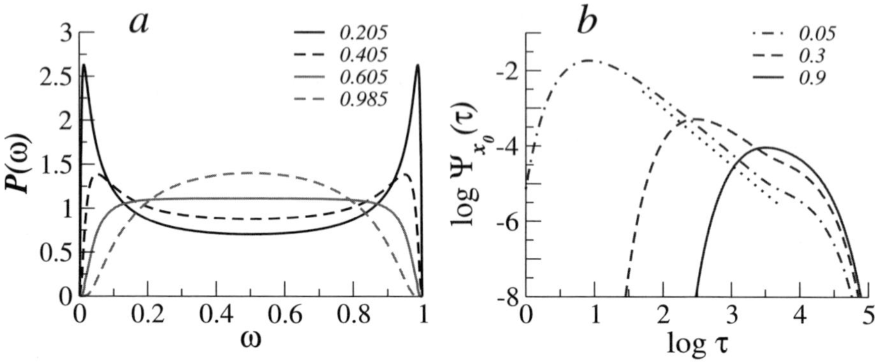

Fig. 2. Two BMs in the interval $[0, L]$. Panel (a): Distribution $P(\omega)$ in Eq. (16) for different values of x_0/L. Panel (b): FPTD $\Psi_{x_0}(\tau)$ in Eq. (15) for $L = 100$, $D = 1/2$ and different values of x_0/L. The dotted curve corresponds to a decay $\sim \tau^{-3/2}$.

FPTD can be obtained exactly as the infinite sum

$$\Psi_{x_0}(\tau) = \frac{2\pi D}{L^2} \sum_{n=0}^{\infty} A_n\left(\frac{x_0}{L}\right) \exp\left(-\frac{\pi^2(n+1/2)^2 D\tau}{L^2}\right), \tag{15}$$

where the coefficients $A_n\left(\frac{x_0}{L}\right) = \left(n + \frac{1}{2}\right)\sin\left(\frac{\pi(n+1/2)x_0}{L}\right)$. Consequently, after some calculations the normalised distribution $P(\omega)$ has the following form [22]

$$P(\omega) = \frac{2}{\pi}\frac{d}{d\omega}\sum_{m=0}^{\infty}\frac{\sin\left(\pi(m+1/2)\frac{x_0}{L}\right)}{m+1/2}\frac{\cosh\left(\pi(m+1/2)\sqrt{\frac{1-\omega}{\omega}}\left(1-\frac{x_0}{L}\right)\right)}{\cosh\left(\pi(m+1/2)\sqrt{\frac{1-\omega}{\omega}}\right)}. \tag{16}$$

Panel (a) of Fig. 2 shows the distribution $P(\omega)$ in Eq. (16) for different values of the ratio x_0/L, evidencing the transition in the shape of $P(\omega)$ at the critical value ≈ 0.605.

Furthermore, the moments of arbitrary order of $\Psi(\tau)$ in Eq. (15) can be straightforwardly calculated. One obtains

$$\langle\tau^m\rangle \equiv \int_0^{\infty} \tau^m \, d\tau \, \Psi(\tau) = \frac{(-1)^m\sqrt{\pi}}{\Gamma(m+1/2)}\left(\frac{L^2}{D}\right)^m E_{2m}\left(\frac{x_0}{2L}\right), \tag{17}$$

where $E_{2k}(\cdot)$ are the Euler polynomials. Consequently, the mean and the variance are given by

$$\langle\tau\rangle = -2\frac{L^2}{D}\, E_2\left(\frac{x_0}{2L}\right), \tag{18}$$

and

$$k_2 = \langle\tau^2\rangle - \langle\tau\rangle^2 = \frac{4}{3}\frac{L^4}{D^2}\left(E_4\left(\frac{x_0}{2L}\right) - 3E_2^2\left(\frac{x_0}{2L}\right)\right) \tag{19}$$

respectively. One can readily check that both statistical measures are monotonically increasing functions of x_0/L and do not show any sign of a particular behaviour at $x_0/L \approx 0.605$. The same occurs for higher order cumulants of the distribution $\Psi(\tau)$ [22].

This puts in evidence the superior sensitivity of the statistics of the uniformity index ω. As we show in sections 4.2 and 5.2, the unimodal or bimodal shape of $P(\omega)$ classifies the parameter space into regions inside which the MFPT is either *meaningful* or *meaningless.*

3.3. *Arcsine law mechanism*

We found that any two BMs arrive to the target at progressively distinct times the closer they are initially to its location, and should most probably arrive together when they are far from it. Notwithstanding this counter-intuitive result, such a behavior is the same as the one behind the famous arcsine law for the distribution of the fraction of time spent by a random walker on a positive half-axis [60]:

Once one of the BMs goes away from the target, it finds it more difficult to return than to keep on going away.

How does this relate to the properties of the distribution of the first passage time? In terms of the FPTD of Eq. (15) in the previous section, x_0 ($a^{1/2}$) and l ($b^{1/2}$) define the effective size of the region in which the decay of the FPTD is governed by the intermediate power-law tail. The larger this region is, the larger the fluctuations in the first passage time become [22]. This is shown in panel (*b*) of Fig. 2. All three curves show an exponential behavior for small and large values of τ, as in Eq. (2). For large τ all three curves merge which signifies that at such values of τ the characteristic decay time is dependent only on l. The lower cut-off is clearly dependent only on the starting point x_0. Evidently the power-law region characterized by a decay $\sim t^{-3/2}$, grows with decreasing x_0/l, *i.e.*, when the BMs start closer to the target.

4. Two-dimensional Brownian motions

In this section we analyze the statistics of ω for BM in different two-dimensional bounded domains.

4.1. *BM in a disc with a reflecting boundary*

We first consider BM in a disc of radius L, centered at the origin $\mathcal{S} = \{\mathbf{r} : |\mathbf{r}| < L\}$, with $\partial\mathcal{S}_a = \{\mathbf{r} : |\mathbf{r}| = l\}$ and $\partial\mathcal{S}_r = \{\mathbf{r} : |\mathbf{r}| = L\}$, where $\mathbf{r} \in \mathbb{R}^2$ and $l < L$. Therefore, the target consist in a concentric disc of smaller radius l.

Suppose next that a BM starts at some point at distance r_0 from the origin and hits the target for the first time at time moment τ. The FPTD $\Psi_{r_0}(\tau)$ is explicitly given by [22]

$$\Psi_{r_0}(\tau) = \frac{lD}{Z} \sum_{n=0}^{\infty} A_n(l, r_0, L) \exp\left(-\lambda_n^2 D\tau\right), \tag{20}$$

where Z is the normalization,

$$A_n(l, r_0, L) = \frac{U_0(\lambda_n r_0) U_0'(\lambda_n l)}{\lambda_n^2 L^2 U_0^2(\lambda_n L) - l^2 \left(U_0'(\lambda_n l)\right)^2},$$

$$U_0(x) = Y_0(\lambda_n l) J_0(x) - J_0(\lambda_n l) Y_0(x); \quad U_0'(\lambda_n l) = \left.\frac{dU_0(\lambda_n x)}{dx}\right|_{x=l}$$

while λ_n are the roots of the function $Y_0(\lambda_n l) J_1(\lambda_n L) - J_0(\lambda_n l) Y_1(\lambda_n L)$, arranged in an ascending order, and $J_n(\cdot)$ and $Y_n(\cdot)$ are Bessel functions of the first and second kind respectively. Note that λ_n depends on L and l.

In this case the distribution of the uniformity index ω can be written as [22]

$$P(\omega) = \frac{l}{Z} \frac{d}{d\omega} \sum_{m=0}^{\infty} \frac{A_m(l, r_0, L)}{\lambda_m^2} \Phi\left(\lambda = \frac{1-\omega}{\omega} D\lambda_m^2\right), \tag{21}$$

where $\Phi(\lambda)$ is the characteristic function of the first passage time distribution in Eq. (20) defined by [39]

$$\Phi(\lambda) = \frac{I_0\left(\sqrt{\frac{\lambda}{D}} r_0\right) K_1\left(\sqrt{\frac{\lambda}{D}} L\right) + K_0\left(\sqrt{\frac{\lambda}{D}} r_0\right) I_1\left(\sqrt{\frac{\lambda}{D}} L\right)}{I_0\left(\sqrt{\frac{\lambda}{D}} l\right) K_1\left(\sqrt{\frac{\lambda}{D}} L\right) + K_0\left(\sqrt{\frac{\lambda}{D}} l\right) I_1\left(\sqrt{\frac{\lambda}{D}} L\right)}, \tag{22}$$

where $I_n(\cdot)$ and $K_n(\cdot)$ are the modified Bessel functions of the first and second type respectively.

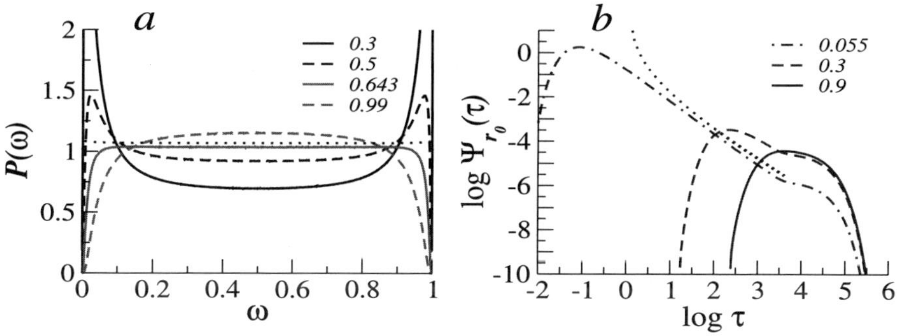

Fig. 3. Two BMs in the two-dimensional disc of radius $L = 100$ with a reflecting boundary and concentric absorbing disc of radius $l = 5$, and $D = 1/2$. Panel (*a*): Distribution $P(\omega)$ in Eq. (21) for different values of r_0/L. The dotted curve corresponds to the numerically computed $P_{\rm av}(\omega)$. Panel (*b*): FPTD $\Psi_{r_0}(\tau)$ in Eq. (20) for different values of r_0/L. The dotted curve corresponds to a decay $\sim 1/\tau \log^2 \tau$.

Fig. 3 shows the FPTD in Eq. (20) and $P(\omega)$ in Eq. (21) for fixed L, D and r, and several values of r_0. Here we observe the same trend as in the previous one-dimensional example, namely the FPTD becomes broader when the starting point of the BMs is closer to $\partial\mathcal{S}_a$, with an intermediate power-law behavior with a logarithmic correction $\sim 1/\tau \log^2(\tau)$ (see the dotted curve in Fig. 3-*b*). The critical value for the transition between bell- and M-shaped $P(\omega)$ is numerically found to be $r_0/L \sim 0.643$. Moreover, if the starting point of two BM is averaged over the domain $\mathcal{S}$, one readily finds, by averaging the result in Eq. (21), that $P_{av}(\omega) \equiv 1$.

4.2. *BM in a disc with aperture*

For other simple geometries but mixed absorbing/reflecting boundary conditions, it is not possible in general to obtain analytical closed expressions for the FPTD. However, the examples discussed previously suggest that the arcsine law mechanism in section 3.3, underlying the appearance of large trajectory-to-trajectory fluctuations of the first passage time, is generic in bounded domains of arbitrary shape.

For instance, consider a circular domain of unit radius (in polar coordinates $\mathbf{r} = (r, \theta)$), $\mathcal{S} = \{\mathbf{r} : r < 1\}$, and the following boundary conditions: the segment with $|\theta| < \Theta/2$ is absorbing while the remaining part of the outer circle is reflective. The aperture of the circular domain is thus an arc of length Θ.

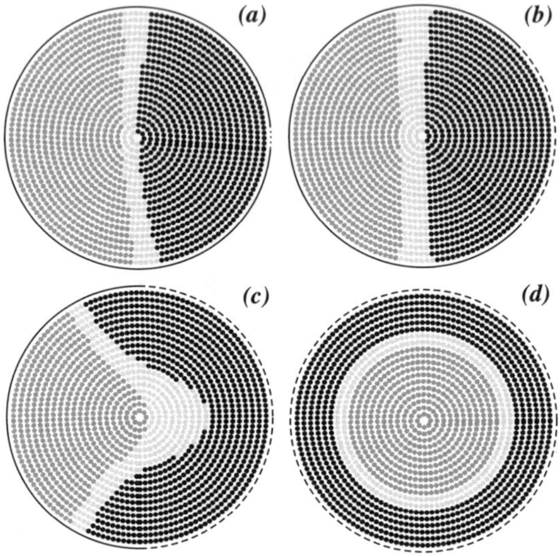

Fig. 4. (Color online) Phase-chart for the shape of the uniformity distribution $P(\omega)$ for a BM in the unit disc with reflective boundary (solid line) and an aperture of size Θ corresponding to the absorbing boundary (dashed line). (a) $\Theta = \pi/18$, (b) $\Theta = \pi/2$, (c) $\Theta = \pi$, and (d) $\Theta = 2\pi$. Starting locations are coloured light blue if $\chi < -\chi_\star$, dark blue if $\chi > \chi_\star$ and beige if $|\chi| < \chi_\star$, where $\chi_\star = 1$.

Obtaining an expression for the FPTD involves the cumbersome problem of finding the solution of the diffusion equation with mixed Dirichlet-Neumann boundary conditions. Here we content ourselves with the numerical analysis of the problem. Let each BM commence at position (ρ_0, θ_0) inside the unit circle and determine the set $\{\tau_i\}$ of first passage times to the location of the aperture. From these data one obtains $P(\omega)$. To easily grasp the geometric dependence of the first passage time fluctuations, we considered the parameter χ that quantifies the shape of the distribution of ω [23]. Performing a fit of the numerically obtained $P(\omega)$ to a quadratic polynomial of ω in the domain $0.05 < \omega < 0.95$, we define χ as the coefficient of the quadratic term, so that the sign of χ determines the shape of $P(\omega)$: $\chi < 0$ corresponds to the unimodal, bell-shaped distribution, $\chi > 0$ signifies that the distribution is bimodal, M-shaped, and a zero value of $\chi = 0$ means that $P(\omega)$ is uniform.

In Fig. 4 we show the phase-chart for the shape of $P(\omega)$ for different sizes of the aperture. As in all previous examples, the closer the starting position of the BM is to the absorbing boundary the larger the fluctuations of the first passage time become. One thus expects that very near the absorbing boundary the standard deviation of the first passage time becomes much larger than its mean. We note that this result is generic irrespectively of the aperture size.

The same analysis was recently carried out for other different shapes, such as pie-wedges or triangles in [23], with the same conclusions.

5. Three-dimensional Brownian motions

5.1. *BM in a sphere with a reflecting boundary*

Consider a BM in a sphere of radius L (in spherical coordinates $\mathbf{r} = (\rho, \theta, \phi)$), $\mathcal{S} = \{\mathbf{r} : \rho < L\}$ with reflecting boundary $\partial\mathcal{S}_r = \{\mathbf{r} : \rho = L\}$, and a concentric absorbing sphere of radius $l < L$, $\partial\mathcal{S}_a = \{\mathbf{r} : \rho = l\}$.

For this geometry, the distribution of the first passage time τ of a BM starting at a distance ρ_0 from the origin to the surface of the target is given explicitly by [22]

$$\Psi_{\rho_0}(\tau) = \frac{D}{Z}\sum_{n=0}^{\infty} A_n(l, \rho_0, L)\exp\left(-\lambda_n^2 D\tau\right) , \tag{23}$$

where the coefficients

$$A_n(l, \rho_0, L) = \frac{2u_0(\lambda_n\rho_0)u_0'(\lambda_n l)}{G(l, L, \lambda_n)} ,$$

with $u_0(x) = y_0(\lambda_n l)j_0(x) - j_0(\lambda_n l)y_0(x)$, $u_0'(\lambda_n l) = du_0(\lambda_n x)/dx|_{x=l}$, and

$$\begin{aligned} G(l, L, \lambda_n) &= Rj_0(\lambda_n L)u_0(\lambda_n L)(\lambda_n L j_0(\lambda_n l) - y_0(\lambda_n l)) \\ &\quad - \frac{j_0(\lambda_n l)y_0(\lambda_n l)}{\lambda_n} + \frac{L-l}{l^2\lambda_n^4} . \end{aligned}$$

The set $\{\lambda_n\}$ are the roots of $y_0(\lambda_n l)j_1(\lambda_n L) - j_0(\lambda_n l)y_1(\lambda_n L)$, arranged in an ascending order, while $j_n(\cdot)$ and $y_n(\cdot)$ are the spherical Bessel functions of the first and of the second kind, respectively.

The corresponding distribution $P(\omega)$ for two independent BMs starting at a distance ρ_0 from the origin is [22]

$$\begin{aligned} P(\omega) &= \frac{1}{Z^2}\sum_{n,m=0}^{\infty}\frac{A_n(l, \rho_0, L)A_m(l, \rho_0, L)}{(\omega\lambda_n^2 + (1-\omega)\lambda_m^2)^2} \\ &= \frac{1}{Z}\sum_{m=0}^{\infty}\frac{A_m(l, \rho_0, L)}{\lambda_m^2}\,\Phi\left(\lambda = \frac{1-\omega}{\omega}D\lambda_m^2\right) , \end{aligned} \tag{24}$$

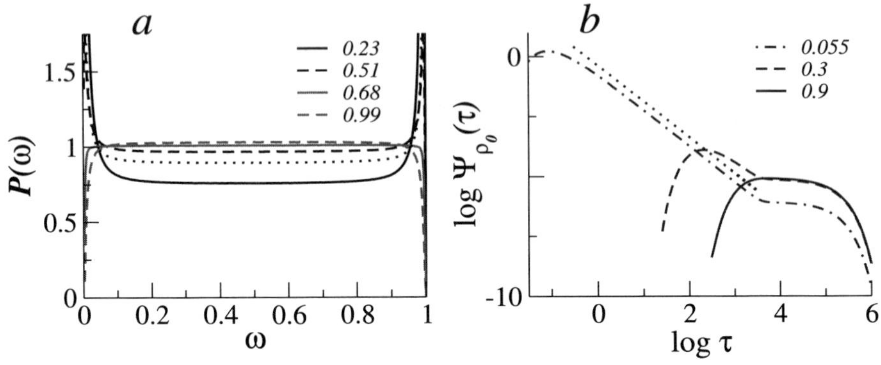

Fig. 5. BM in a sphere of radius $L = 100$ with a reflecting boundary and concentric absorbing sphere of radius $l = 5$, and $D = 1/2$. Panel (a): Distribution $P(\omega)$ in Eq. (24) for different values of ρ_0/L. The dotted curve corresponds to the numerically computed $P_{\rm av}(\omega)$. Panel (b): FPTD $\Psi_{\rho_0}(\tau)$ in Eq. (23) for different values of ρ_0/L. The dotted curve corresponds to a decay $\tau^{-3/2}$.

where $\Phi(\lambda)$ is the characteristic function of $\Psi(\tau)$ in Eq. (23), defined by [39]

$$\Phi(\lambda) = \frac{l}{\rho_0} \frac{\sinh\left(\sqrt{\frac{\lambda}{D}}\,(L-\rho_0)\right) - \sqrt{\frac{\lambda}{D}}\,R\cosh\left(\sqrt{\frac{\lambda}{D}}\,(L-\rho_0)\right)}{\sinh\left(\sqrt{\frac{\lambda}{D}}\,(L-l)\right) - \sqrt{\frac{\lambda}{D}}\,R\cosh\left(\sqrt{\frac{\lambda}{D}}\,(L-l)\right)}. \tag{25}$$

The distributions in Eqs. (23) and (24) are shown in Fig. 5 for fixed L, l and D, and several values of ρ_0. An intermediate power-law $\sim \tau^{-3/2}$ is apparent for $\rho_0/L = 0.055$, persists for one decade for $\rho_0/L = 0.3$ and is entirely absent for $\rho_0/L = 0.9$, *i.e.*, when the BM initiates close to the reflecting boundary (see Fig. 5-b). Correspondingly we find that $P(\omega)$ has a different modality depending on the value of the ratio ρ_0/L with a critical value of $\rho_0/L \sim 0.68$ (see Fig. 5-a). For $\rho_0/L > 0.68$, the distribution has, in principle, a maximum at $\omega = 1/2$ but this maximum is almost invisible so that visually the distribution looks more like a uniform one, as compared to the 1D case in which the maximum is more apparent. When the starting point ρ_0 is uniformly distributed in $\mathcal{S}$, we again find $P_{av}(\omega) \equiv 1$.

5.2. *Narrow Escape Problem*

As a last example we consider the problem of the so-called Narrow Escape Time (NET). This problem, generic in cellular biochemistry, consists of estimating the time that a randomly moving particle spends inside a bounded

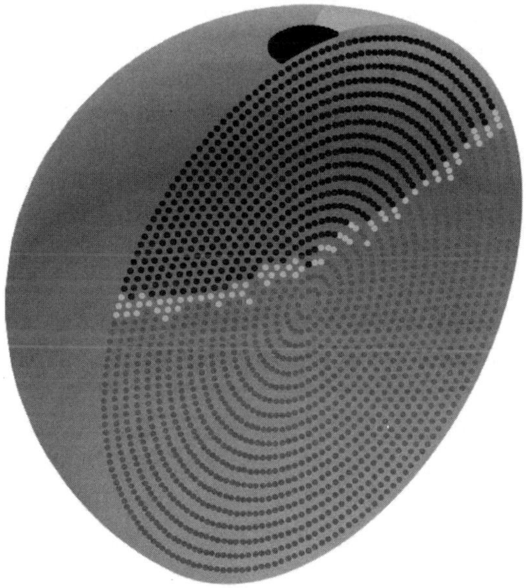

Fig. 6. (Color online) Cut of the phase-chart for the shape of $P(\omega)$ for the NET problem with $p_a = 0$, $\sigma = 1/4\pi$ and $D_V = 0.000416 = D_S = 0.000625$. Starting locations are coloured light blue if $\chi < -\chi_\star$, dark blue if $\chi > \chi_\star$ and beige if $|\chi| < \chi_\star$, where $\chi_\star = 0.1$. The red region on the sphere indicates the position of the narrow aperture. The phase-chart has rotation symmetry with respect to the position of the aperture.

domain before it escapes through a narrow aperture on the boundary of the domain. The particle can be an ion, a ligand or a molecule diffusing inside a cell, a micro-vesicle, a compartment, an endosome, a caveola, a dendritic spine, etc., and a variety of processes in which the importance of the NET problem is striking have been discussed in the past [61–65].

In section 4.2 we discussed a two-dimensional version of this problem. In three dimensions, for a spherical domain of radius L, reflecting everywhere except at a narrow circular aperture of radius $\sigma \ll L$ in which the boundary is absorbing, the survival probability at large times is $\mathscr{S}_{\mathbf{r}_0}(t) \sim \exp(-t/\tau_{3D})$, where the characteristic time decay $\tau_{3D} = \pi L^3/3D\sigma$ corresponds to the MFPT [66].

Here we discuss the statistics of the uniformity index for the NET problem. Consider a sphere of unit radius $\mathcal{S} = \{\mathbf{r} : \rho < 1\}$ (in spherical coordinates $\mathbf{r} = (\rho, \theta, \phi)$), and the following boundary conditions: a solid angle $|\Omega| < \sigma$ is absorbing while the remaining part of the sphere is reflective. Furthermore, consider a BM inside $\mathcal{S}$ with a diffusion coefficient D_V, and initially located at $(\rho_0, \theta_0, \phi_0)$. When the Brownian particle hits the sphere $\partial\mathcal{S}$ it becomes weakly absorbed with probability p_a and starts to diffuse, with diffusion coefficient D_S, along the surface of the sphere. During a time interval Δt, the particle detaches from the sphere with probability $p_a \Delta t$, and continue its motions until it hits the narrow aperture at time τ.

Considering couples of BMs one obtains $P(\omega)$ as before. Fig. 6 shows the phase-chart for the shape of $P(\omega)$, as defined in section 4.2, for $p_a = 0$, *i.e.*, when the BM never gets attached to the sphere. The results qualitatively resemble those obtained in section 4.2 for the two-dimensional disc for narrow apertures (see Fig. 4-a). Note that the statistics of the first passage time inherit the polar symmetry of the geometry with respect to the position of the aperture.

Our results indicate that for BMs starting near the location of the aperture, the MFPT has little significance to determine the NET. We have also considered finite values of the attachment probability p_a (finite particle-surface affinity), obtaining qualitatively similar results than the ones shown in Fig. 6 for $p_a = 0$.

6. Conclusions

In this Chapter we discussed a novel diagnostic method which allows to quantify straightforwardly and meaningfully the effects of the trajectory-to-trajectory fluctuations in first passage phenomena characterised by "narrow" distributions $\Psi(\tau)$ possessing moments of arbitrary order. This diagnostic is based on the simultaneity concept of the first passage events. To this purpose, instead of the original random variable τ characterising the first passage event, we introduced a variable $\omega = \tau_1/(\tau_1 + \tau_2)$, where τ_1 and τ_2 are identical independent random variables with the same distribution $\Psi(\tau)$. Physically, the uniformity index ω shows us the likelihood of the event that two different trajectories arrive to the prescribed location for the first time simultaneously.

We demonstrated that the characteristic shapes of the associated distribution of the uniformity index ω are very sensitive to the trajectory-to-trajectory fluctuation. In some cases, when such fluctuations have only little effect so that the MFPT can be considered as a meaningful property, the distribution $P(\omega)$ is unimodal with a maximum at $\omega = 1/2$, such that $\tau_1 = \tau_2$ is the most probable event. In other cases, $P(\omega)$ is shown to have a bimodal form with two maxima close to $\omega = 0$ and $\omega = 1$ and a local minimum at $\omega = 1/2$. This means that in such situations two different first passage events will be most likely characterised by very different τ_1 and τ_2, so that the trajectory-to-trajectory fluctuations will be very important and the MFPT cannot be considered as a meaningful characteristic measure of the process. In this regard, the underlying "narrow" distributions $\Psi(\tau)$ may show a behaviour specific to broad distributions which do not possess all moments.

To illustrate this concept, we considered several examples of Brownian motion in bounded domains, in different spatial dimensions and with different locations of the target. We demonstrated that, remarkably, within the same domain the very shape of the distribution $P(\omega)$ depends crucially on the location of the target and on the starting point of the Brownian motion — trajectories starting from one point may exhibit considerable fluctuations with respect to the first passage to the target, while trajectories starting from some other point may arrive to the target location almost simultaneously. For some geometries, we evaluated the "phase charts" for the distribution $P(\omega)$ showing which shape — unimodal or bimodal — this distribution will have for different starting point within each domain.

We expect that a similar behaviour will be observed in bounded one-dimensional geometries for fractional BM with arbitrary Hurst index H or for α-stable Lévy flights with $0 < \alpha < 1$, and more generally, for bounded geometries with compact or non compact exploration.

References

1. C. Loverdo, O. Bénichou, M. Moreau, and R. Voituriez, Enhanced reaction kinetics in biological cells, *Nature Phys.* **4**, 134 (2008).
2. C. Loverdo, O. Bénichou, M. Moreau, and R. Voituriez, Reaction kinetics in active media, *J. Stat. Mech.* **02**, P02045 (2009).
3. O. Bénichou, C. Chevalier, J. Klafter, B. Meyer, and R. Voituriez, Geometry-controlled kinetics, *Nature Chemistry* **2**, 472 (2010).
4. O. Bénichou, M. Moreau, and G. Oshanin, Kinetics of stochastically gated diffusion-limited reactions and geometry of random walk trajectories, *Phys. Rev. E.* **61**, 3388 (2000).
5. T. G. Mattos and F. D. A. A. Reis, Phase transitions and crossovers in reaction-diffusion models with catalyst deactivation, *J. Chem. Phys.* **131**, 014505 (2009).
6. M. Smoluchowski, Versuch einer mathematischen Theorie der Koagulations-kinetik kolloidaler Lösungen, *Z. Phys. Chem.* **92**, 129 (1917).
7. G. L. Gerstein and B. B. Mandelbrot, Random walk models for the spike activity of a single neuron, *Biophys. J.* **4**, 41 (1964).
8. A. N. Burkitt, A review of the integrate-and-fire neuron model: I. Homogeneous synaptic input, *Biol. Cybern.* **95**, 1 (2006).

9. G. M. Viswanathan, S. V. Buldyrev, S. Havlin, M. G. E. Da Luz, E. P. Raposo, and H. E. Stanley, Optimizing the success of random searches, *Nature* **401**, 911 (1999).
10. G. M. Viswanathan, M. G. E. da Luz, E. P. Raposo, and H. E. Stanley, *The Physics of Foraging: An Introduction to Random Searches and Biological Encounters.* Cambridge, Cambridge University Press (2011).
11. O. Bénichou, M. Coppey, M. Moreau, P. H. Suet, and R. Voituriez, Optimal Search Strategies for Hidden Targets, *Phys. Rev. Lett.* **94**, 198101 (2005).
12. C. Loverdo, O. Bénichou, M. Moreau, and R. Voituriez, Robustness of optimal intermittent search strategies in one, two, and three dimensions, *Phys. Rev. E* **80**, 031146 (2009).
13. O. Bénichou, C. Loverdo, M. Moreau, and R. Voituriez, Intermittent search strategies, *Rev. Mod. Phys.* **83**, 81 (2011).
14. M. A. Lomholt, T. Ambjörnsson, and R. Metzler, Optimal Target Search on a Fast-Folding Polymer Chain with Volume Exchange, *Phys. Rev. Lett.* **95**, 260603 (2005).
15. M. A. Lomholt, T. Koren, R. Metzler, and J. Klafter, Lévy strategies in intermittent search processes are advantageous, *Proc. Natl. Acad. Sci. USA* **105**, 11055 (2008).
16. M. A. Lomholt, B. van den Broek, S.-M. J. Kalisch, G. J. L. Wuite, and R. Metzler, Facilitated diffusion with DNA coiling, *Proc. Natl. Acad. Sci. USA* **106**, 8204 (2009).
17. G. Oshanin, H. S. Wio, K. Lindenberg, and S. F. Burlatsky, Intermittent random walks for an optimal search strategy: one-dimensional case, *J. Phys.: Cond. Mat.* **19**, 065142 (2007).
18. G. Oshanin, H. S. Wio, K. Lindenberg, and S. F. Burlatsky, Efficient search by optimized intermittent random walks, *J. Phys. A: Math. Theor.* **42**, 434008 (2009).
19. G. Oshanin, O. Vasilyev, P. Krapivsky, and J. Klafter, Survival of an evasive prey, *Proc. Natl. Acad. Sci. USA* **106**, 13696 (2009).
20. F. Rojo, C. E. Budde, and H. S. Wio, Optimal intermittent search strategies, *J. Phys. A: Math. Theor.* **42**, 125002 (2009).
21. F. Rojo, C. E. Budde, H. S. Wio, G. Oshanin, and K. Lindenberg, Intermittent search strategies revisited: effect of the jump length and biased motion, *J. Phys. A: Math. Theor.* **43**, 345001 (2009).
22. C. Mejía-Monasterio, G. Oshanin, and G. Schehr, First passages for a search by a swarm of independent random searchers, *J. Stat. Mech.* **06**, P06022 (2011).
23. T. G. Mattos, C. Mejía-Monasterio, R. Metzler, and G. Oshanin, First passages in bounded domains: When is the mean first passage time meaningful?, *Phys. Rev. E* **86**, 031143 (2012).
24. J. M. Newby and P. C. Bressloff, Random intermittent search and the tug-of-war model of motor-driven transport, *J. Stat. Mech.* **04**, P04014 (2010).
25. I. G. Portillo, D. Campos, and V. Méndez, Intermittent random walks: transport regimes and implications on search strategies, *J. Stat. Mech.* **02**, P02033 (2011).

26. M. R. Evans and S. N. Majumdar, Diffusion with Stochastic Resetting, *Phys. Rev. Lett.* **106**, 160601 (2011).
27. M. R. Evans and S. N. Majumdar, Diffusion with optimal resetting, *J. Phys. A: Math. Theor.* **44**, 435001 (2011).
28. E. Gelenbe, Search in unknown random environments, *Phys. Rev. E* **82**, 061112 (2010).
29. A. L. Lloyd and R. M. May, Epidemiology - How viruses spread among computers and people, *Science* **292**, 1316 (2001).
30. A. Hanke and R. Metzler, Bubble dynamics in DNA, *J. Phys. A: Math. Theor.* **36**, L473 (2003).
31. H. C. Fogedby and R. Metzler, DNA Bubble Dynamics as a Quantum Coulomb Problem, *Phys. Rev. Lett.* **98**, 070601 (2007).
32. G. Oshanin, J. Klafter, and M. Urbakh, Molecular motor with a built-in escapement device, *Europhys. Lett.* **68**, 26 (2004).
33. G. Oshanin, J. Klafter, and M. Urbakh, Saltatory drift in a randomly driven two-wave potential, *J. Phys.: Condens. Matter* **17**, S3697 (2005).
34. V. Palyulin and R. Metzler, How a finite potential barrier decreases the mean first-passage time, *J. Stat. Mech.* **03**, L03001 (2012).
35. J. P. Bouchaud and M. Potters, *Theory of financial risk and derivative pricing: from statistical physics to risk management.* Cambridge, Cambridge University Press (2003).
36. A. V. Chechkin, R. Metzler, V. Y. Gonchar, J. Klafter, and L. V. Tanatarov, First passage and arrival time densities for Lévy flights and the failure of the method of images, *J. Phys. A: Math. Gen.* **36**, L537 (2003).
37. T. Koren, M. A. Lomholt, A. V. Chechkin, J. Klafter, and R. Metzler, Leapover Lengths and First Passage Time Statistics for Lévy Flights, *Phys. Rev. Lett.* **99**, 160602 (2007).
38. K. Lindenberg and B. J. West, The first, the biggest, and other such considerations, *J. Stat. Phys.* **42**, 201 (1986).
39. S. Redner, *A Guide to First-Passage Processes.* Cambridge, Cambridge University Press (2001).
40. R. Metzler and J. Klafter, Boundary value problems for fractional diffusion equations, *Physica A* **278**, 107 (2000).
41. H. Scher, G. Margolin, R. Metzler, J. Klafter, and B. Berkowitz, The dynamical foundation of fractal stream chemistry: The origin of extremely long retention times, *Geophys. Res. Lett.* **29**, 1061 (2002).
42. S. Condamin, O. Bénichou, V. Tejedor, R. Voituriez, and J. Klafter, First-passage times in complex scale-invariant media, *Nature* **450**, 77 (2007).
43. S. Condamin, O. Bénichou, and M. Moreau, First-Passage Times for Random Walks in Bounded Domains, *Phys. Rev. Lett.* **95**, 260601 (2005).
44. S. Condamin, O. Bénichou, and M. Moreau, Random walks and Brownian motion: A method of computation for first-passage times and related quantities in confined geometries, *Phys. Rev. E* **75**, 021111 (2007).
45. B. Meyer, C. Chevalier, R. Voituriez, and O. Bénichou, Universality classes of first-passage-time distribution in confined media, *Phys. Rev. E* **83**, 051116 (2011).

46. C. Chevalier, O. Bénichou, B. Meyer, and R. Voituriez, First-passage quantities of Brownian motion in a bounded domain with multiple targets: a unified approach, *J. Phys. A.: Math. Theor.* **44**, 025002 (2011).
47. J. H. Jeon, A. V. Chechkin, and R. Metzler, First passage behaviour of fractional Brownian motion in two-dimensional wedge domains, *Europhys. Lett.* **94**(2), 20008 (2011).
48. I. Eliazar, On selfsimilar Lévy random probabilities, *Physica A* **356**, 207 (2005).
49. I. Eliazar and I. M. Sokolov, On the fractal characterization of Paretian Poisson processes, *Physica A* **391**, 3043 (2012).
50. I. M. Sokolov and I. Eliazar, The matchmaking paradox: a statistical explanation, *J. Phys. A.: Math. Theor.* **43**, 055001 (2010).
51. I. M. Sokolov and I. Eliazar, Sampling from scale-free networks and the matchmaking paradox, *Phys. Rev. E* **81**, 026107 (2010).
52. G. Oshanin and G. Schehr, Two stock options at the races: Black-Scholes forecasts, *Quantitative Finance* **12**, 1325 (2012).
53. G. Oshanin, Y. Holovatch, and G. Schehr, Proportionate vs disproportionate distribution of wealth of two individuals in a tempered Paretian ensemble, *Physica A* **390**, 4340 (2011).
54. C. Mejía-Monasterio, G. Oshanin, and G. Schehr, Symmetry breaking between statistically equivalent, independent channels in few-channel chaotic scattering, *Phys. Rev. E* **84**, 035203 (2011).
55. D. Boyer, D. S. Dean, C. Mejía-Monasterio, and G. Oshanin, Optimal estimates of the diffusion coefficient of a single Brownian trajectory, *Phys. Rev. E* **85**, 031136 (2012).
56. D. Boyer, D. S. Dean, C. Mejía-Monasterio, and G. Oshanin, Distribution of the least-squares estimators of a single Brownian trajectory diffusion coefficient, *J. Stat. Mech.* **04**, P04017 (2013).
57. S. N. Majumdar, Persistence in Nonequilibrium Systems, *Curr. Sci.* **77**, 370 (1999).
58. A. J. Bray, S. N. Majumdar, and G. Schehr, Persistence and first-passage properties in non-equilibrium systems, *Adv. Phys.* **62**, 225 (2013).
59. H. S. Carslaw and J. C. Jaeger, *Conduction of Heat in Solids.* Oxford University Press, Oxford (1959).
60. P. Lévy, Sur certains processus stochastiques homogènes, *Comp. Math.* **7**, 283 (1939).
61. J. J. Linderman and D. A. Lauffenburger, Analysis of intracellular receptor/ligand sorting, *Biophys. J.* **50**, 295 (1986).
62. I. V. Grigoriev, Y. A. Makhnovskii, A. M. Bereshkovskii, and V. Y. Zitserman, Kinetics of escape through a small hole, *J. Chem. Phys.* **116**, 9574 (2002).
63. Z. Schuss, A. Singer, and D. Holcman, The narrow escape problem for diffusion in cellular microdomains, *Proc. Natl. Acad. Sci. USA* **104**, 16098 (2007).
64. O. Bénichou and R. Voituriez, Narrow-Escape Time Problem: Time Needed for a Particle to Exit a Confining Domain through a Small Window, *Phys. Rev. Lett.* **100**, 168105 (2008).

65. G. Oshanin, M. Tamm, and O. Vasilyev, Narrow-escape times for diffusion in microdomains with a particle-surface affinity: Mean-field results, *J. Chem. Phys.* **132**, 235101 (2010).
66. H. C. Berg and E. M. Purcell, Physics of chemoreception, *Biophys. J.* **20**, 193 (1977).

Chapter 10

Exact Record and Order Statistics of Random Walks via First-Passage Ideas

Grégory Schehr and Satya N. Majumdar

Laboratoire de Physique Théorique et Modèles Statistiques (LPTMS), Univ. Paris-Sud, CNRS, 91405 Orsay Cedex, France

While records and order statistics of independent and identically distributed (i.i.d.) random variables $X_1, \cdots, X_N$ are fully understood, much less is known for strongly correlated random variables, which is often the situation encountered in statistical physics. Recently, it was shown, in a series of works, that one-dimensional random walk (RW) is an interesting laboratory where the influence of strong correlations on records and order statistics can be studied in detail. We review here recent exact results which have been obtained for these questions about RW, using techniques borrowed from the study of first-passage problems. We also present a brief review of the well known (and not so well known) results for records and order statistics of i.i.d. variables.

1. Introduction

Records and order statistics are by now a longstanding issue in the fields of engineering [1], finance [2] or environmental sciences [3] where extreme events might have drastic consequences. Indeed, in these contexts, the statistics of extremes have practical applications which include the prediction of probability distributions of extreme floods, the amounts of large insurance losses, equity risk, the size of freak waves, mutation events during evolution, extreme statistics of time series, etc. These notions are very popular in our societies as, for instance, one always hears and reads, in the media, about record breaking events. This is especially true for sports, where world records are always special and noteworthy [4].

More recently, it was realized that records and order statistics play a crucial role in statistical physics. Hence, there has been a surge of interest for these questions in the physics literature. If one considers a discrete time series $X_1, \cdots, X_N$, where X_i's might represent daily temperatures in a given city or the stock prices of a company, a record happens at time k if the k-th entry is larger than all previous entries $X_1, \cdots, X_{k-1}$ (see Fig. 1 left). One is naturally led to ask the following questions: (a) how many records occur in time N ? (b) how long does a record survive ? what is the longest or shortest age of a record ? Such questions and related ones have found applications in various physical situations ranging from domain wall dynamics [5], spin-glasses [6] and random walks [7–11] to avalanches [12], models of stock prices [10, 13] or the study of global warming [14, 15] and also in evolutionary biology [16, 17] (see [18] for a recent review).

Another interesting question about this sequence concerns the fluctuations of the ordered sequence (so called order statistics) obtained by arranging the values of X_i by decreasing order of magnitude, $M_{1,N} > M_{2,N} > \cdots > M_{N,N}$, $M_{k,N}$ being the k-th maximum of this sequence. Questions related to the statistics of the first maximum, $X_{\max} = M_{1,N}$ have emerged in various areas of physics ranging from disordered systems [19–21] and fluctuating interfaces [22–26] to stochastic processes [27], random matrices [28] and many others. While the statistics of the extremum $X_{\max}$ is important another natural question is: is this extremal value isolated, i.e., far away from the others, or is there many other events close to them? Such questions have led to the study of the density of states of near-extreme events [29, 30]. Order statistics is a natural way to characterize this phenomenon of crowding of near-extreme events. A set of useful observables that are naturally sensitive to the crowding of extremum are the gaps between the consecutive ordered maxima: $d_{k,N} = M_{k,N} - M_{k+1,N}$ denoting the k-th gap. Such questions came up in several physical contexts, in particular in the study of the branching Brownian motion [31, 32] and also for $1/f^\alpha$ signal [33], and more recently for random walks [34, 35].

Records and order statistics of i.i.d. random variables are now perfectly well understood [1, 36–38], and we shall briefly review below the main results in this case. The record statistics when the entries X_i's have a non-identical distributions but still retain their independence were also studied in [39–41], in the so called Linear Drift Model. On the other hand the order statistics of weakly correlated random variables reduce, to a large extent, to the case of i.i.d. random variables. However, much less is known for the difficult case where X_i's are strongly correlated, which turns out to be the case of interest in many problems of statistical physics. Recently, it was shown that one-dimensional random walk (RW) is a non-trivial instance of a set of strongly correlated variables for which exact results for records

[7, 10, 11] and order statistics [34, 35] can be obtained. In this paper, we review the main body of these results, which have been obtained, to a large extent, by methods and ideas stemming from first passage problems (for a review on this topic see [42–44]).

The paper is organized as follows. In section 2, we first focus on records statistics while we focus on order statistics in section 3. In each section, we first present a brief overview of well known, and not so well know, results for i.i.d. random variables. This is then followed by the review of results recently obtained for RW. We finally conclude in section 4 with a summary and discussion of open problems.

2. Record statistics

2.1. *Record statistics of i.i.d. random variables*

We start by a short review on standard results for record statistics of i.i.d. random variables. We denote by $X_1, X_2, \cdots, X_N$ a collection of N i.i.d. random variables, distributed according to a continuous probability density function (pdf) $p(x)$. An entry X_k is an upper record if it is larger than all previous entries (see Fig. 1 left):

$$X_k > \max\{X_1, \cdots, X_{k-1}\} \; , \; k \leq N \; . \tag{1}$$

One can similarly define a lower record which is such that $X_k < \min\{X_1, \cdots, X_{k-1}\}$.

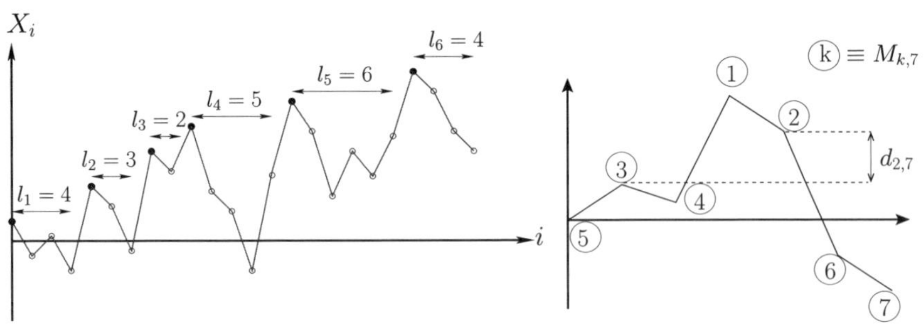

Fig. 1. **Left**: One realization of $N = 24$ random variables X_i's, for which the number of records (the black dots) is $R_{24} = 6$. **Right:** Order statistics of $N = 7$ random variables. $M_{k,7}$ denotes the k-th maximum of the sequence.

In the following, we will focus on upper records (1), which we will simply call "records". Let R_N be the number of records (1) among these N

random variables. We first discuss a straightforward method, based on indicator variables, to investigate the statistics of R_N. Then we discuss more complicated joint probability distributions of the number and the ages of the records. This second method is not only useful to investigate the age of the largest and smallest record but can be generalized, with some appropriate modifications, to the study of the records of random walks.

2.1.1. *Distribution of the number of records*

To study this quantity it is useful to introduce indicator variables σ_k's which take the value 0 or 1.

$$\sigma_k = \begin{cases} 1 \text{ if } X_k \text{ is a record} \;, \\ 0 \text{ if } X_k \text{ is NOT a record} \;, \end{cases} \qquad R_N = \sum_{k=1}^{N} \sigma_k \;. \tag{2}$$

For i.i.d. random variables, these indicator functions σ_k's are independent [see Eq. (8) below]. We define

$$\langle \sigma_k \rangle = r_k \;, \tag{3}$$

where the average is taken over the different realizations of the random variables $X_1, \cdots, X_N$: r_k is thus the probability at which a record is broken, at "time" k. In the following, this probability r_k is called the "record rate". For i.i.d. random variables, it is straightforward to compute the record rate as it is precisely the probability that the event in Eq. (1) happens. This yields

$$r_k = \int_{-\infty}^{\infty} p(x) \left[\int_{-\infty}^{x} p(y) dy \right]^{k-1} dx = \int_0^1 u^{k-1} du = \frac{1}{k} \;, \tag{4}$$

where we have used the change of variable $u = \int_{-\infty}^{x} p(y) dy$. This result $r_k = 1/k$ (4), independently of the parent distribution, can be easily understood: the probability that X_k is the maximum among $X_1, \cdots, X_k$ is indeed $1/k$ as the maximal value can be realized with equal probability by any of these k i.i.d. random variables. From (4), we get the mean number of records as

$$\langle R_N \rangle = \sum_{k=1}^{N} r_k = \sum_{k=1}^{N} \frac{1}{k} = H_N \;, \tag{5}$$

where H_N denotes the N-th Harmonic number. For large N, it behaves as

$$\langle R_N \rangle = \log N + \gamma_E + \mathcal{O}(N^{-1}) \;, \tag{6}$$

where $\gamma_E = 0.57721 \cdots$ is the Euler constant. By a similar calculation, one can compute the variance of the number of records, $\langle R_N^2 \rangle - \langle R_N \rangle^2$. This computation involves the two-point correlations $\langle \sigma_j \sigma_k \rangle$ which, for $j < k$, can be written as

$$\langle \sigma_j \sigma_k \rangle = \int_{-\infty}^{\infty} \left[\int_{-\infty}^{x} \left(\int_{-\infty}^{y} p(y')dy' \right)^{j-1} p(y)dy \right] \times \left[\int_{-\infty}^{x} p(x')dx' \right]^{k-j-1} p(x)\, dx \;, \tag{7}$$

where x denotes the value of the kth record and $y < x$ the value of the jth record. By performing the changes of variables $u = \int_{-\infty}^{x} p(x')dx'$ and $v = \int_{-\infty}^{y} p(y')dy'$, the expression in Eq. (7) can be simply rewritten as

$$\langle \sigma_j \sigma_k \rangle = \int_0^1 u^{k-j-1} \int_0^u v^{j-1} dv\, du = \frac{1}{jk} = \langle \sigma_j \rangle \langle \sigma_k \rangle \;,\; j \neq k \;, \tag{8}$$

which shows that σ_j and σ_k, for $j \neq k$, are indeed independent, as stated above. Using this result (8), one obtains

$$\langle R_N^2 \rangle - \langle R_N \rangle^2 = \sum_{k=1}^{N} \left[\langle \sigma_k^2 \rangle - \langle \sigma_k \rangle^2 \right] = \sum_{k=1}^{N} \left[\frac{1}{k} - \frac{1}{k^2} \right] = \log N + \gamma_E - \frac{\pi^2}{6} + \mathcal{O}(1/N) \;. \tag{9}$$

Similarly, one can compute the generating function (GF) of the probability distribution $P(M|N) = \mathbb{P}(R_N = M)$, where $\mathbb{P}(A = B)$ denotes the probability that A equals B, using (for $N \geq 1$)

$$\sum_{M=1}^{\infty} P(M|N) x^M = \langle x^{R_N} \rangle = \prod_{k=1}^{N} \langle x^{\sigma_k} \rangle = \prod_{k=1}^{N} \left(\frac{1}{k} x + 1 - \frac{1}{k} \right) = \frac{x(x+1)\cdots(x+N-1)}{N!} \;. \tag{10}$$

One recognizes that the rising factorial appearing in (10) is the GF of the unsigned Stirling numbers of the first kind [48]

$$x(x+1)\cdots(x+N-1) = \sum_{M=1}^{N} \begin{bmatrix} N \\ M \end{bmatrix} x^M \;, \tag{11}$$

where the unsigned Stirling numbers $\left[{N \atop M}\right]$ enumerate the number of permutations of N elements with M disjoint cycles exactly. Hence one has

$$P(M|N) = \frac{\left[{N \atop M}\right]}{N!} \;, \tag{12}$$

which thus shows that the number of records of N i.i.d. random variables is distributed like the number of cycles in random permutations of N objects with uniform measure. We will come back later, in section 2.1.5, to this connection with random permutations. Finally, using the asymptotic behaviors of Stirling numbers, one can show that the distribution of R_N approaches, when $N \to \infty$, a Gaussian distribution

$$P(M|N) \sim \frac{1}{\sqrt{2\pi \log N}} \exp\left(-\frac{(M-\log N)^2}{2\log N}\right) . \tag{13}$$

Here we have discussed the case where the random variables X_i's are continuous random variables. We refer the reader to [49] for a discussion of the effects of discreteness, in particular when continuous random variables are subsequently discretized by rounding to integer multiples of a discretization scale.

2.1.2. *Joint distribution of the ages of records and of their number*

Let us consider a realization of these N i.i.d. random variables X_i's, which we consider as a time series, the index i playing the role of discrete time. Let M be the number records in this realization. We denote by $\vec{l} = (l_1, l_2, \cdots, l_M)$ the time intervals between successive records as depicted in Fig. 1. Thus l_i is the age of the i-th record, i. e. it denotes the time up to which the i-th record survives. Note that the last record, the M-th record in this sequence, still stays a record at "time" N. We first compute the joint probability distribution $P(\vec{l}, M|N)$ of the ages $\vec{l}$ and the number M of records, given the length N of the sequence. This joint pdf can be written as

$$P(\vec{l}, M|N) = \int_{-\infty}^{\infty} dy_M p(y_M) \left[\int_{-\infty}^{y_M} p(x)dx\right]^{l_M - 1}$$
$$\times \prod_{k=1}^{M-1} \int_{-\infty}^{y_{k+1}} dy_k p(y_k) \left[\int_{-\infty}^{y_k} p(x)dx\right]^{l_k - 1} \delta_{\sum_{k=1}^M l_k, N} , \tag{14}$$

where the delta function in (14) ensures that the size of the sample is N. If one performs the change of variables $u_k = \int_{-\infty}^{y_k} p(x)dx$, the pdf $P(\vec{l}, M|N)$ in (14) can be written as

$$P(\vec{l}, M|N) = \int_0^1 du_M u_M^{l_M - 1} \prod_{k=1}^{M-1} \int_0^{u_{k+1}} du_k u_k^{l_k - 1} \delta_{\sum_{k=1}^M l_k, N} . \tag{15}$$

This multiple integral in (15) can be performed straightforwardly to obtain

$$P(\vec{l}, M|N) = \frac{1}{l_1(l_1+l_2)\cdots(l_1+l_2+\cdots+l_M)}\delta_{\sum_{k=1}^M l_k, N} \, . \tag{16}$$

Eq. (16) carries more information than just the number of records R_N. Indeed, by summing over $\vec{l}$, one recovers the distribution of R_N given in Eq. (12). In fact, as we show below, this result in (16) can be conveniently used to compute the statistics of the age of longest and shortest records.

2.1.3. *Distribution of the age of the longest record*

We now focus on the age of the longest record, denoted by $l_{\max,N}$, which is defined as

$$l_{\max,N} = \max\{l_1, l_2, \cdots, l_M\} \, . \tag{17}$$

Its cumulative distribution $F(l|N) = \mathbb{P}(l_{\max,N} \leq l)$, $l \leq 1$, is obtained from the full joint pdf (16) by summing over M and $l_1, \cdots, l_M$ with the constraint that $l_1 \leq l$, $\cdots$, $l_M \leq l$. It reads

$$F(l|N) = \sum_{M=1}^{\infty}\sum_{l_1=1}^{l}\cdots\sum_{l_M=1}^{l}\frac{\delta_{\sum_{k=1}^M l_k, N}}{l_1(l_1+l_2)\cdots(l_1+l_2+\cdots+l_M)} \, , \tag{18}$$

while $F(l|0) = 1$. The GF of $F(l|N)$ with respect to (wrt) N is conveniently written using the integral representation of the pdf in (15) as

$$\sum_{N=0} z^N F(l|N) = 1 + \sum_{M=1}^{\infty}\int_0^1 du_M f(u_M)\prod_{k=1}^{M-1}\int_0^{u_{k+1}} du_k f(u_k) \, , \tag{19}$$

$$f(u) = z\sum_{m=1}^{l}(zu)^{m-1} \, . \tag{20}$$

The multiple integral in (19) can be performed by induction in terms of the integral of $f(u)$

$$g(u) = \int_0^u f(v)dv = \sum_{m=1}^{l}\frac{(zu)^m}{m} \, , \tag{21}$$

yielding finally

$$\sum_{N=0}^{\infty} z^N F(l|N) = 1 + \sum_{M=1}^{\infty}\frac{[g(1)]^M}{M!} = \exp\left(\sum_{k=1}^{l}\frac{z^k}{k}\right) \, . \tag{22}$$

From the GF of the full distribution of $l_{\max,N}$ (22) one obtains the GF of the average value $\langle l_{\max,N}\rangle = \sum_{l=1}^{\infty}(1-F(l|N))$ as

$$\sum_{N=0}^{\infty}\langle l_{\max,N}\rangle z^N = \frac{1}{1-z}\sum_{l=1}^{\infty}\left[1-\exp\left(-\sum_{k=l+1}^{\infty}\frac{z^k}{k}\right)\right] . \tag{23}$$

By analysing this expression (23) in the limit $z \to 1$, where the discrete sums can be replaced by integrals (setting $z = e^{-s}$) one obtains the large N behavior of $\langle l_{\max,N}\rangle$ as

$$\langle l_{\max,N}\rangle = c_1 N + \mathcal{O}(1) \ , \ c_1 = \int_0^{\infty} dx(1-e^{-\int_x^{\infty} e^{-y}\frac{dy}{y}}) = 0.62432... \ , \tag{24}$$

where c_1 is the Golomb-Dickman or Goncharov constant [50]. This constant c_1 also describes the linear growth of the longest cycle of a random permutation [50]. The same constant also appeared in a model of growing network [52] and in a one dimensional ballistic aggregation model [47].

2.1.4. *Distribution of the age of the shortest record*

We now focus on the age of the shortest record, denoted as $l_{\min,N}$, which is defined as

$$l_{\min,N} = \min\{l_1, l_2, \cdots, l_M\} . \tag{25}$$

We define $G(l|N) = \mathbb{P}(l_{\min,N} \geq l)$, $l \geq 1$, and using the same reasoning as above for $l_{\max,N}$ we find the GF of $G(l|N) = \mathbb{P}(l_{\min,N} \geq l)$ wrt N as

$$\sum_{N=0}^{\infty} G(l|N) z^N = \exp\left[\sum_{k=l}^{\infty}\frac{z^k}{k}\right] - 1 . \tag{26}$$

The GF of the average value $\langle l_{\min,N}\rangle = \sum_{l=1}^{\infty} G(l|N)$ can be obtained from (26) which yields the asymptotic result for large N [51]

$$\langle l_{\min,N}\rangle = e^{-\gamma_E}\log N + o(\log N) , \tag{27}$$

with the numerical value $e^{-\gamma_E} = 0.5614594835....$

2.1.5. *Connection with random permutations*

As we have seen repeatedly in this section, records statistics bear strong similarities with the statistics of random permutations. The existence of connections between the two fields is well known [53, 54] and they recently showed up in various problems of statistical physics [47, 52]. One of the main manifestations of this connection is that the number of records R_N

for N i.i.d. random variables is distributed like the number of cycles in random permutations of N objects with uniform measure (12). We refer the interested reader to [55] for a more complete discussion of this connection.

2.2. *Record statistics of random walks*

We now study the record statistics of a discrete-time random walker (RW) moving on a continuous line. The position of the RW x_k after k steps evolves via

$$x_k = x_{k-1} + \eta_k \ , \tag{28}$$

starting from $x_0 = 0$ and where the jump variables η_k's are i.i.d. variables, drawn from a distribution $\phi(\eta)$. Here we study the record statistics of a realization of this RW (28) after N steps, $x_0, x_1 \cdots, x_N$ (there are thus $N+1$ random variables in this sequence). As before (1), a record is broken after k steps if $x_k > \max\{x_0, x_1, \cdots, x_{k-1}\}$, with the convention that x_0 is counted as a record. As in the case of i.i.d. variables, we will focus on the number of records R_N as well as on the age of the largest, $l_{\max,N}$, and shortest $l_{\min,N}$ record. As discussed before in the case of i.i.d. random variables, the statistics of these quantities are conveniently obtained from the joint probability distribution $P(\vec{l}, M|N)$ of the ages and the number of records after N time steps. The ages l_i's are thus defined as the number of steps between two records, hence as in the i.i.d. case (see Fig. 1) except that $l_R \to l_R - 1$ (in Fig. 1 one would thus have $l_6 = 5$ for a random walk).

To compute this joint distribution $P(\vec{l}, M|N)$ we need two quantities as inputs [7]. The first one is the probability $q_-(l)$ that a RW, starting in x_0, stays *below* x_0 after l time steps:

$$q_-(l) = \mathbb{P}[x_k < x_0 \ , \ \forall 1 \leq k \leq l] \ . \tag{29}$$

Due to translational invariance, this probability does not depend on x_0 and we can thus set $x_0 = 0$. Its GF, $\tilde{q}_-(z)$, is given by the generalized Sparre Andersen (SA) theorem [46]:

$$\tilde{q}_-(z) = \sum_{k=0}^{\infty} q_-(k) z^k = \exp\left[\sum_{k=1}^{\infty} \frac{z^k}{k} \mathbb{P}(x_k < 0)\right] \ . \tag{30}$$

The second quantity is the first passage probability $f_-(l)$ that the RW crosses its starting point x_0 between steps $(l-1)$ and l from below x_0. Again, $f_-(l)$ is independent of x_0 and one can set $x_0 = 0$. It follows from its definition that $f_-(l) = q_-(l-1) - q_-(l)$ so that its GF can be expressed as

$$\tilde{f}_-(z) = \sum_{l=1}^{\infty} f_-(l) z^l = 1 - (1-z)\tilde{q}_-(z) \; . \tag{31}$$

Armed with these two quantities $q_-(l)$ and $f_-(l)$ we can then write down explicitly the joint distribution of the ages $\vec{l}$ and the number of records M, $P(\vec{l}, M|N)$:

$$P(\vec{l}, M|N) = f_-(l_1) f_-(l_2) \cdots f_-(l_{M-1}) q_-(l_M) \, \delta_{\sum_{k=1}^{M} l_k, N} \; , \tag{32}$$

where we have used the Markov property of the RW which implies that the intervals l_k's are statistically independent, except for an overall global constraint that total length of the interval is N, which is incorporated by the delta function. Note that since the number of records is M, the last interval l_M is not terminated and its pdf is thus $q_-(l_M)$ instead of $f_-(l_M)$. This exact expression (32), together with (30) and (31) is the starting point of the analysis of record statistics of RW [7]. It is the analogue of the expression in (16) obtained in the i.i.d. case.

2.2.1. *Record statistics of a single symmetric random walk*

Continous jump distribution. We first consider the case of symmetric jump distributions, such that $\phi(\eta) = \phi(-\eta)$ and focus, for the moment, on the case where $\phi(\eta)$ is continuous (the case of lattice RW, with discrete jump distribution, will be discussed below). In this case the SA result (30) becomes, thanks to the fact that $\mathbb{P}(x_k < 0) = 1/2$ for any $k \geq 1$:

$$\tilde{q}_-(s) = \tilde{q}(s) = \frac{1}{\sqrt{1-s}} \Longrightarrow q_-(l) = q(l) = \binom{2l}{l} \frac{1}{2^{2l}} \; , \tag{33}$$

independently of the jump distribution $\phi(\eta)$. For large l, one has from (33)

$$q(l) \underset{l \to \infty}{\sim} \frac{1}{\sqrt{\pi l}} \; . \tag{34}$$

One can show that the record rate r_k is given precisely by $r_k = q(k)$ [10]. Therefore we obtain the mean record number $\langle R_N \rangle$ as:

$$\langle R_N \rangle = \sum_{k=0}^{N} r_k = \sum_{k=0}^{N} q(k)$$

$$\langle R_N \rangle = \sum_{k=0}^{N} \binom{2k}{k} \frac{1}{2^{2k}} = (2N+1) \binom{2N}{N} 2^{-2N} \sim \frac{2}{\sqrt{\pi}} \sqrt{N} + \mathcal{O}(N^{-\frac{1}{2}}). \tag{35}$$

It was demonstrated recently that this square root growth $\propto \sqrt{N}$ is robust and remains the same in presence of measurements errors and noise [56]. Note that from the SA theorem (33), $q(l)$ and $f_-(l) = f(l) = q(l-1) - q(l)$ are universal, i.e. independent of the jump distribution: hence the full joint distribution $P(\vec{l}, M|N)$ in (32) and any of its marginals are also universal.

Let us first consider the probability distribution of the number of records $P(M|N) = \mathbb{P}(R_N = M) = \sum_{\vec{l}} P(\vec{l}, M|N)$. From (32) one obtains straightforwardly [7]

$$\sum_{N=M-1}^{\infty} P(M|N) z^N = [\tilde{f}(z)]^{M-1} \tilde{q}(z) = \frac{(1-\sqrt{1-z})^{M-1}}{\sqrt{1-z}} \,, \tag{36}$$

where we have used (31) and (33). From (36) it is possible to obtain the full distribution [7]:

$$P(M|N) = \binom{2N - M + 1}{N} 2^{-2N+M-1} \,, \; M \leq N+1 \,. \tag{37}$$

From (37) we can obtain the mean as in (35) and the variance, which for large N behaves like

$$\langle R_N^2 \rangle - \langle R_N \rangle^2 = 2\left(1 - \frac{2}{\pi}\right) N + \mathcal{O}(\sqrt{N}) \,. \tag{38}$$

It is interesting to compare these results for the RW sequence with that of i.i.d. random variables studied above. In particular, for i.i.d. variables, the fluctuations of R_N (9) are small compared to the mean (6) for large N. In contrast, for the RW sequence, it follows from (35) and (38) that both the mean and the standard deviation grow as $\sqrt{N}$ for $N \gg 1$: thus the fluctuations are large and actually comparable to the mean. This suggests that in the random walk case, at variance with the case of i.i.d. random variables (13), the probability distribution $F(M|N)$ takes the scaling form $F(M|N) \sim (\sqrt{N})^{-1} g_0(M/\sqrt{N})$. This can actually be shown from the analysis of (37) for large N, which yields [7]

$$P(M|N) \sim \frac{1}{\sqrt{N}} g_0\left(\frac{M}{\sqrt{N}}\right) \,, \; g_0(x) = \frac{1}{\sqrt{\pi}} e^{-\frac{x^2}{4}} \,, \; x > 0 \,. \tag{39}$$

What can be said about the statistics of the ages of the records ? The typical age of record $l_{\rm typ}$ can be estimated as $l_{\rm typ} \sim N/\langle R_N \rangle$, which, from (35), thus grows like $l_{\rm typ} \sim \sqrt{4/\pi}\sqrt{N}$. There are however rare records whose age behaves quite differently with N. As was done before in the case of i.i.d. variables we consider the longest lasting record $l_{{\rm max},N}$ in (17) and the shortest duration $l_{{\rm min},N}$ in (25).

We first consider the statistics of $l_{\max,N} = \max\{l_1, \cdots, l_M\}$ and compute the cumulative distribution $F(l|N) = \mathbb{P}(l_{\max,N} \leq l)$. As was done before, it can be computed from the full joint pdf $P(\vec{l}, M|N)$ in (32) by summing up over $l_i \leq l$ and summing up over M. One can thus compute the GF of $F(l|N)$ wrt to N as [7]

$$\sum_{N=0}^{\infty} F(l|N) z^N = \frac{\sum_{k=1}^{l} q(k) z^k}{1 - \sum_{k=1}^{l} f(k) z^k} \ . \tag{40}$$

One can extract, in principle, the expression of $F(l|N)$ from this formula (40). In particular, the asymptotic large N behavior of the average $\langle l_{\max,N} \rangle = \sum_{l=1}^{\infty}[1 - F(l|N)]$ can be extracted explicitly

$$\langle l_{\max,N} \rangle \sim c_2 N \ , \ c_2 = 2 \int_0^{\infty} dy \log\left[1 + \frac{1}{2\sqrt{\pi}}\Gamma(-1/2, y)\right] = 0.626508... \tag{41}$$

Thus the age of the longest record ($\propto N$) is much larger than the typical age ($\propto \sqrt{N}$). Interestingly, the constant c_2, for symmetric random walks (41) is quite close to the Golomb-Dickman or Goncharov's constant c_1 (24) which characterizes the age of the longest record of a i.i.d. sequence. Note however that the origin of universality is quite different in the two problems. Interestingly, the same constant c_2 also appears in the excursion theory of Brownian motion [57]. The precise link between these two problems was shown in [58].

For the shortest lasting record $l_{\min,N} = \min\{l_1, \cdots, l_M\}$, it is also useful to consider the cumulative distribution $G(l|N) = \mathbb{P}(l_{\min,N} \geq l)$. Its GF wrt N is easily obtained from the joint pdf (32) as:

$$\sum_{N=0}^{\infty} G(l|N) z^N = \frac{\sum_{k=l}^{\infty} q(k) z^k}{1 - \sum_{k=l}^{\infty} f(k) z^k} \ . \tag{42}$$

In particular, one can extract from (42) the large N behavior of $\langle l_{\min,N} \rangle$ as [7]

$$\langle l_{\min,N} \rangle \sim \sqrt{\frac{N}{\pi}} \ , \tag{43}$$

which grows in a similar way as that of the typical record, albeit with a smaller prefactor $1/\sqrt{\pi} = 0.56419\cdots$ compared with $\sqrt{\pi/4} = 0.88629\cdots$. Notice also that it grows much faster ($\propto \sqrt{N}$) than in the case of i.i.d. random variables ($\propto \log N$) (27).

Discrete lattice random walks. The above analysis can also be performed for discrete lattice random walks, corresponding to $\phi(\eta) = \frac{1}{2}\delta(\eta +$

$1) + \frac{1}{2}\delta(\eta - 1)$ in (28), except that in this case the expression of $\tilde{q}(z)$ is different from (33) for symmetric jump distributions [this can be seen from Eq. (30) as $\mathbb{P}(x_k = 0) \neq 0$ in this case]. One has then

$$\tilde{q}(z) = \tilde{q}_-(z) = \frac{1}{1-z} - \frac{1-\sqrt{1-z^2}}{z(1-z)} \Longrightarrow q(l) \underset{l\to\infty}{\sim} \frac{\sqrt{2}}{\sqrt{\pi l}} , \tag{44}$$

which differs, by a factor of $\sqrt{2}$ from (34) for the continuous case. In [7], it was shown that $\langle R_N \rangle \sim \sqrt{2N/\pi}$, which is $1/\sqrt{2}$ of the expression for the mean in the continuous case (35). As shown in [10], the number of records R_N is, in this discrete case, directly related to the maximum of the sequence up to step N, $M_N = \max(x_0, \cdots, x_N)$, via the relation $R_N = M_N + 1$. This allows to compute the full distribution of R_N and show [10] that for large N, it takes the scaling form as in Eq. (39), with $g_0(x) = \sqrt{2/\pi} e^{-x^2/2}$. Finally, in [7], it was also found that $\langle l_{\max,N} \rangle \sim c_2 N$ and $\langle l_{\min,N} \rangle \sim \sqrt{2N/\pi}$ which are respectively equal to, and $\sqrt{2}$ times, the corresponding expressions for the continuous case.

2.2.2. *Record statistics of a single random walk with a drift*

Up to now, we have discussed the case of symmetric RWs, where the jump length distribution $\phi(\eta)$ is continuous and symmetric, $\phi(\eta) = \phi(-\eta)$. However, the renewal equation for the joint pdf $P(\vec{l}, M|N)$ in (32), as well as the generalized SA result (30), are still valid for continuous but asymmetric jump distribution. The only difference is that we have to use the appropriate expressions for $q_-(l)$ (30) and $f_-(l)$ (31) instead of $q(l)$ and $f(l)$ in the above expressions for $\langle R_N \rangle$ (35) and for the distributions of $l_{\max,N}$ (40) and $l_{\min,N}$ (42).

In particular, one can study the case of a biased random walk which is constructed from the symmetric random walk x_k in (28) as

$$y_k = x_k + c\,k \Longrightarrow y_k = y_{k-1} + c + \eta_k , \tag{45}$$

where η_k's are i.i.d. variables, drawn from a distribution $\phi(\eta)$: y_k thus represents the position of a discrete-time random walker at step k in presence of a constant drift c. In [12], the authors studied in detail the special case of the Cauchy jump density, $\phi_{\text{Cauchy}}(\eta) = 1/[\pi(1+\eta^2)]$ with arbitrary drift c. In particular it was found that the mean number of records depends algebraically on N with a continuously varying exponent $\theta(c)$ [12]

$$\langle R_N \rangle \sim N^{\theta(c)} , \; \theta(c) = \frac{1}{2} + \frac{1}{\pi}\arctan(c) . \tag{46}$$

On the other hand, the mean number of records $\langle R_N \rangle$ for jump densities with a finite second moment σ^2 and positive drift $c > 0$ was studied in [13], using various approximation schemes. In [11] the authors studied the record statistics of such a biased random walk (45) for arbitrary continuous jump distribution $\phi(\eta)$ such that its Fourier transform $\hat{\phi}(q) = \int_{-\infty}^{\infty} \phi(\eta) e^{iq\eta} d\eta$ behaves, for small q, as

$$\hat{\phi}(q) = 1 - |l_\mu q|^\mu + o(|q|^\mu) , \tag{47}$$

where $0 < \mu < 2$ and l_μ is a typical length scale of the jump. The exponent μ controls the large $|\eta|$ tail of $\phi(\eta)$. For jump density with a well defined second moment $\sigma^2 = \int_{-\infty}^{\infty} d\eta \eta^2 \phi(\eta)$ one has evidently $\mu = 2$, while for $\mu \in (0, 2)$, $\phi(\eta) \sim |\eta|^{-1-\mu}$ for large $|\eta|$. The record statistics of such RW (45) for any value of $\mu \in (0, 2]$ (47) and any drift c was studied exactly in [11]. The analysis performed relied on a detailed study of the behavior of the persistence probability $q_-(l)$ which was found to be very sensitive to these parameters μ and c. This study [11] revealed the existence of five distinct regions in the $(c, 0 < \mu \leq 2)$ strip where R_N, $\langle l_{\max,N} \rangle$ and $\langle l_{\min,N} \rangle$ exhibit very different behaviors. These results are summarized in Table 1.

Table 1. Summary of the main results for the records of RWs with a drift (45), from [11]. The constant c_2 is given in (41) and the exponent $\theta(c)$ is given (46), while the constants $0 < a_\mu(c) < 1$ and $0 < C_{\rm II} < 1$ are non-universal constants given in [11].

	I $\mu \in (0,1)$ $c \in \mathbb{R}$	**II** $\mu = 1$ $c \in \mathbb{R}$	**III** $\mu \in (1,2)$ $c \in \mathbb{R}^+$	**IV** $\mu = 2$ $c \in \mathbb{R}^+$	**V** $\mu \in (1,2]$ $c \in \mathbb{R}^-$
$q_-(N)$	$\propto N^{-\frac{1}{2}}$	$\propto N^{-\theta(c)}$	$\propto N^{-\mu}$	$\propto N^{-\frac{3}{2}} e^{-\frac{c^2 N}{2\sigma^2}}$	$\sim a_\mu(-c)$
$\langle R_N \rangle$	$\propto N^{\frac{1}{2}}$	$\propto N^{\theta(c)}$	$\sim a_\mu(c) N$	$\sim a_2(c) N$	$\sim [a_\mu(-c)]^{-1}$
$\langle l_{\max,N} \rangle$	$\sim c_2\, N$	$\sim C_{\rm II}\, N$	$\propto N^{\frac{1}{\mu}}$	$\propto \log N$	$\sim N$
$\langle l_{\min,N} \rangle$	$\propto N^{\frac{1}{2}}$	$\propto N^{1-\theta(c)}$	$\sim 1 - a_\mu(c)$	$\sim 1 - a_2(c)$	$\sim a_\mu(-c) N$

2.2.3. *Record statistics for multiple random walks*

We conclude this section on records by mentioning results for the records of n symmetric independent RW's each of N steps, which were obtained in [10]. At each time step, each walker jumps by a random length drawn independently from a symmetric and continuous distribution, as in (28). Two cases were considered: (I) when the variance σ^2 of the jump distribution is finite and (II) when σ^2 is divergent as in the case of Lévy flights with index $0 < \mu < 2$ (47). In both cases it was found that the mean record number $\langle R_{N,n} \rangle$ grows universally as $\sim \alpha_n \sqrt{N}$ for large N, but with a very different behavior of the amplitude α_n for $n > 1$ in the two cases. Indeed it was shown that, for large n, $\alpha_n \approx 2\sqrt{\log n}$ independently of σ^2 in case I while, in case II, the amplitude approaches to an n-independent constant for large n, $\alpha_n \approx 4/\sqrt{\pi}$, independently of $0 < \mu < 2$. For finite σ^2 it was argued, and this was confirmed by numerical simulations, that the full distribution of $(R_{N,n}/\sqrt{N} - 2\sqrt{\log n})\sqrt{\log n}$ converges to a Gumbel law [as in Eq. (57) below] as $N \to \infty$ and $n \to \infty$. In case II, numerical simulations indicated that the distribution of $R_{N,n}/\sqrt{N}$ converges, for $N \to \infty$ and $n \to \infty$, to a universal nontrivial distribution, independently of μ, the computation of which remains an open problem. [10] also discussed the applications of these results on records for multiple random walks to the study of the record statistics of 366 daily stock prices from the Standard & Poors 500 index.

3. Order statistics

3.1. *Order statistics of i.i.d. random variables*

Let us first review the standard results for order statistics of i.i.d. random variables. We refer the reader to classical textbooks [37, 38] on the subject for more details (see also [59] for a review). We denote by $X_1, X_2, \cdots, X_N$ a collection of N i.i.d. random variables, distributed according to a probability density function (pdf) $p(x)$. We denote their common cumulative distribution by $P(x) = \int_{-\infty}^{x} p(y)dy$. We define the N order statistics of this sequence by arranging the values of X_i by decreasing order of magnitude (see Fig. 1 right)

$$X_{\max} = M_{1,N} > M_{2,N} > \cdots > M_{N,N} = X_{\min} , \tag{48}$$

where we denote by $X_{\max}$ and $X_{\min}$ the maximum and the minimum among the X_i's.

For i.i.d. random variables, it is possible to write down explicitly the full joint distribution $p_N(m_1, \cdots, m_N)$ of $M_{1,N}, \cdots, M_{N,N}$. To compute it, we first note that, given the realizations of the N order statistics to be $m_1 > m_2 \cdots > m_N$, the original variables X_i's are constrained to take on the values m_i $(i = 1, 2, \cdots, N)$. On the other hand, by symmetry, each of the $N!$ permutations of the X_i's are assigned the same weight. Hence we have [37, 38]

$$p_N(m_1, \cdots, m_N) = N! \prod_{i=1}^{N} p(m_i) \prod_{i=1}^{N-1} \theta(m_i - m_{i+1}) \; , \tag{49}$$

where the product of θ functions ensures the ordering of the variables (we remind that $\theta(x) = 1$ if $x \geq 0$ while $\theta(x) = 0$ if $x < 0$). From this expression (49) one can get, in principle, any characteristic of order statistics of i.i.d. random variables. Here, in addition to the distribution of $M_{k,N}$ we study the gap between two successive maxima (see Fig. 1 right)

$$d_{k,N} = M_{k,N} - M_{k+1,N} \; , \tag{50}$$

which is an interesting observable characterizing the crowding near extreme events [29, 30].

3.1.1. *Finite sample*

We first focus on the pdf $f_{k,N}(m) = \partial_m \mathbb{P}(M_{k,N} \leq m)$ which can be obtained from the full joint pdf (49) by integrating over $m_1 \cdots, m_{k-1}, m_{k+1}, \cdots, m_N$ such that $m_1 > \cdots > m_{k-1} > m > m_{k+1} > \cdots > m_N$:

$$\begin{aligned} f_{k,N}(m) = N!\, p(m) \int_m^\infty dm_{k-1} p(m_{k-1}) \prod_{j=1}^{k-2} \int_{m_{j+1}}^\infty dm_j p(m_j) \\ \times \int_{-\infty}^m dm_{k+1} \prod_{j=k+2}^{N} \int_{-\infty}^{m_{j-1}} dm_j p(m_j) \; , \end{aligned} \tag{51}$$

where $P(m) = \int_{-\infty}^m p(m')dm'$. It is then straightforward to check (for instance by induction) that $f_{k,N}$ in (51) can be written as:

$$f_{k,N}(m) = \frac{N!}{(k-1)!(N-k)!} p(m) \left[P(m)\right]^{N-k} \left[1 - P(m)\right]^{k-1} \; . \tag{52}$$

Note that this formula (52) can also be directly obtained by noticing that the event that $m < M_{k,n} < m + dm$ is the same as the following event where $X_i \geq m + dm$ for $(k-1)$ of the X_i's, $m < X_i < m + dm$ for exactly one of the X_i's and $X_i \leq m$ for the remaining $N-k$ X_i's. In particular for $k=1$ one obtains from (52) the pdf of $X_{\max}$ as

$$f_{1,N} = \frac{\partial}{\partial m}\mathbb{P}(X_{\max} \leq m) = Np(m)\left[P(m)\right]^{N-1} . \tag{53}$$

Similarly the pdf of $X_{\min}$ is obtained by setting $k=N$ in (52).

From the full joint pdf in (49) it is also possible to obtain the joint pdf of $M_{j,N}$ and $M_{k,N}$ and eventually the pdf $p_{k,N}(d)$ of the gap $d_{k,N} = M_{k,N} - M_{k+1,N}$ as

$$p_{k,N}(d) = \theta(d)\frac{N!}{(N-k-1)!(k-1)!}\int_{-\infty}^{\infty} dm_k p(m_k)p(m_k-d) \times \left[P(m_k-d)\right]^{N-k-1}\left[1-P(m_k)\right]^{k-1} . \tag{54}$$

As an example, for the case of exponential i.i.d. random variables X_i's, such that $p(x) = \theta(x)e^{-x}$, one finds from (54)

$$p_{k,N}(d) = \theta(d)ke^{-kd} , \tag{55}$$

independently of N.

3.1.2. *Asymptotic results for large samples*

We now turn to the analysis of these results for i.i.d. variables in the limit of large samples, where N is large. We first focus on $F_{1,N}(m) = \mathbb{P}(X_{\max} \leq m)$ which is known to exhibit a universal behavior when $N \to \infty$. Indeed, one can show that there exist constants a_N and b_N and three distinct families of distributions $G_\rho(z)$, $\rho = \mathrm{I}, \mathrm{II}, \mathrm{III}$ such that

$$\lim_{N\to\infty} F_{1,N}(a_N + b_N z) \to G_\rho(z) \;,\; \rho = \mathrm{I}, \mathrm{II}, \mathrm{III} , \tag{56}$$

where the limiting distribution $G_\rho(z)$, depends only on the large argument of the parent distribution $p(x)$. The large N behavior of extreme value statistics (EVS) of i.i.d. variables is thus characterized by three distinct universality classes: (I) Gumbel, (II) Fréchet and (III) Weibull.

The Gumbel universality class. In this case, the support of $p(x)$ might be bounded or unbounded – though the later is the most commonly encountered. In that case, the Gumbel universality class corresponds to the case where $p(x)$ decays faster than any power law, $p(x) \ll x^{-\eta}$, for any value of $\eta > 0$ and $G_{\mathrm{I}}(z)$ is given by a double exponential

$$G_{\mathrm{I}}(z) = \exp\left[-\exp\left(-z\right)\right] , \tag{57}$$

the so-called Gumbel distribution. The constant a_N is given by the standard relation of EVS

$$1 - P(a_N) = \int_{a_N}^{\infty} p(x)dx = \frac{1}{N} \,, \tag{58}$$

which simply says that there is typically one single variable, the maximum, in the interval $[a_N, +\infty)$. On the other hand, b_N is given by the relation

$$b_N = \frac{\int_{a_N}^{\infty} (x - a_N)p(x)dx}{\int_{a_N}^{\infty} p(x)\, dx} \,, \tag{59}$$

which can be interpreted as the typical distance between $X_{\max}$ and a_N, conditioned to the fact that there is a single variable in $[a_N, +\infty)$. The Gumbel universality class corresponds to the case where, for instance, $p(x)$ is an exponential or a Gaussian distribution. But this also corresponds to the case where $p(x)$ is defined on a bounded support, for instance $x \in [0, 1)$ where $p(x)$ exhibits an essential singularity in $x = 1$, $p(x) \sim \exp\left[-1/(1-x)^{\nu}\right]$, with $\nu > 0$.

The Fréchet universality class. This class corresponds to the case where the support of $p(x)$ is unbounded and where $p(x)$ has a power law tail $p(x) \propto x^{-1-\alpha}$, with $\alpha > 0$. In this case the limiting distribution $G_{\rm II}(z)$ is given by

$$G_{\rm II}(z) = \theta(z) \exp\left[-z^{-\alpha}\right] . \tag{60}$$

Besides, one has in this case $a_N = 0$ while b_N is given by

$$1 - P(b_N) = \int_{b_N}^{\infty} p(x)dx = \frac{1}{N} \,, \tag{61}$$

from which one gets in particular that $b_N \propto N^{\frac{1}{\alpha}}$. This situation corresponds to the case where $p(x)$ is, for instance, a Cauchy distribution or a Pareto distribution.

The Weibull universality class. This corresponds to the situation where the support of $p(x)$ is bounded from above, such that $p(x) = 0$ if $x > x^*$ and $p(x)$ behaves when x approaches x^* as $p(x) \propto (x^* - x)^{\alpha-1}$, $\alpha > 0$. In this case the limiting distribution $G_{\rm III}(z)$ is given by

$$G_{\rm III}(z) = \begin{cases} 1 \,, \; z > 0 \,, \\ \exp\left[-|z|^{\alpha}\right] \,, \; z < 0 \,. \end{cases} \tag{62}$$

In this third case, one naturally has $a_N = x^*$ while b_N is given by

$$\int_{x^*-b_N}^{x*} p(x)dx = \frac{1}{N} \,, \tag{63}$$

from which one gets that $b_N \propto N^{-\frac{1}{\alpha}}$. This universality class includes, for instance, the case where $p(x)$ is a uniform distribution, $p(x) = \theta(x)\theta(1-x)$ (and in this case $\alpha = 1$).

One can now study the limiting behavior of the distribution of the k-th maximum $F_{k,n}(m)$. In this case, depending on the parent distribution $p(x)$, which might belong to one of the three aforementioned universality classes, $\rho = 1, 2, 3$, one can show that [37, 38]

$$\lim_{N\to\infty} F_{k,N}(a_N + b_N z) = G_\rho(z) \sum_{j=0}^{k-1} \frac{[-\log G_\rho(z)]^j}{j!} , \tag{64}$$

$$= \frac{1}{(k-1)!} \int_{[-\ln G_\rho(z)]}^{\infty} e^{-t} t^{k-1} dt , \tag{65}$$

where G_ρ, with $\rho = 1, 2, 3$, is one of the three limiting distributions mentioned above in Eqs. (57), (60) or (62).

A more complete result can also be obtained for the full asymptotic distribution of the vector of the k first maxima [37, 38]

$$\left(\frac{M_{1,N} - a_N}{b_N}, \frac{M_{2,N} - a_N}{b_N}, \cdots, \frac{M_{k,N} - a_N}{b_N}\right) \underset{N\to\infty}{\longrightarrow} (W_1, \cdots, W_k) \tag{66}$$

where the joint pdf of $W_1, \cdots, W_k$ is given by

$$p(w_1, \cdots, w_k) = G_\rho(w_k) \prod_{i=1}^{k} \frac{g_\rho(w_i)}{G_\rho(w_i)} , \quad w_1 > \cdots > w_k , \tag{67}$$

where $g_\rho(z) = G'_\rho(z)$. This expression (67) is already well known. From it we derive the expression for the limiting distribution of the k-th gap $d_{k,N} = M_{k,N} - M_{k+1,N}$, which we have not seen in the literature before. It reads:

$$p_{k,N}(d) \sim \frac{1}{b_N} p_{\text{gap},\rho}\left(\frac{d}{b_N}\right) , \tag{68}$$

where $p_{\text{gap},\rho}(d)$ is given by

$$p_{\text{gap},\rho}(d) = \frac{\theta(d)}{(k-1)!} \int_{-\infty}^{\infty} g_\rho(x) \frac{g_\rho(d+x)}{G_\rho(d+x)} [-\log G_\rho(d+x)]^{k-1} \, dx . \tag{69}$$

In particular, for the Gumbel universality class, one finds simply

$$p_{\text{gap,I}}(d) = \theta(d) k e^{-kd} . \tag{70}$$

For the Fréchet universality class, i.e. $\rho = 2$, one finds from (69):

$$p_{\text{gap,II}}(d) = \theta(d) \frac{\alpha^2}{(k-1)!} \int_0^\infty e^{-x^{-\alpha}} x^{-\alpha-1} (x+d)^{-\alpha k - 1} dx . \tag{71}$$

In particular for large d, it behaves like

$$p_{\mathrm{gap,II}}(d) \underset{d\to\infty}{\sim} \frac{\alpha}{(k-1)!} d^{-\alpha k-1} \; . \tag{72}$$

For $\alpha = 1$, the above integral (71) can be explicitly evaluated

$$p_{\mathrm{gap,II}}(d) = \theta(d) k(k+1) d^{-1-k} \mathrm{U}(k+1, 0, 1/d) \; , \tag{73}$$

where $\mathrm{U}(a,b,z)$ is the confluent (Tricomi) hypergeometric function, which is consistent, for large d with (72) for $\alpha = 1$.

Finally, for the Weibull universality class, one finds

$$p_{\mathrm{gap,III}}(d) = \theta(d) \frac{\alpha^2}{(k-1)!} \int_0^\infty dx (x+d)^{\alpha-1} e^{-(x+d)^\alpha} x^{\alpha k-1} \; , \tag{74}$$

which for large d behaves like

$$p_{\mathrm{gap,III}}(d) \underset{d\to\infty}{\sim} \frac{\alpha^2}{(k-1)!} \alpha^{-k\alpha} \Gamma(\alpha k) d^{(1-\alpha)(\alpha k-2)} e^{-d^\alpha} \; . \tag{75}$$

For $\alpha = 1$, this expression (74) simplifies to yield simply

$$p_{\mathrm{gap,III}}(d) = \theta(d) e^{-d} \; . \tag{76}$$

3.2. *Order statistics of random walks*

As we have seen, the order statistics of i.i.d. random variables is fully understood, thanks in particular to the identification of three different universality classes. In this section, we present recent results for the order statistics of random walks, which offer a non-trivial instance of a set of strongly correlated variables where exact results can be obtained. We will see that the results are quite different from the i.i.d. case.

We thus consider a RW which starts at $x_0 = 0$ at time 0 and evolves via (28) where the η_k's are i.i.d. random jumps each drawn from a symmetric distribution $\phi(\eta)$. We study the fluctuations of the ordered sequence $M_{1,N} > M_{2,N} > \cdots > M_{N+1,N}$ where $M_{k,N}$ is the k-th maximum of the RW after N time steps, hence $k = 1, \cdots, N+1$. The study of order statistics for random walks, beyond the first maximum $X_{\max} = M_{1,N}$, was initiated recently in [34] for the case where the jump distribution $\phi(\eta)$ has a well defined second moment σ^2. In this case, the RW converges, in the limit of a large number of steps N, to the Brownian motion. In [34], it was shown in this case that when $N \to \infty$

$$\frac{\langle M_{k,N} \rangle}{\sigma} = \sqrt{\frac{2N}{\pi}} + \mathcal{O}(1) \; , \tag{77}$$

independently of k. Thus the property of the crowding of extremum (k-dependence) is not captured by the statistics of the maxima $M_{k,N}$ themselves, at least to leading order for large N. The simplest observable that is sensitive to the crowding phenomenon is thus the gap, $d_{k,N} = M_{k,N} - M_{k+1,N}$. The main result of [34] is to show that the statistics of the scaled gap $d_{k,N}/\sigma$ becomes stationary, i.e., independent of N for large N, but retains a rich, nontrivial k dependence which becomes *universal* for large k, i.e. independent of the details of the jump distribution $\phi(\eta)$.

In particular, using the so called Pollaczek-Wendel identity [60, 61], the stationary mean gap $\bar{d}_k = \langle d_{k,\infty} \rangle$ was computed exactly for all k and for arbitrary $\phi(\eta)$ [whose Fourier transform is denoted by $\hat{\phi}(q)$] [34]

$$\bar{d}_k = \langle d_{k,\infty} \rangle = \frac{\sigma}{\sqrt{2\pi}} \frac{\Gamma(k+\frac{1}{2})}{\Gamma(k+1)} - \frac{1}{\pi k} \int_0^\infty \frac{dq}{q^2} \left[[\hat{\phi}(q)]^k - \frac{1}{(1+\frac{\sigma^2}{2}q^2)^k} \right]. \quad (78)$$

In the limit of large k, one finds from (78) that

$$\frac{\bar{d}_k}{\sigma} \sim \frac{1}{\sqrt{2\pi k}}, \quad (79)$$

independently of $\phi(\eta)$. This $k^{-1/2}$ dependence in $\bar{d}_k$ (79) was actually noticed in the numerical study of periodic random walks in [33] and was also conjectured to be exact, based on scaling arguments.

It is natural to wonder about the full distribution of the stationary gap, not only its first moment (78). In [34], this full pdf $p_k(\delta)d\delta = \mathbb{P}(d_{k,\infty} \in [\delta, \delta + d\delta])$ was computed exactly, using backward Fokker-Planck techniques [62], for one particular case of jump variables with an exponential distribution $\phi(\eta) = b^{-1}\exp(-|\eta|/b)$. In the limit of large k, it was shown that there is a scaling regime when $\delta \sim \langle d_{k,\infty} \rangle \simeq \sigma/\sqrt{2\pi k}$ where the pdf scales as, $p_k(\delta) \simeq (\sqrt{k}/\sigma) P(\delta\sqrt{k}/\sigma)$, with a nontrivial scaling function

$$P(x) = 4\Big[\sqrt{\frac{2}{\pi}}(1+2x^2) - e^{2x^2} x(4x^2+3)\mathrm{erfc}(\sqrt{2}x)\Big], \quad (80)$$

where $\mathrm{erfc}(z) = (2/\sqrt{\pi})\int_z^\infty e^{-t^2}\,dt$ is the complementary error function. While it was not possible to compute the gap pdf for arbitrary $\phi(\eta)$, numerical simulations [34] provided strong evidence that the scaling function $P(x)$ in Eq. (80) is actually universal, i.e., independent of $\phi(\eta)$. Somewhat unexpectedly, we find that this universal scaling function has an algebraic tail $P(x) \sim x^{-4}$ for large x. For $\delta \gg \langle d_{k,\infty} \rangle \simeq \sigma/\sqrt{2\pi k}$, the pdf gets cut-off in a nonuniversal fashion. Thus there are two scales associated to $d_{k,\infty}$: a typical fluctuation which is universal and large fluctuations which are non-universal. This is shown to have interesting consequences for the moments of the stationary gap: $\langle d_{k,N}^p \rangle \sim k^{-\frac{p}{2}}$ for $p < 3$, while $\langle d_{k,N}^p \rangle \sim k^{-\frac{3}{2}}$ for $p > 3$.

We end up this section on order statistics of RW by mentioning that exact results have been recently obtained, using first-passage techniques, for the joint distribution $P_N(g, l)$ of the first gap $d_{1,N} = G_N = M_{1,N} - M_{2,N}$ and the time $L_N = n_1 - n_2$ between the occurrence of these first two maxima [35]. This analysis was carried out for any value of the Lévy index $0 < \mu \leq 2$ (47). In particular, it was shown that $P_N(g, l)$ converges to a stationary distribution, i.e. independent of N for large N, which displays a very rich behavior as a function of g and l as μ is varied.

4. Conclusion

To conclude, after a brief review on records and order statistics for i.i.d. random variables, we have presented the main results which were recently obtained for records and order statistics of RW, using first-passage concepts. A striking feature of these statistics for N i.i.d. random variables is their universality with respect to their common parent distribution. For records, universality shows up, to a large extent, already for any finite N. This can be seen, for instance, through their connection with the statistics of random permutations. For extreme and order statistics, universality only appears in the (thermodynamical) limit of large N, thanks to the existence of three distinct universality classes (Gumbel, Fréchet and Weibull). What is left of this universal behavior in the presence of strong correlations is an important question. Quite interestingly, for RW with symmetric and continuous jump distribution $\phi(\eta)$, the records statistics do not depend on the details of $\phi(\eta)$ (including Lévy RW such that $\phi(\eta) \sim |\eta|^{-1-\mu}$ with $0 < \mu < 2$), even for a finite number of steps. This universality is due here to the Sparre Andersen theorem. In the presence of a drift c, universal behavior also emerges but only in the limit of a large number of steps $N \to \infty$. However in this case, this asymptotic behavior depends on both c and the Lévy index μ (see Table 1).

On the other hand, order statistics of RW is quite sensitive to the jump distribution $\phi(\eta)$. For instance, the distribution of the gap $d_{k,N}$ for finite k and N is generically quite sensitive on $\phi(\eta)$. However, for large k and large N, a scaling regime was identified when $d_{k,N} \sim 1/\sqrt{k}$ where the fluctuations are universal and described by a universal scaling function (80), at least in the case where the jump distribution $\phi(\eta)$ has a finite second moment. The statement of the universality of this scaling regime is based on (i) exact calculation for the case of exponential jumps, (ii) numerical simulations. It will be interesting to prove for arbitrary distribution this universal behavior on firmer grounds. In the case where the jump distribution $\phi(\eta)$ does not have a second moment, like for Lévy flights, exact results for the order statistics beyond the first gap [35] is an open question. Finally, it will be interesting to extend this study of records and order statistics to other stochastic processes, in particular non-Markovian ones.

References

1. E. J. Gumbel, *Statistics of Extremes* (Dover), (1958).
2. P. Embrecht, C. Klüppelberg and T. Mikosh, *Modelling Extremal events for insurance and finance* (Springer), Berlin (1997).
3. R. W. Katz, M. P. Parlange and P. Naveau, Statistics of extremes in hydrology, *Adv. Water Resour.* **25**, 1287–1304 (2002).
4. D. Gembris, J. G. Taylor and D. Suter, Sports statistics: Trends and random fluctuations in athletics, *Nature* **417**, 506 (1pp.) (2002).
5. B. Alessandro, C. Beatrice, G. Bertotti and A. Montorsi, Domain-wall dynamics and Barkhausen effect in metallic ferromagnetic materials. I. Theory, *J. Appl. Phys.* **68**, 2901–2907 (1990).
6. P. Sibani, Linear response in aging glassy systems, intermittency and the Poisson statistics of record fluctuations, *Eur. Phys. J. B.* **58**(4), 438–491 (2007).
7. S. N. Majumdar and R. M. Ziff, Universal record statistics of random walks and Lévy flights, *Phys. Rev. Lett.* **101**, 050601 (4pp.) (2008).
8. S. N. Majumdar, Universal first-passage properties of discrete-time random walks and Lévy flights on a line: Statistics of the global maximum and records, *Physica A* **389**, 4299–4316 (2010).
9. S. Sabhapandit, Record Statistics of Continuous Time Random Walk, *Europhys. Lett.* **94**, 20003 (5pp.) (2011).
10. G. Wergen, S. N. Majumdar, G. Schehr, Record statistics for multiple random walks, *Phys. Rev. E* **86**, 011119 (18pp.) (2012).
11. S. N. Majumdar, G. Schehr and G. Wergen, Record statistics and persistence for a random walk with a drift, *J. Phys. A: Math. Theor.* **45**, 355002 (42 pp.) (2012).
12. P. Le Doussal and K. J. Wiese, Driven particle in a random landscape: Disorder correlator, avalanche distribution, and extreme value statistics of records, *Phys. Rev. E* **79**, 051105 (40 pp.) (2009).
13. G. Wergen, M. Bogner and J. Krug, Record statistics for biased random walks, with an application to financial data, *Phys. Rev. E* **83**, 051109 (6pp.) (2011).

14. R. Redner and M.R. Petersen, Role of global warming on the statistics of record-breaking temperatures, *Phys. Rev. E* **74**, 061114 (14pp.) (2006).
15. G. Wergen, J. Krug, Record-breaking temperatures reveal a warming climate, *Europhys. Lett.* **92**, 30008 (6pp.) (2010).
16. J. Krug and K. Jain, Breaking records in the evolutionary race, *Physica A* 358, 1–9 (2005).
17. J. Franke, A. Klözer, J. Arjan G. M. de Visser, J. Krug, Evolutionary accessibility of mutational pathways, *PLos Comp. Biol.* **7**, e1002134 (9pp.) (2011).
18. G. Wergen, Records in stochastic processes – theory and applications, *J. Phys. A: Math. Theor.* **46**, 223001 (35pp.) (2013).
19. J. P. Bouchaud and M. Mézard, Universality classes for extreme value statistics, *J. Phys. A* **30**, 7997–8015 (1997).
20. P. Le Doussal and C. Monthus, Exact solutions for the statistics of extrema of some random 1D landscapes, application to the equilibrium and the dynamics of the toy model, *Physica A* **317**, 140–198 (2003).
21. M. Leblanc, L. Angheluta, K. Dahmen, and N. Goldenfeld, Universal fluctuations and extreme statistics of avalanches near depinning transitions, *Phys. Rev. E* **87**, 022126 (2013).
22. S. Raychaudhuri, M. Cranston, C. Przybla and Y. Shapir, Maximal height scaling of kinetically growing surfaces, *Phys. Rev. Lett.* **87**, 136101 (4pp.) (2001).
23. G. Gyorgyi, P. C. Holdsworth, B. Portelli and Z. Racz, Statistics of extremal intensities for Gaussian interfaces, *Phys. Rev. E* **68**, 056116 (2003).
24. S. N. Majumdar and A. Comtet, Exact maximal height distribution of fluctuating interfaces, *Phys. Rev. Lett.* **92**, 225501 (4pp.) (2004).
25. S. N. Majumdar and A. Comtet, Airy distribution function: from the area under a brownian excursion to the maximal height of fluctuating interfaces, *J. Stat. Phys.* **119**, pp. 777–826, (2005).
26. G. Schehr, S. N. Majumdar, Universal asymptotic statistics of maximal relative height in one-dimensional solid-on-solid models, *Phys. Rev. E* **73**, 056103 (10 pp.) (2006).
27. C. Sire, Probability distribution of the maximum of a smooth temporal signal, *Phys. Rev. Lett.* **98**, 020601 (4pp.) (2007).
28. C. A. Tracy and H. Widom, Level spacing distributions and the Airy kernel, *Commun. Math. Phys.* **159**, 151–174 (1994).
29. S. Sabhapandhit and S. N. Majumdar, Density of near-extreme events, *Phys. Rev. Lett.* **98**, 140201 (4pp.) (2007).
30. S. Sabhapandit, S. N. Majumdar and S. Redner, Crowding at the front of marathon packs, *J. Stat. Mech.* L03001 (2008).
31. E. Brunet, B. Derrida, Statistics at the tip of a branching random walk and the delay of traveling waves, *Europhys. Lett.* **87**, 60010 (5pp.) (2009).
32. E. Brunet, B. Derrida, A branching random walk seen from the tip, *J. Stat. Phys.* **143**, pp. 420-446 (2011).
33. N. R. Moloney, K. Ozogány, Z. Rácz, Order statistics of $1/f^{\alpha}$ signals, *Phys. Rev. E* **84**, 061101 (8pp.) (2011).
34. G. Schehr and S. N. Majumdar, Universal order statistics of random walks,

Phys. Rev. Lett. **108**, 040601 (4pp.) (2012).
35. S. N. Majumdar, Ph. Mounaix and G. Schehr, Exact statistics of the gap and time interval between the first two maxima of random walks and Lévy flights, *Phys. Rev. Lett.* **111**, 070601 (5pp.) (2013).
36. V. B. Nevzorov, *Records: Mathematical Theory*, Am. Math. Soc. (2004).
37. B. C. Arnold, N. Balakrishnan and H. N. Nagaraja, *A first course in order statistics*, Wiley, New York (1992).
38. H. N. Nagaraja, H. A. David, *Order statistics* (third ed.), Wiley, New Jersey (2003).
39. R. Ballerini, S. Resnick, Records from improving populations, *J. Appl. Probab.* **22**, 487–502 (1985).
40. J. Krug, Records in a changing world, *J. Stat. Mech.* P07001 (13pp.) (2007).
41. I. Eliazar and J. Klafter, Record events in growing populations: universality, correlation, and aging, *Phys. Rev. E* **80**, 061117 (7pp.) (2009).
42. S. Redner, *A guide to first-passage processes*, Cambridge University Press, Cambridge, (2001).
43. S. N. Majumdar, Persistence in nonequilibrium systems, *Curr. Sci.* **77**, pp. 370–375 (1999).
44. A. J. Bray, S. N. Majumdar and G. Schehr, Persistence and first-passage properties in nonequilibrium systems, *Adv. Phys.* **62**(3), pp 225-361 (2013).
45. E. Sparre Andersen, On the fluctuations of sums of random variables I, *Math. Scand.* **1**, pp. 263–285 (1953).
46. E. Sparre Andersen, On the fluctuations of sums of random variables II, *Math. Scand.* **2**, pp. 195–233 (1954).
47. S. N. Majumdar, K. Mallick and S. Sabhapandit, Statistical properties of the final state in one-dimensional ballistic aggregation, *Phys. Rev. E* **79**, 021109 (14pp.) (2009).
48. J. Riordan, *Introduction to combinatorial analysis*, Dover, New-York (2002).
49. G. Wergen, D. Volovik, S. Redner and J. Krug, Rounding Effects in Record Statistics, *Phys. Rev. Lett.* 109, 164102 (5pp.) (2012).
50. S. R. Finch, *Mathematical constants*, Cambridge University Press, pp. 284–292 (2003).
51. L. A. Shepp and S. P. Lloyd, Ordered cycle lengths in a random permutation, *Trans. Amer. Math. Soc.* **121**, pp. 340–357 (1966).
52. C. Godrèche and J.-M. Luck, A record-driven growth process, *J. Stat. Mech.*, P11006, (30 pp.) (2008).
53. A. Renyi, Théorie des éléments saillants d'une suite d'observations, *Colloquium on Combinatorial Methods in Probability Theory*, (Math. Inst. Aarhus Univ., Aarhus, Denmark), pp. 104117 (1962).
54. C. M. Goldie, Records, permutations and greatest convex minorants, *Math. Proc. Camb. Phil. Soc.* **106**, pp. 169-177 (1989).
55. P. Flajolet and R. Sedgewick, *Analytic combinatorics*, Cambridge University Press, Cambridge, (2009).
56. Y. Edery, A. B. Kostinski, S. N. Majumdar and B. Berkowitz, Record-breaking statistics for random walks in the presence of measurement error and noise, *Phys. Rev. Lett.* **110**, 180602 (2013).

57. J. Pitman and M. Yor, The two-parameter Poisson-Dirichlet distribution derived from a stable subordinator, *Ann. Probab.* **25**, 855–900 (1997).
58. C. Godrèche, S. N. Majumdar and G. Schehr, The longest excursion of stochastic processes in nonequilibrium systems, *Phys. Rev. Lett.* **102**, 240602 (4pp.) (2009).
59. E. Bertin and M. Clusel, Generalised extreme value statistics and sum of correlated variables, *J. Phys. A* **39**, 7607–7620 (2006).
60. F. Pollaczeck, Fonctions caractéristiques de certaines répartitions définies au moyen de la notion d'ordre. Application à la théorie des attentes, *C. R. Acad. Sci. Paris* **234**, 2334–2336 (1952).
61. J. G. Wendel, Order statistics of partial sums, *Ann. Math. Statist.* **31**, 1034–1044 (1960).
62. A. Comtet, S. N. Majumdar, Precise asymptotics for a random walker's maximum, *J. Stat. Mech.* P06013, (2005).

Chapter 11

First Passage in Conical Geometry and Ordering of Brownian Particles

E. Ben-Naim[1] and P. L. Krapivsky[2]

[1] *Theoretical Division and Center for Nonlinear Studies, Los Alamos National Laboratory, Los Alamos, New Mexico 87545, USA*

[2] *Department of Physics, Boston University, Boston, Massachusetts 02215, USA*

We survey recent results on first-passage processes in unbounded cones and their applications to ordering of particles undergoing Brownian motion in one dimension. We first discuss the survival probability $S(t)$ that a diffusing particle, in arbitrary spatial dimension, remains inside a conical domain up to time t. In general, this quantity decays algebraically $S \sim t^{-\beta}$ in the long-time limit. The exponent β depends on the opening angle of the cone and the spatial dimension, and it is root of a transcendental equation involving the associated Legendre functions. The exponent becomes a function of a single scaling variable in the limit of large spatial dimension. We then describe two first-passage problems involving the order of N independent Brownian particles in one dimension where survival probabilities decay algebraically as well. To analyze these problems, we identify the trajectories of the N particles with the trajectory of one particle in N dimensions, confined to within a certain boundary, and we use a circular cone with matching solid angle as a replacement for the confining boundary. For $N = 3$, the confining boundary is a wedge and the approach is exact. In general, this "cone approximation" gives strict lower bounds as well as useful estimates for the first-passage exponents. Interestingly, the cone approximation becomes asymptotically exact when $N \to \infty$ as it predicts the exact scaling function that governs the spectrum of first-passage exponents.

1. Introduction

Triggering of earthquakes as stress surpasses a threshold [1], onset of avalanches as mass accumulates beyond stability [2], outbreak of an epidemic once the number infected individuals exceeds a threshold [3], and execution of financial transactions once stock prices reach a prescribed value [4]: All are vivid examples of first-passage processes. Indeed, first-passage processes are ubiquitous and are becoming increasingly relevant for understanding stochastic processes in the physical and the natural sciences [5–7].

A first-passage process characterizes how a stochastic variable reaches a fixed threshold for the *first time*. The survival probability $S(t)$ that the threshold is not reached up to time t is a central quantity in the theory of first-passage processes [6, 7]. Basic characteristics follow directly from the survival probability, e.g., $\langle t\rangle = -\int_0^\infty dt\, t\,(dS/dt)$ measures the average duration of the process; this quantity may be finite or infinite. The first-passage probability that the threshold is ever reached equals $1 - S(\infty)$; when $S(\infty) > 0$, the threshold may never be reached.

The widely-known gambler ruin problem which involves the location of a simple random walk on a line is a classic first-passage problem. This process ends when the wealth of the gambler reaches either zero or a high threshold [6, 7]. First-passage properties of multiple random walks [8, 9] and particles undergoing Brownian motion [10–12] in one dimension are the subject of ongoing research and underlie dynamics of spin chains [13–15], voting processes [16–18], urn model [19], and unraveling of knots [20]. Many of these first-passage processes are closely related to the ordering of a set of independent Brownian particles on a line [21–32].

When the number of random walks is no larger than three, first-passage processes involving the order of the particles map onto diffusion in a two-dimensional wedge [7, 21, 27, 28]. For an arbitrary number of particles, a similar mapping remains useful. We recently showed that first-passage processes in high-dimensional cones are connected with a host of problems involving the order of N Brownian particles on a line [33–35]. In this article, we review these results.

We first discuss the survival probability $S(t)$ that a particle diffusing in an infinite circular $d-$dimensional cone with opening angle α does not cross the cone boundary during the time interval $(0, t)$ [36–43]. The survival probability decays algebraically with time

$$S \sim t^{-\beta} \quad \text{as} \quad t \to \infty. \tag{1}$$

Since $S(\infty) = 0$, the surface of the cone is eventually reached. The exponent β turns out to be a root of a transcendental equation involving the associated Legendre functions. We also compute the mean first-passage time which is finite when $\beta > 1$.

In principle, the first-passage exponent β depends on the opening angle α and the spatial dimension d. For $d \gg 1$, the exponent becomes a function of a single scaling variable [44] which is a special combination of α and d. The exponent β is root of the equation [33]

$$D_{2\beta}(y) = 0 \quad \text{with} \quad y = (\cos\alpha)\sqrt{d}, \tag{2}$$

involving the Parabolic cylinder function D_ν [51]. This scaling behavior shows that β is of order unity only when the cone is nearly flat (close to a half space). We also discuss a variety of asymptotic properties, e.g., the behavior in the interior and the exterior of very thin cones.

Next, we discuss first-passage processes involving the order N independent Brownian particles in one dimension. We tag the particles according to initial position. Regardless of the initial conditions, the order of the particles eventually gets "scrambled" and since diffusion is an ergodic process, all N permutations of particle order become equally probable. We show that there is a host of interesting first-passage questions that characterize how the initial order unravels.

For example, we rank the particles from right to left according to initial position with rank $n = 1$ corresponding to the rightmost particle (the "leader") and rank $n = N$ to the leftmost particle (the "laggard"). We ask: what is the survival probability $R_n(t)$ that the rank of the initial leader at time $t = 0$ does not fall below n until time t. The quantity R_1 gives the probability that the leader kept the lead, while R_{N-1} is the probability that the leader is never last [24, 26, 28]. In general, the survival probabilities decay algebraically,

$$R_n \sim t^{-\gamma_n}, \tag{3}$$

in the long-time limit. Interestingly, there is a family of $N - 1$ distinct first-passage exponents with $\gamma_1 > \gamma_2 > \cdots > \gamma_{N-1}$.

The trajectories of the N one-dimensional Brownian particles can be identified with that of a single compound particle undergoing ordinary Brownian motion in N dimensions. Then R_n is the probability that this particle remains within some domain in this higher-dimensional space. When $N \leq 3$, this domain is always a wedge, and the first-passage exponents can be obtained analytically. For arbitrary N, we approximate this domain using an unbounded cone that encloses an identical solid angle. This approach gives a useful approximation for the decay exponents γ_n. When the number of particles is very large, we exploit the scaling rela-

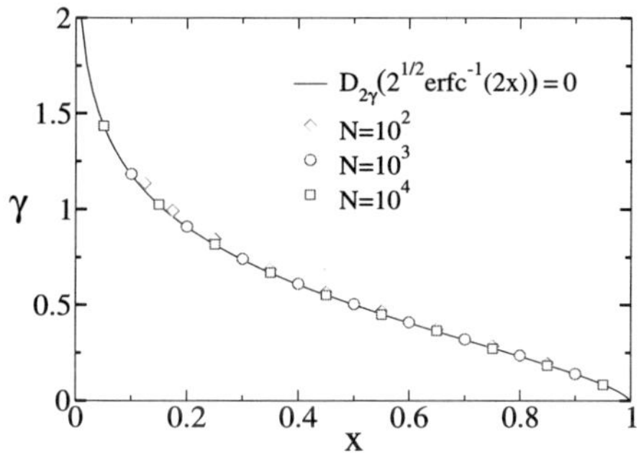

Fig. 1. The exponent γ versus the scaling variable x. Results of Monte Carlo simulations with $N = 10^2$, $N = 10^3$, and $N = 10^4$ particles, are compared with the theoretical prediction (4).

tionship (2) and find that the exponents γ_n, which depend on the threshold n and the number of particles N, become a function of a single scaling variable n/N. The exponent γ is root of the transcendental equation [34]

$$D_{2\gamma}\left(\sqrt{2}\,\mathrm{erfc}^{-1}(2x)\right) = 0 \qquad \text{with} \qquad x = n/N, \tag{4}$$

involving the parabolic cylinder function and the inverse complementary error function. As shown in figure 1, numerical simulations reveal that this result is asymptotically exact: The scaling exponent γ_n is a function of the scaling variable $x = n/N$, and the scaling function is given by (4).

The rest of this paper is organized as follows. In section 2, we provide a pedagogical discussion of first-passage in a two-dimensional wedge. We derive the first-passage exponent directly from the diffusion equation governing the survival probability and show that problems involving particle order map onto a two dimensional wedge when $N = 3$. We then generalize this theoretical description to arbitrary spatial dimensions and obtain the first-passage exponent in section 3. Relevant asymptotic relations including in particular the scaling function underlying β in the infinite-dimension limit are obtained in section 4. In section 5, we develop the cone approximation, using a three-dimensional corner geometry as an illustrative example. We then discuss rank statistics and use the cone approximation to estimate the first-passage exponents and to derive their asymptotic behavior (section 6). In section 7, we introduce the number of pair inversions as a measure of order. Again, we find that the cone approximation yields useful estimates for the first-passage exponents and that it becomes asymptotically exact when the number of particles diverges. We conclude in section 8.

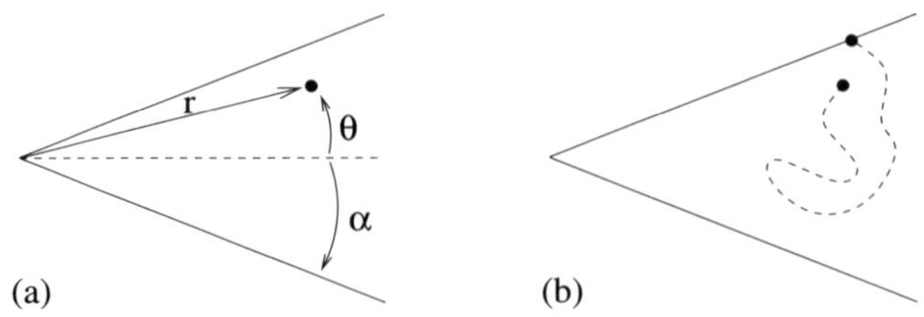

Fig. 2. (a) A wedge with opening angle α. The radial coordinate r and the polar coordinate θ specify the initial location of the Brownian particle. (b) Illustration of the first-passage process: the trajectory of a Brownian particle is indicated by a dashed line while the start and end points are indicated by bullets.

2. Wedges

Consider a particle undergoing simple diffusion in a wedge (Fig. 2a) with an opening angle α. The angle α is measured from the center of the wedge and hence $\alpha \leq \pi$. The first-passage process ends when the particle reaches the boundary of the wedge (Fig. 2b).

The survival probability that the particle avoids the wedge boundary up to time t depends on its initial location (r, θ) of the particle: $S \equiv S(r, \theta, t)$ where r is the distance to the cone apex and θ is the polar angle (Fig. 2a). Since the particle undergoes simple Brownian motion, the survival probability obeys the diffusion equation [8]

$$\frac{\partial S(r, \theta, t)}{\partial t} = D\nabla^2 S(r, \theta, t), \tag{5}$$

where D is the diffusion coefficient. The boundary condition $S(r, \alpha, t) = 0$ reflects that the first-passage process ends when the particle reaches the boundary, and the initial condition is $S(r, \theta, t = 0) = 1$ for all $\theta < \alpha$.

The survival probability decays algebraically

$$S(r, \theta, t) \simeq \Phi(r, \theta)\, t^{-\beta}, \tag{6}$$

in the long-time limit. The algebraic dependence on time is well-known as for $\alpha = \frac{\pi}{2}$ the problem is effectively one-dimensional and $\beta = \frac{1}{2}$. Here, we show how the decay exponent β can be conveniently derived from (5). By substituting the algebraic decay (6) into the diffusion equation (5), we conclude that the amplitude Φ obeys Laplace's equation

$$\nabla^2 \Phi(r, \theta) = 0. \tag{7}$$

The boundary conditions are $\Phi(r, \alpha) = \Phi(0, \theta) = 0$. The amplitude must be positive inside the wedge, $\Phi(r, \theta) > 0$ for $\theta < \alpha$. The Laplace equation

(7) manifests the direct connection between diffusive first-passage processes and electrostatics [9, 45].

Since the quantity S is dimensionless, and since the only dimensionless combination of the quantities r, D, and t is Dt/r^2, the amplitude necessarily has the form

$$\Phi(r,\theta) = \left(\frac{r^2}{D}\right)^{\beta}\psi(\theta). \tag{8}$$

The function $\psi \equiv \psi(\theta)$ depends on the polar angle θ alone. This function is positive inside the wedge, and it vanishes on the boundary, $\psi(\alpha) = 0$. We now substitute (8) into (7) and use

$$\nabla^2 \equiv \frac{\partial^2}{\partial r^2} + \frac{1}{r}\frac{\partial}{\partial r} + \frac{1}{r^2}\frac{\partial^2}{\partial \theta^2}$$

to find that $\psi(\theta)$ obeys the simple eigenvalue equation

$$\frac{\partial^2\psi}{\partial\theta^2} + (2\beta)^2\psi = 0. \tag{9}$$

Equation (9) involves only the angular component of the Laplacian. By symmetry, we expect that the function $\psi(\theta)$ is even, $\psi(\theta) = \psi(-\theta)$. The solution to Eq. (9) is $\psi(\theta) = \cos(2\beta\theta)$, and the first-passage exponent [46]

$$\beta = \frac{\pi}{4\alpha} \tag{10}$$

follows immediately from boundary condition $\psi(\alpha) = 0$. We see that the asymptotic behavior (6) with the exponent (10) holds for all initial conditions. The initial location affects only the amplitude $\Phi(r,\theta)$. Furthermore, using (6) and (8) we conclude that $S \sim r^{2\beta}$.

The exponent is minimal for a needle, $\beta(\pi) = 1/4$ [47], and hence, in two dimensions the particle is bound to encounter a semi-infinite needle [47]. The exponent $\beta(\alpha)$ diverges for infinitely thin wedges, and the duration of the first-passage process is infinitesimal in the limit $\alpha \to 0$. Finally, since $\beta = 1$ when $\alpha = \pi/4$, the mean first-passage time is finite if and only if the wedge is smaller than a square corner.

First-passage problems involving the order of three particles in one-dimension map onto a wedge. Let $x_i(t)$ be the location of the ith particle at time t. Without loss of generality, we label the particles according to the initial positions, $x_1(0) > x_2(0) > x_3(0)$. Furthermore, we can view the three coordinates (x_1, x_2, x_3) as those of a "composite" particle in three dimensions. Since each particle undergoes independent Brownian motion, the composite particle undergoes Brownian motion in three dimensions.

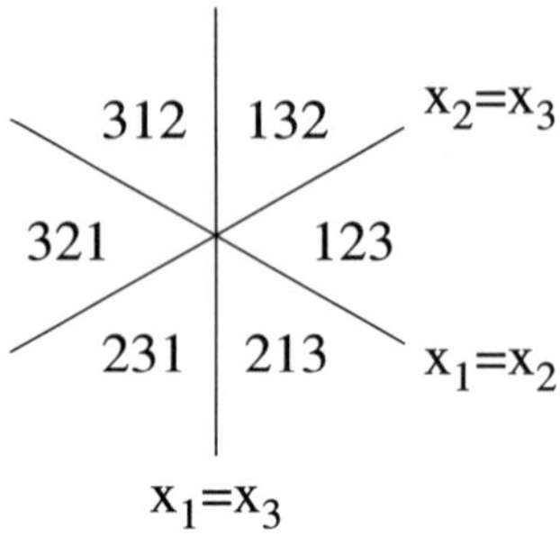

Fig. 3. A projection of three dimensional space onto the plane perpendicular to the diagonal line $x_1 = x_2 = x_3$ divides space into six equal wedges, each with distinct particle order.

The three planes $x_1 = x_2$, $x_1 = x_3$, and $x_2 = x_3$, divide space into six equal wedges [28, 48] with opening angle $\alpha = \pi/6$ (figure 3), and in each wedge the particle order is distinct.

Consider the rank of the initial rightmost particle. Let R_n be the survival probability that this rank does not fall below n up to time t. Figure 4a shows the rank of the leader in each of the six wedges. In two of the wedges $n \leq 1$ and in four of them $n \leq 2$. Hence, the survival probabilities R_1 and R_2 are equivalent to the probability that a diffusing particle remains inside a wedge with opening angles $\alpha = \pi/3$ and $\alpha = 2\pi/3$ respectively. Immediately, we obtain the asymptotic behaviors

$$R_1 \sim t^{-3/4} \qquad \text{and} \qquad R_2 \sim t^{-3/8}. \tag{11}$$

Using the definition (3), the set of exponents is $\{\gamma_1, \gamma_2\} = \{3/4, 3/8\}$. Of course, if there are only two particles, the problem is trivially equivalent to a wedge with opening angle $\alpha = \pi/2$ and the first passage exponent is $\gamma_1 = 1/2$. Based on these two examples, we expect that generally there are $N-1$ exponents $\gamma_n(N)$ that obey $\gamma_1 > \gamma_2 > \cdots > \gamma_{N-1}$.

Figure 3 shows that there is a set of six first-passage exponents

$$\beta_l = \frac{3}{2\,l}, \tag{12}$$

where $l = 1, 2, \ldots, 6$ is the number of wedges. Specifically, the set is $\{1/4, 3/10, 3/8, 1/2, 3/4, 3/2\}$, although for the rank problem, only two of these values are realized.

There are additional measures of particle order, each with a different first-passage process. The rank characterizes only the location of one particle with respect to the rest of the particles. The number of pair inversions,

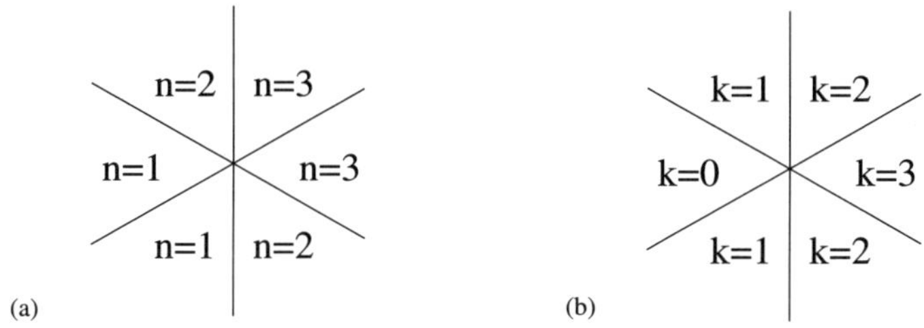

Fig. 4. (a) The rank n of the initial rightmost particle in each of the six wedges shown in figure 3. (b) The number of inversion k in each of the six wedges.

k, measures the number of pairs of particles that are inverted with respect to the initial configuration [49, 50]. For example, if the order changes from 321 initially to 312 later, then number of inversions increases from $k = 0$ to $k = 1$ because the pair 12 is inverted with respect to the initial configuration. The number of inversions is maximal, $k = 3$, when the particle order is completely reversed ($321 \to 123$) and hence $0 \leq k \leq 3$.

What is the survival probability M_k that the number of pair inversions remains smaller than k up to time t? When $k = 0$, the trajectory of the composite particle is confined to a single wedge, $l = 1$. Similarly, the condition $k \leq 1$ corresponds to $l = 3$, and the condition $k \leq 2$ corresponds to $l = 5$. Using equation (12), we find

$$M_1 \sim t^{-3/2}, \qquad M_2 \sim t^{-1/2}, \qquad M_3 \sim t^{-3/10}. \tag{13}$$

In section 7, we show that the number of distinct exponents grows *quadratically* with the number of particles.

3. Circular cones

An infinite circular cone in arbitrary spatial dimension $d > 1$ is specified by the opening angle $\alpha < \pi$, that is, the angle between the cone surface and its axis (see figure 5). The exterior of a cone with opening angle α is a cone with opening angle $\pi - \alpha$. In all dimensions, a cone with opening angle $\alpha = \pi/2$ is a half space.

The survival probability of a diffusing particle again depends only on two variables: the radial coordinate r and the polar angle θ (see figure 5). The survival probability obeys the diffusion equation (5) with the Laplacian

$$\nabla^2 \equiv \frac{\partial^2}{\partial r^2} + \frac{d-1}{r}\frac{\partial}{\partial r} + \frac{1}{r^2(\sin\theta)^{d-2}}\frac{\partial}{\partial \theta}(\sin\theta)^{d-2}\frac{\partial}{\partial \theta}. \tag{14}$$

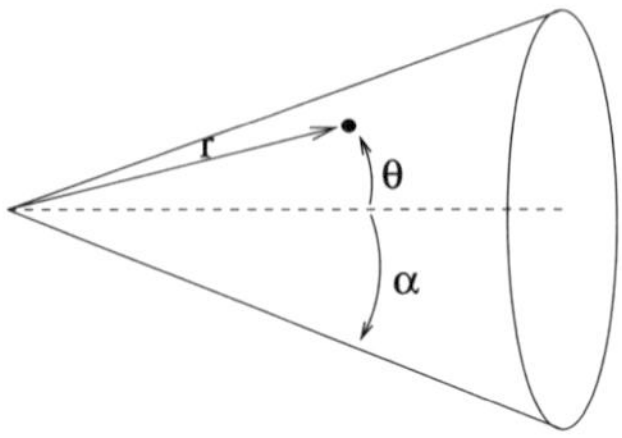

Fig. 5. A cone with opening angle α.

In general, the survival probability decays algebraically as in (6) and the amplitude obeys the Laplace equation (7). By following the steps leading to (9) we find that the angular function ψ satisfies the eigenvalue equation

$$\frac{1}{(\sin\theta)^{d-2}}\frac{d}{d\theta}\left[(\sin\theta)^{d-2}\frac{d\psi}{d\theta}\right]+2\beta(2\beta+d-2)\psi=0. \tag{15}$$

To solve this equation, we introduce the variable $\mu=\cos\theta$, and with this transformation, the function $\psi\equiv\psi(\mu)$ satisfies

$$(1-\mu^2)\frac{d^2\psi}{d\mu^2}-(d-1)\,\mu\,\frac{d\psi}{d\mu}+2\beta(2\beta+d-2)\psi=0. \tag{16}$$

Again, the boundary condition is $\psi(\cos\alpha)=0$.

For $d=3$, the solution to Eq. (16) is given by the Legendre function, $\psi_3(\theta)=P_{2\beta}(\cos\theta)$ [51]. The boundary condition implies that the first-passage exponent is the smallest root of the transcendental equation

$$P_{2\beta}(\cos\alpha)=0. \tag{17}$$

Henceforth, we always choose the smallest root (or equivalently, eigenvalue) because the function $\psi(\theta)$ must remain positive. Further, equation (16) shows that there is a direct connection between the first-passage exponent β and the lowest eigenvalue of the *angular* part of the Laplace operator.

When $d=4$, we can obtain the first-passage exponent explicitly. Using the transformation $\psi_4(\theta)=(\sin\theta)^{-1}u(\theta)$, the eigenvalue equation (15) becomes $u_{\theta\theta}+(2\beta+1)^2u=0$. There are two independent solutions, $\sin[(2\beta+1)\theta]$ and $\cos[(2\beta+1)\theta]$ but since the function ψ must remain finite, only the former is physical. Therefore $\psi_4=\sin\left[(2\beta+1)\theta\right]/\sin\theta$ and the boundary condition $\psi(\alpha)=0$ gives the exponent

$$\beta_4(\alpha)=\frac{\pi-\alpha}{2\alpha}\,. \tag{18}$$

In contrast with equation (10), the exponent now vanishes when $\alpha=\pi$. Hence, in dimensions three and higher, a needle is not reached (Fig. 6). In

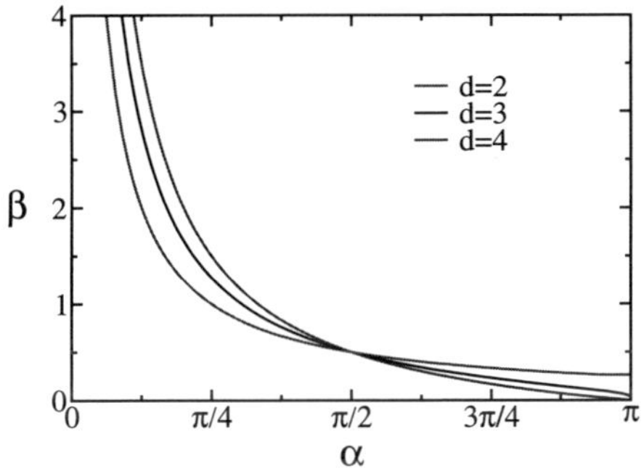

Fig. 6. The first-passage exponent β versus the opening angle α in dimensions $d = 2$, 3, and 4.

the limit of very thin cones, the exponent diverges and is inversely proportional to α as in (10). Equations (16)–(17) were recently used for analyzing entropic forces for polymers in the vicinity of a conical tip [42].

To solve the eigenvalue equation (16), we make the transformation

$$\psi(\mu) = \left(1-\mu^2\right)^{-\delta/2}\Psi(\mu) \qquad \text{with} \qquad \delta = \frac{d-3}{2}. \tag{19}$$

The auxiliary function $\Psi(\mu)$ obeys

$$(1-\mu^2)\frac{d^2\Psi}{d\mu^2} - 2\,\mu\,\frac{d\Psi}{d\mu} + \left[(2\beta+\delta)(2\beta+\delta+1) - \frac{\delta^2}{1-\mu^2}\right]\Psi = 0. \tag{20}$$

This equation specifies the associated Legendre functions $P^{\delta}_{2\beta+\delta}(\mu)$ and $Q^{\delta}_{2\beta+\delta}(\mu)$ of degree $2\beta+\delta$ and order δ [51]. Since ψ must remain finite when $\alpha \to 0$, the first (second) solution is physical when d is odd (even):

$$\psi_d(\theta) = \begin{cases} (\sin\theta)^{-\delta} P^{\delta}_{2\beta+\delta}(\cos\theta) & d \text{ odd}, \\ (\sin\theta)^{-\delta} Q^{\delta}_{2\beta+\delta}(\cos\theta) & d \text{ even}. \end{cases} \tag{21}$$

Using the fact that the survival probability vanishes on the surface of the cone, we find that β is related to the smallest root of the associated Legendre function

$$\begin{aligned} P^{\delta}_{2\beta+\delta}(\cos\alpha) &= 0 \qquad d \text{ odd}, \\ Q^{\delta}_{2\beta+\delta}(\cos\alpha) &= 0 \qquad d \text{ even}. \end{aligned} \tag{22}$$

In general, the first-passage exponent $\beta \equiv \beta_d(\alpha)$ depends on the opening angle α and the dimension d. The exponent α decreases monotonically as the opening angle α increases (figure 6).

The solution $\psi_d(\mu)$ is polynomial in μ when the first-passage exponent is half integer. For example, $\psi_d(\mu) = \mu$ when $\beta = 1/2$ and and $\psi_d(\mu) = d\mu^2 - 1$ when $\beta = 1$. Hence, the mean-first passage time is finite if and only if the opening angle is small enough $\cos\theta > 1/\sqrt{d}$.

In general, the mean first-passage time $T(r,\theta)$ for a particle with initial coordinates (r,θ) obeys the Poisson equation [7, 52]

$$D\nabla^2 T(r,\theta) = -1, \tag{23}$$

and satisfies the boundary conditions $T(r,\alpha) = T(0,\theta) = 0$. Dimensional analysis implies that $T(r,\theta) = (r^2/D)U(\theta)$ with U a dimensionless function of the angle θ. The first-passage time can be obtained explicitly

$$T(r,\theta) = \frac{r^2}{2D}\,\frac{\cos^2\theta - \cos^2\alpha}{d\cos^2\alpha - 1}. \tag{24}$$

This equation holds for $\cos\theta > 1/\sqrt{d}$ where $\beta > 1$. As expected, the duration of the first-passage process is a monotonic function of both distance r and angle θ.

4. Asymptotic properties

Our goal is to apply the first-passage results in unbounded cones to ordering of Brownian particles. We are especially interested in the behavior of a large number of particles, which requires asymptotic properties of (22) when $d \to \infty$. These asymptotics are summarized in this section.

Taking the limit $d \to \infty$ while keeping the angle fixed shows that

$$\lim_{d\to\infty} \beta_d(\alpha) = \begin{cases} \infty & \alpha < \pi/2, \\ 1/2 & \alpha = \pi/2, \\ 0 & \alpha > \pi/2. \end{cases} \tag{25}$$

Hence, the first-passage process is infinitesimally short when the opening angle is acute, and conversely, it lasts forever when the angle is obtuse. For a right-angle, $\alpha = \pi/2$, the problem is effectively one-dimensional. Further, one can also show [33] that the exponent grows linearly $\beta \sim d$ for acute angles, $\alpha < \pi/2$, while in the complementary case, the exponent decays exponentially, $\beta \sim (\sin\alpha)^d$.

The limiting behavior (25) suggests that we should focus on a shrinking region near $\alpha = \pi/2$. Moreover, the fact that $\beta = 1$ when $\sqrt{d}\cos\alpha = 1$ indicates that the first-passage exponent has the scaling form

$$\beta_d(\alpha) \to \beta(y) \qquad \text{with} \qquad y = (\cos\alpha)\sqrt{d}, \tag{26}$$

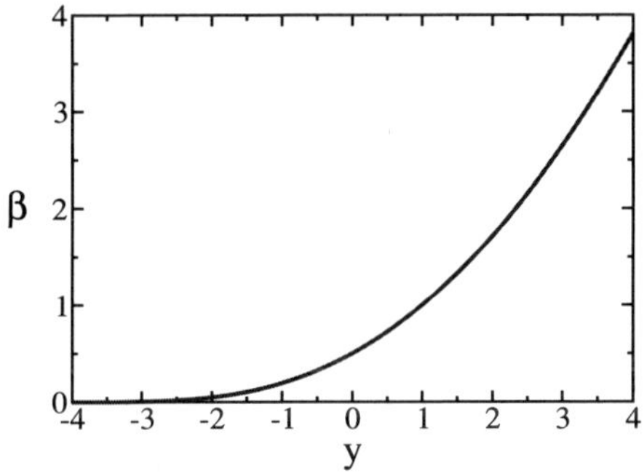

Fig. 7. The exponent β versus the scaling variable $y = (\cos\alpha)\sqrt{d}$.

when $d \to \infty$. If we take the limits $\mu \to 0$ and $d \to \infty$ in such a way that the scaling variable $z = \mu\sqrt{d}$ is finite, then with this scaling transformation equation (16) becomes

$$\psi_{zz} - z\,\psi_z + 2\beta\psi = 0. \tag{27}$$

The boundary condition is $\psi(y) = 0$. We now make the transformation $\psi(z) = \exp(z^2/4)u(z)$ and arrive at the parabolic cylinder equation

$$u_{zz} + \left(2\beta + \frac{1}{2} - \frac{z^2}{4}\right) u = 0. \tag{28}$$

The two independent solutions are the parabolic cylinder functions $D_\nu(z)$ and $D_\nu(-z)$ [51]. Using the asymptotic behavior $D_\nu(-z) \sim \exp(z^2/4)$ as $z \to \infty$ and the fact that the survival probability must be finite on the cone axis, we eliminate the latter solution and hence,

$$\psi(z) = e^{z^2/4} D_{2\beta}(z). \tag{29}$$

By invoking the boundary condition $\psi(\mu) = 0$ we arrive at our main result (2). We choose the largest root of the parabolic cylinder function because ψ is positive. The scaling form (26) implies that β is finite in the narrow region of opening angles

$$\alpha - \pi/2 \sim d^{-1/2}. \tag{30}$$

Hence, the first-passage exponent is finite only when the cone is sufficiently close to a half space.

Asymptotic analysis of Eq. (29) reveals the limiting behaviors (Fig. 7)

$$\beta(y) \simeq \begin{cases} \sqrt{y^2/8\pi}\, \exp\left(-y^2/2\right) & y \to -\infty, \\ y^2/8 & y \to \infty. \end{cases} \tag{31}$$

We also mention the asymptotic behaviors in the interior and the exterior of a slender cone (the dimension d is fixed)

$$\beta_d(\alpha) \simeq \begin{cases} A_d\,\alpha^{-1} & \alpha \to 0, \\ B_d\,(\pi-\alpha)^{d-3} & \alpha \to \pi. \end{cases} \tag{32}$$

Inside a very thin cone, the first-passage process is very fast and β diverges. The proportionality constant is $A_d = \frac{1}{2}\zeta(\delta)$ where $\zeta(\delta)$ is the first positive zero of the Bessel [53] function J_δ of order $\delta = \frac{d-3}{2}$. For example, $A_3 \cong 1.202412$, $A_4 = \frac{\pi}{2} \cong 1.570796$, and $A_5 \cong 1.915852$. As $\alpha \to \pi$, the exponent β vanishes. The coefficient is $B_d = \Gamma\big(\frac{d-2}{2}\big)/\big[2\sqrt{\pi}\,\Gamma\big(\frac{d-3}{2}\big)\big]$. The corresponding behavior (32) holds only when $d > 3$. For $d = 3$, the exponent vanishes logarithmically, $\beta_3 \simeq 1/[4\ln\frac{2}{\pi-\alpha}]$ (Fig. 6).

5. The cone approximation

The above results for diffusion inside a circular cone can be used to approximate diffusion in more complex geometries for which the eigenvalues of the Laplace operator are not known analytically. We now apply this approach to the exterior of the three-dimensional corner (figure 8). We are interested in the survival probability $S(t)$ of a particle that is diffusing in the exterior of an unbounded corner [54] in dimension three, and we expect that the powerlaw decay (1) holds, regardless of initial position.

The corner geometry underlies a first-passage problem involving three particles in one dimension. Let's assume that all particles start in the positive half line, $0 < x_3(0) < x_2(0) < x_1(0)$. Then, the probability that the rightmost particle always remains in the positive half line requires that the composite particle remains outside the corner [34]. The complementary problem that the leftmost particle remains in the positive half line is trivial: since all three particles must remain in the positive half space, then $S(t) = [s(t)]^3$ with $s(t) \sim t^{-1/2}$ the survival probability in one dimension. Hence, for diffusion inside the corner $\beta = 3/2$.

We now replace the corner with an unbounded cone that has identical solid angle. In d dimensions, a cone with opening angle α occupies a fraction V of the total solid angle,

$$V(\alpha) = \frac{\int_0^\alpha d\theta\,(\sin\theta)^{d-2}}{\int_0^\pi d\theta\,(\sin\theta)^{d-2}}. \tag{33}$$

The normalization is such that $V(\pi) = 1$, and of course, $V(\pi/2) = 1/2$. Equation (33) follows from the Jacobian $d\Omega/d\theta \propto (\sin\theta)^{N-2}$ with Ω being

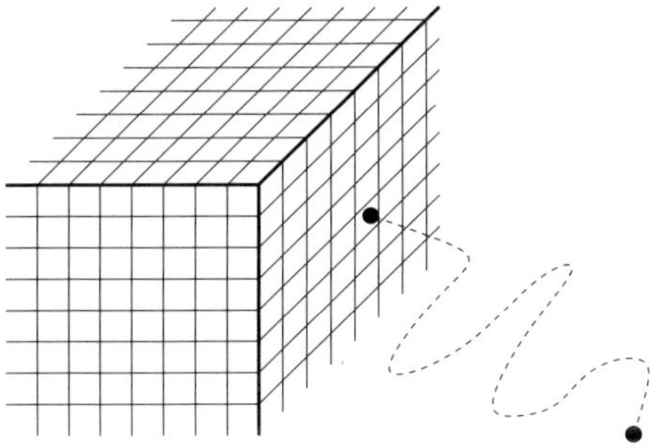

Fig. 8. First-passage process outside a three-dimensional corner.

the solid angle and θ the polar angle.

By substituting $d = 3$ and $V = 7/8$ into (33), we find that the equivalent cone has opening angle $\alpha = 2.418858$ and from equation (17) we obtain

$$\beta^{\text{cone}} = 0.216785 \qquad \text{whereas} \qquad \beta = 0.228. \tag{34}$$

The latter value was obtained from numerical simulations (figure 9). Clearly, the cone approximation produces a useful estimate for β.

The value β^{cone} is a lower bound. This assertion follows from the fact that a sphere has the smallest eigenvalue of the Laplace operator amongst the domains with the same volume. In two dimensions, this result was conjectured by Rayleigh [55] and later proved by Faber and Krahn [56]. In higher dimensions, the Rayleigh-Faber-Krahn theorem remains valid [57]. This result implies for example that the ground state of a quantum particle in a box is lowest when the box is spherical; another consequence is that the lifetime of a Brownian particle inside an absorbing box of a fixed volume is highest if the box is spherical. In our situation, we study a composite particle in a general unbounded conical domain in d dimensions. (By definition, any cone $\mathcal{R}$ with an apex at the origin has the property that every ray from the origin to any point inside $\mathcal{R}$ belongs to $\mathcal{R}$.) The eigenvalue problem generalizing Eqs. (9) and (15) arising for circular cones is specified by

$$\nabla^2\psi = -2\beta(2\beta + d - 2)\psi, \qquad \psi|_{\partial(S^{d-1}\cap\mathcal{R})} = 0. \tag{35}$$

Here ∇^2 is the *angular* component of the Laplacian. The choice of the smallest eigenvalue ensures that the eigenfunction ψ is positive inside $\mathcal{R}$. We need the smallest eigenvalue in the "spherical cap" $S^{d-1} \cap \mathcal{R}$, the intersection of the cone $\mathcal{R}$ with unit sphere S^{d-1}, and ∇^2 is actually the Laplace-Beltrami

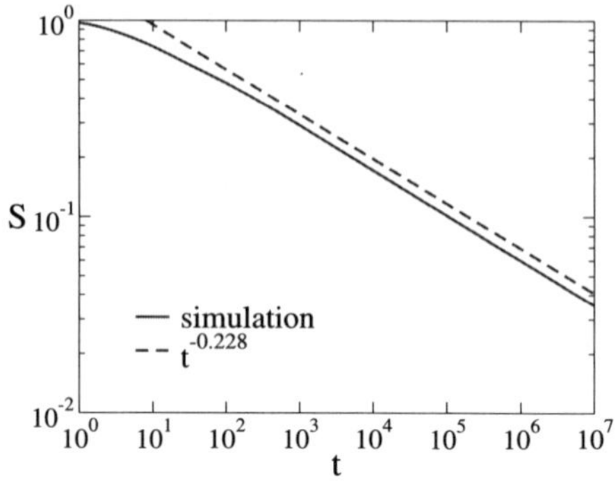

Fig. 9. The survival probability in the exterior of a three-dimensional corner. The solid line indicate results of numerical simulations and the dashed line, a best-fit.

operator on the spherical cap. Fortunately, the Rayleigh-Faber-Krahn theorem generalizes to the smallest eigenvalue of the Laplace-Beltrami operator on Riemannian manifolds [57] and it proves that the smallest eigenvalue $2\beta(2\beta + d - 2)$, and therefore the smallest β, amongst all spherical caps with the same volume (i.e., the same solid angle) corresponds to the circular spherical cap. Hence, the cone approximation produces a lower bound.

Equation (26) shows that for $d \gg 1$, there is universal scaling behavior in terms of the variable $y = (\cos\alpha)\sqrt{d}$. The same scaling variable also underlies the fraction of space V enclosed by the cone. The infinitesimal solid volume element $d\Omega \propto (\sin\theta)^{d-2} d\theta$ in (33) becomes Gaussian, $d\Omega \propto \exp(-y^2/2)dy$. Consequently, the volume which depends on the opening angle and the dimension, $V \equiv V(\alpha, d)$, becomes a function of a single variable y

$$V(\alpha, d) = \frac{1}{2}\,\mathrm{erfc}\left(\frac{y}{2}\right). \tag{36}$$

Here $\mathrm{erfc}(\xi) = \left(2/\sqrt{\pi}\right)\int_\xi^\infty e^{-u^2}du$ is the complementary error function. Combining (36) and (4), we find β as an implicit function of volume V:

$$D_{2\beta}\left(\sqrt{2}\,\mathrm{erfc}^{-1}(2V)\right) = 0. \tag{37}$$

Remarkably, this expression provides an exact description for the two ordering problems we are interested in.

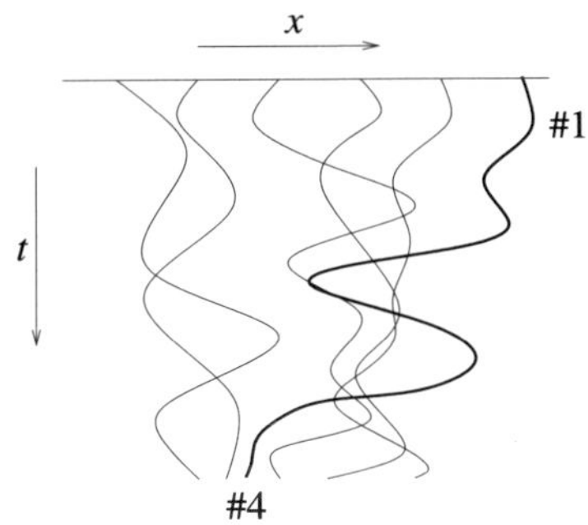

Fig. 10. Illustration of the rank problem. Shown is the spacetime diagram of six Brownian particles in one dimension. The particles are ranked according position, from right to left, and in this example, the rank of the leader falls from one to four.

6. Rank statistics

We now consider N particles, each undergoing independent Brownian motion in one dimension. The particles are ordered from right to left with the rightmost particle at time $t = 0$ regarded as the original leader. As described in the introduction, we are interested in the probability $R_n(t)$ that the rank of the original leader does not fall below n up to time t (figure 10). The extremal cases $n = 1$ and $n = N - 1$ were considered previously [19, 24, 26, 28, 32], yet the long-time kinetics have been established [12] only for $N \leq 3$.

Based on the behavior for $N = 3$ where the set $\{\gamma_1, \gamma_2\} = \{3/4, 3/8\}$ characterizes the decay $R_n \sim t^{-\gamma_n}$, we expect $N - 1$ distinct exponents,

$$\gamma_1 > \gamma_2 > \cdots > \gamma_{N-1}. \tag{38}$$

Simulation results for $N = 4$ confirm this behavior and give (Fig. 11):

$$\gamma_1 = 0.913, \qquad \gamma_2 = 0.556, \qquad \gamma_3 = 0.306. \tag{39}$$

For $N = 4$, the largest exponent has been calculated using electrostatics $\gamma_1 = 0.91342$ [28].

As discussed above, the trajectories of N particles in one dimension can be represented by that of a compound particle in N dimensions. The compound particle is confined to a volume V_n of space in which the position of the initial leader is larger than those of at least $N - n$ of the particles. Figure 3 shows that $V_1 = 1/3$ and $V_2 = 2/3$ for $N = 3$. Generally

$$V_n = \frac{n}{N}. \tag{40}$$

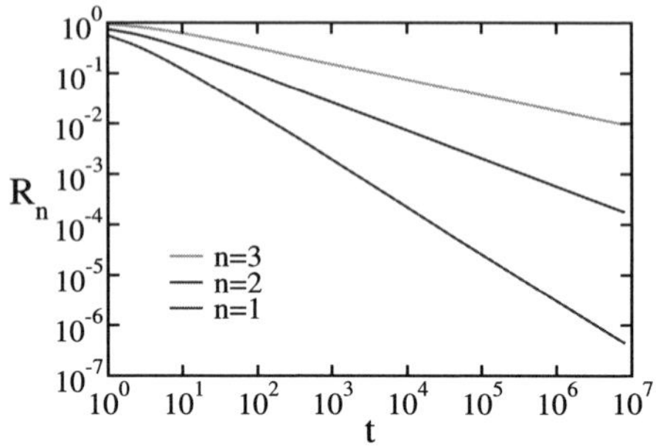

Fig. 11. The survival probability R_n versus rank n for four particles. The results are obtained from 10^{11}, 10^9, and 10^7 independent Monte Carlo runs for $n = 1$, $n = 2$, and $n = 3$ respectively.

In N dimensions, the $N(N-1)/2$ hyperplanes $x_i = x_j$ with $i \neq j$ divide space into $N!$ equal "chambers" each with distinct particle order (figure 2). For example, the region $x_1 > x_2 > \cdots > x_{N-1} > x_N$ is one such chamber where the initial particle order is preserved, and this chamber occupies a fraction $1/N!$ of the total solid angle. Since there are $(N-1)!$ chambers in which the rank of the original leader equals 1, then $V_1 = 1/N$. There are $(N-1)!$ additional chambers in which the rank of the leader equals 2, and hence $V_2 = 2/N$. This argument establishes (40).

According to (40), the compound particle is confined to a region of space that is the union of $n(N-1)!$ chambers. To apply the cone approximation, we replace this complex region with an unbounded spherical cone that encloses a fraction V_n of space. We set $d = N-1$ so that this approximation is exact when $N = 3$. By substituting (40) and $d = N-1$ into (33), we obtain the opening angle α. The first-passage exponent is obtained as root of the associated Legendre function specified in (22) with $d = N-1$. For example, when $N = 3$, we obtain

$$\gamma_1^{\text{cone}} = 0.888644, \qquad \gamma_2 = 1/2, \qquad \gamma_3 = 0.300754. \tag{41}$$

These values, which represent formal lower bounds, provide useful estimates for the numerically measured values (39). Table 1 lists the largest and the smallest first-passage exponents for $3 \leq N \leq 10$. Although cone approximation slightly deteriorates as the number of particle increases, overall the estimates remain faithful.

The predictions of the cone approximation in the limit $N \to \infty$ follow

Table 1. The largest and the smallest exponents γ_1 and γ_{N-1} for $3 \le N \le 10$. Simulation results are compared with the outcome of the cone approximation.

N	γ_1^{cone}	γ_1	$\gamma_{N-1}^{\text{cone}}$	γ_{N-1}
3	3/4	3/4	3/8	3/8
4	0.888644	0.91	0.300754	0.306
5	0.986694	1.02	0.253371	0.265
6	1.062297	1.11	0.220490	0.234
7	1.123652	1.19	0.196216	0.212
8	1.175189	1.27	0.177469	0.190
9	1.219569	1.33	0.162496	0.178
10	1.258510	1.37	0.150221	0.165

immediately from equations (40) and (37),

$$D_{2\gamma}\left(\sqrt{2}\,\mathrm{erfc}^{-1}(2n/N)\right) = 0. \tag{42}$$

The exponent γ which depends on two variables, n and N, becomes a universal function of the scaling variable $x = n/N$ as stated in equation (4). Remarkably, the scaling function specified by equation (42) is exact! Our numerical simulations which use a remarkably large number of particles (up to $N = 10^4$) reveal that: (i) the first-passage exponent is a function of the scaling variable x, and (ii) the scaling function (42) is exact (figure 1).

This finding may indicate that in very high dimensions, the complex boundary confining the composite particle approaches a limiting shape, and that this limiting shape is conical. Another possible scenario is that the complex boundary, which is formed by the intersection of multiple hyperplanes, does not necessarily approach a limiting conical shape, yet the deviations do not affect lowest eigenvalue of the Laplace operator.

The scaling function (42) manifests that there is a continuous spectrum of first-passage exponents $0 < \gamma < \infty$ when $N \to \infty$. The smallest exponent vanishes as N^{-1}, while the largest diverges logarithmically. More precisely,

$$\gamma_n \simeq \begin{cases} \frac{1}{4}\ln N & n = 1, \\ \frac{1}{N}\ln N & n = N-1. \end{cases} \tag{43}$$

These limiting behaviors follow from asymptotic analysis of (42) with $n = 1$ and $n = N-1$, and in both cases, we recover results obtained using heuristic scaling arguments [26, 28]. We also note that median exponent, which characterizes the probability that the leader always ranks above the median particle, approaches the simple limit $\gamma_{N/2} \to 1/2$.

Knowledge of the scaling form (42), merely the fact that γ is function of the variable $x = n/N$ is quite powerful. This realization enables measurement of first-passage exponents for up to $N = 10^4$ particles, corresponding to diffusion in a staggering spatial dimension $d = 10^4$. Whereas accurate measurement of the (vanishing) smallest or the (diverging) largest exponent become prohibitive [24] already at $N \approx 10$, we are able to compute γ by measuring the probability that the rank of the original leader remains in the *first percentile* rather than strictly remaining *first*. The numerical simulation also show that not only does the first-passage exponent becomes a function of the scaling variable $x = n/N$, but so does the entire survival probability: $R_n(N,t) \to \Phi(x,t)$ with $x = n/N$ in the limit $N \to \infty$.

7. Inversion statistics

Rank compares the location of a single particle, the initial leader, with those of the rest of the particles. Generally, there are $N(N-1)/2$ distinct pairs, and the number of pair inversion k treats all particles equally. If initially the particle locations are such that $x_N(0) < x_{N-1}(0) < \cdots < x_1(0)$, then the number of pair inversions is

$$k(t) = \sum_{i \neq j} \Theta\big(x_i(t) - x_j(t)\big), \tag{44}$$

with $\Theta(x)$ the step function: $\Theta(x) = 1$ when $x > 0$ and $\Theta(x) = 0$ otherwise. In figure 12 the initial order is 4321 and later, when the order becomes 2341, the number of inversions equals $k = 3$ as the pairs 23, 24, and 34 are inverted with respect to the initial configuration. The number of inversions (also known as the Mahonian number) is a basic measure of a permutation, and it is routinely used in sorting and ranking algorithms [6, 49, 58, 59]. The number of inversions lies between 0 and $N(N-1)/2$.

There are $N!$ different permutations of particle order and since the diffusion process is ergodic, all of these permutations become equally probable. It is easy to show that if initially the particles are regularly spaced, the "equilibration" time is quadratic in N [35]. Consequently, the equilibrium distribution of the inversion number is given by the well-known Mahonian probability distribution $P_k(N)$ that a random permutation of N integers has k inversions. The generating function reads [50]

$$\sum_k P_k(N) s^k = \frac{1}{N!} \prod_{n=1}^{N} (1 + s + s^2 + \cdots + s^{n-1}). \tag{45}$$

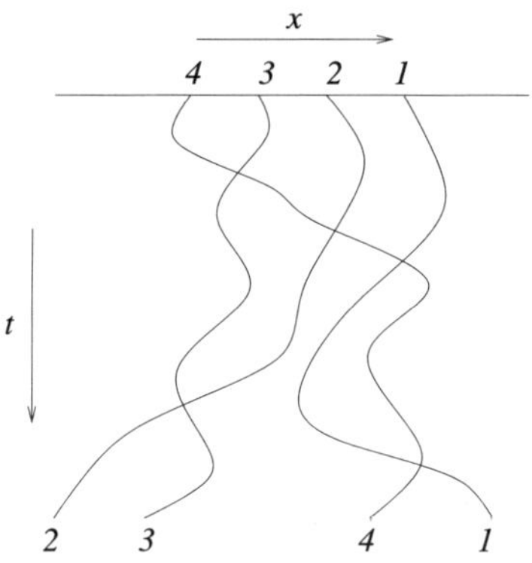

Fig. 12. Illustration of the inversion problem. Shown is the spacetime diagram of four particles in one dimension. The number of inversions grows from $k = 0$ to $k = 3$.

From this generating function, we can obtain the distribution function. For example, for two particles, $P_0(2) = P_1(2) = 1/2$, and for three particles, $P_0(3) = P_2(3) = 1/6$ while $P_1(3) = P_2(3) = 1/3$ (figure 4b). We are particularly interested in the cumulative distribution $V_k(N)$ that the number of inversions is smaller than k

$$V_k(N) = \sum_{j<k} P_j(N). \tag{46}$$

From the generating function (45), we can obtain the average number of inversions $\langle k \rangle$ and the standard deviation σ:

$$\langle k \rangle = \frac{N(N-1)}{4} \qquad \text{and} \qquad \sigma^2 = \frac{N(N-1)(2N+5)}{72}. \tag{47}$$

Notably, the standard deviation is large, $\sigma \sim N^{3/2}$. When the number of particles is large, the distribution P_k becomes normal and is characterized by the mean and the standard deviation [6, 60]. Consequently, the cumulative probability $V_k(N)$ that the inversion number is $\leq k$ approaches the complementary error function in the $N \to \infty$ limit:

$$V_k(N) \to \text{erfc}\left(-\frac{w}{\sqrt{2}}\right) \qquad \text{with} \qquad w = \frac{k - \langle k \rangle}{\sigma}. \tag{48}$$

Let $M_k(t)$ be the survival probability that the number of inversions remains smaller than k up to time t. We anticipate

$$M_k \sim t^{-\nu_k} \qquad \text{with} \qquad \nu_1 > \cdots > \nu_{N(N-1)/2}. \tag{49}$$

Equation (13) shows that $\{\nu_1, \nu_2, \nu_3\} = \{3/2, 1/2, 3/10\}$ when $N = 3$. Numerical simulations confirm that the number of distinct exponents equals

Table 2. First-passage exponents for four particles. Results of Monte Carlo simulations, ν_k versus the outcome of the cone approximation, $\nu_k^{\rm cone}$.

k	1	2	3	4	5	6
$\nu_k^{\rm cone}$	2.67100	1.17208	0.64975	0.39047	0.24517	0.14988
ν_k	3	1.39	0.839	0.455	0.275	0.160

$N(N-1)/2$. In contrast with the rank statistics above, the number of distinct exponents is quadratic in N.

The quantity M_1 equals the probability that the particle order remains intact, and the respective exponent ν_1 is known analytically [21–23, 25, 29, 61]

$$\nu_1 = \frac{N(N-1)}{4}. \tag{50}$$

When $k=1$, the trajectory of the compound Brownian particle is confined to one chamber, $x_1 > x_2 > \cdots > x_N$.

We again use the cone approximation to estimate the first-passage exponents. The fraction (46) calculated from the generating function (45) is substituted into (33) to give the opening angle α. Using this angle and $d = N-1$, we deduce the exponent from equation (22). Results of the cone approximation for the case $N=4$ are listed in table 2. Again, we obtain useful estimates for the first-passage exponents.

To obtain the limiting behavior when the number of particles diverges, $N \to \infty$, we substitute (48) into the scaling relation (4). The scaling exponent ν becomes a function of the scaling variable w defined in (48) and is root of the parabolic cylinder equation,

$$D_{2\nu}(-w) = 0. \tag{51}$$

Hence, if we focus on the probability that the number of inversions does not exceed a fixed number w of standard deviations from the mean, the first-passage process becomes independent of the number of particles. This realization is especially useful in the case of inversion statistics: the family of exponents is quadratic in N, and furthermore, the scaling variable w guides us to measure ν_k in a scaling window of size $N^{3/2}$ around the average, $\langle k\rangle \simeq N^2/4$. Indeed, by using a finite number of measurements, we can verify that $\nu \equiv \nu_k(N)$ becomes a function of the scaling variable w. Again, the cone approximation predicts the exact scaling function governing ν (figure 13).

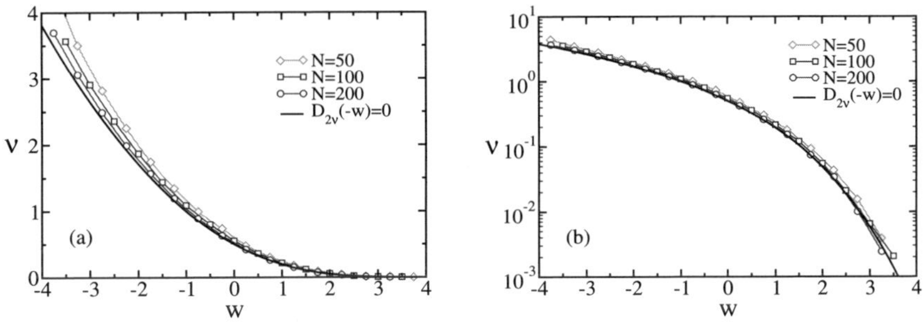

Fig. 13. The exponent ν versus the scaling variable w. Predictions of the cone approximation are compared with results of Monte Carlo simulations with $N = 50$, $N = 100$, and $N = 200$ particles.

8. Conclusions

We studied first-passage properties of a Brownian particle in unbounded circular cones. The first-passage exponent characterizing the decay of the survival probability is root of a transcendental equation involving the associated Legendre equation. Explicit expressions for this exponent are feasible only in two and four dimensions. Another simplification occurs when $d \gg 1$ where the exponent, which generally depends on the cone opening angle α and d, becomes a function of a single variable $\sqrt{d}\cos\alpha$. This universal function is obtained analytically as a root of the parabolic cylinder function.

We analyzed two first-passage problems involving the order of an ensemble of independent Brownian particles in one dimension. The trajectories of the N Brownian particles define a compound trajectory of a single Brownian particle in N dimensions, more precisely in an infinite cone with numerous plane boundaries. As an approximation, we replaced this cone with complex boundary by a circular cone having an identical solid angle. We showed that the cone approximation yields a lower bound and surprisingly good approximations for the first-passage exponents.

There are families of first-passage exponents that govern kinetics of ordering of Brownian particles or equivalently, kinetics of first-passage in high dimensions. We discussed only two out of many possible measures for particle order: the rank of the original leader and the number of pair inversions. In one case, the number of first-passage exponents is linear in the number of particles; in the other case, it is quadratic. The discrete spectrum of exponents becomes continuous when the number of particles diverges.

Surprisingly, the cone approximation predicts this spectrum exactly.

An important challenge is to establish when does the confining geometry becomes equivalent to a cone in a high-dimensional space and the conditions for the cone approximation to be asymptotically exact. We presented two cases for which the scaling function is predicted exactly, but there are counterexamples where the cone approximation correctly predicts the scaling form but fails to predict the actual scaling function [34].

The exponents exhibit scaling when the number of particles diverges. This realization is relevant for numerical simulations [62, 63] where it is difficult to measure exponents that are either much smaller than or much larger than unity. The scaling variable specifies the region where the first-passage exponents are of order one. For example, for the rank problem the scaling behavior implies that the rank should be measured in percentiles.

First-passage processes are inherently nonequilibrium. By definition the particle trajectory is restricted to certain regions in space whereas other regions are strictly excluded. For example, in the corner geometry (8) an unrestricted Brownian particle visits all eight "quadrants" with equal probabilities $p = 1/8$. Yet, for first-passage processes inside a three-dimensional corner, we have $p = 1$ inside the corner and $p = 0$ otherwise. By exploiting the cone geometry, we have shown that the equilibrium probabilities (40) and (48) translate into knowledge of nonequilibrium properties, and in particular, first-passage exponents. Interestingly, geometrical properties allow us to deduce exact nonequilibrium properties of a particle system from its equilibrium properties. It will be interesting to extend these results to other geometries [64, 65] and to establish connections with other problems involving ordering of Brownian particles.

Acknowledgment

We acknowledge DOE grant DE-AC52-06NA25396 for support.

References

1. M. Ohnakam, *The Physics of Rock Failure and Earthquakes* (Cambridge University Press, Cambridge 2013).
2. P. Bak, *How nature works: the science of self-organized criticality* (Copernicus, New York, 1999).
3. J. D. Murray, *Mathematical Biology* (Springer, Berlin 1989)
4. J.-P. Bouchaud and M. Potters, *Theory of Financial Risk and Derivative Pricing* (Cambridge University Press, Cambridge 2003).

5. N. G. van Kampen, *Stochastic Processes in Physics and Chemistry* (North Holland, Amsterdam, 2003).
6. W. Feller, *An Introduction to Probability Theory and Its Applications* (Wiley, New York, 1968).
7. S. Redner, *A Guide to First-Passage Processes* (Cambridge University Press, New York, 2001).
8. G. H. Weiss, *Aspects and Applications of the Random Walk* (North-Holland, Amsterdam, 1994).
9. H. C. Berg, *Random Walks in Biology* (Princeton University Press, Princeton, 1983).
10. B. Duplantier, in: *Einstein, 1905-2005*, eds. Th. Damour, O. Darrigol, B. Duplantier and V. Rivasseau (Birkhäuser Verlag, Basel, 2006); arXiv:0705.1951.
11. P. Mörders and Y. Peres, *Brownian Motion* (Cambridge University Press, Cambridge, 2010).
12. P. L. Krapivsky, S. Redner, and E. Ben-Naim, *A Kinetic View of Statistical Physics* (Cambridge University Press, Cambridge, 2010).
13. B. Derrida, V. Hakim, and V. Pasquier, Phys. Rev. Lett. **75**, 751 (1995).
14. S. N. Majumdar, C. Sire, A. J. Bray, and S. J. Cornell, Phys. Rev. Lett. **77**, 2867 (1996); B. Derrida, V. Hakim, and R. Zeitak, Phys. Rev. Lett. **77**, 2871 (1996).
15. A. J. Bray, S. N. Majumdar, and G. Schehr, arXiv:1304.1195.
16. P. L. Krapivsky, E. Ben-Naim, and S. Redner, Phys. Rev. E **50**, 2474 (1994).
17. M. Howard and C. Godreche, J. Phys. A **31**, L209 (1998).
18. H. Niederhausen, Eur. J. Combinatorics **4**, 161 (1983).
19. T. Antal, E. Ben-Naim, and P. L. Krapivsky, J. Stat. Mech. P07009 (2010).
20. E. Ben-Naim, Z. A. Daya, P. Vorobieff, and R. E. Ecke, Phys. Rev. Lett. **86**, 1414 (2001).
21. M. E. Fisher, J. Stat. Phys. **34**, 667 (1984).
22. D. A. Huse and M. E. Fisher, Phys. Rev. B **29**, 239 (1984).
23. M. E. Fisher and M. P. Gelfand, J. Stat. Phys. **53**, 175 (1988).
24. M. Bramson and D. Griffeath, in: *Random Walks, Brownian Motion, and Interacting Particle Systems: A Festshrift in Honor of Frank Spitzer*, eds. R. Durrett and H. Kesten (Birkhäuser, Boston, 1991).
25. D. J. Grabiner, Ann. Inst. Poincare: Prob. Stat. **35**, 177 (1999).
26. P. L. Krapivsky and S. Redner, J. Phys. A **29**, 5347 (1996); S. Redner and P. L. Krapivsky, Amer. J. Phys. **67**, 1277 (1999).
27. D. ben-Avraham, J. Chem. Phys. **88**, 15 (1988).
28. D. ben-Avraham, B. M. Johnson, C. A. Monaco, P. L. Krapivsky, and S. Redner, J. Phys. A **36**, 1789 (2003).
29. J. Cardy and M. Katori, J. Phys. A **36**, 609 (2003).
30. D. L. Burkholder, Adv. Math. **26**, 182 (1977).
31. S. B. Yuste, L. Acedo, and K. Lindenberg, Phys. Rev. E **64**, 052102 (2001).
32. D. ben-Avraham, S. N. Majumdar, and S. Redner, J. Stat. Mech. L04002 (2007).
33. E. Ben-Naim and P. L. Krapivsky, J. Phys. A **43**, 495007 (2010).
34. E. Ben-Naim and P. L. Krapivsky, J. Phys. A **43**, 495008 (2010).

35. E. Ben-Naim, Phys. Rev. E **82**, 061103 (2010).
36. R. D. DeBlassie, Probab. Theory Relat. Fields **74**, 1 (1987); Probab. Theory Relat. Fields **79**, 95 (1988).
37. B. Davis and B. Zhang, Proc. AMS **121**, 925 (1994).
38. R. Bañuelos and R. G. Smiths, Probab. Theory Relat. Fields **108**, 299 (1997).
39. R. Bañuelos, R. D. DeBlassie, and R. G. Smiths, Ann. Probab. **29**, 882 (2001).
40. N. Th. Varopoulos, Math. Proc. Camb. Phi. Soc. **125**, 335 (1999); Math. Proc. Camb. Phi. Soc. **129**, 301 (1999).
41. J. H. Jeon, A. V. Chechkin, and R. Metzler, EPL **94**, 20008 (2011).
42. M. F. Maghrebi, Y. Kantor, and M. Kardar, EPL **96**, 66002 (2011); Phys. Rev. E **86**, 061801 (2012).
43. E. K. Lenzi, L. R. Evangelista, M. K. Lenzi, and L. R. da Silva, Phys. Rev. E **80**, 021131 (2011).
44. G. I. Barenblatt, *Scaling, Self-Similarity, and Intermediate Asymptotics* (Cambridge University Press, Cambridge, 1996).
45. J. D. Jackson, *Classical Electrodynamics* (Wiley, New York, 1998).
46. F. Spitzer, Trans. AMS **87**, 187 (1958).
47. D. Considine and S. Redner, J. Phys. A **22**, 1621 (1988).
48. D.L. L. Dy and J. P. Esguerra, Phys. Rev. E **78**, 062101 (2008).
49. M. Bóna, *Combinatorics of Permutations* (Chapman and Hall, Boca Raton, 2004).
50. D. E. Knuth, *The Art of Computer Programming, vol. 3: Sorting and Searching* (Addison-Wesley, New York, 1998).
51. *NIST Handbook of Mathematical Functions*, ed. F. W. J. Olver, D. M. Lozier, et al. (Cambridge University Press, Cambridge, 2010).
52. T. G. Mattos, C. Mejia-Monasterio, R. Metzler, and G. Oshanin, Phys. Rev. E **86**, 031143 (2012).
53. J. D. Watson, *A Treatise on the Theory of Bessel Functions* (Cambridge University Press, Cambridge, 1995).
54. L. C. Davis and J. Reitz, J. Math. Phys. **16**, 1219 (1975).
55. J. W. S. Rayleigh, *The Theory of Sound* (Macmillan, New York, 1877; reprinted Dover, New York, 1945).
56. R. Courant and D. Hilbert, *Methods of Mathematical Physics*, vol. I (Wiley, New York, 1953).
57. I. Chavel, *Eigenvalues in Riemannian geometry* (Academic Press, Orlando, 1984).
58. P. A. McMahon, Amer. J. Math. **35**, 281 (1913).
59. G. E. Andrews, *The Theory of Partitions* (Addison-Wesley, Reading, 1976).
60. R. E. Canfield, S. Janson, and D. Zeilberger, Adv. Appl. Math. **49**, 77 (2010).
61. I. M. Gessel and D. Zeilberger, Proc. Amer. Math. Soc. **115**, 27 (1992).
62. P. Grassberger, Computer Phys. Comm. **147**, 64 (2002).
63. T. Oppelstrup, V. V. Bulatov, A. Donev, M. H. Kalos, G. H. Gilmer, and B. Sadigh, Phys. Rev. E **80**, 066701 (2009).
64. M. Lifshits and Z. Shi, Bernoulli **8**, 745 (2002).
65. P. L. Krapivsky and S. Redner, J. Stat. Mech. P11028 (2010).

Chapter 12

First Passage Time Problems in Biophysical Jump Processes with Fast Kinetics

Paul C. Bressloff[1] and Jay M. Newby[2]

[1] *Department of Mathematics, University of Utah, 155 South 1400 East, Salt Lake City UT 84112, USA*

[2] *Math Bioscience Institute, Ohio State University, 1735 Neil Ave. Columbus, OH 43210, USA*

We review recent work on the analysis of first-passage time problems in stochastic hybrid systems. Typically, one or more continuous variables x evolve according to a piecewise deterministic equation of the form $\dot{x} = F_n(x)$, where n labels a set of discrete states. Transitions between these states are governed by a jump Markov process with transition rates that may depend on x. For example, x could represent the membrane voltage of a neuron and n could be the number of open sodium ion channels. Alternatively, x could represent the position of a molecular motor complex on a microtubular track and n could be the internal velocity state of the complex. In the case of a jump process with fast kinetics, a quasi-steady-state diffusion approximation can be performed. However, such an approximation breaks down when considering the escape from a metastable state, and it is necessary to use WKB methods and singular perturbation theory in order to calculate the mean first passage time. The theory is illustrated using the examples of stochastic ion channels and motor-driven intracellular transport.

1. Introduction

There are a growing number of problems in biophysics that involve the coupling between a piecewise deterministic dynamical system and a jump Markov process. Examples include dynamical instabilities of microtubules [1, 2], motor-driven intracellular transport [3–10], stochastic ion channels [11–18], stochastic gene networks [19], and stochastic neural networks [20]. All of these systems can be modeled in terms of an equation having the form

$$\frac{dx}{dt} = \frac{1}{\tau_x} F(x, n) \tag{1}$$

where $x \in \mathbb{R}$ is a continuous variable, $n \in I \subseteq \mathbb{Z}$ is a discrete internal state variable, and the latter evolves according to a jump Markov process $n' \to n$ with transition rates $W(n, n'; x)/\tau_n$.[a] The jump propagator $W(n, n'; x)dt/\tau_n$ is the probability that the system at x switches from the discrete internal state n' at time t to the state n at time $t + dt$. The resulting stochastic process is an example of a stochastic hybrid system based on a piecewise deterministic process. That is, the transition rates generally depend on x, with the latter coupled to the associated jump Markov process according to equation (1), which is only defined between jumps, during which $x(t)$ evolves deterministically. Note that a special case of equation (1) is a deterministic system driven by so-called dichotomous noise [21, 22]:

$$\frac{dx}{dt} = F(x) + G(x)\xi(t), \tag{2}$$

where $\xi(t)$ is a discrete Markov variable that switches between the fixed values $\pm a$, say, at some rate λ. Denote the random state of the full model (1) at time t by $(x(t), n(t))$, and introduce the corresponding probability density

$$\text{Prob}\{x(t) \in (x, x + dx), n(t) = n\} = p(x, n, t|x_0, n_0, 0)dx. \tag{3}$$

The probability density evolves according to the differential Chapman-Kolmogorov (CK) equation (dropping the explicit dependence on initial conditions)

$$\begin{aligned} &\frac{\partial p}{\partial t} + \frac{\partial [F(x, n)p(x, n, t)]}{\partial x} \\ &\quad = \frac{1}{\tau_n} \sum_{n' \in I} [W(n, n'; x)p(\mathbf{x}, n', t) - W(n', n; x)p(x, n, t)] . \end{aligned} \tag{4}$$

We have introduced two time-scales in the system, a relaxation time-scale τ_x for the x-dynamics and a transition time-scale τ_n for the jump process.

In many of the listed biophysical applications, the kinetics associated with the jump process are much faster than the relaxation dynamics of x, that is, $\tau_n \ll \tau_x$. Let us fix the time units by setting $\tau_x = 1$ and

[a]The continuous dynamics may be higher dimensional but we will restrict our discussion to the one–dimensional case. It is also possible to have a set of discrete variables, but one can always relabel the internal states so that they are effectively indexed by a single integer.

introduce the small parameter $\epsilon = \tau_n/\tau_x$. We can then rewrite (4) in the more compact form

$$\frac{\partial p}{\partial t} = -\frac{\partial[F(x,n)p(x,n,t)]}{\partial x} + \frac{1}{\epsilon}\sum_{n'\in I} A(n,n';x)p(x,n',t), \tag{5}$$

with

$$A(n,n';x) = W(n,n';x) - \sum_{m\in I} W(m,n;x)\delta_{n',n}.$$

In the limit $\epsilon \to 0$, equation (1) reduces to the deterministic or mean-field equation

$$\frac{dx}{dt} = \mathcal{F}(x) \equiv \sum_{n\in I} F(x,n)\rho(x,n), \tag{6}$$

where $\rho(x,n)$ is the unique steady-state density satisfying the equation $\sum_{m\in I} A(n,m;x)\rho(x,m) = 0$. We are assuming that for fixed x, the matrix $A(n,m;x)$ is irreducible and has a simple zero eigenvalue with corresponding left eigenvector $\mathbf{1}$ whose components are all unity, that is, $\sum_{n\in I} A(n,m;x) = 0$ for all m. [The latter follows immediately from the definition of A; we are assuming that there are no other eigenvectors with a zero eigenvalue.] The Perron-Frobenius Theorem then ensures that all other eigenvalues are negative and the continuous-time Markov process for fixed x,

$$\frac{dp(x,n,t)}{dt} = \frac{1}{\epsilon}\sum_{m\in I} A(n,m;x)p(x,m,t),$$

has a globally attracting steady-state $\rho(x,n)$ such that $p(x,n,t) \to \rho(x,n)$ as $t \to \infty$.

Now suppose that the system operates in the regime $0 < \epsilon \ll 1$, for which there are typically a large number of transitions between different internal states n while the variable x hardly changes at all. This suggests that the system rapidly converges to the (quasi) steady state $\rho(x,n)$, which will then be perturbed as x slowly evolves. The resulting perturbations can be analyzed using a quasi-steady-state (QSS) diffusion or adiabatic approximation [23–25], in which the CK equation (5) is approximated by a Fokker-Planck (FP) equation. This method was first developed from a probabilistic perspective by Papanicolaou [23], see also [24]. It has subsequently been applied to a wide range of problems in biology, including cell movement [26, 27], traveling-wavelike behavior in models of slow axonal transport [3, 4, 28], and molecular motor-based models of random intermittent search [10, 25, 29, 30]. (Adiabatic elimination has also been applied to chemical master equations, such as one describing CO oxidation on nanoscale surfaces [31].) The basic idea of the QSS reduction is to decompose the probability density as

$$p(x,n,t) = C(x,t)\rho(x,n) + \epsilon w(x,n,t), \tag{7}$$

where $\sum_{n\in I} p(x,n,t) = C(x,t)$ and $\sum_{n\in I} w(x,n,t) = 0$. Carrying out an asymptotic expansion in ϵ, it can be shown that C evolves according to the FP equation

$$\frac{\partial C}{\partial t} = -\frac{\partial}{\partial x}(\mathcal{F}C) + \epsilon\frac{\partial}{\partial x}\left(\mathcal{D}\frac{\partial C}{\partial x}\right), \tag{8}$$

with the drift term given by equation (6), and diffusion coefficient

$$\mathcal{D}(x) = \sum_{m,n\in I} Z(x,n)F(x,n), \tag{9}$$

where $Z(x,n)$ is the unique solution to

$$\sum_{m\in I} A(n,m;x)Z(x,m) = [\mathcal{F}(x) - F(x,n)]\rho(x,n) \tag{10}$$

with $\sum_m Z(x,m) = 0$. Note that we have dropped $\mathcal{O}(\epsilon)$ contributions to the drift term; if these were included then (8) would be of the standard Stratonovich form [24]. The FP equation captures the Gaussian-like fluctuations within the basin of attraction of a fixed point of the mean-field equations. However, the diffusion approximation for small ϵ breaks down when considering rare event transitions between metastable states. (A similar problem arises in approximating chemical master equations by a Fokker–Planck equation in the large N limit [32, 33].) In this chapter, we review asymptotic methods that have been developed to study first passage time problems in stochastic models of the form (5), and then describe a few concrete applications in biophysics.

2. First passage time problem and the quasistationary approximation

Suppose that the mean-field equation (6) has a stable fixed point x_0, $\mathcal{F}(x_0) = 0$, and a basin of attraction given by the semi-infinite interval $\Omega = (-\infty, x_*)$; the point x_* corresponds to an unstable fixed point. For small but finite ϵ the fluctuations about the steady state $\rho(x,n)$ can induce rare transitions out of the basin of attraction due to a metastable trajectory crossing the point x_*. Assume that the stochastic system is initially at x_0. On short time scales ($t \ll 1/\epsilon$) the system rapidly converges to a quasistationary solution within Ω, which can be approximated by a Gaussian solution of the reduced FP equation (8). However, on a longer time scale, probability slowly leaks out of the boundary $\partial\Omega$ due to rare transitions whose exponentially small rates cannot be calculated accurately using the QSS diffusion approximation. One thus has to work with the full CK equation (5) supplemented by absorbing boundary conditions at x_*:

$$p(x_*, n, t) = 0, \text{ for all } n \in \Sigma, \tag{11}$$

where $\Sigma \subset I$ is the set of internal states n for which $F(x_*, n) < 0$. The initial condition is taken to be

$$p(x, n, 0) = \delta(x - x_0)\delta_{n,n_0}. \tag{12}$$

Let T denote the (stochastic) first passage time for which the system first reaches x_*, given that it started at x_0. The distribution of first passage times is related to the survival probability that the system hasn't yet reached x_*:

$$S(t) \equiv \int_\Omega \sum_{n\in I} p(x, n, t)dx. \tag{13}$$

That is, $\text{Prob}\{t > T\} = S(t)$ and the first passage time density is

$$f(t) = -\frac{dS}{dt} = -\int_\Omega \sum_{n\in I} \frac{\partial p}{\partial t}(x, n, t)dx. \tag{14}$$

Substituting for $\partial p_n/\partial t$ using the CK equation (5) shows that

$$f(t) = \int_\Omega \left[\sum_{n\in I} \frac{\partial[F(x,n)p(x,n,t)]}{\partial x}\right] dx = \sum_{n\in I} p(x_*, n, t)F(x_*, n). \tag{15}$$

We have used $\sum_{n\in I} A(n, m; x) = 0$ and $\lim_{x\to-\infty} F(x, n)p(x, n, t) = 0$. The first passage time density can thus be interpreted as the probability flux $J(x, t)$ at the absorbing boundary, since we have the conservation law

$$\sum_{n\in I} \frac{\partial p(x, n, t)}{\partial t} = -\frac{\partial J(x, t)}{\partial x}, \quad J(x, t) = \sum_{n\in I} F(x, n)p(x, n, t). \tag{16}$$

The first passage time problem in the weak noise limit ($\epsilon \ll 1$) has been well studied in the case of FP equations and master equations, see for example [32–38]. (For an FP equation ϵ would represent the noise amplitude, whereas for a master equation $\epsilon = 1/N$, where N is the number of discrete states). One of the characteristic features of the weak noise limit is that the flux through the absorbing boundary and the inverse of the mean first passage time (MFPT) $\langle T \rangle$ are exponentially small, that is, $\langle T \rangle \sim \mathrm{e}^{-C/\epsilon}$ for some constant C. This means that standard singular perturbation theory cannot be used to solve the resulting boundary value problem, in which one matches inner and outer solutions of a boundary layer around the point $x = x_*$. Instead, one proceeds by finding a quasistationary solution using a Wentzel-Kramers-Brillouin (WKB) approximation. The WKB method can be extended to the CK equation (5) using a so-called spectral projection method [14, 15, 39]. In order to apply the projection method, it is necessary to assume certain properties of the non self-adjoint linear operator $-\widehat{L}$ on the right-hand side of (5) with respect to the Hilbert space of functions $h(x,n)$ with inner product defined according to

$$\langle h, g \rangle = \sum_{n \in I} \int_{\Omega} h(x,n) g(x,n) dx. \tag{17}$$

(i) $\widehat{L}$ has a complete set of eigenfunctions ϕ_r with

$$\widehat{L}\phi_r(x,n) \equiv \frac{d}{dx}(F(x,n)\phi_r(x,n)) - \frac{1}{\epsilon}\sum_{m \in I} A(n,m;x)\phi_r(x,m) = \lambda_r \phi_r(x,n), \tag{18}$$

together with the boundary conditions

$$\phi_r(x_*, n) = 0, \text{ for } n \in \Sigma. \tag{19}$$

(ii) The eigenvalues λ_r all have positive definite real parts and the smallest eigenvalue λ_0 is real and simple. Thus we can introduce the ordering $0 < \lambda_0 < \mathrm{Re}[\lambda_1] \le \mathrm{Re}[\lambda_2] \le \ldots$.

(iii) λ_0 is exponentially small, $\lambda_0 \sim \mathrm{e}^{-C/\epsilon}$, whereas $\mathrm{Re}[\lambda_r] = \mathcal{O}(1)$ for $r \ge 1$. In particular, $\lim_{\epsilon \to 0} \lambda_0 = 0$ and $\lim_{\epsilon \to 0} \phi_0(x,n) = \rho(x,n)$.

Under the above assumptions, we can introduce the eigenfunction expansion

$$p(x,n,t) = \sum_{r=0}^{M-1} C_r \mathrm{e}^{-\lambda_r t} \phi_r(x,n), \tag{20}$$

where $M = |I|$ is the number of internal states, and $0 < \lambda_0 \ll \mathrm{Re}[\lambda_r]$ for all $r \ge 1$. Thus, at large times we have the quasistationary approximation

$$p(x,n,t) \sim C_0 \mathrm{e}^{-\lambda_0 t} \phi_0(x,n). \tag{21}$$

Substituting such an approximation into equation (15) gives

$$f(t) \sim C_0 e^{-\lambda_0 t} \sum_{n\in I} F(x_*, n)\phi_0(x_*, n), \quad \lambda_1 t \gg 1. \tag{22}$$

Equation (18) implies that

$$\begin{aligned}\sum_{n\in I} \int_\Omega \widehat{L}\phi_0(x,n)dx &\equiv \sum_{n\in I} F(x_*, n)\phi_0(x_*, n, t)\\ &= \lambda_0 \sum_{n\in I} \phi_0(x,n)dx.\end{aligned}$$

In othor wordo,

$$\lambda_0 = \frac{\sum_{n\in I} F(x_*, n)\phi_0(x_*, n)}{\langle 1, \phi_0\rangle}. \tag{23}$$

Combining equations (23) and the quasistationary approximation (22) shows that the (normalized) first passage time density reduces to

$$f(t) \sim \lambda_0 e^{-\lambda_0 t} \tag{24}$$

and, hence, $\langle T\rangle = \int_0^\infty t f(t)dt \sim 1/\lambda_0$.

It remains to obtain an approximation ϕ_ϵ of the principal eigenfunction ϕ_0, which can be achieved using the WKB method as described in §3.2. This yields a quasistationary density that approximates ϕ_0 up to exponentially small terms at the boundary, that is,

$$\widehat{L}\phi_\epsilon = 0, \quad \phi_\epsilon(x_*, n) = \mathcal{O}(e^{-C/\epsilon}). \tag{25}$$

In order to express λ_0 in terms of the quasistationary density ϕ_ϵ, we consider the eigenfunctions of the adjoint operator, which satisfy the equation

$$\widehat{L}^\dagger \xi_s(x,n) \equiv -F(x,n)\frac{d\xi_s(x,n)}{dx} - \frac{1}{\epsilon}\sum_{m\in I} A(m,n;u)\xi_s(x,m) = \lambda_s \xi_s(x,n) \tag{26}$$

and the boundary conditions

$$\xi_s(x_*, n) = 0, \quad n \notin \Sigma. \tag{27}$$

Given our assumptions regarding the spectral properties of $\widehat{L}$, the two sets of eigenfunctions form a biorthonormal set with

$$\langle \phi_r, \xi_s\rangle = \delta_{r,s}. \tag{28}$$

Now consider the identity

$$\langle \phi_\epsilon, \widehat{L}^\dagger \xi_0\rangle = \lambda_0 \langle \phi_\epsilon, \xi_0\rangle. \tag{29}$$

Integrating by parts the left-hand side of equation (29) picks up a boundary term so that

$$\lambda_0 = -\frac{\sum_{n\in I} \phi_\epsilon(x_*, n)F(x_*, n)\xi_0(x_*, n)}{\langle \phi_\epsilon, \xi_0\rangle}. \tag{30}$$

The calculation of the principal eigenvalue λ_0 thus reduces to the problem of determining the quasistationary density ϕ_ϵ and the adjoint eigenfunction ξ_0 using perturbation methods (see below). Once λ_0 has been evaluated, we can then identify the mean first passage time $\langle T\rangle$ with λ_0^{-1}.

2.1. *WKB method and the quasistationary density*

We now show how the Wentzel-Kramers-Brillouin (WKB) method [32, 33, 36–38] can be used to compute the quasi stationary density ϕ_ϵ of ϕ_0. We thus seek a solution of the WKB form

$$\phi_\epsilon(x, n) \sim R(x, n) \exp\left(-\frac{\Phi(x)}{\epsilon}\right), \tag{31}$$

where $\Phi(x)$ is a scalar potential. Substituting into the equation $\widehat{L}\phi_\epsilon = 0$,

$$\begin{aligned}\sum_{m\in I} (A(n, m; x) + \Phi'(x)\delta_{n,m}F(x, m))\, R(x, m) \\ = \epsilon\frac{dF(x, n)R(x, n)}{dx} + \lambda_0 R(x, n),\end{aligned} \tag{32}$$

where $\Phi' = d\Phi/dx$. Introducing the asymptotic expansions $R \sim R^{(0)}+\epsilon R^{(1)}$ and $\Phi \sim \Phi_0+\epsilon\Phi_1$, and using the fact that $\lambda_0 = \mathcal{O}(\mathrm{e}^{-L/\epsilon})$, the leading order equation is

$$\sum_{m\in I} A(n, m; x)R^{(0)}(x, m) = -\Phi_0'(x)F(x, n)R^{(0)}(x, n). \tag{33}$$

[Note that since $F(x, n)$ is non-zero almost everywhere for $x \in \Omega$, we can identify $-\Phi_0'$ and $R^{(0)}$ as an eigenpair of the matrix operator $\widehat{A}(n, m; x) = A(n, m; x)/F(x, n)$ for fixed x.] Positivity of the probability density ϕ_ϵ requires positivity of the corresponding solution $R^{(0)}$. One positive solution is $R^{(0)} = \rho$, for which $\Phi_0' = 0$. However, such a solution is not admissible since Φ_0 = constant. It can be proven using linear algebra [15], that if $F(x, n)$ for fixed $x \in \Omega$ changes sign as n increases from zero, then there exists one other positive solution, which has the correct behavior to describe the potential well. That is, $\Phi_0'(x)$ has the correct sign and vanishes at the fixed points. Hence, it can be identified as the appropriate WKB solution.

In §3 we will show how $R^{(0)}$ can be calculated explicitly in a particular application.

Proceeding to the next order in the asymptotic expansion of equation (32), we have

$$\sum_{m\in I} \left(A(n,m;x) + \Phi_0'(x)\delta_{n,m}F(x,m)\right) R^{(1)}(x,m) = \frac{dF(x,n)R^{(0)}(x,n)}{dx} - \Phi_1'(x)F(x,n)R^{(0)}(x,n). \tag{34}$$

For fixed x and WKB potential Φ_0, the matrix operator $\bar{A}(n,m;x) = A(n,m;x) + \Phi_0'(x)\delta_{n,m}F(x,m)$ on the left-hand side of this equation has a one-dimensional null space spanned by the positive WKB solution R_0. The Fredholm Alternative Theorem[b] then implies that the right-hand side of (34) is orthogonal to the left null vector S of $\bar{A}$. That is, we have the solvability condition

$$\sum_{n\in I} S(x,n)\left[\frac{dF(x,n)R^{(0)}(x,n)}{dx} - \Phi_1'(x)F(x,n)R^{(0)}(x,n)\right] = 0, \tag{35}$$

with S satisfying

$$\sum_{n\in I} S(x,n)\left(A(n,m;x) + \Phi_0'(x)\delta_{n,m}F(x,m)\right) = 0. \tag{36}$$

Given $R^{(0)}$, S and Φ_0, the solvability condition yields the following equation for Φ_1:

$$\Phi_1'(x) = \frac{\sum_{n\in I} S(x,n)[F(x,n)R^{(0)}(x,n)]'}{\sum_{n\in I} S(x,n)F(x,n)R^{(0)}(x,n)}. \tag{37}$$

Combining the various results, and defining

$$k(x) = \exp\left(-\int_{x_0}^{x} \Phi_1'(y)dy\right), \tag{38}$$

gives to leading order in ϵ,

$$\phi_\epsilon(x,n) \sim \mathcal{N}k(x)\exp\left(-\frac{\Phi_0(x)}{\epsilon}\right)R^{(0)}(x,n), \tag{39}$$

---

[b]Consider an M-dimensional linear inhomogeneous system $\mathbf{Ax} = \mathbf{b}$ with $\mathbf{x}, \mathbf{b} \in \mathbb{R}^M$. Suppose that the $M \times M$ matrix $\mathbf{A}$ has a nontrivial null-space and let $\mathbf{v}$ be a null vector of the adjoint matrix $\mathbf{A}^\dagger$, that is, $\mathbf{A}^\dagger\mathbf{v} = 0$. The Fredholm Alternative Theorem states that the inhomogeneous equation has a (non-unique) solution if and only if $\mathbf{v}\cdot\mathbf{b} = 0$ for all null vectors $\mathbf{v}$.

where we choose $\sum_{n\in I} R^{(0)}(x,n) = 1$ for all x and $\mathcal{N}$ is the normalization factor,

$$\mathcal{N} = \left[\int_\Omega k(x)\exp\left(-\frac{\Phi_0(x)}{\epsilon}\right)\right]^{-1}. \tag{40}$$

The latter can be approximated using Laplace's method to give

$$\mathcal{N} \sim \frac{1}{k(x_0)}\sqrt{\frac{|\Phi_0''(x_0)|}{2\pi\epsilon}}\exp\left(\frac{\Phi_0(x_0)}{\epsilon}\right). \tag{41}$$

2.2. *Perturbation analysis of the adjoint eigenfunction*

Following [14, 15, 40], the adjoint eigenfunction $\xi_0(x,n)$ can be approximated using singular perturbation methods. Since λ_0 is exponentially small in ϵ, equation (26) yields the leading order equation

$$\epsilon F(x,n)\frac{d\xi_0(x,n)}{dx} + \sum_{m\in I} A(m,n;x)\xi_0(x,m) = 0, \tag{42}$$

supplemented by the absorbing boundary condition

$$\xi_0(x_*,n) = 0, \quad n \notin \Sigma. \tag{43}$$

A first attempt at obtaining an approximate solution that also satisfies the boundary conditions is to construct a boundary layer in a neighborhood of the unstable fixed point x_* by performing the change of variables $x = x_* - \epsilon z$ and setting $Q(z,n) = \xi_0(x_* - \epsilon z)$. Equation (42) then becomes

$$F(x_*,n)\frac{dQ(z,n)}{dz} + \sum_{m\in I} A(m,n;x_*)Q(z,m) = 0. \tag{44}$$

This inner solution has to be matched with the outer solution $\boldsymbol{\xi}_0 = \mathbf{1}$, which means that

$$\lim_{z\to\infty} Q(z,n) = 1 \tag{45}$$

for all $n \in I$. Consider the eigenvalue equation

$$\sum_{n\in I} (A(n,m;x) - \mu_r(x)\delta_{n,m}F(x,m))\, S_r(x,n) = 0, \tag{46}$$

with $r = 0,\ldots,M-1$. We take $S_0(x,n) = 1$ so that $\mu_0 = 0$ and set $S_1(x,n) = S(x,n), \mu_1(x) = -\Phi_0'(x)$, where S satisfies equation (36). We then introduce the eigenfunction expansion

$$Q(z,n) = c_0 + \sum_{r=1}^{M-1} c_r S_r(x_*,n)\mathrm{e}^{-\mu_r(x_*)z}. \tag{47}$$

In order that the solution remains bounded as $z \to \infty$ we require that $c_r = 0$ if $\mathrm{Re}[\mu_r(x_*)] < 0$. The boundary conditions (43) generate a system of linear equations for the coefficients c_r with codimension $K = |\Sigma|$. One of the unknowns is determined by matching the outer solution, which suggests that there are $K-1$ eigenvalues with negative real part. The eigenvalues are ordered so that $\mathrm{Re}[\mu_r(x_*)] < 0$ for $r \geq M-K$, and we set the corresponding coefficients c_r to zero.

There is, however, one problem with the above eigenfunction expansion, namely, that $\mu_1(x_*) \equiv -\Phi_0'(x_*) = 0$ so that the zero eigenvalue is degenerate at $x = x_*$. Hence, the solution needs to include a secular term involving the generalized eigenvector $\widehat{S}$,

$$\sum_{n\in I} A(n, m; x_*)\widehat{S}(x_*, n) = -F(x_*, m). \tag{48}$$

The Fredholm Alternative Theorem ensures that $\widehat{S}$ exists and is unique, since the stationary density $\rho(x_*, m)$ is the right null vector of $A(n, m; x_*)$ and $\sum_{n\in I} \rho(x_*, n)F(x_*, n) \equiv \mathcal{F}(x_*) = 0$, see equation (6). The solution for $\mathbf{Q}(z)$ is now

$$Q(z, n) = c_0 + c_1(\widehat{S}(x_*, n) - z) + \sum_{r=2}^{M-K} c_r S_r(x_*, n)\mathrm{e}^{-\mu_r(x_*)z}. \tag{49}$$

The presence of the secular term means that the solution is unbounded in the limit $z \to \infty$, which means that the inner solution cannot be matched with the outer solution. One way to remedy this situation is to introduce an alternative scaling in the boundary layer of the form $x = x_* + \epsilon^{1/2}z$, as detailed in [40]. One can then eliminate the secular term $-c_1 z$ and show that

$$c_1 \sim \sqrt{\frac{2|\Phi_0''(x_*)|}{\pi}} + \mathcal{O}(\epsilon^{1/2}), \quad c_r = \mathcal{O}(\epsilon^{1/2}) \text{ for } r \geq 2 \tag{50}$$

2.3. *Principal eigenvalue*

It turns out that we only require the first coefficient c_1 in order to evaluate the principal eigenvalue λ_0 using equation (30). This follows from equations (33) and (46), and the observation that the left and right eigenvectors of the matrix $\widehat{A}(n, m; x) = A(n, m; x)/F(x, n)$ are biorthogonal. In particular, since the quasistationary approximation ϕ_ϵ is proportional to $R^{(0)}$, see equation (39), it follows that ϕ_ϵ is orthogonal to all eigenvectors S_r, $r \neq 1$. Simplifying the denominator of equation (30) by using the outer solution

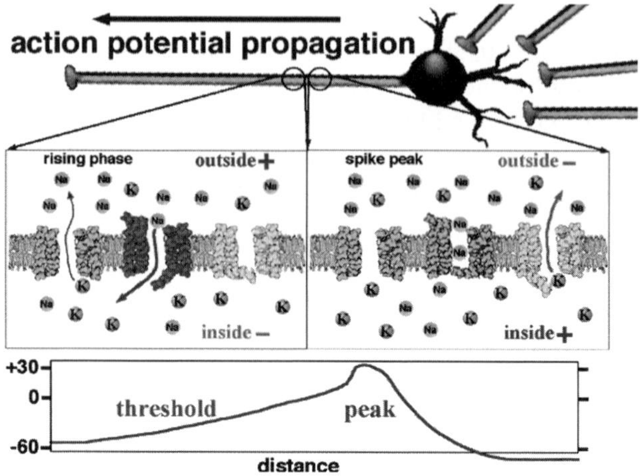

Fig. 1. Schematic diagram illustrating the propagation of an action potential along an axon, which is mediated by the voltage-gated opening and closing of Na^+ and K^+ ion channels. [Public domain figure from Wikimedia Commons].

$\xi_0 \sim \mathbf{1}$, we obtain

$$\lambda_0 \sim -\frac{\sum_{n\in I} \xi_0(x_*,n)F(x_*,n)\phi_\epsilon(x_*,n)}{\langle \phi_\epsilon, \mathbf{1}\rangle}$$

$$\sim c_1 \frac{k(x_*)B(x_*)}{k(x_0)} \sqrt{\frac{|\Phi''(x_0)|}{2\pi}} \exp\left(-\frac{\Phi_0(x_*)-\Phi_0(x_0)}{\epsilon}\right), \qquad (51)$$

with

$$B(x_*) = -\sum_{n\in I} \widehat{S}(x_*,n)F(x_*,n)\rho(x_*,n). \qquad (52)$$

Substituting for c_1

$$\lambda_0 \sim \frac{1}{\pi}\frac{k(x_*)B(x_*)}{k(x_0)} \sqrt{\Phi_0''(x_0)|\Phi_0''(x_*)|} \exp\left(-\frac{\Phi_0(x_*)-\Phi_0(x_0)}{\epsilon}\right). \quad (53)$$

Finally, comparison of equations (48) and (52) with (10) and (9) establishes that $B(x_*) \equiv \mathcal{D}(x_*)$.

3. Biophysical applications

3.1. *Stochastic ion channels*

The generation and propagation of a neuronal action potential arises from nonlinearities associated with active membrane conductances, see Fig. 1.

Ions can diffuse in and out of the cell through ion specific channels embedded in the cell membrane. Ion pumps within the cell membrane maintain concentration gradients, such that there is a higher concentration of Na^+ and Ca^{2+} outside the cell and a higher concentration of K^+ inside the cell. The membrane current through a specific channel varies approximately linearly with changes in the voltage x relative to some equilibrium or reversal potential, which is the potential at which there is a balance between the opposing effects of diffusion and electrical forces. [We will focus on a space-clamped model of a neuron whose cell body is taken to be an iso-potential.] Summing over all channel types, the total membrane current (flow of positive ions) leaving the cell through the cell membrane is

$$I_{\text{con}} = \sum_s g_s(x - V_s), \tag{54}$$

where g_s is the conductance due to channels of type s and V_s is the corresponding reversal potential.

Recordings of the current flowing through single channels indicate that channels fluctuate rapidly between open and closed states in a stochastic fashion. Nevertheless, most models of a neuron use deterministic descriptions of conductance changes, under the assumption that there are a large number of approximately independent channels of each type. It then follows from the law of large numbers that the fraction of channels open at any given time is approximately equal to the probability that any one channel is in an open state. The conductance g_s for ion channels of type s is thus taken to be the product $g_s = \bar{g}_s P_s$ where $\bar{g}_s$ is equal to the density of channels in the membrane multiplied by the conductance of a single channel and P_s is the fraction of open channels. The voltage-dependence of the probabilities P_s in the case of a delayed-rectifier K^+ current and a fast Na^+ current were originally obtained by Hodgkin and Huxley [41] as part of their Nobel prize winning work on the generation of action potentials in the squid giant axon. The delayed-rectifier K^+ current is responsible for terminating an action potential by repolarizing a neuron. One finds that opening of the K^+ channel requires structural changes in 4 identical and independent subunits so that $P_K = n^4$ where n is the probability that any one gate subunit has opened. In the case of the fast Na^+ current, which is responsible for the rapid depolarization of a cell leading to action potential generation, the probability of an open channel takes the form $P_{Na} = m^3 h$ where m^3 is the probability that an activating gate is open and h is the probability that an inactivating gate is open. Depolarization causes m to increase and h to decrease, whereas hyperpolarization has the opposite effect.

The dynamics of the gating variables m, n, h are usually formulated in terms of a simple kinetic scheme that describes voltage-dependent transitions of each gating subunit between open and closed states. More specifically, for each $Y \in \{m, n, h\}$

$$\frac{dY}{dt} = \alpha_Y(x)(1 - Y) - \beta_Y(x)Y, \tag{55}$$

where $\alpha_Y(x)$ is the rate of the transition *closed* $\rightarrow$ *open* and $\beta_Y(x)$ is the rate of the reverse transition *open* $\rightarrow$ *closed*. From basic thermodynamic arguments, the opening and closing rates are expected to be exponential functions of the voltage x

$$\alpha_Y(x) = A_Y \mathrm{e}^{-B_Y x}, \quad \beta_Y(x) = A'_Y \mathrm{e}^{-B'_Y x}.$$

Hodgkin and Huxley originally fitted exponential-like functions to the experimental data obtained from the squid axon. The corresponding conductance-based model (in the absence of synaptic inputs) can then be written in the form

$$C\frac{dx}{dt} = f(x, m, n, h) + I_{\text{ext}}, \tag{56}$$

with

$$f(x, m, n, h) = -\bar{g}_{\text{Na}} m^3 h(x - V_{\text{Na}}) - \bar{g}_{\text{K}} n^4 (x - V_{\text{K}}) - \bar{g}_{\text{L}}(x - V_{\text{L}}). \tag{57}$$

Here $I_L = g_L(x - V_L)$ is called a leak current, which represents the passive flow of ions through nongated channels.

The stochastic opening and closing of ion channels is thought to be one of the main sources of intrinsic noise at the cellular level. Almost all studies of ion channel fluctuations are based on a system size expansion with respect to the number N of ion channels, in which one reduces the dynamics to an effective Langevin equation [11, 16, 42, 43]. However, such a reduction breaks down if the number of ion channels is small. Moreover, it can lead to exponentially large errors when estimating the rate of spontaneous action potential generation. Recently, Keener and Newby [14, 15] have used the WKB method for fast kinetics outlined in §2 in order to analyze spontaneous action potential generation, which holds for small N. They consider the simplified problem of how ion channel fluctuations affect the initiation of an action potential due to the opening of a finite number of Na^+ channels. The slow K^+ channels are assumed to be frozen, so that they effectively act as a leak current, and each sodium channel is treated as a single activating subunit. The stochastic membrane voltage x then evolves according to the piecewise deterministic equation

$$\frac{dx}{dt} = F(x,n) \equiv \frac{1}{N} f(x)n - g(x), \tag{58}$$

where $f(x) = g_{\rm Na}(V_{\rm Na} - x)$ represents the gated sodium currents, $g(x) = -g_{\rm eff}[V_{\rm eff} - x] - I_{\rm ext}$ represents the sum of effective leakage currents and external inputs, and n is the number of open sodium channels. Note that the right-hand side of (58) is negative for large x and positive for small x. This implies that the stochastic voltage x is confined to some bounded domain $[x_1, x_2]$. The opening and closing of the ion channels is described by a birth-death process according to $n \to n+1$ at a rate $\omega_+(x,n) = \alpha(x)(N-n)$ and $n \to n-1$ at a rate $\omega_-(n) = \beta n$ with

$$\alpha(x) = \beta \exp\left(\frac{2(x-v_1)}{v_2}\right)$$

for constants β, v_1, v_2.

The associated probability density $p(x,n,t)$ satisfies the differential Chapman-Kolmogorov (CK) equation

$$\begin{aligned}\frac{\partial p}{\partial t} &= -\frac{\partial[F(x,n)p(x,n,t)]}{\partial v} \\ &\quad + \frac{1}{\epsilon}[\omega_+(x,n-1)p(x,n-1,t) + \omega_-(n+1)p(x,n+1,t) \\ &\qquad -(\omega_+(x,n) + \omega_-(n))p(x,n,t)],\end{aligned} \tag{59}$$

with boundary condition $p(x,-1,t) \equiv 0, p(x,N+1,t) = 0$. The equation for the steady–state distribution $\rho(x,n)$ can be written as [24] $0 = J(x,n+1) - J(x,n)$ with $J(x,n)$ the probability current,

$$J(x,n) = \omega_-(n)\rho(x,n) - \omega_+(x,n-1)\rho(x,n-1).$$

Since $\omega_-(0) = 0$ and $\rho(x,-1) = 0$, it follows that $J(x,0) = 0$ and $J(x,n) = 0$ for all $n \geq 0$. Hence,

$$\rho(x,n) = \frac{\omega_+(x,n-1)}{\omega_-(n)}\rho(x,n-1) = \rho(x,0)\prod_{m=1}^{n}\frac{\omega_+(x,m-1)}{\omega_-(m)} \tag{60}$$

with $\rho(x,0) = 1 - \sum_{m=1}^{N}\rho(x,m)$. Substituting the explicit expressions for the transition rates, we have

$$\rho(x,n) = \rho(x,0)\prod_{m=1}^{n}\frac{(N-m+1)\alpha(x)}{m\beta} = \rho(x,0)\left(\frac{\alpha(x)}{\beta}\right)^n \frac{N!}{(N-n)!n!}. \tag{61}$$

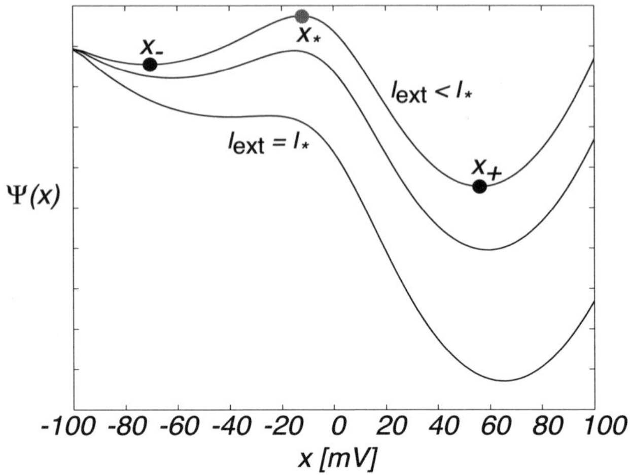

Fig. 2. Plot of deterministic potential $\Psi(x)$ as a function of voltage x for different values of the external stimulus current I_{ext}. Parameter values are $N = 10$, $V_{\text{Na}} = 120$ mV, $V_{\text{eff}} = -62.3$ mV, $g_{\text{Na}} = 4.4$ mS/cm^2, $g_{\text{eff}} = 2.2$ mS/cm^2, $\beta = 0.8$ s^{-1}, and $\alpha(x) = \beta \exp[(x + 1.2)/9]$.

It follows that the normalized density is

$$\rho(x, n) = \frac{N!}{(N-n)!n!} a(x)^n b(x)^{N-n}, \quad a(x) = \frac{\alpha(x)}{\alpha(x) + \beta}, \, b(x) = \frac{\beta}{\alpha(x) + \beta}. \tag{62}$$

The mean number of open channels is then

$$\langle n \rangle = \sum_{n=1}^{N} n\rho(x, n) = Na(x). \tag{63}$$

It follows that the mean-field equation obtained in the $\epsilon \to 0$ limit is

$$\frac{dx}{dt} = a(x)f(x) - g(x) \equiv -\frac{d\Psi}{dx}. \tag{64}$$

We have introduced the effective potential $\Psi(x)$ whose minima and maxima correspond to stable and unstable fixed points of the mean-field equation. By plotting the potential Ψ, it is straightforward to show that equation (64) exhibits bistability for a range of stimuli I_{ext}. That is, there exist two stable fixed points $x_{\pm}$ separated by an unstable fixed point x_*, see Fig. 2.

The rate of initiation of spontaneous action potentials for small ϵ can be reformulated as an FPT problem, in which we calculate the rate of rare transitions from the metastable resting state x_- to the active state x_+. In order to use the machinery of §2, we first rewrite the CK equation (59) in the form (5) with F given by (58), the discrete set $\mathcal{I} = 0, 1, \ldots, N$, and A a tridiagonal matrix (for fixed x):

$$A_{n,n-1;x} = \omega_+(x, n-1),\ A_{n,n;x} = -\omega_+(x,n)-\omega_-(n),\ A_{n,n+1;x} = \omega_-(n+1). \tag{65}$$

Equation (65) is supplemented by absorbing boundary conditions at the unstable fixed point x_*,

$$p_n(x_*, t) = 0, \text{ for } n \in \Sigma = \{0, \ldots, k-1\}, \tag{66}$$

where $0 < k < N$ is the number of negative components of the drift. [The subsequent time to travel from x_* to the high activity fixed point x_+ is insignificant, and can be neglected]. There are three basic steps needed in order to evaluate the escape rate formula (53) with $x_0 \to x_-$.

(1) Find the unique nontrivial positive eigenfunction $\psi_n(x) = R^{(0)}(x, n)$ and associated eigenvalue $\mu(x) = -\Phi_0'(x)$. In the case of the stochastic ion channel model, equation (33) takes the explicit form

$$\begin{aligned}(N-n+1)\alpha\psi_{n-1} - [n\beta + (N-n)\alpha]\psi_n + (n+1)\beta\psi_{n+1} \\ = \mu\left(\frac{n}{N}f(x) - g(x)\right)\psi_n.\end{aligned} \tag{67}$$

Motivated by the form of the stationary density $\rho(x, n)$, equation (61), we try the solution

$$\psi_n(x) = \frac{\Lambda(x)^n}{(N-n)!n!}, \tag{68}$$

which yields the following equation relating Λ and μ:

$$\frac{n\alpha}{\Lambda} + \Lambda\beta(N-n) - n\beta - (N-n)\alpha = \mu\left(\frac{n}{N}f(x) - g(x)\right).$$

We now fix μ in terms of Λ so that the terms linear in n vanish:

$$\mu = \frac{N}{f}\left[\alpha\left(\frac{1}{\Lambda} + 1\right) - \beta(\Lambda + 1)\right].$$

Eliminating μ then shows that

$$\Lambda(x) = \frac{g(x)}{f(x) - g(x)}.$$

We deduce that

$$\mu(x) = N\frac{\alpha(x)f(x) - (\alpha(x) + \beta)g(x)}{g(x)(f(x) - g(x))}, \tag{69}$$

and the normalized eigenvector is

$$\psi_n(x) = \frac{N!}{(N-n)!n!}\frac{(f(x)-g(x))^{N-n}g(x)^n}{f(x)^N}. \quad (70)$$

Note that $\mu(x)$ vanishes at the fixed points x_-, x_* of the mean-field equation (64) with $\mu(x) > 0$ for $0 < x < x_-$ and $\mu_1(x) > 0$ for $x_- < x < x_*$. Moreover, comparing equation (61) with (70) establishes that $\psi_n(x) = \rho(x,n)$ at the fixed points $x_*, x_\pm$. In conclusion $\mathbf{R}^{(0)}(x,n) = \psi_n(x)$ and the effective potential Φ_0 is given by

$$\Phi_0(x) = -\int_{x_-}^{x} \mu_1(y)dy. \quad (71)$$

The effective potential is defined up to an arbitrary constant, which has been fixed by setting $\Phi_0(x_-) = 0$.

(2) Determine the null eigenfunction $\eta_n(x) = S(x,n)$ of equation (36), which becomes

$$\begin{aligned}&(N-m)\alpha\eta_{m+1} - [(N-m)\alpha + m\beta]\eta_m + m\beta\eta_{m-1}\\ &= \mu\left(\frac{m}{N}f(x) - g(x)\right)\eta_m. \end{aligned} \quad (72)$$

Trying a solution of the form $\eta_m(x) = \Gamma(x)^m$ yields

$$(N-m)\alpha\Gamma - ((N-m)\alpha + m\beta) + m\beta\Gamma^{-1} = \mu_1\left(\frac{m}{N}f(x) - g(x)\right). \quad (73)$$

Γ is then determined by canceling terms independent of m, which gives

$$\eta_n(x) = \left(\frac{b(x)g(x)}{a(x)(f(x)-g(x))}\right)^n. \quad (74)$$

The prefactor $k(x)$ may now be determined using equations (37) and (38).

(3) Calculate the generalized eigenvector $\zeta_n = \widehat{S}(x^*, n)$ of equation (48), which reduces to

$$(N-n)\alpha(x_*)\zeta_{n+1} + n\beta\zeta_{n-1} - ((N-n)\alpha(x_*) + n\beta)\zeta_n = g(x_*) - \frac{n}{N}f(x_*). \quad (75)$$

It is straightforward to show that this has the solution

$$\zeta_n = \frac{f(x_*)}{N(\alpha(x_*)+\beta)}n. \quad (76)$$

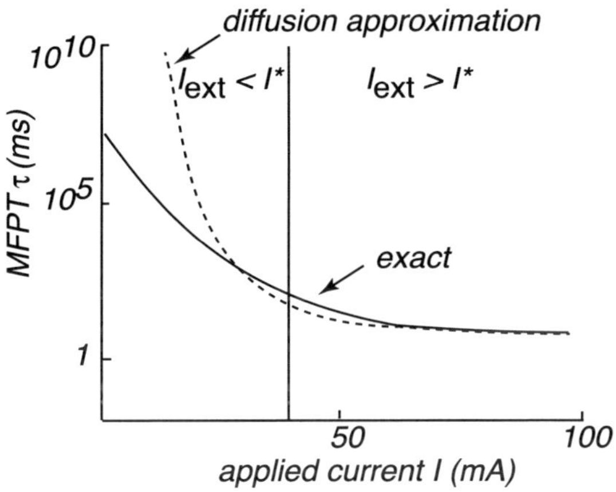

Fig. 3. Schematic diagram comparing MFPT calculated using the diffusion approximation with the MFPT of the full system. The scales of the axes are based on numerical results carried out in [14] for $N = 10$. Other parameter values as in Fig. 2.

It follows from equations (61) and (76) that the factor $B(x*)$ defined by (52) is

$$\begin{aligned} B(x_*) &= -\frac{f(x_*)}{N(\alpha(x_*) + \beta)} \sum_{n=0}^{N} \rho(x_*, n) \left[-g(x_*)n + \frac{f(x_*)}{N} n^2 \right] \\ &= -\frac{f(x_*)}{N(\alpha(x_*) + \beta)} \left[-g(x_*)\langle n \rangle + \frac{f(x_*)}{N} \langle n^2 \rangle \right] \\ &= \frac{f(x_*)^2 \alpha(x_*) \beta}{N(\alpha(x_*) + \beta)^3}, \end{aligned} \tag{77}$$

where the fixed point condition $g(x_*) = f(x_*)a(x_*)$ has been used.

Keener and Newby [14] calculated the MFPT ($\tau = 1/\lambda_0$) using equation (53) and showed that their results agreed very well with Monte-Carlo simulations of the full system, whose probability density evolves according to the CK equation (59). A summary of their findings is shown schematically in Fig. 3, together with the corresponding MFPT obtained using a quasi-steady-state diffusion approximation of (59) given by the FP equation (8). The main observation is that although the Gaussian-like diffusion approximation does well in the superthreshold regime ($I_{\text{ext}} > I_*$), it deviates significantly from the full model results in the subthreshold regime ($I_{\text{ext}} < I_*$), where it overestimates the mean time to spike. This is related to the fact that the effective potential of the steady-state density under the diffusion approximation generates exponentially large errors in the MFPT.

One of the major simplifications of the above analysis by Keener and Newby [14] is to assume that the slow potassium channels are frozen. If one now incorporates the slow opening and closing of these channels, then the underlying deterministic system becomes excitable rather than bistable. That is, there is a single stable fixed point such that for small stimuli the voltage returns directly to rest, whereas for stronger stimuli the voltage makes a large detour before returning to rest, which corresponds to an action potential. There is no longer a well-defined, unique firing threshold. A simple deterministic model of neural excitability is the Morris-Lecar (ML) model [44]:

$$\dot{v} = a(v)f_{\mathrm{Na}}(v) + wf_{\mathrm{K}}(v) - g(v) \tag{78}$$

$$\dot{w} = \frac{w_\infty(v) - w}{\tau_w(v)}, \tag{79}$$

where v is voltage and w represents the fraction of open K^+ channels. The dynamics of this system can be explored using phase-plane analysis as illustrated in Fig. 4. A slow/fast analysis of the deterministic system suggests that the initiation of an action potential occurs without any change in w, thus motivating the approach of Keener and Newby [14]. However, it turns out that this adiabatic approximation breaks down when stochastic fluctuations in the opening and closing of K^+ channels are taken into account.

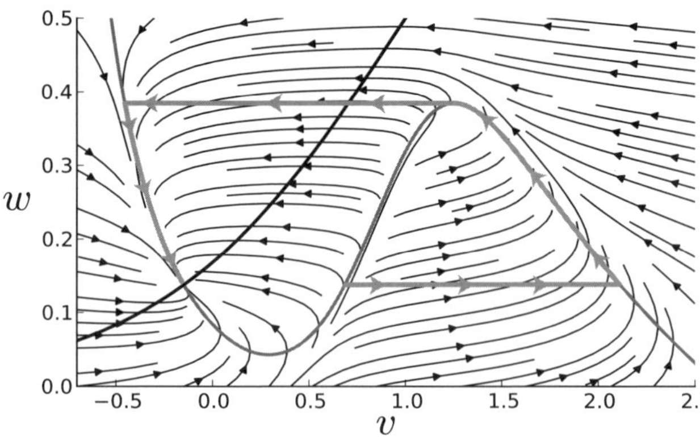

Fig. 4. (Color online) Deterministic phase plane dynamics. Thick curves show the nullclines: $\dot{v} = 0$ as grey and $\dot{w} = 0$ as black. Black stream lines represent deterministic trajectories. Green/blue curves represent an action potential trajectory in the limit of slow w.

This can be established by extending the WKB analysis outlined above to a stochastic version of the ML model [18]. Since one now has two continuous variables x and w in the deterministic limit, it follows that stochastic trajectories in the phase plane correspond to characteristic projections of an underlying Hamiltonian dynamical system. In general, it is difficult to solve FPT problems in more than one dimension. In the case of a metastable state with a well-defined basin of attraction, one has to calculate the MFPT to cross the separatrices forming the boundary of the basin of attraction. There is an additional level of complexity for an excitable system, due to the fact that there is no well-defined deterministic separatrix. Interestingly, we find that the stochastic ML model has an effective separatrix that any stochastic trajectory has to cross in order to generate a stochastic action potential [18], see also [45].

3.2. *Motor-driven intracellular transport*

When considering the active transport of intracellular cargo over relatively long distances, it is often convenient to ignore the microscopic details of how a motor performs a single step (as described by Brownian ratchet models), and to focus instead on the transitions between different types of motion (eg. anterograde vs. retrograde active transport, diffusion vs. active transport). This has motivated a class of mesoscopic models that take the form of a system of PDEs [3, 5–8, 25, 46, 47]. For the sake of illustration, consider an N-state model of a motor-cargo particle moving on a 1D track of length L, see Fig. 5. Such a track could represent a single microtubular filament. Within the interior of the track, $0 < x < L$, the particle is taken to be in one of N internal states labeled by $n = 1, \ldots, N$ with corresponding velocities v_n. The probability density is assumed to evolve according to a particular version of the differential CK equation (5):

$$\frac{\partial p}{\partial t} = -v_n \frac{\partial [p(x,n,t)]}{\partial x} + \frac{1}{\epsilon} \sum_{n' \in I} A(n,n';x) p(x,n',t), \tag{80}$$

where $A(n, n', x)$ specifies the transition rates between the N internal motor states. We assume that the transition rates are fast compared to v/L where $v = \max_n |v_n|$ as indicated by the small parameter ϵ. We also assume that $A(n,m;x)$ is an irreducible matrix for all x with a unique stationary density (right eigenvector) $\rho(x,n)$. In the limit $\epsilon \to 0$, the particle moves deterministically according to the mean-field equation

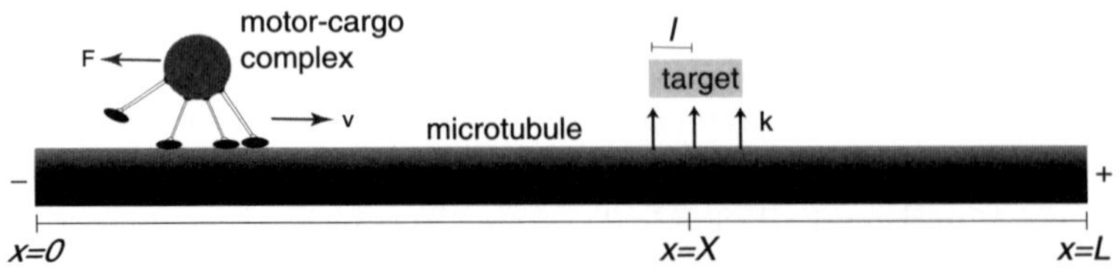

Fig. 5. Schematic diagram illustrating a model of a motor–driven cargo complex moving along a one-dimensional track of length L. The particle can transition between N different internal states with speeds v_n, $n = 1, \ldots, N$. A hidden target of half-width l is located at some unknown position X on the track. The motor-complex can be absorbed by the target at a rate k_n when in internal state n.

$$\frac{dx}{dt} = \mathcal{F}(x) \equiv \sum_{n=1}^{N} v_n \rho(x, n). \tag{81}$$

We will take $v_n > 0$ for $n = 1, \ldots, m$ and $v_n \leq 0$ for $n = m+1, \ldots, N$ with $m > 0$. A simple example is the 3-state model consisting of a forward-moving (anterograde) state ($n = 1, v_1 = v$), a backward-moving (retrograde) state ($n = 2, v_2 = -v$) and a stationary state ($n = 3, v_3 = 0$). If it is assumed that transitions can only occur either into or out of the stationary state, then $A(n, m; x) = A_{nm}(x)$ with the 3×3 matrix

$$\mathbf{A} = \begin{pmatrix} -\beta_+ & 0 & \alpha_+ \\ 0 & -\beta_- & \alpha_- \\ \beta_+ & \beta_- & -\alpha_+ - \alpha_- \end{pmatrix}. \tag{82}$$

The associated stationary density is $\rho(x, n) = \rho_n(x)$ with $\boldsymbol{\rho} = \gamma^{-1}(\beta_+^{-1}, \beta_-^{-1}, \alpha^{-1})^T$ and $\gamma = \beta_+^{-1} + \beta_-^{-1} + \alpha^{-1}$. An example of a more biophysically realistic model is the so-called "tug–of-war model" [48, 49], which identifies an internal state with the number of kinesin and dynein motors bound to the microtubule and exerting a force on the cargo[c].

First suppose that we have a homogeneous track in which the transitions rates are independent of spatial position x. In this case, the steady-state density ρ is independent of x. This means that $\mathcal{F}$ is also independent of x so we have a flat deterministic potential and no fixed points. It is then possible to solve FPT problems using the FP equation (8) under a quasi-steady-state diffusion approximation. In a series of papers [8–10, 25, 29, 30], we have

[c]Microtubules are polarized polymeric filaments with biophysically distinct (+) and (−) ends, and this polarity determines the preferred direction in which an individual molecular motor moves. For example, kinesin moves towards the (+) end whereas dynein moves towards the (−) end.

considered the particular problem of a motor-cargo complex searching for a hidden synaptic target on the dendrite of a neuron, which is formulated as a random intermittent search process, see also [6, 50]. It is straightforward to incorporate a hidden target into the general N-state PDE model (80) by adding the term $k_n\chi(x)p(x,n,t)$ to the right-hand side. The additional term takes into account the probability flux into the target with k_n the rate of target detetction in internal state n and χ is the indicator function

$$\chi(x) = \begin{cases} 1, \text{ if } |x-X| < l \\ 0, \quad \text{otherwise,} \end{cases} \tag{83}$$

where l is the size of the target centered at X. In general k_n will only be non-zero for a subset of states. For example, in the 3-state model, $k_n = k\delta_{j,0}$, that is, the unbound stationary state is identified as the search state. The CK equation (80) is supplemented by absorbing boundary conditions at $x = L$,

$$p(L,n,t) = 0 \text{ for } n = m+1, \ldots N, \tag{84}$$

and a reflecting boundary condition at $x = 0$,

$$\sum_{n=1}^{N} v_n p(0,n,t) = 0. \tag{85}$$

The efficiency of a given search process can be characterized in terms of two important quantities. The first is the hitting probability Π that a particle starting at $x = 0$ at time $t = 0$ finds the target, that is, the particle is absorbed somewhere within the domain $X - l \leq x \leq X + l$. The second is the conditional MFPT $\mathcal{T}$ for the particle to find the target given that it is eventually absorbed by the target. Let $J(t)$ denote the probability flux due to absorption by the target at X,

$$J(t) = \sum_{n=1}^{N} k_n \int_{X-l}^{X+l} p(x,n,t)dx. \tag{86}$$

It follows that

$$\Pi = \int_0^\infty J(t)dt, \quad \mathcal{T} = \frac{\int_0^\infty tJ(t)dt}{\int_0^\infty J(t)dt}. \tag{87}$$

The target detection terms then lead to an inhomogeneous term in the reduced FP equation (8):

$$\frac{\partial C}{\partial t} = -\frac{\partial \mathcal{F}C}{\partial x} + \frac{\partial}{\partial x}\left(\mathcal{D}\frac{\partial C}{\partial x}\right) - \lambda\chi(x-X)u, \tag{88}$$

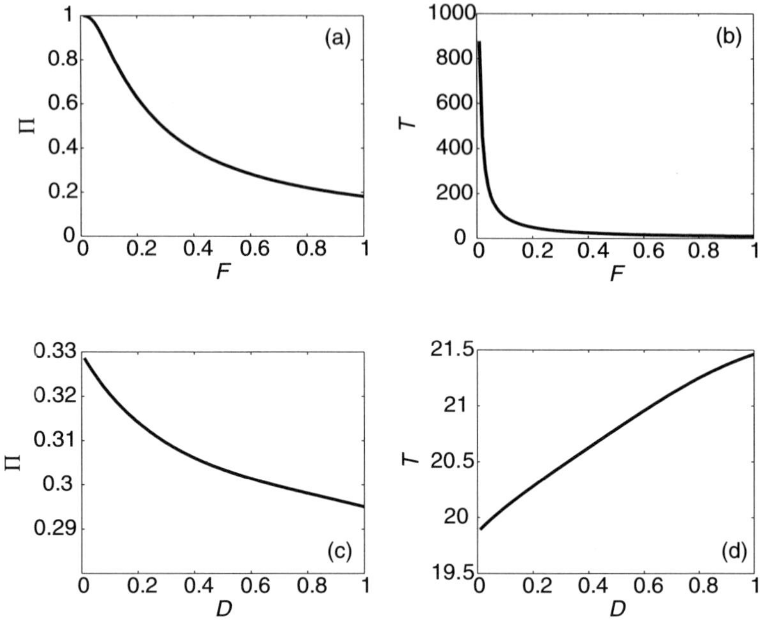

Fig. 6. (a,b) Plots of hitting probability Π and MFPT $\mathcal{T}$ as a function of the drift $\mathcal{F}$ for fixed diffusivity $\mathcal{D} = 0.005$. (c,d) Corresponding plots as a function of D for fixed $V = 0.5$. Other parameter values are $X = 10$, $L = 20$, $\lambda = 0.1$. All parameters are non-dimensionalized by taking the length–scale to be the typical size l of the target and the time–scale to be l/v where v is the typical speed of a molecular motor.

with $\lambda = \sum_{n=1}^{N} k_n \rho(x, n)$. There are then three effective parameters that describe the random search process: the drift $\mathcal{F}$, the diffusivity $\mathcal{D}$, and the target detection rate λ. Each of these parameters are themselves functions of the various cargo velocities, transition rates, and target detection rates contained in the full model. The hitting probability and MFPT are still given by Eqs. (87) except that now the flux is

$$J(t) = \lambda \int_{X-l}^{X+l} u(x,t)dx. \tag{89}$$

In the case of constant drift and diffusion, it is straightforward to solve the FPT problem using the FP equation (88), and the results are summarized in Fig. 6. It can be seen that both Π and $\mathcal{T}$ are monotonically decreasing functions of $\mathcal{F}$ for fixed $\mathcal{D}$. On the other hand, Π is a decreasing function of $\mathcal{D}$, whereas $\mathcal{T}$ is an increasing function of $\mathcal{D}$ for fixed $\mathcal{F}$. This suggests that it is advantageous to reduce the diffusivity as much as possible, since this generates a random search that has a high hitting probability and a relatively small MFPT. However, it should be noted that $\mathcal{D}$ and $\mathcal{F}$ have a complicated dependence on the specific parameters of the underlying biophysical motor transport model, so that changes in $\mathcal{D}$ and $\mathcal{F}$ are generally correlated. Indeed, one finds that it is not possible to simultaneously increase Π and decrease $\mathcal{T}$ by varying a biophysical parameter. That is, there is a playoff between minimizing the MFPT and maximizing the hitting probability. Eleswhere, we have shown that by fixing Π, the MFPT can be minimized by making transport as biased as possible in the anterograde direction. On the other-hand, bidirectional transport is preferable when searching on a dendritic tree structure [29].

One way to enhance the efficiency of the search process would be for the target to generate a local chemical signal that increases the probability of finding the target without a significant increase in the MFPT. This issue has recently been explored using the QSS reduction of the tug-of-war model [9, 25]. One possible signaling mechanism involves microtubule associated proteins (MAPs). These molecules bind to microtubules and effectively modify the free energy landscape of motor-microtubule interactions [51]. For example, tau is a MAP found in the axon of neurons and is known to be a key player in Alzheimer's disease [52]. Another important MAP, called MAP2, is similar in structure and function to tau, but is present in dendrites; MAP2 has been shown to affect dendritic cargo transport [53]. Experiments have shown that the presence of tau or MAP2 on the microtubule can significantly alter the dynamics of kinesin; specifically, by reducing the rate at which kinesin binds to the microtubule [54]. In the simple 3-state model, this could be implemented by taking the binding rate α_+ in the anterograde direction to decrease within the target domain. This means that the effective drift $\mathcal{F}(x)$ in the mean-field equation (81) becomes x-dependent. Suppose, for example, that $\mathcal{F}(x) = \bar{v} > 0$ for $x \notin [X-l, X+l]$ and $\mathcal{F}(x)$ a unimodal function for $x \in [X-l, X+l]$ with a negative minimum at $x = X$. Writing $\mathcal{F}(x) = -\Psi'(x - X)$, the corresponding deterministic potential has the form shown in Fig. 7. Since the drift switches sign within the target domain, it follows that there exists one stable fixed point x_0 and an unstable fixed point x_*.

One interesting effect of a local increase in MAPs is that it can generate stochastic oscillations in the motion of the motor-complex [9]. As a kinesin driven cargo encounters the MAP-coated trapping region the motors unbind at their usual rate and can't rebind. Once the dynein motors are strong enough to pull the remaining kinesin motors off the microtubule, the

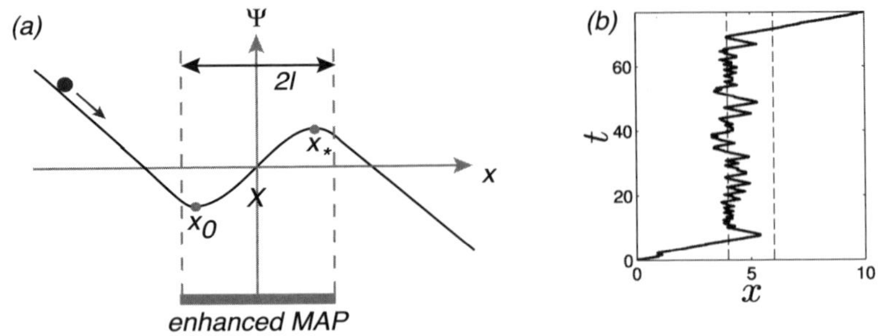

Fig. 7. Diagram showing (a) the effective potential well created by a region of tau coating a MT, and (b) a representative trajectory showing random oscillations within the well.

motor-complex quickly transitions to (−) end directed transport. After the dynein-driven cargo leaves the MAP-coated region, kinesin motors can then re-establish (+) end directed transport until the motor-complex returns to the MAP-coated region. This process repeats until the motor-complex is able to move forward past the MAP-coated region. Interestingly, particle tracking experiments have observed oscillatory behavior during mRNA transport in dendrites [55, 56]. In these experiments, motor-driven mRNA granules move rapidly until encountering a fixed location along the dendrite where they slightly overshoot then stop, move backwards, and begin to randomly oscillate back and forth. After a period of time, lasting on the order of minutes, the motor-driven mRNA stops oscillating and resumes fast ballistic motion. Calculating the mean time to escape the target can be formulated as a FPT problem, in which the particle starts at $x = x_0$ and has to make a rare transition to the unstable fixed point at $x = x_*$. As in the analogous problem of stochastic action potential generation the quasi-steady-state diffusion approximation breaks down for small ϵ, and one has to use the asymptotic methods of §2. The details in the case of the 3-state model can be found elsewhere [15].

References

1. M. Dogterom and S. Leibler, Physical aspects of the growth and regulation of microtubule structures, *Phys. Rev. Lett.* **70**, 1347–1350 (1993).
2. D. J. Bicout, Green's functions and first passage time distributions for dynamic instability of microtubules, *Phys. Rev. E.* **56**, 6656–6667 (1997).
3. M. C. Reed, S. Venakides, and J. J. Blum, Approximate traveling waves in

linear reaction-hyperbolic equations, *SIAM J. Appl. Math.* **50**(1), 167–180 (1990).

4. A. Friedman and G. Craciun, A model of intracellular transport of particles in an axon, *J. Math. Biol.* **51**, 217–246 (2005).
5. A. Friedman and G. Craciun, Approximate traveling waves in linear reaction-hyperbolic equations, *SIAM J. Math. Anal.* **38**, 741–758 (2006).
6. C. Loverdo, O. Benichou, M. Moreau, and R. Voituriez, Enhanced reaction kinetics in biological cells, *Nat. Phys.* **4**, 134–137 (2008).
7. P. Jung and A. Brown, Modeling the slowing of neurofilament transport along the mouse sciatic nerve, *Phys. Biol.* **6**, 046002 (2009).
8. P. C. Bressloff and J. M. Newby, Directed intermittent search for hidden targets, *New J. Phys.* **11**, 023033 (2009).
9. J. M. Newby and P. C. Bressloff, Random intermittent search and the tug-of-war model of motor-driven transport., *J. Stat. Mech.* **P04014** (2010).
10. J. M. Newby and P. C. Bressloff, Local synaptic signalling enhances the stochastic transport of motor-driven cargo in neurons, *Phys. Biol.* **7**, 036004 (2010).
11. R. F. Fox and Y. N. Lu, Emergent collective behavior in large numbers of globally coupled independent stochastic ion channels, *Phys. Rev. E.* **49**, 3421–3431 (1994).
12. C. C. Chow and J. A. White, Spontaneous action potentials due to channel fluctuations, *Biophys. J.* **71**, 3013–3021 (1996).
13. J. R. Groff, H. DeRemigio, and G. D. Smith, *Stochastic methods in neuroscience*, chapter 2. Markov chain models of ion channels and calcium release sites, pp. 29–64. Oxford University Press, Oxford (2009).
14. J. P. Keener and J. M. Newby, Perturbation analysis of spontaneous action potential initiation by stochastic ion channels, *Phys. Rev. E.* **84**, 011918 (2011).
15. J. M. Newby and J. P. Keener, An asymptotic analysis of the spatially inhomogeneous velocity-jump process, *Multiscale Model. Simul.* **9**, 735–765 (2011).
16. J. H. Goldwyn and E. Shea-Brown, The what and where of adding channel noise to the Hodgkin-Huxley equations, *PLoS Comp. Biol.* **7 (11)**, e1002247 (2011).
17. E. Buckwar and M. G. Riedler, An exact stochastic hybrid model of excitable membranes including spatio-temporal evolution., *J. Math. Biol.* **63**, 1051–1093 (2011).
18. J. M. Newby, P. C. Bressloff, and J. P. Keener, The effect of potassium channels on spontaneous action potential initiation by stochastic ion channels, *Phys. Rev. Lett.* **111**, 128101 (2013).
19. J. M. Newby, Isolating intrinsic noise sources in a stochastic genetic switch, *Phys. Biol.* **9**, 026002 (2012).
20. P. C. Bressloff and J. M. Newby, Metastability in a stochastic neural network modeled as a velocity jump Markov process, *SIAM J. Appl. Dyn. Sys.* **12**, 1394–1435 (2013).
21. J. M. Sancho, External dichotomous noise: The problem of the mean first-

passage time, *Phys. Rev. A.* **31**, 3523–3525 (1985).

22. J. M. Porra, J. Masoliver, and K. Lindenberg, Bistability driven by dichotomous noise, *Phys. Rev. A.* **44**, 4866–4875 (1991).
23. G. C. Papanicolaou, Asymptotic analysis of transport processes, *Bull. Amer. Math. Soc.* **81**, 330–392 (1975).
24. C. W. Gardiner, *Handbook of stochastic methods, 4th edition.* Springer, Berlin (2009).
25. J. M. Newby and P. C. Bressloff, Quasi-steady state reduction of molecular-based models of directed intermittent search, *Bull Math Biol.* **72**, 1840–1866 (2010).
26. H. Othmer, S. Dunbar, and W. Alt, Models of dispersal in biological systems, *J. Math. Biol.* **26**, 263–298 (1988).
27. T. Hillen and H. Othmer, The diffusion limit of transport equations derived from velocity-jump processes., *SIAM J. Appl. Math.* **61**, 751–775 (2000).
28. A. Friedman and B. Hu, Uniform convergence for approximate traveling waves in linear reaction-hyperbolic systems, *Indiana University Mathematics Journal.* **56**, 2133–2158 (2007).
29. J. M. Newby and P. C. Bressloff, Directed intermittent search for a hidden target on a dendritic tree, *Phys. Rev. E.* **80**, 021913 (2009).
30. P. C. Bressloff and J. M. Newby, Quasi-steady state analysis of motor-driven transport on a two-dimensional microtubular network, *Phys. Rev. E.* **83**, 061139 (2011).
31. M. Pineda, R. Imbihl, L. Schimansky-Geier, and C. Zulicke, Theoretical analysis of internal fluctuations and bistability in CO oxidation on nanoscale surfaces, *J. Chem. Phys.* **124**, 044701 (2006).
32. P. Hanggi, H. Grabert, P. Talkner, and H. Thomas, Bistable systems: master equation versus fokker–planck modeling, *Z. Physik B.* **28**, 135 (1984).
33. M. I. Dykman, E. Mori, J. Ross, and P. M. Hunt, Large fluctuations and optimal paths in chemical kinetics, *J. Chem. Phys. A.* **100**, 5735–5750 (1994).
34. D. Ludwig, Persistence of dynamical systems under random perturbations, *SIAM Review.* **17**(4), pp. 605–640 (1975). ISSN 00361445.
35. B. J. Matkowsky and Z. Schuss, The exit problem for randomly perturbed dynamical systems, *SIAM J. Appl. Math.* **33**, 365–382 (1977).
36. T. Naeh, M. M. Klosek, B. J. Matkowsky, and Z. Schuss, A direct approach to the exit problem, *SIAM J. Appl. Math.* **50**, 595–627 (1990).
37. R. S. Maier and D. L. Stein, Limiting exit location distribution in the stochastic exit problem, *SIAM J. Appl. Math.* **57**, 752–790 (1997).
38. Z. Schuss, *Theory and applications of stochastic processes: an analytical approach.* vol. v. 170, *Applied mathematical sciences*, Springer, New York (2010).
39. M. J. Ward and J. Lee, On the asymptotic and numerical-analyses of exponentially ill-conditioned singularly perturbed boundary value problems, *Stud. Appl. Math.* **94**, 271–326 (1995).
40. J. Newby and J. Chapman, Metastable behavior in Markov processes with internal states: breakdown of model reduction techniques, *Submitted* (2012).
41. A. L. Hodgkin and A. F. Huxley, A quantitative description of membrane

and its application to conduction and excitation in nerve, *J. Physiol.* **117**, 500–544 (1952).
42. J. A. White, J. T. Rubinstein, and A. R. Kay, Channel noise in neurons, *Trends. Neurosci.* **23**, 131–137 (2000).
43. G. D. Smith. Modeling the stochastic gating of ion channels. In eds. C. Fall, E. S. Marland, J. M. wagner, and J. J. Tyson, *Computational Cell Biology*, chapter 11. Springer-Verlag, New York (2002).
44. C. Morris and H. Lecar, Voltage oscillations in the barnacle giant muscle fiber, *Biophys. J.* **35**, 193–213 (1981).
45. I. A. Khovanov, A. V. Polovinkin, D. G. Luchinsky, and P. V. E. McClintock, Noise-induced escape in an excitable system, *Phys. Rev. E.* **87**, 032116 (2013).
46. D. A. Smith and R. M. Simmons, Models of motor-assisted tranoport of intracellular particles, *Biophys. J.* **80**, 45–68 (2001).
47. A. V. Kuznetsov and A. A. Avramenko, The method of separation of variables for solving equations describing molecular-motor-assisted transport of intracellular particles in a dendrite or axon, *Proc. Roy. Soc. A.* **464**, 2867–2886 (2008).
48. S. Klumpp and R. Lipowsky, Cooperative cargo transport by several molecular motors, *Proc. Natl. Acad. Sci. USA.* **102**, 17284–17289 (2005).
49. M. J. I. Muller, S. Klumpp, and R. Lipowsky, Tug-of-war as a cooperative mechanism for bidirectional cargo transport by molecular motors, *Proc. Natl. Acad. Sci. U. S. A.* **105**, 4609–4614 (2008).
50. O. Benichou, C. Loverdo, M. Moreau, and R. Voituriez, Intermittent search strategies, *Rev. Mod. Phys.* **83**, 81– (2011).
51. I. A. Telley, P. Bieling, and T. Surrey, Obstacles on the microtubule reduce the processivity of kinesin-1 in a minimal in vitro system and in cell extract., *Biophys. J.* **96**, 3341–3353 (2009).
52. K. S. Kosik, C. L. Joachim, and D. J. Selkoe, Microtubule-associated protein tau (tau) is a major antigenic component of paired helical filaments in alzheimer disease, *Proc. Natl. Acad. Sci. U.S.A.* **83**, 4044–4048 (1986).
53. C. Maas, D. Belgardt, H. K. Lee, F. F. Heisler, C. Lappe-Siefke, M. M. Magiera, J. van Dijk, T. J. Hausrat, C. Janke, and M. Kneussel, Synaptic activation modifies microtubules underlying transport of postsynaptic cargo, *Proc. Natl. Acad. Sci. U.S.A.* **106**, 8731–8736 (2009).
54. M. Vershinin, B. C. Carter, D. S. Razafsky, S. J. King, and S. P. Gross, Multiple-motor based transport and its regulation by tau, *Proc. Natl. Acad. Sci. U.S.A.* **104**, 87–92 (2007).
55. J. Dynes and O. Steward, Dynamics of bidirectional transport of Arc mRNA in neuronal dendrites, *The Journal of Comparative Neurology.* **500**(3), 433 (2007).
56. M. S. Rook, M. Lu, and K. S. Kosik, CaMKiiα 3' untranslated regions-directed mRNA translocation in living neurons: Visualization by GFP linkage, *The Journal of Neuroscience.* **20**(17), 6385–6393 (2000).

Chapter 13

First Passage Problems in Biology

T. Chou* and M. R. D'Orsogna†

*Depts. of Biomathematics and Mathematics, UCLA, Los Angeles, CA 90095-1766, USA
tomchou@ucla.edu

†Dept. of Mathematics, CalState-Northridge, Los Angeles, CA 91330-8313, USA

Applications of first passage times in stochastic processes arise across a wide range of length and time scales in biological settings. After an initial technical overview, we survey representative applications and their corresponding models. Within models that are effectively Markovian, we discuss canonical examples of first passage problems spanning applications to molecular dissociation and self-assembly, molecular search, transcription and translation, neuronal spiking, cellular mutation and disease, and organismic evolution and population dynamics. In this last application, a simple model for stem-cell aging is presented and some results derived. Various approximation methods and the physical and mathematical subtleties that arise in the chosen applications are also discussed.

1. Introduction and mathematical preliminaries

Although mainly studied in physical systems, first passage problems [1] arise in many biological contexts, including biomolecular kinetics, cellular function, and population dynamics. First passage problems can be most simply described as finding the distribution of times according to which a random process first exceeds a prescribed threshold or reaches a specified configuration, as described in Fig. 1. While expectations of moments of the random variable are often qualitatively captured by using straightforward approximation methods, other observable quantities such as first passage times may not be, and stochastic approaches must be used.

The probability distribution $P(X,t)$ of a stochastic process $X(t)$ may obey a discrete master equation or a Fokker-Planck or Smoluchowski equation for continuous variables. Other approaches such as the direct analysis of stochastic differential equations (SDEs) for the random variable $X(t)$ or analysis of the branching process [2, 3] describing the evolution of the probability generating function are also often employed. If the system does not harbour long-lived metastable configurations, simple mean-field or closure methods that approximate correlations can be used to analytically find expected trajectories $\langle X(t)\rangle = \sum\!\!\!\!\!\!\int XP(X,t)\mathrm{d}X$ that are often in qualitative agreement with exact results or trajectories derived from approximate, deterministic models.

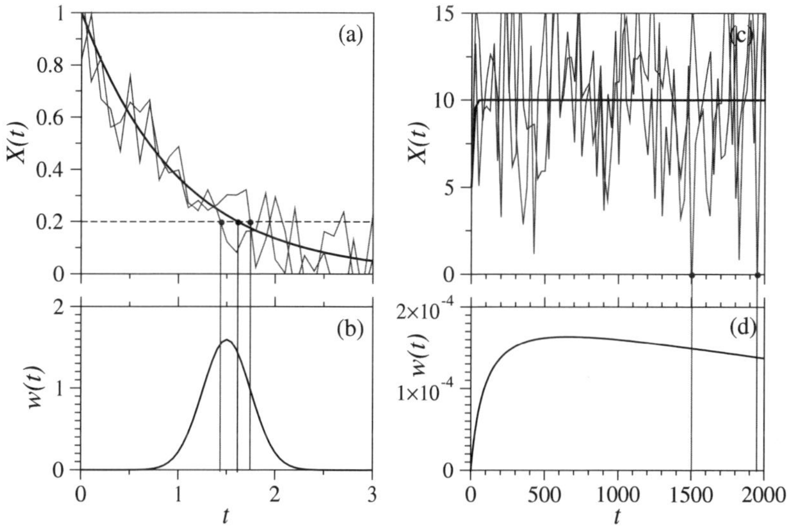

Fig. 1. Trajectories of a random variable $X(t)$ illustrating typical first passage problems. (a) The deterministic or expected trajectory $\langle X(t)\rangle$ (solid black curve) crosses the specified threshold $X^* = 0.2$ at a specific time $T \approx 1.6$; however, when fluctuations are explicitly included, the random variable $X(t)$ can cross $X = 0.2$ at different times $T \approx 1.45$ and $T \approx 1.7$, as shown by the red and blue trajectories, respectively. (b) The distribution of first passage times to $X = 0.2$. (c) Trajectories corresponding to a birth-death process with carrying-capacity (see Eq. (45) in Section 6). In the deterministic model, $X = 10$ (the carrying-capacity in this example) is a stable fixed point while $X = 0$ is an unstable one. With an initial condition $X(0) > 0$, the deterministic model never becomes extinct ($X^* = 0$), but in a stochastic model a random (possibly very rare) fluctuation can extinguish the system. The distribution of first extinction times is schematically shown in (d).

For example, consider the trajectories depicted in Fig. 1. Some deterministic trajectories $\langle X(t)\rangle$ cross a threshold value ($X^* = 0.2$ in Fig. 1(a)) at a unique time T, which then can be used as a qualitatively good estimate of the first passage time for the full stochastic process. In other cases, the deterministic trajectory may never cross a predefined "absorbing" threshold so that $T = \infty$. This is illustrated in Fig. 1(c) where $X(t)$ never reaches the threshold value $X^* = 0$. However, in a stochastic model, fluctuations can bring $X(t)$ to the threshold $X^* = 0$ in finite time. For such cases, there is a clear divergence between the exit times predicted from a deterministic model ($T = \infty$) and that predicted from a stochastic one ($T < \infty$).

To be more concrete, consider a discrete Markov process for a system of N states that can be described by the "forward" master equation

$$\frac{\partial P_{ki}}{\partial t} = M_{kj} P_{ji}, \tag{1}$$

where P_{ki} is the $N \times N$ matrix of probabilities that the system is in configuration k at time t, given that the system started in state i at $t = 0$. The $N \times N$ transition matrix composed of transition rates that take state j to state k is defined by M_{kj}. Note that (k, j) indexes all accessible configurations, including absorbing ones $\mathcal{A}$ from which probability density cannot re-emerge. Transition rates out of configurations $\mathcal{A}$ are defined to be zero while global probability conservation requires $\sum_{k=1}^{N} M_{kj} = 0$. As the dynamics evolve, the flow of probability entering absorbing states $\mathcal{A}$ cannot exit. Eventually, the survival probability defined as $S_i(t) \equiv \sum_{k\notin\mathcal{A}} P_{ki}(t)$ will vanish as $t \to \infty$. The survival probability $S_i(t)$ defines the probability that the system has not reached any absorbing configuration up to time t, given that it started in configuration i at $t = 0$.

Since the first passage time distribution can be derived from $S_i(t)$, it is convenient to consider the adjoint equation that is also obeyed by $P_{ki}(t)$:

$$\frac{\partial P_{ki}}{\partial t} = P_{kj} M_{ji}. \tag{2}$$

This "backward" equation does not operate on the final configurations k so one can perform the sum $\sum_{k\notin\mathcal{A}}$ to find an equation for the survival probability

$$\frac{\partial S_i(t)}{\partial t} = S_j(t)M_{ji}, \tag{3}$$

along with the initial condition $S_k(t = 0) = 1$ for $k \notin \mathcal{A}$ and "boundary condition" $S_k(t) = 0$ for $k \in \mathcal{A}$.

A physical interpretation of Eq. (3) can be easily obtained by considering the *lifetime distribution function* which is a sum over the absorbed states: $F_i(t) \equiv \sum_{k\in\mathcal{A}} P_{ki}(t) = 1 - S_i(t)$. We can now identify the rate of change of F_i as the probability flux into the absorbing state $\mathcal{A}$, so that $\partial_t F_i \equiv J_i^{\mathcal{A}}(t)$. Using $F_j \equiv 1 - S_j$, we can rewrite Eq. (3) as

$$\frac{\partial S_i(t)}{\partial t} = S_j(t)M_{ji} = -J_i^{\mathcal{A}}(t). \tag{4}$$

The latter is also a statement that the probability of survival against entering absorbing configurations decreases in time according to the probability flux into the absorbing states.

From the lifetime distribution $F_i(t)$, one can find the probability that the system reached any absorbing configuration between time t and $t + \mathrm{d}t$ as $F_i(t + \mathrm{d}t) - F_i(t) = S_i(t) - S_i(t + \mathrm{d}t)$. Hence, the first passage time distribution $w_i(t)$ can be found from

$$w_i(t)\mathrm{d}t \equiv \frac{\partial F_i(t)}{\partial t}\mathrm{d}t = -\frac{\partial S_i(t)}{\partial t}\mathrm{d}t, \tag{5}$$

allowing calculation of all moments n of the first passage time

$$\langle T_i^n \rangle = \int_0^\infty w_i(t)t^n \mathrm{d}t. \tag{6}$$

Upon using integration by parts for $n = 1$, the mean first passage time is simply $\langle T_i \rangle = \int_0^\infty S_i(t)\mathrm{d}t$. Integrating Eq. (4) directly, we find an explicit equation for the moments of the first passage time into an absorbing state

$$\langle T_j^n \rangle M_{ji} = -n\langle T_i^{n-1} \rangle, \tag{7}$$

where $\langle T_j^0 \rangle \equiv 1$. Equations (6) and (7) have been used to study moments of first exit times for a random walker to hit either one or two ends of a discrete one-dimensional lattice [4, 5].

A commonly used approximation to Eq. (4) (see Sections 2 and 4) is to assume

$$\frac{\partial S_i(t)}{\partial t} \approx -J_i^{\mathcal{A}}(t)S_i(t), \tag{8}$$

which is motivated by a mass-action argument of the decay of probability of being in the initial surviving state i. Here, $J_i^{\mathcal{A}}(t)$ is the probability current from state i to $\mathcal{A}$. However, the RHS of the exact relationship in Eq. (4) contains the transition matrix M_{ji} which mixes states i with j. Since the approximation in Eq. (8) does not resolve the different surviving states, Eq. (8) is exact only when there is a single surviving state i that directly transitions into $\mathcal{A}$ without any intermediate states. Another limit where Eq. (8) is accurate is if the system mixes quickly among all surviving states well before being absorbed. In this case, the single surviving state i is a lumped average over all the microscopic states j, and first passage can be thought of as slow degradation of a quasi-steady-state configuration. Equation (8) and the associated assumptions have been widely used in practice, particularly to describe bond rupturing in dynamic force spectroscopy of biomolecules (see Section 2).

Another common representation of stochastic processes that is useful for modeling biophysical systems is based on continuous variables. This "Lagrangian" representation is particularly suitable for tracking stochastically-moving, identifiable particles. Starting from Eq. (1), a continuum formulation can be heuristically developed by assuming that each configuration is connected to only a few others. In this case, indices can be chosen such that the transition matrix is banded. For example, a particle at position i on a one-dimensional lattice is allowed to jump only to neighboring positions $i \pm 1$ with probability proportional to an infinitesimal increment of time. If the indices label lattice site positions, the transition matrix will be tridiagonal. Furthermore, if the transition rates vary slowly from site to site, and the system size N is large, we can take a continuum limit where the position of a particle $y = i/N$ and the tridiagonal transition matrix represents a stencil of a differentiation operator.

Upon defining $P(\{\mathbf{y}_j\}, t | \{\mathbf{x}_j\}, 0)$ as the probability that all particles j are located between $\mathbf{y}_j$ and $\mathbf{y}_j + \mathrm{d}\mathbf{y}_j$ at time t given that they were at positions $\{\mathbf{x}_j\}$ at $t = 0$, one can Taylor-expand a discrete master equation in a "diffusion approximation" to find the governing Fokker-Planck or Smoluchowski equation

$$\begin{aligned}\frac{\partial P(\{\mathbf{y}_j\},t|\{\mathbf{x}_j\},0)}{\partial t} &= -\sum_{k=1}^N \nabla_k\cdot(\mathbf{V}_k P) + \sum_{k=1}^N \nabla_k^2(D(\{\mathbf{y}_k\})P) \\ &\equiv \mathcal{L}P(\{\mathbf{y}_j\},t|\{\mathbf{x}_j\},0).\end{aligned} \tag{9}$$

In Eq. (9), N is the total number of particles, $\mathbf{V}_k$ is the drift velocity of the kth particle, and the gradient ∇_k is taken with respect to the coordinates of the kth particle. The density $P(\{\mathbf{y}_j\},t|\{\mathbf{x}_j\},0)$ also obeys the Backward Kolmogorov Equation (BKE) which is simply

$$\partial_t P(\{\mathbf{y}_j\},t|\{\mathbf{x}_j\},0) = \mathcal{L}^\dagger P(\{\mathbf{y}_j\},t|\{\mathbf{x}_j\},0), \tag{10}$$

where $\mathcal{L}^\dagger = \sum_k^N \mathbf{V}_k\cdot\nabla_k + \sum_{k=1}^N D(\{\mathbf{x}_j\})\nabla_k^2$ is the operator adjoint of $\mathcal{L}$. Since $\mathcal{L}^\dagger$ operates on the initial positions $\mathbf{x}_j$, Eq. (10) can be integrated over coordinates $\mathbf{y}_j$ within the domain, excluding the absorbing surfaces. The resulting equation for the survival probability analogous to Eq. (4) is $\partial_t S(\{\mathbf{x}_j\};t) = \mathcal{L}^\dagger S(\{x_j\};t)$, with $S(\{x_j\};t=0) = 1$ for all $\mathbf{x}_j \neq \partial\Omega_{\mathcal{A}}$, and $S(\forall \mathbf{x}_j = \partial\Omega_{\mathcal{A}};t) = 0$. From this survival probability, all moments of the first times any particle hits an absorbing boundary $\partial\Omega_{\mathcal{A}}$ can be derived. Namely, in analogy with Eq. (7), the mean hitting time obeys

$$\mathcal{L}^\dagger\langle T^n(\{x_j\})\rangle = -n\langle T^{n-1}(\{x_j\})\rangle. \tag{11}$$

Both the discrete and continuum stochastic formulations are commonly applied to physical systems; however, care should be exercised in using a continuum description as an approximation for a discrete system where first passage times are sought. Although the continuum diffusion approximation may be accurate in describing probability densities of large discrete systems, it often provides a poor approximation to first passage times of discrete processes. Indeed, using a birth-death process with carrying-capacity (see Section 6), Doering, Sagsyan, and Sander [6] show that the effective potential of a discrete system and its corresponding continuum diffusion approximation differ, leading to different mean first population extinction times. The discrepancy is small only when the convective term in the Fokker-Planck equation is small across all relevant population levels. Thus, depending on the application, continuum diffusion approximations and their numerical discretization should be applied judiciously when first passage times are being analyzed.

The first passage problems defined above assume that one is interested in the distribution of times of the systems arriving at *any* absorbing configuration. However, there may well be states which are physically absorbing (into which probability flux enters irreversibly) but that are not relevant to the biological process. For example, one may be interested in the times it takes for a diffusing protein to first reach a certain target site (see Section 4 below), but the protein may degrade before reaching it. Since decay is irreversible, the system reaches an "unintended" absorbing state through degradation of the protein. If one defines $\mathcal{A}$ to be only the biologically-relevant absorbing configurations, the corresponding survival probability $S_i(t)$ does not vanish in the $t \to \infty$ limit because there are other "irrelevant" absorbing states that absorb some of the probability. In other words, if there are other physical absorbing states competing for probability, the integrated probability flux $J_i^{\mathcal{A}}(t)$ into the relevant absorbing states $\mathcal{A}$ obeys $\int_0^\infty J_i^{\mathcal{A}}(t)\mathrm{d}t < 1$. Also note that since $S_i(t \to \infty) > 0$, the mean first passage time $\langle T_i \rangle = \int_0^\infty S_i(t)\mathrm{d}t = \infty$. All moments also diverge. Provided a measurable fraction of trajectories reach the irrelevant absorbing state, the mean time to arrive at the relevant absorbing state diverges because these "wasted" trajectories will never reach the relevant states.

A more appropriate measure in cases with "interfering" absorbing states is the distribution of first arrival times *conditioned* on arriving at the relevant absorbing configurations $\mathcal{A}$. In other words, we restrict ourselves to the arrival time statistics of only those trajectories that are *not* absorbed by the irrelevant states. The conditioning is a simple statement of Bayes rule: $J_i^{\mathcal{A}}(t) = J_i(t|\mathcal{A}) \times \mathrm{Prob}(\text{exiting through}\, \mathcal{A})$, where $J_i^{\mathcal{A}}(t)$ is the overall probability flux from i into $\mathcal{A}$, and $J_i(t|\mathcal{A})$ is the probability flux of annihilation counting those trajectories that annihilate through the *relevant* absorbing states $\mathcal{A}$. Since the probability of exiting through $\mathcal{A}$ is $\int_0^\infty J_i^{\mathcal{A}}(t)\mathrm{d}t$, the conditional first passage time distribution is

$$J_i(t|\mathcal{A})\mathrm{d}t \equiv w_i(t|\mathcal{A})\mathrm{d}t = \frac{J_i^{\mathcal{A}}(t)\mathrm{d}t}{\int_0^\infty J_i^{\mathcal{A}}(t')\mathrm{d}t'}. \tag{12}$$

Analogous expressions for the continuum representation (Eq. (9)) can be found provided a suitable continuum expression for the probability flux is used. As a simple example, consider a single Brownian particle with diffusivity D in one dimension with absorbing boundaries at $x = \pm 1$. The probability flux through the ends are

$$\mp D \frac{\partial P(y,t|x,0)}{\partial y}\bigg|_{y=\pm 1} \equiv J_x(t|\pm 1). \tag{13}$$

The first passage time distributions sampled over only those trajectories that exit, say, $y = +1$ is thus

$$w_x(t|+1)\mathrm{d}t = \frac{J_x(t|+1)\mathrm{d}t}{\int_0^\infty J_x(t'|+1)\mathrm{d}t'}, \tag{14}$$

which can be explicitly calculated given the solution to the diffusion equation (Eq. (9)) for $P(y,t|x,0)$.

The mathematical approaches presented above, along with many extensions, have been used to model a diverse set of first passage problems arising in biological systems. In the following sections, we survey some illustrative examples of such first passage problems that span length scales ranging from the molecular, to the cellular, to that of populations.

2. Molecular rupture

The times over which molecules dissociate play an important role in chemical biology. For example, ligand-receptor complexes have finite lifetimes that are important determinants of whether signalling is initiated. Cell-substrate and cell-cell adhesion are also mediated by molecules such as glycoproteins [7], and knowing the "strength" of these macromolecular bonds can reveal insight into the biological function of macromolecules.

For a simple single-barrier free energy profile, one simple approximation is to assume a quadratic energy profile and compute the first passage time distribution to a particular displacement, reducing the calculation to that of finding the first crossing time of an over-damped Ornstein-Ulhenbeck process [8, 9]. Another more refined approximation concatenates two harmonic potentials (one of positive curvature, one of negative curvature) together to form an approximate potential. Upon using steepest descents, a simple expression for the *mean* bond rupturing time starting from the energetic minimum ξ_0 can be found in the high barrier (rare crossing) limit:

$$\langle T(\xi_0)\rangle \approx \frac{e^{-(U(\xi^*)-U(\xi_0))}}{2\pi|\kappa_0\kappa_*|}. \tag{15}$$

Here, κ_0 and κ_* are the curvatures of the potential at the local minimum and at the top of the barrier, respectively. Since the barrier is high, and dissociation is a rare event, the distribution of rupturing times can be well-approximated by a single exponential with a dissociation rate $k_{\rm d} \equiv 1/\langle T\rangle$. In addition to the barrier height, Eq. (15) encodes the shape of the bond potential through the curvatures κ_0 and κ^*. However, typical bonds are so strong that their rupture times are too large to be experimentally accessible. Therefore, bonds are typically pulled by external forces in "dynamic force spectroscopy" (DFS) experiments.

Ideally, in a DFS experiment, the applied force on the bond is typically linearly ramped up (in time) until the bond ruptures, and some properties of the bond trajectories or forces sampled [10, 11]. From these data, one may seek to reconstruct properties of the underlying pre-pulled potential. Therefore, under an assumption of no rebinding, analysis of DFS can be reduced to a first passage problem with a time-dependent potential. Nearly all approaches to this problem have included the pulling into a time-dependent free energy barrier $U(\xi,t)$, giving rise to a time-dependent dissociation rate $k_{\rm d}(t)$, which is then used in the mean-field equation (Eq. (8)) for the bond survival probability $\dot{S}(t) \approx -k_{\rm d}(t)S(t)$. As it stands, this rate equation does not provide information about the bond other than the effective barrier height. In order to model finer effects of the bond energy profiles, shape properties need to be incorporated into the analysis. The simplest way to do this is to model how $k_{\rm d}(t)$ depends on the shape of the bond energy, while still retaining the mean-field assumption (Eq. (8)) for the survival probability [11].

One simple approach is to assume the bond potential contains a barrier at bond coordinate ξ^*, beyond which the bond is irreversibly dissociated. To approximate the distribution of times for a bond to spontaneously rupture, one calculates the time it takes for a random walker to reach the "absorbing boundary" ξ^*, given that it started from an initial position ξ_0. The standard calculation proceeds by solving the Fokker-Planck equation for the probability density $P(\xi,t|\xi_0,0)$ and constructing the corresponding survival probability $S(\xi_0;t) = \int_0^{\xi_*} P(\xi,t|\xi_0,0)\mathrm{d}\xi$, or, alternatively, directly solving the Backward Kolmogorov Equation for $S(\xi_0;t)$. The probability density, survival probability, and rupture time distribution are all easily solved numerically. In the over-damped limit of diffusive dynamics, the *mean* bond rupturing time $\langle T\rangle \equiv \int_0^\infty S(\xi_0;t)\mathrm{d}t$ can be found in exact closed form for any general free energy profile $U(\xi)$ [12].

The simplest way to incorporate a time-varying applied force problem in the one-dimensional continuum limit is to define an auxiliary time variable

τ such that $\partial_t \tau = 1$. In the backward equation corresponding to Eq. (11), τ is an independent variable [13]

$$\left(\frac{\partial}{\partial \tau} + F(\tau)\frac{\partial}{\partial \xi} + \mathcal{L}^{\dagger}\right)\langle T(\xi,\tau)\rangle = -1, \tag{16}$$

where ξ is the initial starting coordinate of the bond and $F(\tau) = \gamma\tau$ describes a pulling force that is increased linearly with rate γ. With suitable boundary conditions $\langle T(\xi_*,\tau)\rangle = \langle T(\xi,\infty)\rangle = 0$, one can find the expected rupture time $\langle T(\xi,0)\rangle$ numerically.

Two analytical approximations can be made by assuming the pulling force F is fixed. In this case, the solution to $\left(F\partial_\xi + \mathcal{L}^{\dagger}\right)\langle T(\xi,F)\rangle = -1$ is [13]

$$\langle T(\xi,F)\rangle = Q[\exp\left(-U(\xi) + F\xi\right)], \tag{17}$$

where $Q[...]$ is a complicated, but explicit integral functional [13]. In a first approximation Shillcock and Seifert [13] assumed that the typical rupturing force is determined self-consistently from $F^* \approx \gamma\langle T(\xi,F^*)\rangle$.

A self-consistent approach to estimate the rupture force *distribution* is to solve the mean-field equation $\dot{S}(t;\xi_0) = -k_{\mathrm{d}}(t)S(t;\xi_0)$ and use Eq. (5) to find

$$w(\xi,t)\mathrm{d}t = k_{\mathrm{d}}(\xi,t)\exp\left[-\int_0^t k_{\mathrm{d}}(\xi,t')\mathrm{d}t'\right]\mathrm{d}t, \tag{18}$$

where $k_{\mathrm{d}}(\xi,t)$ is the time-dependent rate of dissociation. Upon using $F(t) = \gamma t$ to convert this distribution to a rupture force distribution yields

$$\begin{aligned} w(\xi,F^*)\mathrm{d}F^* &= \frac{1}{\gamma}k_{\mathrm{d}}(\xi,F^*)\exp\left[-\frac{1}{\gamma}\int_0^{F^*} k_{\mathrm{d}}(\xi,F)\mathrm{d}F\right]\mathrm{d}F^* \\ &= \frac{1}{\gamma}k_{\mathrm{d}}(\xi,F^*)\exp\left[-\frac{1}{\gamma}\int_0^{F^*}\frac{\mathrm{d}F}{Q[\exp(-U(\xi)+F\xi)]}\right]\mathrm{d}F^*, \end{aligned} \tag{19}$$

where for the last equality, $k_{\mathrm{d}}(\xi,F) \approx 1/\langle T(\xi,F)\rangle$ and Eq. (17) were used. These and other mean-field approaches using Eq. (8) typically lead to a most probable rupture force F^* that is proportional to $\ln\gamma$, with proportionality factors related to the spatial width and energetic depth of the underlying bond. Therefore, the rupture force distribution measured as a function of

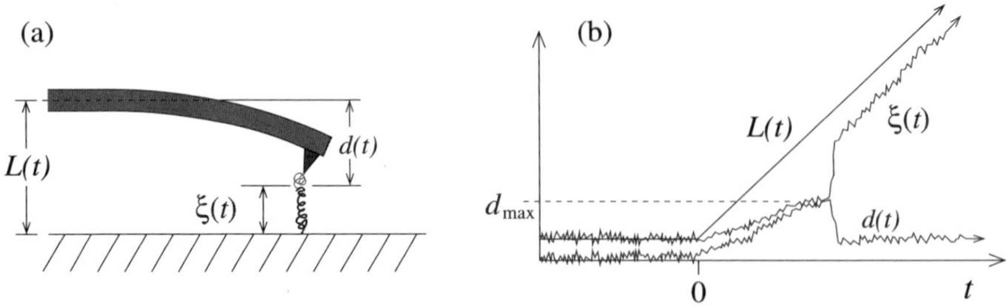

Fig. 2. Schematic of an AFM-controlled bond-rupturing experiment. The AFM is lifted from the rigid substrate with prescribed displacement $L(t)$. The deflection $d(t)$ and the bond coordinate $\xi(t)$ sum to the total AFM displacement: $L(t) = d(t) + \xi(t)$. The deflection $d(t)$ is a measure of the bond-separating force through the relation $F(t) = k_s d(t)$, where k_s is the known spring constant of the AFM cantilever.

loading rate has been widely used as a quick measure of bond strength.

While mathematically well-defined, the analyses above neglects a physical constraint encountered in bond pulling experiments. As noted by Qian and others [14–16], the mechanics of pulling a bond required the introduction of a mechanical spring force, whether manifested through an atomic force microscope (AFM) tip, an optical trap, or a pulled magnetic bead. If these devices are pulled with constant velocity V, the actual pulling potential is of the form $k_s(\xi - Vt)^2/2$, where k_s is the spring constant of the AFM cantilever. The experimental setup depicted in Fig. 2 shows how the pulling force, including an estimate of the maximum force, can be measured through the deflection $d(t)$ of the cantilever.

This and related approximations are used in combination with specific bond energy profiles by many authors to derive expressions for rupture force distributions [17–24]. For example, Dudko *et al.* [18] treat the ensemble where the pulling velocity V is specified. They use a mean-field approximation for the bond survival probability (described in more detail in Section 4) and assume that the total potential is being shifted at a constant velocity V. For rather general potentials, they find a mean rupture force $\langle F^* \rangle \sim (\ln V)^{2/3}$, as well as an expression for the rupture force distribution. These results, however, rely on the use of a soft (small k_s) pulling device. As shown in Fig. 3, a stiff puller (large k_s) results in a single-well effective potential and a distinct rupture event is precluded [14–16]. In this case, it is not fruitful to analyze the problem within a first passage time framework, and a more careful analysis of the force distribution measured during the entire pulling protocol should be used.

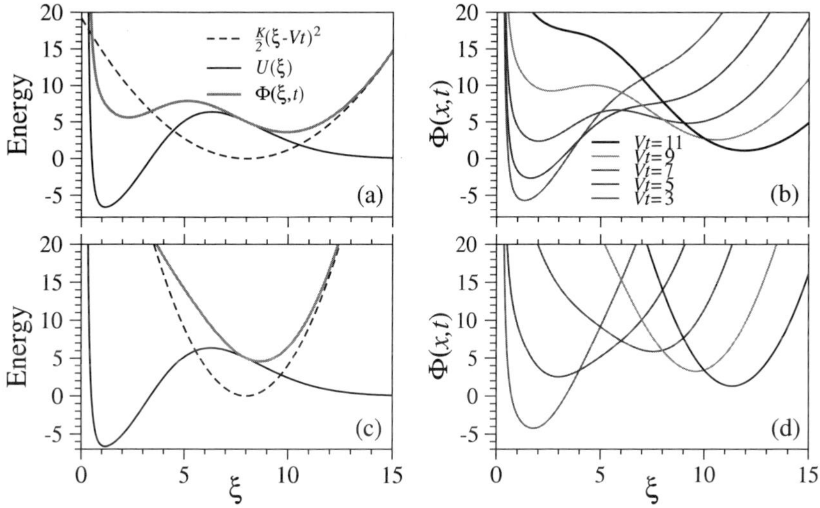

Fig. 3. (Color online) (a) Schematics of free energies as a function of an effective, one-dimensional bond coordinate ξ. The intrinsic molecular potential (black), harmonic potential of the pulling device (dashed), and the total potential (brown) are shown. (b) Only if the device potential is soft does the total potential $\Phi(\xi, t)$ form a barrier as the device is pulled. (c) If the pulling device is stiff, then only a single translating minimum arises, (d). Therefore, only for pulling devices with soft springs does a rupture event and corresponding rupture force qualitatively arise.

In general, the problem, as with many inverse problems, is ill-posed [25]. The reconstruction of a potential from a single rupture time (or rupture force) distribution starting from a single bond coordinate is not unique [25]. However, additional experiments (such as multiple loading forces and multiple starting bond positions) can give rise to multiple rupture time distributions that allow for reconstruction of potentials defined by many more parameters [26]. The extension of these inverse problems to those using rupture force distributions derived from different force loading rates could provide insight into the reconstruction of potentials more complex than simple harmonic, Lennard-Jones, or Morse type potentials.

3. Nucleation and self-assembly

A process complementary to dissociation is self-assembly, which also arises in many biological contexts. The polymerization of actin filaments [27–31] and amyloid fibrils [32], the assembly of virus capsids [33–35] and of antimicrobial peptides into transmembrane pores [36, 37], the assembly of ligands and receptors [38, 39], and the self-assembly of clathrin-coated pits [40–42] are all important processes at the cellular level that can be cast as self-assembly problems. Generally, in biological settings, there exists a maximum cluster size which signals the completion of the assembly process. For example, virus capsids, clathrin coated pits, and antimicrobial peptide pores typically consist of $N \sim 100 - 1000, N \sim 10 - 20$, and $N \sim 5 - 8$ molecular subunits, respectively. Furthermore, in confined spaces such as cellular compartments, the total mass is a conserved quantity. Figure 4 depicts a homogeneous nucleation process where monomers spontaneously bind and detach to clusters one at a time.

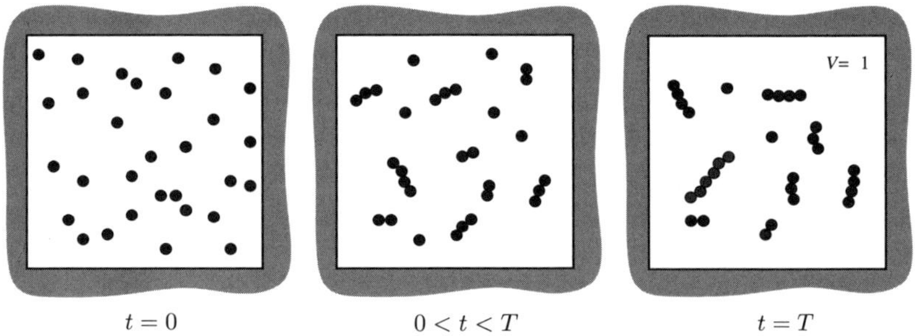

Fig. 4. (Color online) Homogeneous nucleation and growth in the slow detachment ($q \to 0^+$) limit in a closed unit volume initiated with $M = 30$ monomers. If the constant monomer detachment rate q is small, monomers will be nearly exhausted in the long time limit. In this example, we assume that $N = 6$ is the maximum cluster size and that the first maximum cluster is formed at time T (depicted in blue).

The classical description of self-assembly or homogeneous nucleation is a set of mass-action equations (such as the Becker-Döring equations) describing the concentration $c_k(t)$ of clusters of each size k at time t:

$$\begin{aligned}
\dot{c}_1(t) &= -pc_1^2 - pc_1 \textstyle\sum_{j=2}^{N-1} c_j + 2qc_2 + q\sum_{j=3}^{N} c_j \\
\dot{c}_2(t) &= -pc_1c_2 + \tfrac{p}{2}c_1^2 - qc_2 + qc_3 \\
\dot{c}_k(t) &= -pc_1c_k + c_1c_{k-1} - qc_k + qc_{k+1} \\
\dot{c}_N(t) &= pc_1c_{N-1} - qc_N,
\end{aligned} \tag{20}$$

where for simplicity, we have assumed cluster size-independent attachment and detachment rates p and q, respectively. These equations can readily be integrated to provide a mean-field approximation to the numbers of clusters of each possible size k [43].

Given a total number of monomers M one may be interested in the time it takes for the system to first assemble a complete cluster of size N. To address such a first passage problem, a stochastic model for the homogeneous nucleation process must be developed. Consider an N-dimensional probability density $P(n_1, n_2, \ldots, n_N; t)$ for the system exhibiting at time t, n_1 free monomers, n_2 dimers, n_3 trimers...and n_N completed clusters. The forward master equation obeyed by $P(n_1, n_2, \ldots, n_N; t)$ is [43]:

$$\begin{aligned}\dot{P}(\{n\};t) = -\Lambda(\{n\})P(\{n\};t) &+ \frac{1}{2}(n_1+2)(n_1+1)W_1^+W_1^+W_2^-P(\{n\};t)\\ &+ \sum_{i=2}^{N-1}(n_1+1)(n_i+1)W_1^+W_i^+W_{i+1}^-P(\{n\};t)\\ &+q(n_2+1)W_2^+W_1^-W_1^-P(\{n\};t)\\ &+q\sum_{i=3}^{N}(n_i+1)W_1^-W_{i-1}^-W_i^+P(\{n\};t),\end{aligned} \tag{21}$$

where we have rescaled time to p^{-1}. Here, $P(\{n\}, t) = 0$ if any $n_i < 0$, $\Lambda(\{n\}) = \frac{1}{2}n_1(n_1-1) + \sum_{i=2}^{N-1} n_1 n_i + q\sum_{i=2}^{N} n_i$ is total rate out of configuration $\{n\}$, and $W_j^{\pm}$ are the unit raising/lowering operators on the number of clusters of size j. For example,

$$W_1^+W_i^+W_{i+1}^-P(\{n\};t) \equiv P(n_1+1, \ldots, n_i+1, n_{i+1}-1, \ldots; t). \tag{22}$$

The process associated with this master equation has been analyzed using Kinetic Monte-Carlo simulations as well as asymptotic approximations for the mean cluster numbers in limits of small and large q [43, 44].

The first passage problem is to determine the distribution of times for complete assembly of the largest cluster, $n_N = 0 \Rightarrow n_N = 1$. For the purpose of illustration, consider a small system with $M = 7$ or 8, and $N = 3$. Since state-space is small, we can visualize all possible configurations as shown in Fig. 5. The first passage time to a maximum cluster, starting from the all-monomer state ($P(\{n_i\}; t = 0) = \delta_{n_1,M}\prod_{i=2}^{N}\delta_{n_i,0}$) is the time the system takes to reach any of the states highlighted in blue, to the right of the red line.

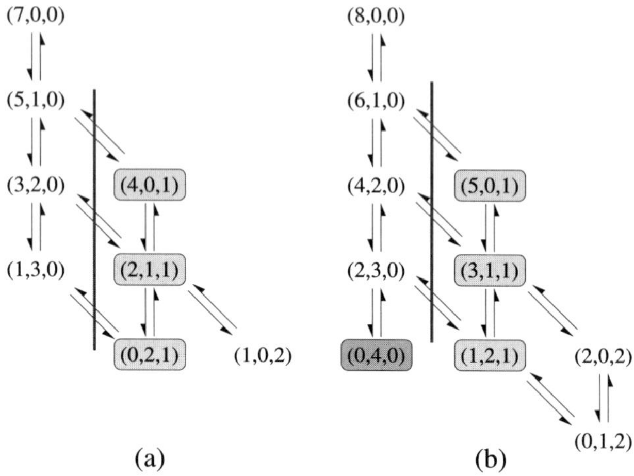

Fig. 5. (Color online) Allowed transitions in stochastic self-assembly starting from an all-monomer initial condition. In this simple example, the maximum cluster size $N = 3$. (a) Allowed transitions for a system with $M = 7$. Since we are interested in the first maximum cluster assembly time, states with $n_3 = 1$ constitute absorbing states. The process is stopped once the system crosses the vertical red line. (b) Allowable transitions when $M = 8$. Note that if monomer detachment is prohibited ($q = 0$), the configuration $(0, 4, 0)$ (yellow) is a trapped state. Since a finite number of trajectories will arrive at this trapped state and never reach a state where $n_3 = 1$, the mean first assembly time $T_3(8, 0, 0) \to \infty$ when $q = 0$.

In the strong binding limit, when $0 < q \ll 1$ and for M even, one can find the dominant pathways to a largest cluster and surmise the leading order behavior $\langle T(q \ll 1)\rangle \sim 1/q$, with a prefactor that depends nontrivially on M and N [44]. This diverging assembly time arises from trapped states as highlighted in yellow in Fig. 5(b). As q is increased, the likelihood of more paths coming out of the trapped states is higher, thereby decreasing the expected time to cluster completion. Only for the special case of $N = 3$ and M odd, where no such traps exist, is $\langle T(q)\rangle$ a nondivergent ratio of polynomials in q, as illustrated in Fig. 6(a).

In the weak binding, $q \gg 1$, maximum cluster formation is a rare event and $\langle T(q \gg 1)\rangle \sim q^{N-2}$. Because of these asymptotic relations, we expect at least a single minimum in the mean first assembly time as a function of the detachment rate q [44]. Figure 6 shows $\langle T(q)\rangle$ as a function of q for $M = 7$ and $M = 8$, clearly indicating a shortest expected maximum cluster formation time at intermediate detachment rates q. As long as M is even or $N \geq 4$, traps states arise and the expected cluster completion

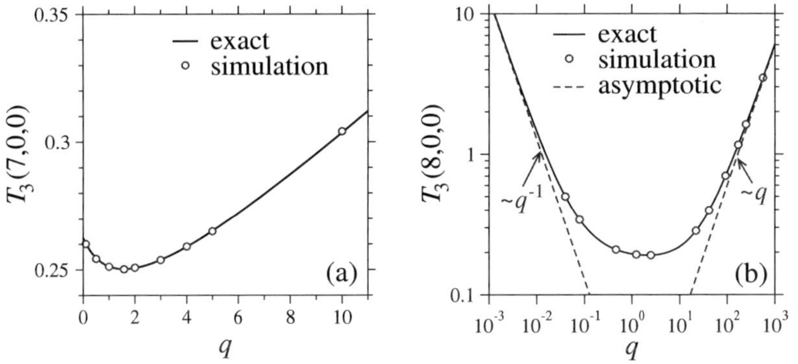

Fig. 6. (Color online) Mean first assembly times for $M = 7$ and $N = 3$ in panel (a) and $M = 8$ and $N = 3$ in panel (b). The notation $T_N(M, 0, 0)$ denotes the mean first maximum cluster (of size N) assembly time $\langle T \rangle$ starting from an initial condition of M monomers. Exact results are plotted as black solid lines, while red circles are obtained by averaging over 10^5 KMC simulation trajectories. The dashed blue lines show the $q \to 0$ and $q \to \infty$ asymptotic approximations.

time diverges as $q \to 0$. Thus, in this limit, it may be physically more meaningful to define the expected assembly time of a maximum cluster, *conditioned* on trajectories yielding complete clusters. The above results can also be extended to first assembly times of the stochastic *heterogeneous* nucleation process [45].

Ideas of self assembly have also been applied to a structurally more specific application of linear filament and microtubule growth [46, 47]. The cell cytoskeleton is a dynamically growing and shrinking assembly of microtubules and filaments that regulate cell migration, internal reorganization such as organelle transport, and mitosis. The assembly and disassembly of microtubules is a key microscopic process for these vital higher order cell functions. The molecular players involved in these processes are numerous and their interaction are biochemically and geometrically complex. However, one basic feature is that the tips of growing filaments can exist in a state that promotes elongation, or one that promotes disassembly. By switching between these two states, the filament can be biased to shrink or grow. A first passage problem that has been studied in this context has been to derive a model for the first disassembly time of a filament starting from a specific length. Using a discrete stochastic model describing the probability density for the number of monomers in a single microtubule, as well as transitions between growing and shrinking states, Rubin calculated its disassembly time distribution in terms of modified Bessel's functions [46].

In later work, Bicout [47] used a semi-Markov model to describe single filament dynamics. During the growth or shrinking phases, the length of the filament was assumed to be continuous variable that increased or decreased according to deterministic velocities $v_{\pm}$. However, the switching between growing and shrinking states was assumed to be Markovian with exponentially distributed times, with rates $f_{\pm}$. For this "Broadwell" model we introduce $P_{\pm}(x,t|x_0,0)$ as the probability that the tip of the filament is moving with velocity $v_{\pm}$ and is located between position x and $x+\mathrm{d}x$ at time t, given that it was at position x_0 at $t=0$. Conservation of probability yields

$$\frac{\partial}{\partial t}\begin{pmatrix} P_+ \\ P_- \end{pmatrix} = \mathcal{L}\begin{pmatrix} P_+ \\ P_- \end{pmatrix}, \tag{23}$$

where

$$\mathcal{L} = \begin{pmatrix} -v_+\frac{\partial}{\partial x} - f_+ & f_- \\ f_+ & v_-\frac{\partial}{\partial x} - f_- \end{pmatrix}, \tag{24}$$

which is also known as the "telegraphers" equation. The ballistic intervals of motion introduces an overall memory into the dynamics. This can be seen by combining $P_+ + P_- = P$ to find an equation for the total probability $P(x,t)$ containing terms of the form $\partial^2 P/\partial t^2$.

By using the associated Green's function, Bicout [47, 48] found explicit solutions for the distribution of lifetimes of a microtubule that started off at a fixed length x_0:

$$w(t;x_0)\mathrm{d}t \sim t^{3/2}\exp\left[-t/\tau_{\mathrm{c}}\right]\mathrm{d}t. \tag{25}$$

The Broadwell model and telegrapher's equation have been used in many other applications, including gas kinetics [49, 50] and photon transport [51]. In the next section, we present another example of a first passage problem from molecular biophysics that involves electron transport and that is also described by equations similar to Eq. (23).

4. Molecular transport and search

A molecular setting in which first passage problems arise in biology is the so called "narrow escape problem", which is simply a higher dimensional generalization of a high-barrier bond-rupturing problem. In cellular environments, numerous confined spaces arise in which molecules diffuse and react. Typically, a small section of the surface of the confined space is "reactive", *i.e.,* contains receptors that bind diffusing molecules, or is a hole that allows escape into a much larger volume. Examples include synaptic clefts connecting neurons, nuclear envelopes and their associated nuclear pore complexes.

Mathematically, the problem is described by Fig. 7(a) in which a particle diffuses in the domain Ω, bounded by $\partial\Omega$. The boundary $\partial\Omega$ is made of two regions, a reflecting boundary $\partial\Omega_{\rm r}$, and an absorbing one $\partial\Omega_{\rm a}$, representing a hole or irreversibly binding surface. Asymptotic results for mean first passage times have been derived for $\varepsilon = \partial\Omega_{\rm a}/\partial\Omega \ll 1$. Since escaping is a rare event in this limit, we expect that the escape time will be insensitive to the starting position.

A number of asymptotic results for the mean escape time of particles in confined geometries have been determined by Singer, Schuss, and Holcman [52], as well as Ward *et al.* [53].

Figure 7(a) shows the diffusing particle in a volume Ω that can escape from a small hole of size $\sim \varepsilon\Omega^{1/d}$, where d is the spatial dimensionality of Ω. If $\varepsilon \ll 1$, estimates of the mean first exit times have been derived using asymptotic analysis of Eq. (11) and conformal mapping. Specifically, in 2D and 3D, for escape from a small hole punched through a smooth boundary as shown in Fig. 7(a), we find

$$\begin{aligned}\langle T\rangle &\sim \frac{\Omega}{4\pi D}\left[\log\frac{1}{\varepsilon} + O(1)\right], \qquad \text{2D}\\ \langle T\rangle &\sim \frac{\Omega^{2/3}}{\varepsilon D}\left[1 + \frac{\varepsilon}{\pi}\log\frac{1}{\varepsilon} + \ldots\right]. \qquad \text{3D}\end{aligned} \tag{26}$$

Analogous results were obtained for the constriction escape problem, where a narrow bottleneck is formed by circles or spheres of radius $R_{\rm c}$ approaching each other or revolved to form a three-dimensional bottleneck:

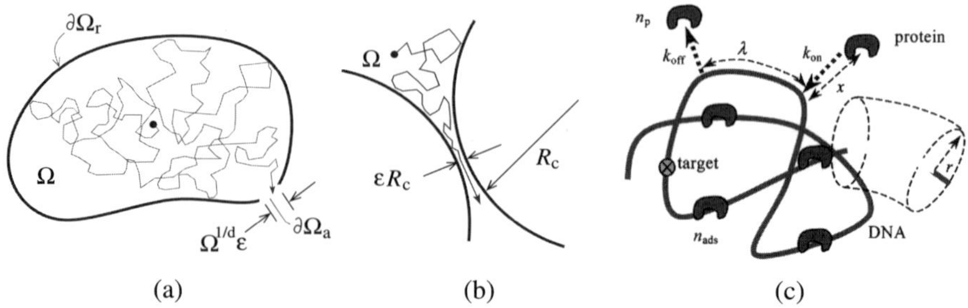

Fig. 7. (a) The canonical narrow escape problem. A particle with diffusivity D can escape from an asymptotically small aperture. The mean time $\langle T(\mathbf{x})\rangle$ to escape Ω, as a function of initial position $\mathbf{x}$ can be calculated in the asymptotic limit $\varepsilon \to 0$. (b) An escape problem where the escape hatch is at a cusp. (c) DNA target site search problem. Search is facilitated by 1D diffusion along the DNA chain.

$$\begin{aligned} \langle T\rangle &\sim \frac{\pi\Omega}{2D\sqrt{\varepsilon}}, \qquad && \text{2D} \\ \langle T\rangle &\sim \frac{\Omega}{\sqrt{2}R_c D}\frac{1}{\varepsilon^{3/2}}, \qquad && \text{3D}. \end{aligned} \tag{27}$$

Similar results have been derived for different geometries such as diffusion to the tip of a corner, and first passage to the end of a long neck. Due to the chosen geometries, escape is a rare event, and the particle reaches a quasi-steady-state distribution before any escape has occurred. Since the time to reach the quasi-steady-state distribution starting from a specific position is negligible compared to the mean escape time, all the above results are independent of the particle's initial position $\mathbf{x}$.

Another related and biologically important example of first passage is the search of molecules for their target sites, such as the binding of transcription factors (sequence-specific DNA-binding proteins) to their corresponding binding sites along DNA [54–58] (see Fig. 7(c)). These sites are often proximal to the genes they regulate, although in reality, numerous transcription factors, including basal factors, RNA polymerase, coactivators, and activators must assemble before transcription of a specific gene is initiated. The search problem is of theoretical interest because experimental search times are much shorter than those estimated from simple 3D diffusion alone, leading to the idea of facilitated diffusion, a mechanism whereby more than one transport path is available. Since DNA is a linear, often compacted polymer, sections many bases away from the target may nonetheless be spatially proximal to it. These physical features have been incorporated into transport models to estimate the time it takes for an enzyme to bind its intended target along DNA of arclength L. The original phenomenological model assumes an effective absorbing sphere of radius λ around the target, where λ is the typical contiguous length traveled along the DNA. A simple heuristic expression for this "antennae effect" on the search time was derived: $\langle T \rangle \approx (L/\lambda)(\tau_1 + \tau_3)$, where τ_1 and τ_3 are the typical times spent on the DNA and in the bulk, respectively. To obtain realistic search times using this expression requires that the enzyme spend approximately an equal amount of time on DNA as in the bulk. However, in reality, enzymes spend an overwhelming majority of time associated with DNA. Moreover, this expression breaks down in certain singular limits such as when the one-dimensional diffusivity $D_1 \to 0$, leading to $\tau_1 \to \infty$. An improved expression for the mean search time has been recently derived [59],

$$\langle T \rangle \approx \frac{Lr}{2D_3 n_\mathrm{p}} \left(\frac{r}{\lambda n_\mathrm{ad}} + \frac{\lambda D_3 n_\mathrm{p}}{D_1 r n_\mathrm{ads}} + \frac{2D_3 k_\mathrm{off}}{k_\mathrm{on} D_1 \sqrt{n_\mathrm{p}}} \right), \tag{28}$$

where L is the arclength of the DNA, r is its effective thickness, n_p and n_ads are the number of bulk and adsorbed proteins, and k_on and k_off are the attachment and detachment rates of protein. Note that in this treatment, k_on was defined using a reference protein concentration of one molecule per search volume. The typical arclength a protein stays within r of the DNA before dissociating is thus estimated to be

$$\lambda \approx \frac{r\sqrt{k_\mathrm{on} D_1}}{\sqrt{k_\mathrm{off} D_3 n_\mathrm{ads}}}. \tag{29}$$

The result (29) is able to resolve a number of quantitative kinetic issues. In particular, Cherstvy, Kolomeisky, and Korynyshev [59] were able to find optimal binding energies that minimize the search time. Within a realistic parameter regime, the reduction in search time relative to 3D diffusion alone can be obtained even for small D_1/D_3. Additional details and references are found in Kolomeisky [60]. Note that all results on this problem are independent of the initial starting position of the searching enzyme since an initial uniform distribution of enzyme positions is implicitly assumed.

The molecular search problem is also intimately related to the filament growth described in the previous section. During mitosis, the ends of growing and shrinking microtubules emanating from centrosomal bodies form a party in search of kinetochores that hold together chromosomes [61, 62]. Using the Green's function approach of Bicout [47] for a single microtubule as a starting point, Gopalakrishnan and Govindan [63] found estimates for the search time to one kinetochore

$$\langle T\rangle \approx \frac{e^{\Delta d}}{p}\left(1+\frac{f_-(1-e^{-\Delta d})}{v_-\Delta}\right)\left(\frac{v_+ + v_-}{\Delta v_- v_+}+\frac{1}{f}\right), \tag{30}$$

where $\Delta \equiv (v_- f_+ - v_+ f_-)/(v_+ v_-)$, and f is the frequency of nucleation of new microtubules from the centrosome that is located a distance d from the kinetochore target. The probability that any new microtubule is pointed in the right direction and within the capture cone is $p \ll 1$. The microtubule velocities $v_\pm$ and flipping rates $f_\pm$ take on the same meaning as in Eq. (23) used by Bicout to study the lifetime of a single microtubule. Equation (30) holds only when the cell radius $R \gg d$. This and related formulae allow for an easy determination of optimal parameters that minimize the mean search time. The topic of capture of multiple kinetochores associated with multiple chromosomes has also been treated by Wollman *et al.* [62].

Besides the filament growth and search problems described in Section 3 and above, two other examples of cellular transport involving first passage times have been recently discussed: optimal microtubule transport of virus material to a host cell nucleus [64], and localization of DNA damage repair enzymes to DNA lesions [65, 66].

When a virus first enters a mammalian host cell its genetic material needs to be processed and transported into the host cell nucleus before productive infection can occur. The transport is often mediated by molecular motors that carry viral RNA or DNA towards the nucleus. This process was modeled by a unidirectional convection of cargo in multiple stages, while detachment of the motor and degradation of the viral cargo was implemented by a decay term. Nuclear entry probabilities and conditional first arrival times for cargo starting at the cell periphery and ending at the nucleus were calculated [64]. These were found to depend on parameters describing convection, decay, and transformation in nontrivial ways which suggested new strategies for drug intervention of the transport process.

Another biophysical example where finding first passage times is important is the localization of proteins to certain sites on DNA using an electron ejection mechanism [65]. A redox mechanism for certain DNA repair enzymes to localize near DNA damage sites has been proposed [67–69], as depicted in Fig. 8(a). Here, a recently deposited repair enzyme oxidizes by releasing an electron that can either scatter or absorb at guanine bases and damaged DNA sites. The oxidized repair enzyme has a higher binding affinity to DNA. However, if the electron returns, the reduced enzyme will dissociate from the DNA.

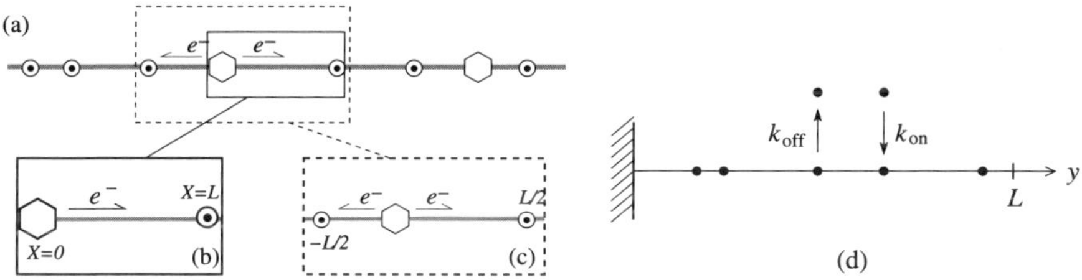

Fig. 8. (a) A repair enzyme (hexagon) adsorbs onto a DNA which is initially populated by guanine radicals (circled dots) with a density ρ. An electron is emitted to the left or right with equal probability. The emitted electron has flip rates $f_\pm$, rightward/leftward velocity $v_\pm$ and decay rate M. (b) The one-sided Broadwell problem. An electron is emitted from position $X = 0$ with probability 1 toward a guanine radical at $X = L$. (c) The two-sided Broadwell problem. An enzyme is deposited between two guanine radicals which are at a distance L apart. Immediately after landing inside this segment, an electron is emitted to the left or right with equal probability. (d) First passage time to a boundary position $y = L$ in the presence of multiple particles undergoing Langmuir kinetics.

Within this overall mechanism, the problems of the first electron return time, conditioned on its returning arises. The model equation for this subproblem is identical to Eq. (23) except that x, $v_\pm$, and $f_\pm$ are the position, speeds, and flip rates of an electron along the DNA, and decay terms are added to describe the absorption of electrons "off" the DNA. The effective desorption rate was calculated from the probability and time of electron return. For repair enzymes that land far from electron absorbing lesions, and if other electron-absorbing mechanisms are negligible, return of the emitted electron is likely and the enzyme will detach before it can diffuse sufficiently far. However, in a finite cell volume, the detached enzyme reenters the bulk pool and can reattach to the DNA, potentially closer to the lesion. Deposition near a lesion will likely be longer-lived because the ejected electron will be more likely absorbed rather than returning and dissociating the enzyme. In this way, Fok and Chou [65, 66] were able to find conditions under which the repair enzymes statistically localize near electron-absorbing damage sites on DNA.

Finally, search problems can involve multiple diffusing particles. In this case, it is still reasonable to define the state-space in terms of the positions $\{x_j\}$, $1 \leq j \leq N$ for each of, say, N particles. In one-dimension, the first hitting time for any particle to reach an absorbing point of a line segment has been examined by Sokolov *et al.* [70] who considered noninteracting particles that diffuse and undergo Langmuir kinetics as shown in Fig. 8(b). In their study, the authors employ a mean-field assumption for Eq. (4) where the probability current $J(t)$ is *conditioned* on no other particle having exited the interval previous to time t. The mean-field assumption arises by expressing this conditioning as $J_{\text{conditioned}}(t) = J_{\text{unconditioned}}S(t)$. The mean-field solution to the probability $S(t)$ that no particle has hit the target site up to time t is

$$S(t) = J(t) \exp\left[-\int_0^t J(t')\mathrm{d}t'\right], \tag{31}$$

where $J(t)$ is the unconditioned probability flux. Note that for this approximation to yield physical results, we require

$$\lim_{t\to\infty} tJ(t) > 0 \tag{32}$$

in order for $\int_0^t J(t')\mathrm{d}t'$ to diverge and $S(t) \to 0$ as $t \to \infty$. In this problem, the flux was approximated by $J(t) = -D\partial_y n(y,t)|_{y=L}$, where $n(y,t)$ is the particle density at position y that is found from

$$\frac{\partial n(x,t)}{\partial t} = D\frac{\partial^2 n(x,t)}{\partial x^2} - k_{\text{off}}n + k_{\text{on}}, \tag{33}$$

where D is the one-dimensional diffusivity, and k_{on} and k_{off} are the particle adsorption and desorption rates. Because of the implied infinite bulk reservoir (through rate k_{on}) the mean-field flux satisfies Eq. (32). Even in the case $k_{\text{off}} = k_{\text{on}} = 0$, if an infinite system size is assumed, the condition in Eq. (32) is also satisfied. In fact, when the system is infinite, the mean-field assumption in Eq. (31) is exact.

A more general approach that does not initially rely on the mean-field assumption, and can be used for finite-sized systems, is to note that if the particles are noninteracting, the survival probability $S(t;\{x_i\}) = \prod_{i=1}^{N} S_1(t;x_i)$ is a product of the survival probabilities of each particle with initial position x_i. We assume a finite segment and assume N total of particles, including those in the bulk. In this way, we can compute the single particle probability flux $J_1(t) = -D\partial_y P_1(y,t|x,0)|_{y=L}$, and use the exact relation

$$\frac{\partial S_1(t;x)}{\partial t} = -J_1(t;x) = D\partial_y P_1(y,t|x,0)|_{y=L}. \tag{34}$$

Using conservation of probability, $\int_0^\infty J_1(t';x)\mathrm{d}t' = 1$ and, assuming the initial positions (including the possibility of being detached from the lattice) of all particles are identical, we find

$$S(t;x) = \left[1 - \int_0^t J_1(t';x)\mathrm{d}t'\right]^N. \tag{35}$$

A direct comparison can be made with the mean field result in the case $k_{\text{off}} = k_{\text{on}} = 0$. Upon solving Eq. (33), we can find the the Laplace transform of the single-particle probability flux, assuming a uniformly distributed initial condition

$$\tilde{J}_1(s) = -D\frac{\partial\tilde{n}(x,s)}{\partial x}\bigg|_{x=L} = \frac{\tanh L\sqrt{s/D}}{L\sqrt{s/D}}. \tag{36}$$

Upon inverse Laplace-transforming, and using the result in Eq. (35), we can find the exact survival probability. Note that this result is different from using $NJ_1(t)$ for $J(t)$ in the mean-field approximation Eq. (31). Only in the infinite system size limit of $L, N \to \infty$, but $N/L = n_0$ constant do the mean-field and exact result $S(t) = \exp\left[-2n_0\sqrt{Dt/\pi}\right]$ coincide. This can be shown mathematically by using $L = N/n_0$ in Eq. (36), inverse Laplace transforming, substituting the result in Eq. (35), and taking the $N \to \infty$ limit. The discrepancy can be most easily seen by assuming all particles start at x and

$$\frac{\partial S(t;x)}{\partial t} = NS_1^{N-1}(t;x)\frac{\partial S_1(t;x)}{\partial t} = -NJ_1(t;x)S_1^{N-1}(t;x). \tag{37}$$

For noninteracting particles, the total annihilation flux $J(t;x) = NJ_1(t;x)$, and

$$\frac{\partial S(t;x)}{\partial t} = -J(t;x)S_1^{N-1}(t;x) = -J(t;x)\frac{S(t;x)}{S_1(t;x)}. \tag{38}$$

The relative effect of the extra factor $S_1(t;x) < 1$ on $S(t;x)$ decreases as

$N \to \infty$. It should be stressed that independence of the diffusing particles allow for the exact analysis above. However, certain approximate results for interacting particles have also been obtained [71].

Multiple particle first passage problems also illustrate the concept of order statistics. Although Eq. (35) provides the survival probability of a boundary untouched by *any* one of the diffusing particles, one might be interested in the statistics of the first, second, third, etc., particle to leave the interval, as well as the complete clearing time distribution. These order statistics and asymptotic expressions for the first two moments of the j exit times have been derived for independent particles diffusing in one-dimension [72] and d–dimensions [73].

5. Neuronal spike trains

An important first passage problem within a living, functioning nerve cell, or group of nerve cells arises in the study of the timing of electrical spike trains. While modeling the stochastic dynamics of the membrane potential of a neuron requires taking into account a large number of detailed microscopic processes, such as nonlinear ion channel gating and membrane capacitance and leakage, the overall phenomena of spike trains can be effectively described by a stochastic process with a threshold membrane potential V^*. When the voltage of a neuron reaches V^*, highly nonlinear processes take over, the voltage quickly spikes, and returns to a reset voltage, as shown in Fig. 9(a). The interspike times are distributed according to the time that the transmembrane potential first reaches V^* after the previous resetting.

A simple one-dimensional stochastic model for predicting interspike times for a single neuron has been proposed by Stein [74]. Here, the transmembrane voltage is assumed to dissipate through a "leak" current, while other connected neurons impart noise to the neuron of interest. The model implicitly relies on a mean field assumption in the sense that none of the other neurons are affected by the behavior of the neuron in question. The "bath" neurons provide random excitatory and inhibitory signals through unspecified physical connections with the isolated neuron. Starting from a stochastic differential equation (SDE) formulation, increments of the transmembrane voltage V can be expressed as

$$\mathrm{d}V = -\frac{V}{\tau}\mathrm{d}t + a_{\mathrm{e}}\mathrm{d}\pi_{\mathrm{e}}(r_{\mathrm{e}}, t) - a_{\mathrm{i}}\mathrm{d}\pi_{\mathrm{i}}(r_{\mathrm{i}}, t), \tag{39}$$

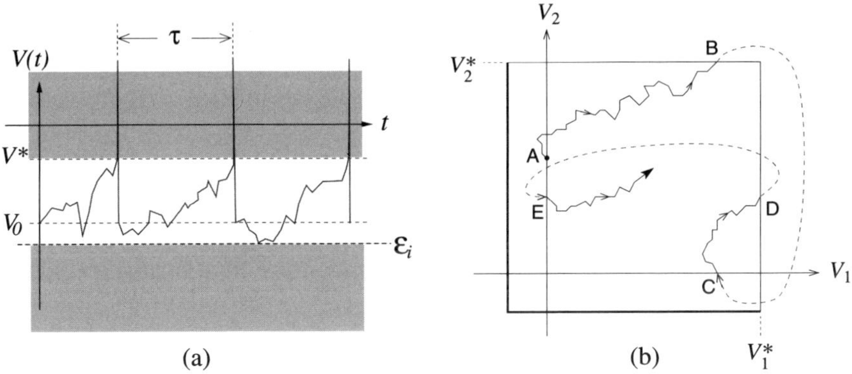

Fig. 9. First exit times in simple neuronal firing models. (a) A schematic time trace of the transmembrane potential showing voltage spikes triggered at V^* and resetting back to V_0. The subthreshold voltage dynamics is a stochastic processes with the interspike time distribution measuring the statistics of the first passage time to the threshold voltage. (b) Voltage trajectories for two coupled neurons, with transmembrane voltage V_1 and V_2. If neuron 2 spikes first at point (B), V_2 spikes and quickly resets to point (C). In this example, neuron 1 spikes next at point (D), and V_1 resets to point (E).

where $a_{\rm e}$ and $a_{\rm i}$ are the fixed amplitudes of the excitatory and inhibitory spikes feeding into the neuron, and $\pi_{\rm e}(r_{\rm e},t)$ and $\pi_{\rm i}(r_{\rm i},t)$ are possibly time-varying unit excitatory and inhibitory Poisson processes with rates $r_{\rm e}$ and $r_{\rm i}$, respectively. Suppose the voltage starts at $V(t=0)=X$ and that the threshold for spiking is V_*. The recursion equations for the moments $M_n(X;V_*)\equiv\langle T^n(X;V_*)\rangle$ of the interspike times are [75]

$$\frac{X}{\tau}\frac{\mathrm{d}M_n}{\mathrm{d}X}-r_{\rm e}M_n(X+a_{\rm e})-r_{\rm i}M_n(X-a_{\rm i})+(r_{\rm e}+r_{\rm i})M_n(X)=nM_{n-1}(X), \quad (40)$$

where $M_0(X)\equiv 1$. The mean interspike times $M_1(X,V_*)\equiv\langle T(X)\rangle$ were analyzed by Cope and Tuckwell [76] using asymptotic analysis for large negative reset voltages, and continuing the solutions to the threshold V_*. Assuming $a_{\rm e}=a_{\rm i}$, their result for the mean first time $T(V)$ to spiking starting from an initial voltage V can be expressed in the form

$$\langle T(X,V_*)\rangle\approx\frac{1}{r_{\rm e}}\left[\frac{1}{\tau r_{\rm e}}\log\left(\frac{V}{a_{\rm e}}\right)+C\left(\frac{V_*}{a_{\rm e}}\right)+\sum_{n=1}^{\infty}A_n\left(\frac{a_{\rm e}}{V}\right)\right], \quad (41)$$

where the function $C(V_*/a_{\rm e})$ and the coefficients A_n were numerically found from recursion relations of a set of linear equations. However, note that the associated equation for the voltage probability density $P(V,t|V_0,0)\mathrm{d}V$ is

$$\frac{\partial P}{\partial t} = \frac{1}{\tau}\frac{\partial (VP)}{\partial V} + r_{\mathrm{e}}P(V-a_{\mathrm{e}},t) + r_{\mathrm{i}}P(V+a_{\mathrm{i}},t) - (r_{\mathrm{e}}+r_{\mathrm{i}})P, \tag{42}$$

where only arguments of P that are different from $(V,t|V_0,0)$ are explicitly written. A further simplification can be taken by assuming the noise amplitudes $a_{\mathrm{e,i}}$ are small and Taylor expanding the probability densities to second order in $a_{\mathrm{e,i}}$ (a "diffusion" approximation). The Fokker-Planck or Smoluchowski equation now takes the form

$$\frac{\partial P}{\partial t} = \frac{\partial}{\partial V}\left[\left(\frac{V}{\tau} - r_{\mathrm{e}}a_{\mathrm{e}} + r_{\mathrm{i}}a_{\mathrm{i}}\right)P\right] + \frac{1}{2}\left(r_{\mathrm{e}}a_{\mathrm{e}}^2 + r_{\mathrm{i}}a_{\mathrm{i}}^2\right)\frac{\partial^2 P}{\partial V^2}, \tag{43}$$

with $P(V,t|V_0,0) = \delta(t)$ when $V = V_*$. This model for subthreshold neuron voltage is simply a first passage problem of the Ornstein-Ulhenbeck process that has been used to describe particle escape from a quadratic potential or rupturing of a harmonic bond. Recasting the problem using a Backward Kolmogorov Equation, the survival probability (the probability that no spike has occurred) as well as the moments of the interspike times can be expressed in terms of special functions [75]. Tuckwell and Cope [75] also provide a careful analysis of the accuracy of the diffusion approximation in approximating the "exact" results from Eq. (40). As expected the diffusion approximation is accurate in the limit of large excitatory and inhibitory spike noise rates r_{e} and r_{i}, and when the threshold voltage V_* is far from the reset voltage.

Besides simple one-dimensional models, higher dimensional models that include more mechanistic details of a single neuron have also been studied. In particular, stochastic first passage problems for Fitzhugh-Nagumo [77] and Hodgkin-Huxley models [78] have been developed. These more complex models still focus on the voltage dynamics of a single neuron, with the voltage dynamics of other connected neurons subsumed into the "noise" felt by the neuron. Typically, the multiple neuron voltages can be simultaneously measured using multielectrode recordings, allowing for the quantification of the correlations between the spiking times of connected neurons. A first approach for modeling these higher dimensional data is to treat the stochastic dynamics of a small number of interacting neurons. For the two neuron problem illustrated in Fig. 9(b), the dynamics of the subthreshold voltages of neurons 1 and 2, V_1 and V_2, respectively, are independent of each other, and the probabilities factorize: $P(V_1,V_2)\mathrm{d}V_1\mathrm{d}V_2 = P_1(V_1)P_2(V_2)\mathrm{d}V_1\mathrm{d}V_2$.

Interactions between the two neurons occur when either voltage spikes. A neuron connected to one that spikes can suffer a small voltage displacement. Rather than treating each neuron as subject to independent noise, the spiking time statistics of the neurons provide one component of the random noise of the other neuron. The full spiking time statistics must be computed self-consistently. Trajectories in the state space shown in Fig. 9(b) can be described moving along a torus with jumps in the orthogonal direction each time it crosses circumferentially or axially. Mathematically, the probability densities for the two subthreshold voltages obey

$$\frac{\partial P_i(V_i,t)}{\partial t} = \frac{\partial}{\partial V_i}\left[U_i(V_i)P_i\right] + D_i\frac{\partial^2 P_i}{\partial V_i^2}, \tag{44}$$

where D_i is the voltage diffusivity in neuron i. However, as soon as one V_i reaches V_i^*, not only does it reset, but $V_{j\neq i} \to V_{j\neq i} + \delta_j$ is shifted by δ_j.

6. Cellular and organismic population dynamics

The simplest nonspatial deterministic population model, describing growth limitations due to a carrying-capacity, centers on the logistic equation

$$\frac{\mathrm{d}n(t)}{\mathrm{d}t} = rn(t)\left(1 - \frac{n(t)}{K}\right), \tag{45}$$

where $n(t)$ is the population density and K is the carrying-capacity. This deterministic model has stable fixed points at $n = 0$ and $n = K$. There are multiple ways to define stochastic birth-death models that in the mean field limit reduce to Eq. (45) [79]. Nonetheless, all of these models require at least one existing organism for proliferation to take place. Therefore, these models contain an absorbing state at $n = 0$, where the population is extinct. Although the deterministic equation predicts, at long times, a permanent population $n = K$, a stochastic model predicts a finite extinction time T after which $n(t \geq T) = 0$. Approximations to this extinction time have been analyzed by Kessler and Shnerb [80] using a WKB approximation and Assaf and Meerson [81] using a generating function approach and properties of the associated Sturm-Liouville equation. Both methods use the approximation $K \gg 1$, for which extinction is rare, and a near equilibrium number distribution is first achieved before an extinction event occurs. This approximation is analogous to that of assuming "local thermodynamic equilibrium" (as opposed to kinetic theory) for transport calculations [82]. The probability flux is then constructed from the rate of transport into an absorbing state from this near equilibrium density. The distribution of times for the rare extinction events are nearly exponential

$$w(t)\mathrm{d}t \approx \Gamma e^{-\Gamma t}\mathrm{d}t, \tag{46}$$

where to leading order the extinction rate is of the form

$$\Gamma \sim K^{3/2}e^{-K}. \tag{47}$$

Note that these results, as with those of the narrow escape problem (Section 4), do not depend on the initial number $n_0 = n(t=0)$ because equilibration to a quasi-stationary state occurs on a time scale much faster than Γ^{-1}.

Other classic population models, such as models for cell genotype/phenotype populations, Lotka-Volterra type models [83], and disease models (such as SIS and SIR) [84, 85] have also been extended into the stochastic realm, and the corresponding exit times into absorbing configurations analyzed (see Ovaskainen and Meerson [86] for a review). Here, the total organism number is a random variable determined by the dynamical rules of the model, which may include "interacting" effects such as carrying-capacity. The simplest model for heterogeneity in a birth-death process is the Wright-Fisher model or, in continuous-time, the Moran model. The latter is a stochastic model for two-competing species with numbers n_1 and n_2, where the total population $n_1 + n_2 \equiv N$ is fixed. Since $n_2 = N - n_1$, the problem state-space reduces to one-dimension. The transition rules in the Moran model are defined by randomly selecting an individual for annihilation, but instantaneously replacing it with either one of the same type (so that the system configuration does not change), or one of the opposite type. The transition probability in time interval dt for converting an n_1 individual to an n_2 individual is thus $r_1 n_1 n_2 \mathrm{d}t = r_1 n_1(N - n_1)\mathrm{d}t$, while conversion of n_2 to n_1 occurs with probability $r_2 n_2(N - n_2)\mathrm{d}t$. By defining $P(n,t|m,0)$ as the probability that there are $n = n_1$ type 1 individuals at time t, given that there were initially m type 1 individuals, the BKE is simply

$$\begin{aligned}\frac{\partial P(n,t|m,0)}{\partial t} = m(N-m)\big[r_1 P(n,t|m+1,0) + r_2 P(n,t|m-1,0) \\ -(r_1+r_2)P(n,t|m,0)\big].\end{aligned} \tag{48}$$

Note that $n = 0$ and $n = N$ are absorbing states corresponding to the entire population being fixed to either type 1 or type 2 individuals. Upon summing $\sum_{n=1}^{N-1} P(n,t|m,0) \equiv S(t;m)$, we can find the corresponding BKE for the probability of survival against fixation at either $n = 0$ or $n = N$. The mean time to fixation can then be found from inverting the matrix equation

$$m(N-m)\left[r_1\langle T(m+1)\rangle + r_2\langle T(m-1)\rangle - (r_1+r_2)\langle T(m)\rangle\right] = -1, \quad (49)$$

with $\langle T(0)\rangle = \langle T(N)\rangle = 0$, to give the well-known result

$$\langle T(m)\rangle = N\sum_{k=1}^{m}\frac{N-m}{N-k} + N\sum_{k=m+1}^{N-1}\frac{m}{k}. \quad (50)$$

If spontaneous mutations are included in the model, there is strictly no fixation since the states $n = 0, N$ are no longer absorbing. Many generalizations of the Moran model have been investigated, including extensions to include more species, fluctuating population sizes, and time-dependent parameters such as the rates $r_1(t), r_2(t)$ [87, 88]. These extended models are not typically amenable to closed form solutions such as Eq. (50). Nonetheless, it is often possible to employ asymptotic analysis in the large N limit and derive a corresponding PDE for either the probability density or its generating function. For example, if one assumes $N \to \infty$ and takes $x = m/N$ one finds the diffusion approximation for the BKE

$$\frac{\partial S(t;x)}{\partial t} = D_{\text{eff}}x(1-x)\frac{\partial^2 S(t;x)}{\partial x^2}, \quad 0 \le x \le 1. \quad (51)$$

Here, we have introduced $D_{\text{eff}} = r_1 N^2 = r_2 N^2$. The corresponding PDEs for more complex Moran-type models are often amenable to analysis, making the Moran model one of the paradigmatic theories in population biology and ecology. However, recall from Section 1 the discrepancy between the first passage times derived from discrete and corresponding continuum theories [6]. For Eq. (51), there is no selection or mutation giving rise to a convection term, so the corresponding mean first passage time asymptotically approaches the discrete result in Eq. (50) as $N \to \infty$. However, care should be exercised for more complex models that include effective convection terms.

Higher dimensional generalizations of these types of discrete models can also be readily applied to problems in cell population biology such as cancer modeling and stem-cell proliferation. When the total population size constraint is relaxed, a linear, multiple state model shares many mathematical features with the Zero-Range Process (ZRP) [89], as shown in Fig. 10. The multiple sites in such a ZRP might represent the number of cells in a tissue at a particular mutation stage as the cells progress towards a cancerous state. Of interest is the first time that a certain number of cells arrive at the final, "fully cancerous" state [a].

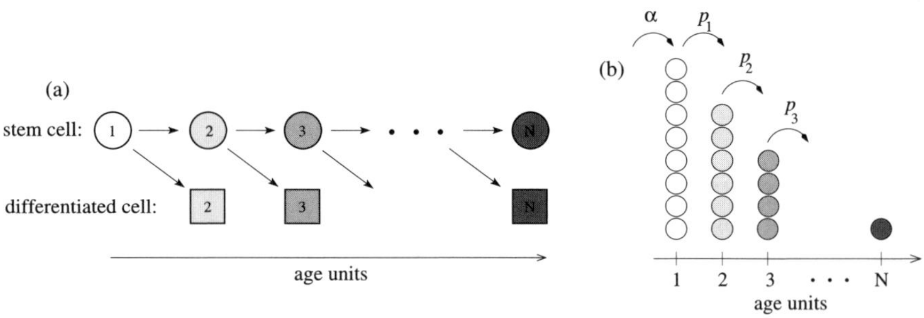

Fig. 10. Schematic of a reduced model of stem-cell aging. (a) Asymmetric division of aging stem-cells. The circles represent stem cells, while the squares represent differentiated cells. The numerical index represents the age of the cell and is assumed to be inversely related to the telomere length. (b) A lattice representation of the stem-cell aging model. The rate of asymmetric differentiation are shown as p_k, while the death rates μ_k at each age k are not indicated.

Besides multi-hit models of cancer and evolution, the Zero-range process can also be adapted to model aging in a stem-cell population. Consider stem-cells that have a limited number of divisions due to shortening telomeres, ends of their DNA that are shortened at each division. Without telomerase to rebuild these ends, cells will generally be programmed for death. As shown in Fig. 10(a), our model assumes that each division leads to one stem-cell and one differentiated cell, both aged by one unit (or both with shortened telomeres). Since all cell divisions are asymmetric, yielding one stem-cell and one differentiated cell, one only needs to keep track of the number of stem-cells. The forward master equation for the process has been derived in Shargel, D'Orsogna, and Chou [90], as well as

[a]In other contexts, such as individual survival probabilities against death from cancer are called Kaplan-Meier curves which represent the fraction of a population alive as a function of time after the initial diagnosis of cancer.

the associated equation for the generating function:

$$\frac{\partial G}{\partial t} = -\sum_{j=1}^{N-1}(\mu_j + p_j)z_j\frac{\partial G}{\partial z_j} + \sum_{j=1}^{N}\mu_j\frac{\partial G}{\partial z_j} + \sum_{j=1}^{N-1}p_j z_j\frac{\partial G}{\partial z_j} - \mu_N z_N\frac{\partial G}{\partial z_N}, \quad (52)$$

where

$$G(z_1, \cdots, z_N; t) = \sum_{n_j} P(n_1, \cdots, n_N; t) z_1^{n_1} \cdots z_N^{n_N} \quad (53)$$

and $P(\{n\}; t)$ is the probability that there are exactly n_k stem-cells of age k at time t. If we do not assume an immigration of new stem-cells defined as having age $k = 1$ (as was done in Shargel, D'Orsogna, and Chou [90]), Eq. (52) can be expressed in the form $\mathrm{d}G/\mathrm{d}t = 0$ and solved using the method of characteristics. The vector of characteristic trajectories $\mathbf{Z} = (z_1, z_2, \ldots, z_N)^T$ can be found by solving $\dot{\mathbf{Z}} = \mathbf{PZ} - \mathbf{M}$, where

$$\mathbf{P} = \begin{bmatrix} \mu_1 + p_1 & -p_1 & 0 & \cdots & 0 \\ 0 & \mu_2 + p_2 & -p_2 & \cdots & 0 \\ 0 & \cdots & \mu_j + p_j & -p_j & 0 \\ 0 & \cdots & 0 & \mu_{N-1} + p_{N-1} & -p_{N-1} \\ \cdots & \cdots & \cdots & \cdots & \cdots \\ 0 & \cdots & 0 & 0 & \mu_N \end{bmatrix} \quad (54)$$

and $\mathbf{M} = (\mu_1, \cdots, \mu_j, \cdots, \mu_N)^T$. For an initial condition of one stem-cell of age $k = 1$, these trajectories can be inverted and expressed in terms of the initial values $z_i(t = 0)$, which form the independent variable in the generating function:

$$G(\mathbf{Z}; t) = z_1 e^{-\Delta_1 t} + \sum_{i=2}^{N}\left[z_i(-1)^{i+1}\left(\prod_{\ell=1}^{i-1} p_\ell\right)\sum_{j=1}^{i-1}\frac{e^{-\Delta_j t} - e^{-\Delta_i t}}{\prod_{k\neq j}^{i}(\Delta_j - \Delta_k)}\right] + \quad (55)$$
$$1 - e^{-\Delta_1 t} + \sum_{i=2}^{N}\left[(-1)^i\left(\prod_{\ell=1}^{i} p_\ell\right)\sum_{j=1}^{i-1}\frac{e^{-\Delta_j t} - e^{-\Delta_i t}}{\prod_{k\neq j}^{i}(\Delta_j - \Delta_k)}\right],$$

where $\Delta_j \equiv p_j + \mu_j$, for $1 \leq j \leq N - 1$ and $\Delta_N = \mu_N$. From the generating function in Eq. (55) we can derive the probability $P(n_1 = 0, ..., n_j = 1, ..., n_N = 0; t)$ that a certain age by the descendants of single cell can be found at a given age j:

$$P(n_1 = 1, n_2 = 0, \cdots, n_N = 0) = e^{-\Delta_1 t}, \tag{56}$$

while for all other ages $1 < j < N$ we find

$$\begin{aligned} P(0, \cdots, n_j = 1, \cdots, 0; t) &= (-1)^j \left(\prod_{\ell=1}^{j-1} p_\ell\right) \sum_{k=1}^{j-1} \frac{e^{-\Delta_j t} - e^{-\Delta_k t}}{\prod_{i \neq k}^{j} (\Delta_k - \Delta_i)} \\ &= \frac{(pt)^{j-1}}{(j-1)!} e^{-(\mu+p)t}, \end{aligned} \tag{57}$$

where the last equality holds in the case where all $p_i = p$ and $\mu_i = \mu$ are age-independent. Finally, the probability for complete extinction of the lineage is given by

$$P(\{n\} = 0; t) = 1 - e^{-\Delta_1 t} + \sum_{i=2}^{N} \left[(-1)^i \left(\prod_{\ell=1}^{i-1} p_\ell\right) \sum_{j=1}^{i-1} \frac{e^{-\Delta_j t} - e^{-\Delta_i t}}{\prod_{k \neq j}^{i} (\Delta_j - \Delta_k)} \right]. \tag{58}$$

It can be easily verified that the sum of the probabilities in Eqs. (56), (57) and (58) add to unity, and that $P(n_1 = 0, \cdots, n_N = 0; t \to \infty) \to 1$, indicating that a single cell will eventually age and that its lineage will go extinct with certainty.

From these probabilities we can construct the probability that the oldest age reached by a lineage is Q_k:

$$Q_k = \int_0^\infty [p_{k-1} P(0, \ldots, n_{k-1} = 1, \ldots, 0; t) - p_k P(0, \ldots, n_k = 1, \ldots, 0; t)] \, \mathrm{d}t. \tag{59}$$

Equation (59) is derived by considering the difference between the probability flux into age k and the flux out of age k into age $k+1$ (excluding death). The time-integrated result Q_k is thus the probability that the lineage died at age k. For the constant rate case $p_i = p$ and $\mu_i = \mu$, we find explicitly

$$Q_1 = \frac{\mu}{\mu + p}, \quad Q_k = \frac{\mu p^{k-1}}{(\mu + p)^k}, \text{ and } Q_N = \frac{p^{N-1}}{(\mu + p)^{N-1}}. \tag{60}$$

From these probabilities, we can define the first passage time to age k conditioned on the system reaching at least age k. Since the decay at all ages preceding k are "interfering" absorbing states, we can use $J_1^k(t) = pP(0, ..., n_{k-1} = 1, ..., 0; t)$ in Eq. (12) to find

$$w_1(t|k) \equiv J_1(t|k) = \frac{(\mu + p)((\mu + p)t)^{k-2}}{(k-2)!} e^{-(\mu+p)t}, \quad k \geq 2, \tag{61}$$

with a corresponding conditional mean arrival time to age k: $\langle T_1(k) \rangle = (k-1)/(\mu + p)$. Note that if the decay rate μ is high, the *conditional* mean arrival time is small because only fast trajectories will survive to state k.

Our simple stem-cell aging model assumes all divisions are asymmetric at all ages. Nonetheless, this model serves as an illustrative example of an application of a simple Markov process to cell biology. Indeed, since aging only increases, our model can also be represented by a simple asymmetric, decaying random walk of a single "particle" in one-dimension, with the position of the particle representing the age of the single stem-cell in the system at any given time. The more complicated approach we have illustrated above allows our model to be generalized to include effects of multiple initial stem-cells and symmetric stem-cell division, as well as a more complete analysis of differentiated cell populations.

7. Summary

We have surveyed only a few mathematical and physical models wherein first passage problems play a central role in the quantitative understanding of biological observations and experiments. These applications span all scales from molecular to cellular to populations. Most applications thus far have been concerned with low dimensional models with few degrees of freedom. As measurements improve and more complex systems can be quantitatively studied, first passage time problems should become increasingly important in higher dimensional settings where additional analytic and numerical insights will be desired. Furthermore, first passage problems provide a new framework with which to fit experimental data, model biological processes, and develop inverse problems of model determination.

Acknowledgments

The authors are grateful to B. van Koten, J. Newby, and J. J. Dong for incisive discussions and comments. This work was supported by the NSF through grants DMS-1021818 (TC) and DMS-1021850 (MD). TC is supported by the Army Research Office through grant 58386MA. MD was also supported by an Army Research Office MURI grant W911NF-11-1-0332.

References

1. S. Redner, *A Guide to First-Passage Processes.* Cambridge University Press, Cambridge, UK (2001).
2. K. B. Athreya and P. E. Ney, *Branching Processes.* Dover, Mineola, NY (2000).
3. T. E. Harris, *The Theory of Branching Processes.* Dover, Mineola, NY (1989).
4. G. H. Weiss, First passage time problems for one-dimensional random walks, *Journal of Statistical Physics.* **24**(4), 587–594 (1981).
5. P. A. Pury and M. O. Caceres, Mean first-passage and residence times of random walks on asymmetric disordered chains, *Journal of Physics A: Mathematical and General.* **36**, 2695 (2003).
6. C. R. Doering, K. V. Sargsyan, and L. M. Sander, Extinction times for birth-death processes: Exact results, continuum asymptotics, and the failure of the Fokker–Planck approximation, *Multiscale Model. Simul.* **3**, 283–299 (2005).
7. G. I. Bell, Models for the specific adhesion of cells to cells, *Science.* **200**, 618–627 (1978).
8. A. G. Nobile, L. M. Ricciardi, and L. Sacerdote, Exponential trends of Ornstein-Ulhenbeck first passage-time densities, *Journal of Applied Probability.* **22**, 360–369 (1988).
9. L. M. Ricciardi and S. Sato, First-passage-time density and moments of the Ornstein-Uhlenbeck process, *Journal of Applied Probability.* **25**(1), 43–57 (1988).
10. E. Evans, Probing the relation between force-lifetime-and chemistry in single molecular bonds, *Annual Reviews in Biophysics and Biomolecular Structure.* **30**, 105–128 (2001).
11. E. A. R. Bizzarri and S. Cannistraro, *Dynamic Force Spectroscopy and Biomolecular Recognition.* CRC Press, Taylor and Francis Grp, LLC, Boca Raton, FL (2012).
12. C. W. Gardiner, *Handbook of Stochastic Methods: For Physics, Chemistry, and the Natural Sciences.* Springer, Berlin (2004).
13. J. Shillcock and U. Seifert, Escape from a metastable well under a time-ramped force, *Phys. Rev. E.* **57**, 7301–7304 (1998).
14. B. E. Shapiro and H. Qian, A quantitative analysis of sinhle protein-ligand complex separation with the atomic force microscope, *Biophysical Chemistry.* **67**, 211–219 (1997).
15. B. E. Shapiro and H. Qian, Hysteresis in force probe measurements: a dynamical systems perspective, *J. theor. Biol.* **194**, 551–559 (1998).
16. H. Qian and B. E. Shapiro, Graphical method for force analysis: Macromolecular mechanics with atomic force microscopy, *Proteins: Structure, Function, and Genetics.* **37**, 576–581 (1999).

17. G. Hummer and A. Szabo, Kinetics from nonequilibrium single-molecule pulling experiments, *Biophysical Journal* **85**, 5–15 (2003).
18. O. K. Dudko, A. E. Filippov, J. Klafter, and M. Urbach, Beyond the conventional description of dynamics force spectroscopy of adhesion bonds, *Proc. Natl. Acad. Sci. USA.* **100**, 11378–11381 (2003).
19. A. Fuhrmann, D. Anselmetti, and R. Ros, Refined procedure of evaluating experimental single-molecule force spectroscopy data, *Physical Review E.* **77**, 031912 (2008).
20. S. Getfert, M. Evstigneev, and P. Reimann, Single-molecule force spectroscopy: practical limitations beyond Bell's model, *Physica A.* **388**, 1120–1132 (2009).
21. L. B. Freund, Characterizing the resistance generated by a molecular bond as it is forcibly separated, *Proc. Natl. Acad. Sci. USA.* **106**, 8818–8823 (2009).
22. A. Fuhrmann and R. Ros, Single-molecule force spectroscopy: a method for quantitative analysis of ligand-receptor interactions, *Nanomedicine.* **5**, 657–666 (2010).
23. V. K. Gupta and C. D. Eggleton, A theoretical method to determine unstressed off-rate from multiple bond force spectroscopy, *Colloids and Surfaces B: Biointerfaces.* **95**, 50–56 (2012).
24. V. K. Gupta, Rupture of single receptor-ligand bonds: A new insight into probability distribution function, *Colloids and Surfaces B: Biointerfaces.* **101**, 501–509 (2013).
25. G. Bal and T. Chou, On the reconstruction of diffusions using a single first-exit time distribution, *Inverse Problems.* **20**, 1053–1065 (2004).
26. P.-W. Fok and T. Chou, Reconstruction of bond energy profiles from multiple first passage time distributions, *Proc. Roy. Soc. A.* **466**, 3479–3499 (2010).
27. T. P. J. Knowles, C. A. Waudby, G. L. Devlin, S. I. A. Cohen, A. Aguzzi, M. Vendruscolo, E. M. Terentjev, M. E. Welland, and C. M. Dobson, An analytical solution to the kinetics of breakable filament assembly, *Science.* **326**, 1533–1537 (2009).
28. D. Sept and A. J. McCammon, Thermodynamics and kinetics of actin filament nucleation, *Biophysical Journal.* **81**, 667–674 (2001).
29. J. Miné, L. Disseau, M. Takahashi, G. Cappello, M. Dutreix, and J.-L. Viovy, Real-time measurements of the nucleation, growth and dissociation of single Rad51-DNA nucleoprotein filaments, *Nucleic Acids Res.* **35**, 7171–7187 (2007).
30. L. Edelstein-Keshet and G. Ermentrout, Models for the length distributions of actin filaments: I. simple polymerization and fragmentation, *Bulletin of Mathematical Biology.* **60**, 449–475 (1998).
31. M. F. Bishop and F. A. Ferrone, Kinetics of nucleation-controlled polymerization. a perturbation treatment for use with a secondary pathway, *Biophysical Journal.* **46**, 631–644 (1984).
32. E. T. Powers and D. L. Powers, The kinetics of nucleated polymerizations at high concentrations: amyloid fibril formation near and above the "supercritical concentration, *Biophysical Journal.* **91**, 122–132 (2006).
33. D. Endres and Z. A., Model-based analysis of assembly kinetics for virus cap-

sids or other spherical polymers, *Biophysical Journal.* **83**, 1217–1230 (2002).
34. A. Zlotnick, Distinguishing reversible from irreversible virus capsid assembly, *Journal of Molecular Biology.* **366**, 14–18 (2007).
35. A. Y. Morozov, R. F. Bruinsma, and R. J., Assembly of viruses and the pseudo-law of mass action, *Journal of Chemical Physics.* **131**, 155101 (2009).
36. K. A. Brogden, Antimicrobial peptides: pore formers or metabolic inhibitors in bacteria?, *Nature Reviews Microbiology.* **3**, 238–250 (2005).
37. G. L. Ryan and A. D. Rutenberg, Clocking out: Modeling phage-induced lysis of *escherichia coli*, *J. Bacteriology.* **189**, 4749–4755 (2007).
38. M. R. D'Orsogna and T. Chou, First passage and cooperativity of queueing kinetics, *Physical Review Letters.* **95**, 170603 (2005).
39. G. Bel, B. Munsky, and I. Nemenman, The simplicity of completion time distributions for common complex biochemical processes, *Physical Biology.* **7**, 016003 (2010).
40. M. Ehrlich, W. Boll, A. van Oijen, K. Hariharan R., Chandran, M. L. Nibert, and T. Kirchhausen, Endocytosis by random initiation and stabilization of clathrin-coasted pits, *Cell.* **118**, 591–605 (2004).
41. B. I. Shraiman, On the role of assembly kinetics in determining the structure of clathrin cages, *Biophysical Journal.* **72**, 953–957 (1997).
42. L. Foret and P. Sens, Kinetic regulation of coated vesicle secretion, *Proc. Natl. Acad. Sci. USA.* **105**, 14763–14768 (2008).
43. M. R. D'Orsogna, G. Lakatos, and T. Chou, Stochastic self-assembly of incommensurate clusters, *Journal of Chemical Physics.* **126**, 084110 (2012).
44. R. Yvinec, M. R. D'Orsogna, and T. Chou, First passage times in homogeneous nucleation and self-assembly, *Journal of Chemical Physics.* **137**, 244107 (2012).
45. M. R. D'Orsogna, B. Zhao, B. Berenji, and C. T., Combinatory analysis of heterogeneous stochastic self-assembly, *Journal of Chemical Physics.* **139**, 121918 (2013).
46. R. J. Rubin, Mean lifetime of microtubules attached to nucleating sites, *Proc. Natl. Acad. Sci. USA.* **85**, 446–448 (1988).
47. D. J. Bicout, Green's functions and first passage time distributions for dynamic instability of microtubules, *Phys. Rev. E.* **56**, 6656–6667 (1997).
48. D. J. Bicout and R. J. Rubin, Classification of microtubule histories, *Phys. Rev. E.* **59**, 913–920 (1999).
49. J. E. Broadwell, Shock structure in a simple discrete velocity gas, *Physics of Fluids.* **7**, 1243–1247 (1964).
50. T. Platkowski and R. Illner, Discrete velocity models of the boltzmann equation: A survey on the mathematical aspects of the theory, *SIAM Review.* **30**, 213–255 (1988).
51. D. J. Durian and J. Rudnick, Photon migration at short times and distances and in cases of strong adsorption, *J. Opt. Soc. Am. A.* **14**, 235–245 (1997).
52. A. Singer, Z. Schuss, and D. Holcman, Narrow escape and leakage of brownian particles, *Phys. Rev. E.* **78**, 051111 (2008).
53. M. J. Ward, S. Pillay, A. Peirce, and T. Kolokolnikov, An asymptotic analysis of the mean first passage time for narrow escape problems: Part I: Two-

dimensional domains, *Multiscale modeling and Simulation.* **8**, 803–835 (2010).
54. S. E. Halford and J. F. Marko, How do site-specific DNA-binding proteins find their targets?, *Nucleic Acids Research.* **32**, 3040–3052 (2004).
55. T. Hu, A. Y. Grosberg, and B. I. Shklovskii, How proteins search for their specific sites on DNA: The role of DNA conformation, *Biophysical Journal.* **90**, 2731–2744 (2006).
56. L. Mirny, M. Slutsky, Z. Wunderlich, A. Tafvizi, J. Leith, and A. Kosmrlj, How a protein searches for its site on DNA: the mechanism of facilitated diffusion, *Journal of Physics A: Mathematical and Theoretical.* **42**(43), 434013 (2009).
57. A. B. Kolomeisky and A. Veksler, How to accelerate protein search on DNA: Location and dissociation, *Journal of Chemical Physics.* **136**, 125101 (2012).
58. A. Veksler and A. B. Kolomeisky, Speed-selectivity paradox in the protein search for targets on DNA: Is it real or not?, *The Journal of Physical Chemistry B.* p. In press (2013).
59. A. G. Cherstvy, A. B. Kolomeisky, and A. A. Korynyshev, Protein-DNA interactions: reaching and recognizing the targets, *Journal of Physical Chemistry B.* **112**, 4741–4750 (2008).
60. A. B. Kolomeisky, Physics of protein-DNA interactions: mechanisms of facilitated target search, *Physical Chemistry and Chemical Physics.* **13**, 2088–2095 (2011).
61. T. E. Holy and S. Leibler, Dynamic instability of microtubules as an efficient way to search in space, *Proc. Natl. Acad. Sci. USA.* **91**, 5682–5685 (1994).
62. R. Wollman, E. N. Cytrynbaum, J. T. Jones, T. Meyer, J. M. Scholey, and A. Mogilner, Efficient chromosome capture requires a bias in the "search-and-capture" process during mitotic spindle assembly, *Current Biology.* **15**, 828–832 (2005).
63. M. Gopalakrishnan and B. S. Govindan, A first-passage-time theory for search and capture of chromosomes by microtubules in mitosis, *Bulletin of Mathematical Biology.* **73**, 2483–2506 (2011).
64. M. R. D'Orsogna and T. Chou, Optimal cytoplasmic transport in viral infections, *PLoS One.* **4**, e8165 (2009).
65. P. W. Fok, C. L. Guo, and T. Chou, Charge-transport-mediated recruitment of DNA repair enzymes, *Journal of Chemical Physics.* **129**, 235101 (2008).
66. P. W. Fok and T. Chou, Accelerated search kinetics mediated by redox reactions of DNA repair enzymes, *Biophysical Journal.* **96**, 3949–3958 (2009).
67. S. D. Bruner, D. P. G. Norman, and G. L. Verdine, Structural basis for recognition and repair of the endogenous mutagen 8-oxoguanine in DNA, *Nature.* **403**, 859–866 (2000).
68. E. M. Boon, A. L. Livingston, M. H. Chmiel, S. S. David, and J. K. Barton, DNA-mediated charge transport for DNA repair, *Proceedings of the National Academy of Science.* **100**, 12543–12547 (2003).
69. K. A. Eriksen, Location of DNA damage by charge exchanging repair enzymes: effects of cooperativity on location time, *Theoretical Biology and Medical Modelling.* **2**, 15 (2005).
70. I. M. Sokolov, R. Metzler, K. Pant, and M. C. Williams, First passage time of

n excluded-volume particles on a line, *Physical Review E.* **72**, 041102 (2005).
71. A. Zilman and G. Bel, Crowding effects in non-equilibrium transport through nano-channels, *Journal of Physics: Condensed Matter.* **22**, 454130 (2010).
72. S. B. Yuste and K. Lindenberg, Order statistics for first passage times in one-dimensional diffusion processes, *J. Stat. Phys.* **85**, 501–512 (1996).
73. S. B. Yuste, L. Acedo, and K. Lindenberg, Order statistics for d-dimensional diffusion processes, *Physical Review E.* **64**, 052102 (2001).
74. R. B. Stein, Some models of neuronal variability, *Biophysical Journal.* **7**, 38–68 (1967).
75. H. C. Tuckwell and D. K. Cope, Accuracy of neuronal interspike times calculated from a diffusion approximation, *Journal of Theoretical Biology.* **83**, 377–387 (1980).
76. D. K. Cope and H. C. Tuckwell, Firing rates of neurons with random excitation and inhibition, *Journal of Theoretical Biology.* **80**, 1–14 (1979).
77. H. C. Tuckwell, R. Rodriguez, and W. F. Y. M., Determination of firing times for the stochastic Fitzhugh-Nagumo neuronal model, *Neural Computation.* **15**, 143–159 (2003).
78. H. C. Tuckwell and F. Y. M. Wan, Time to first spike in stochastic Hodgkin-Huxley systems, *Physica A: Statistical Mechanics and its Applications.* **351**, 427–438 (2005).
79. L. J. S. Allen, *An Introduction to Stochastic Processes with Applications to Biology, Second Edition.* Chapman & Hall/CRC Mathematical & Computational Biology, London, UK (2010).
80. D. A. Kessler and N. M. Shnerb, Extinction rates for fluctuation-induced metastabilities: a real-space WKB approach, *Journal of Statistical Physics.* **127**, 861–886 (2007).
81. M. Assaf and B. Meerson, Extinction of metastable stochastic populations, *Phys. Rev. E.* **81**, 021116 (2010).
82. L. E. Reichl, *A Modern Course in Statistical Physics.* Wiley-VCH, Berlin (2009).
83. A. Dobrinevski and E. Frey, Extinction in neutrally stable stochastic lotka-volterra models, *Phys. Rev. E.* **85**, 051903 (2012).
84. M. I. Dykman, I. B. Schwartz, and A. S. Landsman, Disease extinction in the presence of random vaccination, *Physical Review Letters.* **101**, 078101 (2008).
85. J. R. Artalejo, A. Economou, and M. J. Lopez-Herrero, On the number of recovered individuals in the SIS and SIR stochastic epidemic models, *Mathematical Biosciences.* **228**, 45–55 (2010).
86. O. Ovaskainen and B. Meerson, Stochastic models of population extinction, *Trends in Ecology & Evolution.* **25**, 643–652 (2010).
87. S. Hod, Time-dependent random walks and the theory of complex adaptive systems, *Physical Review Letters.* **90**, 128701 (2003).
88. K. E. Emmert and L. J. Allen, Population extinction in deterministic and stochastic discrete-time epidemic models with periodic coefficients with applications to amphibian populations, *Natural Resource Modeling.* **19**(2), 117–164 (2006).

89. M. R. Evans and T. Hanney, Nonequilibrium statistical mechanics of the zero-range process and related models, *Journal of Physics A: Mathematical and General.* **38**, R195–R240 (2005).
90. B. H. Shargel, M. R. D'Orsogna, and T. Chou, Arrival times in a zero-range process with injection and decay, *Journal of Physics A: Math. Theor.* **43**, 305003 (2010).

Chapter 14

The Effect of Detection Mechanisms on Spatial Search and Foraging

Daniel Campos* and Vicenç Méndez

Universitat Autònoma de Barcelona, Departament de Física, Facultat de Ciències, 08193 Bellaterra, Spain

**daniel.campos@uab.es*

1. Introduction

First-passage problems have been traditionally applied to a wide range of areas in physics ranging from optics to quantum or polymer physics [1], and have also found important applications for example in areas of finance [2] and molecular biology [3]. However, its use is still marginal in ecology as a measure of the efficiency of a foraging process or, more in general, the random search of living organisms. While it is true that there is a large biological literature in *random encounter* models [4, 5] that are used to measure the probability and/or intensity of interactions between individuals, most of the approaches within this field do not take spatial effects explicitly into account. It has been only recently that new approaches have started to emphasize the possible evolutionary interrelations between the specific patterns of motions followed by animals and their search efficiency [6, 7].

The main reasons why first-passages have not traditionally found its place in ecological research, we presume, are basically the lack of an interdisciplinary communication among researchers (aggravated by the mathematical background necessary to understand the details of these problems) and a number of technical problems that make applicability to this specific field difficult. A summary of the most relevant difficulties and obstacles could be the following:

(i) Research on animal foraging and search has been traditionally dominated by the idea that individuals have perfect knowledge of the target (e.g.

food) locations. Within this context, the search efficiency question fits better a path optimization problem [7] (as the traveling salesman's) or an energy optimization process [8], while aspects about stochastic motion of the individuals are considered as residual or unimportant. More in general, animals and humans tend to combine deterministic or systematic strategies (according to their capacity to use and/or acquire information for decision-making) with random elements when searching, so these processes involve a partial information use difficult to implement mathematically.

(ii) Most works and results from the area of first-passage problems have usually been focused on Brownian walkers or classical diffusion, while there is a common knowledge that the stochastic patterns of motion of biological organisms resemble very much a correlated or persistent walk [9], something which is based on experimental evidence ranging from cell tracking [10, 11] to telemetry experiments in large predators [12].

(iii) First-passage problems assume a perfect detection efficiency upon encounter with the target, while nature is full of examples that contradict this. For example, target detection can depend in a non-trivial way on aspects like the distance to the target [13], the speed of the searcher [14], time, etc.

While the first of these three points is definitely the most fundamental and the one to be faced in order to develop in the future a theory of search for living organisms, in the present contribution we show that the other two points can be adequately addressed under rather general conditions. In particular, note that if non-perfect detection is assumed the search problem then becomes a n-passage problem, where n itself is a random variable. In the following, we show how to derive the formal equations that describe formally this non-perfect detection process. After that, we will explore their application to a general class of persistent random walks, with the focus put on their use in animal random searches.

2. The non-perfect detection problem

For the remaining of the present contribution, we will restrict our discussion to the problem of a particle moving within a finite interval of size $2L$ centered on the origin of the x-axis, at which a single point target is located, with periodic boundary conditions at $x = \pm L$ (Figure 1). The corresponding solution of the first-passage problem (i.e. with perfect detection upon encounter with the target) could be properly derived using well-known methods as backward Fokker-Planck equations [15] or alternative techniques [1]. Here, to study the non-perfect detection problem we will employ the renewal approximation, which has been used (with some variations) in several classical papers and books [16–18]. We will adopt a similar approach to that from van Kampen's book (albeit adpated to continuous media), so we define $q(t; x_0)$ as the probability density that the particle is detected at time t having started its walk from x_0, and $f(t; x_0)$ as the corresponding first-detection time distribution. Then one can write

$$q(t; x_0) = z(t; x_0) + \int_0^t dt' q(t - t'; 0) f(t'; x_0) \tag{1}$$

so the paths contributing to the density $q(t; x_0)$ are divided into those that have never detected the target (these are denoted by $z(t; x_0)$) and those coming from a first-detection occurrence at time t' followed by a return-and-detection to the target at time t. Within this framework, note that the particle is assumed to lose the memory of its previous history at the instant t' (this is the renewal approximation) so this means that Equation (1) is strictly valid for Markovian motion processes only.

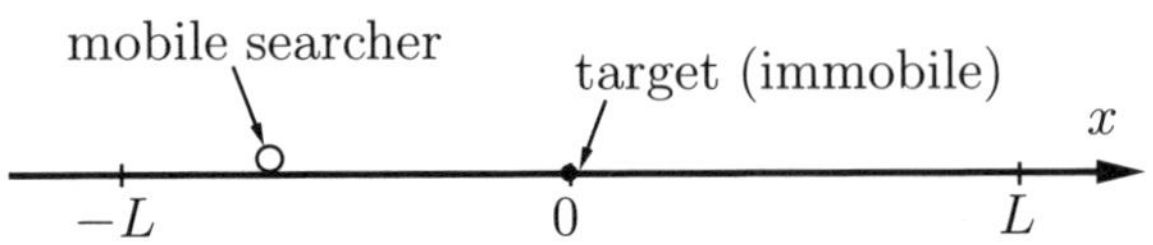

Fig. 1. Schematic representation of the non-perfect detection problem with a single target in 1D.

Next step in most approaches in the literature, including van Kampen's, is to consider $z = 0$, since by definition z gives the contribution from paths never detecting the target. Then one can easily find a closed expression for the first-passage probability $f(t; x_0)$. This is more conveniently written by transforming (1) from the real time space to the Laplace space with argument s, so using the time convolution theorem one gets

$$f(s; x_0) = \frac{q(s; x_0)}{q(s; 0)}. \tag{2}$$

We note that this expression, however, presents a serious problem since the limit $x_0 \to 0$ leads immediately to $f(s; x_0) \to 1$ or, equivalently, $f(t; x_0) \to \delta(t)$. In consequence, the Mean Detection Time (MDT) will vanish with certainty as long as the particle starts its walk at an arbitrarily small distance from the target. This is not true, however, for many patterns of motion as for example in persistent walks, as has been noted by Weiss and others before [19, 20]. Then, in order to adress properly those situations, we think that a more natural choice for z is to consider $z(t;x_0) \equiv f(t;x_0)$, so this term represents the contribution from all paths that have never detected the target *prior to t*. Using this alternative definition, the first-detection probability reads (again in the Laplace space)

$$f(s;x_0) = \frac{q(s;x_0)}{1+q(s;0)} \tag{3}$$

and then the problem at $x_0 \to 0$ is overcome. The MDT $\langle T \rangle$ can be now computed, since by definition

$$\langle T \rangle = -\lim_{s\to 0} \frac{df(s;x_0)}{ds} = -\lim_{s\to 0} \frac{\frac{\partial q(s;x_0)}{\partial s}(1+q(s;0)) - \frac{\partial q(s;0)}{\partial s} q(s;x_0)}{(1+q(s;0))^2}. \tag{4}$$

Finally, this expression can be conveniently simplified if we take into account that the limit $s \to 0$ in Laplace space corresponds to the long-time regime, at which the probability density of occupation within the finite interval $(-L, L)$ will tend to be uniform. Then q (which should depend explicitly on that occupation probability) is expected to approach asymptotically a constant value, here denoted as q_∞. This allows us to introduce the expansion $\lim_{t\to\infty} q(t;x_0) \equiv q_\infty + q_1(t;x_0)$, where the function $q_1(t;x_0)$ should decay to zero asymptotically with time. In Laplace space we can write, equivalently, $\lim_{s\to 0} q(s;x_0) \equiv q_\infty/s + q_*(x_0)$, where note that $q_*(x_0) \equiv \lim_{s\to 0} q_1(s;x_0)$ will be independent of s at the leading order.

Introducing the previous expansion into (4) and rearranging terms yields the general expression for the MDT

$$\langle T \rangle = \frac{q_*(0) - q_*(x_0)}{q_\infty} + \frac{1}{q_\infty} \equiv \langle T_1 \rangle + \langle T_2 \rangle. \tag{5}$$

The distinction made here between $\langle T_1 \rangle$ and $\langle T_2 \rangle$ is not arbitrary. Note that the first contribution to the MDT ($\langle T_1 \rangle$) vanishes for $x_0 = 0$; this means that the second contribution $\langle T_2 \rangle$ represents the MDT for a search that starts at a point arbitrarily close to the target (this second term would be lost if the expression (2) was used for the first-detection distribution instead of (3)). Hence, $\langle T_1 \rangle$ can be properly interpreted as the initial time required to reach the target position starting from x_0 (Figure 1). It is important to stress that $\langle T_2 \rangle$ is not necessarily small albeit it implies a search time starting from the target location; this is because target detection is considered as non-perfect and moreover the particle motion, if it exhibits long-time correlations, could lead again the particle far away from the target.

2.1. *Classical diffusion in the non-perfect detection problem*

From the previous discussion one can infer that classical or Fickian diffusion, where relaxation times or persistence of the individual particles are considered as negligible, introduces strong simplifications in the scheme presented. To illustrate this idea, let us use the image method, which is one of the most powerful and intuitive ways to explore first-passage problems for diffusive particles. To simplify things, let us consider that $x_0 \ll L$ so for small times we can neglect the effects from the boundaries. The image method allows one to determine the probability density that the particle has not crossed through the origin up to time t by summing the contributions from the real particle starting its walk at x_0 and the contribution of an image "antiparticle" starting its walk at $-x_0$ (further details can be found in Chapter 3 from Redner's book [1]). Then, the corresponding probability of not having crossed the origin by t reads

$$S(x,t) = \frac{1}{\sqrt{4\pi Dt}} \left[e^{-(x-x_0)^2/4Dt} - e^{-(x+x_0)^2/4Dt} \right]. \tag{6}$$

In the limit $x_0 \to 0$, which is the focus of interest here, this expression can be simplified to [1]

$$S(x,t) \approx \frac{1}{\sqrt{4\pi Dt}} \frac{x x_0}{Dt} e^{-(x^2 - x_0^2)/4Dt}. \tag{7}$$

In consequence, the probability vanishes as the initial position gets close to the target, and so does the corresponding mean first-passage time. By extension, the probability of not having crossed twice, three times, four times... the origin by time t will also vanish for x_0 small enough, and so in general one can conclude that the MDT will tend to zero in the limit $x_0 \to 0$. Using the notation from the previous section, this implies necessarily that $\langle T_2 \rangle \to 0$ for diffusive particles, so the only contribution to the MDT is that given by the time necessary to approach the target, i.e. $\langle T_1 \rangle$. The consequence of this reasoning is that *the non-perfect detection problem is meaningless for the case of diffusive particles*, since for that case it just reduces to a first-passage problem.

An intuitive way to understand these results consists of explicitly assuming that the particles have a characteristic persistence or relaxation time denoted by τ. This is a measure of the characteristic time that the particle travels without changing direction, as illustrated in Figure 2. If $v\tau$ is of the order of x_0 (where v is the characteristic speed of the particle) or larger then the particle has a finite probability to depart from the target initially so the MDT cannot vanish regardless of how small x_0 is taken. This is the reason why mean first-passage times (and in consequence MDTs) do not vanish for persistent walkers even if the motion starts from arbitrarily close to the target [19]. Instead, in classical diffusion (or in a Wiener process) by definition $\tau \to 0$ and $v\tau \to 0$ (while the diffusion coefficient $D \sim v^2\tau$ is kept constant); so that, diffusive particles lack this capacity of persistently departing from the origin, and the MDT vanishes for x_0 small.

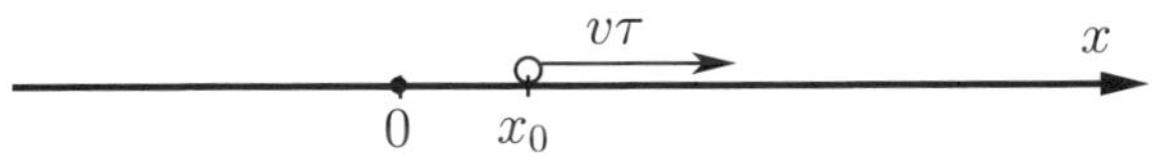

Fig. 2. Visualization of the non-perfect detection problem for a persistent walker starting its search very close to the target, showing the explicit role that the persistence length $v\tau$ will play in departing the particle from the origin.

3. Detection mechanisms in velocity models

Velocity models are motion models where the speed or the velocity of the particle are considered explicitly as a variable. By assuming a finite v for the particles, a characteristic persistence appears straightforward, so velocity models offer a suitable framework to study non-perfect detection, according to our discussion in the previous Section. Moreover, many of the detection constraints one can find in real systems are closely related to the speed of the particle. For example, particles trying to escape from a well potential (e.g. a square well) may need to reach the boundaries of the well with a speed large enough to overcome the corresponding barrier (this picture could serve as a toy model for many problems in reaction kinetics). Also, going back to searches in living organisms, it is known that increasing the speed of an animal can reduce its perception abilities so reducing the detection probability [14]. Finally, the speed of individuals can be often easily related to energy consumption, so a velocity framework allows us also to explore energy restrictions in motion, as we shall briefly illustrate in Section 3.3.

The practical outcome one should obtain from any velocity model is a closed expression for the probability density $p(x, v, t; x_0)$, which determines the probability that the particle is at time t at position x moving with velocity v, provided that it was at position x_0 at $t = 0$. The detection probability $q(t; x_0)$ defined in Section 2, which is the one we need to find in order to compute the MDT, will be directly related to $p(x, v, t; x_0)$ by means of the specific detection mechanism chosen. To illustrate this, we give a few practical examples:

(i) *First-passage detection:* In this case detection will occur with certainty once the particle crosses over the target position; then we have

$$q(t;x_0) = \int_{-\infty}^{\infty} dv|v|p(0,v,t;x_0). \tag{8}$$

(ii) *Partially reactive detection:* In this case detection will occur with a certain probability Ω_0 at the passage over the target position, so we simply use

$$q(t;x_0) = \Omega_0 \int_{-\infty}^{\infty} dv|v|p(0,v,t;x_0). \tag{9}$$

(iii) *Speed-selective detection:* This case would be similar to the previous one but now the detection probability (denoted by $\Omega(v)$) depends explicitly on the speed of the particle, so

$$q(t;x_0) = \int_{-\infty}^{\infty} dv|v|\Omega(v)p(0,v,t;x_0). \tag{10}$$

Obviously, alternative and more complicated detection mechanism could be proposed. More in general, the detection probability could be written for instance as $q(t;x_0) = \mathbb{G}\left[p(x,v,t;x_0)\right]$, where $\mathbb{G}$ represents an integrodifferential operator which could be acting on any variable of the function $p(x,v,t;x_0)$. So the detection could be assumed to depend explicitly not just on the searcher speed, but also on its specific location, on time, etc. In the remaining of our work, however, we will rather focus on the speed-selective mechanism, since it is a simple case but general enough to explore several aspects of the non-perfect detection problem; moreover, it allows us to recover the *first-passage* and the *partially reactive* mechanisms as particular cases.

3.1. *A working example: walks with random velocities*

We now propose a rather general velocity model in which particles are assumed to move with constant velocity v during a characteristic (persistence) time τ; after that time, the particle chooses a new velocity and goes on moving according to this new value until the next velocity switching. The variable v will be assumed to be randomly chosen from an arbitrary probability distribution function (pdf) $h(v)$ while the sojourn duration τ is here considered as constant for simplicity (the case where τ can also be a random variable independent of v has been considered in other works [21]). Note that this model is specially suitable to analyse experimental trajectories obtained from tracking experiments in animals as well as other moving entities, since the outcome from this kind of experiments is usually the position of every individual or particle computed at fixed time intervals (e.g. in case of data obtained from snapshots or from movies, the frequency of images or frames determines the duration of this interval).

Next, we will determine the corresponding expression for the probability density $p(x, v, t; x_0)$ in the random velocity model described; to do this, first we need to introduce the function $j(x, t; x_0)$ to denote the probability density that this particle chooses a new velocity value at position x at time t. From this, one can write the balance equations

$$j(x, t; x_0) = j(x_0)\delta(t) + \int_{-\infty}^{\infty} dvh(v)j(x - v\tau, t - \tau; x_0) \tag{11}$$

$$p(x, v, t; x_0) = h(v) \int_0^t dt'\Theta(t')j(x - vt', t - t'; x_0). \tag{12}$$

The first term on the rhs of (11) accounts for the initial condition, so the particle is assumed to choose its initial velocity at $t = 0$ at a location determined by the function $j(x_0)$. The last term in that equation takes into account the arrival of the particles to x after completing a sojourn time τ moving with random velocity v. Likewise, Equation (12) indicates that the particles present at time t at x will be those which chose its present velocity at position $x - vt'$ at time $t - t'$ and have not completed yet that sojourn; here $\Theta(t')$ denotes a step function which equals 1 in the time interval $t' < \tau$, and it is zero otherwise.

The first problem we need to address in order to adapt this set of equations to our specific situation depicted in Figure 1 is that of how to implement the boundary conditions in this scheme. To do this, we simply use periodic summation techniques (which is a straightforward extension of the image method aforementioned) and consider that, if the actual initial position of the particle is x_0, then the expression of the probability density $p(x, v, t; x_0)$ within the interval $(-L, L)$ will follow from summing the contributions of infinite particles initially located at positions $x_0 - 2mL$, where m can take any integer value. This at practice implies that we have to choose

$$j(x_0) = \sum_{m=-\infty}^{\infty} \delta(x - x_0 + 2mL) = \sum_{m=-\infty}^{\infty} \frac{e^{-i\pi m(x - x_0)/L}}{2L} \tag{13}$$

for periodic boundary conditions, where in the last identity Fourier expansion of the Dirac delta function has been used. Now, since the corresponding probability distributions $p(x,v,t;x_0)$ and $j(x,t;x_0)$ will be periodic functions in the whole real axis (albeit we are only interested in the solution for $-L<x<L$) we can use Fourier analysis and assume a solution of the type

$$j(x,t;x_0) = \sum_{m=-\infty}^{\infty} b_m\left(t,x_0\right)e^{-i\pi mx/L} \tag{14}$$

$$p(x,v,t;x_0) = \sum_{m=-\infty}^{\infty} c_m\left(v,t,x_0\right)e^{-i\pi mx/L}. \tag{15}$$

This can be now introduced into (11,12). After Laplace transform, this substitution yields

$$b_m(s,x_0) = \frac{e^{i\pi mx_0/L}}{2L} + b_m(s,x_0)\widehat{h}\left(\frac{\pi m\tau}{L}\right)e^{-\tau s} \tag{16}$$

$$c_m(v,s,x_0) = \frac{1-e^{-\tau(s+i\pi mv/L)}}{s+i\pi mv/L}h(v)b_m(s,x_0) \tag{17}$$

where the hat $\hat{}$ denotes the Fourier transform with respect to the velocity variable v. From the previous expressions we find the explicit form of the Fourier coefficients for $p(x,v,t;x_0)$:

$$c_m(v,s,x_0) = \frac{h(v)}{2L}\frac{1-e^{-\tau(s+i\pi mv/L)}}{s+i\pi mv/L}\frac{e^{i\pi mx_0/L}}{1-\widehat{h}\left(\frac{\pi m\tau}{L}\right)e^{-\tau s}}. \tag{18}$$

Our main interest is in the limit $s\to 0$ of the previous result; this leads to

$$\begin{aligned}\lim_{s\to 0} p(x,v,s;x_0) &= \lim_{s\to 0}\sum_{m=-\infty}^{\infty} c_m\left(v,s,x_0\right)e^{-i\pi mx/L} = \\ &= \frac{h(v)}{2L}\left(\frac{1}{s}+\sum_{m\neq 0}\frac{1-e^{-i\tau\pi mv/L}}{i\pi mv/L}\frac{e^{-i\pi m(x-x_0)/L}}{1-\widehat{h}\left(\frac{\pi m\tau}{L}\right)}\right)\end{aligned} \tag{19}$$

where the term $m=0$ is treated separately to avoid the singularity, taking into account the normalization condition $\widehat{h}(0)=1$.

One can finally compute the detection probability $q(s;x_0)$ in the limit $s\to 0$ by using the speed-selective mechanism from expression (10). The corresponding expression reads

$$\lim_{s\to 0} q(s;x_0) = \frac{\langle |v|\Omega(v)\rangle}{2Ls} + \sum_{m\neq 0}\int_{\infty}^{\infty} dv|v|\Omega(v)h(v)\frac{1-e^{-i\tau\pi mv/L}}{i\pi mv/L}\frac{e^{i\pi m x_0/L}}{1-\widehat{h}\left(\frac{\pi m\tau}{L}\right)} \tag{20}$$

where the brackets denote the weighted integral over $h(v)$, i.e. $\langle f(v)\rangle \equiv \int_{-\infty}^{\infty} dvh(v)f(v)$.

A direct comparison to the expansion $\lim_{s\to 0} q(s;x_0) = q_\infty/s + q_*(x_0)$ used in Section 2 allows us to identify the terms:

$$q_\infty = \frac{\langle |v|\Omega(v)\rangle}{2L} \tag{21}$$

$$q_*(x_0) = \sum_{m\neq 0}\frac{e^{i\pi m x_0/L}}{1-\widehat{h}\left(\frac{\pi m\tau}{L}\right)}\int_{\infty}^{\infty} dv|v|\Omega(v)h(v)\frac{1-e^{-i\tau\pi mv/L}}{i\pi mv/L}. \tag{22}$$

This lengthy derivation has then leaded us to determine the details of the asymptotic behavior of the detection probability. Now, to get a more simplified expression we can focus in the limit case $L\to\infty$; by doing this we get

$$\lim_{L\to\infty} q_*(x_0) = \frac{\langle |v|\Omega(v)\rangle L}{\langle v^2\rangle \tau\pi^2}\sum_{m\neq 0}\frac{e^{i\pi m x_0/L}}{m^2}. \tag{23}$$

Finally, by replacing this expression into the MDT expression (4) we obtain

$$\langle T_1\rangle = \frac{2L^2}{\langle v^2\rangle\tau\pi^2}\sum_{m\neq 0}\frac{1-e^{i\pi m x_0/L}}{m^2} \quad , \quad \langle T_2\rangle = \frac{2L}{\langle |v|\Omega(v)\rangle}. \tag{24}$$

So, we can easily observe the scaling of the MDT with the main parameters of the problem for the specific limit considered. Let us observe, since we will use these results later, that the case $x_0 = L$ leads to

$$\langle T_1\rangle = \frac{L^2}{\langle v^2\rangle\tau} \tag{25}$$

while for x_0 uniformly distributed in the interval $(-L, L)$ one finds

$$\langle T_1\rangle = \frac{2L^2}{3\langle v^2\rangle\tau}. \tag{26}$$

We now present some direct comparison to Monte Carlo simulations in order to illustrate the goodness of the analytical results obtained. For this purpose, we need to assign explicit expressions to the velocity distribution $h(v)$ and the probability $\Omega(v)$ for non-perfect detection. For the moment,

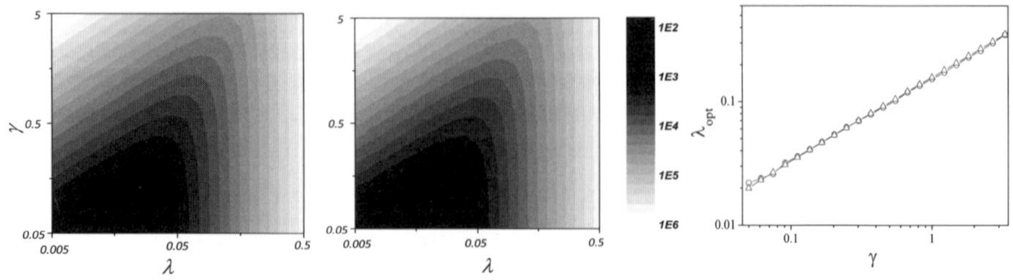

Fig. 3. MDTs as a function of the inverse mean speed $\lambda = 1/\langle v \rangle$ and the parameter γ assuming an exponential decay of the speed pdf $h(v)$ and the probability $\Omega(v)$. A comparison is shown between MDTs obtained from Monte Carlo simulations (left panel) and those obtained from expression (25) (middle panel) corresponding to $x_0 = L$. To facilitate comparison we include (right panel) explicitly the results for the optimal values of λ for which the MDT gets minimized as a function of γ.

let us show the results when both functions are taken as exponential, i.e. for $h(v) = \lambda e^{-\lambda|v|}$, $\Omega(v) = e^{-\gamma|v|}$. In Figure 3 the values of the MDT $\langle T \rangle$ from Monte Carlo simulations are compared to our analytical results (left and middle panels) for $x_0 = L$. As can be observed the agreement found between both is excellent so confirming the validity of the renewal approximation used in our derivation. As an interesting consequence from the non-perfect detection criteria used, we find that there appears an optimal value of the mean speed (note that $\langle |v| \rangle = \lambda^{-1}$ in the exponential case considered here) that minimizes the MDT. This is because the speed-selective constraints imposed make the detection process extremely difficult if the cruise velocity of the particle is very large, while searching at very low speed will be neither effective. Put in a biological context, this result reproduces a speed-accuracy trade-off that appears in search or foraging process in which moving too fast leads to a reduction in the perception abilities [21]. In accordance with that, we find (Figure 3, right panel) that as long as the detection constraints are assumed to be more restrictive (i.e. for γ large) the optimal value of λ increases (this is, the optimal mean speed of the searcher decreases). In particular, for exponential expressions for $h(v)$ and $\Omega(v)$ one could try to determine the optimization condition by differentiating the MDT with respect to λ. This leads to a third order equation

$$\frac{L^2}{\pi^2 \tau} \left(\sum_{m \neq 0} \frac{e^{i\pi m x_0/L} - 1}{m^2} \right) \lambda^3 + 2L(\lambda^2 - \gamma^2) = 0 \tag{27}$$

whose solution yields the optimal value of λ as a function of γ. From the solution of this third order equation one expects the optimal value to satisfy $\lambda_{opt} \sim \gamma^{2/3}$, in agreement with the results found in Figure 3. For the sake of completeness, we also show the results for a uniform initial condition x_0 (Figure 4), which leads qualitatively to the same results.

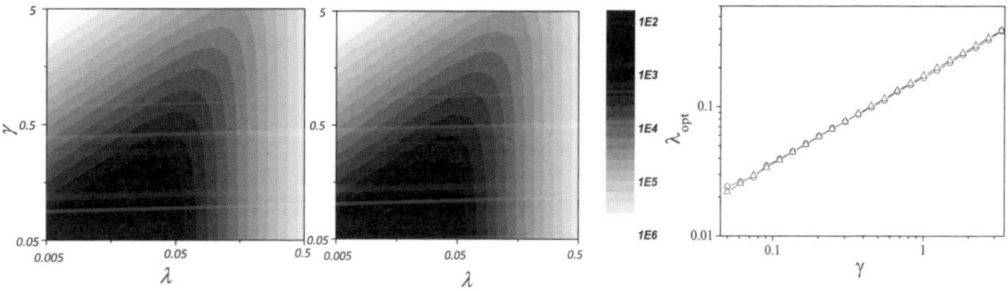

Fig. 4. MDTs as a function of the inverse mean speed $\lambda = 1/\langle v \rangle$ and the parameter γ assuming an exponential decay of the speed pdf $h(v)$ and the probability $\Omega(v)$. A comparison is shown between MDTs obtained from Monte Carlo simulations (left panel) and those obtained from expression (25) (middle panel) corresponding to uniform x_0 in the interval $(-L, L)$. To facilitate comparison we include (right panel) explicitly the results for the optimal values of λ for which the MDT gets minimized as a function on γ.

Next, we move our attention to the case where $h(v)$ does not decay exponentially but rather as a power-law. Then, since we are considering that single sojourns have constant duration, this is equivalent to have a power-law distribution of sojourn lengths, i.e. the searcher is performing Lévy flights. This situation is of special interest due to the intensive work published in the last years around the *Lévy flight hypothesis*. This hypothesis arises from the result that Lévy flights represent an optimal strategy for random search under particular situations [22], since these strategies involve an adequate compromise between intensive search (in order to exhaustively explore particular areas) and relocation events to new unexplored areas. This has lead to the claiming (specially among several ecologists) than Lévy flights should be expected to be favored by selective pressures in species whose survival and basic functions largely depend on random search (e.g. for foraging purposes).

So, we will consider now $h(v) = \alpha\lambda/(1 + \lambda|v|)^{1+\alpha}$ (with $|v| < v_{max}$ for some v_{max}, so at practice we consider truncated Lévy flights). The equivalent to Figures 3 and 4 for the case of truncated Lévy flights (plots not shown here) would exhibit the same qualitative properties than the exponential case, this is, an optimal mean speed exist for which the benefits from fast search get compensated with the speed detection constraints. For the case of a first-passage detection, it is known that if the initial condition is far away from the target then the optimal solution (i.e. that which minimizes the mean first-passage time) is found in the limit $\alpha \to 0$, while for an initial condition very close to the target an intermediate optimal value $\alpha = 1$ is found [7]. So that, it is interesting to check if these results still appear for the non-perfect detection problem considered here.

From Figure 5 one can check that the speed-selective detection strongly modifies the picture expected. Since we have used a uniform distribution of x_0 for building the plots, the case of a first-passage detection mechanism should lead in the limit $\alpha \to 0$ to smaller MDTs. However, what we observe in Figure 5 is that this only happens for λ large, while for λ small the MDT decreases as α decreases. This is because the combination of small λ and small α correspond to enormous values of the mean speed, and so the speed-selective mechanism punishes those strategies. If comparing the two panels in Figure 5 (which correspond to $\gamma = 0.05$ and $\gamma = 2$, respectively) we observe that the region where this punishment to low α-values occur increases as the speed-selective mechanism gets more restrictive, in agreement with our reasoning.

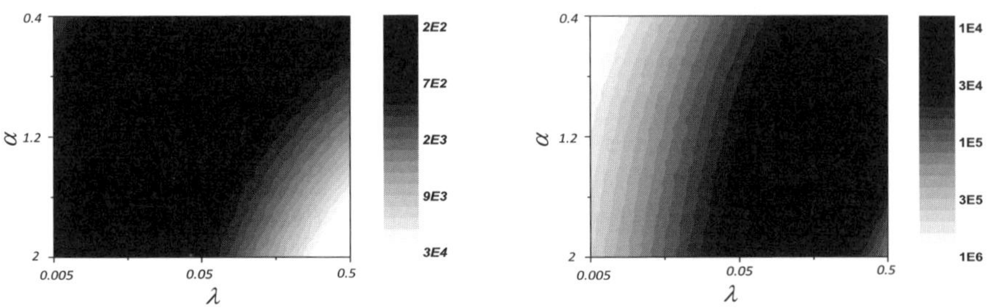

Fig. 5. MDTs as a function of the parameter λ and the power-law exponent α by assuming a power-law decay of the speed pdf $h(v)$, and an exponential decay $\Omega(v) = e^{-\gamma|v|}$. Left and right panels correspond to the cases $\gamma = 0.05$ and $\gamma = 2$, respectively.

3.2. *Non-perfect detection in random media*

If the non-perfect detection problem stated in the previous Sections is to be applied in real searches in animals and humans, one of the ingredients that need to be improved in order to make the situation more realistic is the fact that the problem described in Figure 1 is just amenable to searches in periodic media, while most real landscapes present random/patchy properties. As a simple attempt to amend this problem we now show how to extend our results to the case where one has a set of targets randomly (and homogeneously) distributed in a one-dimensional media (Figure 6). Then, the distances $2L_1, 2L_2, ...$ between consecutive targets must be considered as random variables with average value $2L$.

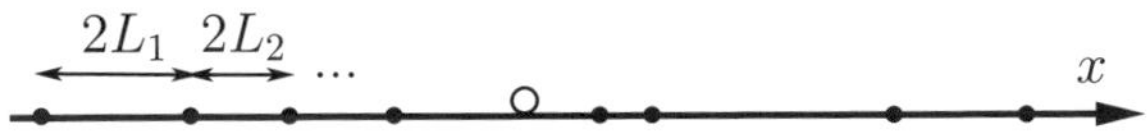

Fig. 6. The non-perfect detection problem in the case of randomly distributed targets in 1D media.

In the case of a first-passage detection mechanism the solution to this problem would be rather trivial. Given that the particle initially falls within an interval of length $2L_i$ and the mean-first passage time $\langle T_{FP} \rangle$ within an interval of that length is known (with x_0 uniformly distributed within that interval), then the mean time to detection is

$$\int_0^\infty dL_i \langle T_{FP} \rangle \rho(L_i), \tag{28}$$

where $\rho(L_i)$ is the probability density that x_0 lies within an interval of size $2L_i$. For the case of targets randomly distributed throughout the horizontal axis we know that the distribution of size intervals decays exponentially as $\sim e^{-L_i/L}$; hence the probability that x_0 lies within that interval is proportional to its size, i.e.

$$\rho(L_i) \sim L_i e^{-L_i/L} \to \rho(L_i) = \frac{L_i}{L^2} e^{-L_i/L}.$$

So that, the expression for the detection time could be computed easily.

For the non-perfect detection problem, instead, we have an additional difficulty since the particle, during its search process, can pass over many different intervals of different sizes. This makes the analytical treatment rather complicated unless we consider some limit situations, in particular:

(i) If the effect of the speed-selective restriction is very small, then $\Omega(v) \approx 1$ and one can approximate the solution to that of the first-passage detection problem. By replacing (24,26) into $\int_0^\infty dL_i \langle T \rangle \rho(L_i)$ one obtains then

$$\langle T \rangle_r = \frac{4L^2}{\langle v^2 \rangle \tau} + \frac{4L}{\langle |v| \Omega \rangle} \tag{29}$$

where the r stands for *random*.

(ii) On the contrary, for very restrictive detection constraints one can assume that the searcher needs to visit so many different intervals that the final distance covered will be approximately the same than that it would have covered in a periodic media. For that case the effect of the fluctuations in the size intervals will vanish and so the MDT will coincide with the case of periodic media (i.e. with (24,26)).

The previous discussion is corroborated from our Monte Carlo simulations, as shown in Figure 7. There one can observe that, again for $\Omega(v) = e^{-\gamma|v|}$, the results for restrictive detection (γ large) coincide with those from (24,26), while the expression (29) fits well the results obtained for the smaller values of γ.

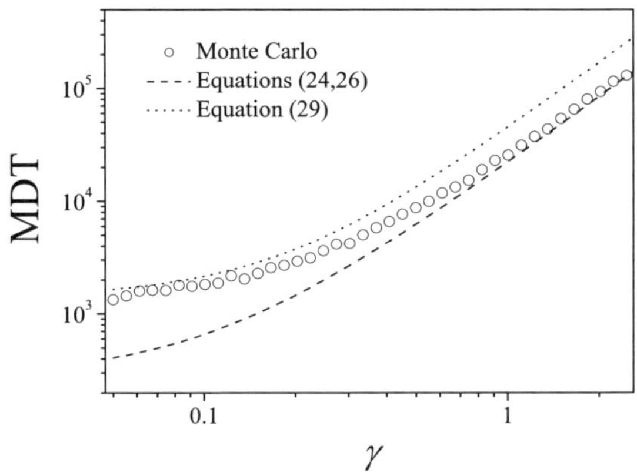

Fig. 7. Comparison between the MDT obtained numerically from simulations and the analytical approximations discussed for the case of random media. Equations (24,26) fit the numerical results in the limit of large γ and the expression (29) is able to explain the results for $\gamma \to 0$.

3.3. *Time vs energy constraints*

When first-passage (or similar) statistics is applied to understand optimization principles in random search of living organisms, one is implicitly assuming that the mean time to detect the target is the essential parameter of interest. Alternatively, however, one can find that a more convenient optimization criteria is that based on the probability that the search finishes successfully before a threshold time [23] t_s as a way to guarantee survival through a more or less regular nutrient income. Actually, the MDT can be of scarce utility in situations where the distribution of detection times shows a great dispersion around their average value, as has been studied in recent works for the case of first-passages [24].

More in general, optimization criteria for living entities should be rather based on energy arguments, so the more energy consumed in the search process the less efficient it becomes. If we define ε as the energy rate consumption, then the mean energy invested by the organism in searching up to detection at the random time T, i.e.

$$\langle E \rangle = \left\langle \int_0^T \varepsilon dt \right\rangle, \tag{30}$$

is the parameter to be minimized. In case ε is approximately constant then we find that energy minimization and time minimization are equivalent. However, this will not be realistic for many situations; for example, energy consumption related to movement will largely depend on the searcher speed, and then search processes carried out at large speeds will be punished according to this criteria.

Energy costs of transport are expected to be more dramatic for microorganisms or small organisms. This is because, while mechanic energy necessary for movement (kinetic plus potential) is essentially proportional to body mass m, the metabolic energy devoted to movement is empirically observed [8] to scale just as $m^{0.68}$. This means that larger organisms are in general more efficient at converting chemical energy into mechanical work, and in consequence energy limitations due to movement should be less restrictive for them. This is one of the reasons why models based on the coupled dynamics between particle motion and its energy depot (as the well-known model of active particles [25, 26]) have been traditionally focused on applications at the microscopic level.

In general, one cannot provide a universal explicit dependence of ε on the speed v, but we can assume here just for illustration purposes that the energy required by microorganisms to move basically comes from the effects of the viscous drag forces they need to overcome in order to advance. Under this assumption we can take $\varepsilon = \varepsilon_0 + \varepsilon_v v^2$, where ε_0 represents the basal (constant) metabolic rate of the organism, while ε_v measures the role of the energy consumption rate due to viscous drag, which is proportional to v^2 (see e.g. the Stokes' law). With all this in mind, we now present (Figures 8 and 9) again the results from Figure 4 but now representing the values for $\langle E \rangle / \varepsilon_0$ instead of those for the MDT, and for different values of the ratio $\varepsilon_v / \varepsilon_0$. The results obtained make evident, as expected, that search strategies based on too large mean speeds are additionally punished

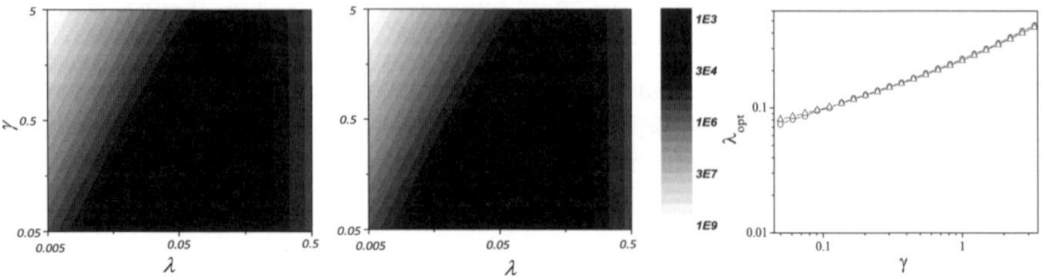

Fig. 8. Normalized energy consume $\langle E \rangle/\varepsilon_0$ as a function of λ and γ for $\varepsilon_v/\varepsilon_0 = 0.01$. A comparison is shown between results obtained from Monte Carlo simulations (left panel) and those obtained from the adiabatic approximation (middle panel) for uniform $x_0 = L$. To facilitate comparison we include (right panel) explicitly the results for the optimal values of λ for which the energy consumed gets minimized as a function of γ.

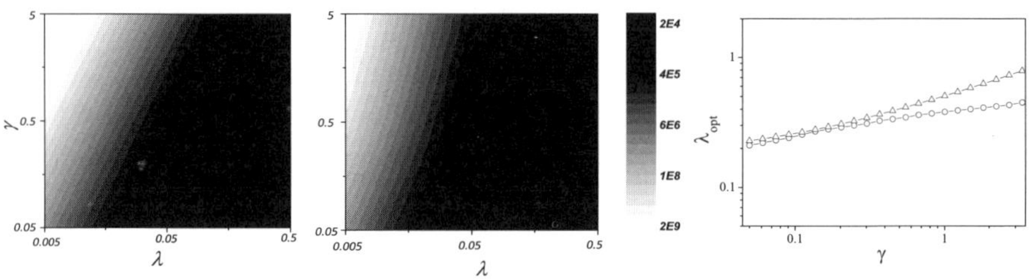

Fig. 9. Normalized energy consume $\langle E \rangle/\varepsilon_0$ as a function of λ and γ for $\varepsilon_v/\varepsilon_0 = 0.5$. A comparison is shown between results obtained from Monte Carlo simulations (left panel) and those obtained from the adiabatic approximation (middle panel) for uniform $x_0 = L$. To facilitate comparison we include (right panel) explicitly the results for the optimal values of λ for which the energy consumed gets minimized as a function of γ.

now. For small values of $\varepsilon_0/\varepsilon_v$ that effect is small and so we find that an optimal mean speed (this is, an optimal value of λ) still appears and show similar features to those found from minimizing the MDT. However, as the ratio $\varepsilon_0/\varepsilon_v$ increases (Figure 9) that optimal value rapidly increases, so making evident that strategies with very small speeds should be selected in those cases. This provides, for example, a rough justification for *ambush* search strategies [27] observed in many animals. These strategies basically consist of waiting most of the time (so minimizing energy consumption) and only moving from time to time or when the prey is perceived in a certain range (albeit this certainly would involve target motion, which has not been considered in our model).

Finally, in order to compare these simulations with analytical expressions we should be able to compute the integral in (30), which is in general impossible since v depends explicitly on time within our random-walk approach. However, if we assume that the search process takes long enough then we can simply use an adiabatic approximation which consists in replacing the actual speed of the searcher by $\langle v^2 \rangle$, so assuming that the speed variable changes very fast if compared to the timescale T. In consequence, the mean energy becomes simply $\langle E \rangle = \langle T \rangle \left(\varepsilon_0 + \varepsilon_v \langle v^2 \rangle \right)$. The corresponding analytical results, using our expression for $\langle T \rangle$ found in Section 3.1, are presented in the middle panels from Figures 8 and 9. From these panels, we can verify the goodness of the adiabatic approximation used as long as speed constraints are not too strong (note that for the case $\varepsilon_v/\varepsilon_0 = 0.5$ and large γ the approximation fails).

4. Conclusions

As stated in the introduction, understanding search processes in living organisms essentially requires to know how individuals select and acquire information from their environment, how they process this information and bring it together with their previous knowledge (what psychologists call working and reference memories), and how this translates into action through decision-making mechanisms. All this represents a formidable task which requires at least a joint collaboration between behavioral psychologists, neuroscientists, anthropologists, ecologists, and many other. Moreover, the complexity of all the molecular processes involved in these tasks makes it necessary to deal with a certain level of stochasticity, so even if these mechanisms are understood in detail one day, it will be possible only through a probabilistic description.

So that, in the present contribution we have tried to illustrate how one could use such a probabilistic or statistical description to study mathematically and predict the properties (e.g. the efficiency) of random search processes. Though we have focused here on the speed-selective mechanism for the sake of simplicity, we stress that the approach presented in Section 2 allows for much more general situations to be considered. Anyway, even for the speed-selective case we have been able to explore and reproduce several features of realistic random searches in ecology, as the existence of an optimal mean speed to solve the compromise between speed and accuracy in detection, or the role that energy constraints may play on selecting different search strategies. Overall, we hope the approach provided here can represent a promising start in order to develop in the future a general search theory of living organisms.

Acknowledgements

We thank Frederic Bartumeus for fruitful discussions on the non-perfect detection problem. This research has been partially supported by Grants No. FIS 2012-32334 (VM, DC), SGR 2009-00164 (VM, DC).

References

1. S. Redner. *A Guide to First-Passage Processes.* Cambridge Univ. Press, Cambridge (2001).
2. M. Nardon. First Passage and Excursion Time Models for Valuing Defaultable Bonds: A Review with Some Insights,*Front. Fin. Econ.* **5**, 1–25 (2008).
3. O. Bénichou, C. Loverdo, M. Moreau and R. Voituriez. Intermittent search strategies, *Rev. Mod. Phys.* **83** 81–129 (2011).
4. J.A. Byers. An encounter rate model of bark beetle populations searching at random for susceptible host trees. *Ecological Modelling* **91**, 57–66 (1996).
5. A.W. Visser and T. Kiorboe. Plankton motility patterns and encounter rates. *Oecologia* **148**, 538–546 (2006).
6. G.M. Viswanathan, E.P. Raposo and M.G.E. da Luz. Levy flights and superdiffusion in the context of biological encounters and random searches. *Phys. Life Rev.* **5**, 133–150 (2008).
7. F. Bartumeus and J. Catalán. Optimal search behavior and classic foraging theory. *J. Phys. A: Math. Theor.* **42**, 434002 (2009).
8. R.M. Alexander. *Energy for animal life.* Oxford Univ. Press, Oxford (1999).
9. E.A. Codling, M.J. Planck and S. Benhamou. Random walk models in biology. *J. Roy. Soc. Interf.* **5**, 813–834 (2008).
10. D. Selmeczi, S. Mosler, P.H. Hagedom, N.B. Larsen and H. Flyvbjerg. Cell Motility as Persistent Random Motion: Theories from Experiments. *Biophys. J.* **89** 912–931 (2005).
11. D. Campos, V. Méndez, I. Llopis. Persistent random motion: Uncovering cell migration dynamics. *J. Theor. Biol.* **267**, 526–534 (2010).
12. N.E. Humphries et al. Environmental context explains Lvy and Brownian movement patterns of marine predators. *Nature* **465**, 1066–1069 (2010).
13. B.O. Koopman. *Search and screening: general principles with historical applications.* Pergamon Press, New York (1980).
14. D.L. Kramer and R.L. McLaughlin. The behavioral ecology of intermittent locomotion. *Am. Zool.* **41**, 137–153 (2001).
15. C.W. Gardiner. *Handbook of Stochastic Methods for Physics, Chemistry and the Natural Sciences*, 3rd Ed. Springer-Verlag, Berlin Heidelberg (2004).
16. E. Montroll and G.H. Weiss. Random walks on lattices. II. *J. Math. Phys.* **6**, 167–181 (1965).
17. R.L. Stratonovich. *Topics in the Theory of Random Noise.* Gordon and Breach Pub, New York (1967).
18. N.G. van Kampen. *Stochastic processes in physics and chemistry*, 3rd Ed. Elsevier, Amsterdam (2007).

19. G.H. Weiss. First Passage Times for Correlated Random Walks and Some Generalizations. *J. Stat. Phys.* **37** 325–330 (1984).
20. G.H. Weiss. *Aspects and Applications of the Random Walk.* North-Holland, Amsterdam (1994).
21. D. Campos, V. Méndez and F. Bartumeus. Optimal Intermittence in Search Strategies under Speed-Selective Target Detection. *Phys. Rev. Lett.* **108**, 028102 (2012).
22. G.M. Viswanathan, S. Buldyrev, S. Havlin, M.G.E. da Luz, E.P. Raposo and H.E. Stanley. Optimizing the success of random searches. *Nature* **401**, 911–914 (1999).
23. B.M. Friedrich. Search along persistent random walks. *Phys. Biol.* **5**, 026007 (2008)
24. T.G. Mattos, C. Mejía-Monasterio, R. Metzler and G. Oshanin. First passages in bounded domains: When is the mean first passage time meaningful? *Phys. Rev. E* **86**, 031143 (2012).
25. W. Ebeling, F. Schweitzer and B. Tilch. Active Brownian particles with energy depots modelling animal mobility. *BioSystems* **49**, 17–29 (1999).
26. F. Schweitzer. *Brownian Agents and Active Particles.* Springer-Verlag, Berlin Heidelberg (2003).
27. W.J. Brien, H.I. Browman and B.I. Evans. Search strategies of foraging animals. *Am. Sci.* **78**, 152–160 (1990).

Chapter 15

Search in Random Media with Lévy Flights

Erol Gelenbe and Omer H. Abdelrahman

Department of Electrical & Electronic Engineering,
Imperial College London, SW7 2BT, UK
{e.gelenbe,o.abd06}@imperial.ac.uk

1. Introduction and motivation

Our work on the subject of search has been motivated by a variety of applications in engineering and technology, and it differs from the physicists' motivation for such problems, e.g. [1–3], or the motivation provided by biochemistry [4, 5], and in theoretical biology [6–9].

One of our reasons for a more fundamental approach to search comes from the area of traffic routing in wired networks [10–12] where each end user may send out many search or Smart Packets to seek out paths for the traffic it needs to forward, so that the paths that are used will satisfy useful quality of service (QoS) properties such as minimising delay or packet loss. A similar question arises, with even more acuity, when the networks are ad-hoc and wireless, and the need to reduce uncertainty regarding possible paths is even greater [13–15]. A distinct but related motivation for our work has been the search for network paths that avoid ongoing Cyberattacks [16].

Another important motivation for our work comes from the field of explosive mine detection and removal on which one of us worked in the past [17, 18], and in such circumstances it is sensible to see how the natural world deals with such challenges [19, 20]. The related fields of tactical operations [21], and emergency operations and management [22–26] also raise many interesting problems related to the search for targets [17], objects such as landmines [18], and victims, as part of an organised response to an incident or an emergency.

On the other hand, one of us first introduced Lévy flights in diffusion models in the 1970's, motivated by the performance analysis of computer systems [27] to deal with continuous approximations to discrete queueing models under light traffic conditions, whereas the usual approach to such approximations for queues have focused on heavy-traffic conditions. Here, the instantaneous jump or Lévy flight was used to represent a jump from the origin (the empty state) of the diffusion process, to the value +1 representing the arrival of a customer to an empty queue. When the diffusion model was used to model the waiting time process of the queue, the instantaneous jump represented the service time of the first customer arriving to an empty queue [28] including non-Markovian multiple-phase models. This analysis then gave rise to a call admission control (CAC) algorithm for asynchronous transfer model (ATM) networks [29] which resulted in a patented CAC scheme for ATM networks [30].

In the sequel we review some of our work regarding search, focusing on the Lévy flight models, including the issues of time, energy, inhomogeneous media and security. In Section 2 we develop the basic mathematical model representing N searchers that proceed independently, and which can be affected by destruction or loss, and time-outs. A time-out is a length of time such that if a searcher has not been successful after the time-out elapses, the organiser of the search simply aborts that particular searcher and some time later replaces it with a new one. In related approaches [31], the authors study the fluctuations, from one trajectory to another, of the first passage time of N independent searchers.

In Section 2.1 we present the system of N coupled diffusion equations, together with the discrete probability equations representing losses and time-outs, and subsequent restarts, and their analytical solution in steady-state is obtained in Section 2.2.

An approach based on Laplace-Stieltjes transform of the search time is then developed in Section 3 for the case where the success of the search requires that k out of the N searchers be successful. Section 3.1 then considers the problem of estimating the number N of serachers that are needed so that k out of these N searchers will be successful in time B, provided B is large. Then, computing the amount of energy needed to conduct a successful search is discussed in Section 3.2.

All the previous analysis assumed that the search space has homogeneous and time-invariant characteristics. Our work on non-homogeneous media is actually motivated by security models where, for instance, as the searchers get closer to the object they seek, the search can become more difficult and searchers may be more frequently destroyed. This is certainly the case where countermeasures are in place to protect or defend the immediate surroundings of (say) a network node that is being defended against Cyberattacks. Thus in Section 4 we present an iterative numerical solution approach that allows us to analyse the search time and the energy needed to find an object when the search space is non-homogeneous. We then apply this approach in Section 4.1 to the analysis of a *region of protection* around an object that is sought out by a searcher. Finally in Section 4.2 we show that there is a form of phase transition that occurs when the speed of approach is varied in relation to the rate at which searchers are lost or destroyed; thus it is possible to reach a situation where the searcher (or attacker) will always be successful against the defender that destroys the attacker in the proximity of the object being sought out (and defended). Some suggestions for further research are made in the final section of this paper.

2. Modelling the search process

Consider N independent searchers sent out simultaneously in the quest for the same object. Let $Z_i(t)$ denote the i-th searcher's *distance* from its destination at time $t \geq 0$ with $Z_i(0) = D$, and denote its search time by $T_i = \inf\{t : Z_i(t) = 0\}$. Let $T_{1,N} \leq T_{2,N} \leq \cdots \leq T_{N,N}$ be the variables T_i rearranged in ascending order, i.e. the corresponding order statistics; this section deals with the computation of $E[T_{1,N}]$ [32]. The state of the searcher at time $t \geq 0$ is $s_i(t)$ which can take one of the values $\{\mathbf{S_i}, \mathbf{L_i}, \mathbf{W_i}, \mathbf{P}\}$ defined as follows:

• $\mathbf{S_i}$: If the i-th searcher is searching and its distance from the destination is $Z_i(t) > 0$. We denote the probability density function (pdf) of the distance $Z_i(t)$ by $f_i(z_i, t)dz_i = P[z_i < Z_i(t) \leq z_i + dz_i,\ s_i(t) = \mathbf{S_i}]$.

• $\mathbf{W_i}$: The i-th searcher's life-span has ended, and so has its search. Note that this may have happened because it was destroyed or became lost, but this becomes known to the source via the time-out which is exponentially distributed with parameter r. After an additional exponentially distributed delay of parameter μ, it is replaced at the source by a new searcher with the same identity. We write $W_i(t) = P[s_i(t) = \mathbf{W_i}]$.

• $\mathbf{L_i}$: The i-th searcher has been destroyed or lost, and its search is ended; for small Δt and $Z_i(t) = z > 0$, this happens with a probability $\lambda(z)\Delta t + o(\Delta t)$, where $\lambda(z) \geq 0$ is the destruction rate at distance z. The time spent in this state is exponentially distributed with the same parameter r as the

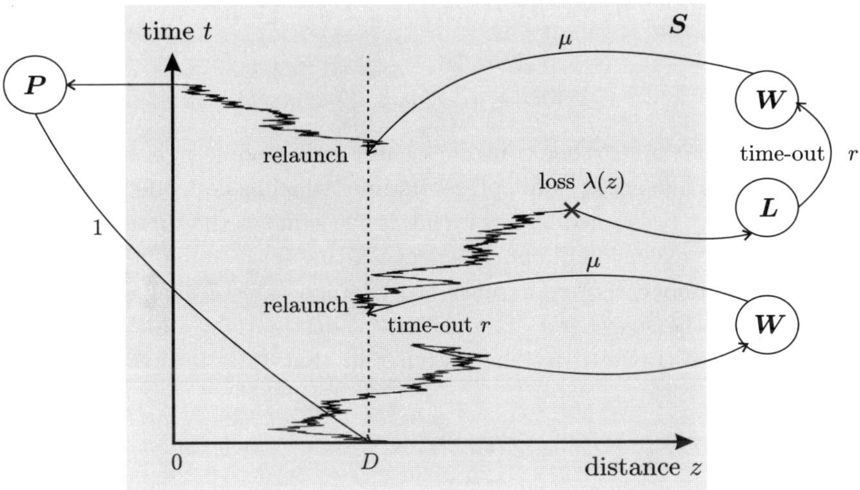

Fig. 1. An instance of the search process with $N = 1$ where the search is relaunched twice, due to timeout and destruction, before the object is finally found.

life-span since the source realizes that the searcher is lost or destroyed via the life-span effect. At the end of this exponentially distributed time, the searcher is handled just as if it has "died". We write $L_i(t) = P[s_i(t) = \mathbf{L_i}]$.

• **P**: One of the searchers has found the object being sought; the search process stops for all searchers, including the ones who are lost or dead. Notice that **P** is a synchronised state for all of the searchers. After one time unit, the search process starts again as before at the source with N searchers being sent out. We write $P(t) = P[s_i(t) = \mathbf{P}]$.

Notice that the above process repeats itself indefinitely, and $E[T_{1,N}]$ is the average time that it takes from any successive start of the search until the first instance when state **P** is reached again. Strictly speaking, the Lévy flight takes place only in this last instance, since diffusion process is reset, or jumps to the distance D for all of the searchers. However in a more abstract sense, whenever the search process of any one of the searchers leaves the diffusion in order to join another discrete probability variable, i.e. such as $L_i(t)$ or $W_i(t)$, we have a form of Lévy flight as an instantaneous jump away from the diffusion and into a discrete set of states, before the process restarts at a later time. A schematic representation of the search process for $N = 1$ is presented in Fig. 1.

Let $P(t)$ be the probability that the model we have just described is in state $\mathbf{P}$ at time $t \geq 0$, and let $P = \lim_{t\to\infty} P(t)$. Then:

$$P = \frac{1}{1 + E[T_{1,N}]}, \quad E[T_{1,N}] = \frac{1-P}{P}. \tag{1}$$

We assume that a searcher consumes energy only while it is *moving* (i.e., in state $\mathbf{S_i}$) at one energy unit per unit time, which is equivalent to stating that the energy consumed by a searcher is the same as the time it spends in motion, to the exclusion of epochs spent waiting for a time-out to relaunch it. A real number $b(z)$ represents the average rate of change over time $[t, t + \Delta t)$ of the searcher's distance to the destination, and the variance of the distance travelled by the searcher in that time interval is $c(z)\Delta t$, $c(z) \geq 0$:

$$b(z) = \lim_{\Delta t\to 0} \frac{E[Z(t+\Delta t) - Z(t)|Z(t) = z]}{\Delta t},$$

$$c(z) = \lim_{\Delta t\to 0} \frac{E[(Z(t+\Delta t) - Z(t))^2] - (E[Z(t+\Delta t) - Z(t)])^2|Z(t) = z]}{\Delta t}.$$

While $b(z) < 0$ is the favourable case where the searcher on average gets closer to the destination with time, we may also have cases of interest with $b(z) > 0$, which means that the searcher on average moves *away* from the object of interest, for instance because intermediate locations provide wrong information on average, or it lacks information altogether when $b(z) = 0$. It was shown that [13, 14] in a homogeneous medium, even if $b \geq 0$ it is possible to have a travel time to destination which is finite on average.

2.1. *The system of coupled equations*

We now express the process $\{s_i(t) : t \geq 0\}$ in terms of a system of equations describing a mixed continuous space (diffusion) supplemented by a discrete space random process [27–29, 33], where the discrete part in this case describes the reset process. Note that in [27–29] the discrete portion is related to transitions to internal states where the transition time is non-exponential so that the process is non-Markovian. We first write the equations that the probability density function $f_i(z_i, t)$, $z_i > 0$, and the probability masses $L_i(t)$, $W_i(t)$ and $P(t)$, $t \geq 0$ will satisfy. We represent the interaction between the diffusion processes using a parameter $a_i(t)$, $1 \leq i \leq N$ in the following manner: $a_i(t)$ is the total rate of attraction exerted at time t by all other diffusion processes, on the i-th diffusion due to the fact that one of the other diffusions may have reached its absorbing barrier. The system of coupled differential and partial differential equations representing the search is:

$$\frac{\partial f_i(z_i,t)}{\partial t} = \frac{1}{2}\frac{\partial^2[c(z_i)f_i(z_i,t)]}{\partial {z_i}^2} - \frac{\partial[b(z_i)f_i(z_i,t)]}{\partial z_i} - (\lambda(z_i)+r+a_i(t))f_i(z_i,t) + [P(t)+\mu W_i(t)]\delta(z_i - D), \quad (2)$$

while

$$\frac{dL_i(t)}{dt} = -(r+a_i(t))L_i(t) + \int_{0+}^{\infty} \lambda(z_i)f_i(z_i,t)dz_i, \quad (3)$$

$$\frac{dW_i(t)}{dt} = -(\mu+a_i(t))W_i(t) + r(L_i(t) + \int_{0+}^{\infty} f_i(z_i,t)dz_i), \quad (4)$$

$$\frac{dP(t)}{dt} = -P(t) + \sum_{i=1}^{N} \lim_{z_i\to 0^+}[\frac{1}{2}\frac{\partial[c(z_i)f_i(z_i,t)]}{\partial z_i} - b(z_i)f_i(z_i,t)], \quad (5)$$

and

$$a_i(t) = \sum_{j=1,j\neq i}^{N} \lim_{z_j\to 0^+}[\frac{1}{2}\frac{\partial[c(z_j)f_j(z_j,t)]}{\partial z_j} - b(z_j)f_j(z_j,t)], \quad (6)$$

which is the rate at which the i-th searcher is attracted to the origin, i.e. to finish its search, because any one of the *other* $N-1$ searchers has found the object being sought. This is reflected both in (2) and in the equations (3) and (4) where the searcher can be forced into the rest state from the "lost" state and the "time-out before retransmission" state, as well. We also see that we enter the loss state from any position $z_i > 0$, and that a time-out can occur for a searcher that is in the lost state. Since the behavior of all searchers when they are not in the rest state are independent, it follows that the event that triggers the jump of searcher i into the rest state does not depend on the prior state of searcher i but on the state of the *other* searchers. We also have that the sum of the probabilities is one:

$$1 = L_i(t) + W_i(t) + P(t) + \int_{0+}^{\infty} f_i(z_i,t)dz_i. \quad (7)$$

Note again that these equations represent the system where, whenever any one searcher has reached the destination, all other searchers' progress is artificially stopped and re-started from the rest state. The purpose here is to compute $E[T_{1,N}]$ by constructing a synthetic ergodic process. In this section we assume that the medium in which the searchers move is homogeneous, i.e. $b(z) = b$, $c(z) = c$ and $\lambda(z) = \lambda$, while in Section 4 we deal with spatially non-homogeneous environments.

2.2. The stationary solution

Dropping the dependence on i because all searchers are statistically identical, and writing the system of differential equations in steady-state yield:

$$\begin{aligned} 0 &= \frac{c}{2}\frac{d^2f(z)}{dz^2} - b\frac{df(z)}{dz} - (\lambda + r + a)f(z) + [P + \mu W]\delta(z - D), \\ 1 &= P + \frac{\mu + r + a}{r}W, \\ P &= N \lim_{z\to 0^+}[\frac{c}{2}\frac{df(z)}{dz} - bf(z)], \\ a &= (N-1)\lim_{z\to 0^+}[\frac{c}{2}\frac{\partial f(z)}{\partial z} - bf(z)]. \end{aligned} \tag{8}$$

The solution for $f(z)$ can be shown to take the following form:

$$f(z) = \begin{cases} A[e^{uz} - e^{vz}], & z \le D, \\ A[e^{(u-v)D} - 1]e^{vz}, & z \ge D, \end{cases} \tag{9}$$

where A depends on a, and u, v are, respectively, the positive and negative real roots of the characteristic polynomial of the differential equation (2):

$$u, v = \frac{b \pm \sqrt{b^2 + 2c(\lambda + r + a)}}{c}.$$

Solving the system of equations in (8) with (9) and (1) we obtain:

Result 1. The average time for one of N searchers to find the object being sought is:

$$E[T_{1,N}] = \frac{\mu + r + a}{(\mu + a)(r + a)}\frac{e^{uD} - 1}{N}, \tag{10}$$

where a is given by the solution of the non-linear equation:

$$a = \frac{N-1}{N + \frac{\mu+r+a}{(\mu+a)(r+a)}[e^{uD} - 1]}. \tag{11}$$

The average energy consumed *until one of the searchers is successful* is:

$$E[J^-_{1,N}] = N(1 + E[T_{1,N}])\int_{0+}^{\infty} f(z)dz = \frac{e^{uD} - 1}{\lambda + r + a}. \tag{12}$$

3. A transform approach when k out of N searchers must be successful

In this section we focus on the case where at least k out of N searchers must be successful [34], and obtain approximate and asymptotic estimates for the search times $T_{k,N}$. Indeed, in hunting or foraging, search may take place as a group effort [19], and k out of N members of a team must be successful in reaching the object. Another important motivation comes from communication networks [35] where packets can be encoded so that the information transmitted is correctly received when at least k out of N transmitted packets arrive at the destination, in the presence of possible packet loss. The energy consumed is computed (i) when the search is stopped as soon as the first k searchers find the object, and (ii) when the remaining searchers (other than the first k successful ones) continue their search till successful completion or until they are destroyed or stopped by their time-out. The latter case is relevant when the source cannot communicate with searchers or when successful searchers cannot communicate with their peers. We use a Laplace transform (LT) approach to compute the distribution of the search time and of the total energy consumption in the two aforementioned cases. Thus the state of a searcher at time t is again $s(t) \in \{\mathbf{S}, \mathbf{W}, \mathbf{L}, \mathbf{P}\}$ but unlike the recurrent approach of Section 2, $\mathbf{P}$ is now an absorbing state.

The probability that a single searcher has reached the object by time t is denoted by $G(t) \equiv \Pr[T \leq t]$ and its pdf $g(t)$. The time until return to the origin of a pure diffusion process starting at distance D is [36]:

$$\begin{aligned} g_0(t) &= \frac{D}{\sqrt{2\pi c t^3}}\, e^{-\frac{(D+bt)^2}{2ct}}, \\ G_0(t) &= \frac{1}{2}\left[\operatorname{erfc}\left(\frac{D+bt}{\sqrt{2ct}}\right) + e^{-2bD/c}\operatorname{erfc}\left(\frac{D-bt}{\sqrt{2ct}}\right)\right], \end{aligned} \tag{13}$$

where $\operatorname{erfc}(x) = \frac{2}{\sqrt{\pi}}\int_x^\infty e^{-y^2}dy$ and the subscript 0 indicates that the quantity refers to a pure diffusion. The pdf of the searcher's distance z to the destination at time t is:

$$f_0(z,t) = \frac{e^{-\frac{b^2 t}{2c}}\, e^{-\frac{b}{c}(D-z)}}{\sqrt{2\pi c t}}\left[e^{-\frac{(z-D)^2}{2ct}} - e^{-\frac{(z+D)^2}{2ct}}\right]. \tag{14}$$

Writing the LT of any $y(t)$ as $\bar{y}(s) = \int_0^\infty y(t)e^{-st}dt$, the above quantities yield:

$$\begin{aligned}\bar{g}_0(s) =& e^{-(b+\sqrt{b^2+2cs})D/c},\\ \bar{f}_0(z,s) =& \frac{e^{b(z-D)/c}}{\sqrt{b^2+2cs}}\left[e^{-\frac{\sqrt{b^2+2cs}}{c}|z-D|} - e^{-\frac{\sqrt{b^2+2cs}}{c}(z+D)}\right].\end{aligned} \tag{15}$$

In the model with loss and time-out, let X and Y be the mutually independent random variables representing the time to the next loss and the time to the next time-out, respectively, which are exponentially distributed with parameters λ and r. Then $\gamma_\iota(t)$, the pdf of the duration of a search time until its first interruption is $\gamma_\iota(t)dt = \Pr[t \le \min(X,Y) \le t+dt,\ T_0 > t]$ since T_0 is the total search time if there is no interruption and its pdf is given in (13). Therefore:

$$\gamma_\iota(t) = (\lambda+r)e^{-(\lambda+r)t}[1-G_0(t)], \bar{\gamma}_\iota(s) = \frac{\lambda+r}{s+\lambda+r}\,[1-\bar{g}_0(s+\lambda+r)]. \tag{16}$$

Search is interrupted randomly several times in this manner, and after each interruption it starts again at the origin after a further delay whose pdf is:

$$\begin{aligned}\psi(t) =& \frac{r}{\lambda+r}\,\mu e^{-\mu t} + \frac{\lambda}{\lambda+r}\int_0^t re^{-ry}\mu e^{-\mu(t-y)}dy,\\ \bar{\psi}(s) =& \frac{s+\lambda+r}{\lambda+r}\,\frac{\mu r}{(s+\mu)(s+r)}.\end{aligned}$$

The last and hence successful attempt at reaching the destination has a duration whose pdf $\gamma_d(t)dt = \Pr[t \le T_0 \le t+dt,\ \min(X,Y) > t]$ or:

$$\gamma_d(t) = g_0(t)e^{-(\lambda+r)t}, \quad \bar{\gamma}_d(s) = \bar{g}_0(s+\lambda+r). \tag{17}$$

If the searcher is successful in locating the object being sought in its first attempt then the search time T and energy consumption J are equivalent. On the other hand, if the search is interrupted at least once then T will exceed J by the amount of time spent in states **L** and **W**. Therefore, the joint density of T and J can be obtained by accounting for the possibilities of locating the object in $1, 2, \cdots$ attempts while including the time spent in states **L** and **W** in T but not in J. Let $\phi(x,t)$ be the joint probability density of search time and energy consumption: $\phi(x,t)dxdt = \Pr[x \le J \le x+dx,\ t \le T \le t+dt]$. Since each attempt is independent of its predecessors, we have for $t \ge x$:

$$\phi(x,t) = \gamma_d(t)\delta(t-x) + \int_0^x \gamma_\iota(y)\psi(t-x)\gamma_d(x-y)dy + \cdots$$

and $\phi(x,t)=0$ for $t<x$. Evaluating the 2D LT of $\phi(x,t)$ yields:

$$\hat{\phi}(\xi,s)=\int_0^\infty\int_x^\infty \phi(x,t)e^{-st-\xi x}dtdx=\frac{\bar{\gamma}_d(s+\xi)}{1-\bar{\psi}(s)\bar{\gamma}_\iota(s+\xi)}. \tag{18}$$

The LT of the pdf of the total search time is $\bar{g}(s)=\hat{\phi}(0,s)$. Inversion of the LT for small and large values of the real variable is performed by taking the limit of the corresponding Laplace variable as it tends to ∞ and 0 yielding:

$$g(t)\sim\begin{cases}\gamma_d(t)+\frac{\mu r}{\mu+r}\left[\Gamma_d(t)-\int_0^t\gamma_d(t-\tau)e^{-(\mu+r)\tau}d\tau\right], & t \text{ small},\\ \beta_T e^{-\beta_T t}, & t \text{ large},\end{cases} \tag{19}$$

where $\Gamma_d(t)=\int_0^t\gamma_d(\tau)d\tau$ and $\beta_T^{-1}=\left[\frac{1}{r}+\frac{1}{\mu}\right]e^{(b+\sqrt{b^2+2c(\lambda+r)})D/c}$. Similarly the pdf of the energy consumption is $\bar{h}(\xi)=\hat{\phi}(\xi,0)$ which can be asymptotically inverted to:

$$h(x)\sim\begin{cases}\gamma_d(x)+(\lambda+r)\Gamma_d(x), & x \text{ small},\\ \beta_J e^{-\beta_J x}, & x \text{ large},\end{cases} \tag{20}$$

where $\beta_J^{-1}=\frac{e^{(b+\sqrt{b^2+2c(\lambda+r)})D/c}}{\lambda+r}$. Since it may not be possible to analytically invert $\bar{g}(s)$ and $\bar{h}(\xi)$ for all values of t and x, we perform numerical inversion [37] using MATLAB [38]. Fig. 2 shows that the pdf of the search time and energy consumption using asymptotic and numerical inversion [37] agree well for a wide range of delay and energy values.

3.1. *Asymptotic formula for k out of N to be successful in large time B*

The probability that k out of N independent searchers will be successful by time t is:

$$G_{k,N}(t)\equiv \Pr[T_{k,N}\le t]=\binom{N}{k}G(t)^k[1-G(t)]^{N-k}. \tag{21}$$

Define $G^{-1}(p)=\inf\{t:\ G(t)\ge p\}$, $0<p<1$, the *quantile function* of the distribution of the search time for a single searcher. When N is large, it is known that $T_{\lceil pN\rceil,N}$, the p-th sample quantile, is asymptotically normally distributed [39]:

$$T_{\lceil pN\rceil,N}\sim\mathcal{N}\left(G^{-1}(p),\frac{p(1-p)}{N\ [g(G^{-1}(p))]^2}\right). \tag{22}$$

Thus for large N the distribution of the time for k out of N searchers to be successful tends to a constant equal to the $p\approx k/N$-th quantile of $G(t)$. As

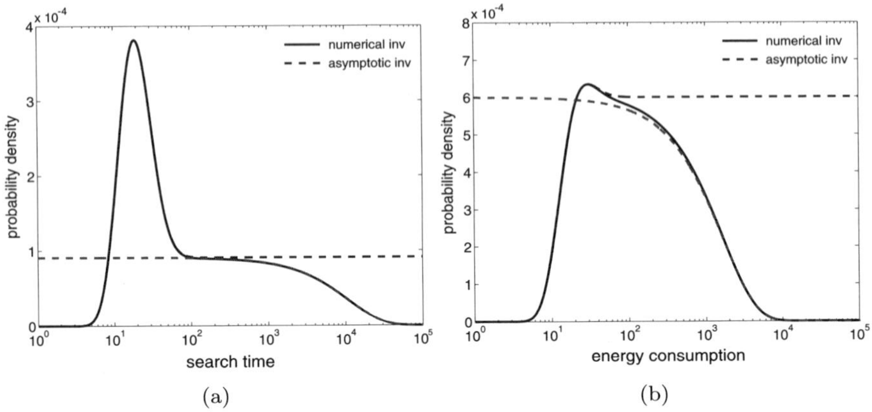

Fig. 2. The pdf of (a) search time $g(t)$ and (b) energy consumption $h(t)$ when $\lambda = 0.05$, $b = 0.1$, $c = 1$, $r = 0.01$, $\mu = 0.1$, and $D = 10$. For this example that has a small positive value of b and hence relatively high uncertainty in search direction, the pdf of the search time has a long tail which is apparent from the logarithmic scale on the horizontal axis.

a consequence, the number of searchers $N(B, k)$ required to find the object in time B when N is large is given approximately by:

$$N(B, k) \cong \left\lceil \frac{k}{G(B)} \right\rceil. \tag{23}$$

Since convergence to the normal distribution (22) is fast, the expression (23) provides a good approximation even for relatively small $N(B, k)$. The good agreement between the asymptotic approximation of (23) and the detailed analysis for $G_{k,N}(B)$ using the formula (21) and the numerical inversion technique [37] is illustrated in Fig. 3.

3.2. *Energy consumption*

In order to derive the energy required for k out of N independent searchers to locate the object, we need to evaluate the time-dependent pdf for the energy expended by a searcher. If we define $J(t)$ as the energy consumed by a searcher up to time t, then the joint pdf of $J(t)$ and the state of the searcher $s(t)$ can be evaluated in three distinct cases:

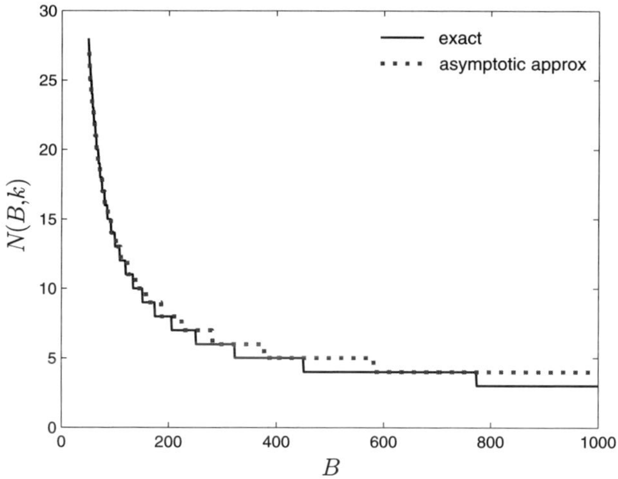

Fig. 3. Comparison of the asymptotic approximation with exact analysis for the total number of searchers $N(B,k)$ that are required so that $k = 3$ of them find the object within time B. Here $b = 0$, $c = 1$, $\lambda = 0.0025$, $r^{-1} = 78$, $\mu^{-1} = 10$ and $D = 10$.

• The searcher reached the *destination* at some time $\tau \le t$:

$$h_d(x,t) \equiv \frac{d}{dx}\Pr[J(t) \le x, s(t) = \mathbf{P}] = \int_x^t \phi(x,\tau)d\tau, \quad \hat{h}_d(\xi,s) = \frac{\hat{\phi}(\xi,s)}{s}. \tag{24}$$

• The searcher is *idle* at time t due to a blocking or time-out that has not yet ended:

$$\begin{aligned} h_\iota(x,t) \equiv \frac{d}{dx}\Pr[J(t) \le x, s(t) = \mathbf{W} \textit{ or } \mathbf{L}] &= \gamma_\iota(x)[1 - \Psi(t-x)] \\ &+ \int_0^x \int_0^{t-x} \gamma_\iota(y_1)\psi(y_2)\gamma_\iota(x-y_1)[1-\Psi(t-x-y_2)]dy_2dy_1 + \cdots, \end{aligned}$$

where $\Psi(t) = \int_0^t \psi(\tau)d\tau$. Here the i-th term denotes the case where the time instant t occurs after the search process is suspended i times but before restarting the $(i+1)$-th search attempt. Taking the double LT of the above equation yields:

$$\hat{h}_\iota(\xi,s) = \frac{[1-\bar{\psi}(s)]\bar{\gamma}_\iota(s+\xi)}{s[1-\bar{\psi}(s)\bar{\gamma}_\iota(s+\xi)]} = \frac{1}{s}\left[1 - \frac{1-\bar{\gamma}_\iota(s+\xi)}{1-\bar{\psi}(s)\bar{\gamma}_\iota(s+\xi)}\right]. \tag{25}$$

- The searcher is *active* at time t:

$$h_a(x,t) \equiv \frac{d}{dx}\Pr[J(t) \le x, s(t) = \mathbf{S}] = e^{-(\lambda+r)t}[1 - G_0(t)]\delta(x-t) + \int_0^x \gamma_\iota(y)\psi(t-x)e^{-(\lambda+r)(x-y)}[1 - G_0(x-y)]dy + \cdots .$$

In the first term, no time-out or blocking has occurred up to t and consequently the total energy consumption is equal to t. The i-th term corresponds to the case where at time t the search is ongoing after it was restarted $i-1$ times so that the pdf of the energy utilisation up to t is given by the convolution of the pdf of $i-1$ interrupted search periods (each followed by an idle period in which energy is not consumed) and a single search period which does not end before the time instant t. We then end up with:

$$\hat{h}_a(\xi,s) = \frac{1 - \bar{g}_0(s+\xi+\lambda+r)}{[s+\xi+\lambda+r][1 - \bar{\psi}(s)\bar{\gamma}_\iota(s+\xi)]} = \frac{1}{\lambda+r}\frac{\bar{\gamma}_\iota(s+\xi)}{1 - \bar{\psi}(s)\bar{\gamma}_\iota(s+\xi)}. \tag{26}$$

Note that $\sum_{p=\{d,\iota,a\}} h_p(x,t) \equiv h(x,t)$ is the pdf of the energy consumed up to t by the searcher $\hat{h}(\xi,s) = \frac{1}{s}[1 - \frac{\xi}{\lambda+r}\frac{\bar{\gamma}_\iota(s+\xi)}{1-\bar{\psi}(s)\bar{\gamma}_\iota(s+\xi)}]$ and $\lim_{t\to\infty} h(x,t) = h(x)$ as expected.

On the other hand, if all active but unsuccessful searchers continue searching after the first k successful ones complete their search, we need to know the energy expended by an active searcher up to t, and its distance to the object being sought so as to compute the additional energy consumed before it actually stops moving. Define $f(z,x,t)dzdx = \Pr[x \le J(t) \le x + dx, z \le Z(t) \le z + dz]$ where $x \le t, z > 0$, which can be derived using $f_0(z,t)$ from (14) as:

$$f(z,x,t) = e^{-(\lambda+r)t} f_0(z,t)\delta(x-t) + \int_0^x \gamma_\iota(\tau)\psi(t-x) f_0(z, x-\tau) e^{-(\lambda+r)(x-\tau)} d\tau + \ldots,$$

where the first term is the probability that the searcher reaches distance z in time t without being interrupted so that $x = t$; the second term is the probability that the search is stopped at some time $\tau \in [0,x]$, it is restarted after $t-x$ time units, and distance z is reached during the next search attempt in a time interval $x-\tau$ in which no interruption occurs. More generally, the i-th term represents the case where the time instant t lies in the i-th attempt to locate the object, when the searcher is at distance z

and has consumed x units of energy. The 2D LT of $f(z,x,t)$ is:

$$\hat{f}(z,\xi,s) = \frac{\bar{f}_0(z, s+\xi+\lambda+r)}{1-\bar{\psi}(s)\bar{\gamma}_\iota(s+\xi)}. \tag{27}$$

Note that $f(z,x,t)$ satisfies the equalities $\int_0^t f(z,x,t)dx = f(z,t)$ and $\int_0^\infty f(z,x,t)dz = h_a(x,t)$ so that $\bar{f}(z,s) = \hat{f}(z,0,s) = \frac{\bar{f}_0(z,s+\lambda+r)}{1-\bar{\psi}(s)\bar{\gamma}_\iota(s)}$.

Let $J^-_{k,N}$ and $J^+_{k,N}$ be the total energy consumption when search is suspended immediately after the object being sought is found by k searchers, and when there is no mechanism to immediately stop active searchers after the completion of the search, respectively. We then have:

Result 2. The LT of the pdf for the total energy consumption is:

$$\begin{aligned}\bar{h}^{\pm}_{k,N}(\xi) &\equiv \frac{d}{dx}\Pr[J^{\pm}_{k,N} \le x] \\ &= \frac{N!}{(k-1)!(N-k)!}\int_0^\infty \bar{h}_d(\xi,t)^{k-1}\bar{\phi}(\xi,t)[\bar{h}_\iota(\xi,t)+\bar{h}_*(\xi,t)]^{N-k}dt,\end{aligned} \tag{28}$$

where h_* is substituted by h_a in the case of $h^-_{k,N}$ while for $h^+_{k,N}$ it is replaced by the pdf of the total energy expended by an unsuccessful active searcher until it stops moving after the search ends at t:

$$\begin{aligned}h_c(x,t) &= \int_0^\infty \int_0^x f(z,x-u,t)\big[\gamma_\iota(u|z)+\gamma_d(u|z)\big]dudz, \\ \hat{h}_c(\xi,s) &= \frac{1}{s(\xi+\lambda+r)}\,\frac{s\bar{\gamma}_\iota(s+\xi)+\xi[\bar{\gamma}_d(\xi)-\bar{\gamma}_d(s+\xi)]}{1-\bar{\psi}(s)\bar{\gamma}_\iota(s+\xi)},\end{aligned}$$

and $\gamma_\iota(.|z)$ and $\gamma_d(.|z)$ are computed as in (16) and (17), respectively, but with initial distance z instead of D.

Proof. For $J^-_{k,N}$ to be equal to x with a search time t, it is necessary that exactly $k-1$, 1 and $N-k$ searchers locate the object being sought in the intervals $[0,t]$, $[t,t+dt]$ and $[t+dt,\infty]$ respectively, and that the energy expended by each individual searcher is at most t while their sum is x. The probabilities that a search succeeds in the three respective intervals while consuming w units of energy up to t are $h_d(w,t)dw$, $\phi(w,t)dwdt$ and $[h_\iota(w,t)+h_a(w,t)]dw$. $h^-_{k,N}(x)$ then follows by accounting for all possible combinations, convolving with respect to the energy variable (which is equivalent to multiplication in the ξ-domain), and integrating over all possible values of t. The pdf $h^+_{k,N}$ is evaluated in the same manner as $h^-_{k,N}$ except that we use h_c instead of h_a to take account of the additional energy consumed by active searchers upon the completion of the search. Specifically, if the object being sought is found at some time t while a searcher is at distance $z > 0$ and has consumed $x - u$ units of energy, then the searcher will continue to move and consume additional u units of energy with probability $\gamma_\iota(u|z)du$ if it is interrupted before reaching the destination or $\gamma_d(u|z)du$ otherwise. $h_c(y,t)$ is obtained by integrating over all possible values of the distance $z > 0$ and energy $u \in [0, x]$ at the time instant t, and its double LT follows directly. □

The total average energy consumption is then $E[J^{\pm}_{k,N}] = -\lim_{\xi\to 0} \frac{d\bar{h}^{\pm}_{k,N}(\xi)}{d\xi}$ yielding:

$$E[J^{\pm}_{k,N}] = -\lim_{\xi\to 0} \frac{N!}{(k-1)!(N-k)!} \int_0^\infty dt \Bigg[G(t)^{k-1}[1-G(t)]^{N-k} \frac{\partial\bar{\phi}(\xi,t)}{\partial\xi} + (k-1)G(t)^{k-2} g(t)[1-G(t)]^{N-k} \frac{\partial\bar{h}_d(\xi,t)}{\partial\xi} + (N-k) \times G(t)^{k-1} g(t)[1-G(t)]^{N-k-1} [\frac{\partial\bar{h}_\iota(\xi,t)}{\partial\xi} + \frac{\partial\bar{h}_*(\xi,t)}{\partial\xi}]\Bigg], \tag{29}$$

which, unlike $h^{\pm}_{k,N}(x)$, requires only 1D LT numerical inversions for $\frac{\partial\bar{\phi}(\xi,t)}{\partial\xi}$, $\frac{\partial\bar{h}_d(\xi,t)}{\partial\xi}$ and $\frac{\partial}{\partial\xi}[\bar{h}_\iota(\xi,t) + \bar{h}_*(\xi,t)]$ when $\xi \to 0$. Fig. 4 shows $E[J^{-}_{k,N}]$ and $E[J^{+}_{k,N}]$ versus the time-out $1/r$ for different values of N. One sees that the minimum energy consumed until the object is found (i.e. $E[J^{-}_{k,N}]$) does not vary much with the number of searchers N. However, in the absence of a stopping mechanism the minimum energy consumed by the search *increases* with N while the "optimum" time-out decreases in order to reduce the additional energy wasted by active searchers after the completion of the search. Of course the intuitive but interesting observation is that when N increases, the *energy consumed increases if there is no stopping mechanism*, while the opposite occurs with the stopping mechanism.

4. Search in a non-homogeneous medium

This section considers a *single* searcher in an infinite random non-homogeneous medium, with spatially non-homogeneous events that may stop or impede the current search [40, 41]. A motivation for this work is the case where the object being sought is protected from the searcher: as the searcher approaches, its progress becomes more frequently blocked

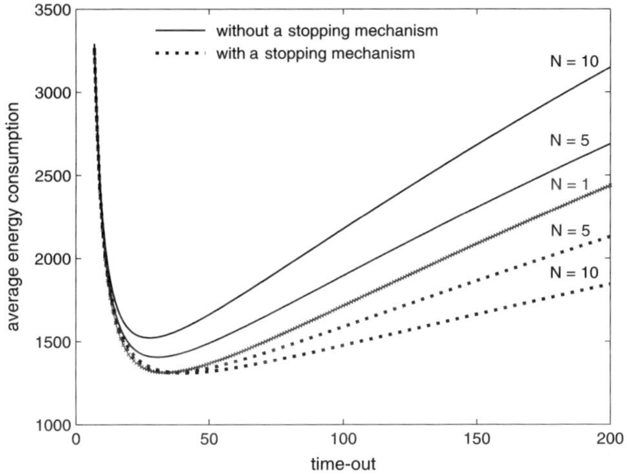

Fig. 4. Average energy consumption (with and without a stopping mechanism) versus time-out $1/r$ for $k = 1, b = 0.15, c = 1.25, \lambda = 0.001, \mu = 0.1, D = 10$ and different values of N. When N increases, the *energy consumed increases if there is no stopping mechanism*, and the opposite is true with the stopping mechanism.

or destroyed, and a new searcher has to be sent out to replace it. Another example of a non-homogeneous search space occurs when the search progresses faster as the searcher approaches the object, for instance when directional information (the smell of food for the forager or a radio signature in a wireless network) becomes more readily available in the proximity of the object. We develop an analytical solution technique based on a finite but unbounded number of internally homogeneous segments, yielding the average search time and the energy expended. The results are illustrated by several examples.

We simplify the model of a non-homogeneous search space, described in (2)–(5) with $N = 1$ and $a = 0$, by considering a finite but unbounded number of "segments" that have different parameters for the Brownian motion describing the searcher's movement as a function of its distance to the object being sought, while within each segment the parameters are the same. The first segment is in the immediate proximity of the object being sought, starting at distance $z = 0$. Each segment may have a different size, and we assume that there are a total of $m < \infty$ segments. By choosing as many segments as we wish, and letting each segment be as small as we wish (all segments need not be of the same length), we can approximate as closely as needed any physical situation that arises where the searcher's

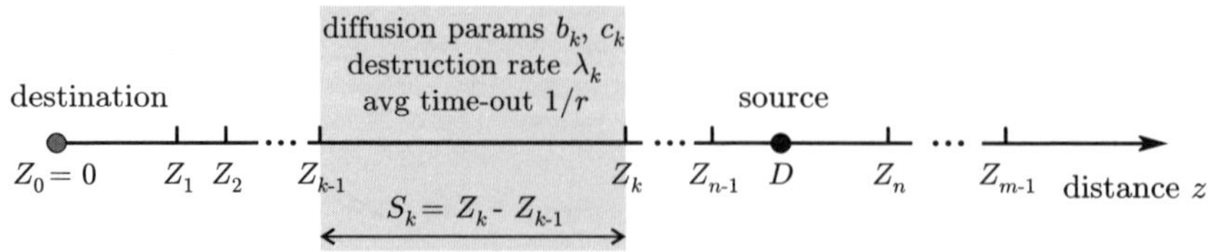

Fig. 5. Piece-wise approximation of a non-homogeneous medium.

motion characteristics vary over the distance of the searcher to the object being sought. We also show that this discrete representation leads to a neat algebraic "product form" representation of the average search time, and that it thus provides a useful analytic form that offers a more intuitive representation of the analytical results.

We denote by $0 \leq Z_k < \infty$ the boundary between the k-th and $(k+1)$-th segments with $Z_0 = 0$. The last segment goes from Z_{m-1} to $+\infty$, and we assume that both m and Z_{m-1} are finite but unbounded. Thus for greater accuracy in representing the search we can take as many segments as we wish, and they may be as small as needed, but they are all finite except the last segment. Thus for $0 \leq k \leq m$, the k-th segment represents the range of distances $Z_{k-1} \leq z < Z_k$, and let $S_k = Z_k - Z_{k-1}$ denote its size. We use n to denote the segment number in which the source point of the search is located, i.e. $Z_{n-1} \leq D < Z_n$. The piece-wise approximation is illustrated in Figure 5. If we write the parameters of the diffusion model for the k-th segment as:

$$\{f(z,t),\ b(z),\ c(z),\ \lambda(z)\} = \{f_k(z,t),\ b_k,\ c_k,\ \lambda_k\}, \quad Z_{k-1} \leq z < Z_k,$$

then the differential equation for the *stationary* solution of the distance dependent diffusion equation for $k \neq n$ is:

$$0 = \frac{c_k}{2}\frac{d^2 f_k(z)}{dz^2} - b_k \frac{df_k(z)}{dz} - (\lambda_k + r) f_k(z),$$

while the equation for the segment where the source is located is:

$$-[P + \mu W]\delta(z - D) = \frac{c_n}{2}\frac{d^2 f_n(z)}{dz^2} - b_n \frac{df_n(z)}{dz} - (\lambda_n + r) f_n(z).$$

We will also have:

$$rL = \sum_{k=1}^{m} \lambda_k \int_{Z_{k-1}}^{Z_k} f_k(z)dz,$$

$$\mu W = r[L + \sum_{k=1}^{m} \int_{Z_{k-1}}^{Z_k} f_k(z)dz],$$

$$P = \lim_{z \to 0^+} [\frac{c_1}{2} \frac{df_1(z)}{dz} - b_1 f_1(z)],$$

and the normalization condition:

$$1 = P + W + L + \sum_{k=1}^{m} \int_{Z_{k-1}}^{Z_k} f_k(z)dz.$$

Result 3. The total average search time, which is obtained by solving for P so that $E[T] = P^{-1} - 1$, is given by:

$$E[T] = (\frac{1}{r} + \frac{1}{\mu})[\sqrt{\frac{b_n^2 + 2c_n(\lambda_n + r)}{b_1^2 + 2c_1(\lambda_1 + r)}} \frac{A_n E_n e^{u_n S_n} - B_n C_n e^{v_n S_n}}{E_n e^{u_n (Z_n - D)} + C_n e^{v_n (Z_n - D)}} - 1], \tag{30}$$

where $u_k, v_k = \frac{b_k \pm \sqrt{b_k^2 + 2c_k(\lambda_k + r)}}{c_k}$ and the remaining parameters in (30) are computed as follows. Define:

$$\alpha_k^- = \frac{c_k u_k - c_{k-1} v_{k-1}}{c_k(u_k - v_k)}, \quad \beta_k^- = \frac{c_k u_k - c_{k-1} u_{k-1}}{c_k(u_k - v_k)},$$
$$\alpha_k^+ = \frac{c_k u_k - c_{k+1} v_{k+1}}{c_k(u_k - v_k)}, \quad \beta_k^+ = \frac{c_k u_k - c_{k+1} u_{k+1}}{c_k(u_k - v_k)}.$$

Then set $A_1 = 1$ and $B_1 = -1$ and for $2 \le k \le n$ compute:

$$\begin{bmatrix} A_k \\ B_k \end{bmatrix} = \begin{bmatrix} \alpha_k^- & \beta_k^- \\ 1 - \alpha_k^- & 1 - \beta_k^- \end{bmatrix} \begin{bmatrix} e^{u_{k-1} S_{k-1}} & 0 \\ 0 & e^{v_{k-1} S_{k-1}} \end{bmatrix} \begin{bmatrix} A_{k-1} \\ B_{k-1} \end{bmatrix}.$$

Then set $C_m = 0$ and $E_m = e^{v_m Z_m}$, and start another computation at $k = m - 1$ for $n \le k \le m - 1$ with:

$$\begin{bmatrix} C_k \\ E_k \end{bmatrix} = \begin{bmatrix} \alpha_k^+ & \beta_k^+ \\ 1 - \alpha_k^+ & 1 - \beta_k^+ \end{bmatrix} \begin{bmatrix} e^{-u_{k+1} S_{k+1}} & 0 \\ 0 & e^{-v_{k+1} S_{k+1}} \end{bmatrix} \begin{bmatrix} C_{k+1} \\ E_{k+1} \end{bmatrix}.$$

This completes all terms in $E[T]$, and the proof [41] consists in showing that the pdf $f_k(z)$ has the form:

$$f_k(z) = \begin{cases} \eta[A_k e^{u_k(z-Z_{k-1})} + B_k e^{v_k(z-Z_{k-1})}], \ Z_{k-1} \le z \le \min(D, Z_k), \\ \sigma[C_k e^{-u_k(Z_k-z)} + E_k e^{-v_k(Z_k-z)}], \ \ \max(D, Z_{k-1}) \le z \le Z_k, \end{cases}$$

where $\eta = \omega[E_n e^{u_n(Z_n-D)} + C_n e^{v_n(Z_n-D)}]$, $\sigma = \omega[A_n e^{u_n S_n} e^{v_n(Z_n-D)} + B_n e^{v_n S_n} e^{u_n(Z_n-D)}]$ and

$$\omega = \frac{r\mu/(r+\mu)}{\sqrt{b_n^2 + 2c_n(\lambda_n + r)}} \{A_n E_n e^{u_n S_n} - B_n C_n e^{v_n S_n} - [1 - \frac{r\mu}{r+\mu}] \sqrt{\frac{b_1^2 + 2c_1(\lambda_1 + r)}{b_n^2 + 2c_n(\lambda_n + r)}} [E_n e^{u_n(Z_n-D)} + C_n e^{v_n(Z_n-D)}]\}^{-1}.$$

Note that if the last segment includes the starting point $z = D$ (i.e. $m = n$) then $E[T]$ takes the much simpler form:

$$E[T] = \frac{r+\mu}{r\mu}[\sqrt{\frac{b_n^2 + 2c_n(\lambda_n + r)}{b_1^2 + 2c_1(\lambda_1 + r)}} A_n e^{u_n(D-Z_{n-1})} - 1], \tag{31}$$

and with a homogeneous search space $m = n = 1$ so that $E[T] = \frac{r+\mu}{r\mu}[e^{u_1 D} - 1]$ as we would expect from (10) when $N = 1$.

4.1. *Search in a protected neighborhood*

We consider the case where the neighborhood of the object being sought, up to a distance S, is protected by randomly located traps that destroy the searcher. In the rest of the search space accidental destruction of the searcher may occur, but at much lower rate. Thus we take $m = n = 2$, so that $E[T]$ is obtained from (31) with $\lambda_1 >> \lambda_2$. We choose the time-out parameter r so as to minimize $E[T]$ when $D = 100$, $b_k = 0.25$, $c_k = 1$, $\lambda_k = 0$, $\mu = 0.1$ and $S = 10$. In Fig. 6 we raise the question about how to select S and λ_1 together in order to maximize the protection offered to the object being sought. Thus we take λ_1 to be inversely proportional to S in Fig. 6a so that the average number of sources of protection, placed at rate λ_1, remains constant in proportion to the protection space of size S. The mapping of time rate to spatial rate will remain constant for any fixed value of b_1 which is the speed of motion. In this context, we examine whether there is a size S^* of the protected neighborhood which maximizes protection, i.e. that maximizes the average time to locate the object. Fig. 6a shows that there is indeed an optimum S^* that varies with the speed b_1 of the searcher inside the protected neighborhood. As the speed increases, the optimum size of the neighborhood gets smaller: a smaller size implies

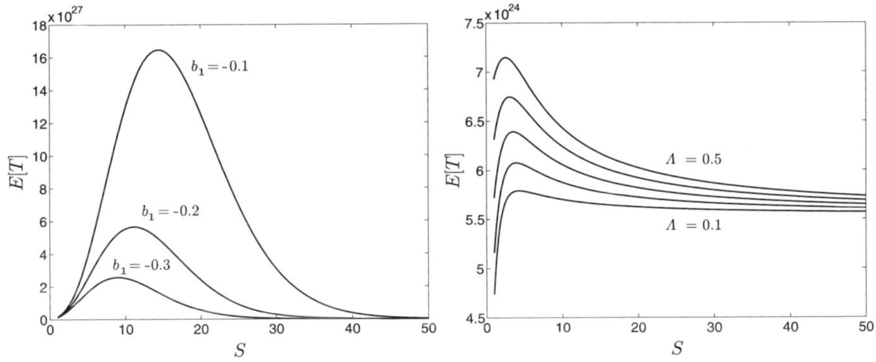

(a) (b)

Fig. 6. Average search time $E[T]$ versus size of the protected neighborhood S when (a) $\lambda_1 = 10/S$ for different values of b_1: the optimum protection area needed becomes smaller so that λ_1 increases when the search speed increases (b) $\lambda_1 = \Lambda/S^2$ for $\Lambda = 0.1$ to 0.5 with a step of size 0.1 and $b_1 = b_2 = 0.25$: the protection area needed to maximize the search time decreases as Λ increases.

a higher "rate of protection" and hence more frequently occurring destructions of the searcher which compensate for the higher speed of the searcher. However the corresponding maximum values of $E[T]$ do become smaller as the searcher's speed increases. In Fig. 6b we set $b_1 = b_2 = 0.25$ and Λ is varied in $\lambda_1 = \Lambda/S^2$. The results are similar to the previous ones.

4.2. *A phase transition effect*

The destruction of the searcher and the time-out will both relaunch the search process allowing the searcher to improve its chances to attain the object, we suspect that if the object being sought is heavily defended when the searcher gets very close to it, then the searcher may never attain the object. This is confirmed in Fig. 7 where we observe that if $\lambda_k = e^{\frac{1}{k\rho}}$ and $b_k = -e^{\frac{1+\epsilon}{k\rho}}$ with $\epsilon \geq 0$, then as ρ becomes very small, $E[T]$ tends to infinity despite the fact that near the origin the search speed is greater and its randomness is smaller. However it is interesting to see that if the searcher's speed of approach to the object grows faster than the rate at which the searcher may be destroyed, then $E[T]$ remains finite and may tend to zero, while in the opposite case it will tend to infinity presenting a form of phase transition.

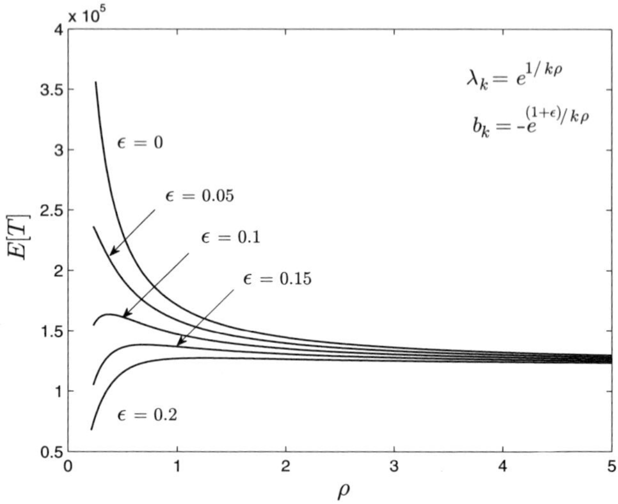

Fig. 7. Average search time $E[T]$ versus ρ when $\lambda_k = e^{\frac{1}{k\rho}}$ and $b_k = -e^{\frac{1+\epsilon}{k\rho}}$ for different values of $\epsilon \geq 0$; $c_k = 1$, $D = 10$, $r = 0.05$, $\mu = 0.025$ and $S_k = 1$ for $k < m = 20$.

5. Future work

In future work, we expect to address issues of load sharing or balancing so as to achieve overall better performance in search activities [42]. For instance, it would be reasonable to subdivide the search space among multiple searchers so that each search space is covered by a subset of the searchers. Also, it may be interesting to investigate how it would be possible to exploit the distinct performance capacities of multiple classes of searchers [43] so as to improve overall performance with respect to both time and energy needed for a search.

Another interesting area of research concerns the analysis of searchers that learn from each other so that they take advantage of their more successful colleagues. Similarly we can evaluate situations where searchers may try to conceal their knowledge of the location of an object, or even deceive the other searchers so as to impede their success. Thus we see that this area of research still reserves many opportunities for interesting research problems that are of value to different areas of application.

References

1. M. Coppey, O. Bénichou, R. Voituriez, and M. Moreau, Kinetics of target site localization of a protein on DNA: A stochastic approach, *Biophys. J.* **87** (3), 1640 – 1649 (2004). doi: 10.1529/biophysj.104.045773.
2. O. Bénichou, M. Moreau, P.-H. Suet, and R. Voituriez, Intermittent search process and teleportation, *J. Chem. Phys.* **126**(23), 234109 (JUN 21, 2007). doi: 10.1063/1.2741516.
3. M. R. Evans and S. N. Majumdar, Diffusion with stochastic resetting, *Phys. Rev. Lett.* **106**(16), 160601 (Apr, 2011). doi: 10.1103/PhysRevLett.106. 160601.
4. O. Berg, R. Winter, and P. Von Hippel, Diffusion-driven mechanisms of protein translocation on nucleic acids. 1. models and theory, *Biochemistry.* **20** (24), 6929–6948 (1981). doi: 10.1021/bi00527a028.
5. I. Eliazar, T. Koren, and J. Klafter, Parallel search of long circular strands: Modeling, analysis, and optimization, *J. Phys. Chem. B.* **112**(19), 5905–5909 (2008). doi: 10.1021/jp075113k.
6. B. Tilch, F. Schweitzer, and W. Ebeling, Directed motion of Brownian particles with internal energy depot, *Physica A.* **273**, 294–314 (1999). doi: 10.1016/S0378-4371(99)00247-2.
7. G. Oshanin, O. Vasilyev, P. L. Krapivsky, and J. Klafter, Survival of an evasive prey, *Proc. Natl. Acad. Sci.* **106**(33), 13696–13701 (2009). doi: 10. 1073/pnas.0904354106.
8. F. Rojo, J. Revelli, C. E. Budd, H. S. Wio, G. Oshanin, and K. Lindenberg, Intermittent search strategies revisited: effect of the jump length and biased motion, *J. Phys. A: Math. Theor.* **43**(34), 345001 (2010). doi: 10.1088/1751-8113/43/34/345001.
9. A. Gabel, S. N. Majumdar, N. K. Panduranga, and S. Redner, Can a lamb reach a haven before being eaten by diffusing lions?, *J. Stat. Mech.* **2012** (05), P05011 (2012). doi: 10.1088/1742-5468/2012/05/P05011.
10. E. Gelenbe, Z. Xu, and E. Seref. Cognitive packet networks. In *Proc. 11th Int'l Conf. on Tools for Artificial Intelligence (ICTAI'99)*, pp. 47–54, IEEE Computer Society (1999).
11. E. Gelenbe, Cognitive packet network (CPN), *US Patent 6,804,20* (Oct., 2004).
12. E. Gelenbe, Steps towards self-aware networks, *Commun. ACM.* **52**(7), 66–75 (Jul., 2009). doi: 10.1145/1538788.1538809.
13. E. Gelenbe. Travel delay in a large wireless ad hoc network. In *Proc. 2nd W'shop on Spatial Stochastic Models for Wireless Networks (SpaSWiN'06)*, pp. 1–6, IEEE, Boston, MA (3–6 Apr., 2006). doi: 10.1109/WIOPT.2006. 1666483.
14. E. Gelenbe, A diffusion model for packet travel time in a random multi-hop medium, *ACM Trans. Sen. Netw.* **3**(2), 1–19 (Jun., 2007). doi: 10.1145/ 1240226.1240230.
15. S. Shakkottai, Asymptotics of search strategies over a sensor network, *IEEE Trans. Automat. Contr.* **50**(5), 594–606 (May, 2005). doi: 10.1109/TAC.2005.

847061.
16. E. Gelenbe and G. Loukas, A self-aware approach to denial of service defence, *Comput. Netw.* **51**(5), 1299–1314 (2007). doi: 10.1016/j.comnet.2006.09.009.
17. E. Gelenbe and Y. Cao, Autonomous search for mines, *Eur. J. Oper. Res.* **108**(2), 319–333 (1998). doi: 10.1016/S0377-2217(97)00373-1.
18. E. Gelenbe and T. Koçak, Area-based results for mine detection, *IEEE Trans. Geosci. Remote Sens.* **38**(1), 12–24 (2000). doi: 10.1109/36.823897.
19. E. Gelenbe, N. Schmajuk, J. Staddon, and J. Reif, Autonomous search by robots and animals: A survey, *Robot. Auton. Syst.* **22**(1), 23 – 34 (1997). doi: 10.1016/S0921-8890(97)00014-6.
20. E. Gelenbe, Natural computation, *Comput. J.* **55**(7), 848–851 (2012). doi: 10.1093/comjnl/bxs077.
21. E. Gelenbe, K. Hussain, and V. Kaptan, Simulating autonomous agents in augmented reality, *J. Syst. Softw.* **74**(3), 255–268 (Feb., 2005). doi: 10.1016/j.jss.2004.01.016.
22. A. Filippoupolitis, L. Hey, G. Loukas, E. Gelenbe, and S. Timotheou. Emergency response simulation using wireless sensor networks. In *Proc. 1st Int'l Conf. on Ambient Media and Systems (AMBI-SYS'08)*, pp. 21:1–21:7, ICST, Brussels, Belgium (2008).
23. N. Dimakis, A. Filippoupolitis, and E. Gelenbe, Distributed building evacuation simulator for smart emergency management, *Comput. J.* **53**(9), 1384–1400 (2010). doi: 10.1093/comjnl/bxq012.
24. E. Gelenbe, G. Gorbil, and F.-J. Wu. Emergency cyber-physical-human systems. In *Proc. 21st Int'l Conf. on Computer Communications and Networks (ICCCN'12)*, pp. 1–7, IEEE Communications Society, Munich, Germany (Aug., 2012). doi: 10.1109/ICCCN.2012.6289183.
25. E. Gelenbe and F.-J. Wu, Large scale simulation for human evacuation and rescue, *Comput. Math. Appl.* **64**(2), 3869–3880 (2012). doi: 10.1016/j.camwa.2012.03.056.
26. A. Desmet and E. Gelenbe, Graph and analytical models for emergency evacuation, *Future Internet.* **5**(1), 46–55 (2013). doi: 10.3390/fi5010046.
27. E. Gelenbe, On approximate computer system models, *J. ACM.* **22**(2), 261–269 (Apr., 1975). doi: 10.1145/321879.321888.
28. E. Gelenbe, Probabilistic models of computer systems. part II: Diffusion approximations, waiting times and batch arrivals, *Acta Inform.* **12**, 285–303 (1979). doi: 10.1007/BF00268317.
29. E. Gelenbe, X. Mang, and R. Onvural, Diffusion based statistical call admission control in ATM, *Perform. Eval.* **27** & **28**, 411–436 (1996). doi: 10.1016/S0166-5316(96)90038-3.
30. E. Gelenbe, X. Mang, G. A. Marin, and R. O. Onvural, Statistical call admission control, *US Patent* (Apr., 2001).
31. C.Mejia-Monasterio, G. Oshanin, and G. Schehr, First passages for a search by a swarm of independent random searchers, *J. Stat. Mech.* **6**, P06022 (2011). doi: 10.1088/1742-5468/2011/06/P06022.
32. E. Gelenbe, Search in unknown random environments, *Phys. Rev. E.* **82**(6), 061112 (Dec., 2010). doi: 10.1103/PhysRevE.82.061112.

33. A. Einstein, *Investigations on the Theory of Brownian Motion.* Dutton & Dover, New York, NY (1926).
34. O. H. Abdelrahman and E. Gelenbe, Time and energy in team-based search, *Phys. Rev. E.* **87**(3), 032125 (Mar, 2013). doi: 10.1103/PhysRevE.87.032125.
35. Y. Wang, S. Jain, M. Martonosi, and K. Fall. Erasure-coding based routing for opportunistic networks. In *Proc. SIGCOMM W'shop on Delay-tolerant Networking (WDTN'05)*, pp. 229–236, ACM, Philadelphia, PA (Aug., 2005). doi: 10.1145/1080139.1080140.
36. S. Redner, *A Guide to First-Passage Processes.* Cambridge University Press, Cambridge, UK (2001).
37. F. R. de Hoog, J. H. Knight, and A. N. Stokes, An improved method for numerical inversion of Laplace transforms, *SIAM J. Sci. and Stat. Comput.* **3**(3), 357–366 (Sep., 1982). doi: 10.1137/0903022.
38. K. J. Hollenbeck. Invlap.m: A MATLAB function for numerical inversion of Laplace transforms by the de Hoog algorithm. `http://cambridge.org/us/engineering/author/nellisandklein/software.htm` (1998).
39. R. J. Serfling, *Approximation Theorems of Mathematical Statistics.* John Wiley & Sons, Inc., New York (1980).
40. O. H. Abdelrahman and E. Gelenbe, Search in non-homogenous random environments, *SIGMETRICS Perform. Eval. Rev.* **39**(3), 37–39 (Dec., 2011). doi: 10.1145/2160803.2160853. Special issue on MAMA'11.
41. O. H. Abdelrahman and E. Gelenbe, Packet delay and energy consumption in non-homogeneous networks, *Comput. J.* **55**(8), 950–964 (2012). doi: 10.1093/comjnl/bxs056.
42. J. Aguilar and E. Gelenbe, Task assignment and transaction clustering heuristics for distributed systems, *Information Sciences.* **97**(1-2), 199–219 (1997). doi: 10.1016/S0020-0255(96)00178-8.
43. E. Gelenbe and J.-M. Fourneau, Random neural networks with multiple classes of signals, *Neural Comput.* **11**(4), 953–963 (1999). doi: 10.1162/089976699300016520.

Chapter 16

Exit Strategies: Visual Search and the Quitting Time Problem

Todd S. Horowitz

Basic Biobehavioral and Psychological Sciences Branch
National Cancer Institute, Rockville, MD 20850, USA
todd.horowitz@nih.gov

Visual search is the study of how humans find desired objects in their visual environment. I provide brief outline of work on visual search, and then describe work on specific topics such as the effects of target prevalence and visual foraging.

1. Introduction

In the morning, you look for your keys before leaving the house, or you try to find a pen on your cluttered desk. A radiologist scrutinizes a mammogram for a tumor. At the airport, a screener examines an x-ray image of your bag for potential weapons. All of these are examples of *visual search.* Visual search is ubiquitous in everyday life; we conduct hundreds of visual searches each day. But visual search also crops up in a wide array of socially critical venues, from an intelligence analyst searching for signs of missile silos in North Korea to a mechanic searching for signs of stress on an airplane's wing.

Psychologists have studied visual search since the 1920s [1, 2]. Study of this task became central to the study of attention following seminal papers by Schneider and Shiffrin [3, 4] and Treisman and Gelade [5]. Since then, visual search has become one of the most prominent topics in vision science and cognitive psychology. I will provide a brief overview of the field. For a more thorough primer, see Wolfe [6]. For reviews, see Eckstein [7]. Then I will discuss in detail the topic of quitting rules in search.

2. Review of visual search

2.1. *Methodologies*

2.1.1. *Psychophysics*

The modal visual search experiment involves a human searcher presented with an image and asked to report, as quickly and accurately as possible, whether a pre-specified *target* is present or absent in the image. The searcher is variously referred to as the "subject" or "participant"; here we will use the term *observer.* The key dependent variables are the reaction time (RT) between the onset of the stimulus and the observer's response, and the accuracy of that response. The image is highly controlled, comprising an array of non-target items (*distractors*) in addition to the target (if present, usually on half the trials). The number of distractors (*set size*) is typically a key independent variable, along with the similarity between targets and distractors, and among distractors. The observer will perform many trials (the total number will depend on the number of different stimulus conditions, but typically is on the order of 100) in order to provide reliable data.

Figure 1(a) shows an example of a *spatial configuration* search task, in which targets and distractors are composed of the same elements in different arrangements. The target is in the shape of a "5", and the distractors are "2"s. Figure 1(b) plots RT as a function of set size for target present and target absent trials from this kind of experiment. RTs increase linearly with set size, and the slope of the RT $\times$ set size regression (in time/item) is an index of the efficiency of the search; less time per item (flatter slopes) indicates more efficient search.

In RT experiments, the stimulus is generally available until the observer responds. In accuracy experiments, stimulus exposure is controlled and typically brief (i.e., $<$500 ms). Visual processing is then terminated by a specially designed *mask.* Performance is measured as a function of stimulus duration. Accuracy measures may be transformed according to Signal Detection Theory (SDT) [8] into parameters estimating sensitivity (ability to discriminate between target presence or absence) and response bias.

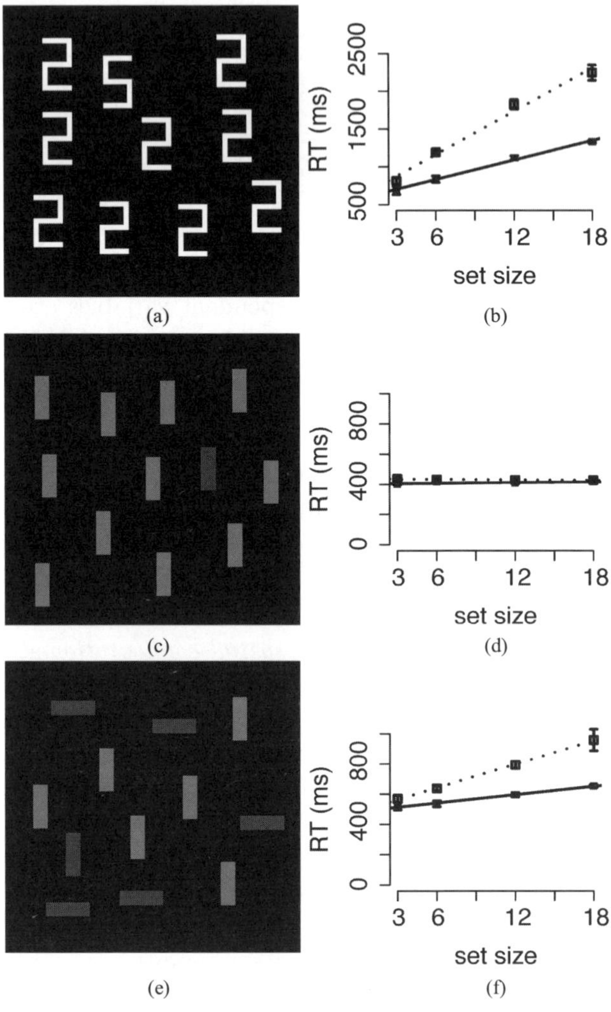

Fig. 1. (Color online) Typical visual search stimuli and results. Panel (a) illustrates a typical spatial configuration search. The target is a "5" among distractor "2"s. Results from such a task are illustrated in panel (b). Mean RT for 9 observers are plotted against set size. Panel (c) illustrates a feature search task. The target is a red bar among green bar distractors. Mean RTs for 9 observers in this task are plotted against set size in panel (d). Panel (e) illustrates a conjunction search task. The target is a red vertical bar among green vertical and red horizontal bar distractors. Mean RTs for 10 observers in this task are plotted against set size in panel (f). Note that the data in (b) are plotted on a different scale from the data in (d) and (f). Error bars in all figures denote the within-subject confidence interval [9, 10]. These data have been previously published [11, 12].

2.2. *Phenomena*

2.2.1. *Pop-out and conjunction search*

Treisman and Gelade [5] proposed a two-stage theory of visual perception. In the first stage, the visual system analyzes the visual stimulus into a set of elementary or basic features, such as color, orientation, and size. This analysis is carried out in parallel across the visual field. Each feature is represented in a separate feature map. In the second stage, information from different feature maps can be combined via focal selective attention to a particular location. However, the presence of activity in a feature map can be detected without focal attention. Thus, a target defined by such a basic feature should be easily detected, while a target defined by a combination of features should require effortful search. This prediction was operationalized via the slope of the RT × set size function. Figure 1(c) shows a *feature search* task, in which the target (a red square) differs from the distractors (green bars) on a putative basic feature, color. Introspectively, the search feels effortless; the target seems to "pop out" of the display. Accordingly, this kind of search task is typically described as "pop-out search". Quantitatively, pop-out is diagnosed by the near-zero slope on the RT × search size function in Fig. 1(d). Figure 1(e) illustrates a conjunction search task, in which the target is defined as the combination of red and vertical, among distractors which can be red and horizontal or green and vertical. This search is noticeably more effortful, as reflected in the steeper RT × set size slope in Fig. 1(f).

The method of proposing candidate basic features and testing whether they produce flat RT × set size slopes has produced a cottage industry [13]. Well-established features include color, orientation, size, motion, depth, luminance polarity, closure, and curvature. Other putative features include lighting direction, looming/expansion, and aspect ratio. There is substantial argument over proposed features such as novelty [14, 15], angry faces [16, 17], and alphanumeric category (i.e., letters vs. digits) [18–20].

2.2.2. *Search asymmetries*

Some searches are asymmetrical, in that search for A among B is efficient, while search for B among A is not. In general, search for the presence of a feature is easy, search for absence is difficult [21]. For example, a "Q" among "O"s will pop-out, because the Q has a crossbar that the Os lack. In contrast, search for O among Q is typically inefficient. Such search asymmetries can also tell us something about how some feature dimensions are represented. Generalizing from the present/absent asymmetry, if the target and distractors differ along a single feature dimension, targets that produce more activation on the dimension than distractors should pop-out, while target that produce less activation than distractors will require inefficient search. For example, long lines pop-out among short lines, but short lines are difficult to detect among long lines; similarly, curved lines are easy to detect among straight lines, but the reverse search is inefficient. One way of accounting for these results is to suggest that search relies on detectors which respond with increasing intensity to length and curvature, respectively [22]. Alternatively, we might assume that the response to short lines or straight lines is more variable than the response to long or curved lines [23].

2.3. *Theories*

Treisman and Gelade's Feature Integration Theory (FIT) [5] remains the most influential approach to visual search. As noted above, FIT proposes that the visual scene is analyzed in parallel into a set of independent feature maps. In order to combine information across maps, focal attention must be directed to a particular location on a master location map. While the theory has been modified several times in response to empirical challenges [5, 24–26], the basic insight, that focal attention serves to bind disparate features, has remained intact.

Wolfe's Guided Search Theory (GST) [27–30] adopts the basic two-stage (parallel, then serial) architecture put forth by Treisman but, as the name suggests, puts the emphasis on how attention is guided to likely targets. In GST, information from the various feature maps is combined into a unified guidance map, which is used to guide attention to objects likely to be worthy of scrutiny. The guidance map combines two types of information, *bottom-up* and *top-down*. Some things, such as bright spots or edges, are simply going to attract your attention, regardless of what you're looking for. These items have bottom-up salience, which typically reflects locations where feature values are rapidly changing. However, the content of the guidance map also changes as a function of what you're looking for. If you're searching for a friend in a red shirt, for example, input from the red feature map will be more heavily weighted, so that red things will appear more salient. Attention is then directed to objects or regions in proportion to the activity they generate on the guidance map. It is important to note that the guidance map only highlights what is likely to be important, it doesn't indicate why; bottom-up and top-down sources of guidance are indistinguishable.

Another influential theory built on the FIT platform is the Reverse Hierarchy Theory (RHT) [31]. RHT begins from the observation that many of the visual features which guide search do not seem to match the properties of the early visual system. Neurons in the early visual system are characterized by precise tuning curves and small receptive fields, while human performance on search tasks implies large visual fields with coarse tuning functions [32, 33]. Search also seems to use fairly high level features, such as 3-dimensional form cues [34, 35], that are not computed by the early visual system. In RHT, "basic" features are computed late in the visual system in higher cortical areas, such as inferotemporal cortex. The large receptive fields in these areas mediate the phenomenon of pop-out, while their placement late in the visual stream allows them to compute complex properties. The function of attention is then to "drill down" the hierarchy to extract the precise, location-specific information we may require. Thus, when we are attending to an object, we have very fine-grained information, computed in the early visual system, but outside the focus of attention, we have only coarse, categorical feature information and very little information about where things might be. One controversial corollary of RHT is the claim that very complex, almost semantic visual properties, such as "faceness" or "houseness" can generate pop-out and function as a basic visual feature [36, 37] (but see [38]).

One noted competitor to the two-stage models discussed above is the Attentional Engagement Theory (AET) [39, 40]. This theory is built around another set of parameters which are known to modulate search performance: similarity between targets and distractors (TD similarity) and similarity among distractors (DD). As common sense would indicate, as targets become dissimilar from their surrounds, they are easier to find. Targets also become easier to find as the distractors become more similar to one another. In AET theory, all of the visual properties of objects are computed prior to the deployment of attention. The role of attention is not to improve or finalize the processing of object identities, as in two-stage theories, but to provide access to response mechanisms. Thus, two-stage theories are also termed "early selection" theories, because selection occurs before higher visual processing, while AET is an example of a "late selection" account.

According to AET, objects compete for access to attention via two avenues: activation and suppression. When you decide to look for your red-shirted friend in a crowd, activation spreads among objects that share the target feature (red), while suppression spreads among objects that do not. If your friend is in a crowd with pink or purple shirts, these distractors will receive some of the activation intended for the target, making him harder to find. Similarly, the more disparate colors there are in the crowd, the slower suppression will spread among the distractors, again camouflaging your friend.

While AET dispenses with the two-stage architecture, it does share a common assumption that objects must compete for access to a limited resource. The signal detection theory (SDT) approach dispenses with that assumption. According to SDT accounts of search, the limiting factor on search performance is not attention, but noise. In this approach, the target is conceived of as a signal embedded in noise generated by the distractors. As the number of distractors increases, the probability that a noise response will exceed the signal response increases, leading to the set size effect. This approach has been successful in explaining performance in a wide variety of accuracy experiments [41–43] (but see [44]).

3. Knowing when to quit

An important problem currently being studied is how humans know when to quit searching. If we consider a search task where the number of targets is well-defined, say [0, 1], and each possibility is mapped to a different reponse ["absent", "present"], the decision rule for when to respond "present" is fairly simple. When you find something you think is a target, you respond "present". But if there is a possibility that there may be no target, how do you know when to respond "absent"?

One straightforward answer is to simply examine everything and then answer "absent" if you find no targets. This "exhaustive search" strategy assumes that attention samples the array without replacement. This assumption is built into many models of search [5, 45–48], often implemented with the help of an "inhibitory tagging" mechanism, in which the neural representations of examined items are suppressed [49].

Over the last 15 years, there has been a vigorous debate over whether search employs sampling without replacement. Horowitz and Wolfe [50–52] proposed that attention samples with replacement in search tasks such as those depicted in Fig. 1(a). Note that in the general case, search is almost certainly not memoryless; surely you remember that you already looked under the bed for your shoes. The claim here is that attentional sampling of locations at 20 Hz is not influenced by which items have been previously selected. This claim has been disputed [53–55]; for a review of this issue, see Horowitz [56]. Recently, Hulleman [57–59] has resolved several of the methodological issues that dogged earlier studies, and concluded that individual items are not tagged unless the search task is very difficult. The question is somewhat more tractable when studying eye movements, because you get a more direct measure of which items have been fixated. Peterson *et al.* demonstrated that it is possible for observers to employ a near-perfect sampling without replacement strategy [60]. However, this is probably too strong as a general claim. Tasks which require each item to be individually fixated proceed at roughly 4 Hz, which is slow enough for deliberate strategies to be implemented. The appearance of sampling without replacement can be created by prospective strategies [61], such as scanning left-to-right and top-to-bottom [62], which do not imply any inhibitory tagging mechanism. The memory capacity for previously fixated locations has been estimated at 3-4 [63].

Even setting this debate aside, however, the data do not support exhaustive search in most cases. Exhaustive search predicts that the RT $\times$ set size slope for target absent trials should be twice that of target present trials, but the actual ratio is typically greater than 2:1 [64]. Furthermore, it is clear that searchers change their quitting behavior in response to feedback, speeding up after a series of correct "absent" decisions, but slowing down dramatically after a miss [65]. Thus, it seems highly unlikely that observers simply quit after they've looked at everything; we need a more nuanced account of search termination behavior.

3.1. *The prevalence effect*

As previously noted, the typical laboratory visual search experiment takes the form of a present/absent discrimination; there is either one target in a display, or zero targets. The research emphasis is usually on the ability to find targets under various situations, and target-absent trials are included mainly as catch trials, to ensure that observers are actually doing the task. Thus, the typical experiment has 50% targets, or more if it can be justified.

However, in many, many real-world search and search-like tasks, the *prevalence* of targets is much lower. Consider a lifeguard's search for drowning victims [66], a security officer's search for threats in carry-on baggage at the airport [67], a cytologist's search for an abnormal cell [68], or a liquor store clerk checking whether a driver's license photo fails to match the customer [69]. Prevalence in these cases is typically less than 1%.

Wolfe, Horowitz, and Kenner [70] asked observers to search for tools (e.g., hammers, wrenches) among other objects. These stimuli were constructed from commercially available clip-art. In an effort to roughly approximate conditions in airport baggage screening, the objects were semi-transparent and could overlap one another, and were presented in visual noise. They tested three levels of prevalence: 50% to 10% to 1%. They found that miss rates substantially increased as prevalence was reduced; while observers missed fewer than 1 in 10 targets on the high (50%) prevalence trials, they missed nearly 1 in 3 targets on the low (1% prevalence) trials; a result that would be quite disturbing if it proved to generalize to the previously mentioned real-world tasks.

The changes in error rates were accompanied by equally dramatic changes in the RT $\times$ set size functions. At high prevalence, target absent RTs were substantially slower than target present RTs, as is typical. RTs on miss trials were slightly faster than correct rejection RTs, suggesting that high prevalence misses occured when the observer had been searching for a while, gave up a little too early, and guessed wrong. On low prevalence trials, however, target absent trials were actually slower than target present responses, and miss RTs tracked correct rejection RTs more closely. One way of understanding this pattern is to imagine that observers have an estimate of how long they should have to search in order to be sure that there is no target is present. On a given trial, the actual maximum search time the observer is willing to tolerate is distributed around this estimate; on some trials observers are more patient, on others less so. At high prevalence, this estimate is fairly long, because the high proportion of targets means that if you give up early, you will miss a substantial proportion of targets. Thus, while the mean target present RT is well below the threshold, behavior is governed by the right tail, because there are enough trials where waiting longer is rewarded by finding a target (or where giving up and responding "absent" is punished). This leads to target absent trials much longer than target present trials. At low prevalence, however, since there are fewer targets, there are correspondingly fewer trials out in the right tail of the distribution to govern behavior, and observers become more and more confident declaring the target to be absent after shorter and shorter search times. Hit RTs then are drawn disproportionately from those trials on which they happened to search a little longer than usual.

Given that observers in these experiments were making simple button-press responses, perhaps a simpler explanation is that observers get so used to hitting the "absent" button that it is hard to suppress that response even after successfully locating the rare targets. Fleck and Mitroff [71] ran a similar experiment, in which observers searched for tools among other objects, at 2%, 10%, and 50% prevalence. Critically, half of the observers were given the option to "correct" their responses; after the next trial had begun, they could press the "escape" key to indicate that their reponse on the previous trial should be reversed. They found that the observers who could not correct their responses replicated Wolfe *et al.* [70]. The observers in the correction condition also replicated the original Wolfe *et al.* results before correction; however, once the corrections were taken into account, the increase in misses with decreased prevalence was substantially reduced.

If the prevalence effect were merely an artifact of hasty button presses, it would be a mere laboratory curiosity, and not a problem for socially critical search tasks. Unfortunately, using more realistic artificial baggage stimuli assembled from real x-ray images provided by the TSA, Van Wert, Horowitz, and Wolfe [72] showed that the opportunity to correct their responses did not alleviate the prevalence effect, suggesting that low prevalence can affect behavior through more than one pathway. Converging evidence was provided by Godwin *et al.* [73], who asked observers to search for either of two targets. Half of the trials always contained a target, but the relative prevalence of the two targets was varied. Here, response priming or the strength of the "target present" response was constant and unrelated to prevalence, but decreased prevalence still led to increased misses for the low prevalence target.

An important characteristic of performance of the "clip-art" object search tasks used by both Wolfe *et al.* [70] and Fleck and Mitroff [71] is that the false alarm rates were extremely low (less than 0.01 in the Fleck and Mitroff study, even lower in the Wolfe, *et al.* study). In contrast, the more naturalistic baggage stimuli of van Wert and Wolfe [72] produced false alarm rates an order of magnitude greater. This suggests that errors in the clip-art search tasks come from failure to attend or fixate the target, rather than from noisy target identification processes. That is, if you attend to the target, you recognize it; the trick is getting your attention to the right place in the image. This is referred to as a "high threshold theory" task. In contrast target identification in naturalistic baggage search appears to be a noisy process governed by signal detection theory.

Kundel, Nodine, and Carmody [74] identified three types of errors observers could make in search tasks: search errors (the observer quits the search before attending the target); identification errors (the observer attends the target, but fails to recognize it as the target); and response errors (the observer correctly identifies the target, but inadvertently makes the incorrect response). Each error can be identified by a characteristic eye movement behavior. If the observer did not fixate the target before responding "absent", that would reflect early quitting. If the observer fixated the target, then continued to search before responding "absent", that can be inferred to reflect a failure to identify the target. Finally, if the observer was fixating the target at the time of response, but still reported "absent", that was assumed to reflect a motor error. Rich *et al.* [75] applied these classifications to low- and high-prevalence search tasks in order to determine the source of the additional errors that observers commit at low prevalence. Note that the stimuli for this experiment (small Ts and Ls) were specifically designed to ensure that observers needed to fixate each item in order to identify it as target or distractor. Figure 2 shows the results. Dark gray bars indicate the rate of miss trials in each category at high prevalence, and light gray bars indicate the rates at low prevalence. Some of the additional errors at low prevalence do seem to reflect motor errors, à la Fleck and Mitroff [71], and there is a small fraction of identification errors. However, in Rich *et al.*'s data, the majority of the additional misses at low prevalence could be attributed to premature search termination.

In the accuracy domain, increasing prevalence generally shifts errors from the miss column to the false alarm column. In signal detection theory terms, prevalence affects "criterion" rather than "sensitivity"[a]; it does not reduce the target's visibility, but rather changes the amount of evidence the observer requires in order to decide that a target is present [76]. In the RT domain, the effect of prevalence is much stronger on target absent (i.e., "no") responses than on target present ("yes") responses [76, 77]. While I have so far been discussing comparisons of "high" (~50%) and "low" (~1–2%) prevalence, this phenomenon can be observed across the entire range of prevalence. Wolfe and van Wert [78] varied prevalence sinusoidally from 0% to 100% and back again across 1000 trials, and found that criterion and target absent RTs varied accordingly, following prevalence with a lag of around 50 trials, while sensitivity and target present RTs were roughly constant.

[a]Here we are using the term "sensitivity" as it is used in the psychophysical literature, to denote the observer's ability to detect the target; or, more specifically, the distance between the signal and noise distributions in SDT. In the medical literature, "sensitivity" is the percentage of targets that are found (referred to here as "hit rate"). This can be confusing.

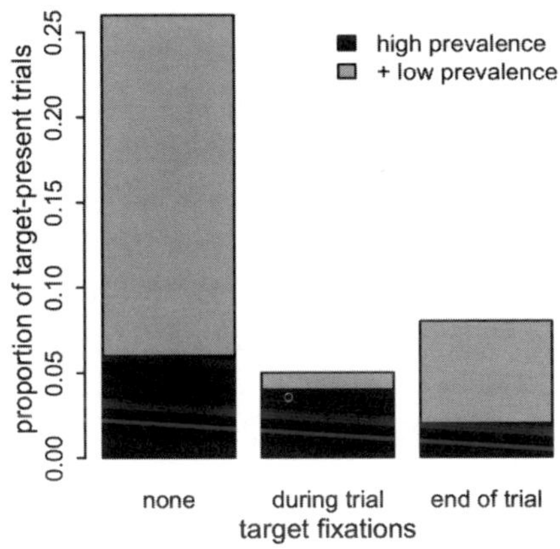

Fig. 2. Miss rates at high and low prevalence, classified into three different categories (see text). Dark gray denotes high prevalence misses, light gray denotes low prevalence misses. "None" indicates that the target was not fixated before response, implicating premature search temination. "During trial" indicates that the target was fixated during the trial, but the observer continued searching before responding, implicating identifcation errors. "End of trial" indicates that the target was the last item fixated before response, likely implicating motor errors. Data taken from [75].

In order to explain this pattern, Wolfe and van Wert [78] proposed a two-criterion model of prevalence effects in visual search. They modeled visual search as a series of decisions. On each cycle of the model, one stimulus is selected, and a decision is made as to whether or not it is a target. This decision is noisy: both targets and distractors generate some neural activity, the strength which is assumed to be normally distributed. The distributions of both targets and distractors have the same variance, but there is a difference between the means such that targets will, on average, yield larger activations than distractors. The larger the difference between the means, d, the easier the discrimination. Note that the activation level does not, in itself, convey the answer as to whether the item is a target or distractor. Targets generate larger values on average, but distractors will sometimes generate large values as well. The observer makes a decision by setting some criterion, c, such that if the observed activity exceeds c, the item is declared to be a target; if the activation does not exceed c, the item is assumed to be a distractor. So far, I have simply described the basic rules for a 2-alternative forced-choice (2AFC) decision in Signal Detection Theory [8]. Now, if the outcome of the discrimination is "target", the observer should respond by pressing the "present" button (or whatever response is appropriate for detecting a target, such as opening your suitcase or calling you back for a biopsy). If the outcome is "distractor", however, she has another choice to make. Should she give up on this trial, assume that there is no target and respond "absent" (wave you on to your plane, give you a clean bill of health), or should she select another item and repeat the cycle? Wolfe and van Wert model this decision, which they call the "search termination decision", as diffusion towards a threshold [79–81], Q. The longer the trial goes on without a target being detected, the closer the diffusion process will get to Q. As long as the diffusion variable is below Q, the observer keeps searching. When Q is exceeded, the observer declares the target to be absent.

The model therefore has two criteria which can be adjusted by the observer in response to changing prevalence. As prevalence goes down, c becomes more conservative; it takes a stronger signal to convince the observer that the item is really a target. Simultaneously, Q goes down; the observer quits the search more readily. Increasing prevalence has the opposite effect: the observer adopts a more liberal criterion in the 2AFC judgment, and is willing to wait around longer to see if a target is present. Changing c does not affect either sensitivity (indexed by d) or RT (the decision is assumed to take a fixed amount of time regardless of prevalence), while Q affects only the quitting decision. The model therefore explains the known pattern of data quite well. The core architecture of this model has been incorporated into a general search model called "competitive guided search" [82].

Some supporting evidence for the link between the quitting threshold and hit rates comes from work by Levin, Angelone and Beck [83], using highly cluttered naturalistic scenes (e.g., desk drawers, refrigerators). They demonstrated that hit rates in rare target search were predicted by what they term the "search termination (ST) correlation": the correlation between normalized individual RTs for target-absent images and the group average RTs on those same images. Levin *et al.* hypothesized that the ST correlation reflected an individual observer's ability at *distractor-tuning*; judging how difficult a particular image would be to search through and adjusting their quitting time (i.e., Q) accordingly. Additionally, the Levin *et al.* data provided converging evidence for the conclusion that the prevalence effect is not an artifact of a speed accuracy tradeoff. If such a tradeoff were in action, then the relationship between ST correlation and hit rate should have disappeared when they controlled for the overall mean or standard deviation of RT, but that was not the case.

The prevalence effect is a robust phenomenon in the laboratory, with several different components pushing human performance toward misses at low prevalence and false alarms at high prevalence. This has important implications for many socially important search tasks, such as the interpretation of medical images or airport baggage screening. In many such tasks, prevalence is very low. Prevalence of abnormalities in cervical cancer screening, for example, is estimated to be below 1% [84]. These tasks are also characteristically difficult from a perceptual point of view, which makes them more vulnerable to prevalence.

Performance feedback is a complicating factor. In laboratory search tasks, observers are routinely given full feedback about their performance, such that both false alarms and misses are flagged, in order to encourage observers to perform at optimal levels. In field search tasks, such feedback is rarely available. Consider a screener at the airport inspecting hand luggage. If she declares a threat in the bag, she receives immediate feedback as to whether she has made a hit or a false alarm. If she lets the bag go, she will never find out if it contained a threat or not. Similarly, a radiologist who calls back a patient after reading a mammogram will get fairly rapid hit vs. false alarm feedback, but may not find out about a miss until the woman comes back for her next annual screening, if then. This lack of feedback may actually be helpful; Van Wert, Horowitz and Wolfe [72] found that the prevalence effect was weaker when observers did not receive feedback about their performance. Presumably, without feedback, observers' internal estimates of prevalence tend to drift towards their prior assumptions about the state of the world. Indeed, two groups have proposed potential "cures" for the prevalence by exploiting feedback. Wolfe *et al.* [76] designed a protocol in which long stretches of low prevalence trials with no feedback are broken up with brief "bursts" of high prevalence trials with feedback. The bursts provide the visual system with strong evidence for high prevalence, thus moving c and Q (in terms of the Wolfe and Van Wert model [78]) to more neutral settings, while the lack of feedback helps slow down the drift towards more conservative settings during low prevalence. This method reduced the prevalence effect to statistically non-significant levels. Schwark *et al.* [85] provided false feedback to observers on 20% of target-absent trials; telling observers that they had missed targets when they had, in fact, correctly indicated that no target was present. This manipulation again improved target detection at low prevalence by shifting criterion to a more neutral setting. Thus, the prevalence effect can be mitigated by changing *perceived* prevalence.

Another important feature of socially important search tasks is that the consequences attached to misses and false alarms are asymmetrical. For the baggage screener at the airport, for example, false alarms have an immediate consequence in terms of frustrated passengers, but the consequences of a miss may be catastrophic. For the radiologist searching a mammogram, however, the asymmetry is less stark; a miss means that a tumor goes undetected for another year, but a false alarm may mean a risky biopsy. These payoff structures may motivate professional searchers such that prevalence effects are minimized in the field. However, payoff structure is known to be less powerful than prevalence in affecting performance [86]. It has proved difficult in the laboratory to counteract the prevalence effect by changing the costs of different types of errors or correct responses. Wolfe *et al.* [76] reported being unable to ameliorate the prevalence affect at all with payoffs, while Navalpakkam *et al.* [87] were only successful when they framed the experiment as a contest among the observers. Of course, laboratory experiments have only monetary outcomes to dangle in front of observers; they cannot threaten exploding airplanes or metastasizing cancers.

Given this discrepancy in payoff structure and motivation, laboratory studies are not guaranteed to generalize to the field. It is of course more difficult to study practicing radiologists than psychology undergraduates, but there is a literature on low prevalence in the medical decision-making field. The results are rather inconsistent, some showing a prevalence effect and some not, but most of these studies use very small stimulus sets (e.g., 40 images [88]), leading to low power and unreliable results [89, 90]. A large-scale study of fourteen trained radiologists (comprising eight faculty, two fellows, and four residents) and 1632 chest images and two prevalence levels (28% and 2%) [91] reported no effect of prevalence on the ability to detect abnormalities, as measured by the area under the ROC curve (AUC), concluding that there was likely no prevalence effect for radiologists. However, this result is actually consistent with laboratory-based prevalence effects, which result from a change in criterion, rather than sensitivity [76, 92]; a change in AUC would not be predicted. Indeed, a later analysis of data from the same study [93] reported a change in observers' confidence in their responses a function of prevalence, in a direction consistent with the prevalence effect. More recently, Evans and her colleagues have reported two studies indicating that medical imaging experts are subject to the prevalence effect when searching in their domain of expertise. In one study [68], they tested two samples of professional cytologists, one in the US and one in the UK. Each sample saw 2000 images in a low prevalence condition

(2%, or 40 target present images with abnormalities for the US sample; 5%, or 100 target present images for the UK sample) and 100 images in a high prevalence condition (50%, 50 target present images). The US cytotechnologists, with an average of 18 years of experience, missed 17% of the abnormalities at high prevalence, and 28% at low prevalence. The UK cytology screeners, with an average of 10 years of experience, missed 27% of the abnormalities at high prevalence, and 42% at low prevalence. Consistent with laboratory results on naïve observers, Evans *et al.* found that sensitivity (d') was not impaired at low prevalence, but that the increased miss rate was due to a change in criterion. In a later study [94], they tested professional radiologists at a large US research hospital (5-30 years of experience) using 100 test images. The test images comprised 50 positive cases (i.e., with tumors) and 50 negative cases. In the high prevalence condition, the radiologists saw just the 100 test images, so prevalence was 50%. For the low prevalence condition, the same 100 test images were inserted into the normal workflow of the radiology clinic, over the course of 9 months. Prevalence in this arm of the study, including both test images and real tumors discovered during the normal course of events, was estimated at 1%. While the radiologists missed only 12% of the cancers in the high prevalence condition, they missed 30% of those very same cancers in the low prevalence condition; false alarms moved in the opposite direction, indicating a criterion effect. Since the low prevalence arm of the study was carried out in the clinic, this is fairly strong evidence that the prevalence effect is a real phenomenon for expert observers working under field conditions. Of course, the high prevalence data point comes from an artificial setting, but it is difficult to ascribe the superior performance in this condition to non-prevalence factors, unless one assumes that, for example, radiologists are more motivated to search for cancers in a laboratory setting than when real patients' lives are at risk.

3.2. *Visual foraging*

In the prevalence studies described above, the number of targets in a given trial is always well-defined. That is, while the observer may be uncertain as to whether or not there is a target, there are always either 0 or 1 targets. However, in many practical search tasks, the number of targets is ill-defined. For example, a medical x-ray may contain an unknown number of abnormalities. Or consider the examples from the start of the article. The intelligence analyst scrutinizing satellite photos does not know how many missile silos to look for (if any), but needs to find as many as possible. Similarly, the number of stress fractures on an airplane wing is not known in advance, but it is important not to miss any.

These sorts of tasks are beginning to be understood in terms of optimal foraging theory. While foraging theory was originally developed in population biology to explain the behavior of organisms seeking food, such as a green lacewing's hunt for mites or aphids [95], insights from foraging theory have recently been applied to human behavior in a variety of contexts, from simulated fishing tasks [96] to word puzzles [97] to foraging for information [98, 99].

Assume that potential food sources are distributed in a patchy fashion across the terrain, with variable yield. A good example might be to pretend you are an animal foraging for berries. Berries are not uniformly distributed on the ground; they grow on bushes, which are irregularly sited. Each bush has a varying number of berries on it. If you are currently picking berries from bush n, when is it time to give up and move on to bush $n + 1$? Let's assume that your task is to maximize your intake of berries in a given time frame, before the sun goes down and you have to return to your burrow. If you wait until you are sure that you have picked every single berry on bush n before moving on, you will probably not maximize your nutrition. Since you eat each berry once you find it, berry-picking is sampling without replacement. Thus, berries will become sparser over time and the rate of acquiring berries from the current bush will tend to go down. That last berry may take a long time to find. A fundamental problem in foraging theory, then, is to compute optimal patch (bush) leaving times [100].

One durable principle to come out of the literature on optimal patch leaving times is the *Marginal Value Theorem* (MVT) [101]. MVT states that the organism should leave a patch when the local (or marginal) capture rate in the patch drops to the average global capture rate for the habitat. When you first arrive at a given berry bush, you acquire berries at a rapid rate. As you pick the berries, the berry distribution becomes sparser, and the inter-berry time increases, reducing your rate. When this rate is lower than the average rate for the environment, it becomes highly likely that the rate would be higher if you switched to another bush, so you are wasting your time. Note, however, that the global rate will be affected by the travel time between patches. If bushes are widely separated, the average rate will go down, and it will make sense to remain longer at a given bush. While MVT has been criticized for its simplifying assumptions [102], it remains a powerful tool for explaining patch leaving times [100, 103–105].

As noted earlier, search termination is a problem that the field of visual search has not yet solved, even for well-defined present/absent decisions. Insights from foraging theory are likely to be important for the development of an adequate account of search termination, as well as for generalization of search theories into the realm of undefined target sets.

In the visual search context, a trial or an image serves as a patch. Consider our hard-working radiologist with a stack of x-rays to get through. She has to balance finding as many tumors or fractures in a given image with the fact that she has to get through a large number of images in the work day. Her quitting time for a particular image will, according to MVT, be inversely proportional to the average rate at which she finds abnormalities in the average image. Similarly, an observer in a laboratory visual search experiment is faced with a series of trials, and must decide how much time to devote to "harvesting" a particular trial before moving on to the next trial. Note that MVT-style behavior could be suboptimal in contexts such as radiology, where it may be important to find the last "berry" on the current bush before moving to the next.

Studies of optimal search behavior are not new [87, 106], but only very recently has foraging theory been applied to the visual search context. Cain *et al.* [107] studied how quitting behavior changed as they varied the distribution of targets across trials. For each observer, overall target prevalence was fixed across the experiment, but different groups of observers experienced 25%, 50%, or 75% prevalence. However, the number of targets on a particular trial was drawn from a geometric distribution, adjusted such that the mean number of targets per trial was always 1. So in the 75% condition, for example, 25% of the trials had 0 targets, and 75% had 1 or more targets, up to a maximum for 4 targets on a single trial. In the 25% condition, 75% of the trials had 0 targets, but a single trial could have up to 12 targets. Critically, instead of merely discriminating target present trials from target absent trials, observers were required to localize targets with the computer mouse, and were rewarded with 15 points for each target they found. The experiment was over when the observer reached the goal of 2000 points, so observers had an incentive to acquire targets as efficiently as possible. Using a variant of MVT adapted for discrete cases [108], they found that observers in the 50% prevalence condition followed the MVT. Behavior in the lower and higher prevalence conditions departed from MVT, but much of this departure could be accounted for by assuming that observers had a prior expectation of 50% prevalence, which was corrected only slowly during the course of the experiment.

A set of experiments by Wolfe [109] aimed to closely approximate the berry foraging scenario I outlined above. Observers were presented with a simulated aerial view comprising a large set of "patches". They had to click on this view to bring up an individual patch, which comprised red and blue disks (berries) and green ovals (leaves). As in the Cain *et al.* [107] study, the task was to click on the target berries. The redder a berry was, the more likely it was to be a "good" berry; points were only given for good berries. After foraging in a given patch, observers could click on the aerial view to move on to the next patch. The global rate of acquiring targets for the environment was manipulated by varying the time to travel between patches and the difficulty of moving the mouse within a patch. Under these conditions, observers' patch leaving behavior was well-described by MVT. However, when Wolfe introduced variability into patch quality, such that target prevalence could vary from 20% to 80%, he found that only the average behavior fit MVT; when the data were broken down by prevalence, MVT no longer predicted quitting times. Instead, observers seemed to be adopting a "redness criterion": they would pick all the berries down to a particular red level, a strategy which mimicked MVT when averaged across prevalence levels. In a follow-up experiment, Wolfe eliminated the color information, so that this strategy was unavailable. Observers again did not follow MVT; instead, they appeared to be probability-matching, picking 20% of the berries in the 20% prevalence patches, and so on, though the mechanism by which they determined probability was not evident. As noted above, there is still considerable work to be done on this quitting problem.

4. Concluding remarks: Visual search and search as a stochastic process

Visual search is a varied and interesting field of enquiry. Studying visual search can provide insights into basic cognitive processes, as well as many problems of practical, social importance. However, theoretical accounts of visual search have been drawn primarily from the computer science field (e.g. [45]) or from engineering sources such as SDT [110]. The reader may have noticed that there have been no references to topics covered elsewhere in this volume on search (with the exception of foraging, see Campos chapter). The closest point of contact between the visual search literature and search as covered elsewhere in this volume (see Gelenbe chapter) is a small body of work on modeling eye movements in terms of Lévy flights [111–113].

I believe that recent developments in the field, however, point to an opening for researchers working in stochastic processes to bring their conceptual tools to bear on the theoretical problems of visual search. Optimal foraging theory [107, 109] and diffusion models [78, 80] (see Grebenkov; Rosso and Zoia) represent points of contact where the theoretical language of stochastic processing has already been translated or naturalized into the context of visual search. Therefore, I believe we could be at the beginning of a new era of collaboration between these two disciplines.

Acknowledgments

I would like to thank Carlos Mejia-Monasterio, Raul Toral, and Horacio Wio for inviting me to the workshop on "Search and Stochastic Phenomena", and Sidney Redner, Gleb Oshanin, and Ralf Metzler for inviting me to contribute to this volume. I would also like to thank Jeremy Wolfe for valuable comments on the manuscript.

References

1. Kingsley, H. Search; a function intermediate between perception and thinking. *Psychol Monogr* **35**, 16-55 (1926).
2. Ruckmick, C. A. On overlooking familiar objects. *Am J Psychol* **37**, 631-632 (1926).
3. Schneider, W. & Shiffrin, R. Controlled and automatic human information processing: I. Detection, search, and attention. *Psychol Rev* **84**, 1-66 (1977).
4. Shiffrin, R. & Schneider, W. Controlled and automatic human information processing: II. Perceptual learning, automatic attending, and a general theory. *Psychol Rev* **84**, 127-190 (1977).
5. Treisman, A. & Gelade, G. A feature-integration theory of attention. *Cogn Psychol* **12**, 97-136 (1980).
6. Wolfe, J. M. Visual search. *Curr. Biol.* **20**, R346 (2010).
7. Eckstein, M. P. Visual search: A retrospective. *J Vis* **11**, 14-14 (2011).
8. Macmillan, N. & Creelman, C. *Detection theory: A user's guide* (Lawrence Erlbaum Associates, 2004).
9. Cousineau, D. Confidence intervals in within-subject designs: A simpler solution to Loftus and Masson's method. *Tutorial in Quantitative Methods for Psychology* **1**, 42-45 (2005).
10. Morey, R. D. Confidence Intervals from Normalized Data: A correction to Cousineau (2005). *Tutorials in Quantitative Methods for Psychology* **4**, 61-64 (2008).
11. Wolfe, J. M., Palmer, E. M. & Horowitz, T. S. Reaction time distributions constrain models of visual search. *Vis Res* **50**, 1304-1311 (2010).

12. Palmer, E. M., Horowitz, T. S., Torralba, A. & Wolfe, J. M. What are the shapes of response time distributions in visual search? *J Exp Psychol Hum Percept Perform* **37**, 58-71 (2011).
13. Wolfe, J. M. & Horowitz, T. S. What attributes guide the deployment of visual attention and how do they do it? *Nat Rev Neurosci* **5**, 495-501 (2004).
14. Wang, Q., Cavanagh, P. & Green, M. Familiarity and pop-out in visual search. *Percept Psychophys* **56**, 495-500 (1994).
15. Frith, U. A curious effect with reversed letters explained by a theory of schema. *Percept Psychophys* **16**, 113-116 (1974).
16. Nothdurft, H. Faces and facial expressions do not pop out. *Perception* **22**, 1287-98 (1993).
17. Suzuki, S. & Cavanagh, P. Facial organization blocks access to low-level features: An object inferiority effect. *J Exp Psychol Hum Percept Perform* **21**, 901-913 (1995).
18. Jonides, J. & Gleitman, H. A conceptual category effect in visual search: O as letter or as digit. *Percept Psychophys* **12**, 457-460 (1972).
19. Duncan, J. Category effects in visual search: a failure to replicate the 'oh-zero' phenomenon. *Percept Psychophys* **34**, 221-32 (1983).
20. Krueger, L. The category effect in visual search depends on physical rather than conceptual differences. *Percept Psychophys* **35**, 558-64 (1984).
21. Treisman, A. & Souther, J. Search asymmetry: a diagnostic for preattentive processing of separable features. *J Exp Psychol Gen* **114**, 285-310 (1985).
22. Treisman, A. & Gormican, S. Feature analysis in early vision: evidence from search asymmetries. *Psychol Rev* (1988).
23. Rosenholtz, R. Search asymmetries? What search asymmetries? *Percept Psychophys* **63**, 476-489 (2001).
24. Treisman, A. & Sato, S. Conjunction search revisited. *J Exp Psychol Hum Percept Perform* **16**, 459-78 (1990).
25. Treisman, A. How the deployment of attention determines what we see. *Vis Cogn* **14**, 411-443 (2006).
26. Treisman, A. Search, similarity, and integration of features between and within dimensions. *J Exp Psychol Hum Percept Perform* **17**, 652-676 (1991).
27. Wolfe, J. M. "Effortless" texture segmentation and 'parallel' visual search are not the same thing. *Vis Res* **32**, 757-763 (1992).
28. Wolfe, J. M. & Gancarz, G. Guided Search 3.0: A model of visual search catches up with Jay Enoch 40 years later. In *Basic and Clinical Applications of Vision Science.* Ed. V. Lakshminarayanan (Springer, Netherlands, 1997), pp. 189-192.
29. Wolfe, J. M. Guided Search 4.0: Current progress with a model of visual search. In *Integrated models of cognitive systems*, Ed. W. D. Gray (Oxford University Press, New York, 2007), pp. 99-119.
30. Wolfe, J. M., Cave, K. & Franzel, S. Guided search: an alternative to the feature integration model for visual search. *J Exp Psychol Hum Percept Perform* **15**, 419-33 (1989).
31. Hochstein, S. & Ahissar, M. View from the Top Hierarchies and Reverse Hierarchies in the Visual System. *Neuron* **36**, 791-804 (2002).

32. Reijnen, E., Wolfe, J. M. & Krummenacher, J. Coarse guidance by numerosity in visual search. *Atten Percept Psychophys* **75**, 16-28 (2013).
33. Wolfe, J. M., Friedman-Hill, S. R., Stewart, M. I. & O'Connell, K. M. The role of categorization in visual search for orientation. *J Exp Psychol Hum Percept Perform* **18**, 34-49 (1992).
34. Aks, D. J. & Enns, J. T. Visual search for size is influenced by a background texture gradient. *J Exp Psychol Hum Percept Perform* **22**, 1467-1481 (1996).
35. Enns, J. T. & Rensink, R. Sensitivity to three-dimensional orientation in visual search. *Psychol Sci* **1**, 323-326 (1990).
36. Hershler, O. & Hochstein, S. At first sight: A high-level pop out effect for faces. *Vis Res* **45**, 1707-1724 (2005).
37. Hershler, O. & Hochstein, S. With a careful look: still no low-level confound to face pop-out. *Vis Res* **46**, 3028-35 (2006).
38. Van Rullen, R. On second glance: Still no high-level pop-out effect for faces. *Vis Res* **46**, 3017-3027 (2006).
39. Duncan, J. & Humphreys, G. W. Beyond the search surface: visual search and attentional engagement. *J Exp Psychol Hum Percept Perform* **18**, 578-88; discussion 589-93 (1992).
40. Duncan, J. & Humphreys, G. W. Visual search and stimulus similarity. *Psychol Rev* **96**, 433-58 (1989).
41. Eckstein, M. P., Thomas, J. P., Palmer, J. & Shimozaki, S. S. A signal detection model predicts the effects of set size on visual search accuracy for feature, conjunction, triple conjunction, and disjunction displays. *Percept Psychophys* **62**, 425-451 (2000).
42. Palmer, J. Set-size effects in visual search: the effect of attention is independent of the stimulus for simple tasks. *Vis Res* **34**, 1703-1721 (1994).
43. Baldassi, S. & Verghese, P. Comparing integration rules in visual search. *J Vis* **2**, 559-570 (2002).
44. Palmer, E. M., Fencsik, D. E., Flusberg, S., Horowitz, T. S. & Wolfe, J. M. Signal detection evidence for limited capacity in visual search. *Atten Percept Psychophys* **73**, 2413-2424 (2011).
45. Cave, K. R. & Wolfe, J. M. Modeling the role of parallel processing in visual search. *Cogn Psychol* **22**, 225-71 (1990).
46. Grossberg, S., Mingolla, E. & Ross, W. D. A neural theory of attentive visual search: interactions of boundary, surface, spatial, and object representations. *Psychol Rev* **101**, 470-489 (1994).
47. Koch, C. & Ullman, S. Shifts in selective visual attention: towards the underlying neural circuitry. *Hum Neurobiol* **4**, 219-27 (1985).
48. Itti, L. & Koch, C. A saliency-based search mechanism for overt and covert shifts of visual attention. *Vis Res* **40**, 1489-506 (2000).
49. Klein, R. Inhibitory tagging system facilitates visual search. *Nature* **334**, 430-431 (1988).
50. Horowitz, T. S. & Wolfe, J. M. Visual search has no memory. *Nature* **394**, 575-7 (1998).
51. Horowitz, T. S. & Wolfe, J. M. Memory for rejected distractors in visual search? *Vis Cogn* **10**, 257-298 (2003).

52. Horowitz, T. S. & Wolfe, J. M. Search for multiple targets: remember the targets, forget the search. *Percept Psychophys* **63**, 272-285 (2001).
53. Takeda, Y. Search for multiple targets: evidence for memory-based control of attention. *Psychon Bull Rev* **11**, 71-76 (2004).
54. Kristjánsson, A. In search of remembrance: evidence for memory in visual search. *Psychol Sci* **11**, 328-332 (2000).
55. von Mühlenen, A., Müller, H. & Muller, D. Sit-and-wait strategies in dynamic visual search. *Psychol Sci* (2003).
56. Horowitz, T. S. & Wolfe, J. M. Visual Search: The role of memory for rejected distractors. In *Neurobiology of Attention*, Eds. L. Itti, G. Rees & J. K. Tsotsos (Elsevier, San Diego, 2005) pp. 264-268.
57. Hulleman, J. No need for inhibitory tagging of locations in visual search. *Psychon Bull Rev* **16**, 116-120 (2009).
58. Hulleman, J. Inhibitory tagging in visual search: only in difficult search are items tagged individually. *Vis Res* **50**, 2069-2079 (2010).
59. Hulleman, J. Inhibitory tagging in visual search: only in difficult search are items tagged individually. *Vis Res* **50**, 2069-2079 (2010).
60. Peterson, M. S., Kramer, A., Wang, R. F., Irwin, D. E. & McCarley, J. S. Visual search has memory. *Psychol Sci* **12**, 287-292 (2001).
61. Peterson, M. S., Beck, M. R. & Vomela, M. Visual search is guided by prospective and retrospective memory. *Percept Psychophys* **69**, 123-35 (2007).
62. Gilchrist, I. D. & Harvey, M. Evidence for a systematic component within scan paths in visual search. *Vis Cogn* (2006).
63. McCarley, J. S., Wang, R. F., Kramer, A. & Irwin, D. How much memory does oculomotor search have? *Psychol Sci* **9**, 33-39 (2003).
64. Wolfe, J. M. What can 1 million trials tell us about visual search? *Psychol Sci* **9**, 33-39 (1998).
65. Chun, M. M. & Wolfe, J. M. Just say no: how are visual searches terminated when there is no target present? *Cogn Psychol* **30**, 39-78 (1996).
66. Schwebel, D. C., Lindsay, S. & Simpson, J. Brief report: a brief intervention to improve lifeguard surveillance at a public swimming pool. *J Pediatr Psychol* **32**, 862-8 (2007).
67. McCarley, J. S., Kramer, A. F., Wickens, C. D., Vidoni, E. D. & Boot, W. R. Visual Skills in Airport-Security Screening. *Psychol Sci* **15**, 302-306 (2004).
68. Evans, K. K., Tambouret, R. H., Evered, A., Wilbur, D. C. & Wolfe, J. M. Prevalence of Abnormalities Influences Cytologists' Error Rates in Screening for Cervical Cancer. *Arch. Pathol. Lab. Med.* **135**, 1557-1560 (2011).
69. Bindemann, M., Avetisyan, M. & Blackwell, K. Finding needles in haystacks: identity mismatch frequency and facial identity verification. *J Exp Psychol Appl* **16**, 378-86 (2010).
70. Wolfe, J. M., Horowitz, T. S. & Kenner, N. M. Rare items often missed in visual searches. *Nature* **435**, 439-40 (2005).
71. Fleck, M. S. & Mitroff, S. R. Rare targets are rarely missed in correctable search. *Psychol Sci* **18**, 943-947 (2007).

72. Van Wert, M. J., Horowitz, T. S. & Wolfe, J. M. Even in correctable search, some types of rare targets are frequently missed. *Atten Percept Psychophys.* **71**, 541-553 (2009).
73. Godwin, H. J. et al. The impact of Relative Prevalence on dual-target search for threat items from airport X-ray screening. *Acta Psychologica* **134**, 79-84 (2010).
74. Kundel, H. L., Nodine, C. F. & Carmody, D. Visual scanning, pattern recognition and decision-making in pulmonary nodule detection. *Invest Radiol* **13**, 175-81 (1978).
75. Rich, A. N. et al. Why do we miss rare targets? Exploring the boundaries of the low prevalence effect. *J Vis* **8**, 15.1-17 (2008).
76. Wolfe, J. M. et al. Low target prevalence is a stubborn source of errors in visual search tasks. *J Exp Psychol Gen* **136**, 623-638 (2007).
77. Ishibashi, K., Kita, S. & Wolfe, J. M. The effects of local prevalence and explicit expectations on search termination times. *Atten Percept Psychophys* **74**, 115-123 (2012).
78. Wolfe, J. M. & Van Wert, M. J. Varying target prevalence reveals two dissociable decision criteria in visual search. *Curr. Biol.* **20**, 121-124 (2010).
79. Reeves, A., Santhi, N. & Decaro, S. A random-ray model for speed and accuracy in perceptual experiments. *Spat Vis* **18**, 73-83 (2005).
80. Palmer, J., Huk, A. C. & Shadlen, M. N. The effect of stimulus strength on the speed and accuracy of a perceptual decision. *J Vis* **5**, 376-404 (2005).
81. Ratcliff, R. Modeling response signal and response time data? *Cogn Psychol* **53**, 195-237 (2006).
82. Moran, R., Zehetleitner, M., Müller, H. J. & Usher, M. Competitive guided search: Meeting the challenge of benchmark RT distributions. *J Vis* **13**, 1-31 (2013).
83. Levin, D. T., Angelone, B. L. & Beck, M. R. Visual search for rare targets: Distracter tuning as a mechanism for learning from repeated target-absent searches. *Br J Psychol* **102**, 313-327 (2011).
84. Benard, V. B. *et al.* Cervical Screening in the National Breast and Cervical Cancer Early Detection Program, 1995-2001. *Obstetrics & Gynecology* **103**, 564-571 (2004).
85. Schwark, J., Sandry, J., MacDonald, J. & Dolgov, I. False feedback increases detection of low-prevalence targets in visual search. *Atten Percept Psychophys* **74**, 1583-1589 (2012).
86. Maddox, W. T. Toward a unified theory of decision criterion learning in perceptual categorization. *J Exp Anal Behav* **78**, 567-595 (2002).
87. Navalpakkam, V., Koch, C. & Perona, P. Homo economicus in visual search. *J Vis* **9**, 31.1-16 (2009).
88. Egglin, T. K. P. & Feinstein, A. R. Context bias: A problem in diagnostic radiology. **276**, 1752-1755 (1996).
89. Kundel, H. L. Disease prevalence and radiological decision making. *Invest Radiol* **17**, 107-9 (1982).
90. Swensson, R. G., Hessel, S. J. & Herman, P. G. The value of searching films without specific preconceptions. *Invest Radiol* **20**, 100-14 (1985).

91. Gur, D. et al. Prevalence effect in a laboratory environment. *Radiology* **228**, 10-14 (2003).
92. Schwark, J., Sandry, J., Macdonald, J. & Dolgov, I. False feedback increases detection of low-prevalence targets in visual search. **74**, 1583-1589 (2012).
93. Gur, D. et al. The Prevalence Effect in a Laboratory Environment. *Acad Radiol* **14**, 49-53 (2007).
94. Evans, K. K., Birdwell, R. L. & Wolfe, J. M. If You Don't Find It Often, You Often Don't Find It: Why Some Cancers Are Missed in Breast Cancer Screening. *PLoS ONE* **8**, e64366 (2013).
95. Bond, A. B. Giving-up as a Poisson process: the departure decision of the green lacewing. *Anim Behav* **29**, 2, 629-630 (1981).
96. Hutchinson, J. M. C., Wilke, A. & Todd, P. M. Patch leaving in humans: can a generalist adapt its rules to dispersal of items across patches? *Anim Behav* **75**, 1331-1349 (2008).
97. Wilke, A., Hutchinson, J. M. C., Todd, P. M. & Czienskowski, U. Fishing for the Right Words: Decision Rules for Human Foraging Behavior in Internal Search Tasks. *Cognitive Science* **33**, 497-529 (2009).
98. Pirolli, P. & Card, S. Information foraging. *Psychol Rev* **106**, 643 (1999).
99. Pirolli, P. Rational analyses of information foraging on the web. *Cogn Sci* **29**, 343-73 (2005).
100. Pyke, G. H., Pulliam, H. R. & Charnov, E. L. Optimal foraging: a selective review of theory and tests. *Q Rev Biol* 137-154 (1977).
101. Charnov, E. L. Optimal foraging, the marginal value theorem. *Theor Popul Biol* **9**, 129-136 (1976).
102. Nonacs, P. State dependent behavior and the marginal value theorem. *Behav Ecol* **12**, 71-83 (2001).
103. Marshall, H. H., Carter, A. J., Ashford, A., Rowcliffe, J. M. & Cowlishaw, G. How do foragers decide when to leave a patch? A test of alternative models under natural and experimental conditions. *J Anim Ecol* **82**, 894-902 (2013).
104. Wajnberg, E., Fauvergue, X. & Pons, O. Patch leaving decision rules and the Marginal Value Theorem: an experimental analysis and a simulation model. *Behavioral Ecology* **11**, 577-586 (2000).
105. Hayden, B. Y., Pearson, J. M. & Platt, M. L. Neuronal basis of sequential foraging decisions in a patchy environment. *Nat. Neurosci.* **14**, 933-939 (2011).
106. Najemnik, J. & Geisler, W. S. Optimal eye movement strategies in visual search. *Nature* **434**, 387-91 (2005).
107. Cain, M. S., Vul, E., Clark, K. & Mitroff, S. R. A Bayesian optimal foraging model of human visual search. *Psychol. Sci.* 1-7 (2012).
108. McNamara, J. M., Green, R. F. & Olsson, O. Bayes' theorem and its applications in animal behaviour. *Oikos* **112**, 243-251 (2006).
109. Wolfe, J. M. When is it time to move to the next raspberry bush? Foraging rules in human visual search. *J Vis* **13**, 10-10 (2013).
110. Green, D. & Swets, J. *Signal detection theory and psychophysics.* (Wiley, New York, 1966).

111. Brockmann, D. & Geisel, T. The ecology of gaze shifts. *Neurocomputing*, **32**, 643-650 (2000).
112. Boccignone, G. & Ferraro, M. Modelling gaze shift as a constrained random walk. *Physica A: Statistical Mechanics and its Applications* **331**, 207-218 (2004).
113. Credidio, H. F., Teixeira, E. N., Reis, S. D., Moreira, A. A. & Andrade, J. S., Jr. Statistical patterns of visual search for hidden objects. *Scientific Reports* **2**, 1-6 (2012).

Chapter 17

Statistical Physics of Evolutionary Trajectories on Fitness Landscapes

Michael Manhart and Alexandre V. Morozov*

Department of Physics and Astronomy &
BioMaPS Institute for Quantitative Biology,
Rutgers University, Piscataway, NJ 08854, USA

Random walks on multidimensional landscapes are important to many areas of science and engineering. In particular, properties of adaptive first-passage trajectories on fitness landscapes determine population fates and thus play a central role in evolutionary biology. The topography of fitness landscapes and its effect on evolutionary dynamics have been extensively studied in the literature. We will survey the current knowledge in this field, focusing on a recently developed systematic approach to characterizing path lengths, mean times, and other statistics of the first-passage path ensemble. This approach, based on general techniques from statistical physics, is applicable to landscapes of arbitrary complexity and structure. It is especially well-suited to quantifying the diversity of stochastic trajectories and repeatability of evolutionary events. We demonstrate this methodology using a biophysical model of protein evolution that describes how proteins maintain folding stability while evolving new functions.

1. Introduction

Random walks on networks are ubiquitous in nature. As an example, consider proteins, macromolecules that carry out a myriad of chemical and mechanical functions inside cells [1]. Each protein is a chain of amino acids chemically bonded to make a linear polypeptide [2], and the sequence of amino acids determines the protein fold — a compact 3D conformation which has the minimum free energy. Unlike random heteropolymers, naturally-occuring proteins have unique folds that they achieve robustly and, in many cases, rapidly (on the time scales of micro- or milliseconds),

*Corresponding author: morozov@physics.rutgers.edu

starting from arbitrary unfolded conformations [3].

Proteins are produced in the unfolded state inside the cell and have to fold before they can function. Protein conformations are often represented by sets of dihedral angles (torsion angles defined by three chemical bonds connecting four atoms [3]). Although in general the values of dihedral angles are continuous, they are typically discretized in protein structure prediction algorithms. In this case, protein folding can be viewed as a random walk on a network of conformation states with connectivity defined by the move set — a set of instructions for changing the dihedral angles in each step.

The network is very high-dimensional. For example, for a relatively small protein with $L = 100$ amino acids, 2 dihedral angles per amino acid, and 10° dihedral angle increments, there are 36^{200} possible conformations. With a simple move set that updates one angle at a time, each node is connected to $200 \times (36 - 1) = 7000$ neighbors. In such a large space, how can a protein reach its unique folded shape on reasonable time scales? This problem is known as the Levinthal paradox [4], and key to its resolution is the idea of the protein folding landscape [5, 6]. Each protein conformation has a free energy which is a function of the 200 dihedral angles, forming a free energy landscape over the network. The free energy values at a node and its neighbors determine rates of transition between nodes, e.g., according to the Metropolis algorithm [7]. This landscape is believed to have a global funnel shape, which allows the protein to find its folded structure efficiently through incremental moves, without searching the entire space [6].

This picture generalizes to many other search problems on networks in which each node, corresponding to a discrete (or discretized) state of the system, can be assigned a value of the objective function which sets the transition rates. As with protein folding, a major question is how the landscape topography and the move set determine the dynamics, especially first-passage processes. For example, an important quantity of interest is the mean first-passage time (e.g., to the global minimum on the protein folding landscape), which should be minimal in optimized algorithms [8].

The effect of landscape topography on dynamics is of particular importance in evolutionary theory, the study of how populations of organisms change over time through mutation and natural selection [9]. The genotype (genetic state) of an organism is represented by a sequence σ of letters drawn from an alphabet of size k. The sequence may represent nucleotides in genomic DNA ($\{\mathsf{A}, \mathsf{C}, \mathsf{G}, \mathsf{T}\}$, $k = 4$), amino acids in a protein ($k = 20$), or the binary presence/absence of a mutation at several genes across the genome ($k = 2$). Assuming a fixed number L of sites in each sequence, the space of all k^L possible sequences represents a network with sequence nodes connected to each other if they differ by a mutation at a single site [10]. For simplicity we neglect recombination between sequences and insertion/deletion of sites which would redefine network connectivity and in some cases the total number of nodes.

Each sequence σ can be assigned a fitness value $\mathcal{F}(\sigma)$ that characterizes the reproductive success of an organism with that sequence. The exact definition of fitness can vary widely across different contexts, often depending on the construction of a model or what is observable in an experiment [11, 12]. Here we use a general theoretical definition of fitness as the relative probability an individual with that sequence will survive to reproduce [13]. This probabilistic definition of fitness is sometimes known as *multiplicative fitness*. In some circumstances it is more useful to consider $\log \mathcal{F}$, or *additive fitness*, which is related to growth rate.

Either case defines a fitness landscape or, more precisely, a genotypic fitness landscape [14]. Just as the folding landscape's structure is key to a protein's ability to reach its folded state efficiently, the fitness landscape is key to understanding how complex biological structures, such as bacterial flagella or the human eye, can arise through random, incremental mutations [10, 15]. Evolutionary adaptation, therefore, is represented by first-passage trajectories leading to local or global maxima on the fitness landscape. Characterizing the statistical properties of these first-passage trajectories is a major goal for evolutionary theory.

1.1. *Evolutionary dynamics*

In general, individuals in a population will have different sequences, occupying a distribution of points on the fitness landscape. However, in the limit $u \ll (LN \log N)^{-1}$ [16, 17], where u is the mutation rate (defined as the probability of mutation per site per generation) and N is an effective population size [13, 18], new mutations arise individually and either fix in the population or disappear from it on time scales that are short compared with the times between successive mutations [19–21]. Thus the population is monomorphic and, apart from short transition periods, occupies a single point in sequence space. The stochastic process of a new mutant appearing and fixing in the population is known as substitution, and the substitution rate from sequence σ to σ' is given by [19]

$$\langle \sigma' | \mathbf{W} | \sigma \rangle = N u \phi(s), \tag{1}$$

where Nu is the total number of new mutations in the population per site per generation, and $\phi(s)$ is the probability of a single σ' mutant fixing in a population of σ when the selection coefficient is $s = \mathcal{F}(\sigma')/\mathcal{F}(\sigma) - 1$ ($s > 0$ for beneficial mutations, $s < 0$ for deleterious ones).

The exact form of the fixation probability $\phi(s)$ depends on the underlying population dynamics. However, there are some common approximations valid in different asymptotic regimes [22]. For example, when $N \gg 1$ and $|s| \gg 1$, all beneficial mutations are essentially guaranteed to fix, while deleterious ones are guaranteed to be eliminated. Similar to zero-temperature Monte Carlo, the population can only undergo substitutions that increase fitness, and all allowed substitutions occur with the same rate Nu (since for $|s| \gg 1$, $\phi(s) \approx 1$ when $s > 0$ and $\phi(s) \approx 0$ when $s < 0$). Thus adaptation follows trajectories on the landscape along which fitness increases monotonically. This approximation is common for studying dynamics on model landscapes (see Sec. 1.5).

A different approximation holds when $N \gg 1$ but $N^{-1} \ll |s| \ll 1$. In this case, $\phi(s) \approx s$ for $s > 0$ and $\phi(s) \approx 0$ for $s < 0$ [13]. Thus deleterious mutations always get eliminated as before, but beneficial mutations fix at the rate Nus [23]. The true dynamics of real populations may not always fall into these special cases, instead involving more complex dynamics such as interference between multiple simultaneous mutations [24]. However, the simplified evolutionary dynamics described here are useful when our main objective is to qualitatively understand the role of the fitness landscape in constraining evolution over long time scales, rather than fine-grained details of short-time population dynamics.

1.2. *Epistasis*

The most basic aspect of fitness landscape topography is known as *epistasis*. Let the sequence σ be $\sigma^1\sigma^2\ldots\sigma^L$, where σ^μ is the letter at site $\mu \in \{1, \ldots, L\}$. In general the multiplicative fitness function $\mathcal{F}(\sigma)$ cannot be decomposed into a product of independent contributions from each site μ or, equivalently, the additive fitness $\log \mathcal{F}(\sigma)$ cannot be decomposed into a sum. This means that the fitness effect of a mutation at a given site may depend on the state of other sites. If this is true, the sites will be correlated, which can be thought of as a coupling between the sites. Mathematically this is reminiscent of a Hamiltonian for a system of interacting particles.

Epistasis is precisely this interactive coupling in the context of genotypic sequences. Following convention, we use additive fitness here ($\log \mathcal{F}$), and

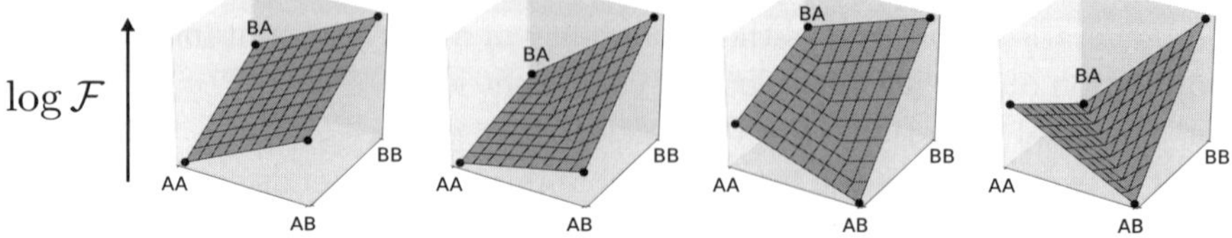

Fig. 1. The four qualitative types of epistasis for a two-letter, two-site model. From left to right: *no epistasis*, where each mutation has the same effect on additive fitness regardless of the state of the other site, yielding a linear landscape; *magnitude epistasis*, where the magnitude (but not the sign) of the additive fitness effect of a mutation depends on the other site; *sign epistasis*, where the sign of a mutation's fitness effect (beneficial or deleterious) depends on the other site; *reciprocal sign epistasis*, where multiple instances of sign epistasis can lead to local fitness maxima.

categorize types of epistasis according to the qualitative differences in the additive fitness effects of mutations. We summarize the four possible cases using a two-letter, two-site model in Fig. 1 in which sequence AA evolves into sequence BB, which has the highest fitness. In the first case on the left of Fig. 1, there is no epistasis: the fitness effect of A $\to$ B substitution at site 2 is the same regardless of the state of site 1, and vice versa. Thus the additive fitness can be decomposed into a sum of contributions from each site: $\log \mathcal{F}(\sigma) = \log \mathcal{F}_1(\sigma^1) + \log \mathcal{F}_2(\sigma^2)$, i.e., the additive fitness landscape is linear in sequence space. Under the aforementioned strong-selection evolutionary dynamics, both trajectories from AA to the global maximum at BB are accessible.

In the second case of Fig. 1, the additive fitness effect of A $\to$ B at site 2 differs in magnitude but not in sign depending on whether site 1 has A or B. This situation is known as *magnitude epistasis* [25, 26]. Magnitude epistasis does not completely block any trajectories, but affects quantitative aspects of dynamics such as adaptation times. Note that there are two kinds of magnitude epistasis: one in which the fitness benefit of a mutation is enhanced by the presence of other mutations (as shown in the second panel on the left of Fig. 1), and the "diminishing returns" case in which the fitness benefit is decreased by other mutations.

The third case of Fig. 1 shows how the A $\to$ B substitution at site 2 can have opposite effects on fitness depending on the state of site 1: it is deleterious if σ^1 = A, but beneficial if σ^1 = B. Since the sign of the fitness effect depends on the other site, the situation is known as *sign epistasis*. Sign epistasis can significantly affect accessibility of genotypes on the landscape by blocking trajectories: under strong-selection dynamics, the trajectory AA → AB → BB is unavailable since it requires a deleterious substitution. When sign epistasis exists at multiple sites, it is known as *reciprocal sign epistasis*, as shown in the fourth case of Fig. 1. Reciprocal sign epistasis is a necessary condition for the existence of multiple local maxima [27]. As the examples in Fig. 1 show, epistasis underlies landscape ruggedness that can constrain evolutionary trajectories. Thus the existence and nature of epistasis is of prime interest in evolution.

1.3. *Measures of landscape ruggedness and accessibility*

Numerous measures have been proposed to quantify epistatic ruggedness of fitness landscapes and accessibility of evolutionary trajectories (summarized in [12]). One commonly-used measure is the number of local fitness maxima, which is indicative of the presence and type of epistasis [27]: more local maxima indicates a more rugged or epistatic landscape. For binary alphabets, deviations of the additive fitness function from linearity can be quantified by fitting a linear function and calculating the sum of squares of residuals, known as a roughness parameter [12]. A more local measure of ruggedness can be obtained by considering all pairs of sites and all pairs of possible letters at those sites, and then determining the sub-landscape for each combination like those shown in Fig. 1. Each sub-landscape can then be classified into one of the four categories of epistasis.

Other measures consider accessibility and other properties of the first-passage trajectories themselves, especially those leading to the global fitness maximum. For example, without epistasis all first-passage trajectories to the global maximum from any other point on the landscape are accessible under strong-selection dynamics, but with sign epistasis some trajectories become blocked. The distributions of first-passage trajectory times and lengths are important for understanding the effect of landscape ruggedness on adaptation [28–32].

1.4. *Repeatability of evolution and diversity of evolutionary trajectories*

Landscape ruggedness is especially relevant in its effect on the repeatability of evolution, a question of paramount importance in biology [33]. If "life's tape" could be replayed, would we see a completely different outcome because evolution is a largely stochastic phenomenon, or are accessible evolutionary trajectories so constrained that the outcome would have been the same or recognizably similar [15]?

Discussing this question in general entails many issues, including environmental conditions, initial conditions, and details of population dynamics. One specific approach focuses on the diversity of first-passage trajectories leading from an ancestral state to a particular descendant state or set of states. One assessment of this diversity is simply counting the number of such trajectories. For example, Weinreich and co-workers found that only a small fraction of all possible trajectories from wild-type *E. coli* to a strain resistant to antibiotics was accessible to adaptation [34, 35]. In another approach, Koonin and co-workers devised a measure called mean path divergence [33, 36]:

$$\mathcal{D} = \sum_{\varphi_1 \neq \varphi_2} \Delta(\varphi_1, \varphi_2) p(\varphi_1) p(\varphi_2), \tag{2}$$

where the sum is over all pairs of distinct paths in an ensemble, $p(\varphi)$ is the probability of path φ, and $\Delta(\varphi_1, \varphi_2)$ is the path distance between φ_1 and φ_2. The path distance is defined as the average of the shortest Hamming distances between each sequence σ_1 on path φ_1 and all sequences on path φ_2, and vice versa:

$$\Delta(\varphi_1, \varphi_2) = \frac{1}{\mathcal{L}[\varphi_1] + \mathcal{L}[\varphi_2]} \left(\sum_{\sigma_1 \in \varphi_1} \operatorname*{argmin}_{\sigma_2 \in \varphi_2} d(\sigma_1, \sigma_2) + \sum_{\sigma_2 \in \varphi_2} \operatorname*{argmin}_{\sigma_1 \in \varphi_1} d(\sigma_2, \sigma_1) \right), \tag{3}$$

where $\mathcal{L}[\varphi]$ is the length (number of jumps) of path φ and $d(\sigma_1, \sigma_2)$ is the Hamming distance between σ_1 and σ_2. The divergence therefore captures not only how many paths are available, but weighs them by their proximity in sequence space. Other measures of path diversity, such as path entropy and the distribution of path lengths and times [32], are discussed in Secs. 2 and 3.

1.5. *Model and empirical landscapes*

A few simple models have traditionally dominated theoretical studies of fitness landscapes and evolutionary trajectories. These models serve as useful null hypotheses or limits of more complex scenarios; they generally consider sequences with binary alphabets ($k = 2$), in which case sequence space is a unit hypercube. Without attempting to account for every landscape proposed in the literature, we will discuss and motivate several popular choices. In Kauffman's NK (or "LK") model [31, 37, 38], each of the L sites in the gene (or genes in the genome) interacts with K other sites chosen by random sampling. The additive fitness of genotype σ is given by

$$\log \mathcal{F}(\sigma) = \sum_{\mu=1}^{L} \log \mathcal{F}_\mu(\sigma^\mu, \sigma^{n_1(\mu)}, \ldots, \sigma^{n_K(\mu)}), \tag{4}$$

where $n_1(\mu), \ldots, n_K(\mu)$ are the randomly-chosen interaction partners of site μ. The single-site fitnesses $\mathcal{F}_\mu$ are obtained by sampling from a continuous distribution; each combination of 2^{K+1} possible states of the argument corresponds to an independent sampling. When $K = 0$, the NK landscape becomes fully additive and thus non-epistatic. Because in this limit the landscape is smooth and has a single maximum, it is sometimes called the "Mount Fuji" model [39]. The amount of epistasis, or landscape ruggedness, can be tuned by increasing K to the maximum value of $L - 1$. With $K = L - 1$, the fitnesses of different sequences are uncorrelated; this model is called the "House of Cards" [40] due to the unpredictable fitness effects of mutations. Realistically, closely-related genotypes should have correlated fitnesses, so this limit serves mainly as a null model. Many results are known for the NK landscape and its adaptive first-passage trajectories [28, 30, 37, 41, 42]. For example, in the $K = L - 1$ limit the average number of local maxima is $k^L/(L(k-1)+1)$ for any alphabet size k [12].

Another class of models starts from a non-epistatic landscape and adds a tunable amount of ruggedness to it. For example, in the "rough Mount Fuji" model [39], sequence σ_0 is arbitrarily picked as the global maximum and the fitness of sequence σ is given by

$$\log \mathcal{F}(\sigma) = \eta(\sigma) - \theta d(\sigma, \sigma_0), \tag{5}$$

where $d(\sigma, \sigma_0)$ is the Hamming distance between sequences σ and σ_0, θ is the parameter which controls the slope of the smooth part of the landscape, and $\eta(\sigma)$ is a zero-mean random variable sampled independently for each sequence σ. The ruggedness of the landscape is controlled by the ratio of θ and the standard deviation of the distribution from which the random variables $\eta(\sigma)$ are sampled.

The landscapes described above are dominated by selection. Another approach to evolution is based on the neutral theory, which postulates that the majority of mutations have either no phenotypic effect (i.e., are selectively neutral) or are strongly deleterious and thus rapidly removed from the population [19]. This picture leads to evolution on a neutral network where all viable nodes have the same fitness [10, 43]. With some probability, a viable individual can acquire a lethal mutation from which it cannot recover, and will disappear from the population. Thus the population as a whole can only make transitions between viable, selectively-neutral states. Evolution on such a landscape is reminiscent of the percolation problem [44]: each node is assigned fitness 1 with probability p and fitness zero with probability $1 - p$, independent of the other nodes [31].

Due to the enormous number of sequences involved, the structure of fitness landscapes is difficult to probe experimentally. Typically, only a small number of sites is studied (approximately 4 to 9, summarized in [12]), and at those sites only a subset of all possible mutations is introduced, resulting in fitness measurements for tens or hundreds of different genotypes. In addition, because genotype survivability is not directly accessible in experiments, proxy measures of fitness are employed, such as growth rates and antibiotic resistance. Although these empirical studies can be used to probe the local structure of the landscapes, they are insufficient for analyzing the global properties of adaptive trajectories because adaptation may involve mutations outside of the experimentally-probed subset.

Nevertheless, many studies have attempted to characterize empirical landscapes in terms of their epistatic features, accessibility, and correspondence to theoretical models [12, 25, 31, 34–36, 45–47]. For example, magnitude and sign epistasis have been observed, as well as significant constraints on evolutionary trajectories [34, 35, 46, 47]. One general finding of such studies is that empirical landscapes include some epistasis, but are far from the House of Cards regime in which all fitness values are completely uncorrelated [25]. The emerging picture is closer to the rough Mount Fuji model, which includes a limited amount of epistasis around a mostly linear landscape [12].

2. Statistical physics of stochastic paths

Analytical treatments of evolutionary dynamics on fitness landscapes are typically restricted to uncorrelated or highly symmetric models, such as those described in Sec. 1.5. Simulations, meanwhile, can suffer from numerical inaccuracy and may be computationally expensive when rare events are considered. More systematic tools are necessary, especially tools that directly address statistical properties of stochastic paths that are relevant for understanding the diversity of evolutionary pathways.

Physics and chemistry have long grappled with similar problems in the field of reaction rate theory [48], which studies rare transitions between metastable states that model phenomena ranging from protein folding [3] to chemical reactions [49]. In these systems, quantities of interest include not only mean first-passage times and reaction rates but also the spatial distribution of transition paths and identification of kinetic bottlenecks.

Transition state theory is a well-known approach to these problems; however, it relies on the existence and *a priori* identification of key transition states [48]. A more recent development has been transition path sampling [49–54], in which paths are directly sampled via Monte Carlo to estimate their statistical properties. Similar methods have been used in phylogenetic analysis of protein sequences [55–59]. These techniques are based on a finite sample of paths and do not provide natural cutoffs for the size of the sample, which may lead to noisy estimates of various path statistics. Another technique, called transition path theory [60–64], relies on explicit solutions to the Kolmogorov backward equation. This approach, though more systematic, does not directly address the diversity of paths.

Here we discuss a general statistical physics treatment of stochastic paths that provides many useful tools for analyzing evolutionary models and other stochastic processes [32]. A semi-Markov process (i.e., continuous-time random walk [65]) on the discrete state space $\mathcal{S}$ consists of jumps between states and continuous-time waiting within states; the jump process is memoryless, but the waiting process need not be. Thus the process is defined by a set of jump probabilities, $\langle\sigma'|\mathbf{Q}|\sigma\rangle$ for the jump $\sigma \to \sigma'$ ($\sigma, \sigma' \in \mathcal{S}$), and waiting time distributions $\psi_\sigma(t)$, which is the probability density of waiting exactly time t in state σ before jumping out. We assume that $\psi_\sigma(t)$ has finite mean $w(\sigma)$ for all $\sigma \in \mathcal{S}$. For fully Markov processes with memoryless waiting, $\psi_\sigma(t) = e^{-t/w(\sigma)}/w(\sigma)$; non-exponential $\psi_\sigma(t)$ may arise due to coarse-graining a Markov process [66–68]. Note that the space $\mathcal{S}$ equipped with the jump matrix $\mathbf{Q}$ defines a network with directed and weighted edges.

Define a trajectory as a path through state space $\varphi = \{\sigma_0, \sigma_1, \ldots, \sigma_\ell\}$ combined with a set of intermediate waiting times $\{t_0, t_1, \ldots, t_{\ell-1}\}$, where t_i is the waiting time in σ_i. We consider the trajectory finished once it reaches the final state σ_ℓ, and thus do not count the waiting time in that state. The probability functional of starting in the initial state σ_0 and completing the path φ no later than time t is

$$\Pi[\varphi, t] = \pi(\sigma_0) \left(\prod_{i=0}^{\ell-1} \langle \sigma_{i+1} | \mathbf{Q} | \sigma_j \rangle \right) \times \left(\prod_{i=0}^{\ell-1} \int_0^\infty dt_i \ \psi_{\sigma_i}(t_i) \ \Theta \left(t - \sum_{i=0}^{\ell-1} t_i \right) \right), \tag{6}$$

where the first factor is the initial state probability $\pi(\sigma_0)$, the second is the product of jump probabilities, the third integrates over waiting times, and the fourth constrains the total waiting time to be less than t (Θ is the Heaviside step function). In the $t \to \infty$ limit we obtain the probability of the path φ for any duration,

$$\Pi_\infty[\varphi] = \pi(\sigma_0) \prod_{i=0}^{\ell-1} \langle \sigma_{i+1} | \mathbf{Q} | \sigma_i \rangle, \tag{7}$$

which is just the product of jump probabilities. In the time-dependent case, the Laplace transform of Eq. (6) results in a simpler expression through deconvolution [69]:

$$\tilde{\Pi}[\varphi, s] = \frac{\pi(\sigma_0)}{s} \prod_{i=0}^{\ell-1} \langle \sigma_{i+1} | \mathbf{Q} | \sigma_i \rangle \ \tilde{\psi}_{\sigma_i}(s), \tag{8}$$

where $\tilde{\psi}_{\sigma_i}(s)$ is the Laplace transform of $\psi_{\sigma_i}(t)$. For Markov processes, Eq. (8) becomes [70]

$$\tilde{\Pi}[\varphi, s] = \frac{\pi(\sigma_0)}{s} \prod_{i=0}^{\ell-1} \frac{\langle \sigma_{i+1} | \mathbf{Q} | \sigma_i \rangle}{1 + s w(\sigma_i)}. \tag{9}$$

2.1. *Path ensemble averages*

The distribution $\Pi[\varphi, t]$ in principle contains all statistical information on a set of paths. However, direct analysis of this distribution is typically prohibitive due to the high dimensionality of path space. The simplest alternative entails taking averages of various path properties over this distribution. Let Φ be an ensemble of paths that defines some dynamical process; for example, this may be all first-passage paths from a set of initial states $\mathcal{S}_i$ to a set of final states $\mathcal{S}_f$. The partition function for this ensemble is

$$\mathcal{Z}_\Phi(t) = \sum_{\varphi \in \Phi} \Pi[\varphi, t], \tag{10}$$

which represents the total probability of reaching $\mathcal{S}_f$ from $\mathcal{S}_i$ by time t via paths in Φ.

We define the following path functionals:

$$\mathcal{L}[\varphi] = \text{length (number of jumps) of } \varphi, \quad \mathcal{I}_\sigma[\varphi] = \begin{cases} 1 \text{ if } \sigma \in \varphi, \\ 0 \text{ otherwise,} \end{cases}$$

$$\mathcal{T}[\varphi] = \sum_{i=0}^{\ell-1} w(\sigma_i), \qquad \mathcal{T}_\sigma[\varphi] = \sum_{i=0}^{\ell-1} \delta_{\sigma,\sigma_i} w(\sigma_i), \tag{11}$$

where δ is the Kronecker delta. We can now express various path statistics as averages of these functionals over the ensemble, conditioned on completing the process by time t. For example, the average time of paths is given by [53]

$$\bar{\tau}_\Phi(t) = \langle \mathcal{T}(t) \rangle_\Phi = \frac{1}{\mathcal{Z}_\Phi(t)} \sum_{\varphi \in \Phi} \mathcal{T}[\varphi] \Pi[\varphi, t]. \tag{12}$$

The distribution of path lengths is given by

$$\rho_\Phi(\ell, t) = \frac{1}{\mathcal{Z}_\Phi(t)} \sum_{\varphi \in \Phi} \delta_{\ell, \mathcal{L}[\varphi]} \Pi[\varphi, t], \tag{13}$$

from which the average length $\bar{\ell}_\Phi(t) = \langle \mathcal{L}(t) \rangle_\Phi$ and standard deviation of length $\ell_\Phi^{\text{sd}}(t)$ are readily obtained.

Averages over state-dependent functionals can be used to characterize the spatial structure of paths. For example, the fraction of time paths spend in a state σ can be expressed as $\langle \mathcal{T}_\sigma(t) \rangle_\Phi / \bar{\tau}_\Phi(t)$; this is a normalized distribution over all states $\sigma \in \mathcal{S}$ and therefore it represents the density of states on the paths in the ensemble Φ. The quantity $\langle \mathcal{T}_\sigma(t) \rangle_\Phi / w(\sigma)$ gives the average number of visits to state σ. The probability that a path will visit a state σ at all is given by $\langle \mathcal{I}_\sigma(t) \rangle_\Phi$, which we will refer to as the density of paths in the ensemble Φ. We can also construct the two-point correlation function $\langle \mathcal{I}_{\sigma'}(t) \mathcal{I}_\sigma(t) \rangle_\Phi$, which gives the probability of paths passing through both σ and σ'.

In many cases we are interested in the time-independent versions of these quantities, i.e., statistical properties of paths taking any amount of time to finish. These can be obtained as the $t \to \infty$ limit of the above expressions, which amounts to replacing $\Pi[\varphi, t]$ (Eq. (6)) with $\Pi_\infty[\varphi]$ (Eq. (7)). We will denote these time-independent properties by simply omitting the time dependence, e.g., $\lim_{t\to\infty} \bar{\tau}_\Phi(t) = \bar{\tau}_\Phi$. The convergence of these limits depends on the path ensemble Φ. For example, if Φ includes all possible paths connecting the initial and final states, including those that visit the final state multiple times, then these limits generally diverge: typically, there is no finite average time or length for these paths. However, restriction to first-passage paths in Φ, as is our focus here, guarantees convergence.

This formalism also allows for development of path thermodynamics. The entropy of the path ensemble is given by

$$\begin{aligned} S_\Phi(t) &= -\frac{1}{\mathcal{Z}_\Phi(t)} \sum_{\varphi\in\Phi} \Pi[\varphi, t] \log\left(\frac{\Pi[\varphi, t]}{\mathcal{Z}_\Phi(t)}\right) \\ &= -\langle \log \Pi(t)\rangle_\Phi + \log \mathcal{Z}_\Phi(t). \end{aligned} \tag{14}$$

Indeed, if we define the path Hamiltonian to be

$$\mathcal{H}[\varphi, t] = -\log(\Pi[\varphi, t]), \tag{15}$$

(so that $\Pi[\varphi, t] = e^{-\mathcal{H}[\varphi,t]}$), we can express the path ensemble free energy as

$$F_\Phi(t) = \langle \mathcal{H}(t)\rangle_\Phi - S_\Phi(t) = -\log \mathcal{Z}_\Phi(t). \tag{16}$$

The partition function $\mathcal{Z}_\Phi(t)$ monotonically increases with time. Therefore the free energy $F_\Phi(t)$ monotonically decreases as $t \to \infty$, corresponding to equilibration of the path ensemble.

For recurrent processes (i.e., where the system will almost surely reach the final states eventually [67]), $\lim_{t\to\infty} \mathcal{Z}_\Phi(t) = \mathcal{Z}_\Phi = 1$, and hence equilibrium free energy is zero. In these cases, equilibrium path entropy is equal to the average Hamiltonian. If the ensemble Φ consists of only a single path with nonzero probability, its entropy is $S_\Phi = 0$. This situation may arise if a landscape is so constrained that only a single viable pathway exists between the initial and final states. In contrast, consider a purely random walk on a homogeneous network with γ nearest neighbors per node. The jump probability between any pair of neighboring nodes is thus γ^{-1}, so any path φ has probability $\Pi_\infty[\varphi] = \gamma^{-\mathcal{L}[\varphi]}$, and the entropy of the ensemble is given by

$$S_\Phi = -\langle \log \Pi_\infty \rangle_\Phi = \bar{\ell}_\Phi \log \gamma. \tag{17}$$

Note that path entropy and the average path Hamiltonian scale with the average path length, which defines a notion of extensivity in the path ensemble. This is sensible if we think of a path as a gas of particles, where each jump in the path corresponds to a particle. The path ensemble, which includes paths of many lengths, therefore is equivalent to the grand canonical ensemble of the gas. In the case of the gas, extensive quantities like entropy and energy scale with the number of particles, and hence these quantities here scale with the path length.

2.2. *Numerical algorithm*

The factorized form of the path probability distribution functional (Eqs. (7)–(9)) permits efficient calculation of path ensemble averages via a recursive algorithm [32]. Here for simplicity we consider the time-independent case, and thus assume that Φ consists of first-passage paths to guarantee convergence of path averages. Let $|\sigma\rangle$ be the vector with 1 at position σ and zero otherwise, and let $|\pi\rangle = \sum_\sigma \pi(\sigma)|\sigma\rangle$ be the vector of initial state probabilities. For each jump ℓ and intermediate state σ, we calculate $P_\ell(\sigma) = \langle\sigma|\mathbf{Q}^\ell|\pi\rangle$, the total probability of all paths that end at σ in ℓ jumps; $T_\ell(\sigma)$, the total average time of all such paths; and $\Gamma_\ell(\sigma)$, the total entropy of all such paths. These quantities obey the following recursion relations:

$$\begin{aligned} P_\ell(\sigma') &= \sum_{\text{nn } \sigma \text{ of } \sigma'} \langle\sigma'|\mathbf{Q}|\sigma\rangle P_{\ell-1}(\sigma), \\ T_\ell(\sigma') &= \sum_{\text{nn } \sigma \text{ of } \sigma'} \langle\sigma'|\mathbf{Q}|\sigma\rangle \left[T_{\ell-1}(\sigma) + w(\sigma)P_{\ell-1}(\sigma)\right], \\ \Gamma_\ell(\sigma') &= \sum_{\text{nn } \sigma \text{ of } \sigma'} \langle\sigma'|\mathbf{Q}|\sigma\rangle \left[\Gamma_{\ell-1}(\sigma) - \log\langle\sigma'|\mathbf{Q}|\sigma\rangle P_{\ell-1}(\sigma)\right], \end{aligned} \tag{18}$$

where $P_0(\sigma) = \pi(\sigma)$ and $T_0(\sigma) = \Gamma_0(\sigma) = 0$ for all $\sigma \in \mathcal{S}$, and the sums run over all nearest neighbors (nn) σ of σ'. The final states $\sigma \in \mathcal{S}_f$ are treated as absorbing to ensure that only first-passage paths are counted. This procedure can be considered a generalization of the exact-enumeration algorithm of [71]. Path ensemble averages are then given by

$$\mathcal{Z}_\Phi = \sum_{\ell=1}^{\infty} \sum_{\sigma \in \mathcal{S}_f} P_\ell(\sigma), \qquad \rho_\Phi(\ell) = \frac{1}{\mathcal{Z}_\Phi} \sum_{\sigma \in \mathcal{S}_f} P_\ell(\sigma), \tag{19}$$

$$\bar{\tau}_\Phi = \frac{1}{\mathcal{Z}_\Phi} \sum_{\ell=1}^{\infty} \sum_{\sigma \in \mathcal{S}_f} T_\ell(\sigma), \qquad S_\Phi = \frac{1}{\mathcal{Z}_\Phi} \sum_{\ell=1}^{\infty} \sum_{\sigma \in \mathcal{S}_f} \Gamma_\ell(\sigma).$$

We can similarly calculate state-dependent quantities such as $\langle \mathcal{I}_\sigma \rangle_\Phi$ and $\langle \mathcal{T}_\sigma \rangle_\Phi$. The two quantities to be recursively updated are $P_\ell(\sigma';\sigma)$, the total probability of all paths currently at σ' at jump ℓ that have visited σ at least once previously, and $T_\ell(\sigma';\sigma)$, the total average time that all such paths have spent in σ. These obey the following recursion relations:

$$P_\ell(\sigma';\sigma) = \begin{cases} \sum_{\text{nn } \sigma'' \text{ of } \sigma'} \langle \sigma' | \mathbf{Q} | \sigma'' \rangle P_{\ell-1}(\sigma'';\sigma), & \sigma' \neq \sigma, \\ P_\ell(\sigma), & \sigma' = \sigma, \end{cases} \tag{20}$$

$$T_\ell(\sigma';\sigma) = \sum_{\text{nn } \sigma'' \text{ of } \sigma'} \langle \sigma' | \mathbf{Q} | \sigma'' \rangle [T_{\ell-1}(\sigma'';\sigma) + \delta_{\sigma,\sigma''} w(\sigma'') P_{\ell-1}(\sigma'';\sigma)],$$

with the initial conditions $P_0(\sigma';\sigma) = T_0(\sigma';\sigma) = 0$ for all $\sigma, \sigma' \in \mathcal{S}$, $\sigma \neq \sigma'$ ($P_0(\sigma;\sigma) = \pi(\sigma)$, $T_0(\sigma;\sigma) = 0$). Averages are then expressed as

$$\langle \mathcal{I}_\sigma \rangle_\Phi = \frac{1}{\mathcal{Z}_\Phi} \sum_{\ell=1}^{\infty} \sum_{\sigma' \in \mathcal{S}_f} P_\ell(\sigma';\sigma), \qquad \langle \mathcal{T}_\sigma \rangle_\Phi = \frac{1}{\mathcal{Z}_\Phi} \sum_{\ell=1}^{\infty} \sum_{\sigma' \in \mathcal{S}_f} T_\ell(\sigma';\sigma). \tag{21}$$

Furthermore, we can calculate mean path divergence that characterizes the spatial diversity of the paths in Φ:

$$\mathcal{D}_\Phi = \sum_{\ell=1}^{\infty} \sum_{\sigma,\sigma' \in \mathcal{S}} d(\sigma,\sigma') P_\ell(\sigma) P_\ell(\sigma'), \tag{22}$$

where $d(\sigma,\sigma')$ is a distance metric on $\mathcal{S}$. This definition is distinct from that proposed in [33, 36] (Eq. (2)) in that it dynamically calculates distances between points on paths as they propagate, rather than comparing the minimal distance between complete paths. Thus for a path that revisits some states multiple times, the divergence with a path that travels through the same set of states without revisiting any of them will be zero according to Eq. (2), but nonzero with the definition in Eq. (22).

This algorithm allows for very general definitions of the path ensemble Φ without having to explicitly enumerate all paths. For instance, Φ can include paths that begin and end at arbitrary sets of states, or are prohibited from passing through arbitrary sets of intermediate states. The time complexity of the algorithm is $\mathcal{O}(\gamma N \Lambda)$ for $\mathcal{Z}_\Phi$, $\rho_\Phi(\ell)$, $\bar{\tau}_\Phi$, S_Φ, and $\mathcal{O}(\gamma N^2 \Lambda)$ for $\langle \mathcal{I}_\sigma \rangle_\Phi$, $\langle \mathcal{T}_\sigma \rangle_\Phi$, $\mathcal{D}_\Phi$, where γ is the average number of nearest neighbors, N is the number of states visited by paths in Φ, and Λ is the cutoff path length. The cutoff Λ scales with network size N in the same way as the average path length $\bar{\ell}_\Phi$; for simple random walks, it is known that

$$\Lambda \sim \bar{\ell}_\Phi \sim \begin{cases} N^{d_w/d_f}, & d_w \geq d_f \quad \text{(compact exploration)}, \\ N, & d_w < d_f \quad \text{(non-compact exploration)}, \end{cases} \tag{23}$$

where d_w is the dimension of the walk and d_f is the fractal dimension of the space [72, 73]. Therefore, the algorithm scales as

$$\mathcal{O}(\gamma N \Lambda) = \begin{cases} \mathcal{O}(\gamma N^{1+d_w/d_f}), & d_w \geq d_f, \\ \mathcal{O}(\gamma N^2), & d_w < d_f, \end{cases} \tag{24}$$

automatically accounting for the sparseness of network connections. This scaling compares favorably with standard linear algebra algorithms, which in general require $\mathcal{O}(N^3)$ operations [74] to solve the backward equation [62, 63].

2.3. *Evolution on a neutral network*

As a simple application of this approach, we consider a population evolving on a neutral network [31]. In the space of all sequences of length L and with an alphabet of size k, we assign each sequence fitness 1 with probability p or fitness zero with probability $1-p$. The subset of fit states connected to each other forms a neutral network; there can be several disconnected neutral networks in each landscape realization. All jumps between neighboring fit states occur at the same rate, and waiting times are Markovian. We choose $L = 8$ and a binary alphabet $\{\mathsf{A}, \mathsf{B}\}$ ($k = 2$), which gives $2^8 = 256$ total nodes in the network, and we consider the ensemble Φ of first-passage paths from the sequence AAAAAAAA to the sequence BBBBBBBB.

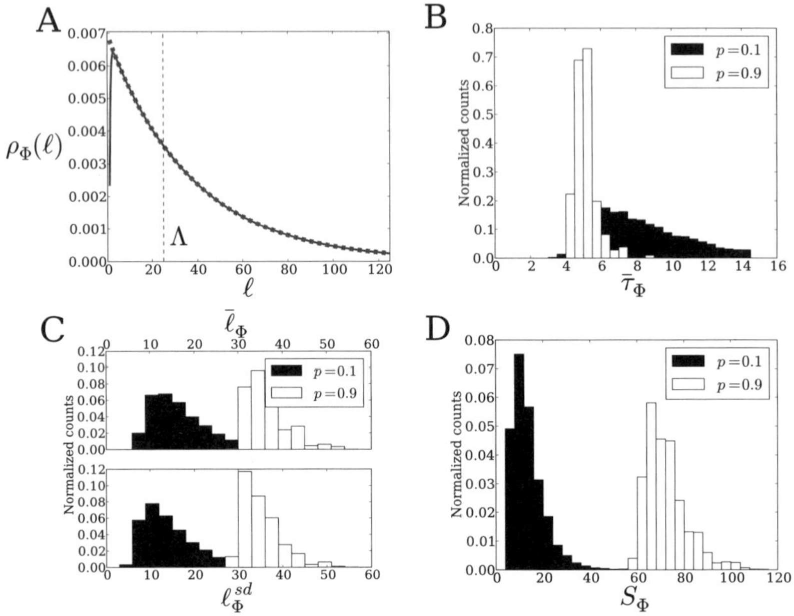

Fig. 2. (Color online) First-passage path ensemble statistics in a neutral network model. (A) The path length distribution $\rho_\Phi(\ell)$ (solid, blue) and exponential fit (dashed, green) in the interval $[\Lambda - 5, \Lambda]$ for $\Lambda = 25$ in a single realization of a neutral network with $p = 0.9$. (B) Distribution of mean path times $\bar{\tau}_\Phi$, (C) distribution of mean path lengths $\bar{\ell}_\Phi$ and standard deviations of path lengths ℓ_Φ^{sd}, and (D) distribution of path entropies S_Φ for $p = 0.1$ and $p = 0.9$. All quantities in (A)-(D) are per site. Histograms in (B)-(D) are generated from 10^4 successful random realizations of the neutral network for each value of p; a realization is considered successful if both initial and final states are included in a single connected network.

Figure 2A shows $\rho_\Phi(\ell)$ for a single realization of this model with $p = 0.9$. The exponential tail of $\rho_\Phi(\ell)$ is a universal feature of first-passage processes on finite spaces [72]; other path statistics, such as the average time $\bar{\tau}_\Phi(\ell)$ of paths up to length ℓ, also show asymptotic behavior that is exponential for long paths. We can use this feature to determine the cutoff path length Λ for the algorithm: Λ is set at a length such that $\rho_\Phi(\ell)$ and $\bar{\tau}_\Phi(\ell)$ are close to exponential in a region around Λ. Then one need only consider paths with $\ell < \Lambda$ and infer the contributions of all longer paths from an exponential fit to the tail, which considerably improves the efficiency of the algorithm. This procedure takes advantage of the fact that information about longer paths is already contained in the structure of shorter paths; the longer paths are built on the shorter paths by adding loops. The maximum length Λ of the shorter paths that must be explicitly calculated depends on the chemical distance between the initial and final states and the lengths over which the landscape is correlated. This essentially implements a numerical renormalization scheme on the ensemble of paths [70].

In Fig. 2B,C,D we show distributions of the mean path time $\bar{\tau}_\Phi$, mean path length $\bar{\ell}_\Phi$, path length standard deviation ℓ_Φ^{sd}, and path entropy S_Φ for multiple realizations of the neutral network with high and low values of p. We see that long paths are likely in these models: dozens of substitutions can occur at each site before the final state is reached. The larger size of the neutral network for $p = 0.9$ allows longer paths on average than for $p = 0.1$. However, the mean time of paths for the larger neutral network is usually smaller (Fig. 2B), since the increased connectivity of the network leads to shorter waiting times at individual nodes. Larger p leads to substantially more diversity of paths and path lengths, as expected due to the increased size and connectivity of the network (Fig. 2C,D). Note that the distributions of $\bar{\ell}_\Phi$ and ℓ_Φ^{sd} in Fig. 2C are very similar, owing to the nearly exponential distribution of $\rho_\Phi(\ell)$ in this model (cf. Fig. 2A).

In an unconstrained sequence space, the number of nearest neighbors is $\gamma = L(k-1)$, and the average path length $\bar{\ell}_\Phi$ scales as $N = k^L$ (Eq. (23)). According to Eq. (17), the entropy of paths in sequence space is

$$S_\Phi = \bar{\ell}_\Phi \log L(k-1) \sim k^L \log L(k-1). \tag{25}$$

When $p = 0.9$ the neutral network is nearly the size of the entire sequence space, and these results hold approximately. Indeed, we see that $\bar{\ell}_\Phi$ and S_Φ differ by roughly a factor of $\log L(k-1) \approx 2.1$ (Fig. 2C,D).

3. Biophysics of protein evolution

We now consider more realistic models of evolution based on protein biophysics, where the fitness landscape depends on protein folding stability and energetics of intermolecular interactions [3]. Many recent studies have focused on how proteins evolve under the constraint of maintaining thermodynamic stability of their folded state [75–82]. Suppose that an organism encodes a particular protein that is folded with probability $1/(1 + e^{\beta E_f})$, where E_f is the free energy of folding (i.e., the free energy difference between folded and unfolded states of the protein) and $\beta = 1.7$ (kcal/mol)$^{-1}$ is inverse room temperature. The protein contributes multiplicative fitness

1 if it is folded and $f_0 < 1$ if it is unfolded. Then the total fitness averaged over all proteins in an organism is given by

$$\mathcal{F}(E_f) = \frac{1 + f_0 e^{\beta E_f}}{1 + e^{\beta E_f}}. \tag{26}$$

Equation (26) states that robust protein folding confers a fitness advantage; in the extreme case of a protein essential to the organism, $f_0 = 0$ and so $\lim_{E_f \to +\infty} \mathcal{F}(E_f) = 0$. Some studies simplify this idea further by assuming that the folding energy E_f need only be below a particular threshold E_f^{thr}; below that threshold all proteins are adequately stable and equivalent in fitness [75, 77, 81]. Mathematically,

$$\mathcal{F}(E_f) = \Theta(E_f^{\text{thr}} - E_f), \tag{27}$$

where Θ is the Heaviside step function. This model is equivalent to the zero-temperature limit of Eq. (26). Similar approaches based on protein-DNA binding energies have also been used to study evolution of gene regulation [20, 83–87].

An extension of this model considers both protein stability and function [32], which we take to be the ability to bind a target molecule such as another protein (e.g., in a signaling pathway). Let E_b be the free energy of binding relative to the chemical potential of the target molecule, so that the probability of binding is $1/(1 + e^{\beta E_b})$. We assume that the protein contributes fitness 1 if it both folds and binds, and $f_0 < 1$ otherwise [88]. Then fitness averaged over all proteins in an organism is given by

$$\mathcal{F}(E_f, E_b) = \frac{1 + f_0(e^{\beta E_f} + e^{\beta E_b} + e^{\beta(E_f + E_b)})}{1 + e^{\beta E_f} + e^{\beta E_b} + e^{\beta(E_f + E_b)}}. \tag{28}$$

The folding and binding energies depend on the amino acid sequence σ. Many proteins have only a small number of residues at the binding interface that contribute the majority of the binding affinity; these are known as "hotspot" residues [89]. We assume that there are L such residues and that they make additive contributions to the total folding and binding free energies [90]:

$$E_f(\sigma) = E_f^0 + \sum_{\mu=1}^{L} \epsilon_f(\mu, \sigma^\mu), \quad E_b(\sigma) = E_b^0 + \sum_{\mu=1}^{L} \epsilon_b(\mu, \sigma^\mu), \tag{29}$$

where E_f^0, E_b^0 are overall offsets and $\epsilon_f(\mu, \sigma^\mu), \epsilon_b(\mu, \sigma^\mu)$ are the folding and binding energy contributions of amino acid σ^μ at position μ. The offset E_f^0 is a fixed contribution to the folding energy from all other residues in the protein, which we assume to be perfectly adapted. We sample ϵ_f's from a Gaussian distribution with mean 1.25 kcal/mol and standard deviation 1.6 kcal/mol, consistent with computational studies showing the mutational effects on stability are universally distributed [91].

We randomly assign a sequence $\sigma_{\rm bb}$ to be the best-binding sequence. Since binding hotspot residues typically have a minimum penalty of 1–3 kcal/mol for mutations away from the wild-type amino acid [92] (this requirement is used to define which residues make up the hotspot), we set $\epsilon_b(\mu, \sigma_{\rm bb}^\mu) = 0$ for all $\mu \in \{1, \ldots, L\}$, and sample the other ϵ_b's from an exponential distribution defined in the range of $(1, \infty)$ kcal/mol, with mean 2 kcal/mol. This distribution is consistent with alanine-scanning experiments which probe energetics of amino acids at the binding interface [93]. Here we consider $L = 5$ hotspot residues and a reduced alphabet of $k = 8$ amino acids grouped by physico-chemical properties, resulting in $8^5 = 32768$ possible sequences. Different choices of these parameters can be considered, but they appear to have little effect on the overall qualitative features of the model.

We consider a population of individuals whose genomes encode a protein of interest. In each individual, the sequence of the protein is perfectly adapted to binding an original target molecule. Then the population is subjected to a selection pressure which favors binding a new target. This situation is common in directed evolution experiments which attempt to evolve new protein functions in a laboratory [94]. To model such experiments, we sample one set of ϵ_f's and two sets of ϵ_b's (one for each target), while E_f^0 and E_b^0 are assumed to be fixed. This procedure defines two fitness landscapes, $\mathcal{F}_1$ and $\mathcal{F}_2$, through Eq. (28); the entire population begins at the sequence with the global maximum on $\mathcal{F}_1$ and proceeds to adapt to a new global or local maximum on $\mathcal{F}_2$. We assume the strong-selection evolutionary dynamics as described in Sec. 1.1: any beneficial mutation that arises is guaranteed to fix in the population. Thus the substitution rate from σ to σ' is $\langle\sigma'|\mathbf{W}|\sigma\rangle = Nu$ if $\mathcal{F}(\sigma') > \mathcal{F}(\sigma)$ and zero otherwise, and the mean waiting time for sequence σ is $w(\sigma) = (\sum_{\text{nn } \sigma' \text{ of } \sigma} \langle\sigma'|\mathbf{W}|\sigma\rangle)^{-1} = (Nub(\sigma))^{-1}$, where $b(\sigma)$ is the number of beneficial mutations available from σ. Therefore the jump probability is $\langle\sigma'|\mathbf{Q}|\sigma\rangle = \langle\sigma'|\mathbf{W}|\sigma\rangle w(\sigma)$, which equals $1/b(\sigma)$ for a beneficial mutation and zero for a deleterious one. Note that in this limit the results are independent of f_0 and the population mutation rate

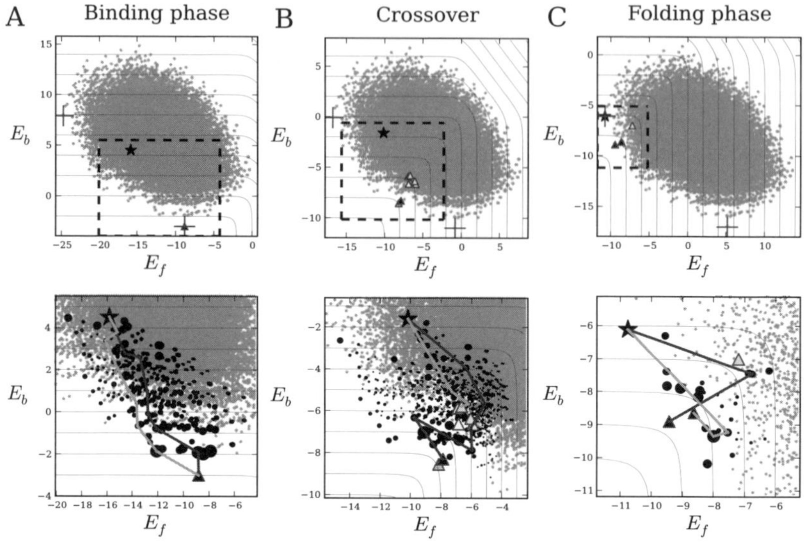

Fig. 3. (Color online) Three realizations of the fitness landscape. The offsets E_f^0 and E_b^0 are different for each realization, but ϵ_f's and the two sets of ϵ_b's (one for $\mathcal{F}_1$ and another for $\mathcal{F}_2$) are the same. (A) Binding phase, with $E_f^0 = -17$ kcal/mol and $E_b^0 = -3$ kcal/mol. (B) Crossover regime, with $E_f^0 = -9$ kcal/mol and $E_b^0 = -11$ kcal/mol. (C) Folding phase, with $E_f^0 = -3$ kcal/mol and $E_b^0 = -17$ kcal/mol. Top panels of (A)-(C) show the global distribution of all $8^5 = 32768$ sequences in energy space according to $\mathcal{F}_2$, where the blue crosses indicate the best-folding (σ_{bf}) and best-binding (σ_{bb}) sequences, red triangles indicate local fitness maxima on $\mathcal{F}_2$ (shaded according to their commitment probabilities), and black stars indicate the initial state for adaptation (sequence with global maximum on $\mathcal{F}_1$). Black lines are contours of constant fitness $\mathcal{F}_2$. In the bottom panels of (A)-(C) only the regions of energy space accessible to APs are shown; these regions are outlined by dashed lines in the top panels. Example APs are shown in blue and green; black circles indicate intermediate states along APs, sized proportional to the AP density $\langle \mathcal{I}_\sigma \rangle_{\mathrm{AP}}$; small gray circles are sequences inaccessible to APs.

Nu only affects the overall time scale. The path ensemble consists of all adaptive paths (APs), which are first-passage paths leading from the initial state to a local maximum on $\mathcal{F}_2$, with fitness monotonically increasing along each path. In Fig. 3 we show three realizations of $\mathcal{F}_2$ with examples of APs.

We focus on the generic properties of these landscapes, averaged over multiple realizations of ϵ_f and ϵ_b (Fig. 4). Varying E_f^0 and E_b^0 reveals two qualitatively different phases of adaptation. When E_f^0 is low and E_b^0 is high (see Fig. 3A for an example), adaptation is in the *binding phase*, i.e., the

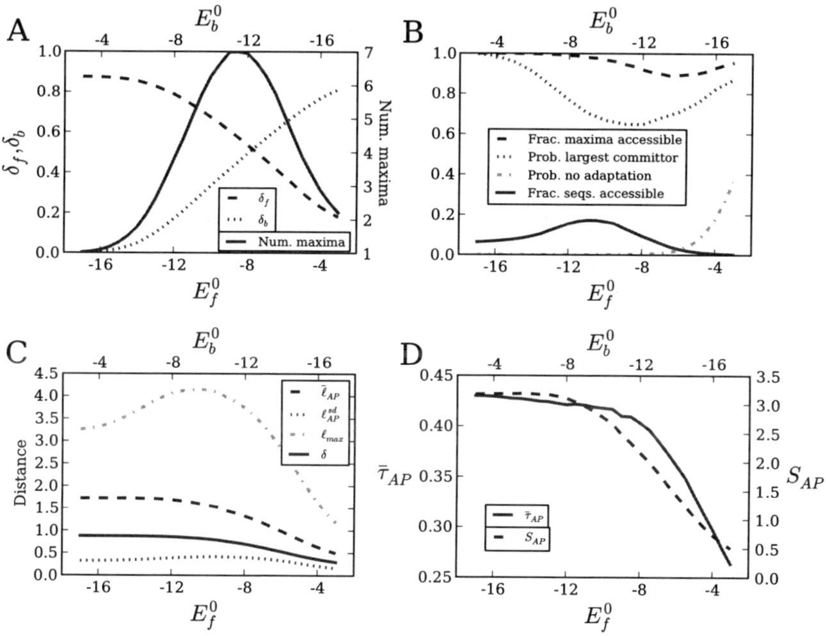

Fig. 4. (Color online) Statistics of fitness landscapes and adaptive paths averaged over multiple landscape realizations. All quantities in (A), (C), and (D) are per-residue (except the number of local maxima). (A) Average number of local fitness maxima (solid, green), average Hamming distance δ_f (number of mutations) between the maxima and the best-folding sequence $\sigma_{\rm bf}$ (dashed, blue), and average Hamming distance δ_b between the maxima and the best-binding sequence $\sigma_{\rm bb}$ (dotted, red) for the parameter subspace $E_f^0 + E_b^0 = -20$ kcal/mol. Note that the average distance between two random sequences is $1 - 1/k = 0.875$, where $k = 8$ is the size of the alphabet. (B) Fraction of local fitness maxima accessible from the initial state (dashed, blue), fraction of all landscape realizations in which the global maximum has the largest commitment probability (committor) among all local maxima (dotted, red), probability that the initial sequence starts at a local maximum resulting in no adaptation (dashed and dotted, cyan), and fraction of all sequences accessible to APs (solid, green). (C) Mean AP length $\bar{\ell}_{\rm AP}$, standard deviation $\ell^{\rm sd}_{\rm AP}$ of APs, average Hamming distance δ between the initial state and the final states, and average length $\ell_{\max}$ of the longest APs connecting the initial state with the final states. (D) Path ensemble entropy $S_{\rm AP}$ (dashed, blue) and the mean time of paths $\bar{\tau}_{\rm AP}$ (solid, green), in units of inverse population mutation rate $(Nu)^{-1}$. The probability of no adaptation in (B) is an average over 2×10^4 landscape realizations; all other data points are averages over 5×10^3 realizations, and realizations with no adaptation are excluded.

need to bind the new target molecule dominates evolutionary dynamics. In this phase, there is typically a single local fitness maximum which coincides with the best-binding sequence $\sigma_{\rm bb}$ (Fig. 4A). In contrast, when E_f^0 is high and E_b^0 is low (see Fig. 3C for an example), adaptation is in the *folding phase*, where evolution is mostly constrained by the need to maintain or increase folding stability. In this case there are also few local maxima and they tend to be close in sequence space to the best-folding sequence σ_{bf} (Fig. 4A). Between these phases there is a crossover regime, where folding and binding compete more equally in shaping the landscape and adaptive dynamics (see Fig. 3B for an example). The crossover regime has the most epistasis, as indicated by the number of local maxima, the accessibility of those maxima, and the fraction of fitness landscape realizations in which the global maximum has the largest commitment probability (Fig. 4A,B). The differences in the landscape structure in the binding and folding phases lead to substantial differences in adaptive dynamics. In particular, APs are longer and take more time in the binding phase compared to the folding phase; they are also more diverse (Fig. 4C,D). Initial and final states in the binding phase are separated by longer Hamming distances (Fig. 4C). In the folding phase, there is an appreciable probability that no adaptation occurs at all, since the initial state may coincide with one of the local maxima (Fig. 4B).

This model reproduces several important features of molecular evolution observed in experimental studies. First of all, we see that adaptive dynamics involve tradeoffs between folding and binding as frequently observed in directed evolution experiments [94–96], even though mutational effects on folding and binding energies are uncorrelated [97]. This evolutionary coupling between folding and binding is introduced through nonlinearities in the fitness function $\mathcal{F}(E_f, E_b)$ (Eq. (28)), which contains both magnitude and sign epistasis. We note that although these landscapes are generated from randomly-drawn parameters ϵ_f and ϵ_b, similar to many classical model landscapes (Sec. 1.5), the protein landscapes studied here are highly correlated [98]: fitness values of k^L sequences are determined by $2Lk$ ϵ_f and ϵ_b parameters. Indeed, the average number of local maxima on a House of Cards landscape for the same sequence space ($L = 5$ and $k = 8$) is ≈ 910, while the protein landscape has on average no more than 7 (Fig. 4A). Thus, this model is far less epistatic than completely uncorrelated landscapes [38]. This more moderate level of epistasis is consistent with previous analyses of empirical fitness landscapes [12, 25, 36].

There are other features of the model that correspond to experimental observations. In the folding phase, APs tend to be short and no adaptation may occur if the old global maximum on $\mathcal{F}_1$ coincides with a new local maximum on $\mathcal{F}_2$ (Fig. 4B). This lack of adaptation is sometimes observed

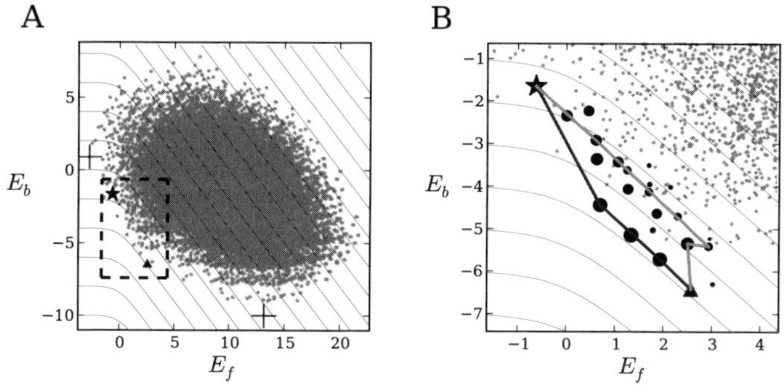

Fig. 5. (Color online) A realization of the fitness landscape for the three-state model (Eq. (30)) in the crossover regime, with $E_f^0 = 5$ kcal/mol and $E_b^0 = -10$ kcal/mol. As in Fig. 3, $k = 8$ and $L = 5$, resulting in 8^5 sequences. Note that ϵ_f's and the two sets of ϵ_b's (one for $\mathcal{F}_1$ and another for $\mathcal{F}_2$) are the same as in Fig. 3. Black lines are contours of constant fitness $\mathcal{F}_2$ (Eq. (30)). All symbols are identical to those in Fig. 3.

in experiments in which a protein already exhibiting some affinity for the new ligand cannot increase it any further [94]. Natural proteins are often found to have only marginal folding stability [76]; the model presented here shows that marginal stability can arise purely from the evolutionary tradeoff between binding and folding in the crossover regime, unlike previous hypotheses that explain it with mutational entropy [78] or a fitness function that explicitly disfavors hyperstable proteins [76].

The basic model described here can be generalized to account for many other aspects of protein evolution. For example, we can incorporate chaperone-assisted folding [99] by modifying E_f^0 or the distribution of ϵ_f's, and include "folding hotspots" away from the binding interface, which may acquire stabilizing mutations as a buffer against destabilizing but function-improving mutations at the interface [94, 95]. Neutral and weakly-selected mutations can be incorporated as well by using substitution rates from more complex population genetics models [18, 21], although we expect non-adaptive substitutions to play little role on short time scales.

Another important case is binding-mediated stability, in which binding stabilizes an otherwise disordered protein [100]. In this case, instead of considering folding and binding to be independent as in Eq. (28), which leads to four possible protein states, we assume that the protein can only bind when it is folded. This results in three possible protein states (excluding the bound-and-unfolded state), yielding a fitness function

$$\mathcal{F}(E_f, E_b) = \frac{1 + f_0(e^{\beta E_b} + e^{\beta(E_f+E_b)})}{1 + e^{\beta E_b} + e^{\beta(E_f+E_b)}}, \tag{30}$$

where E_f, E_b are again defined by Eq. (29). In this model a protein may have high fitness even if $E_f > 0$, as long as the protein is stabilized by binding ($E_f + E_b < 0$). The methodology described here can be straightforwardly applied to this new fitness function. When E_f^0 is low, adaptive dynamics resemble the binding phase of the four-state model (Eq. (28)). However, when $E_f^0 > 0$, the dynamics enter a crossover regime, in which adaptation changes folding and binding energies simultaneously (Fig. 5). Unlike the four-state model (Eq. (28)), there is no folding-dominated phase.

In this chapter we have summarized many aspects of evolutionary trajectories on fitness landscapes, focusing especially on landscape topography and the statistical properties of adaptive first-passage trajectories. We have also described a general statistical physics-based methodology for studying stochastic paths on arbitrary landscapes and networks. This approach can be widely applied to first-passage problems in physics, chemistry, biology, and engineering, including protein folding, transport and search in complex media, stochastic phenotypes, and cell-type differentiation. Here we have emphasized its utility in exploring evolutionary problems, which can often be modeled as random walks on fitness landscapes and where the diversity and reproducibility of evolutionary paths is a central issue. The path-based methodology is well-suited for providing intuitive path statistics in problems whose complexity and high dimensionality make direct visualizations impossible.

References

1. P. Nelson, *Biological Physics: Energy, Information, Life.* W.H. Freeman and Company, New York, USA (2007).
2. T. E. Creighton, *Proteins: Structures and Molecular Properties.* W.H. Freeman and Company, New York, USA (1992).
3. A. V. Finkelstein and O. Ptitsyn, *Protein Physics: A Course of Lectures.* Academic Press, London, UK (2002).
4. C. Levinthal, Are there pathways for protein folding?, *J. Chim. Phys.* **65**, 44 (1968).
5. J. D. Bryngelson, J. N. Onuchic, N. D. Socci, and P. G. Wolynes, Funnels, pathways, and the energy landscape of protein folding: A synthesis, *Proteins: Str. Func. Genet.* **21**, 167–195 (1995).
6. K. A. Dill and J. L. MacCallum, The protein-folding problem, 50 years on, *Science.* **338**, 1042 (2012).

7. N. Metropolis, A. Rosenbluth, M. Rosenbluth, A. Teller, and E. Teller, Equation of state calculations by fast computing machines, *J. Chem. Phys.* **21**, 1087–1092 (1953).
8. D. Tolkunov and A. V. Morozov, Single temperature for Monte Carlo optimization on complex landscapes, *Phys. Rev. Lett.* **108**, 250602 (2012).
9. C. R. Darwin, *The Origin of Species*. J. Murray, London (1859).
10. J. Maynard Smith, Natural selection and the concept of a protein space, *Nature*. **225**, 563–564 (1970).
11. H. A. Orr, Fitness and its role in evolutionary genetics, *Nat. Rev. Genet.* **10**, 531–539 (2009).
12. I. G. Szendro, M. F. Schenk, J. Franke, J. Krug, and J. A. de Visser, Quantitative analyses of empirical fitness landscapes, *J. Stat. Mech.* p. P01005 (2013).
13. J. Gillespie, *Population Genetics: A Concise Guide*. The Johns Hopkins University Press, Baltimore, USA (2004).
14. S. Wright, The roles of mutation, inbreeding, crossbreeding and selection in evolution, *Proc. 6th Int. Cong. Genet.* **1**, 356–366 (1932).
15. S. J. Gould, *Wonderful Life: The Burgess Shale and the Nature of History*. W. W. Norton and Company, New York, USA (1990).
16. N. Champagnat, A microscopic interpretation for adaptive dynamics trait substitution sequence models, *Stochastic Process. Appl.* **116**, 1127–1160 (2006).
17. N. Champagnat, R. Ferrire, and S. Mlard, Unifying evolutionary dynamics: from individual stochastic processes to macroscopic models, *Theor. Popul. Biol.* **69**, 297–321 (2006).
18. J. F. Crow and M. Kimura, *An Introduction to Population Genetics Theory*. Harper and Row, New York (1970).
19. M. Kimura, *The Neutral Theory of Molecular Evolution*. Cambridge University Press, Cambridge, UK (1983).
20. M. Lässig, From biophysics to evolutionary genetics: statistical aspects of gene regulation, *BMC Bioinformatics*. **8**, S7 (2007).
21. M. Manhart, A. Haldane, and A. V. Morozov, A universal scaling law determines time reversibility and steady state of substitutions under selection, *Theor. Popul. Biol.* **82**, 66–76 (2012).
22. J. Wakeley, The limits of theoretical population genetics, *Genetics*. **169**, 1–7 (2005).
23. J. H. Gillespie, Molecular evolution over the mutational landscape, *Evolution*. **38**, 1116–1129 (1984).
24. M. M. Desai and D. S. Fisher, Beneficial mutation-selection balance and the effect of linkage on positive selection, *Genetics*. **176**, 1759–1798 (2007).
25. M. Carneiro and D. L. Hartl, Adaptive landscapes and protein evolution, *Proc. Natl. Acad. Sci. USA*. **107**, 1747–1751 (2010).
26. D. M. Weinreich, R. A. Watson, and L. Chao, Sign epistasis and genetic constraint on evolutionary trajectories, *Evolution*. **59**, 1165–1174 (2005).
27. F. J. Poelwijk, S. Tanase-Nicola, D. J. Kiviet, and S. J. Tans, Reciprocal sign epistasis is a necessary condition for multi-peaked fitness landscapes,

J. Theor. Biol. **272**, 141–144 (2011).
28. H. Flyvbjerg and B. Lautrup, Evolution in a rugged fitness landscape, *Phys. Rev. A.* **46**, 6714–6723 (1992).
29. A. Traulsen, Y. Iwasa, and M. A. Nowak, The fastest evolutionary trajectory, *J. Theor. Biol.* **249**, 617–623 (2007).
30. S. Kryazhimskiy, G. Tkačik, and J. B. Plotkin, The dynamics of adaptation on correlated fitness landscaps, *Proc. Natl. Acad. Sci. USA.* **106**, 18638–18643 (2009).
31. J. Franke, A. Klozer, J. A. de Visser, and J. Krug, Evolutionary accessibility of mutational pathways, *PLoS Comput. Biol.* **7**, e1002134 (2011).
32. M. Manhart and A. V. Morozov, Path-based approach to random walks on networks characterizes how proteins evolve new functions, *Phys. Rev. Lett.* **111**, 088102 (2013).
33. A. E. Lobkovsky and E. V. Koonin, Replaying the tape of life: quantification of the predictability of evolution, *Front. Gene.* **3**, 246 (2012).
34. D. M. Weinreich, N. F. Delaney, M. A. DePristo, and D. L. Hartl, Darwinian evolution can follow only very few mutational paths to fitter proteins, *Science.* **312**, 111–114 (2006).
35. F. J. Poelwijk, D. J. Kiviet, D. M. Weinreich, and S. J. Tans, Empirical fitness landscapes reveal accessible evolutionary paths, *Nature.* **445**, 383–386 (2007).
36. A. E. Lobkovsky, Y. I. Wolf, and E. V. Koonin, Predictability of evolutionary trajectories in fitness landscapes, *PLoS Comput. Biol.* **7**, e1002302 (2011).
37. S. A. Kauffman and E. D. Weinberger, The NK model of rugged fitness landscapes and its application to maturation of the immune response, *J. Theor. Biol.* **141**, 211–245 (1989).
38. S. Kauffman, *The Origins of Order: Self-Organization and Selection in Evolution.* Oxford University Press, New York (1993).
39. T. Aita, H. Uchiyama, T. Inaoka, M. Nakajima, T. Kokubo, and Y. Husimi, Analysis of a local fitness landscape with a model of the rough Mt. Fuji-type landscape: application to prolyl endopeptidase and thermolysin, *Biopolymers.* **54**, 64–79 (2000).
40. J. F. C. Kingman, A simple model for the balance between selection and mutation, *J. Appl. Probab.* **15**, 1–12 (1978).
41. L. Altenberg. NK fitness landscapes. In eds. T. Bäck, D. Fogel, and Z. Michalewicz, *Handbook of Evolutionary Computation*, pp. B2.7:5–B2.7:10. IOP Publishing Ltd and Oxford University Press (1997).
42. D. R. Rokyta, C. J. Beisel, and P. Joyce, Properties of adaptive walks on uncorrelated landscapes under strong selection and weak mutation, *J. Theor. Biol.* **243**, 114–120 (2006).
43. E. van Nimwegen, J. P. Crutchfield, and M. Huynen, Neutral evolution of mutational robustness, *Proc. Natl. Acad. Sci. USA.* **96**, 9716–9720 (1999).
44. D. Stauffer and A. Aharony, *Introduction to Percolation Theory.* Taylor and Francis, London (1994).
45. F. Mammano, V. Trouplin, V. Zennou, and F. Clavel, Retracing the evo-

lutionary pathways of human immunodeficiency virus type 1 resistance to protease inhibitors: virus fitness in the absence and in the presence of drug, *J. Virol.* **74**, 8524–8531 (2000).
46. A. I. Khan, D. M. Dinh, D. Schneider, R. E. Lenski, and T. F. Cooper, Negative epistasis between beneficial mutations in an evolving bacterial population, *Science.* **332**, 1193–1196 (2011).
47. H.-H. Chou, H.-C. Chiu, N. F. Delaney, D. Segrè, and C. J. Marx, Diminishing returns epistasis among beneficial mutations decelerates adaptation, *Science.* **332**, 1190–1192 (2011).
48. P. Hänggi, P. Talkner, and M. Borkovec, Reaction rate theory: fifty years after kramers, *Rev. Mod. Phys.* **62**, 251–341 (1990).
49. P. G. Bolhuis, D. Chandler, C. Dellago, and P. L. Geissler, Transition path sampling: Throwing ropes over rough mountain passes, in the dark, *Ann. Rev. Phys. Chem.* **53**, 291–318 (2002).
50. C. Dellago, P. G. Bolhuis, F. S. Csajka, and D. Chandler, Transition path sampling and the calculation of rate constants, *J. Chem. Phys.* **108**, 1964–1977 (1998).
51. C. Dellago, P. G. Bolhuis, and P. L. Geissler, Transition path sampling, *Adv. Chem. Phys.* **123**, 1 (2003).
52. G. Hummer, From transition paths to transition states and rate coefficients, *J. Chem. Phys.* **120**, 516–523 (2004).
53. B. Harland and S. X. Sun, Path ensembles and path sampling in nonequilibrium stochastic systems, *J. Chem. Phys.* **127**, 104103 (2007).
54. T. More, A. Walczak, and F. Zamponi, Transition path sampling algorithm for discrete many-body systems, *Phys. Rev. E.* **85**, 036710 (2012).
55. D. M. Robinson, D. T. Jones, H. Kishino, N. Goldman, and J. K. Thorne, Protein evolution with dependence among codons due to tertiary structure, *Mol. Biol. Evol.* **20**, 1692–1704 (2003).
56. N. Rodrigue, N. Lartillot, D. Bryant, and H. Philippe, Site interdependence attributed to tertiary structure in amino acid sequence evolution, *Gene.* **347**, 207–217 (2005).
57. N. Rodrigue, H. Philippe, and N. Lartillot, Assessing site-interdependent phylogenetic models of sequence evolution, *Mol. Biol. Evol.* **23**, 1762–1775 (2006).
58. S. C. Choi, A. Hobolth, D. M. Robinson, H. Kishino, and J. L. Thorne, Quantifying the impact of protein tertiary structure on molecular evolution, *Mol. Biol. Evol.* **24**, 1769–1782 (2007).
59. N. Rodrigue, C. L. Kleinman, H. Philippe, and N. Lartillot, Computational methods for evaluating phylogenetic models of coding sequence evolution with dependence between codons, *Mol. Biol. Evol.* **26**, 1663–1676 (2009).
60. W. E and E. Vanden-Eijnden, Toward a theory of transition paths, *J. Stat. Phys.* **123**, 503–523 (2006).
61. P. Metzner, C. Schütte, and E. Vanden-Eijnden, Illustration of transition path theory on a collection of simple examples, *J. Chem. Phys.* **125**, 084110 (2006).
62. P. Metzner, C. Schütte, and E. Vanden-Eijnden, Transition path theory for

Markov jump processes, *Multiscale Model. Simul.* **7**, 1192–1219 (2009).
63. F. Noé, C. Schütte, E. Vanden-Eijnden, L. Reich, and T. R. Weikl, Constructing the equilibrium ensemble of folding pathways from short off-equilibrium simulations, *Proc. Natl. Acad. Sci. USA.* **106**, 19011–19016 (2009).
64. W. E and E. Vanden-Eijnden, Transition-path theory and path-finding algorithms for the study of rare events, *Ann. Rev. Phys. Chem.* **61**, 391–420 (2010).
65. G. H. Weiss, *Aspects and Applications of the Random Walk.* North Holland, Amsterdam (1994).
66. J. Klafter and R. Silbey, Derivation of the continuous-time random-walk equation, *Phys. Rev. Lett.* **44**, 55 (1980).
67. S. Redner, *A Guide to First-Passage Processes.* Cambridge University Press, Cambridge (2001).
68. C. Maes, K. Netočný, and B. Wynants, Dynamical fluctuations for semi-markov processes, *J. Phys. A: Math. Theor.* **42**, 365002 (2009).
69. O. Flomenbom and J. Klafter, Closed-form solutions for continuous time random walks on finite chains, *Phys. Rev. Lett.* **95**, 098105 (2005).
70. S. X. Sun, Path summation formulation of the master equation, *Phys. Rev. Lett.* **96**, 210602 (2006).
71. I. Majid, D. ben-Avraham, S. Havlin, and H. E. Stanley, Exact-enumeration approach to random walks on percolation clusters in two dimensions, *Phys. Rev. B.* **30**, 1626–1628 (1984).
72. E. M. Bollt and D. ben-Avraham, What is special about diffusion on scale-free nets?, *New J. Phys.* **7**, 26–47 (2005).
73. S. Condamin, O. Bénichou, V. Tejedor, R. Voituriez, and J. Klafter, First-passage times in complex scale-invariant media, *Nature.* **450**, 77–80 (2007).
74. W. H. Press, S. A. Teukolsky, W. T. Vetterling, and B. P. Flannery, *Numerical Recipes in C: The Art of Scientific Computing*, second edn. Cambridge, Cambridge (1992).
75. J. D. Bloom, J. J. Silberg, C. O. Wilke, D. A. Drummond, C. Adami, and F. H. Arnold, Thermodynamic prediction of protein neutrality, *Proc. Natl. Acad. Sci. USA.* **102**, 606–611 (2005).
76. M. A. DePristo, D. M. Weinreich, and D. L. Hartl, Missense meanderings in sequence space: a biophysical view of protein evolution, *Nat. Rev. Genet.* **6**, 678–687 (2005).
77. J. D. Bloom, S. T. Labthavikul, C. R. Otey, and F. H. Arnold, Protein stability promotes evolvability, *Proc. Natl. Acad. Sci. USA.* **103**, 5869–5874 (2006).
78. K. B. Zeldovich, P. Chen, and E. I. Shakhnovich, Protein stability imposes limits on organism complexity and speed of molecular evolution, *Proc. Natl. Acad. Sci. USA.* **104**, 16152–16157 (2007).
79. J. D. Bloom, Z. Lu, D. Chen, A. Raval, O. S. Venturelli, and F. H. Arnold, Evolution favors protein mutational robustness in sufficiently large populations, *BMC Biology.* **5**, 29 (2007).
80. J. D. Bloom, P. A. Romero, Z. Lu, and F. H. Arnold, Neutral genetic

drift can alter promiscuous protein functions, potentially aiding functional evolution, *Biology Direct.* **2**, 17 (2007).
81. J. D. Bloom, A. Raval, and C. O. Wilke, Thermodynamics of neutral protein evolution, *Genetics.* **175**, 255–266 (2007).
82. S. Bershtein, K. Goldin, and D. S. Tawfik, Intense neutral drifts yield robust and evolvable consensus proteins, *J. Mol. Biol.* **379**, 1029–1044 (2008).
83. A. M. Sengupta, M. Djordjevic, and B. I. Shraiman, Specificity and robustness in transcription control networks, *Proc. Natl. Acad. Sci. USA.* **99**, 2072–2077 (2002).
84. U. Gerland and T. Hwa, On the selection and evolution of regulatory DNA motifs, *J. Mol. Evol.* **55**, 386–400 (2002).
85. J. Berg and M. Lässig, Stochastic evolution of transcription factor binding sites, *Biophysics (Moscow).* **48**, S36–S44 (2003).
86. J. Berg, S. Willmann, and M. Lässig, Adaptive evolution of transcription factor binding sites, *BMC Evolutionary Biology.* **4**, 42 (2004).
87. V. Mustonen, J. Kinney, C. G. Callan, and M. Lässig, Energy-dependent fitness: A quantitative model for the evolution of yeast transcription factor binding sites, *Proc. Natl. Acad. Sci. USA.* **105**, 12376–12381 (2008).
88. S. Mayer, S. Rüdiger, H. C. Ang, A. C. Joerger, and A. R. Fersht, Correlation of levels of folded recombinant p53 in *escherichia coli* with thermodynamic stability *in vitro*, *J. Mol. Biol.* **372**, 268–276 (2007).
89. T. Clackson and J. A. Wells, A hot spot of binding energy in a hormone-receptor interface, *Science.* **267**, 383–386 (1995).
90. L. Serrano, A. G. Day, and A. R. Fersht, Step-wise mutation of barnase to binase. a procedure for engineering increased stability of proteins and an experimental analysis of the evolution of protein stability, *J. Mol. Biol.* **233**, 305–312 (1993).
91. N. Tokuriki, F. Stricher, J. Schymkowitz, L. Serrano, and D. S. Tawfik, The stability effects of protein mutations appear to be universally distributed, *J. Mol. Biol.* **369**, 1318–1332 (2007).
92. A. A. Bogan and K. S. Thorn, Anatomy of hot spots in protein interfaces, *J. Mol. Biol.* **280**, 1–9 (1998).
93. K. S. Thorn and A. A. Bogan, Asedb: a database of alanine mutations and their effects on the free energy of binding in protein interactions, *Bioinformatics.* **17**, 284–285 (2001).
94. J. D. Bloom and F. H. Arnold, In the light of directed evolution: Pathways of adaptive protein evolution, *Proc. Natl. Acad. Sci. USA.* **106**, 9995–10000 (2009).
95. N. Tokuriki and D. S. Tawfik, Stability effects of mutations and protein evolvability, *Curr. Opin. Struct. Biol.* **19**, 596–604 (2009).
96. X. Wang, G. Minasov, and B. K. Shoichet, Evolution of an antibiotic resistance enzyme constrained by stability and activity tradeoffs, *J. Mol. Biol.* **320**, 85–95 (2002).
97. N. Tokuriki, F. Stricher, L. Serrano, and D. S. Tawfik, How protein stability and new functions trade off, *PLoS Comput. Biol.* **4**, e1000002 (2008).
98. L. D. Bogarad and M. W. Deem, A hierarchical approach to protein molec-

ular evolution, *Proc. Natl. Acad. Sci. USA.* **96**, 2591–2595 (1999).
99. S. L. Rutherford, Between genotype and phenotype: protein chaperones and evolvability, *Nat. Rev. Genet.* **4**, 263–274 (2003).
100. C. J. Brown, A. K. Johnson, A. K. Dunker, and G. W. Daughdrill, Evolution and disorder, *Curr. Opin. Struct. Biol.* **21**, 441–446 (2011).

Chapter 18

Some Applications of First-Passage Ideas to Finance

Rémy Chicheportiche and Jean-Philippe Bouchaud

Capital Fund Management,

23–25, rue de l'Université, 75 007 Paris, France

Many problems in finance are related to first passage times. Among all of them, we chose three on which we contributed personally. Our first example relates Kolmogorov-Smirnov like goodness-of-fit tests, modified in such a way that tail events and core events contribute equally to the test (in the standard Kolmogorov-Smirnov, the tails contribute very little to the measure of goodness-of-fit). We show that this problem can be mapped onto that of a random walk inside moving walls. The second example is the optimal time to sell an asset (modelled as a random walk with drift) such that the sell time is as close as possible to the time at which the asset reaches its maximum value. The last example concerns optimal trading in the presence of transaction costs. In this case, the optimal strategy is to wait until the predictor reaches (plus or minus) a threshold value before buying or selling. The value of this threshold is found by mapping the problem onto that of a random walk between two walls.

1. Introduction

Quantitative finance is the bounty land of statistics and probabilities. Bachelier proposed to model price paths as random walks in 1900, but his amazingly creative work was forgotten until the sixties, when Samuelson, Black & Scholes revived the Brownian motion framework, which is now the cornerstone of modern mathematical finance. This is probably unfortunate, because the continuous time, Gaussian random walk misses most of the important "stylized facts" of financial markets — fat (power-law) tails, intermittency and long memory, etc. Many of the results that hold true for a Gaussian process go awry in reality — for example, the well-known perfect hedge of Black-Scholes, that would enable one to sell option contracts without any risk, is a figment of the very specific assumptions of the model. It is all too easy to get carried away with the beauty of a mathematical model, and forget that it does not bear any relation with reality. This is of special concern in the case of financial markets, where inadequate models can contribute (and have contributed) to systemic risks [1].

This is however not to say that probabilistic methods are useless in this context. Quite on the contrary, empirically motivated, faithful models do help in controlling risks better and pricing derivative contracts more accurately. Many questions that are relevant in practice can be addressed. Some of them are directly related to first passage time problems, which is our topic here. For example, one might be interested to invest in financial markets with a profit objective, that would allow another project to be financed. What is the distribution of the time one should wait until this profit is reached? Conversely, one might be worried about the default of a company or a bank. This is often modeled as the first passage time when a random walk (the value of the asset) goes below a certain threshold (the equity) [2]. Another example is that of "barrier options", which disappear when the price of the underlying hits a certain predefined value (the barrier); a related issue is the early exercise of so-called American options, when the option first reaches a value where it is optimal to exercise and cash the current pay-off rather than let the option run to maturity [3].

We review here three examples of the use of first passage ideas in finance: the design of a goodness-of-fit test whose law is the survival probability of a process before hitting a barrier, the optimal time to sell an asset, and the optimal value of the price threshold at which the expected benefit outweighs linear transaction costs. Extensions and open questions related to these three problems are briefly discussed in the conclusion.

2. Weighted Kolmogorov-Smirnov tests and first passages [4]

Our first example concerns goodness-of-fit (GoF) testing, which is ubiquitous in all fields of science and engineering. This is the problem of testing whether a null-hypothesis theoretical probability distribution is compatible with the empirical probability distribution of a sample of observations. GoF tests are designed to assess quantitatively whether a sample of N observations can statistically be seen as a collection of N realizations of a given probability law, or whether two such samples are drawn from the same hypothetical distribution.

The best known theoretical result is due to Kolmogorov and Smirnov (KS) [5, 6], and has led to the eponymous statistical test for an *univariate* sample of *independent* draws. The major strength of this test lies in the fact that the asymptotic distribution of its test statistics is completely independent of the null-hypothesis cdf.

Several specific extensions have been studied (and/or are still under scrutiny), including: different choices of distance measures, multivariate samples, investigation of different parts of the distribution domain, dependence in the successive draws (which is particularly important for financial applications), etc.

This class of problems has a particular appeal for physicists since the works of Doob [7] and Khmaladze [8], who showed how GoF testing is related to stochastic processes. Finding the law of a test amounts to computing a survival probability in a diffusion system. In a Markovian setting, this is often achieved by treating a Fokker-Planck problem, which in turn maps into a Schrödinger equation for a particle in a certain potential confined by walls.

2.1. *Empirical cumulative distribution and its fluctuations*

Let $\boldsymbol{X}$ be a random vector of N independent and identically distributed variables, with marginal cumulative distribution function (cdf) F. One realization of $\boldsymbol{X}$ consists of a time series $\{x_1, \ldots, x_n, \ldots, x_N\}$ that exhibits no persistence (see Ref. [9] when some non trivial dependence is present). The empirical cumulative distribution function

$$F_N(x) = \frac{1}{N} \sum_{n=1}^{N} \mathbb{1}_{\{X_n \leq x\}} \tag{1}$$

converges to the true CDF F as the sample size N tends to infinity. For finite N, the expected value and fluctuations of $F_N(x)$ are

$$\mathbb{E}[F_N(x)] = F(x),$$
$$\mathrm{Cov}(F_N(x), F_N(x')) = \frac{1}{N}\left[F(\min(x,x')) - F(x)F(x')\right].$$

The rescaled empirical CDF

$$Y_N(u) = \sqrt{N}\,\left[F_N(F^{-1}(u)) - u\right] \tag{2}$$

measures, for a given $u \in [0,1]$, the difference between the empirically determined cdf of the X's and the theoretical one, evaluated at the u-th quantile. It does not shrink to zero as $N \to \infty$, and is therefore the quantity on which any statistics for GoF testing is built.

Limit properties

One now defines the process $Y(u)$ as the limit of $Y_N(u)$ when $N \to \infty$. According to the Central Limit Theorem, it is Gaussian and its covariance function is given by:

$$I(u,v) = \min(u,v) - uv, \tag{3}$$

which characterizes the so-called Brownian bridge, i.e. a Brownian motion $Y(u)$ such that $Y(u{=}0) = Y(u{=}1) = 0$. Interestingly, the function F does not appear in Eq. (3) anymore, so the law of any functional of the limit process Y is independent of the law of the underlying finite size sample. This property is important for the design of *universal* GoF tests.

Norms over processes

In order to measure a limit distance between distributions, a norm $||.||$ over the space of continuous bridges needs to be chosen. Typical such norms are the norm-2 (or 'Cramer-von Mises' distance)

$$||Y||_2 = \int_0^1 Y(u)^2 \mathrm{d}u,$$

as the bridge is always integrable, or the norm-sup (also called the Kolmogorov distance)

$$||Y||_\infty = \sup_{u\in[0,1]} |Y(u)|,$$

as the bridge always reaches an extremal value.

Unfortunately, both these norms mechanically overweight the core values $u \approx 1/2$ and disfavor the tails $u \approx 0, 1$: since the variance of $Y(u)$ is zero at both extremes and maximal in the central value, the major contribution to $||Y||$ indeed comes from the central region and not from the tails. To alleviate this effect, in particular when the GoF test is intended to investigate a specific region of the domain, it is preferable to introduce additional weights and study $||Y\sqrt{\psi}||$ rather than $||Y||$ itself. Anderson and Darling show in Ref. [10] that the solution to the problem with the Cramer-von Mises norm and arbitrary weights ψ is obtained by spectral decomposition of the covariance kernel. They design an eponymous test [11] with the specific choice of $\psi(u) = 1/I(u,u)$ equal to the inverse variance, which equi-weights all quantiles of the distribution to be tested. We analyze here the case of the same weights but with the Kolmogorov distance.

So again $Y(u)$ is a Brownian bridge, i.e. a centered Gaussian process on $u \in [0,1]$ with covariance function $I(u,v)$ given in Eq. (3). In particular, $Y(0) = Y(1) = 0$ with probability equal to 1, no matter how distant F is from the sample cdf around the core values. In order to put more emphasis on specific regions of the domain, let us weight the Brownian bridge as follows: for given $a \in]0,1[$ and $b \in [a,1[$, we define

$$\tilde{y}(u) = y(u) \cdot \begin{cases} \sqrt{\psi(u)} & , a \leq u \leq b \\ 0 & , \text{otherwise.} \end{cases} \tag{4}$$

We will characterize the law of the supremum $K(a,b) \equiv \sup_{u\in[a,b]} |\tilde{y}(u)|$:

$$\mathcal{P}_<(k|a,b) \equiv \mathbb{P}[K(a,b) \leq k] = \mathbb{P}[|\tilde{y}(u)| \leq k, \forall u \in [a,b]] \,.$$

2.2. *The equi-weighted Brownian bridge: Kolmogorov-Smirnov*

In the case of a constant weight, corresponding to the classical KS test, the probability $\mathcal{P}_<(k;0,1)$ is well defined and has the well known KS form [5]:

$$\mathcal{P}_<(k;0,1) = 1 - 2\sum_{n=1}^{\infty}(-1)^{n-1}\mathrm{e}^{-2n^2k^2}, \tag{5}$$

which, as expected, grows from 0 to 1 as k increases. The value k^* such that this probability is 95% is $k^* \approx 1.358$ [6]. This can be interpreted as follows: if, for a data set of size N, the maximum value of $|Y_N(u)|$ is larger than ≈ 1.358, then the hypothesis that the proposed distribution is a "good fit" can be rejected with 95% confidence.

Diffusion in a cage with fixed walls

The Brownian bridge Y is nothing else than a Brownian motion with imposed terminal condition, and can be written as $Y(u) = X(u) - u\,X(1)$ where X is a Brownian motion. The survival probability of Y in a cage with absorbing walls can be found by counting the number of Brownian paths that go from $Y(0) = 0$ to $Y(1) = 0$ without ever hitting the barriers. More precisely, the survival probability of the Brownian bridge in the same stripe can be computed as $f_1(0;k)/f_1(0;\infty)$, where $f_u(y;k)$ is the transition kernel of the Brownian motion within the allowed region $[-k,k]$. It satisfies the simple Fokker-Planck equation

$$\begin{cases} \partial_u f_u(y;k) = \frac{1}{2}\partial_y^2 f_u(y;k) \\ f_u(\pm k;k) = 0 \end{cases}, \quad \forall u \in [0,1].$$

By spectral decomposition of the Laplacian, the solution is found to be

$$f_u(y;k) = \frac{1}{k}\sum_{n\in\mathbb{Z}} \mathrm{e}^{-E_n u}\cos\left(\sqrt{2E_n}\,y\right), \quad \text{where} \quad E_n = \frac{1}{2}\left(\frac{(2n-1)\pi}{2k}\right)^2$$

and the free propagator in the limit $k \to \infty$ is the usual

$$f_u(y;\infty) = \frac{1}{\sqrt{2\pi u}}\,\mathrm{e}^{-\frac{y^2}{2u}},$$

so that the survival probability of the constrained Brownian bridge is

$$\mathcal{P}_<(k;0,1) = \frac{\sqrt{2\pi}}{k}\sum_{n\in\mathbb{Z}}\exp\left(-\frac{(2n-1)^2\pi^2}{8k^2}\right). \tag{6}$$

Although it looks different from Eq. (5), the two expressions can be shown to be exactly identical. But the above proof looks to us way easier than the canonical ones [10].

Diffusion in a cage with moving walls

The problem can be looked at differently. Under the following change of variable and time

$$W(t) = (1+t)\,Y\left(\frac{t}{1+t}\right), \quad t = \frac{u}{1-u} \in \left[\frac{a}{1-a}, \frac{b}{1-b}\right], \tag{7}$$

the problem can be transformed into that of a Brownian diffusion inside a box with walls *moving at constant velocity*. Indeed, one can check that

$$\mathrm{Cov}\big(W(t), W(t')\big) = \min(t,t'),$$

and that $\mathcal{P}_<(k|0,1)$ can be now written as

$$\mathcal{P}_<(k|0,1) = \mathbb{P}[|W(t)| \leq k\,(1+t), \forall t \in [0,\infty[\,]\,.$$

Since the walls expand as $\sim t$ faster than the diffusive particle can move ($\sim \sqrt{t}$), the survival probability converges to a positive value, which is again given by the usual Kolmogorov distribution (5) [12–14].

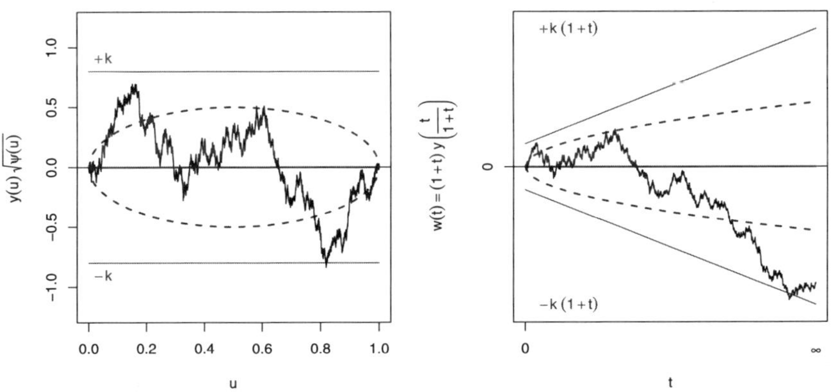

Fig. 1. The equi-weighted Brownian bridge, $\psi(u) = 1$. The time-changed rescaled process lives in a geometry with boundaries receding at constant speed.

2.3. *The variance-weighted Brownian bridge: Accounting for the tails*

As mentioned above, the classical KS test is only weakly sensitive to the quality of the fit in the tails of the distribution, when it is often these tail events (corresponding to centennial floods, devastating earthquakes, financial crashes, etc.) that one is most concerned with (see, e.g., [15]).

A simple and elegant GoF test for the tails *only* can be designed starting with digital weights in the form $\psi(u;a) = \mathbb{1}_{\{u\geq a\}}$ or $\psi(u;b) = \mathbb{1}_{\{u\leq b\}}$ for upper and lower tail, respectively. The corresponding test laws can be read off Eq. (5.9) in [10].[a] Investigation of both tails is attained with $\psi(u;q) = \mathbb{1}_{\{u\leq 1-q\}} + \mathbb{1}_{\{u\geq q\}}$ (where $q > \frac{1}{2}$).

[a]The quantity M appearing there is the volume under the normal bivariate surface between specific bounds, and it takes a very convenient form in the unilateral cases $\frac{1}{2} \leq a \leq u \leq 1$ and $0 \leq u \leq b \leq \frac{1}{2}$. Mind the missing j exponentiating the alternating (-1) factor.

Here we rather focus on a GoF test for a univariate sample of size $N \gg 1$, with the Kolmogorov distance but equi-weighted quantiles, which is equally sensitive to *all regions* of the distribution.[b] We unify two earlier attempts at finding asymptotic solutions, one by Anderson and Darling in 1952 [10] and a more recent, seemingly unrelated one that deals with "life and death of a particle in an expanding cage" by Krapivsky and Redner [12, 16]. We present here the exact asymptotic solution of the corresponding stochastic problem, and deduce from it the precise formulation of the GoF test, which is of a fundamentally different nature than the KS test.

So in order to zoom on the tiny differences in the tails of the Brownian bridge, we weight it as explained earlier, with its variance

$$\psi(u) = \frac{1}{u\,(1-u)}.$$

Solutions for the distributions of such variance-weighted Kolmogorov-Smirnov statistics were studied by Noé, leading to the laws of the one-sided [17] and two-sided [18] finite sample tests. They were later generalized and tabulated numerically by Niederhausen [19, 20]. However, although exact and appropriate for small samples, these solutions rely on recursive relations and are not in closed form. We instead come up with an analytic closed-form solution for large samples that relies on an elegant analogy from statistical physics.

Diffusion in a cage with moving walls

After performing the above change of variable (7) that converts a Brownian bridge into a Brownian motion, $\mathcal{P}_<(k|a,b)$ can be written as

$$\mathcal{P}_<(k|a,b) = \mathbb{P}\Big[|W(t)| \le k\sqrt{t}, \forall t \in [\tfrac{a}{1-a}, \tfrac{b}{1-b}]\Big].$$

The problem with initial time $\frac{a}{1-a} = 0$ and horizon time $\frac{b}{1-b} = T$ has been treated by Krapivsky and Redner in Ref. [12] as the survival probability $S(T; k = \sqrt{\frac{A}{2D}})$ of a Brownian particle diffusing with constant D in a cage with walls expanding as $\sqrt{At}$. Their result is that for large T,

$$S(T;k) \equiv \mathcal{P}_<(k|0, \tfrac{T}{1+T}) \propto T^{-\theta(k)}.$$

They obtain analytical expressions for $\theta(k)$ in both limits $k \to 0$ and $k \to \infty$. The limit solutions of the very same differential problem were

[b] Other choices of ψ generally result in much harder problems.

found earlier by Turban for the critical behavior of the directed self-avoiding walk in parabolic geometries [21].

We take here a slightly different route, suggested (but not finalized) by Anderson and Darling in Ref. [10]. Our specific contributions are: (i) we treat the general case $a > 0$ for *any* k; (ii) we explicitly compute the k-dependence of both the exponent *and* the prefactor of the power-law decay; and (iii) we provide the link with the theory of GoF tests and compute the pre-asymptotic distribution when $]a, b[\to]0, 1[$ of the weighted Kolmogorov-Smirnov test statistics.

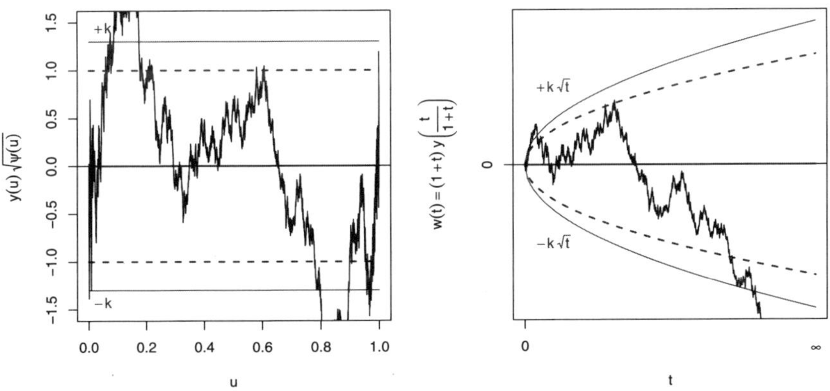

Fig. 2. The variance-weighted Brownian bridge, $\psi(u) = 1/[u(1-u)]$. The time-changed rescaled process lives in a geometry with boundaries receding as $\sim \sqrt{t}$.

Mean-reversion in a cage with fixed walls

Introducing now the new time change $\tau = \ln \sqrt{\frac{1-a}{a}\, t}$, the variable $Z(\tau) = W(t)/\sqrt{t}$ is a stationary Ornstein-Uhlenbeck process on $[0, T]$ where

$$T = \ln \sqrt{\frac{b\,(1-a)}{a\,(1-b)}}, \tag{8}$$

and

$$\mathrm{Cov}\big(Z(\tau), Z(\tau')\big) = \mathrm{e}^{-|\tau-\tau'|}.$$

Its dynamics is described by the stochastic differential equation

$$\mathrm{d}Z(T) = -Z(T)\mathrm{d}T + \sqrt{2}\,\mathrm{d}B(T), \tag{9}$$

with $B(T)$ an independent Wiener process. The initial condition for $T = 0$ (corresponding to $b = a$) is $Z(0) = Y(a)/\sqrt{\mathbb{V}[Y(a)]}$, a random Gaussian variable of zero mean and unit variance. The distribution $\mathcal{P}_<(k|a, b)$ can now be understood as the unconditional survival probability of a mean-reverting particle in a cage with fixed absorbing walls:

$$\begin{aligned}\mathcal{P}_<(k|T) &= \mathbb{P}[-k \le Z(\tau) \le k, \forall \tau \in [0, T]]\\ &= \int_{-k}^{k} f_T(z; k)\,\mathrm{d}z,\end{aligned}$$

where

$$f_T(z; k)\,\mathrm{d}z = \mathbb{P}\big[Z(T) \in [z, z + \mathrm{d}z[\,|\, \{Z(\tau)\}_{\tau<T}\big]$$

is the density probability of the particle being at z at time T, when walls are in $\pm k$. Its dependence on k, although not explicit on the right hand side, is due to the boundary condition associated with the absorbing walls (it will be dropped in the following for the sake of readability)[c].

The Fokker-Planck equation governing the evolution of the density $f_T(z)$ reads

$$\partial_\tau f_\tau(z) = \partial_z\left[z\, f_\tau(z)\right] + \partial_z^2\left[f_\tau(z)\right], \quad 0 < \tau \le T.$$

Calling $\mathcal{H}_{\rm FP}$ the second order differential operator $-\left[\mathbb{1} + z\partial_z + \partial_z^2\right]$, the full problem thus amounts to finding the general solution of

$$\begin{cases} -\partial_\tau f_\tau(z) = \mathcal{H}_{\rm FP}(z) f_\tau(z) \\ f_\tau(\pm k) = 0, \forall \tau \in [0, T] \end{cases}.$$

We have explicitly introduced a minus sign since we expect that the density decays with time in an absorption problem. Because of the term $z\partial_z$, $\mathcal{H}_{\rm FP}$ is not hermitian and thus cannot be diagonalized. However, as is well known, one can define $f_\tau(z) = \mathrm{e}^{-\frac{z^2}{4}}\phi_\tau(z)$ and the Fokker-Planck equation becomes

$$\begin{cases} -\partial_\tau \phi_\tau(z) = \left[-\partial_z^2 + \frac{1}{4}z^2 - \frac{1}{2}\mathbb{1}\right]\phi_\tau(z) \\ \phi_\tau(\pm k) = 0, \forall \tau \in [0, T] \end{cases}.$$

Its Green's function, i.e. the (separable) solution *conditionally on the initial position* $(z_{\rm i}, T_{\rm i})$, is the superposition of all modes

$$G_\phi(z, T \mid z_{\rm i}, T_{\rm i}) = \sum_\nu \mathrm{e}^{-\theta_\nu(T-T_{\rm i})}\widehat{\varphi}_\nu(z)\widehat{\varphi}_\nu(z_{\rm i}),$$

[c]In particular, $\mathcal{P}_<(k|0) = \mathrm{erf}\left(\frac{k}{\sqrt{2}}\right)$.

where $\widehat{\varphi}_\nu$ are the normalized solutions of the stationary Schrödinger equation

$$\begin{cases} \left[-\partial_z^2 + \frac{1}{4}z^2\right]\varphi_\nu(z) = \left(\theta_\nu + \frac{1}{2}\right)\varphi_\nu(z) \\ \varphi_\nu(\pm k) = 0 \end{cases},$$

each decaying with its own energy θ_ν, where ν labels the different solutions with increasing eigenvalues, and the set of eigenfunctions $\{\widehat{\varphi}_\nu\}$ defines an orthonormal basis of the Hilbert space on which $\mathcal{H}_\mathrm{S}(z) = \left[-\partial_z^2 + \frac{1}{4}z^2\right]$ acts. In particular,

$$\sum_\nu \widehat{\varphi}_\nu(z)\widehat{\varphi}_\nu(z') = \delta(z - z'), \tag{10}$$

so that indeed $G_\phi(z, T_\mathrm{i} \mid z_\mathrm{i}, T_\mathrm{i}) = \delta(z - z_\mathrm{i})$, and the general solution writes

$$f_T(z_T; k) = \int_{-k}^{k} \mathrm{e}^{\frac{z_\mathrm{i}^2 - z_T^2}{4}} G_\phi(z_T, T \mid z_\mathrm{i}, T_\mathrm{i})\, f_0(z_\mathrm{i})\, \mathrm{d}z_\mathrm{i},$$

where $T_\mathrm{i} = 0$, which corresponds to the case $b = a$ in Eq. (4), and f_0 is the distribution of the initial value z_i which is here, as noted above, Gaussian with unit variance.

$\mathcal{H}_\mathrm{S}$ describes an harmonic oscillator of mass $\frac{1}{2}$ and frequency $\omega = \frac{1}{\sqrt{2}}$ within an infinitely deep well of width $2k$: its eigenfunctions are parabolic cylinder functions [22, 23]

$$\begin{aligned} y_+(\theta; z) &= \quad \mathrm{e}^{-\frac{z^2}{4}}\, {}_1F_1\left(-\tfrac{\theta}{2}, \tfrac{1}{2}, \tfrac{z^2}{2}\right) \\ y_-(\theta; z) &= z\,\mathrm{e}^{-\frac{z^2}{4}}\, {}_1F_1\left(\tfrac{1-\theta}{2}, \tfrac{3}{2}, \tfrac{z^2}{2}\right) \end{aligned}$$

properly normalized. The only acceptable solutions for a given problem are the linear combinations of y_+ and y_- which satisfy orthonormality (10) and the boundary conditions: for periodic boundary conditions, only the integer values of θ would be allowed, whereas with our Dirichlet boundaries $|\widehat{\varphi}_\nu(k)| = -|\widehat{\varphi}_\nu(-k)| = 0$, real non-integer eigenvalues θ are allowed. For instance, the fundamental level $\nu = 0$ is expected to be the symmetric solution $\widehat{\varphi}_0(z) \propto y_+(\theta_0; z)$ with θ_0 the smallest possible value compatible with the boundary condition:

$$\theta_0(k) = \inf_{\theta > 0} \left\{\theta : y_+(\theta; k) = 0\right\}. \tag{11}$$

In what follows, it will be more convenient to make the k-dependence explicit, and a hat will denote the solution with the normalization relevant to our problem, namely $\widehat{\varphi}_0(z;k) = y_+(\theta_0(k);z)/||y_+||_k$, with the norm

$$||y_+||_k^2 \equiv \int_{-k}^{k} y_+(\theta_0(k);z)^2 \,\mathrm{d}z,$$

so that $\int_{-k}^{k} \widehat{\varphi}_\nu(z;k)^2 \,\mathrm{d}z = 1$.

Asymptotic survival rate

Denoting by $\Delta_\nu(k) \equiv [\theta_\nu(k) - \theta_0(k)]$ the gap between the excited levels and the fundamental, the higher energy modes $\widehat{\varphi}_\nu$ cease to contribute to the Green's function when $\Delta_\nu T \gg 1$, and their contributions to the above sum die out exponentially as T grows. Eventually, only the lowest energy mode $\theta_0(k)$ remains, and the solution tends to

$$f_T(z;k) = A(k)\,\mathrm{e}^{-\frac{z^2}{4}}\,\widehat{\varphi}_0(z;k)\,\mathrm{e}^{-\theta_0(k)T},$$

when $T \gg (\Delta_1)^{-1}$, with

$$A(k) = \int_{-k}^{k} \mathrm{e}^{\frac{z_\mathrm{i}^2}{4}}\,\widehat{\varphi}_0(z_\mathrm{i};k) f_0(z_\mathrm{i})\,\mathrm{d}z_\mathrm{i}. \tag{12}$$

Let us come back to the initial problem of the weighted Brownian bridge reaching its extremal value in $[a,b]$. If we are interested in the limit case where a is arbitrarily close to 0 and b close to 1, then $T \to \infty$ and the solution is thus given by

$$\begin{aligned}\mathcal{P}_<(k|T) &= A(k)\,\mathrm{e}^{-\theta_0(k)T}\int_{-k}^{k} \mathrm{e}^{-\frac{z^2}{4}}\,\widehat{\varphi}_0(z;k)\,\mathrm{d}z\\ &= \widetilde{A}(k)\,\mathrm{e}^{-\theta_0(k)T},\end{aligned}$$

with $\widetilde{A}(k) \equiv \sqrt{2\pi}A(k)^2$.

We now compute explicitly the limit behavior of both $\theta_0(k)$ and $\widetilde{A}(k)$.

$\boxed{\boldsymbol{k \to \infty}}$ As k goes to infinity, the absorption rate $\theta_0(k)$ is expected to converge toward 0: intuitively, an infinitely far barrier will not absorb anything. At the same time, $\mathcal{P}_<(k|T)$ must tend to 1 in that limit. So $\widetilde{A}(k)$ necessarily tends to unity. Indeed,

$$\theta_0(k) \xrightarrow{k\to\infty} \sqrt{\frac{2}{\pi}}k\,\mathrm{e}^{-\frac{k^2}{2}} \to 0, \tag{13}$$

$$\widetilde{A}(k) \xrightarrow{k\to\infty} \left(\int_{-\infty}^{\infty} \widehat{\varphi}_0(z;\infty)^2\,\mathrm{d}z\right)^2 = 1.$$

In principle, we see from Eq. (12) that corrections to the latter arise both (and jointly) from the functional relative difference of the solution $\epsilon(z;k) = y_+(\theta_0(k);z)/y_+(0;z) - 1$, and from the finite integration limits ($\pm k$ instead of $\pm\infty$). However, it turns out that the correction of the first kind is of second order in ϵ, see [4]. The correction to $A(k)$ is thus dominated by the finite integration limits $\pm k$, so that

$$\widetilde{A}(k \to \infty) \approx \operatorname{erf}\left(\frac{k}{\sqrt{2}}\right)^2 . \tag{14}$$

$\boxed{\boldsymbol{k \to 0}}$ For small k, the system behaves like a free particle in a sharp and infinitely deep well, since the quadratic potential is almost flat around 0. The fundamental mode becomes then

$$\widehat{\varphi}_0(z; k \to 0) = \frac{1}{\sqrt{k}} \cos\left(\frac{\pi z}{2k}\right),$$

and consequently

$$\theta_0(k) \xrightarrow{k\to 0} \frac{\pi^2}{4k^2} - \frac{1}{2}, \tag{15}$$

$$\widetilde{A}(k) \xrightarrow{k\to 0} \frac{16}{\pi^2\sqrt{2\pi}} k. \tag{16}$$

We show in Fig. 3 the functions $\theta_0(k)$ and $\widetilde{A}(k)$ computed numerically from the exact solution, together with their asymptotic analytic expressions. In intermediate values of k (roughly between 0.5 and 3) these limit expressions fail to reproduce the exact solution.

Higher modes and validity of the asymptotic ($N \gg 1$) solution

Higher modes $\nu > 0$ with energy gaps $\Delta_\nu \lesssim 1/T$ must in principle be kept in the pre-asymptotic computation. This, however, is irrelevant in practice since the gap $\theta_1 - \theta_0$ is never small. Indeed, $\widehat{\varphi}_1(z;k)$ is proportional to the asymmetric solution $y_-(\theta_1(k);z)$ and its energy

$$\theta_1(k) = \inf_{\theta > \theta_0(k)} \left\{\theta : y_-(\theta;k) = 0\right\}$$

is found numerically to be very close to $1 + 4\theta_0(k)$. In particular, $\Delta_1 > 1$ (as we illustrate in Fig. 4) and thus $T\Delta_1 \gg 1$ will always be satisfied in cases of interest.

2.3.1. *Back to GoF testing*

Let us now come back to GoF testing. In order to convert the above calculations into a meaningful test, one must specify values of a and b. The natural choice would be $a = 1/N$, corresponding to the min of the sample series since $F(\min x_n) \approx F_N(\min x_n) = \frac{1}{N}$. Eq. (8) above motivates a slightly different value of $a = 1/(N+1)$ and $b = 1 - a$, such that the relevant value of T is given correspondingly by

$$T = \ln \sqrt{\frac{b\,(1-a)}{a\,(1-b)}} = \ln N.$$

This leads to our central result for the cdf of the weighted maximal Kolmogorov distance $K(\frac{1}{N+1}, \frac{N}{N+1})$ under the hypothesis that the tested and the true distributions coincide:

$$\boxed{S(N;k) = \mathcal{P}_<(k|\ln N) = \widetilde{A}(k) N^{-\theta_0(k)}}, \tag{17}$$

which is valid whenever $N \gg 1$ since, as we discussed above, the energy gap Δ_1 is greater than unity.

The final cumulative distribution function (the test law) is depicted in Fig. 5 for different values of the sample size N. Contrarily to the standard KS case, this distribution *still depends on* N: as N grows toward infinity, the curve is shifted to the right, and eventually $S(\infty;k)$ is zero for any k. In particular, the threshold value k^* corresponding to a 95% confidence level (represented as a horizontal grey line) increases with N. Since for large N, $k^* \gg 1$ one can use the asymptotic expansion above, which soon becomes quite accurate, as shown in Fig. 5. This leads to:

$$\theta_0(k^*) \approx -\frac{\ln 0.95}{\ln N} \approx \sqrt{\frac{2}{\pi}} k^* \, \mathrm{e}^{-\frac{k^{*2}}{2}},$$

which gives $k^* \approx 3.439, 3.529, 3.597, 3.651$ for, respectively, $N = 10^3, 10^4, 10^5, 10^6$. For exponentially large N and to logarithmic accuracy, one has: $k^* \sim \sqrt{2\ln(\ln N)}$. This variation is very slow, but one sees that as a matter of principle, the "acceptable" maximal value of the weighted distance is much larger (for large N) than in the KS case.

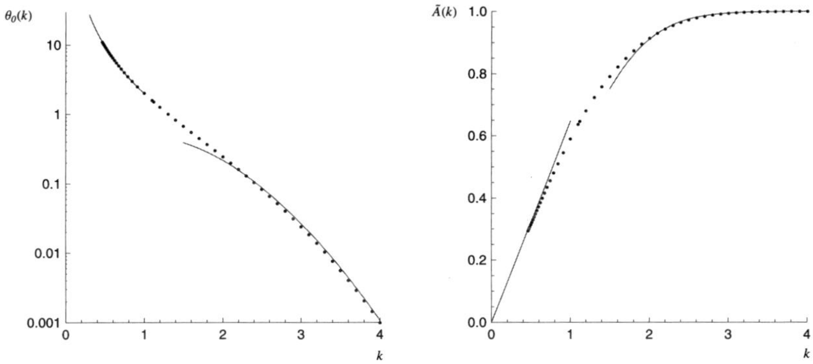

Fig. 3. **Left:** Dependence of the exponent θ_0 on k; similar to Fig. 2 in [12], but in lin-log scale; see in particular Eqs. (9b) and (12) there. **Right:** Dependence of the prefactor $\widetilde{A}$ on k. The red solid lines illustrate the analytical behavior in the limiting cases $k \to 0$ and $k \to \infty$.

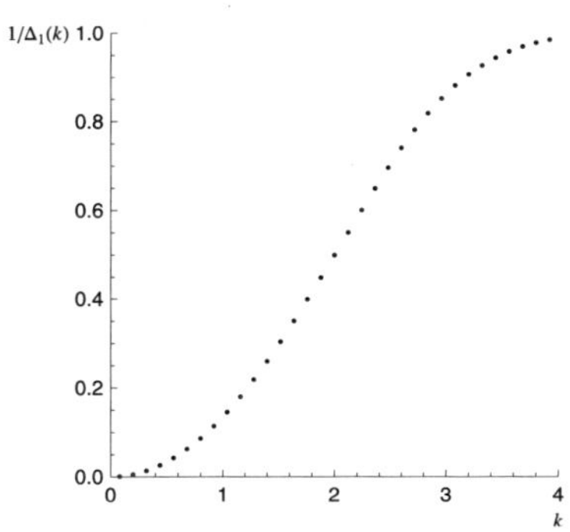

Fig. 4. $1/\Delta_1(k)$ saturates to 1, so that the condition $N \gg \exp[1/\Delta_1(k)]$ is virtually always satisfied.

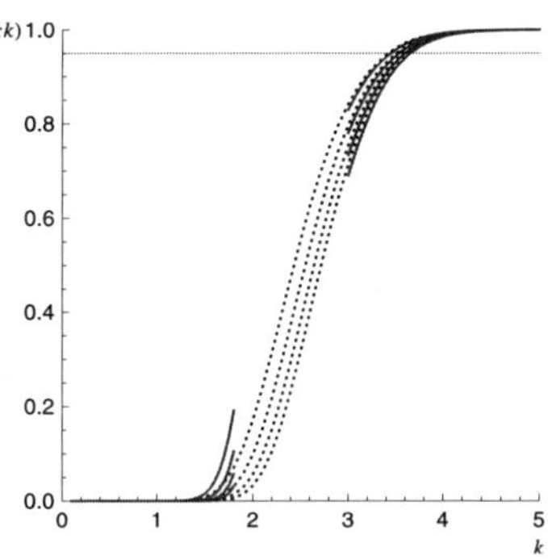

Fig. 5. Dependence of $S(N;k)$ on k for $N = 10^3, 10^4, 10^5, 10^6$ (from left to right). The red solid lines illustrate the analytical behavior in the limiting cases $k \to 0$ and $k \to \infty$.

3. Optimal time to sell a stock[d]

Consider the problem of holding a stock at an initial date $t = 0$, and hoping to sell it back at a time $t = \tau$ before a deadline $t = T$. The goal is to find the *ex ante* optimal selling time τ, i.e. take a decision at $t = 0$ as of when to sell it in the future for optimal profit. If the (log-)price can be modelled as a stationary random walk, this is not a restriction of generality, since at any later time the problem is identical with however a reduced horizon.

3.1. *Minimizing the expected distance to the maximum*

In technical terms, we aim at minimizing the expected (relative) spread

$$S(\tau; T) = \frac{M_T - X_\tau}{M_T}$$

between the instantaneous price X_τ and the *ex post* realized maximum over the allowed horizon $M_T = \max\{X_t, t \in [0, T]\}$.

In order for this forward-looking problem to be handled analytically, we impose that the price process follows a (possibly drifted) geometric Brownian motion $X_t = \mathrm{e}^{x_t}$ with

$$\mathrm{d}x_t = \mu\,\mathrm{d}t + \sigma\,\mathrm{d}B_t \quad \text{or} \quad \dot{x} = \mu + \sigma\eta$$

where B_t in the Stochastic Differential Equation (left) is a Wiener process, and $\eta = \frac{\mathrm{d}B_t}{\mathrm{d}t}$ in the Langevin equation (right) is a standard White Gaussian Noise. We rewrite the maximum value as $M_T = \mathrm{e}^{m_T}$ with obviously $m_T = \max\{x_t, t \in [0, T]\}$. The problem is clearly invariant under a shift of both x_t and m_T, so that we can arbitrarily set $x_0 = 0$.

The optimal time to sell is then defined as the solution of the minimization problem

$$\tau^* = \operatorname*{arg\,min}_{\tau \in [0,T]} \mathbb{E}[\ln S(\tau; T)] = \operatorname*{arg\,min}_{\tau \in [0,T]} \mathbb{E}[s(\tau; T)]$$

where $s(\tau; T) = m_T - x_\tau$. The expectation estimator is clearly inter-temporal, and the probability distribution function $\mathcal{P}_\mu$ of $s(\tau; T)$ can be written in terms of the joint density f_μ of (x_τ, m_T) as

$$\begin{aligned} \mathcal{P}_\mu(s; \tau, T) &= \int_0^\infty \int_{-\infty}^\infty f_\mu(x, m; \tau, T)\,\delta(m - x - s)\,\mathrm{d}x\,\mathrm{d}m \\ &= \int_0^\infty f_\mu(m - s, m; \tau, T)\,\mathrm{d}m. \end{aligned} \tag{18}$$

[d] Joint work with S. Majumdar [24], motivated by a paper by Shiryaev, Xu and Zhou [25].

This is equivalent to writing $\mathcal{P}_\mu(s;\tau,T) = \int \mathrm{d}F_\mu(m-s,m;\tau,T)$, with the (partial) cumulative distribution function

$$F_\mu(x,m;\tau,T) = \mathbb{P}[x_\tau = x, m_T \le m]$$

counting the fraction of the paths arriving in $[x, x+\mathrm{d}x]$ at time τ among all paths never crossing m from below over the whole horizon $[0,T]$. It is expressed in terms of the *causal* propagator G_μ as the probability of arriving in x at time τ without ever hitting m, and then arriving anywhere below m in the remaining time $T-\tau$:

$$F_\mu(x,m;\tau,T) = G_\mu(x,\tau;m) \int_{-\infty}^{m} G_\mu(x',T-\tau;m-x)\,\mathrm{d}x'. \tag{19}$$

The propagator $G_\mu(x,\tau;m)$ describes a μ-drifted diffusion close to a fixed absorbing boundary, or equivalently a pure diffusion close to a boundary moving at constant velocity ("daredevil at the edge of a receding cliff", [12]), see Fig. 6. It can be written in terms of the propagator G_0 of the zero-drift diffusion:

$$G_\mu(x,t;m) = \exp\left(-\frac{\mu^2 - 2\mu x}{2\sigma^2}\right) G_0(x,t;m),$$

where G_0 can be computed in several ways — method of images [16, 26], path-integral method [24, 27], solution of the Fokker-Planck equation. The solution writes (up to a normalizing constant) as the difference $G_0(x,t;m) \propto G_0(x,t;\infty) - G_0(x-2m,t;\infty)$ between the free propagator

$$G_0(x,t;\infty) = \frac{1}{\sqrt{2\pi\sigma^2 t}} \exp\left(-\frac{(x-x_0)^2}{2\sigma^2 t}\right)$$

with initial positions at $x_0 = 0$ and $x_0 = 2m$. As expected, $G_0(m,t;m) = 0$ at all times, saying that the probability of presence *at* the boundary is nil.

Once G_μ is known, the joint distribution f_μ of x_τ and m_T is obtained by differentiating Eq. (19) with respect to m, and the distribution of the spread is found from Eq. (18) to be

$$\boxed{\mathcal{P}_\mu(s;\tau,T) = a_\mu(s;\tau)\,b_{-\mu}(s,T-\tau) + a_{-\mu}(s;T-\tau)\,b_\mu(s,\tau)}\,, \tag{20}$$

where

$$a_\mu(s;\tau) = \frac{\mu}{2\sigma^2}\exp\left(-\frac{2s\mu}{\sigma^2}\right)\mathrm{erfc}\left(\frac{s-\mu\tau}{\sqrt{2\sigma^2\tau}}\right) + \frac{1}{\sqrt{2\pi\sigma^2\tau}}\exp\left(-\frac{(s+\mu\tau)^2}{2\sigma^2\tau}\right)$$

$$b_\mu(s;\tau) = -\exp\left(-\frac{2s\mu}{\sigma^2}\right)\mathrm{erfc}\left(\frac{s-\mu\tau}{\sqrt{2\sigma^2\tau}}\right) + \mathrm{erfc}\left(-\frac{s+\mu\tau}{\sqrt{2\sigma^2\tau}}\right).$$

Finally, the expected spread is

$$\mathbb{E}[s(\tau, T)] = \int_0^\infty s\,\mathcal{P}_\mu(s;\tau,T)\,\mathrm{d}s = \iint_0^\infty s\,f_\mu(m-s,m;\tau,T)\,\mathrm{d}m\,\mathrm{d}s, \quad (21)$$

and the optimal τ^* is found by minimizing this function with respect to τ:

$$\tau^* = \begin{cases} T & \text{if } \mu \geq 0 \\ 0 & \text{if } \mu \leq 0 \end{cases}. \quad (22)$$

It is degenerate at $\mu = 0$ where both $\tau^* = 0$ and $\tau^* = T$ are optimal. This result states in technical terms a very intuitive truism: whenever the log-prices are expected to increase in average ($\mu > 0$) one should keep the stock as long as possible, and conversely if the log-prices are expected to fall ($\mu < 0$) one should sell immediately. It should not be surprising that τ^* is not affected by the value of the so-called "volatility" parameter σ, since the optimization program only focused on "maximizing the gain" without controlling for the encountered risk, and thus the solution applies to a risk-neutral agent only.

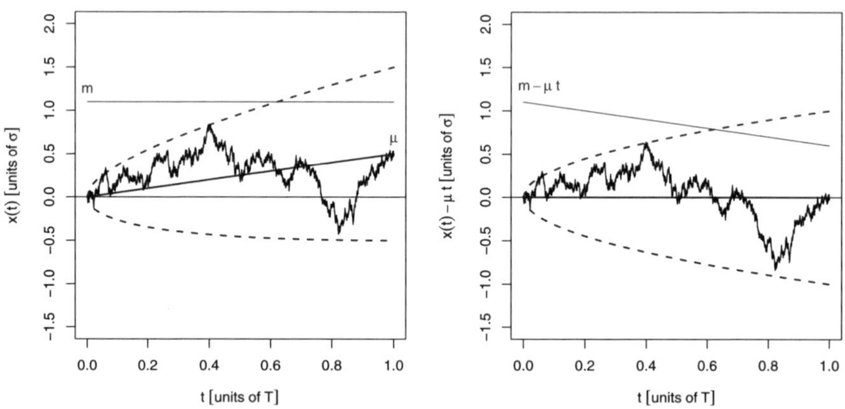

Fig. 6. Illustration for a positive drift ($\mu > 0$). In the case where $\mu < 0$, the upper barrier moves *away* linearly. See Sect. 2.2 page 451 for an application of the bilateral case.

3.2. *Maximizing the occurrence time probability of maximum*

Alternatively to minimizing the spread $s(\tau;T)$ between the expected maximum m_T and the log-price x_τ at the selling time, one can try to maximize the probability $p_\mu(\tau;T)\mathrm{d}\tau = \mathrm{d}\mathbb{P}[x_\tau = m_T]$ that m_T will occur at time τ, whatever value it takes. But the joint probability that the global maximum over $[0,T]$ has a value $m_T = m$, and that this maximum is reached for the first (and in fact only) time at $t=\tau$ is nothing else than $f_\mu(m,m;\tau,T)$.

In order to avoid the issue of infinite crossings of the continuous Brownian motion [24, 28, 29], we allow for an infinitesimal spread $s = m - x$ that we eventually take to 0:

$$p_\mu(\tau;T) \propto \int_0^\infty \lim_{s\to 0} f_\mu(m-s,m;\tau,T)\,\mathrm{d}m = \mathcal{P}_\mu(0^+;\tau,T),$$

with $\mathcal{P}_\mu$ given in Eq. (20). Noticing that a_μ converges to a finite value $a_\mu(0,\tau)$ when $s\to 0$, and that the expansion of b_μ to first order in s around 0 is $b_\mu(s\to 0,\tau) \approx 4s\,a_\mu(0,\tau)$, we have

$$\mathcal{P}_\mu(s\to 0;\tau,T) = 8s\,a_\mu(0,\tau)\,a_{-\mu}(0,T-\tau)$$

and finally, normalizing with $\int_0^T \mathcal{P}_\mu(s;\tau,T)\mathrm{d}\tau$, we get

$$\boxed{p_\mu(\tau;T) = 2\sigma^2\,a_\mu(0,\tau)\,a_{-\mu}(0,T-\tau)}\,. \tag{23}$$

Notice that when $\mu = 0$ the function $a_0(s,\tau)$ is the centered normal distribution with variance $\sigma^2\tau$. In particular $a_0(0,\tau) = 1/\sqrt{2\pi\sigma^2\tau}$ and one recovers Lévy's result [30]:

$$p_0(\tau;T) = \frac{1}{\pi}\frac{1}{\sqrt{\tau\,(T-\tau)}},$$

with two global maxima at $\tau^\star = 0$ and $\tau^\star = T$. For non zero μ's, the distribution still has inverse square-root singularities both at $\tau = 0$ and $\tau = T$, but with unequal amplitudes. For $\mu < 0$, the amplitude of the $\tau = 0$ singularity is larger than that of the $\tau = T$ singularity, and vice-versa when $\mu > 0$. Therefore one concludes that:

$$\tau^\star = \underset{\tau\in[0,T]}{\arg\max}\; p_\mu(\tau;T) = \begin{cases} T & \text{if } \mu \geq 0 \\ 0 & \text{if } \mu \leq 0 \end{cases} \tag{24}$$

and is thus equal to τ^*, see Eq. (22).

Whereas the minimization program of the previous section embedded the information of all the possible gaps s (as revealed by Eq. (21)), maximizing the occurrence time distribution only cares for the infinitesimal proximity of the maximum $s = m - x \to 0$. Nevertheless, although the objective function is not the same (minimize the spread or maximize the probability), the solution of the optimal time is not sensitive to the chosen criterion.

4. Optimal trading with linear costs[e]

The problem addressed in this section is to determine the optimal strategy in the presence of "linear" trading costs (i.e. a fixed cost *per share*, neglecting any price impact) and a constraint on the maximum size of the position (both long and short). This problem is of very significant interest in practice, at least for small sizes. For large sizes, a quadratic cost can be added to mimic price impact; the problem is however not (yet ?) solved in full generality.

We consider an agent who wants to maximize his/her expected gains, by trading a single asset, of current price Price_t, over a long period $[0, T]$ (we will later consider the limit $T \to \infty$). The position (signed number of shares/contracts) of the trader at time t is π_t. We assume that the agent has some signal p_t that predicts the next price change $r_t = \text{Price}_{t+1} - \text{Price}_t$, and is faced with the following constraints:

- His/her risk control system is simply a cap on the absolute size of his/her position : $|\pi_t| \leq M$, with no other risk control.
- He/she has to pay linear costs $\Gamma|\Delta\pi_t|$ whenever he/she trades a quantity $\Delta\pi_t \equiv \pi_{t+1} - \pi_t$

We assume that the predictor has a number of "nice" (but natural) properties; in particular, the predictability $\mathfrak{L}_t(p) = \mathbb{E}[r_t|p_t = p]$ is an odd, continuous and strictly increasing function of p. We also assume that it is Markovian: $\forall \omega_{t+1}$, $\mathbb{P}[\omega_{t+1}|p_t, p_{t-1}, \ldots] = \mathbb{P}[\omega_{t+1}|p_t]$ where ω_{t+1} is any event at $t+1$. In what follows, we will use the notation $P_t(p'|p)\mathrm{d}p' = \mathbb{P}[p_{t+1} = p'|p_t = p]\mathrm{d}p'$. We also define an integrated predictability at $t = \infty$, depending on p_t :

$$p_\infty(p_t) = \mathbb{E}[\text{Price}_\infty - \text{Price}_t|p_t] = \sum_{n=0}^{\infty} \mathbb{E}[r_{t+n}|p_t].$$

This quantity indicates how much one will gain in the future if one keeps a fixed position $\pi_{t' \geq t} = \pi$: the expected gain is then $p_\infty(p_t)\,\pi$.

4.1. *The optimal strategy*

A naïve solution

At first sight, the solution to this problem seems straightforward: if the expected future gain (given by the integrated predictability) exceeds the

[e] Joint work with J. de Lataillade, C. Deremble and M. Potters [31].

trading cost per contract Γ, then one trades in the direction of the signal (if not already at the maximum position), otherwise one does not. This solution obviously generates a positive average gain, but it has no reason to be the optimal solution. Indeed, because the predictor is auto-correlated in time, it might be worthy (and in general it will be) to wait for a larger value of the predictor, in order to grab the opportunities that have the most chances to get realized, and discard the others. As we shall see, the mistake in this naïve reasoning is not to compare the future gain with the cost, but rather comes from a wrong definition of the future gain, which does not include future trading decisions.

The Bellman method: general solution

The framework to attack this problem is Bellman's optimal control theory, or dynamic programming [32], which consists in solving the problem backwards: by assuming one follows the optimal strategy for all future times $t' > t$, one can find the optimal solution at time t. As is usual in dynamic programming, one has a *control variable* π_t, which needs to be optimized, and a *state variable* p_t, which parameterizes the solution. The optimization is done through a *value function* $V_t(\pi, p)$, which gives the maximal expected gains between time t and $+\infty$, considering that the position at $t-1$ is π and the predictor's value at t is p. The optimal solution of the system will be denoted $(\pi_t^*)_{t\in[0,T]}$.

At the last time step $t = T$, the expected future return is really $p_\infty(p)\pi_T$ where $p = p_T$, since no trading is allowed beyond that time. Any trade $\Delta\pi_T$ induces a cost $\Gamma|\Delta\pi_T|$, so:

$$\begin{array}{lllll}
\text{If } p_\infty(p) & \geq +\Gamma & \text{then } \pi_T^* = +M, & \text{and } V_T(\pi,p) = & +p_\infty(p)M - \Gamma\,(M-\pi) \\
\text{If } p_\infty(p) & \leq -\Gamma & \text{then } \pi_T^* = -M, & \text{and } V_T(\pi,p) = & -p_\infty(p)M - \Gamma\,(M+\pi) \\
\text{If } |p_\infty(p)| & < \Gamma & \text{then } \pi_T^* = \pi\ \ , & \text{and } V_T(\pi,p) = & p_\infty(p)\pi.
\end{array}$$

Hence, one recovers exactly the naïve solution in this case, but this is only because there is no trading beyond $t = T$. Now at $t < T$, the quantity to be maximized includes immediate gains, costs and future gains. This leads to the following recursion relation:

$$V_t(\pi, p) = \max_{|\pi'|\leq M} \left\{ \mathfrak{L}_t(p)\cdot\pi' - \Gamma|\pi' - \pi| + \int V_{t+1}(\pi', p')P_t(p'|p)\mathrm{d}p' \right\}, \tag{25}$$

and π_t^* is the value of π' which realises this maximum when $\pi = \pi_{t-1}^*$ and $p = p_t$. The general solution is given by the following construction [31]:

- $\pi_t^* = \begin{cases} \pi_{t-1}^* & \text{if } |p_t| < q_t \\ \text{sign}(p_t) \cdot M & \text{if } |p_t| \geq q_t \end{cases} \qquad \text{(with } \pi_{-1} = 0\text{)}$
- q_t is such that $q_t \geq 0$ and $g(t, q_t) = \Gamma$, where $g(t, p)$ is a continuous, strictly increasing function of p which satisfies, for $t < T$:

$$g(t,p) = \mathfrak{L}_t(p) + \Gamma \left[\int_{q_{t+1}}^{\infty} - \int_{-\infty}^{-q_{t+1}} \right] P_t(p'|p)\mathrm{d}p' \qquad (26)$$
$$+ \quad \int_{-q_{t+1}}^{q_{t+1}} g(t+1, p') P_t(p'|p)\mathrm{d}p'.$$

The stationary solution

The solution provided by Bellman's method above exhibits in general a dependence in t. Let us now consider the case where $T \to \infty$, and suppose that the predictor is stationary, i.e. $P_t(p'|p) = P(p'|p)$ is independent of t. Then we obtain a *telescopic* (self-consistent) equation for the one-variable function g, and the solution for the threshold q^*:

$$g(p) = \mathfrak{L}(p) + \Gamma \left[\int_{q^*}^{\infty} - \int_{-\infty}^{-q^*} \right] P(p'|p)\mathrm{d}p' + \int_{-q^*}^{q^*} g(p') P(p'|p)\mathrm{d}p', \quad (27)$$

$$g(q^*) = \Gamma. \qquad (28)$$

The optimal solution π_t^* to the system is then similar to the general solution, but with a constant threshold q^*. Thus, we obtain a very simple trading system, always saturated at $\pm M$, with a threshold to decide at each step whether we should revert the position or not. This of course looks a lot like the naïve solution of page 466. The only difference lies in the value of the threshold $q^* = g^{-1}(\Gamma)$, defined by Eqs. (27,28), instead of $q_{\text{naïve}} = p_\infty^{-1}(\Gamma)$ for the naïve solution. Intuitively, these equations take our future trading into account, whereas the naïve solution does not.

If we look closely at Eq. (27), its interpretation becomes transparent: $2M\, g(p)$ equals the expected difference in total future profit between the situation where $\pi = +M$ and the situation where $\pi = -M$. This difference is made up of:

- $\Delta\pi\, \mathfrak{L}(p)$ which represents the difference in immediate gain
- $\Delta\pi\, \Gamma\, \mathbb{P}[p_{t+1} > q^* | p_t = p]$ which represents the loss if the current position is $-M$ and in the next time step the predictor goes over the positive threshold q^* (hence π will go to $+M$)

- $\Delta\pi\,\Gamma\,\mathbb{P}[p_{t+1} < -q^* | p_t = p]$ which represents the loss if the current position is $+M$ and in the next time step the predictor goes below the negative threshold $-q^*$ (hence π will go to $-M$)
- $\int_{-q^*}^{q^*} P(p'|p)\ 2Mg(p')\ \mathrm{d}p'$ which is the expected difference in total future profit if, in the next step, the predictor remains between the two thresholds (leaving π unchanged).

Since the change of position between $-M$ and $+M$ costs $2M\Gamma$, it makes sense to compare $2Mg(p)$ with it and only trade when $g(p)$ is greater than Γ. Hence, $g(p)$ can be seen as the "gain per traded lot".

According to Eq. (27), $|g(p)| \geq |\mathfrak{L}(p)|$ and with Eq. (28) this implies in particular that $\mathfrak{L}(q^*) \leq \Gamma$. This property is actually rather intuitive: indeed, if the immediate expected gain was higher than the trading cost, then there would be no reason not to trade the maximal possible amount.

From here on, we only consider a linear predictor, $\mathfrak{L}_t(p) = p$.

Reformulation as a path integral

Although Eq. (27) is easy to interpret, it proves very difficult to solve in concrete cases. It can be rewritten by expanding the function g:

$$
\begin{aligned}
g(p) &= p + \int_{-q^*}^{q^*} p_1 P(p_1|p)\mathrm{d}p_1 + \int_{-q^*}^{q^*}\int_{-q^*}^{q^*} p_2 P(p_2|p_1)P(p_1|p)\mathrm{d}p_1\mathrm{d}p_2 + \dots \\
&\quad + \Gamma\cdot\left[\int_{q^*}^{+\infty} P(p_1|p)\mathrm{d}p_1 + \int_{q^*}^{+\infty}\int_{-q^*}^{q^*} P(p_2|p_1)P(p_1|p)\mathrm{d}p_1\mathrm{d}p_2 + \dots\right] \\
&\quad - \Gamma\cdot\left[\int_{-\infty}^{-q^*} P(p_1|p)\mathrm{d}p_1 + \int_{-\infty}^{-q^*}\int_{-q^*}^{q^*} P(p_2|p_1)P(p_1|p)\mathrm{d}p_1\mathrm{d}p_2 + \dots\right] \\
&= \mathcal{G}(p) + \Gamma\left[\mathcal{P}_+(p) - \mathcal{P}_-(p)\right],
\end{aligned}
\tag{29}
$$

and can thus be understood as a path integral: $\mathcal{G}(p)$ can be interpreted as the average over all possible exit times n of the cumulated predictor $\sum_{i=0}^{n-1} p_i$, where the expectation is taken only over all paths that stay in $[-q^*, q^*]$ until n. Similarly $\mathcal{P}_+(p)$ and $\mathcal{P}_-(p)$ are the probabilities for a path starting at $p_0 = p$ to exit (at any possible later time) above q^* or below $-q^*$, respectively.

Using now the fact that $g(q^*) = \Gamma$, and $\mathcal{P}_+(q^*) + \mathcal{P}_-(q^*) = 1$, we get:

$$\mathcal{G}(q^*) - 2\Gamma\,\mathcal{P}_-(q^*) = 0. \tag{30}$$

In some cases, both terms of this equation will tend to be infinitesimal, so it is rather the ratio $\lim_{p\to q^*} \mathcal{G}(p)/\mathcal{P}_-(p)$ that we will ask to equal 2Γ. Figure 7 illustrates our reformulation of the problem in terms of first passage times properties.

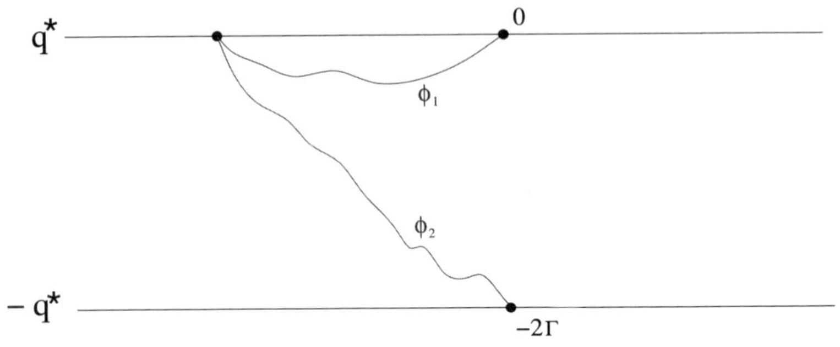

Fig. 7. (Color online) Path integral representation of Eq. (30). The value of q^* is such that the "penalty" 2Γ over all paths exiting through $-q^*$ is equal to the average gain over all paths exiting either through q^* (e.g. ϕ_1, blue) or through $-q^*$ (e.g. ϕ_2, red).

Note that Eq. (30) is completely general provided the assumptions of page 466 are satisfied, it does not rely on any specific statistics of the predictor. In the next section, we will explicitly solve this equation when the predictor is Gaussian and follows an auto-regressive evolution.

4.2. *Application to an auto-regressive linear predictor*

Let us assume that the predictor follows a discrete auto-regressive dynamics:

$$p_{t+1} = \rho \cdot p_t + \beta \cdot \xi_t, \tag{31}$$

where $(\xi_t)_{t\in\mathbb{R}}$ is a set of independent $\mathcal{N}(0,1)$ Gaussian random variables. One classical example of such a predictor is an exponential moving average of price returns:

$$p_t^{\text{EMA}} = K \sum_{t'<t} \rho^{t-1-t'} r_t.$$

If we suppose, as is usual, that the returns r_t are $\mathcal{N}(0,\sigma_r)$ random variables, then p_t^{EMA} follows the dynamics in Eq. (31) with $\beta = K\sigma_r$. Note however that the r_t must have some small correlations in order to be predictable! Therefore, in this case, the discussion in terms of an auto-regressive process is only consistent in the limit $K \ll 1$.

When $\rho = 0$, the predictor is a white noise in time: $\mathbb{E}[p_t p_{t+1}] = 0$. Since we assume a perfect $p \leftrightarrow -p$ symmetry, the self-consistent equation becomes simply $g(p) = p$, which trivially implies that $q^* = \Gamma$ in that case. This threshold is what we expect from such a system: without any auto-correlation, the best strategy is to trade as soon as the instantaneous predictability is above the trading cost. Note that in this case $p_\infty = p_t$, so this threshold also coincides with the naïve solution.

At the other extreme, when $\epsilon \equiv 1-\rho \ll 1$, we have $\mathbb{E}[p_{t+n}|p_t] \approx \mathrm{e}^{-\epsilon n} p_t$, so $\tau = 1/\epsilon$ is the auto-correlation time of the predictor p_t. The standard deviation of the predictor, i.e. its average predictability, is $\sigma_p = \sqrt{\mathbb{E}[p_t^2]} = \beta/\sqrt{2\epsilon}$. The integrated predictability is given by

$$p_\infty(p) = \sum_{n=0}^{\infty} \mathbb{E}[p_{t+n}|p_t] \approx \sum_{n=0}^{\infty} \mathrm{e}^{-\epsilon n} p_t \approx p/\epsilon,$$

what implies that the naïve threshold value is given by $q_{\text{naïve}} = \Gamma\epsilon$, while the integrated average predictability is $\sigma_\infty = \beta/\sqrt{2\epsilon^3}$. In what follows, we will study the problem by distinguishing between two cases:

- $\boxed{\beta \gg \Gamma}$: the predictor can easily beat its transaction costs at every step. This situation (which is not very realistic) requires to keep a *discrete time* approach of the problem.
- $\boxed{\beta \ll \Gamma}$: the predictor needs in general a large number of steps to beat the costs. This will lead to a *continuous* formulation of the problem.

Discrete case: $\beta \gg \Gamma$

We already explained in page 469 that $q^* \leq \Gamma$ with a linear predictor. Consequently, whenever $\beta \gg \Gamma$, we also have $\beta \gg q^*$. This means that, starting at $p = q^*$, one will typically jump beyond q^* or $-q^*$ in just one step. Thus:

$$\mathcal{G}(q^*) = q^* \quad \text{and} \quad \mathcal{P}_-(q^*) = \int_{x^*}^{+\infty} \frac{\mathrm{e}^{-x^2/2}}{\sqrt{2\pi}} \mathrm{d}x,$$

where $x^* = (2-\epsilon)q^*/\beta$. Since $\beta \gg q^*$, one has $x^* \ll 1$ and thus $\mathcal{P}_-(q^*) \approx 1/2$. Equation (30) finally gives: $q^* = \Gamma$. Hence, if the volatility of each predictor change is very large compared to the trading costs, then one needs to be as selective as possible.

Continuous case: $\beta \ll \Gamma$

If the threshold was of the order of the predictor's surprise $q^* \approx \beta$, the predictor would have a significant probability of switching from above q^* to below $-q^*$ in just one step. The optimal strategy would then require to resell everything at cost 2Γ, whereas the immediate gain would only be of the order of magnitude of β. So when $\beta \ll \Gamma$, we necessarily have $q^* \gg \beta$, and many steps are required for the predictor to get from $q^* \gg \beta$ to $-q^* \ll -\beta$. This is effectively the continuum limit, where the variation of the predictor at each time step is infinitesimal compared to q^*. We can then approximate the dynamics of the predictor by the Ornstein-Uhlenbeck drift-diffusion process:

$$\mathrm{d}p_t = -\epsilon\, p_t\, \mathrm{d}t + \beta\, \mathrm{d}X_t, \tag{32}$$

where $(X_t)_t$ is a Wiener process.

In such a continuous setting, the quantities $\mathcal{G}(q^*)$ and $\mathcal{P}_-(q^*)$ are actually ill-defined because the diffusion process starts on an absorbing boundary. This is a classical problem, which is handled by starting infinitesimally close to q^*. Therefore we consider $\mathcal{G}(p)$ and $\mathcal{P}_-(p)$ for $p = q^* - \delta < q^*$. It can be shown that these two functions obey two Kolmogorov backward equations, that read:

$$\frac{1}{2}\beta^2\frac{\partial^2\mathcal{G}}{\partial p^2} - \epsilon p\frac{\partial\mathcal{G}}{\partial p} = -p \quad \text{and} \quad \frac{1}{2}\beta^2\frac{\partial^2\mathcal{P}_-}{\partial p^2} - \epsilon p\frac{\partial\mathcal{P}_-}{\partial p} = 0, \tag{33}$$

with boundary conditions: $\mathcal{G}(\pm q^*) = 0$ and $\mathcal{P}_-(q^*) = 0$, $\mathcal{P}_-(-q^*) = 1$. We therefore encounter again the problem of a Brownian harmonic oscillator confined between two walls, already discussed in Sect. 2.3. The solutions of these equations are

$$\mathcal{G}(p) = \frac{1}{\epsilon}\left(p - q^*\frac{I(p\sqrt{a})}{I(q^*\sqrt{a})}\right) \quad \text{and} \quad \mathcal{P}_-(p) = \frac{1}{2}\left(1 - \frac{I(p\sqrt{a})}{I(q^*\sqrt{a})}\right),$$

with

$$I(x) = \int_0^x \mathrm{e}^{v^2}\,\mathrm{d}v \qquad \text{and} \qquad a = \frac{\epsilon}{\beta^2}.$$

To first order in $\delta \to 0$, Eq. (30) becomes

$$-\frac{\delta}{\epsilon} + \frac{\delta q^*}{\epsilon}\sqrt{a}\cdot\frac{I'(q^*\sqrt{a})}{I(q^*\sqrt{a})} \approx \Gamma\delta\sqrt{a}\frac{I'(q^*\sqrt{a})}{I(q^*\sqrt{a})}.$$

As expected, δ disappears from the equation, to give the following solution for the threshold q^*:

$$\boxed{q^* = \frac{\beta}{\sqrt{\epsilon}} F^{-1}\left(\frac{\Gamma\epsilon^{3/2}}{\beta}\right) \quad \text{where} \quad F(x) = x - I(x)/I'(x)}. \qquad (34)$$

Note that when $\epsilon \ll 1$, this equation can be expressed entirely in terms of the integrated predictability:

$$p_\infty(q^*) = \sigma_\infty\sqrt{2} \cdot F^{-1}\left(\frac{\Gamma}{\sigma_\infty\sqrt{2}}\right).$$

This means that we can find the optimal threshold for a predictor by studying only its total predictive power (if we suppose of course that it satisfies all the required properties).

One can now study the limits of Eq. (34) for large and small values of the only remaining adimensional parameter $\eta = \Gamma\epsilon^{3/2}/\beta$. Interestingly, $\eta \approx 1$ is the regime of practical interest where predictability beats costs whenever the predictor's value is of the order of its rms. The asymptotes of $F(x)$ are as follows:

- If $x \gg 1$, then $\int_0^x \mathrm{e}^{v^2}\mathrm{d}v \ll \mathrm{e}^{x^2}$, so $F(x) \approx x$.
- If $x \ll 1$, then $F(x) \simeq x - (1 - x^2)\int_0^x (1 - v^2)\mathrm{d}v \approx 2x^3/3$.

Therefore when $\eta \gg 1$, the threshold is simply given by $q^* = \Gamma\epsilon$. This result is rather intuitive: if β is very small then the predictability of the predictor is weak, compared to the trading cost. Hence, it makes sense to try to catch any profitable opportunity, without taking future trading into account. That is why we recover the naïve solution of page 466. If on the other hand $\beta \gg \Gamma\epsilon^{3/2}$ then $\eta \ll 1$, and $F^{-1}(\eta) \approx \sqrt[3]{3\eta/2}$, which yields

$$q^* = \sqrt[3]{\frac{3}{2} \cdot \Gamma\beta^2}.$$

This says that if β is large enough, the optimal threshold is independent of the mean-reversion parameter ϵ. The surprise, however, is the rather unexpected dependence of the threshold q^* as the 1/3 power of the trading costs. This result was obtained in the literature before, in the limit considered here of a continuous time random walk, see Refs. [33–35]. Our formulation is however much more general, and would allow one to treat non Gaussian and non stationnary situations as well.

5. Some open problems

We presented three very different examples of "first passage time" problems coming from quantitative finance. Let us discuss some extensions and open issues concerning these three problems.

As far as the Kolmogorov-Smirnov goodness-of-fit test is concerned, we believe that extensions of this test to higher-dimensional, multivariate settings, would be quite interesting. More precisely, the concept of "copulas" (that describe the correlation structure between dependent variables) has become an important one in theoretical finance in the recent years. For pairs of dependent variables, the copula $C(u, v)$ is an increasing function of both its arguments, from $[0, 1] \times [0, 1]$ to $[0, 1]$. The trivial copula, corresponding to independent variables, is such that $C(u, v) = uv$. It turns out that it is always possible to transform an arbitrary copula into the independent one by an appropriate change of variables, $(u, v) \to (s, t)$ [36, 37]. One can then, in the spirit of KS, test the GoF in a *copula independent* manner. The problem boils down to estimating the distribution of the maximum of a pinned "Brownian sheet" that generalizes the Brownian bridge described above. This is still an unsolved problem, but there is a hope that an exact solution can be found. Extensions to weights that emphasize the "tails" of the copula, similar to our one-dimensional problem above, would be quite interesting too.

The second problem, concerning the optimal selling time, is interesting from a mathematical/pedagogical point of view, but the final result turns out to be quite trivial from a financial point of view. A more interesting problem would be to add some correlations in the returns, accounting for trends or mean-reversion, for example with an exponentially decaying correlation function of the lag that would allow to make the problem Markovian.

Finally, the issue of optimal strategies in the presence of transaction costs would deserve much more attention. One particularly relevant endeavor would be to solve the problem in the presence of both linear *and* quadratic costs, i.e. when the cost of a change of position $\Delta\pi$ is of the form $\Gamma|\Delta\pi| + \Gamma'|\Delta\pi|^2$. The case treated in this review corresponds to $\Gamma' = 0$, but in practice *price impact* is very important: prices tend to go up when one buys, and down when one sells.

References

1. E. Derman and P. Wilmott. The financial modelers' manifesto. Online blog.
2. R. C. Merton, On the pricing of corporate debt: The risk structure of interest rates, *The Journal of Finance.* **29**(2), 449–470 (1974).
3. J. C. Hull, *Options, Futures and Other Derivatives.* Prentice Hall finance series, Prentice Hall Higher Education (2009).
4. R. Chicheportiche and J.-P. Bouchaud, Weighted Kolmogorov-Smirnov test: Accounting for the tails, *Physical Review E.* **86**, 041115 (Oct, 2012).
5. A. N. Kolmogorov, Sulla determinazione empirica di una legge di distribuzione, *Giornale dell'Istituto Italiano degli Attuari.* **4**(1), 83–91 (1933).
6. N. Smirnov, Table for estimating the goodness of fit of empirical distributions, *The Annals of Mathematical Statistics.* **19**(2), 279–281 (1948).
7. J. L. Doob, Heuristic approach to the Kolmogorov-Smirnov theorems, *The Annals of Mathematical Statistics.* **20**(3), 393–403 (1949).
8. E. V. Khmaladze, Martingale approach in the theory of goodness-of-fit tests, *Theory of Probability and its Applications.* **26**, 240 (1982).
9. R. Chicheportiche and J.-P. Bouchaud, Goodness-of-fit tests with dependent observations, *Journal of Statistical Mechanics: Theory and Experiment.* **2011** (09), P09003 (2011).
10. T. W. Anderson and D. A. Darling, Asymptotic Theory of Certain "Goodness of Fit" Criteria Based on Stochastic Processes, *The Annals of Mathematical Statistics.* **23**(2), 193–212 (1952).
11. D. A. Darling, The Kolmogorov-Smirnov, Cramer-von Mises Tests, *The Annals of Mathematical Statistics.* **28**(4), 823–838 (1957).
12. P. L. Krapivsky and S. Redner, Life and death in an expanding cage and at the edge of a receding cliff, *American Journal of Physics.* **64**(5), 546–551 (1996).
13. A. J. Bray and R. Smith, The survival probability of a diffusing particle constrained by two moving, absorbing boundaries, *Journal of Physics A: Mathematical and Theoretical.* **40**(10), F235 (2007).
14. A. J. Bray and R. Smith, Survival of a diffusing particle in an expanding cage, *Journal of Physics A: Mathematical and Theoretical.* **40**(36), 10965 (2007).
15. A. Clauset, C. R. Shalizi, and M. E. J. Newman, Power-law distributions in empirical data, *SIAM review.* **51**(4), 661–703 (2009).
16. S. Redner, *A guide to first-passage processes.* Cambridge University Press, Cambridge, UK (2001).
17. M. Noé and G. Vandewiele, The calculation of distributions of Kolmogorov-Smirnov type statistics including a table of significance points for a particular case, *The Annals of Mathematical Statistics.* **39**(1), 233–241 (1968).
18. M. Noé, The calculation of distributions of two-sided Kolmogorov-Smirnov type statistics, *The Annals of Mathematical Statistics.* **43**(1), 58–64 (1972).
19. H. Niederhausen. Tables of significance points for the variance-weighted Kolmogorov-Smirnov statistics. Technical report, Stanford University, Department of Statistics (February, 1981).
20. R. R. Wilcox, Percentage points of a weighted Kolmogorov-Smirnov statistic,

Communications in Statistics - Simulation and Computation. **18**(1), 237–244 (1989).
21. L. Turban, Anisotropic critical phenomena in parabolic geometries: the directed self-avoiding walk, *Journal of Physics A: Mathematical and General.* **25**, L127 (1992).
22. W. N. Mei and Y. C. Lee, Harmonic oscillator with potential barriers — exact solutions and perturbative treatments, *Journal of Physics A: Mathematical and General.* **16**, 1623 (1983).
23. I. S. Gradshteyn and I. M. Ryzhik, *Table of Integrals, Series and Products. Corrected and Enlarged Edition.* Academic Press, New York, London, Toronto and Tokyo (1980).
24. S. N. Majumdar and J.-P. Bouchaud, Optimal time to sell a stock in the Black-Scholes model: comment on Ref. [25], *Quantitative Finance.* **8**(8), 753–760 (2008).
25. A. Shiryaev, Z. Xu, and X. Y. Zhou, Thou shalt buy and hold, *Quantitative Finance.* **8**(8), 765–776 (2008).
26. W. Feller, *An introduction to probability theory and its applications.* vol. 1 and 2, *Wiley Series in Probability and Mathematical Statistics*, John Wiley & Sons, New York (1968).
27. S. N. Majumdar, Brownian functionals in physics and computer science, *Current Science.* **89**(12), 2076 (2005).
28. J. Randon-Furling and S. N. Majumdar, Distribution of the time at which the deviation of a Brownian motion is maximum before its first-passage time, *Journal of Statistical Mechanics: Theory and Experiment.* **2007**(10), P10008 (2007).
29. S. N. Majumdar, J. Randon-Furling, M. J. Kearney, and M. Yor, On the time to reach maximum for a variety of constrained Brownian motions, *Journal of Physics A: Mathematical and Theoretical.* **41**(36), 365005 (2008).
30. P. Lévy, Sur certains processus stochastiques homogènes, *Compositio Mathematica.* **7**, 283–339 (1939).
31. J. De Lataillade, C. Deremble, M. Potters, and J.-P. Bouchaud, Optimal trading with linear costs, *arXiv preprint q-fin.PM/1203.5957* (2012).
32. R. E. Bellman, *Dynamic Programming.* Dover Books on Mathematics, Dover Publications (2003).
33. L. C. G. Rogers. Why is the effect of proportional transaction costs $O(\delta^{2/3})$? In *Contemporary Mathematics*, vol. 351, pp. 303–308, American Mathematical Society, Providence, RI (2004).
34. R. J. Martin and T. Schöneborn, Mean reversion pays, but costs, *Risk Magazine.* pp. 96–101 (Feb, 2011).
35. R. J. Martin, Optimal multifactor trading under proportional transaction costs, *arXiv preprint q-fin.TR/1204.6488* (2012).
36. R. Chicheportiche. *Non linear Dependences in Finance.* PhD thesis, École Centrale Paris (Jun, 2013).
37. R. Chicheportiche and J.-P. Bouchaud. Ironing the copula surface. In preparation (2013).

Chapter 19

First-Passage and Extremes in Socio-Economic Systems

Jaume Masoliver* and Josep Perelló†

Departament de Fisica Fonamental,
Universitat de Barcelona,
Diagonal, 647, 08028 Barcelona, Spain
**jaume.masoliver@ub.edu*
†josep.perello@ub.edu

We present an introduction to level-crossing problems aimed to applications in socio-economic systems. We summarize the general theory of first-passage and exit problems and apply it to diffusions, random diffusions and discontinuous random walks. Some results regarding financial problems, especially related to dangerous fluctuations and risk control are also briefly discussed.

1. Introduction

First-passage and extreme value problems –also referred to under the collective name of level-crossing problems– are a crucial aspect of stochastic methods with a long and standing tradition of applications to physics, biology, chemistry and engineering [1; 2; 3; 4; 5; 6]. In addition, many natural records need also a similar description as are, for instance, floods, very high temperatures and solar flares or earthquakes [7].

In dealing with extreme value statistics of a given time series one wants to learn about the distribution of the extreme events, that is, the maximum and/or the minimum values of the signal within time intervals of fixed duration, and the statistical properties of their sequences. In hydrological engineering, for example, extreme value statistics are commonly applied to decide what building projects are required to protect riverside areas against typical floods that occur once in 100 years [7].

All this effort and knowledge still have appealing spaces for further exploration in many socio-economic contexts. First-passage and extreme values techniques and models, which have mostly remained inside physical sciences and engineering without any great spread out of it, can then be of renewed interest and motivate new challenges. The explosion of what has been received the name of Big Data in recent years has drastically increased the data available in a large list of socio-economic phenomena and this in turn has allowed a better observation and characterization of extreme events. One very popular case nowadays is the occurrence of microcrashes in speculative markets mostly due to the high frequency trading activity which is faster than a second. Within the community mostly coming from statistical physics there have been efforts toward the study of first-passage properties specially in financial markets. See for instance [9; 10; 11]. Our aim in this chapter is to summarize some of these efforts and suggest new lines of research.

In the study of any socio-economic system one inevitably deals with time series. In order to model these series and extract some regularities that may be hidden in them, an useful assumption is to assume that the series correspond to a given realization of some continuous random process, usually a diffusion process. This is, for instance, the case of continuous-time finance where speculative prices are supposed to follow the geometric Brownian motion. In such an approach, if $x_1, x_2, \dots, x_n, \dots$ is a given time series, one assumes the existence of a random process in continuous time (for example, a diffusion process) $X(t)$ such that $X(t_n) = x_n$ at discrete times $t_1, t_2, \dots, t_n, \dots$. Note, however, that in taking this approach one is assuming that between two adjacent data points x_n and x_{n+1} there exist infinitely many registers corresponding to $X(t)$ with $t_n < t < t_{n+1}$; in other words, we are assuming that the actual trajectories of $X(t)$ are continuous.

If, on the other hand, we know with certainty that time is a discrete variable and that between x_n and x_{n+1} there are no intermediate points (for example in tic-by-tic financial time series) then the diffusion approach may not be the most convenient one. In such cases a powerful tool is provided by the Continuous Random Walk formalism [12].

In this report we shall address the first-passage problem using both approaches. We first present a general setting with most challenging issues in reference to socio-economical contexts and afterwards consider both the continuous and discrete time approaches with some models which are enough sophisticated to improve the basic Gaussian dynamics and solve first-passage and extreme value problems.

2. General setting

Let us briefly review the main definitions involved in level-crossing problems where the main objective is to determine statistical information on the length of time that it takes a random process to reach certain boundary points for the first time. If the boundary consists in only one point x_c, usually referred to as the *critical value* or *threshold*, we have a *first-passage* or *hitting* problem. If the boundary consists in two points a and b we have an *escape* or *exit* problem out of the interval (a, b).

2.1. *The first-passage problem*

The hitting problem is completely solved when one knows the first-passage probability to threshold x_c. Let us denote by $W_c(t|x)$ the probability of first reaching x_c, at time t, when the process $X(t)$ starts at $X(t_0) = x$ at some initial time $t_0 \leq t$. In what follows we will usually assume that the process $X(t)$ is time homogeneous so that t_0 can be taken equal to 0.

Closely related to the hitting probability is the so-called *survival probability*, $S_c(t|x)$, which is the probability that the process starting at x has not reached at time t, or during any previous instant of time, the level x_c. Obviously

$$S_c(t|x) = 1 - W_c(t|x).$$

Let us denote by $\tau_c(x)$ the first-passage time for the process, starting at x, to first reach a given threshold x_c. This time is a random variable depending on each realization of the process. Formally

$$\tau_c(x) = \inf\{t|X(t) > x_c; X(0) = x < x_c\}$$

when the initial value is below threshold, and

$$\tau_c(x) = \inf\{t|X(t) < x_c; X(0) = x > x_c\}$$

when the initial value is above threshold.

We easily relate the first-passage time with the hitting probability. Indeed W_c can be defined as

$$W_c(t|x) = \text{Prob}\{\tau_c(x) \leq t\}$$

for if the observation time t is greater than or equal to the first-passage time, the process $X(t)$ has reached the critical value with certainty. Note also that this definition shows that $W_c(t|x)$ is the distribution function of the first-passage time. The corresponding probability density function

$$f_c(t|x)dt = \text{Prob}\{t \leq \tau_c(x) < t + dt\}$$

is, therefore, related to the distribution function W_c by

$$f_c(t|x) = \frac{\partial W_c(t|x)}{\partial t}. \tag{1}$$

The moments of this distribution are

$$T_n(x|x_c) = \int_0^\infty t^n f_c(t|x)dt,$$

$(n = 1, 2, 3, \ldots)$ and the mean first-passage time (MFPT) is the first moment

$$T_c(x) \equiv T_1(x|x_c).$$

In terms of the Laplace transform,

$$\hat{f}_c(s|x) = \int_0^\infty e^{-st} f_c(t|x)dt,$$

the first-passage moments are

$$T_n(x|x_c) = (-1)^n \left. \frac{\partial^n \hat{f}_c(s|x)}{\partial s^n} \right|_{s=0}.$$

This implies that, as long as moments $T_n(x|x_c)$ $(n = 1, 2, 3, \ldots)$ exist, the Laplace transform of the first-passage time density has the following expansion in powers of s

$$\hat{f}_c(s|x) = \sum_{n=0}^{\infty} \frac{(-1)^n}{n!} T_n(x|x_c)s^n. \tag{2}$$

On the other hand the Laplace transform of Eq. (1), along with the initial condition $W_c(0|x) = 0$, yields

$$\hat{W}_c(s|x) = \frac{1}{s}\hat{f}(s|x). \tag{3}$$

By combining Eqs. (2) and (3) we then have

$$\hat{W}_c(s|x) = \sum_{n=0}^{\infty} \frac{(-1)^n}{n!} s^{n-1} T_n(x|x_c), \tag{4}$$

expansion that furnishes the basis for the asymptotic analysis of the hitting probability $W(t|x)$. Indeed, the so-called Tauberian theorems prove that the long-time behavior of a function $g(t)$ is determined by the small s behavior of its Laplace transform $\hat{g}(s)$ [13]. For the case of the first-passage probability we see from Eq. (4) that the small s behavior of $\hat{W}_c$ is

$$\hat{W}_c(s|x) = \frac{1}{s} - T_c(x) + O(s) = \frac{1}{s}[1 - sT_c(x) + O(s^2)], \qquad (5)$$

where $T_c(x) \equiv T_1(x|x_c)$ is the mean first-passage time (MFPT). Note that expansion (5) may also be written, within the same level of approximation, as

$$\hat{W}_c(s|x) = \frac{1}{s[1 + sT_c(x) + O(s^2)]}, \qquad (6)$$

which by the Tauberian theorems implies that the long-time behavior of the first-passage probability $W_c(t|x)$ is given by the Laplace inversion of Eq. (6). That is,

$$W_c(t|x) \simeq 1 - e^{-t/T_c(x)}, \qquad (t \to \infty). \qquad (7)$$

We have thus obtained an asymptotic form for the hitting probability to threshold x_c and see that the MFPT determines the long-time behavior of the first-passage probability.

In terms of $\hat{W}_c(s|x)$ obtaining the MFPT is straightforward. In effect from Eq. (5) we see that

$$T_c(x) = \lim_{s \to 0} \left[\frac{1}{s} - \hat{W}_c(s|x) \right], \qquad (8)$$

which is the basis for obtaining the mean first-passage time when one knows the hitting probability –as long as the limit $s \to 0$ exists, which is not always de case.

2.2. *Extremes*

We now review the question of the extreme values reached by the process during a given interval of time. We will see below that this problem is closely related to the first-passage problem and encompasses the maximum and minimum values attained by the process, as well as the range or span; the latter defined as the difference between maximum and minimum. In this short review we will focus on maximum and minimum values and refer reader to the literature [14] for a more complete account of this important, yet rather forgotten and unexplored, question.

2.2.1. *The maximum*

Let us denote by $M(t)$ the *maximum value* reached by process $X(t)$ over the time span $(0,t)$. Obviously $M(t)$ is a random variable (in fact, a random process) whose value depends on the particular realization of the process $X(t)$. The distribution function of $M(t)$ is defined by

$$F_{\max}(\xi,t|x) = \text{Prob}\big\{M(t) < \xi | X(0) = x\big\}.$$

We shall now see how the problem of characterizing $M(t)$ is related to the first-passage problem of process $X(t)$. Suppose first that the level ξ of the maximum is greater than the initial value, $\xi > x$. Little reflection is needed to realize that in such a case the event $\{M(t) < \xi|x\}$ is equivalent to say that $X(t)$ has not crossed ξ at time t or before. The probability of the latter event being $1 - W_\xi(t|x)$ where $W_\xi(t|x)$ is the first-passage probability to threshold ξ at time t and starting from $X(0) = x$. Therefore

$$F_{\max}(\xi,t|x) = 1 - W_\xi(t|x), \qquad (\xi > x).$$

However, if the level of the maximum is lower than the initial point, $\xi < x$, the event $\{M(t) < \xi|x\}$ is impossible and has zero probability, In other words $F_{\max}(\xi,t|x) = 0$ when $\xi < x$. Therefore,

$$F_{\max}(\xi,t|x) = \big[1 - W_\xi(t|x)\big]\theta(\xi - x), \tag{9}$$

where $\theta(x)$ is the Heaviside step function. The probability density function (pdf) of the maximum,

$$\phi_{\max}(\xi,t|x) = \frac{\partial}{\partial\xi}F_{\max}(\xi,t|x),$$

then reads

$$\phi_{\max}(\xi,t|x) = -\frac{\partial W_\xi(t|x)}{\partial\xi}\theta(\xi - x) + \big[1 - W_\xi(t|x)\big]\delta(\xi - x),$$

where $\delta(x) = d\theta(x)/dx$ is the Dirac delta function. Since $W_\xi(t|x)\delta(\xi - x) = \delta(\xi - x)$ (recall that $W_x(t|x) = 1$) we have

$$\phi_{\max}(\xi,t|x) = -\frac{\partial W_\xi(t|x)}{\partial\xi}\theta(\xi - x). \tag{10}$$

Using this expression we can evaluate the *average maximum value* of the random process in terms of the first-passage probability as

$$\mathrm{E}\big[M(t)|x\big] = -\int_x^\infty \xi\frac{\partial W_\xi(t|x)}{\partial\xi}d\xi, \tag{11}$$

which, after an integration by parts, reads

$$\mathrm{E}\big[M(t)|x\big] = x + \int_x^\infty W_\xi(t|x)d\xi, \tag{12}$$

where we have assumed that $\xi W_\xi \to 0$ as $\xi \to \infty$.

2.2.2. *The minimum*

Let us denote by $m(t)$ the minimum value attained by $X(t)$ during the time interval $(0,t)$, and let $F_{\min}(\xi,t|x) = \mathrm{Prob}\{m(t) < \xi|X(0) = x\}$ be its distribution function.

Note that when (i) $\xi < x$ the event $\{m(t) < \xi|X(0) = x\}$ means that $X(t)$ has crossed the threshold ξ at time t or before. Hence $F_{\min}(\xi,t|x) = W_\xi(t|x)$. Whereas (ii) when $\xi > x$, $\{m(t) < \xi|X(0) = x\}$ is a sure event and $F_{\min}(\xi,t|x) = 1$. Summarizing

$$F_{\min}(\xi,t|x) = \theta(\xi - x) + W_\xi(t|x)\theta(x - \xi). \tag{13}$$

Denote by $\phi_{\min}(\xi|t)$ the pdf of $m(t)$. Taking the derivative with respect to ξ of the distribution function and noting that $W_\xi(t|x)\delta(x-\xi) = \delta(x-\xi)$ we arrive at

$$\phi_{\min}(\xi,t|x) = \frac{\partial W_\xi(t|x)}{\partial \xi}\theta(x - \xi). \tag{14}$$

The *average minimum value* is, therefore, given by

$$\mathrm{E}\big[m(t)|x\big] = \int_{-\infty}^x \xi\frac{\partial W_\xi(t|x)}{\partial \xi}d\xi. \tag{15}$$

Assuming that $\xi W_\xi \to 0$ as $\xi \to 0$, the integration by parts of Eq. (15) yields

$$\mathrm{E}\big[m(t)|x\big] = x - \int_{-\infty}^x W_\xi(t|x)d\xi. \tag{16}$$

2.3. *The escape problem*

We finish this section by briefly summarizing the escape problem which consists in knowing whether or not the process $X(t)$, starting at some point inside the interval (a,b), has exited this interval for the first time. The answer to this question lies in the knowledge of the survival probability $S_{ab}(t|x)$ defined as the probability that, beginning at $x \in (a,b)$, the process at time t has not left the interval at that time or during any previous instant of time. The escape probability, that is, the probability that at time t the process has exited the interval is thus given by

$$W_{ab}(t|x) = 1 - S_{ab}(t|x). \tag{17}$$

The time taken by the process to leave an interval, the so-called *escape* or *exit time*, is the random variable:

$$\tau_{ab}(x) = \inf\{t|a \geq X(t) \geq b; a < x < b\},$$

where $x = X(0)$. Note that the event $\{\tau_{ab}(x) < t\}$ is equal to the event $\{a \geq X(t) \geq b\}$; hence the distribution function of the exit time,

$$\text{Prob}\{\tau_{ab}(x) < t|X(0) = x\} = \text{Prob}\{t|a \geq X(t) \geq b; X(0) = x\},$$

is precisely the escape probability $W_{ab}(t|x)$. Therefore, the moments of the exit time will be defined by

$$T_{ab}^{(n)}(x) = \int_0^\infty t^n dW_{ab}(t|x),$$

$(n = 1, 2, 3, \ldots)$ and the mean escape time (MET) is the first moment:

$$T_{ab}^{(1)}(x) \equiv T_{ab}(x).$$

Proceeding as before one easily sees that the Laplace transform of the escape probability can be written as (cf Eq. (4))

$$\hat{W}_{ab}(s|x) = \sum_{n=0}^{\infty} \frac{(-1)^n}{n!} s^{n-1} T_{ab}^{(n)}(x),$$

from which it follows that (cf Eq. (8))

$$T_{ab}(x) = \lim_{s\to 0} \left[\frac{1}{s} - \hat{W}_{ab}(s|x)\right]. \tag{18}$$

Also, in terms of the Laplace transform of the survival probability (cf. Eq. (17)), we see that

$$T_{ab}(x) = \hat{S}_{ab}(0|x). \tag{19}$$

Moreover, and analogously to the first-passage problem discussed above, we can easily prove that the long-time behavior of the escape probability is also solely determined by the MET (see Eq. (7)):

$$W_{ab}(t|x) \simeq 1 - e^{-t/T_{ab}(x)}, \qquad (t \to \infty). \tag{20}$$

3. Diffusion processes

As explained in Sec. 1, when dealing with socio-economic time series a convenient assumption consists in supposing that time series are realizations of diffusion processes. Let us recall that diffusion processes are Markovian random processes with continuous sample paths. From a mathematical point of view they are characterized, in one dimension, by two functions: the drift, $f(x)$, and a positive-defined diffusion coefficient $D(x) \geq 0$. The sample paths of any diffusion process can thus be pictured as the continuous trajectory resulting from the superposition of a deterministic evolution, governed by $f(x)$, and fluctuations around it, the latter characterized by $D(x)$.

Denoting the process by $X(t)$, the diffusion picture becomes apparent by the fact that the time evolution of $X(t)$ is governed by the stochastic differential equation

$$dX(t) = f(X(t))dt + \sqrt{D(X(t))}dW(t), \tag{21}$$

where $W(t)$ is the Wiener process, that is, a Gaussian process with zero mean, unit variance, and correlation function $\mathrm{E}[W(t_1), W(t_2)] = \min(t_1, t_2)$, where $\mathrm{E}[\cdot]$ is the average over all realization of the Wiener process. In all the stochastic differentials appearing in this paper we use the Itô convention.

Within the framework of financial time series in which $P(t)$ may represent the price of a given asset, the first diffusion model was proposed by Bachelier [15] at the opening of the twentieth century and consisted in assuming that prices are themselves a Brownian motion. That is to say, $P(t)$ is a diffusion process with constant drift and constant diffusion coefficient. This simple model suffers severe shortcomings being the most important that $P(t)$ can be negative with certainty. Attending that prices cannot be negative, Bachelier model became unrealistic and it was replaced during the nineteen fifties by the *geometric Brownian motion* defined by

$$\frac{dP}{P} = adt + kdW(t), \tag{22}$$

where a and $k > 0$ are constants [15]. The drift and diffusion coefficients are now given by $f(x) = ax$ and $D(x) = k^2x^2$. Note that the integration of Eqs. (22) yields (we use Itô convention)

$$P(t) = P(0)\exp\{(a - k^2/2)t + kW(t)\}$$

and the process is always positive.

Another random process that has been found most useful, not only in modeling socio-economic systems, but also in theoretical biology such as population dynamics and neuron firing processes, is the *Feller process* [16; 17]. It has been recently applied to reproduce cholera epidemics as a susceptible-infected-recovered model (SIR) [18]. It is also an important single neuron model where functionals of the first-passage time are utilized to characterize Feller model parameters [19; 20].

The Feller process is a diffusion process with linear drift and linear diffusion coefficient vanishing at the origin. The time evolution of the process is governed by

$$dX(t) = [-\alpha X(t) + \beta]dt + k\sqrt{X(t)}dW(t), \tag{23}$$

where $a > 0, \beta > 0$ and $k > 0$ are constant parameters. Note that the linear drift $f(x) = -\alpha x + \beta$ with $\alpha > 0$ results in a restoring force which, in the absence of noise, makes the process decaying toward the value β. On the other hand, the state dependent coefficient, $D(x) = k^2 x$, for large values of x enhances the effect of noise while when x goes to zero the effect of noise vanishes. Therefore, when the process reaches the origin, the drift drags it toward β. Hence, since $\beta > 0$, the process, starting at some positive value, cannot reach the negative region with the overall result that the Feller process remains always positive.

The fact that $X(t)$ never attains negative values makes the process an ideal candidate for modeling a number of phenomena in natural and social sciences. Thus, for instance, in financial markets the process was introduced in 1985 to represent the term structure of interest rates –under the name of Cox, Ingersoll, and Ross (CIR) model [21]. The process is also being considered to provide a random character for the volatility of a given stock (see Sect. 3.3 below).

3.1. *First-passage and escape for diffusion processes*

Consider a stationary diffusion process $X(t)$ which initially takes the value x. We know from Sec. 2 that the first-passage problem is completely characterized by the hitting probability $W_c(t|x)$ while the exit problem is solved by knowing the escape probability W_{ab}. We shall prove that W_c and W_{ab} satisfy the same type of differential equation with specific boundary conditions.

Considering financial markets, the hitting probability is crucial to calibrate the default probability [22] or to obtain credit-risk estimations [23].

Less dramatic situations, involving the pricing of American and other exotic options in financial derivatives markets, also need to measure statistics related to first-passage time events [24; 25; 26]. Moreover, intraday algorithmic trading strategies [27; 28] of buying or selling a given asset could learn from first-passage statistics.

Let us begin with the hitting problem and let $p(x', \tau'|x, \tau)$ $(\tau' \geq \tau)$ be the (unrestricted) probability density function –also called the propagator– of the diffusion process $X(t)$ described by Eq. (21). The propagator obeys the Fokker-Planck equation which in backward form reads [1]

$$\partial_\tau p(x', \tau'|x, \tau) = -f(x)\partial_x p(x', \tau'|x, \tau) - \frac{1}{2}D(x)\partial^2_{xx} p(x', \tau'|x, \tau),$$

$(\tau' \geq \tau)$. The propagator also obeys the "final" condition $p(x', \tau'|x, \tau') = \delta(x' - x)$. Moreover, the assumed stationary character of the process implies invariance under time translations, in other words

$$p(x', \tau'|x, \tau) = p(x', \tau' - \tau|x).$$

Thus calling $\tau' - \tau = t \geq 0$ the above equation turns into

$$\partial_t p(x', t|x) = f(x)\partial_x p(x', t|x) + \frac{1}{2}D(x)\partial^2_{xx} p(x', t|x), \tag{24}$$

with the "initial" condition

$$p(x', 0|x) = \delta(x' - x). \tag{25}$$

We will see next that knowing the propagator is equivalent to know the first-passage probability, for if we assume the initial value of the process lies below threshold, i.e. $x < x_c$ then both probabilities are related by [a]

$$W_c(t|x) = 1 - \int_{-\infty}^{x_c} p(x', t|x)dx'. \tag{26}$$

Indeed the integral on the right is the probability that $X(t)$ has not crossed x_c (from below), so that the overall expression yields the hitting probability. Integrating Eq. (24) with respect to x' from $-\infty$ to x_c and substituting for Eq. (26) result in the following Fokker-Planck equation for the first-passage probability:

$$\partial_t W_c(t|x) = f(x)\partial_x W_c(t|x) + \frac{1}{2}D(x)\partial^2_{xx} W_c(t|x). \tag{27}$$

[a]When the initial value lies above threshold, $x > x_c$, Eq. (26) is replaced by

$$W_c(t|x) = 1 - \int_{x_c}^{\infty} p(x', t|x)dx'.$$

Note that at $t = 0$ the process cannot have reached the threshold provided that $x \neq x_c$. Hence, the following initial condition holds

$$W_c(0|x) = 0. \tag{28}$$

Moreover, if the starting value happens to be the threshold itself, $x = x_c$, then crossing is a sure event and we have the following boundary condition

$$W_c(t|x_c) = 1. \tag{29}$$

Before proceeding further, let us remark that the same Fokker-Planck equation, and the same initial and boundary conditions than those of Eqs. (27)–(29), are obtained had we assumed the process started *above* threshold, $x > x_c$.

The difficulty of solving the initial-boundary problem (27)–(29) is somewhat diminished by taking the time Laplace transform,

$$\hat{W}_c(s|x) = \int_0^\infty e^{-st} W_c(t|x)dt,$$

for this transformation reduces the original problem to the solution of an ordinary differential equation

$$D(x)\frac{d^2\hat{W}_c}{dx^2} + f(x)\frac{d\hat{W}_c}{dx} - s\hat{W}_c = 0, \tag{30}$$

with boundary condition

$$\hat{W}_c(s|x_c) = \frac{1}{s}. \tag{31}$$

Let us remark that since W_c is a probability, any solution of the problem must be finite and nonnegative, which may restrict the set of the possible solutions.

Let us briefly touch on the escape problem out of an interval (a, b). In such a case, in terms of the propagator $p(x', t|x)$, the escape probability W_{ab} is (compare with Eq. (26))

$$W_{ab}(t|x) = 1 - \int_a^b p(x', t|x)dx'.$$

Following the same reasoning as that of the hitting probability we easily see that W_{ab} follows an equation analogous to (27); that is,

$$\partial_t W_{ab}(t|x) = f(x)\partial_x W_{ab}(t|x) + \frac{1}{2}D(x)\partial^2_{xx} W_{ab}(t|x),$$

with the same initial condition as that of Eq. (28)

$$W_{ab}(0|x) = 0,$$

but two boundary conditions instead of one

$$W_{ab}(t|a) = W_{ab}(t|b) = 1.$$

3.2. *The first-passage problem in the Feller processes*

In the Feller process the drift and the diffusion coefficients are (see Eq. (23)) $f(x) = -(\alpha x - \beta)$ and $D(x) = kx$. The equation for the hitting probability, Eq. (30), then reads

$$\frac{1}{2}kx\frac{d^2\hat{W}_c}{dx^2} - (\alpha x - \beta)\frac{d\hat{W}_c}{dx} - s\hat{W}_c = 0, \tag{32}$$

with boundary condition given by Eq. (31). Equation (32) is the Kummer equation whose general solution is given by a linear combination of confluent hypergeometric functions [29]. Details are found in [30] and we here summarize some of the main findings.

The Laplace transform of first-passage probability is found to be

$$\hat{W}_c(s|x) = \begin{cases} \frac{\alpha F(s/\alpha,\theta,2\alpha x/k^2)}{sF(s/\alpha,\theta,2\alpha x_c/k^2)}, & x \leq x_c, \\ \frac{\alpha U(s/\alpha,\theta,2\alpha x/k^2)}{sU(s/\alpha,\theta,2\alpha x_c/k^2)}, & x \geq x_c, \end{cases} \tag{33}$$

where

$$\theta = 2\beta/k^2$$

is a convenient parameter, and F and U are Kummer functions [29].

In general these expressions for the Laplace transform of the hitting probability cannot be inverted exactly in an analytical fashion and one has to resort to numerical inversion. There are some instances, however, where we can obtain analytical expressions in real time. This is the case of hitting the origin ($x_c = 0$) which has a significant interest in many practical applications as, for example, in financial markets where reaching the origin may be related to bankruptcy or in population dynamics where it means extinction. We denote by $W_0(t|x)$ the first-passage probability to threshold $x_c = 0$, we have shown elsewhere that [30]

$$W_0(t|x) = \begin{cases} \frac{1}{\Gamma(1-\theta)}\Gamma\left(1-\theta, \frac{2\alpha x e^{-\alpha t}}{k^2(1-e^{-\alpha t})}\right), & \theta < 1, \\ 0, & \theta > 1, \end{cases}$$

where $\Gamma(a, z)$ is the incomplete Gamma function [29]. Let us incidentally note that when $\theta > 1$ this expression shows the impossibility for the Feller process to reach the origin. Therefore, parameter estimation would be crucial to determine whether the process hits the threshold $x_c = 0$.

Recall that the knowledge of the hitting probability $\hat{W}_c$ allows us to obtain the expression for the mean first-passage time $T_c(x)$. Indeed, from Eqs. (8) and (33) we have [30]

$$T_c(x) = \frac{1}{\beta}\int_{x_c}^{x} F(1, 1+\theta, 2\alpha z/k^2)dz, \quad (x \leq x_c),$$

and

$$T_c(x) = \frac{2}{k^2}\int_{x_c}^{x} U(1, 1+\theta, 2\alpha z/k^2)dz, \quad (x \geq x_c).$$

For the Feller process one can solve the escape problem as well. In such a case the Laplace transform of the exit probability out of the interval (a, b) also obeys the Kummer equation

$$\frac{1}{2}kx\frac{d^2\hat{W}_{ab}}{dx^2} - (\alpha x - \beta)\frac{d\hat{W}_{ab}}{dx} - s\hat{W}_{ab} = 0,$$

with the boundary conditions

$$\hat{W}_{ab}(s|a) = \hat{W}_{ab}(s|b) = 1/s.$$

The exact solution is again given by combinations of the confluent hypergeometric functions F and U and we refer the reader to [30] for a complete account of either W_{ab} and the mean exit time T_{ab}.

3.3. *Random diffusions*

The dynamics of particles in random media may be used for a large variety of phenomena in statistical mechanics and condensed matter physics [31]. One possible approach to tackle the problem consists in assuming that the diffusion coefficient of the random process under study is itself random. In such an approach the function $D(x)$ appearing in Eq. (21) is a stochastic quantity which in many instances is supposed to be another diffusion process.

In such a case one deals with a two-dimensional random process, $(X(t), Y(t))$, governed by two coupled stochastic differential equations one representing the usual diffusion process and the other "the diffusion of diffusion":

$$dX(t) = f(X(t))dt + Y(t)dW_1(t),$$
$$dY(t) = h(Y(t)) + g(Y(t))dW_2(t),$$

where the Wiener processes W_1 and W_2 may, or may not, be correlated. Even in simple cases this is a rather intricate problem from a mathematical point of view and the first-passage problem is even more complicated since it requires a higher dimensional phase space and higher dimensional Fokker-Planck equations (see below) involving boundary conditions which usually make obtaining the solution to the problem a very difficult task [5].

In socio-economic systems one also encounters random diffusions. This is the case of the so-called stochastic volatility (SV) models appearing in financial analysis for which the volatility $\sigma(t)$, defined as the standard deviation of the return, is a random quantity. There is substantial empirical evidence that supports this assumption [32]. Financial engineering had started to consider this approach after the 1987 crash. The industry in this way improved the geometric Brownian motion description with the aim of providing more precise evaluation of financial option prices [33]. There seems to be empirical evidence, embodied in the so-called stylized facts, that the assumption of constant volatility does not properly account for important features of markets such as fat tails in price fluctuations, negative skewness in price changes, long-range volatility autocorrelation and return-volatility asymetric cross-correlation [32].

Moreover, volatility –which is, by definition, a positive quantity– seems to "revert to some average value". This latter fact suggesting the existence of a drift dragging volatility toward its mean value. The simplest conjecture being that of a linear restoring force as in the harmonic oscillator [33]. This makes the Feller process discussed in the previous section a rather convenient candidate as a model of volatility. However, either the Ornstein-Uhlenbeck process or the exponential Ornstein-Uhlenbeck process are comparable candidates while keeping the formalism as simple as possible.

The combination of the geometric Brownian motion for price, Eq. (22), and the Feller process, Eq. (23), for the volatility results in the so-called Heston model [34]

$$dX(t) = \sqrt{Y(t)}dW_1(t), \tag{34}$$

$$dY(t) = -\alpha\left[Y(t) - m^2\right]dt + k\sqrt{Y(t)}dW_2(t), \tag{35}$$

where $X(t)$ is the detrended log-return[b], $Y(t)$ is the volatility[c] and $W_1(t)$

[b]The actual definition of $X(t)$ is through the stochastic differential (in the Itô sense)

$$dX(t) = \frac{dP(t)}{P(t)} - \mathrm{E}\left[\frac{dP(t)}{P(t)}\right],$$

where $P(t)$ is a price or the value of an index.

[c]Note that the volatility is $\sigma(t) = \sqrt{Y(t)}$ and the diffusion coefficient of the return is $D(t) = Y(t)$, a random quantity.

and $W_2(t)$ are two Wiener processes.

We have addressed the level crossing problems for SV models in a series of papers. In [35] we have analyzed the extreme times not only for the Heston model but for several SV models. In [36] we have solved the complete escape problem –including survival and exit probabilities– for the Heston model and finally in [37] we have fully solved the first-passage problem for the same model assuming $W_1(t)$ and $W_2(t)$ are uncorrelated processes.

We will here summarize the main findings of the first-passage problem for the Heston model and refer to the reader to [37] for details. The problem we want to address is only related to the process $X(t)$ and not to the process $Y(t)$. In other words, we want to know the hitting probability for the return to attain a critical value x_c *regardless the value of the volatility.*

We shall concentrate on the survival probability (SP), $S_c(t|x,y)$ to a certain critical level x_c of the return starting at $X(0) = x$ (usually $x = 0$[d]) with initial volatility $Y(0) = y$. As we have seen, closely related to the SP is the hitting probability $W_c(t|x,y) = 1 - S_c(t|x,y)$. When the critical level, $x_c > x$, is greater than the initial return, W_c gives the probability of first hitting a profit at time t. On the contrary if $x_c < x$, W_c gives the probability of first having a loss quantified by x_c (when $x_c \to -\infty$ this is the default probability).

For the bidimensional diffusion process $(X(t), Y(t))$ described by Eqs. (34)-(35) the survival probability $S_c(t|x,y)$ is the solution to the following higher dimensional Fokker-Planck equation [1]:

$$\partial_t S_c = -\alpha(y - m^2)\partial_y S_c + \frac{1}{2}k^2 y \partial^2_{yy} S_c + \frac{1}{2} y \partial^2_{xx} S_c, \tag{36}$$

with initial and boundary conditions given by[e]

$$S_c(0|x,y) = 1, \qquad S_c(t|x_c,y) = 0. \tag{37}$$

The problem is solved exactly by means of the Fourier-sine integral of the SP and we refer the reader to [37] for details. We have also obtained useful approximations to the exact SP. Thus, in the asymptotic regimes of either long times or large initial volatilities, the SP goes to the following

[d] For most applications x represents today's return and, by definition, $x = 0$. However, to avoid confusion we keep the symbol x.

[e] Also the first-passage probability obeys the same Fokker-Planck equation (36), but now the initial and boundary conditions are, respectively, $W_c(0|x,y) = 0$ and $W_c(t|x_c,y) = 1$.

Gaussian probability

$$S_c(t|x,y) \simeq \mathrm{Erf}\left(\frac{|x-x_c|}{\sqrt{\lambda(t,y)}}\right), \tag{38}$$

where

$$\lambda(t,y) = 2m^2t + 2(y/\alpha)(1-e^{-\alpha t}),$$

and the hitting probability is given by $W_c = 1 - S_c$.

Recall that the approximation given by (38) is valid for long times ($t \gg (k/\alpha m)^{1/2}$) and large volatilities ($y \gg \alpha$) [37]. Therefore, as long as neither the volatility nor the time are small, we can safely use (38) to obtain the behavior of the SP as $|x_c| \to \infty$. This is, perhaps, one of the most interesting aspects of the first-passage problem to look at, because that limit is intimately related to extreme risks and, ultimately, to default. Taking the limit $|x_c| \to \infty$ in Eq. (38) one easily finds for the default probability (and also for the uprising probability when $x_c \to +\infty$) the following fast exponential decay:

$$W_c \sim \frac{e^{-x_c^2/\lambda}}{|x_c|}, \qquad |x_c| \to \infty. \tag{39}$$

Two additional consequences of the asymptotic form above are, for one hand, the non-existence of a mean first-passage time, since the SP falls off as $S(x,y,t) \sim 1/\sqrt{t}$ as $t \to \infty$ and mean first-passage time, $T(x,y) = \infty$, does not exists. On the other hand, when the volatility variable y increases the SP decreases as the power law $S(x,y,t) \sim y^{-1/2}$ ($y \to \infty$).

Let us also discuss other very useful approximations based on the dimensionless parameter $\beta = k/\alpha$. Looking at the Heston model described in Eqs. (34)-(35) we see that β measures the strength of the volatility fluctuations, given by the vol of vol k, in relation to the deterministic pull, characterized by the parameter α, which drives volatility towards the normal level m^2 (with a characteristic time given by α^{-1}). We have termed β as *the strength of normalized volatility fluctuations* because a large value of this parameter implies large values of the noise term in the volatility and small values of α as well. This latter fact (*i.e.*, α small) means a slow relaxation towards the mean. As far as the hitting problem is concerned, two different patterns arise according to whether β is small or large.

In the case of small fluctuations where $\beta \ll 1$, the exact SP is precisely approximated by Eq. (38). Therefore, the case of weak fluctuations leads to the same approximation than that of long times or large volatilities [37].

On the other hand, when fluctuations are strong, $\beta \gg 1$, the exact SP is approximated by the Cauchy distribution [37]

$$S_c(t|x,y) \sim \frac{2}{\pi} \arctan\left(\frac{\beta|x-x_c|}{m^2 t + y/\alpha}\right) \qquad (\beta \gg 1). \tag{40}$$

Let us observe that this asymptotic form verifies the boundary condition $S_c(t|x_c,y) = 0$, but not the initial condition, since $S(0|x,y) \neq 1$ for $y \neq 0$. Consequently, the approximation (40) will work only after an initial period of time, representing transient effects, has elapsed.

The most striking difference between the above asymptotic forms of the first-passage probability, Eqs. (38) and (40), appears when one considers the extreme risk of default. In such a case $x_c \to -\infty$ and the hitting probability, $W_c(t|x,y)$, corresponding to the Gaussian form (38) quickly decays by following the decreasing quadratic exponential shown in Eq. (39). On the other hand, as volatility fluctuations increase, the non-Gaussian approximation given by Eq. (40) implies a much slow decay since now default is ruled by the hyperbolic law:

$$W_c \sim \frac{1}{\beta|x_c|}, \qquad |x_c| \to \infty. \tag{41}$$

In other words, we have *a significant increase of risk when the fluctuations of the volatility soar.*

4. Continuous Time Random Walks

In many practical situations the time series of socio-economic systems represent the complete information on the underlying random process. This is the case in finance of tic-by-tic data which give the microscopic description of the market. In such a situation, diffusion models –which have, by definition, continuous trajectories– only represent a "mesoscopic" approximation to the actual process. A more accurate description is provided by Continuous Time Random Walks (CTRW's) where the inherent discontinuous nature of any microscopic description is taken into account from the start. This kind of random walk was introduced almost five decades ago by Montroll and Weiss [38] and a rather complete account of the formalism and its application to financial problems may be found in [39].

We shall briefly summarize the main traits of CTRW's before addressing the level-crossing problem. In its most common version a given random process $X(t)$ shows a series of random increments or jumps at random times $\ldots, t_{-1}, t_0, t_1, t_2, \ldots, t_n, \ldots$ remaining constant between these jumps.

Thus, after a certain time interval $\tau_n = t_n - t_{n-1}$, the process suffers a random increment $\Delta X_n(\tau_n) = X(t_n) - X(t_{n-1})$ with the result that the trajectory of the process consists of a series of steps. Both waiting times τ_n and jumps $\Delta X_n(\tau_n)$ are assumed to be independent and identically distributed random variables described by their pdf's which we denote by $\psi(\tau)$ and $h(x)$, respectively. We combine these two causes of randomness into a single density $\rho(x,t)$ representing the joint pdf of waiting times and random jumps:

$$\rho(x,t)dxdt = \text{Prob}\big\{x < \Delta X_n \le x + dx; t < \tau_n \le t + dt\big\}.$$

It is implicit in this definition that the probability for any jump ΔX_n to take the value x does not depend on the value of previous jumps and the probability for any time interval between jumps, τ_n, to take the value t does not depend on previous waiting times either (see [40] where this restriction has been relaxed).

It is frequently assumed that $\rho(x,t)$ is an even function of x so that there is no net drift in the evolution of $X(t)$. We observe that if waiting times are independent of jumps then $\rho(x,t) = h(x)\psi(t)$. In any other situation one has to specify a functional form of $\rho(x,t)$ that is compatible with the observed data [39]. Moreover since the jump and waiting time pdf's, $h(x)$ and $\psi(t)$, are the marginal densities of of the joint density, any proposed form of $\rho(x,t)$ must satisfy:

$$h(x) = \int_0^\infty \rho(x,t)dt, \qquad \psi(t) = \int_{-\infty}^\infty \rho(x,t)dx.$$

We now address the level-crossing problem within the framework of CTRW's. In the context of financial time series we have studied the problem in a number of papers [11; 40; 42]. From a mathematical point of view, and contrary to diffusion models where all probabilities related to the first-crossing problem obey differential equations, in CTRW's they satisfy integral equations.

In order to set the formalism it is more convenient to first address the escape problem and recover from it the first-passage problem. Let $S_{ab}(t|x)$ be the survival probability of the random walker within the interval (a,b). Given the discontinuous character of the trajectory, the process $X(t)$ will leave (a,b) as long as a sufficiently large jump occurs. Since exit may take place at the end of the first sojourn or after it, we write the survival probability as the sum

$$S_{ab}(t|x) = A(t|x) + B(t|x),$$

where $A(t|x)$ is the probability of remaining inside the interval and that *no jump has occurred* whereas $B(t|x)$ takes into account that one or more jumps has occurred but the walker is still inside (a, b). Moreover, since the initial position x lies inside the interval, $A(t|x)$ will be given by the probability that no jump has occurred up to time t which is the cumulative distribution of $\psi(t)$:

$$A(t|x) = \int_t^\infty \psi(t')dt' \equiv \Psi(t).$$

On the other hand, suppose the walker has not exited the interval during the first sojourn this means that the process has attained at time t' some value x' still inside (a, b) and from this point the survival probability is given by $S_{ab}(t-t'|x')$. Hence, the probability that the process is still inside the interval and at least one jump occurred reads

$$B(t|x) = \int_0^t dt' \int_a^b \rho(x'-x, t')S_{ab}(t-t'|x')dx'.$$

Summing A and B we find that the survival probability obeys the following renewal equation:

$$S_{ab}(t|x) = \Psi(t) + \int_0^t dt' \int_a^b \rho(x'-x, t')S_{ab}(t-t'|x')dx'. \qquad (42)$$

Although the escape probability $W_{ab}(t|x)$ obeys a similar integral equation, it is more straightforward to calculate it by means of $W_{ab}(t|x) = 1 - S_{ab}(t|x)$ once we know the survival probability.[f]

Let us note that the first-passage probabilities to some threshold x_c are special cases of the escape probabilities. For example, if the threshold is above the starting point, $x_c > x$, the hitting probability W_c is given by the exit probability W_{ab} by setting $a = -\infty$ and $b = x_c$, while when threshold x_c is below x the substitution is $a = x_c$ and $b = \infty$.

The difficulty in solving Eq. (42) is soften by taking the Laplace transform:

$$\hat{S}_{ab}(s|x) = \int_0^\infty e^{-st}S_{ab}(t|x)dt.$$

[f]The integral equation for $W_{ab}(t|x)$ is

$$W_{ab}(t|x) = \phi(t|x) + \int_0^t dt' \int_a^b \rho(x'-x, t')W_{ab}(t-t'|x')dx',$$

and, since $\phi(t|x) = \int_0^t \psi(t')dt' - \int_0^t dt' \int_a^b \rho(x'-x, t')dx'$, this equation can be more difficult to solve than that of S_{ab}.

Equation (42) transforms to

$$\hat{S}_{ab}(s|x) = \hat{\Psi}(s) + \int_a^b \hat{\rho}(x'-x,s)\hat{S}_{ab}(s|x')dx'. \tag{43}$$

In the special case where jumps and waiting times are independent, $\rho(x,t) = h(x)\psi(t)$, Eq. (43) reduces to

$$\hat{S}_{ab}(s|x) = \hat{\Psi}(s) + \hat{\psi}(s)\int_a^b h(x'-x)\hat{S}_{ab}(s|x')dx'. \tag{44}$$

When jumps are exponentially distributed according to the Laplace density,

$$h(x) = \frac{\gamma}{2}e^{-\gamma|x|}, \tag{45}$$

the integral equation (44) can be solved exactly by turning it into an ordinary differential equation (see [40] for details and more results). We write here the expression corresponding to the first-passage probability

$$\hat{W}_c(s|x) = \frac{\hat{\psi}(s)/s}{1+\sqrt{1-\hat{\psi}(s)}}\exp\left\{-\gamma\sqrt{1-\hat{\psi}(s)}\big|x-x_c\big|\right\}.$$

It is instructive to compare this probability with that of the simplest diffusion process: the free Brownian motion. As is well known [1; 2; 40] the hitting probability for the Wiener process of variance σ is

$$\hat{W}_c(s|x) = \frac{1}{s}\exp\left\{-\frac{\sqrt{2s}}{\sigma}\big|x-x_c\big|\right\}. \tag{46}$$

Note that the two expressions coincide for small values of s, that is to say, for large timescales which again constitutes a restatement of the Central Limit Theorem [12].

In finance one relevant application of first-passage probabilities is in risk control [40]. To clarify the concepts we remove model dependent inferences (that is, particular choices of $\psi(t)$ and $h(x)$) and use the Wiener process. This simple Gaussian model is usually employed to evaluate the "Value at Risk" (VaR) level. The VaR gives the worst return one can get at the end of a fixed time interval t for a given confidence level $\alpha \leq 1$. If we assume that the market volatility is σ, the (unrestricted) Gaussian pdf yields

$$\text{VaR}(\alpha) = \sigma\sqrt{t}\text{N}^{-1}(1-\alpha), \tag{47}$$

where $\text{N}(\cdot)$ is the Normal distribution. Note that this measure of risk ignores the instantaneous risk aversion of investors, because it neglects the fact that investors may not assume all paths leading to the same final return (indeed, some paths, while ending at the given point, may have hit any intermediate dangerous threshold).

This would have not been the case had we used the survival probability, instead of the unrestricted probability, for quantifying risk. Indeed, the use of the former ensures, within the desired level of confidence, that a given position has never crossed the risk measure. We call this safer measure of risk "survival probability risk" (SpR).

Since the survival probability to level x_c is $S_c(s|x) = 1 - W_c(s|x)$, we get after Laplace inverting Eq. (46):

$$S_c(t|x) = 1 - 2\mathrm{N}\left(\frac{|x - x_c|}{\sigma\sqrt{t}}\right),$$

whence

$$\mathrm{SpR}(\alpha) = \sigma\sqrt{t}\mathrm{N}^{-1}\left(\frac{1-\alpha}{2}\right). \tag{48}$$

By comparing Eqs. (47) and (48) we have $\mathrm{VaR}(\alpha) > \mathrm{SpR}(\alpha)$ meaning that VaR systematically underestimates risk. This may have significant practical consequences, since VaR is one of the most widely used measures of risk [43].

We will finish this exposition on CTRW methods by noting that the mean exit time, T_{ab}, obeys also an integral equation [11] which is simpler than the renewal equation (42) for $S_{ab}(t|x)$. In effect, recalling that in terms of the Laplace transform of the survival probability the MET is given by $T_{ab}(x) = \hat{S}(s = 0|x)$ (see Eq. (19)) and taking into account, firstly

$$\lim_{s\to 0}\hat{\Psi}(s) = \lim_{s\to 0}\left\{\frac{1}{s}\left[1 - \int_0^t e^{-st}\psi(t)dt\right]\right\} = \int_0^\infty t\psi(t)dt = \mathrm{E}[\tau],$$

where $\mathrm{E}[\tau]$ is the average time between two consecutive jumps; and secondly $\hat{\rho}(x'-x|, s = 0) = h(x'-x)$ (remember that the jump density is the marginal of the joint density), we obtain from Eq. (43) the following renewal equation for the mean exit time:

$$T_{ab}(x) = \mathrm{E}[\tau] + \int_a^b h(x' - x)T_{ab}(x')dx'. \tag{49}$$

Note that a similar equation may be written for the mean first-passage time $T_c(x)$ to threshold x_c, by setting $a = -\infty$ and $b = x_c > x$ or $a = x_c < x$ and $b = \infty$.

In [11] and [42] we have solved Eq. (49) for a number of cases and applied the solutions to financial time series (see also [41] for scaling properties of first-passage probabilities). Thus, for example, in the case of Laplacian jumps, Eq. (45), we obtained the quadratic growth [11]

$$T_{ab}(x) = \frac{1}{2}\mathrm{E}[\tau]\left[1 + \left(1 + \frac{\gamma L}{2}\right)^2 - \gamma^2\left(x - \frac{a+b}{2}\right)^2\right],$$

where $L = b - a$ is the interval length. Let us observe that for Laplacian jumps, the mean-first passage time, $T_c(x)$, does not exist because $T_{ab}(x) \to \infty$ as $L \to \infty$.

5. Concluding remarks

First-passage and extreme value statistics are crucial elements in financial markets trading and start having important impact in a better understanding of social systems such as human decision making or human mobility. In the last few years, Big Data has provided massive amounts of time series from several socio-economic systems and this gives the opportunity to observe and analyze extreme events in these contexts. The approach from physical sciences and engineering can then be of renewed interested and it motivates new challenges. After furnishing the general setting for level crossing, we have here provided some hints and some ways of exploration by considering continuous and discrete time approaches.

We are nonetheless conscious that we have kept aside other relevant approaches. For example, starting from the continuous time approach, there has been, mostly in the mathematics literature, some amount of work on first-passage problems in Lévy processes with applications to financial markets (see [44; 45]).

Another approach that we have not considered here is furnished by a combination of discrete and continuous processes as, for instance, diffusion processes with jumps. This would amount to merge the diffusion theory of level crossings explained in Sec. 3 with that of the CTRW's of Sec. 4. Such an approach, which has many applications in physical and economic sciences –see, for example, Merton's classic work on jump-diffusions in finance [46]– has not been, to our knowledge, extensively studied with few exceptions (see, however, [47] where the MFPT of the Brownian motion with superimposed random jumps was obtained).

Acknowledgment

Partial financial support from Dirección General de Investigación under Contract No. FIS2009-09689 is acknowledged.

References

1. C. W. Gardiner, *Handbook of Stochastic Methods* (Springer, Berlin, 1985).
2. S. Redner, *A Guide to First-Passage Processes* (Cambridge University Press, Cambridge, England, 2001).
3. G. H. Weiss, *First-Passage Time Problems in Chemical Physics*, in Advadces in Chemical Physiscs, Vol. 13, edited by I. Prigogine (J. Wiley, Hoboken, NJ, 2007).
4. S. Condamin, O. Bénichou, V. Tejedor, R. Volturiez and J. Klafter, Nature **450**, 77 (2007).
5. J. Masoliver and J. Porrà, Phys. Rev. Lett. **75**, 189 (1995).
6. M. F. Shlesinger, Nature (London) **450**, 40 (2007).
7. J. Eichner, J. Kantelhardt, A. Bunde, and S. Havlin Phys. Rev. E **73**, 016130 (2006).
8. S. K. Gou and H. Wang, Adv. Appl. Prob. **35**, 504 (2003) (and references therein).
9. G. Bonanno, D. Valenti, and B. Spagnolo, Eur. Phys. J. B **53**, 405–409 (2006); D. Valenti, B. Spagnolo, and G. Bonanno, Physica A **382**, 311-320 (2007).
10. J. Inoue and N. Sazuka, Phys. Rev. E **70**, 021111 (2007); N. Sazuka, J. Inoue and E. Scalas, Physica A **388** 2839 (2009).
11. J. Masoliver, M. Montero and J. Perelló Phys. Rev E **71**, 056130 (2005). See also Refs. [35; 36; 37; 40; 41] and [42].
12. G. H. Weiss *Aspects and Applications of the Random Walk* (North-Holland, Amsterdam, 1994).
13. R. A. Handelsman and J. S. Lew, SIAM J. Math. Anal. **5**, 425 (1974).
14. D. A. Darling and A. F. J. Siegert, Ann. Math. Statist. **24**, 624 (1953); I. F. Blake and W. C. Lindsay, IEEE Transactions on Information Theory, **IT-19**, 295 (1993).
15. P. H. Cootner (Ed.), *The random character of stock market prices* (MIT Press, Cambridge, 1964).
16. L. M. Ricciardi, Diffusion Processes and Related Topics in Biology (Springer-Verlag, Berlin, 1977).
17. W. Gerstner and W. M. Kistler, Spiking Neuron Models (Cambridge University Press, Cambridge, 2002).
18. S. Azaele, A. Maritan, E. Bertuzzo, I. Rodriguez-Iturbe, and A. Rinaldo, Phys. Rev. E 81, 051901 (2010).
19. S. Ditlevsen, P. Lansky, Phys. Rev. E 73, 061910 (2006).
20. E. Bibbona, P. Lansky, and R. Sirovich, Phys. Rev. E 81, 031916 (2010).
21. J. C. Cox, J. E. Ingersoll, and S. A. Ross, Econometrica **53**, 385 (1985).

22. Y. Katz, V. Shokhirev, Phys. Rev. E **82**, 016116 (2010).
23. T. Bielecki and M. Rutkowski, Credit Risk: Modeling, Valuation and Hedging. (Springer, Berlin, 2004).
24. L. Alili and A. E. Kyprianou, Ann. Appl. Probab. 15, 2062 (2005).
25. F. Avram, A. E. Kyprianou, and M. R. Pistorius, Ann. Appl. Probab. 14, 215 (2004).
26. Delbaen, F. and Yor, M. Passport options. Math. Finance 12, 299 (2002).
27. F. Almgren, Appl. Math. Finance 10, 1 (2003).
28. T. Hendershott, Ch.M. Jones, and A. J.Menkveld, J. Finance 66, 1 (2011).
29. W. Magnus, F. Oberhettinger, and R. P. Soni, *Formulas and Theorems for the Special Functions of Mathematical Physics* (Springer-Verlag, New York, 1966).
30. J. Masoliver and J. Perelló, Phys. Rev. E **86**, 041116 (2012).
31. D. Ben-Avraham and S. Havlin, *Diffusion and Reactions in Fractals and Disordered Systems* (Cambridge University Press, Cambridge, 2000).
32. R. Cont, Quant. Fin. **1**, 223 (2001).
33. J-P-. Fouque, G. Papanicolau, and K. R. Sircar, *Derivatives in Financial Markets with Stochastic Volatility* (Cambridge University Press, Cambridge, 2000).
34. S.L. Heston, Rev. Financ. Stud. 6, 327 (1993).
35. J. Masoliver and J. Perelló, Phys. Rev. E **75**, 046110 (2007).
36. J. Masoliver and J. Perelló, Phys. Rev. E **78**, 056104 (2008).
37. J. Masoliver and J. Perelló, Phys. Rev. E **80**, 016108 (2009).
38. E. W. Montroll and G. H. Weiss, J. Math. Phys. **6**, 167 (1965).
39. J. Masoliver, M. Montero, J. Perelló and G. H. Weiss, J. Economic Behavior and Organization **61**, 577 (2006).
40. M. Montero and J. Masoliver, Eur. Phys. J. B **57**, 181 (2007).
41. J. Perelló, M. Gutiérrez-Roig and J. Masoliver Phys. Rev. E **84**, 066110 (2011).
42. M. Montero, J. Perelló, J. Masoliver, F. Lillo, S. Miccichè and R. N. Mantegna, Phys. Rev E **72**,05101 (2005).
43. J-P. Bouchaud, Nature (London) **455**, 1181 (2008).
44. R. Doney, J. London Math. Soc. **44**, 566 (1991).
45. N. Packhama, L. Schloeglb, W.M. Schmidt, Quant. Fin. (2013) published online 27 Feb 2013 Doi: 10.1080/14697688.2012.739729.
46. R. Merton, J. Finan. Econ. **5**, 125 (1976).
47. J. Masoliver and G. H. Weiss, Physica A **149**, 395 (1988).

Chapter 20

Transport and the First Passage Time Problem with Application to Cold Atoms in Optical Traps

Eli Barkai and David A. Kessler

Department of Physics
Institute of Nanotechnology and Advanced Materials
Bar-Ilan University, Ramat-Gan 52900, Israel

Measurements of spatial diffusion of cold atoms in optical lattices have revealed anomalous super-diffusion, which is controlled by the depth of the optical lattice. We use first passage time statistics to derive the diffusion front of the atoms. In particular, the distributions of areas swept under the first passage curve till its first arrival, and of areas under the Bessel excursion are shown to be powerful tools in the analysis of the atomic cloud. A rather general relation between first passage time statistics and diffusivity is discussed, showing that first passage time analysis is a useful tool in the calculation of transport coefficients. A brief introduction to the semi-classical description of Sisyphus cooling is provided which yields a rich phase diagram for the dynamics.

1. Introduction

Fick's second law predicts how diffusion causes concentration to change in time. Instead of concentration we will use a probability density in one dimension $P(x,t)$ and then Fick's law reads

$$\frac{\partial P}{\partial t} = K_2 \frac{\partial^2 P}{\partial x^2} \tag{1}$$

where K_2 is the diffusion coefficient. For an open system, and particles initially localized at the origin, the solution is a spreading Gaussian packet, with a width proportional to time:

$$\langle x^2 \rangle = 2K_2 t. \tag{2}$$

Such normal diffusion is ubiquitous, which is hardly surprising since the diffusion is merely a sum of random displacements with zero mean. Hence, following the central limit theorem we expect to find in experiment a Gaussian distribution for the particle position. The Gaussian universality of diffusion processes leaves us with one difficult task, and that is to compute the diffusion constant K_2. This constant is non-universal, unlike the nearly universal shape of the Gaussian probability packet. Still, non-equilibrium statistical mechanics gives us a recipe (in principle) for the calculation of K_2 which is the famous Green-Kubo formula

$$K_2 = \int_0^\infty \mathrm{d}\tau \langle v(t+\tau)v(t)\rangle. \tag{3}$$

This relation between the velocity correlation function $\langle v(t+\tau)v(t)\rangle$ and the diffusivity is extensively used in the context of hard and soft condensed matter. The correlation function and the underlying process itself is assumed to be stationary so $\langle v(t+\tau)v(t)\rangle$ is independent of t. The Green-Kubo formula is important since once we obtain the diffusivity of a system we also know its mobility. This is made possible via linear response theory and the well-known Einstein relation which connects the diffusivity with the mobility via the thermal scale $k_B T$ where T is temperature.

All this was established a long time ago, so it is not surprising that physicists have turned their focus to problems that violate this normal behavior. One example is the diffusion and transport of atoms in optical molasses driven by counter-propagating laser beams [1]. This system has attracted considerable attention, both for applied purposes like cooling and control of atoms and from a more fundamental point of view, since it exhibits unusual friction and in some cases large deviations from ordinary statistical mechanics. Indeed, laser cooling is today the tool of choice in many laboratories which investigate low temperature physics. All this makes a fundamental understanding of statistical properties of laser cooled atoms a timely subject.

Within the semi-classical picture of Sisyphus cooling, the atoms are subject to a velocity dependent friction force, $F(v)$, defined more precisely below. At low velocities, $F(v) \propto -v$, which is of course normal in the sense that it mimics Stoke's friction for a macroscopic Brownian particle in a viscous medium at room temperatures. Such a friction force is intuitive since it indicates that the faster the particle moves the more energy it dissipates to the surrounding bath. The large v behavior of the atomic friction on the other hand is counterintuitive: it decreases with velocity, $F(v) \sim -1/v$, thus when $v \to \infty$ the system becomes frictionless. Such a system can be called asymptotically dissipationless, and it exhibits physical behaviors very different than standard systems such as non-dissipative onesSisyphus and systems with kinetic dissipation which does not vanish at high speeds.

This behavior of the friction force is due to the fact that the laser cooling is not effective beyond some critical velocity. Briefly, the effect is caused since a fast particle cannot distinguish the uphills from the downhills in the spatially periodic optical lattice; hence the Sisyphus mechanism which favours the loss of energy when the particle is close to the maximum of the periodic lattice is not effective (see Fig. 1 and caption). Since the friction is vanishingly small for large velocities the equilibrium distribution of the velocities can be shown to obey power law statistics (see details below), and so various average quantities are dominated by fast particles (in contrast, for example, to the Maxwellian distribution which is Gaussian). More generally, the kinetics and equilibrium properties of atoms under laser cooling are far from standard: anomalous diffusion, non-Gibbsian states, etc. can be found not only in Sisyphus cooling but also in other cooling approaches like sub-recoil laser cooling [2], and in Doppler cooling as well.

Indeed in the closing paragraph of Cohen-Tannoudji's Nobel lecture [3] he writes "It is clear finally that all the developments which have occurred in the field of laser cooling and trapping are strengthening the connections which can be established between atomic physics and other branches of physics such as condensed matter or statistical physics. The use of Lévy statistics for analyzing sub-recoil cooling is an example of such a fruitful dialogue". Here such a dialogue is extended to the popular Sisyphus cooling mechanism. Lévy statistics roughly refers to power law distributions, which violate the conditions leading to the Gaussian central limit theorem. An example of this is found in a recent experiment performed at the Weizmann Institute.

1.1. *Spatial diffusion of cold atoms*

There, Sagi *et al.* [4] measured the diffusion of ultra-cold ^{87}Rb atoms in a one dimensional optical lattice. Starting with a very narrow atomic cloud they recorded the time evolution of the density of the particles, here denoted $P(x,t)$ (normalized to unity). Their work employed the well-known Sisyphus cooling scheme [5]. As predicted theoretically by Marksteiner *et al.* [6], the diffusion of the atoms was not Gaussian, so that the assumption that the diffusion process obeys the standard central limit theorem is not valid in this case. We recently [7] determined the precise nature of the non-equilibrium spreading of the atoms, in particular the dynamical phase diagram of the various different types of behaviors exhibited as the depth of the optical potential is varied, at least within the semiclassical approach.

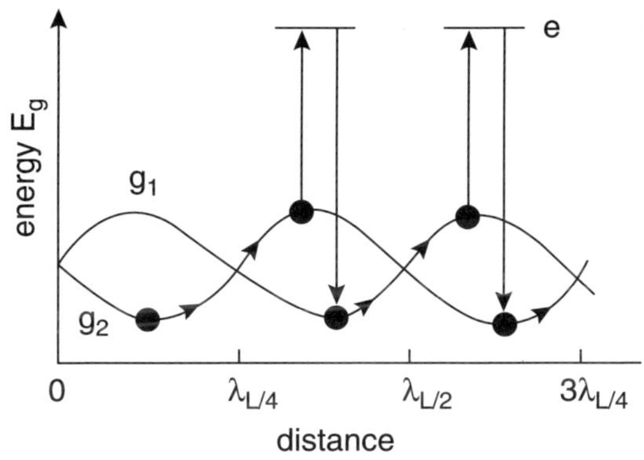

Fig. 1. Sisyphus was punished by the Greek gods to push a heavy stone uphill just to have it roll back to its starting point. In Sisyphus cooling, an atom will go up a potential hill created by laser fields, and then falls down to the local ground state, thus losing energy. Such a scheme is made possible only when transitions from the potential maximum to minimum are statistically preferred, which in turn is made possible by clever quantum mechanics, and polarization fields (which control the transitions) which are periodic functions of distance specially designed to correspond to the spatial modulation of the energy surfaces (the period is determined by the wavelength of the laser field, λ_L). However, if Sisyphus runs like a bolt, over the hills and the valleys he can fool the gods. They cannot push him down from the top of the hills, since they need some finite time to implement his punishment. This corresponds to a fast atom whose transitions are not synchronized with it being at a maximum of the energy surface. So for fast particles, the Sisyphus friction becomes small and decreases, according to the detailed calculations, like $F(v) \propto -1/v$. In the semiclassical approach, an average over the spatial period of the optical lattice is made, hence this periodicity in not explicitly found in our treatment. For more on Sisyphus cooling see [1].

In [4], the anomalous diffusion data was compared to the solutions of the fractional diffusion equation [8–10]

$$\frac{\partial^\beta P(x,t)}{\partial t^\beta} = K_\nu \nabla^\nu P(x,t), \tag{4}$$

with $\beta = 1$, so that the time derivative on the left hand side is a first-order derivative. In [7], we briefly presented the derivation of this equation from the semi-classical theory. The fractional space derivative on the right hand side is a Weyl-Rietz fractional derivative [10], defined below. Here the anomalous diffusion coefficient K_ν has units cm$^\nu$/sec. A fundamental challenge is to derive such fractional equations from a microscopic theory, without invoking power-law statistics in the first place. Furthermore, the solutions of such equations exhibit a diverging mean-square displacement $\langle x^2 \rangle = \infty$, which violates the principle of causality[a], which restricts physical phenomena to spread at finite speeds. So how can fractional equations like Eq. (4) describe physical reality? We will address this paradox in this work.

The solution of Eq. (4) for an initial narrow cloud is given in terms of a Lévy distribution (see details below). The Lévy distribution generalizes the Gaussian distribution in the mathematical problem of the sum of a large number of independent random variables symmetrically distributed about 0, in the case where the variance of the summands diverges, corresponding physically to scale free systems. Here our aim is to derive Lévy statistics and the fractional diffusion equation from the semi-classical picture of Sisyphus cooling. Specifically, we will show that $\beta = 1$ and relate the value of the exponent ν to the depth of the optical lattice U_0, deriving an expression for the constant K_ν. Furthermore, we discuss the limitations of the fractional framework, and show that for a critical value of the depth of the optical lattice, the dynamics switches to a non-Lévy behavior (i.e. a regime where Eq. (4) is not valid); instead it is related to Richardson-Obukhov diffusion found in turbulence. Thus the semiclassical picture predicts a rich phase diagram for the atomistic diffusion process. We will then compare the results of this analysis to the experimental findings, and see that there are still unresolved discrepancies between the experiment and the theory. Reconciling the two thus poses a major challenge for the future.

Since usual approaches to diffusion, namely Fick's second law and the Green-Kubo formula break down for the case of interest, we must take another tack. Here the power of first passage time statistics enters. We will show how analysis of first passage time statistics yields the diffusion front of a packet of particles. In essence the first passage time tool replaces the failing Green Kubo formula, and leads us to the solution of the problem, namely the exponents β and ν as well as the generalized diffusion constant K_ν. But this is a long journey, so first let us briefly review the concept of Brownian excursions, first passage times for simple Brownian motion, and the distribution of the area swept under Brownian motion till its first passage.

[a]See the discussion in [11] where the unphysical nature of Lévy flights is discussed and the resolution in terms of Lévy walks is addressed.

1.2. *Brief survey: Joint PDF for first passage time and the area swept under Brownian motion*

Consider a Brownian motion in one dimension. The time τ_f it takes the particle starting at $x = \epsilon > 0$ to reach $x = 0$ for the first time is the first passage time. The distribution of τ_f has been well investigated both for Brownian motion and for other types of random walks [12–15]. A quantity which until recently was only of mathematical interest is the total area under the Brownian path in the time interval $(0, \tau_f)$, which was treated by Kearney and Majumdar [16]. This area, which we denote χ_f, is obviously positive if we start at $x = \epsilon > 0$. Later we will take $\epsilon \to 0$ which is a subtle point, but for now we keep ϵ finite. The pair of random quantities χ_f and τ_f are clearly correlated since a large τ_f implies statistically a large χ_f. The joint probability density function is denoted as $\psi(\tau_f, \chi_f)$. Using Bayes' law we may write this density as

$$\psi(\tau_f, \chi_f) = g(\tau_f)P(\chi_f|\tau_f) \tag{5}$$

where $g(\tau_f)$ is the first passage time probability density function (PDF), which at least for Brownian motion is well investigated. In particular, $g(\tau_f)$ decays as a power for large τ_f,

$$g(\tau_f) \propto (\tau_f)^{-3/2}, \tag{6}$$

so that the average first passage time, $\langle \tau_f \rangle = \infty$, diverges for one dimensional Brownian motion on the infinite line. $P(\chi_f|\tau_f)$ is the conditional probability density function of χ_f given some fixed τ_f. This leads us to the concept of a Brownian excursion.

A Brownian excursion is a conditioned one dimensional Brownian motion $x(t)$ over the time interval $0 \le t \le \tau_f$ (recall here τ_f is fixed to a specific value). The motion starts at $x(0) = \epsilon > 0$, $\epsilon \ll 1$, and ends at $x(\tau_f) = 0$ and is constrained not to cross the origin $x = 0$ in the observation time $(0, \tau_f)$, namely the Brownian excursion is always positive. The area under the Brownian excursion is the object of interest since that gives the conditional probability density $P(\chi_f|\tau_f)$. Thus the statistics of the area under the Brownian excursion also gives the joint PDF of the first passage time and the total area under the Brownian motion.

The area under the Brownian excursion was treated by mathematicians with applications in computer science and graph theory and more recently in the physical context of fluctuations of interfaces [17–19]. Majumdar and Comtet [18] give a concise introduction and derivation of the solution to this problem in terms of the Airy distribution. We shall soon show that

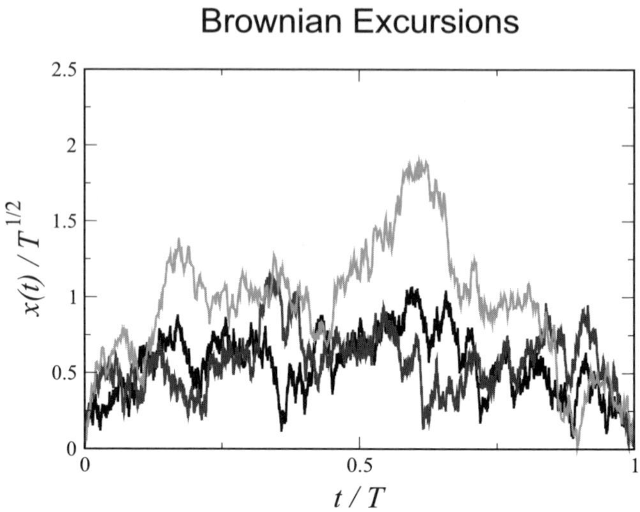

Fig. 2. An example of a Brownian excursion $x(t)$, which in the text is generalized to a Langevin excursion in momentum space, $p(t')$. These are random curves which start and end close to the origin and remain positive for a fixed time. Surprisingly the random area under these curves has started to attract physical attention in diverse problems ranging from fluctuating interfaces and as we show here diffusion of atoms in optical lattices. A related problem is the Tracy-Widom distribution.

the solution of the dynamics of the atomic packet demands the solution of a first passage time problem in momentum space, and the distribution of the area under the random first passage time curve. Thus the random area under a random curve terminated at its first passage turns into a relevant problem, but, fortunately or unfortunately, the apposite curve is not a simple Brownian one.

So, for perspective, we first review what is known about the distribution of A, the random area under the Brownian excursion, $A = \int_0^{\tau_f} x(t) \mathrm{d}t$, in the $\epsilon \to 0$ limit. Since Brownian motion scales like $x \sim t^{1/2}$ the area A scales like $\int_0^{\tau_f} x(t) \mathrm{d}t \propto (\tau_f)^{3/2}$. Thus the distribution has the scaling behavior $P_{\tau_f}(A) = (\tau_f)^{-3/2} f(A/(\tau_f)^{3/2})$, where the $(\tau_f)^{-3/2}$ prefactor stems from the normalization condition. The Laplace $x \to u$ transform of $f(x)$, $\hat{f}(u)$ was computed by Darling and Louchard [20, 21]

$$\hat{f}(u) = u\sqrt{2\pi} \sum_{k=1}^{\infty} \exp\left(-\alpha_k u^{2/3} 2^{-1/3}\right) \tag{7}$$

(here, the diffusion constant of the Brownian motion, arising from the simple random walk used to model the motion, is 1/2). The α_k's are the magnitudes of the zeros of the Airy function $Ai(z)$ on the negative real axis $\alpha_1 = 2.3381..$, $\alpha_2 = 4.0879...$, $\alpha_3 = 5.5205..$ and for that reason the distribution is called the Airy distribution. Takács [22] inverted the Laplace transform to find

$$f(x) = \frac{2\sqrt{6}}{x^{10/3}} \sum_{k=1}^{\infty} e^{-b_k/x^2} b_k^{2/3} U(-5/6, 4/3, b_k/x^2) \tag{8}$$

where $b_k = 2(\alpha_k)^3/27$ and $U(a, b, z)$ is the confluent hypergeometric function [23]. Moments and asymptotic behaviors of the Airy distribution can be found in [18]. The function can be calculated and plotted by truncating the sum at some x dependent value.

From this, we can deduce the power law exponent describing the tail of the PDF of the area swept under the Brownian curve till its first passage event $q(\chi_f)$, averaged over all τ_f. Clearly the marginal PDF $q(\chi_f)$ is found from the joint PDF in the usual way by integration

$$q(\chi_f) = \int_0^{\infty} g(\tau_f) P(\chi_f|\tau_f) \mathrm{d}\tau_f \tag{9}$$

Using Eq. (6) and the scaling behavior of the conditional PDF $P(\chi_f|\tau_f)$, we have

$$q(\chi_f) \propto \int_0^{\infty} (\tau_f)^{-3/2} (\tau_f)^{-3/2} f(\chi_f/(\tau_f)^{3/2}) \mathrm{d}\tau_f. \tag{10}$$

Changing integration variables to $y \equiv \tau_f/(\chi_f)^{2/3}$ we get the asymptotic form

$$q(\chi_f) \propto (\chi_f)^{-4/3}. \tag{11}$$

We see that the scaling exponents 3/2 for $g(\tau)$ and 4/3 for $q(\chi_f)$ are easy to obtain and are both related to the scaling of Brownian motion $x \propto t^{1/2}$. The full distribution of $q(\chi_f)$ was recently found by Kearney and Majumdar for finite ϵ. We will later deal with such a calculation in more rigour, but for now we just focus on the exponents. We now turn to showing how the generalization of these concepts lead to the description of the dynamics of cold atoms.

2. Model and goal

We wish to investigate the spatial density of the atoms, $P(x, t)$. The trajectory of a single particle is $x(t) = \int_0^t p(t) \mathrm{d}t/m$ where $p(t)$ is its momentum.

Within the standard picture [5, 6] of Sisyphus cooling, two competing mechanisms describe the dynamics. The cooling force

$$F(p) = -\overline{\alpha}p/[1 + (p/p_c)^2] \tag{12}$$

acts to restore the momentum to the minimum energy state $p = 0$. Momentum diffusion is governed by a diffusion coefficient which is momentum dependent, $D(p) = D_1 + D_2/[1 + (p/p_c)^2]$. The latter describes momentum fluctuations which lead to heating (due to random emission events). We use dimensionless units, time $t \to t\overline{\alpha}$, momentum $p \to p/p_c$, the momentum diffusion constant $D = D_1/(p_c)^2\overline{\alpha}$ and $x \to xm\overline{\alpha}/p_c$. For simplicity, we set $D_2 = 0$ since it does not modify the asymptotic $|p| \to \infty$ behavior of the diffusive heating term, nor that of the force, and therefore does not modify our main conclusions. The Langevin equations

$$\frac{\mathrm{d}p}{\mathrm{d}t} = F(p) + \sqrt{2D}\xi(t), \tag{13a}$$

$$\frac{\mathrm{d}x}{\mathrm{d}t} = p \tag{13b}$$

describe the dynamics in phase space. Here the noise term is Gaussian, has zero mean and is white $\langle \xi(t)\xi(t')\rangle = \delta(t - t')$. The now dimensionless cooling force is

$$F(p) = -\frac{p}{1 + p^2}. \tag{14}$$

The stochastic Eq. (13) give the trajectories of the standard Kramers picture for the semi-classical dynamics in the optical lattice which in turn was derived from microscopical considerations [5, 6]. From the semiclassical treatment of the interaction of the atoms with the counterpropagating laser beams, we have

$$D = cE_R/U_0, \tag{15}$$

where U_0 is the depth of the optical potential and E_R the recoil energy, and the dimensionless parameter c depends on the atomic transition involved.[b] For $p \ll 1$, the cooling force Eq. (14) is harmonic, $F(p) \sim -p$, while in the opposite limit, $p \gg 1$, $F(p) \sim -1/p$. It is useful, given the correspondence between Eq. (13a) and overdamped motion in space subject to a force F,

[b] For atoms in molasses with a $J_g = 1/2 \to J_e = 3/2$ Zeeman substructure in a lin $\perp$ lin laser configuration $c = 12.3$ [6]. [5, 24, 25] give $c = 22$. As pointed out in [6] different notations are used in the literature.

to define an effective potential

$$V(p) = -\int_0^p F(p)\mathrm{d}p = (1/2)\ln(1+p^2). \tag{16}$$

This potential is asymptotically logarithmic, $V(p) \sim \ln(p)$ when p is large, which has as its consequence various unusual equilibrium and non-equilibrium properties of the momentum distribution [24–29]. For example the steady state momentum distribution is $W_{eq}(p) \propto \exp[-V(p)/D]$ which has a power law form

$$W_{eq}(p) = \mathcal{N}(1+p^2)^{-1/2D} \tag{17}$$

where $\mathcal{N}$ is the normalization. Such non-Gaussian behavior was observed in the experiments of Renzoni's group [25]. For $D < 1/3$, corresponding to shallow optical lattices, the averaged kinetic energy diverges. A similar effect was found experimentally by Walther's group [26]: when the depth of the optical lattice is tuned one may observe a gigantic increase of the energy of the system. Of course the energy of a physical system cannot be infinite and this paradox was recently resolved [27, 28] by treating the problem dynamically, revealing that the energy may increase with time but it is never infinite. Of course, when the energy becomes too large, other experimental effects become relevant as well.

While the velocity distribution of the atomic cloud appears to be well understood, the new experiment [4] demands a theory for the spatial spreading. This will be the focus of this chapter. However, by now the reader might think that laser cooling is an oxymoron. Namely if we get non-Maxwellian distributions in equilibrium, fat tails, and Lévy behavior, where is the thermal state found? The answer, as we understand it, is rather involved. First the equilibrium momentum PDF can be rather narrow in its center, and in that sense the atoms are cold. Further, the divergence of energy happens for low D, so an experimentalist interested in cooling will turn his knobs away from that limit. In fact there is an optimal depth of the optical potential in the sense that it yields the smallest value for the mean kinetic energy of the particles, since the basic energy scale is set by the optical potential depth (see, e.g., Fig. 4 of [26] and Fig. 2 of [30]). Strictly speaking, however, for all D the equilibrium state is not thermal since the momentum distribution is not Gaussian. Experimentalists intuitively recognize this problem, and hence also use a method called evaporation to reach an ultimate cold state. Evaporation simply means that atoms in the tails, i.e. atoms with high velocities, are a problem for cooling, so they are allowed by the experimentalists to leave the system. Another mechanism that may lead to a realistic thermal state is internal collisions among the atoms (which will depend, of course, on the density). It is believed that such effects will lead the system in the long time limit to thermal equilibrium. However, at least in the experiments briefly surveyed here [4, 25, 26], which take long times on atomic scale (e.g. milliseconds) no hint of thermal Maxwellian equilibrium is found, at least for values of D of interest. Hence it is an important task to map the different phases of the dynamics which are controlled by D. Indeed this system is truly wonderful in the sense that the experimentalist can tune D and hence explore different effects both in and out of equilibrium.

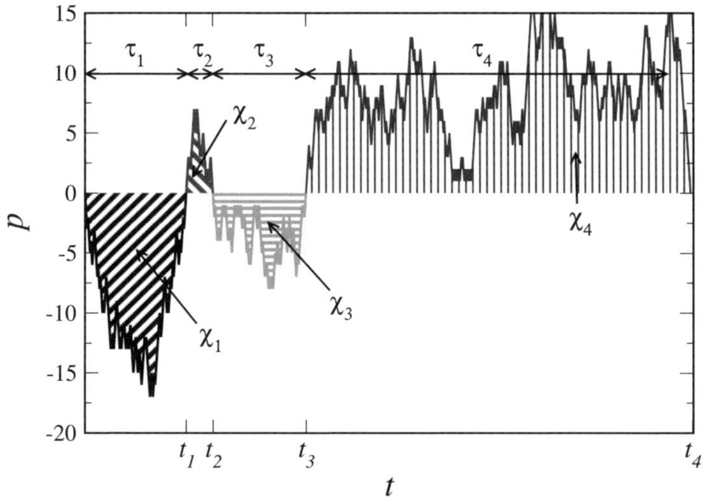

Fig. 3. Schematic presentation of the momentum of the particle versus time. The times between consecutive zero crossings are called the jump durations τ and the shaded area under each excursion are the random flight displacements χ. The τ's and the χ's are correlated, since statistically a long jump duration implies a large displacement.

3. From area under the Bessel excursion to Lévy walks for the packet of atoms

The heart of our analysis is the mapping of the Langevin dynamics to a recurrent set of random walks. The particle along its stochastic path in momentum space crosses $p = 0$ many times when the measurement time is long. Let $\tau > 0$ be the random time between one crossing event to the next crossing event, and let $-\infty < \chi < \infty$ be the random displacement (for the corresponding τ). As schematically shown in Fig. 3, the process starting at the origin with zero momentum is defined by the sequence of jump durations, $\{\tau_1, \tau_2, \ldots\}$ with corresponding displacements $\{\chi_1, \chi_2, \ldots\}$, with $\chi_1 \equiv \int_0^{\tau_1} p(\tau)d\tau$, $\chi_2 \equiv \int_{\tau_1}^{\tau_1+\tau_2} p(\tau)d\tau$, etc. The total displacement x at time t is a sum of the individual displacements χ_i. Since the underlying Langevin dynamics is continuous, we need a more precise definition of this process. Let τ be the time it takes the particle with initial momentum p_i to reach $p_f = 0$ for the first time. So τ is the first passage time for the process in momentum space. Eventually we will take $p_i \to p_f$, see below. Similarly, χ is the displacement of the particle during this flight. Namely by definition χ is the area under the Langevin curve (in momentum space) till the first passage time. The probability density function (PDF) of the displacement χ is denoted $q(\chi)$ and of the jump durations $g(\tau)$. To conclude we see that first passage time statistics, in particular the area swept under the random momentum process $p(\tau)$ until the first passage time, constitute the microscopical random jumps and correlated random waiting times which eventually give the full displacement of the the particle.

As shown by Marksteiner *et al.* and Lutz [6, 24], these PDFs exhibit power law behavior

$$g(\tau) \propto \tau^{-\frac{3}{2}-\frac{1}{2D}}, \tag{18a}$$

$$q(\chi) \propto |\chi|^{-\frac{4}{3}-\frac{1}{3D}}, \tag{18b}$$

as a consequence of the logarithmic potential, which makes the diffusion for large enough p only weakly bounded. It is this power-law behavior, with its divergent second moment of the displacement χ for $D > 1/5$, which gives rise to the anomalous statistics for x (measured in the experiment). Notice that taking the limit $D \to \infty$ in Eq. (18) we get the expected results as given in Eqs. (6) and (11). In this limit, the cooling force $F(p)$ is negligible, the dynamics is governed by the noise in Eq. (13) and so the process $p(t)$ approaches a Brownian motion. This corresponds to atoms emitting a photon in random directions, and hence from the recoil jolts, they are undergoing an unbiased random walk in momentum space (which is the pure heating limit).

Importantly, and previously overlooked, there is a strong correlation between the jump duration τ and the spatial extent of the jumps χ. These correlations are responsible, in particular, for the finiteness of the moments of $P(x,t)$. Physically, such a correlation is obvious, since long jump durations involve large momenta, which in turn induce a large spatial displacement. The theoretical development starts then from the quantity $\psi(\chi,\tau)$, the joint probability density of $|\chi|$ and τ. From this, we construct a Lévy walk scheme [31–33] which relates the microscopic information $\psi(\chi,\tau)$ to the atomic packet $P(x,t)$ for large x and t.

4. Scaling theory for anomalous diffusion

We rewrite the joint PDF

$$\psi(\chi,\tau) = \frac{1}{2} g(\tau) p\left(|\chi| \,|\, \tau\right), \tag{19}$$

where $p(\chi \,|\, \tau)$ is the conditional probability to find a jump length of $0 < \chi < \infty$ for a given jump duration τ. The factor $\frac{1}{2}$ accounts for the walks with $\chi < 0$. Numerically we observed that the conditional probability scales at large times like

$$p(\chi|\tau) \approx 2\tau^{-\gamma} B(|\chi|/\tau^{\gamma}); \qquad \gamma = 3/2. \tag{20}$$

$B(\cdot)$ is a scaling function soon to be determined. To analytically obtain the scaling exponent $\gamma = 3/2$ note that $q(\chi) = \int_0^\infty \mathrm{d}\tau \psi(\chi,\tau)$, giving

$$q(\chi) \sim \int_{\tau_0}^{\infty} \mathrm{d}\tau \tau^{-\frac{3}{2}-\frac{1}{2D}} \tau^{-\gamma} B\left(\frac{|\chi|}{\tau^{\gamma}}\right) \propto |\chi|^{-\left(1+\frac{1+1/D}{2\gamma}\right)}. \tag{21}$$

Here τ_0 is a time scale after which the long time limit in Eqs. (18) holds and is irrelevant for large χ. Comparing Eq. (21) to Eq. (18b) yields the consistency condition $1+(1+1/D)/(2\gamma) = 4/3+1/(3D)$ and hence $\gamma = 3/2$, as we observe in Fig. 4.

We see that the scaling function $B(.)$ in Eq. (20) is similar to the problem of the area under a Brownian excursion. However, now we are confronted with the additional impact of the friction force. The concept of the Brownian excursion needs to be extended to a stochastic curve called the Bessel excursion [34], which corresponds to the friction force $F(p) = -1/p$. The term Bessel excursion stems from the fact that mathematically speaking, diffusion in momentum space, in the non-regularized potential $\ln(p)$, corresponds to a process called the Bessel process [35, 36]. As with the Brownian excursion, the Bessel excursion is the Langevin path $p(t')$ over the time interval $0 \le t' \le \tau$ such that the path starts on $p_i \to 0$ and ends on the origin, but is *constrained* to stay positive (if $p_i > 0$) or negative (if $p_i < 0$). Brownian excursions are just the $D \to \infty$ limit of this more

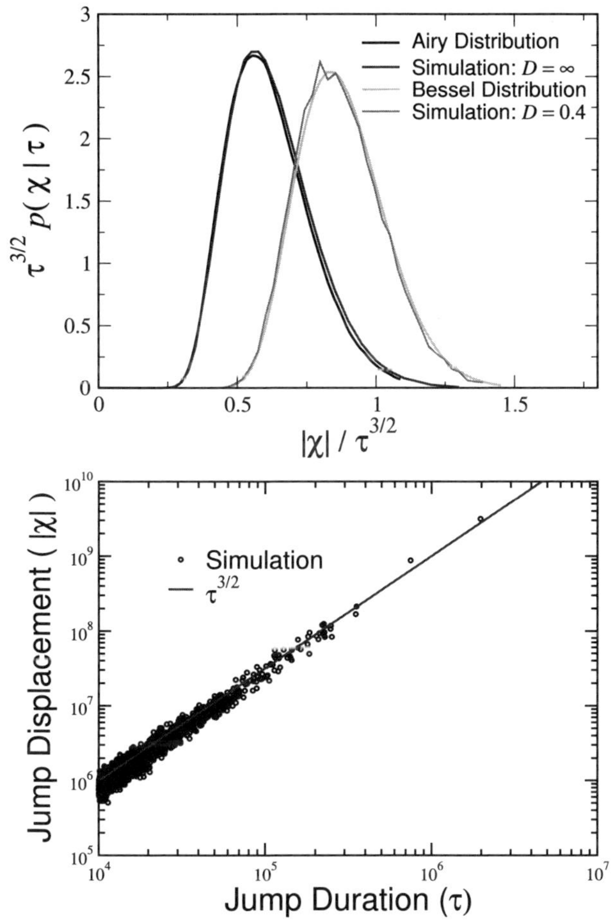

Fig. 4. Lower panel: We plot the flight displacement $|\chi|$ versus the jump duration τ to demonstrate the strong correlations between these two random variables. Here $D = 2/3$ and the line has slope 3/2 reflecting the $\chi \sim \tau^{3/2}$ scaling discussed in the text. Upper panel: The conditional probability $p(\chi|\tau)$ is described by the Airy distribution [17, 18] when $D \to \infty$; otherwise by the distribution derived here based on the Feynman-Kac theory, Eq. (24). They describe areas under constrained Brownian (or Bessel) excursions when the cooling force $F(p) = 0$ (or $F(p) \neq 0$) respectively.

general problem. Schematic excursions $p(t)$ are presented in Fig. 3, albeit there with various excursion times τ. The displacement $\chi = \int_0^\tau p(t')\mathrm{d}t'$ is the area under the excursion as shown in Fig. 3.

4.1. *Area under the Bessel excursion*

We briefly outline how to obtain the distribution of the area under the Bessel excursion, namely we find the rather formidable $p(\chi|\tau)$, Eq. (24) below. The main tool we use is a modified Feynman-Kac theory. The original well-known formalism deals with functionals of Brownian motion, while here we deal with functionals of Langevin motion, hence the modification. The Feynman-Kac equation is essentially a Shrödinger equation in imaginary time and is reviewed in [37]. The calculation involves three steps:

(a) We first find $\widetilde{G}(s,p,\tau)$, the Laplace transform with respect to χ of the joint PDF for the positively constrained path to reach p with area χ at time τ, which we denote by $G(\chi,p,t)$.

(b) Taking the limit of the initial and final momentum $p_i = p \to 0$, we get $\tilde{p}(s|\tau) = \lim_{p,p_i\to 0} \widetilde{G}(s,p,\tau)/\widetilde{G}(s=0,p,\tau)$; the $\widetilde{G}(s=0,p,\tau)$ denominator yields the correct normalization.

(c) Finally, using the inverse Laplace transform, we get $p(\chi\,|\,t)$.

As mentioned, the main tool we use is the modified Feynman-Kac formalism for functionals of over-damped Langevin paths [37, 38]. For step (a) we solve:

$$\frac{\partial \widetilde{G}(s,p,\tau)}{\partial \tau} = \left[\widetilde{L}_{\rm fp} - sp\right] \widetilde{G}(s,p,\tau), \tag{22}$$

where $\widetilde{L}_{\rm fp} = D(\partial_p)^2 - \partial_p F(p)$ is the Fokker Planck operator. When $F(p) = 0$, Eq. (22) is the celebrated Feynman-Kac equation which is an imaginary-time Schrödinger equation in a linear potential (i.e. the $-sp$ term). We do not provide here a derivation of Eq. (33) but in an Appendix provide a derivation of a related equation. The constraint $p > 0$ gives the boundary condition $\widetilde{G}(s,p=0,t) = 0$, i.e., an infinite potential barrier when $p < 0$ in the quantum language. We use the force $F(p) = -1/p$ since we are interested only in the scaling behavior of the problem where large excursions imply that the small p behavior of the force is not important. This assertion can be mathematically justified, but here for the sake of space we will only demonstrate it numerically (see Fig. 4). Solving Eq. (22) and following the recipe (a) - (b) we find

$$\widetilde{p}(s|\tau) = \sum_{k=1}^{\infty} 2^{3\nu+1}\Gamma\left(1+\frac{3}{2}\nu\right)\left(sD^{1/2}\tau^{2/3}\right)^{\nu+2/3}[g_k'(0)]^2 e^{-\tau D^{1/3}\lambda_k s^{2/3}}. \tag{23}$$

Here $Dg_k'' + (g_k/p)' - pg_k = -D\lambda_k g_k$ is the eigenvalue equation which gives the eigenvalues λ_k and eigenstates $g_k(p)$ which are normalized according $\int_0^\infty [g_k(p)]^2 p^{1/D} \mathrm{d}p = 1$. A detailed mathematical derivation will soon be published [39].

It is easy to tabulate the eigenvalues λ_k and the slopes on the origin $g_k'(p = 0)$ with standard numerically exact techniques. With this information and the inverse Laplace transform we find the solution in terms of generalized hypergeometric functions

$$\begin{aligned} p(\chi \,|\, \tau) = & -\frac{\Gamma\left(1+\frac{3\nu}{2}\right)}{4\pi\chi}\left(\frac{4D^{1/3}\tau}{\chi^{2/3}}\right)^{1+3\nu/2} \times \\ & \sum_k [g_k'(0)]^2 \left[{}_2F_2\left(\frac{4}{3}+\frac{\nu}{2}, \frac{5}{6}+\frac{\nu}{2}; \frac{1}{3}, \frac{2}{3}; -\frac{4D\lambda_k^3\tau^3}{27\chi^2}\right) c_0 \right. \\ & - {}_2F_2\left(\frac{7}{6}+\frac{\nu}{2}, \frac{5}{3}+\frac{\nu}{2}; \frac{2}{3}, \frac{4}{3}; -\frac{4D\lambda_k^3\tau^3}{27\chi^2}\right) c_1 \\ & \left. + {}_2F_2\left(2+\frac{\nu}{2}, \frac{3}{2}+\frac{\nu}{2}; \frac{4}{3}, \frac{5}{3}; -\frac{4D\lambda_k^3\tau^3}{27\chi^2}\right) \frac{c_2}{2} \right], \end{aligned} \tag{24}$$

where

$$c_j = \left(\frac{D^{1/3}\lambda_k\tau}{\chi^{2/3}}\right)^j \Gamma\left(\frac{5+2j}{3}+\nu\right) \sin\left(\pi\frac{2+2j+3\nu}{3}\right) \qquad j = 0\ldots 2 \tag{25}$$

and the summation is over the eigenvalues. Here we have introduced the parameter

$$\nu \equiv \frac{1+D}{3D} \tag{26}$$

which will turn out to play a crucial role in the dynamics. Notice the $\chi \sim \tau^{3/2}$ scaling which proves the scaling hypothesis, Eq. (20). To compare with numerical simulations, we evaluate this function truncating the sum beyond some k, since for fixed x the terms decay rapidly in k. The Langevin simulations were performed by mapping the problem onto a biased random walk in momentum space, with momentum spacing $\delta p = 0.1$.

In Fig. 4, we present $p(\chi|\tau)$ obtained from numerical simulations, showing that it perfectly matches our theory on Bessel excursions without fitting. Our theory and simulations show that the areas under the Brownian and Bessel excursions share the same $\chi \sim \tau^{3/2}$ scaling behavior. In both cases, $p(\chi|\tau)$ falls off rapidly for large $|\chi|$, ensuring finite moments of this distribution, which will prove important later. When $D \gg 1$, the Bessel and the Brownian excursions coincide, since then the force $F(p)$ plays a vanishing role, as can be shown directly from Eq. (24). The Bessel excursion PDF for finite D is compared with the limiting Airy distribution in Fig. 4, where it is seen that the Bessel excursions are pushed further away from the origin. The $-1/p$ force implies that trajectories not crossing the origin in $(0, \tau)$ are typically further away from the origin, compared with the free Brownian particle, since trajectories close to the origin are more likely to cross it and thus be excluded, due to the attractive force.

To continue with the analysis we should consider also the area under the Bessel meander. The Brownian meander is a Brownian motion, starting at ϵ that does not cross the origin in the time interval $(0, \tau)$. Unlike the Brownian excursion, the end point of the Brownian meander is not constrained to lie at the origin. Similarly, we can define a Bessel meander, using Eq. (13a) with $F(p) = -1/p$.

The meander is important in the statistical description of the last jump event. Consider a process $p(\tau)$ with n zero crossings of the momentum origin in the time interval $(0, t)$ as shown in Fig. 3. These zero crossings epochs are related to the waiting times $t_1 = \tau_1, t_2 = \tau_1 + \tau_2, \ldots$ etc. so t_n denotes the time for the last zero crossing event before the observation time t. So clearly at time t_n the momentum is zero, while at time t it is not. Since by definition the particle did not cross the momentum origin in (t_n, t) the random momentum curve in that interval is a meander. For normal diffusion, e.g., if the friction force is linear, the last time interval in the sequence, i.e. (t_n, t), sometimes called the backward recurrence time, is of no physical consequence, in the limit of long measurement times. In contrast, for certain anomalous diffusion processes the contribution of the backward recurrence time is crucial for precise statistical modelling. Luckily certain observables and conclusions are not sensitive to the statistics of the meander, so we avoid its discussion here to save space. As we will show in a future publication [39], the analysis of the area under the meander is important if one wishes to investigate, for example, the mean square displacement, when the atomic diffusion is anomalous.

4.2. *Coupled continuous time random walk theory from first passage time statistics*

The next step is to relate the first passage time properties of the underlying Langevin process in momentum space, namely the joint distribution of the area under the Bessel excursion and the first passage time, with the spreading of the density $P(x,t)$. Given our scaling solution for $p(\chi|t)$, and $g(\tau)$ which is considered in detail in [39] (see also Appendix here) and hence $\psi(\chi,t)$, we construct a theory for $P(x,t)$ using tools developed in the random walk community [31]. As soon detailed, one first obtains a Montroll-Weiss [10] type of equation for the Fourier-Laplace transform of $P(x,t)$, $\widetilde{P}(k,u)$, in terms of $\widetilde{\psi}(k,u)$, the Fourier-Laplace transform of the joint PDF $\psi(\chi,\tau)$:

$$\widetilde{P}(k,u) = \frac{\Psi_M(k,u)}{1-\widetilde{\psi}(k,u)}. \tag{27}$$

Here, $\Psi_M(k,u)$ is the Fourier-Laplace transform of $\tau^{-3/2}B_M(|\chi|/\tau^{3/2})[1-\int_0^t \psi(\tau)d\tau]$. Here $B_M(.)$ describes the random area under the Bessel meander, which as mentioned is not presented here [39], concentrating instead on observables which are not sensitive to it. The last step is then to invert Eq. (27) back to the x, t domain.

To derive Eq. (27) we present a formalism which relates the joint distribution

$$\psi(\chi,\tau) = g(\tau)\tau^{-3/2}B\left(|\chi|/\tau^{3/2}\right) \tag{28}$$

with $P(x,t)$ ($B(.)$ is given in Eqs. (20,24)). Define $\eta_s(x,t)\mathrm{d}t\mathrm{d}x$ as the probability that the particle crossed the momentum state $p=0$ for the sth time in the time interval $(t,t+\mathrm{d}t)$ and that the particle's position was $(x,x+\mathrm{d}x)$. This probability is related to the probability of the previous crossing according to

$$\eta_s(x,t) = \int_{-\infty}^{\infty}\mathrm{d}v_{3/2}\int_0^{\infty}\mathrm{d}\tau\,\eta_{s-1}\Big(x-v_{3/2}\tau^{3/2},t-\tau\Big)\,B\left(|v_{3/2}|\right)g\left(\tau\right) \tag{29}$$

where we changed the variable of integration to $v_{3/2}\equiv\chi/\tau^{3/2}$. Now the process is described by a sequence of jump durations $\tau_1,\tau_2,\cdots$ and the corresponding generalized velocities $v_{3/2}(1),v_{3/2}(2),\cdots$ (with units $[x]/[t^{3/2}]$). The displacements in the sth interval is: $\chi_s = v_{3/2}(s)[\tau_s]^{3/2}$. The advantage of the representation of the problem in terms of the pair of microscopic stochastic variables $\tau, v_{3/2}$ (instead of the correlated pair τ,χ) is clear from Eq. (29): we may treat $v_{3/2}$ and τ as independent random variables whose corresponding PDFs are $g(\tau)$ and $B(v_{3/2})$ respectively. The initial condition $x=0$ at time $t=0$ implies $\eta_0(x,t)=\delta(x)\delta(t)$. Let $P(x,t)$ be the probability of finding the particle in $(x,x+\mathrm{d}x)$ at time t, which is found according to

$$P(x,t) = \sum_{s=0}^{\infty} \int_{-\infty}^{\infty} \mathrm{d}v_{3/2} \int_0^{\infty} \mathrm{d}\tau\, \eta_s\Big(x - v_{3/2}\tau^{3/2}, t-\tau\Big)\, B_M\left(|v_{3/2}|\right) W\left(\tau\right). \tag{30}$$

since for a time series with s intervals, the last jump event took place at $t-\tau$ and in the time period $(t-\tau, t)$ the particle did not cross the momentum origin (hence the survival probability $W(\tau) = 1 - \int_0^{\tau} g(\tau)\mathrm{d}\tau$ and the use of the meander). The summation in Eq. (30) is a sum of all possible numbers s of returns to the momentum origin $p_f = 0$.

As usual [10, 31], we consider the problem in Laplace-Fourier space where $t \to u$ and $x \to k$. Eq. (29) then translates to

$$\tilde{\eta}_s(k,u) = \tilde{\eta}_{s-1}(k,u)\, \widetilde{\psi}(k,u). \tag{31}$$

Here the tildes refer to the Fourier-Laplace transforms, and we have used Eq. (20) to write that $\psi(\chi,\tau) = g(\tau)\tau^{-3/2}B(|\chi|/\tau^{3/2})$. Hence, summing the Fourier-Laplace transform of Eq. (30) using Eq. (31), we get the Montroll-Weiss Eq. (27).

The Montroll-Weiss equation is widely used in applications of the random walk. This equation holds since the underlying dynamics is Markovian, namely each time the process $p(t)$ crosses zero, the process is renewed. However, there is a subtle point behind our approach, which makes it different. Since the trajectories are continuous, the number of zero crossings s is infinite within any finite time interval $(0,t)$, for a particle starting with $p=0$. In contrast, in usual continuous time random walk theories the number of renewals is always finite, and it increases with time the size of the time interval, t. The functions entering the right hand side of Eq. (27) still depend on the initial momentum that we assign to the particle $\pm p_i$ immediately after each zero crossing (if we use a grid in a numerical simulation this is handled automatically). We still must take the limit $p_i \to 0$ and show that our final expression will give sensible results in that regularized limit (see also [39]).

5. The Lévy phase

We now explain why Lévy statistics describes the diffusion profile $P(x,t)$ when $1/5 < D < 1$, provided that x is not too large. The key idea is that, for x's which are large, but not extremely large, the problem decouples, and $\widetilde{\psi}(k,u)$ can be expressed as a product of the Fourier transform of $q(\chi)$, $\tilde{q}(k)$ and the Laplace transform of $g(\tau)$, $\tilde{g}(u)$. This is valid as long as $x \ll t^{3/2}$, since otherwise paths where $\chi \sim t^{3/2}$ are relevant, for which the correlations are strong, as we have seen. The long-time, large-x behavior of $P(x,t)$ in the decoupled regime is then governed by the small-k behavior of $\tilde{q}(k)$ and the small u behavior of $\tilde{g}(u)$. When the second moment of $q(\chi)$ diverges, i.e. for $D > 1/5$, the small-k behavior of $\tilde{q}(k)$ is determined by the large-χ asymptotics of $q(\chi)$ as given in Eq. (20), $q(\chi) \sim x_*^\nu/|\chi|^{1+\nu}$. When the first moment of τ is finite, i.e. for $D < 1$, the small-u behavior of $\tilde{g}(u)$ is governed by the first moment, $\langle\tau\rangle$. From these follow the small-k, small-u behavior of $\widetilde{P}(k,u)$:

$$\widetilde{P}(k,u) \sim \frac{1}{u + K_\nu |k|^\nu} \tag{32}$$

where $K_\nu = \pi x_*^\nu/(\langle\tau\rangle\Gamma(1+\nu)\sin\frac{\pi\nu}{2})$ (see some details in Appendix A).

Both x_*^ν and $\langle\tau\rangle$ can be calculated (see Appendix A and [39]) via appropriate backward Fokker-Planck equations. They both vanish as the magnitude of the initial momentum of the walk goes to zero, but their ratio has a finite limit, so that K_ν, upon returning to dimensionfull units, is

$$K_\nu = \frac{\sqrt{\pi}(3\nu-1)^{\nu-1}\Gamma\left(\frac{3\nu-1}{2}\right)}{\Gamma\left(\frac{3\nu-2}{2}\right)3^{2\nu-1}[\Gamma(\nu)]^2\sin\left(\frac{\pi\nu}{2}\right)}\left(\frac{p_c}{m}\right)^\nu(\overline{\alpha})^{-\nu+1}. \tag{33}$$

$\widetilde{P}(k,u)$, as given in Eq. (32), is in fact precisely the symmetric Lévy distribution in Laplace-Fourier space with index ν, whose (x,t) representation is (see Eq. (B17) of [40])

$$P(x,t) \sim \frac{1}{(K_\nu t)^{1/\nu}} L_{\nu,0}\left[\frac{x}{(K_\nu t^{1/\nu})}\right]. \tag{34}$$

The properties of the Lévy function $L_{\nu,0}(.)$ are well-known, and the solution in time for the spatial Fourier transform is $P(k,t) = \exp(-K_\nu t|k|^\nu)$ with $2/3 < \nu < 2$. This distribution is the solution of the fractional diffusion equation, Eq. (4), with $\beta = 1$ and an initial distribution located at the origin. This justifies the use of Eq. (4) in [4] for $1/5 < D < 1$ and provides ν and K_ν in terms of the experimental parameters. We can verify this behavior in simulations, as shown in Fig. 3, where we see excellent agreement to our theoretical prediction, Eqs. (34) and (33), without any fitting.

The fractional diffusion equation, Eq. (4), has the following meaning and connection to the underlying random walk. We re-express Eq. (32) as $u\widetilde{P}(k,u) - 1 = -K_\nu|k|^\nu\widetilde{P}(k,u)$. Recalling [10] that the Fourier space representation of the Weyl-Rietz fractional derivative, ∇^ν, is $-|k|^\nu$ and that

$u\widetilde{P}(k,u) - 1$ is the representation of the time derivative in Laplace space for δ-function initial conditions centered at the origin, we see that in fact $P(x,t)$ satisfies the fractional diffusion equation, Eq. (4), with $\beta = 1$.

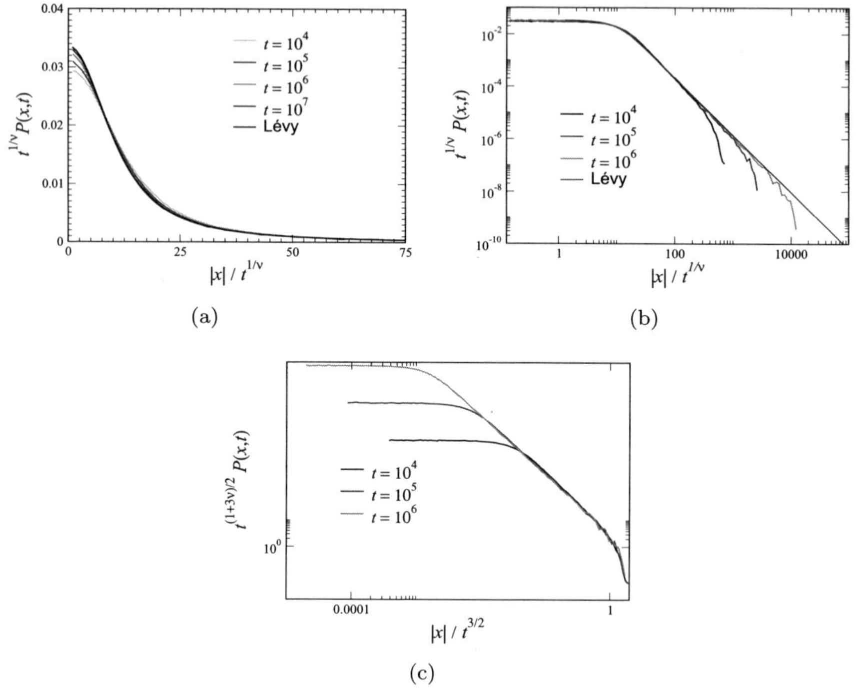

Fig. 5. (a) $t^{1/\nu}P(|x|,t)$ versus $|x|/t^{1/\nu}$ for $D = 2/5$, i.e. $\nu = 7/6$. The theory, i.e., the Lévy PDF Eq. (34) along with K_ν, Eq. (33), perfectly matches simulations without fitting. (b) Same data in log-log scale. Notice the cutoff for large x which is due to the coupling between jump lengths and jump durations, making $x > t^{3/2}$ extremely improbable. (c) $t^{(1+3\nu)/2}P(|x|,t)$ versus $|x|/t^{3/2}$ for $D = 2/5$, showing the crossover from power-law to Gaussian behavior at $|x| \sim t^{3/2}$. This cutoff ensures the finiteness of the mean square displacement of the underlying spreading of the packet of atoms.

Figure 5 (b) also illustrates the cutoff on the Lévy distribution, which is found at distances $x \sim t^{3/2}$, as can be seen also from Fig. 5(c). Beyond this length scale, the density falls off rapidly (Fig. 5(c)). This, as noted above, is the result of the correlation between χ and τ, as there are essentially no walks with a displacement greater than of order $t^{3/2}$. This cutoff ensures the finiteness of the mean square displacement. Using the power law tail of the Lévy PDF $L_\nu(x) \sim x^{-(1+\nu)}$ and the cutoff we get:

$$\langle x^2 \rangle \simeq \int^{t^{3/2}} t^{-(1/\nu)} (x/t^{1/\nu})^{-(1+\nu)} x^2 \mathrm{d}x \sim t^{4-3\nu/2},$$

for $2/3 < \nu < 2$, in agreement with rigorous results of [41] (where the prefactor is also computed). As noted in the introduction, if we rely on the fractional diffusion equation, Eq. (4), naively, we get $\langle x^2 \rangle = \infty$. Thus the fractional equation must be used with care, realizing its limitations in the statistical description of the moments of the distribution and its tails. The exact calculation of $\langle x^2 \rangle$ demands more space, since it involves the calculations of moments of the area under the Bessel excursion and meander.

6. Richardson phase

When the average jump duration, $\langle \tau \rangle$, diverges, i.e., for $D > 1$, the dynamics of $P(x,t)$ enters a new phase. Since the Levy index ν approaches 2/3 as D approaches 1, x scales like $t^{3/2}$ in the limit. Due to the correlations, x cannot grow faster than this, so in this regime, $P(x,t) \sim t^{-3/2} h(x/t^{3/2})$. This scaling is that of free diffusion, namely momentum diffuses like $p \sim t^{1/2}$ and hence the time integral over the momentum scales like $x \sim t^{3/2}$. Indeed, in the absence of the logarithmic potential, namely in the limit $D \gg 1$ Eq. (13) gives

$$P\left(x,t\right) \sim \sqrt{\frac{3}{4\pi D t^3}} \exp\left[-\frac{3x^2}{4Dt^3}\right]. \qquad (35)$$

This limit describes the Obukhov model for a tracer particle path in turbulence, where the velocity follows a simple Brownian motion [42, 43]. These scaling properties are related to Kolmogorov's theory of 1941 (see Eq. (3) in [43]) and to Richardson diffusion [41, 44] $\langle x^2 \rangle \sim t^3$. In the statistical physics community, this large D limit corresponds to the random acceleration process [45, 46].

Equation (35) is valid when the optical potential depth is small since $D \to \infty$ when $U_0 \to 0$. This limit should be taken with care, as the observation time must be made large before considering the limit of weak potential. In the opposite scenario, i.e. $U_0 \to 0$ before $t \to \infty$, we expect ballistic motion, $|x| \sim t$, since then the optical lattice has not had time to make itself felt [4].

7. Relation between first passage times and diffusivity

When the variance of χ is finite, namely $\nu > 2$, we get normal diffusion so that $P(x,t)$ is a Gaussian. In that case the spatio-temporal distribution of jump times and jump lengths decouples $\psi(\chi,\tau) \simeq q(\chi)g(\tau)$. The mean square displacement $\langle x^2 \rangle \sim 2K_2 t$ and the diffusion constant is

$$K_2 = \lim_{\epsilon \to 0} \frac{\langle \chi^2 \rangle}{2\langle \tau \rangle}. \tag{36}$$

This equation relates the first passage time properties of the momentum process and the spatial diffusivity and it has the structure of the famous Einstein relation, relating the variance of jump sizes and the average time to make a jump, with a diffusion constant. However here we are dealing with a special limit, namely $\langle \tau \rangle$ is the average time it takes for a particle with initial momentum $p_i = \pm\epsilon$ to reach $p = 0$ for the first time. This average first passage time vanishes in the limit of $\epsilon \to 0$. The variance of χ, is $\langle \chi^2 \rangle = \int_{-\infty}^{\infty} \chi^2 q(\chi) \mathrm{d}\chi$ namely the variance of the area under the random first passage process $p(t')$ till its first return to $p = 0$, and it too vanishes in the limit $\epsilon \to 0$. Notice that here χ may be either positive or negative (see Fig. 3) and its mean is zero (since the particle may start on $p_i = \pm\epsilon$) with equal probability since the force $F(p)$ is zero on $p = 0$. Of course, from symmetry $F(p) = -F(-p)$, so it suffices to consider χ for say positive areas only.

The toolbox for obtaining $q(\chi)$ and hence $\langle \chi^2 \rangle$ is presented in an Appendix. The calculation of $\langle \tau \rangle$ and $\langle \chi^2 \rangle$ is based on backward Fokker-Planck equations Eqs. (B.1,B.3) (see further details in [39]) and we find

$$K_2 = \frac{1}{D\mathcal{Z}} \int_{-\infty}^{\infty} \mathrm{d}p e^{V(p)/D} \left[\int_p^{\infty} \mathrm{d}p' e^{-V(p')/D} p' \right]^2, \tag{37}$$

and $\mathcal{Z} = \int_{-\infty}^{\infty} \exp[-V(p)/D] \mathrm{d}p$. This equation was derived previously using a different approach in [47]. The coefficient K_2 diverges as $D \to 1/5$ from below indicating the transition to the superdiffusive phase. A sharp increase in the diffusion coefficient of an atomic cloud as the laser intensity reaches a critical value was detected experimentally by Hodapp *et al.* [47], which could be the fingerprint of the transition from Gaussian to the Lévy phase. We see that the Green-Kubo formula, Eq. (3), is not the only way to calculate the diffusion constant. For Markovian processes, with non-linear friction, using Eq. (36) and calculating the first passage time properties is an alternative. In the normal diffusion case, the Green-Kubo formalism is so successful that the existence of an alternate method is only of academic interest. However, as shown throughout this current work, calculations of first passage time properties of the velocity process are very useful for the unraveling of the anomalous spatial diffusion properties of the system, where we cannot use the standard Green-Kubo formalism nor the Gaussian central limit theorem. See further work on a generalized Green-Kubo formalism and non-stationary momentum correlation functions in [30].

8. Discussion of the experiment

Our work shows a rich phase diagram of the dynamics, with two transition points. For deep wells, $D < 1/5$, the diffusion is Gaussian, while for $1/5 < D < 1$ we have Lévy statistics, and for $D > 1$ the Richardson-Obukhov scaling $x \sim t^{3/2}$ is found. So far, experiments detected a Gaussian phase and an anomalous one. A transition to Richardson-Obukhov scaling has not yet been detected. Sagi *et al.* [4] clearly demonstrated that changing the optical potential depth modifies the anomalous exponents in the Lévy spreading packet. However, the experiment showed at most ballistic behavior, with the spreading exponent δ, defined by $x \sim t^{\delta}$, always less than unity. This might be related to our observation that to go beyond ballistic motion, $\delta > 1$, one must take the measurement time to be very long. A more serious problem is that, in the experimental fitting of the diffusion front to the Lévy propagator, an additional exponent was introduced in [4] to describe the time dependence of the full width at half maximum. In contrast, our semi-classical theory shows that a single exponent ν is needed within the Lévy scaling regime $1/5 < D < 1$, with the spreading exponent $\delta = 1/\nu$. This might be related to the cutoff of the tails of Lévy PDF which demands that the fitting be performed in the central part of the atomic cloud. On the other hand we cannot rule out other physical effects not included in the semiclassical model. For example it would be very interesting to simulate the system with quantum Monte Carlo simulations, though we note that these are not trivial in the $|x| \sim t^{3/2}$ regime since the usual simulation procedure introduces a cutoff on momentum, which may give rise to an artificial ballistic motion. Further, the experiment suffers from leakage of particles, which could be crucial, while our theory assumes conservation of the number of particles. Another effect our theory does not address is the Doppler cooling which may serve as a cutoff at long time scales (see some discussion on this in [30]). Thus while there is some tantalizing points of contact between the theory and experiment, full agreement is yet far away.

9. Summary

Usually first passage times are derived from a random diffusive processes. Here our goal was a bit different. We use first passage time statistics to derive the spatial diffusivity of the atomic cloud. We revealed also general relations between first passage times and the normal diffusion constant K_2. Since the latter, through linear response, is in many cases related to the mobility, we see that analysis of first passage time problems in momentum space can be very helpful. For this analysis, we need not only distributions of first passage times (which was a topic thoroughly investigated previously) but also relatively new tools (at least in physics), namely the distributions of areas under the Langevin curve till the random first arrival, and the analysis of the area under positive excursion curves like the Bessel excursion with fixed time interval. The area under the Brownian excursion, e.g. the Airy distribution, is found to be useful in a special limit ($D \to \infty$). It is encouraging that problems of a rather pure mathematical nature, like the distribution of areas under Brownian and non-Brownian excursions, are slowly but surely finding applications in seemingly diverse fields like solid on solid models [19] or as discussed here diffusion of atoms in optical lattices [7]. The rich phase diagram, i.e. Normal, Lévy, Richardson diffusions, together with the divergence of the diffusion constant close to the transition, revealed by the semi-classical approach used here, point to interesting physics, though many open questions and challenges both experimental and theoretical remain open with respect to the dynamics and thermodynamics of laser cooled atoms.

Acknowledgments

This work was supported by the Israel Science Foundation. We thank Andreas Dechant, Eric Lutz, Nir Davidson, Yoav Sagi, and Satja Majumdar for discussions. EB thanks the Free University of Berlin for hospitality during the preparation of the manuscript.

Appendix A.

We here find the equation for the distribution of the area swept by a particle until its first passage. More precisely, we briefly outline the derivation of the equation for the PDF of the area $\chi = \int_0^t p(t')\mathrm{d}t'$ under the stochastic process $p(t) \geq 0$ which starts at $p_i > 0$ until its reaches $p = 0$ for the first time, so that t is the first passage time. The PDF $q(|\chi|)$ has a Laplace $\chi \to s$ transform which we call $Q(s, p_i)$ (here $\chi > 0$ the extension to negative χ being trivial). The equation for this quantity is

$$D\frac{d^2Q}{dp_i^2} + F(p_i)\frac{dQ}{dp_i} - sp_iQ = 0 \ . \tag{A.1}$$

For simplicity and for conciseness we consider the case $F(p) = 0$, so $p(t)$ is undergoing Brownian motion without bias. By definition $\chi_{p_i} = \int_0^t p(t')\mathrm{d}t'$ so

$$Q(s, p_i) = \langle \exp[-s \int_0^t p(t')\mathrm{d}t']\rangle_{p_i} \tag{A.2}$$

where t is the first passage time. The starting point is denoted by the subscript. Now we consider the process $p(t)$ as a random walk on a lattice with probability 1/2 to jump from $p(t) \to p(t) + \delta p$ or $p(t) \to p(t) - \delta p$, so δp is the lattice spacing. The constant times between these jumps is δt and we consider the limit $\delta p \to 0$ and $\delta t \to 0$ while keeping $D = (\delta p)^2/2\delta t$ finite, this being the Brownian limit.

A particle starting with p_i at time $t = 0$ can either have momentum $p_i - \delta p$ at time δt with probability one half, or $p_i + \delta p$ with the same probability. From the definition of χ_{p_i} and for a particle starting with p_i we have

$$\chi_{p_i} = \begin{cases} p_i\delta t + \chi_{p_i - \delta p} & \text{Prob } 1/2 \\ p_i\delta t + \chi_{p_i+\delta p} & \text{Prob } 1/2. \end{cases} \tag{A.3}$$

The first case corresponds to a particle initially jumping down in p, $p_i \to p_i - \delta p$ while the second case is for a particle jumping up. The term $p_i\delta t$ stems from the displacement during the interval δt, and the remaining term (either χ_{p_i} or $\chi_{p_i+2\delta p}$ is the displacement for a new process which starts either with $p_i - \delta_p$ or $p_i + \delta p$). Equation (A.3) is now used to find

$$\langle \exp(-s\chi_{p_i})\rangle = \frac{1}{2}\sum_{\sigma=\pm 1} \langle \exp\left[-s(p_i\delta t - s\chi_{p_i+\sigma\delta p}\right]\rangle. \tag{A.4}$$

Expanding the right hand side of Eq. (A.4) to second order in δp and first order in δt, and inserting these expansions in Eq. (A.4), using $D =$

$(\delta p)^2/2\delta t$, we get Eq. (A.1). For $F(p_i) \neq 0$ a similar approach is used though now weights on the jumping probabilities must be implemented (see similar expansions in [16, 38]). The boundary condition $Q(s,p_i)|_{p_i=0} = 1$ is used since a particle starting on $p_i \to 0$ will immediately reach $p = 0$ and then the PDF of χ is a delta function centered on zero, so the Laplace transform of this delta function $Q(s,0)$ is equal to unity. We use eq. (A.1) in Appendix A (and in more detail in [39]) to derive the long tail of $q(\chi)$ and from it K_ν.

Appendix B.

In this appendix we derive the small k, small u behavior of $\widetilde{\psi}$. We first note [32] that for $u^{3/2} \ll k \ll 1$, $D < 1$, we can approximate $\widetilde{\psi}$ by its decoupled form, $\widetilde{\psi}(k,u) \sim \tilde{g}(u)\tilde{q}(k)$. For small u, since g is a probability distribution, we have $\tilde{g}(u) \sim 1 - u\langle\tau\rangle$, as long as the first moment exists. This first moment satisfies the backward Fokker-Planck equation [48]

$$D\frac{d^2\langle\tau\rangle}{dp_i^2} + F(p_i)\frac{d\langle\tau\rangle}{dp_i} = -1 \tag{B.1}$$

and the small p_i limit of the solution is:

$$\langle\tau\rangle \sim \frac{p_i}{D}\int_0^\infty e^{-V(p)}dp = \frac{p_i\sqrt{\pi}\Gamma\left(\frac{3\nu-2}{2}\right)}{2D\Gamma\left(\frac{3\nu-1}{2}\right)} \tag{B.2}$$

For small k behavior of $\tilde{q}(k)$, given that the second moment diverges, we have $\tilde{q}(k) \sim 1 - \pi x_*^\nu |k|^\nu/(\Gamma(1+\nu)\sin\frac{\pi\nu}{2})$. The length scale x_* governing the decay of $q(\chi)$ can in principle be determined from the scaling function B, but it is easier to get it directly from another backward Fokker-Planck equation for the Laplace transform of $q(\chi)$, $Q(s,p_i)$, where we have made explicit the dependence on the initial momentum $p_i > 0$ (so $\chi > 0$):

$$D\frac{d^2Q}{dp_i^2} + F(p_i)\frac{dQ}{dp_i} - sp_iQ = 0\ . \tag{B.3}$$

From this, we derive the small p_i, small s behavior: $Q(s,p_i) \sim 1 - p_i\Gamma(1-\nu)(3\nu-1)^\nu s^\nu/(\Gamma(\nu)3^{2\nu-1})$, which in turn yields

$$x_*^\nu \sim p_i\frac{\nu(3\nu-1)^\nu}{2\cdot 3^{2\nu-1}\Gamma(\nu)} \tag{B.4}$$

where the $1/2$ prefactor stems from equal consideration of positive and negative excursions. Putting these two results together, we have

$$\widetilde{\psi}(k,u) \approx 1 - \langle\tau\rangle u - x_*^\nu|k|^\nu \cdots\ . \tag{B.5}$$

To leading order, we may approximate $\Psi(k,u) \sim \Psi(0,u) \sim 1/\langle\tau\rangle$, which, together with Eq. (B.5), gives us Eq. (32).

References

1. C. N. Cohen-Tannoudji and W. D. Phillips, New mechanisms for laser cooling, *Phys. Today.* **43**(10), 33–40 (1990).
2. F. Bardou, J. P. Bouchaud, A. Aspect, and C. Cohen-Tannoudji, *Lévy Statistics and Laser Cooling.* Cambridge Univ. Press, Cambridge (2002).
3. C. N. Cohen-Tannoudji, Nobel Lecture: Manipulating atoms with photons, *Rev. Mod. Phys.* **70**(3), 707–719 (1998).
4. Y. Sagi, M. Brook, I. Almog, and N. Davidson, Observation of anomalous diffusion and fractional self-similarity in one dimension, *Phys. Rev. Lett.* **108** (9), 093002 (2012).
5. Y. Castin, J. Dalibard, and C. Cohen-Tannoudji. The limits of Sisyphus cooling. In eds. L. Moi, S. Gozzini, C. Gabbanini, E. Arimondo, and F. Strumia, *Light Induced Kinetic Effects on Atoms, Ions and Molecules*, ETS Editrice, Pisa (1991).
6. S. Marksteiner, K. Ellinger, and P. Zoller, Anomalous diffusion and Lévy walks in optical lattices, *Phys. Rev. A.* **53**(5), 3409–3430 (1996).
7. D. A. Kessler and E. Barkai, Theory of fractional Lévy kinetics for cold atoms diffusing in optical lattices, *Phys. Rev. Lett.* **108**(23), 230602 (2012).
8. A. I. Saichev and G. M. Zaslavsky, Fractional kinetic equations: solutions and applications, *Chaos.* **7**(4), 753–764 (1997).
9. R. Metzler, E. Barkai, and J. Klafter, Deriving fractional Fokker-Planck equations from a generalized master equation, *Europhys. Lett.* **46**(4), 431–436 (1999).
10. R. Metzler and J. Klafter, The random walk's guide to anomalous diffusion: a fractional dynamics approach, *Phys. Rep.* **339**(1), 1–77 (2000).
11. J. Klafter, M. F. Shlesinger, and G. Zumofen, Beyond Brownian motion, *Physics Today.* **49**(2), 33–39 (1996).
12. E. Schrödinger, The theory of drop and rise tests on Brownian motion particles, *Physik. Z.* **16**, 289–295 (1915).
13. S. Redner, *A Guide to First-Passage Processes.* Cambridge Univ. Press, Cambridge (2001).
14. S. Condamin, O. Bénichou, V. Tejedor, R. Voituriez, and J. Klafter, First-passage times in complex scale-invariant media, *Nature.* **450**, 77–80 (2007).
15. A. J. Bray, S. N. Majumdar, and G. Schehr, Persistence and first-passage properties in non-equilibrium system, *Advances in Physics.* **62**(3), 225–361 (2013).
16. S. N. Majumdar and M. J. Kearney, On the area under a continuous time Brownian motion till its first-passage time, *J. Phys. A: Mathematical and General.* **38**(19), 4097–4104 (2005).
17. S. N. Majumdar and A. Comtet, Exact maximal height distribution of fluctuating interfaces, *Phys. Rev. Lett.* **92**(22), 225501 (2004).

18. S. N. Majumdar and A. Comtet, Airy distribution function: From the area under a brownian excursion to the maximal height of fluctuating interfaces, *J. Stat. Phys.* **119**(3–4), 777–826 (2005).
19. G. Schehr and S. N. Majumdar, Airy distribution function: From the area under a brownian excursion to the maximal height of fluctuating interfaces, *Phys. Rev. E.* **73**, 056103 (2006).
20. D. A. Darling, On the supremum of a certain Gaussian process, *Ann. Probab.* **11**(3), 803–806 (1983).
21. G. Louchard, Kac formula, Lévy local time and Brownian excursion, *J. Appl. Probab.* **21**(3), 479–499 (1984).
22. L. Takács, A Bernoulli excursion and its various applications, *Adv. Appl. Probab.* **23**(3), 557–585 (1991).
23. M. Abramowitz and I. A. Stegun, eds., *Handbook of Mathematical Functions.* Dover, New York (1972).
24. E. Lutz, Power-law tail distributions and nonergodicity, *Phys. Rev. Lett.* **93** (19), 190602 (2004).
25. P. Douglas, S. Bergamini, and F. Renzoni, Tunable Tsallis distributions in dissipative optical lattices, *Phys. Rev. Lett.* **96**(11), 110601 (2006).
26. H. Katori, S. Schlipf, and H. Walther, Anomalous dynamics of a single ion in an optical lattice, *Phys. Rev. Lett.* **79**(12), 2221–2224 (1997).
27. D. A. Kessler and E. Barkai, Infinite invariant densities for anomalous diffusion in optical lattices and other logarithmic potentials, *Phys. Rev. Lett.* **105**(12), 120602 (2010).
28. A. Dechant, E. Lutz, E. Barkai, and D. A. Kessler, Solution of the Fokker-Planck equation with a logarithmic potential, *J. Stat. Phys.* **145**(6), 1524–1545 (2011).
29. O. Hirschberg, D. Mukamel, and G. M. Schütz, Approach to equilibrium of diffusion in a logarithmic potential, *Phys. Rev. E.* **84**(4), 041111 (2011).
30. A. Dechant, E. Lutz, D. A. Kessler, and E. Barkai, Scaling Green-Kubo relation and application to three aging systems, *submitted* .
31. J. Klafter, A. Blumen, and M. F. Shlesinger, Stochastic pathway to anomalous diffusion, *Phys. Rev. A.* **35**(7), 3081–3085 (1987).
32. A. Blumen, G. Zumofen, and J. Klafter, Transport aspects in anomalous diffusion: – Lévy walks, *Phys. Rev. A.* **40**(7), 3964–3974 (1989).
33. E. Barkai and J. Klafter. Anomalous diffusion in the strong scattering limit: A Lévy walk approach. In eds. S. Benkadda and G. M. Zaslavsky, *Chaos, Kinetics and Nonlinear Dynamics in Fluids and Plasmas: Proceedings of a workshop held in Carry-Le Rouet, France 16-21 June 1997*, vol. 511, *Lecture Notes in Physics*, pp. 373–394, Springer, Berlin (1998).
34. Y. Hu and Z. Shi, Extreme lengths in Brownian and Bessel excursions, *Bernoulli.* **3**(4), 387–402 (1997).
35. A. J. Bray, Random walks in logarithmic and power-law potentials, nonuniversal persistence, and vortex dynamics in the two-dimensional XY model, *Phys. Rev. E.* **62**(1), 103–112 (2000).
36. G. Schehr and P. Le Doussal, Extreme value statistics from the real space renormalization group: Brownian motion, Bessel processes and continuous

time random walks, *J. Stat. Mech.: Theory and Expt.* **P01009** (2010).
37. S. N. Majumdar, Brownian functionals in physics and computer science, *Current Sci.* **89**(12), 2076–2092 (2005).
38. S. Carmi and E. Barkai, Fractional Feynman-Kac equation for weak ergodicity breaking, *Phys. Rev. E.* **84**(6), 061104 (2011).
39. D. A. Kessler and E. Barkai, From the area under the Bessel excursion to Lévy walks in optical lattices, *(in preparation)* .
40. J. P. Bouchaud and A. Georges, Anomalous diffusion in disordered media: Statistical mechanisms, models and physical applications, *Phys. Rep.* **195** (4–5), 127–293 (1990).
41. A. Dechant, E. Lutz, D. A. Kessler, and E. Barkai, Fluctuations of time averages for Langevin dynamics in a binding force field, *Phys. Rev. Lett.* **107** (24), 240603 (2011).
42. A. M. Obukhov, Description of turbulence in terms of Lagrangian variable, *Adv. Geophys.* **6**, 113–116 (1959).
43. A. Baule and R. Friedrich, Investigation of a generalized Obukhov model for turbulence, *Phys. Lett. A.* **350**(3–4), 167–173 (2006).
44. L. F. Richardson, Atmospheric diffusion shown on a distance-neighbour graph, *Proc. Roy. Soc. Lond. A.* **110**(756), 709–737 (1926).
45. J. Massoliver and J. M. Porra, Exact solution to the exit-time problem for an undamped free particle driven by Gaussian noise, *Phys. Rev. E.* **53**, 2243–2255 (1996).
46. T. W. Burkhardt, The random accelaration process in bounded geometries, *J. Stat. Mech.: Theory and Expt.* **P07004** (2007).
47. T. W. Hodapp, C. Gerz, C. Furtlehner, C. I. Westbrook, W. D. Phillips, and J. Dalibard, Three-dimensional spatial diffusion in optical molasses, *Appl. Phys. B.* **60**(2–3), 135–143 (1995).
48. C. W. Gardiner, *Handbook of Stochastic Methods for Physics, Chemistry and the Natural Sciences*, 3 edn. Springer, Berlin (2003).

Chapter 21

The Excursion Set Theory in Cosmology

Michele Maggiore and Antonio Riotto

*Département de Physique Théorique and Centre for Astroparticle Physics,
24 quai E. Ansermet, CH-1211 Geneva, Switzerland*

Understanding the formation of large-scale structures in the Universe is a central problem of modern cosmology. A key element is the formation of dark matter halos, which provide the gravitational potential wells where galaxies and galaxy clusters will later form. The process of formation of dark matter halos via gravitational instability can be formulated as a first-passage time problem, in which one studies the evolution of the smoothed density fluctuations with respect to the smoothing scale, and halo formation takes place when a given barrier is hit. We review this classic approach to halo formation, as well as recent developments in which this first-passage time problem is reformulated in terms of a path integral, which also allows us to include non-markovianity effect intrinsic to the physics of halo formation. This application also provides a vivid example of the generality and elegance of statistical tools such as first-passage time theory.

1. Introduction

Our current understanding of the evolution of the universe is based upon the Friedmann-Robertson-Walker (FRW) cosmological model. The model is so successful that it has become known as the standard cosmology and that it is possible to make sensible speculations about the universe at times as early as 10^{-43} sec after the Big Bang.

The most important feature of our universe is its large scale homogeneity and isotropy. The best evidence for the isotropy of the observed universe is the uniformity of the temperature of the cosmic microwave background (CMB) radiation: aside from the observed dipole anisotropy, the temperature difference between two antennas separated by angles ranging from about 10 arc seconds to $180°$ is smaller than about one part in 10^5.

Homogeneity and isotropy is of course true if the universe is observed at sufficiently large scales. The observable patch of the universe is of order 3000 Mpc. Redshift surveys suggest that the universe is homogeneous and isotropic only when coarse grained on scales of the order of 100 Mpc. On smaller scales there exist large inhomogeneities, such as galaxies, clusters and superclusters. It is firmly established by observations that our universe

- is homogeneous and isotropic on scales larger than 100 Mpc and has well developed inhomogeneous structure on smaller scales;
- expands according to the Hubble law for which the recession velocity of, say, galaxies is proportional to their distances.

Concerning the matter composition of the universe, we know that [1]

- it is pervaded by thermal microwave background radiation with temperature $T_0 \simeq 2.73$ K;
- there is baryonic matter, roughly one baryon per 10^9 photons, but no substantial amount of antimatter;
- the chemical composition of baryonic matter is about 75% hydrogen, 25% helium, plus trace amounts of heavier elements;
- baryons contribute only a small percentage of the total energy density (about 5%);
- the rest of the content of the universe made of dark components, which appear to be composed of cold Dark Matter (DM) with negligible pressure (about 25% of the energy content) and dark energy with negative pressure (about 70%).

Observations of the fluctuations in the cosmic microwave background radiation suggest that there were only small fluctuations of order 10^{-5} in the energy density distribution when the universe was a thousand times smaller than now and that structures (galaxies, clusters of galaxies, etc.) develop from these small fluctuations under the action of gravity. This process goes under the name of gravitational instability. DM is absolutely necessary to drive the formation of structure in the universe. If DM were not present, and since baryons decouple from the photon fluid too late, there would not be enough time for the structures to form and develop. DM creates gravitational wells where the baryons, once liberated from the photon pressure, can fall in and develop structure formation. Therefore, according to current understanding, DM overdensities collapse to highly nonlinear objects, called DM halos, whose potential wells will make the baryons fall into them.

Larger structures may form because of merging of smaller structures. This is the so-called hierarchical paradigm: big DM halos form from the merging of small DM halos.

The study of the mass function of DM halos, that is how many halos have a given mass, is a central problem in modern cosmology. Various planned large-scale galaxy surveys, both ground-based (DES, PanSTARRS and LSST) and on satellite (EUCLID and ADEPT) have as a target to characterize with large accuracy the properties of DM halos. This is because the halo mass function is a sensitive probe of cosmological parameters and a crucial ingredient when one studies the DM distribution.

As the reader can guess, the formation and evolution of dark matter halos is a highly complex dynamical process, and a detailed understanding of it can only come through large-scale N-body simulations. Some analytical understanding is however also desirable and this is the subject of this chapter. The rest of it is planned as follows. We will begin in Sect. 2 by discussing the classic Press-Schechter theory [2] and its extension known as excursion set theory [3]. In particular, the latter transforms the problem of halo formation into a first-passage time problem [4]. We will then see how to further formulate this first-passage time problem in terms of a path integral with a boundary [5]. The fact that a problem in cosmology such as the formation of DM halos can be formulated as a first-passage time problem, and finally expressed in a path-integral formulation, is quite interesting by itself, and underscores the generality and power of these theoretical tools. Furthermore this path integral reformulation, beside displaying a certain formal elegance, also allows us to tackle problems that had remained unsolved since the classic 1991 paper by Bond et al. [3] that invented excursion set theory. In particular, we will see in sect. 3 that the random walks associated with realistic halo formation are intrinsically non-markovian and this non-markovianity, together with other statistic aspects that are needed for a realistic description of the halo formation process, can indeed be included in the path integral formulation of excursion set theory [6, 7]. Finally, in sect. 4 we will see how these techniques can shed light on some other aspects of the physics of dark matter halo, such as the so-called halo bias, i.e. the relation between the distribution of DM halo and that of galaxies.

2. The Press-Schecther formalism and the excursion set method

To compute the halo mass function dn/dM, that is how many halos per unit volume $dn(M)$ are present in our universe in the mass range M and $(M + dM)$, we start from the famous Press-Schecther (PS) formalism [2]. Press and Schecther observed that the fraction of mass in collapsed objects more massive than some mass M is related to the fraction of volume samples in which the smoothed initial density fluctuations are above some density threshold. This yields a formula for the mass function (distribution of masses) of objects at any given time. In other words the philosophy is the following. We know that the collapse of DM halo is a very complicated phenomenon. Nevertheless we are not interested in describing the dynamics itself, but only in computing the probability that at a given point $\mathbf{x}$ a halo will form. But when does it? Let us simplify the dynamics and assume a spherical collapse. In this approximation one ignores the tidal effects of neighbouring density perturbations upon the evolution of an isolated, homogeneous, spherical density perturbation. To justify this we can appeal to Birkhoff's theorem in General Relativity, or Gauss's law in Newtonian Gravity. Under these simplifying assumptions an exact analytical treatment is possible: an overdensity region in the universe behaves as a separated closed FRW universe (that is with energy density larger than the critical energy density), expands to a maximum radius, turn arounds, and collapses to form a spherical halo. Upon solving exactly the dynamics of the system, one discovers that at the time of collapse, the corresponding linear overdensity $\delta \equiv \delta\rho/\rho$ would have reached a critical value $\delta_c \simeq 1.68$.

PS assumed that the collapse is spherical and argued that a halo of mass M and radius R is formed when the corresponding smoothed linear density contrast at recollapse is larger than the critical value $\delta_c \simeq 1.68$ computed through the spherical collapse model. Of course the real density contrast will be much larger, of the order of 200, but we are not interested in it. Notice that in the PS formalism therefore there is no dynamics in time, it only provides the probability that at a given point a halo of mass M will form. In order to compute such a probability we turn to the excursion set method (see also the review [8]).

The computation of the halo mass function can be formulated in terms of a stochastic process. One considers the density contrast $\delta(\mathbf{x})$ and smooths it on some scale R

$$\delta(\mathbf{x}, R) = \int \frac{d^3k}{(2\pi)^3} \, \delta_{\mathbf{k}} W(k, R) e^{-i\mathbf{k}\cdot\mathbf{x}}. \tag{1}$$

We focus on the evolution of $\delta(\mathbf{x}, R)$ with R at a fixed value of $\mathbf{x}$, that we can choose without loss of generality as $\mathbf{x} = 0$, and we write $\delta(\mathbf{x} = 0, R)$ simply as $\delta(R)$. Taking the derivative of Eq. (1) with respect to R we get

$$\frac{\partial \delta(R)}{\partial R} = \zeta(R), \tag{2}$$

where

$$\zeta(R) \equiv \int \frac{d^3k}{(2\pi)^3} \, \delta_{\mathbf{k}} \frac{\partial W(k, R)}{\partial R}. \tag{3}$$

Since the modes $\delta_{\mathbf{k}}$ are stochastic variables, $\zeta(R)$ is a stochastic variable too, and Eq. (2) has the form of a Langevin equation, with R playing the role of time, and $\zeta(R)$ playing the role of noise. When $\delta(R)$ is a Gaussian variable, only its two-point connected correlator is non-vanishing. The two-point function of δ defines the power spectrum $P(k)$,

$$\langle \delta(\mathbf{k}) \delta_(\mathbf{k}')\rangle = (2\pi)^3 \delta^{(3)}(\mathbf{k} + \mathbf{k}') P(k). \tag{4}$$

From this it follows that

$$\langle \zeta(R_1) \zeta(R_2) \rangle = \int_{-\infty}^{\infty} d(\ln k) \; \Delta^2(k) \frac{\partial W(k, R_1)}{\partial R_1} \frac{\partial W(k, R_2)}{\partial R_2}, \tag{5}$$

where $\Delta^2(k) = k^3 P(k)/2\pi^2$. For a generic filter function the right-hand side is a function of R_1 and R_2, different from a Dirac delta $\delta_{\rm D}(R_1 - R_2)$. Things simplify considerably for a sharp k-space filter $W(k, R) = \theta(k - k_{\rm f})$. Using $k_{\rm f} = 1/R$ instead of R, and defining $Q(k_{\rm f}) = -(1/k_{\rm f})\zeta(k_{\rm f})$, we find

$$\frac{\partial \delta(k_{\rm f})}{\partial \ln k_{\rm f}} = Q(k_{\rm f}), \tag{6}$$

and

$$\langle Q(k_{\rm f1}) Q(k_{\rm f2}) \rangle = \Delta^2(k_{\rm f1}) \delta^{(3)}(\ln k_{\rm f1} - \ln k_{\rm f2}). \tag{7}$$

Therefore, we have a Dirac delta noise. We can write these equations in an even simpler form using as "pseudotime" variable the variance S

$$S(R) = \int_{-\infty}^{\infty} d(\ln k) \, \Delta^2(k) |W(k, R)|^2. \tag{8}$$

For a sharp k-space filter, S becomes

$$S(k_{\rm f}) = \int_{-\infty}^{\ln k_{\rm f}} d(\ln k) \, \Delta^2(k), \tag{9}$$

so

$$\frac{\partial S}{\partial \ln k_{\rm f}} = \Delta^2(k_{\rm f}). \tag{10}$$

Thus, redefining finally $\eta(k_{\rm f}) = Q(k_{\rm f})/\Delta^2(k_{\rm f})$, we get

$$\frac{\partial \delta(S)}{\partial S} = \eta(S), \tag{11}$$

with

$$\langle \eta(S_1)\eta(S_2)\rangle = \delta(S_1 - S_2), \tag{12}$$

which is a the Langevin equation with Dirac-delta noise, with S playing the role of time. In hierarchical power spectra, at $R = \infty$ we have $S = 0$, and S increases monotonically as R decreases. Therefore we can start from $R = \infty$, corresponding to "time" $S = 0$, where $\delta = 0$, and follow the evolution of $\delta(S)$ as we decrease R, *i.e.* as we increase S. The fact that this evolution is governed by the Langevin equation means that $\delta(S)$ performs a random walk, with respect to the "time" variable S. We may refer to the evolution of δ as a function of S as a "trajectory". In the spherical collapse model, a virialized object forms as soon as the trajectory exceeds the threshold $\delta = \delta_c$.

According to the PS formalism an halo is formed when the stochastic quantity $\delta(R)$ in its random walk crosses the barrier at δ_c for the first time. It is therefore a problem known in statistical mechanics as the first-time passage problem. We are not interested in subsequent crossing of the barrier as the halo has formed after the first crossing. We therefore consider an ensemble of trajectories, all starting from the initial value $\delta = 0$ at initial "time" $S = 0$, and have compute the function that gives the probability distribution of reaching a value δ_c at "time" S for the first time. Notice that, for a Gaussian theory, the only non-vanishing connected correlator is then the two-point correlator $\langle \delta(S_1)\delta(S_2)\rangle_c$, where the subscript c stands for connected. We consider an ensemble of trajectories all starting at $S_0 = 0$ from an initial position $\delta(0) = \delta_0$, and we follow them for a time S. We discretize the interval $[0, S]$ in steps $\Delta S = \epsilon$, so $S_k = k\epsilon$ with $k = 1, \ldots n$, and $S_n \equiv S$. A trajectory is defined by the collection of values $\{\delta_1, \ldots, \delta_n\}$, such that $\delta(S_k) = \delta_k$. There is no absorbing barrier, i.e. $\delta(S)$ is allowed to range freely from $-\infty$ to $+\infty$. The probability density in the space of trajectories is

$$W(\delta_0; \delta_1, \ldots, \delta_n; S_n) \equiv \langle \delta_{\rm D}(\delta(S_1) - \delta_1) \ldots \delta_{\rm D}(\delta(S_n) - \delta_n)\rangle. \tag{13}$$

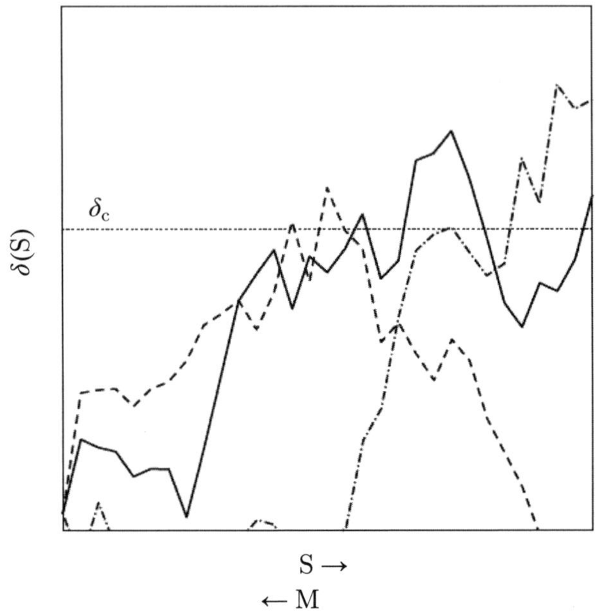

Fig. 1. The stochastic motion of the smoothed density contrast.

In terms of W we define

$$\Pi(\delta_0;\delta_n;S_n) \equiv \int_{-\infty}^{\delta_c} d\delta_1 \ldots \int_{-\infty}^{\delta_c} d\delta_{n-1}\, W(\delta_0;\delta_1,\ldots,\delta_{n-1},\delta_n;S_n), \quad (14)$$

where $S_n = n\epsilon$. So, $\Pi(\delta_0;\delta;S)$ is the probability density of arriving at the "position" δ in a "time" S, starting from δ_0 at time $S_0 = 0$, through trajectories that never exceeded δ_c. Observe that the final point δ ranges over $-\infty < \delta < \infty$. The usefulness of Π is that it allows us to compute the first-crossing rate from first principles, without the need of postulating the existence of an absorbing barrier. Simply, the quantity

$$F(S) = 1 - \int_{-\infty}^{\delta_c} d\delta\, \Pi(\delta_0;\delta;S) \quad (15)$$

gives the fraction of trajectories that crossed the barrier at time S. The rate of change of this quantity is therefore equal to minus the rate at which trajectories cross for the first time the barrier, so the first-crossing rate is

$$\mathcal{F}(S) = -\frac{dF}{dS} = -\int_{-\infty}^{\delta_c} d\delta\, \frac{\partial}{\partial S}\Pi(\delta_0;\delta;S). \quad (16)$$

The halo mass function follows if one has a relation $M = M(R)$ that gives the mass associated to the smoothing of δ over a region of radius R. Once $M(R)$ is given, we can consider F as a function of M rather than of $S(R)$. Then $|dF/dM|dM$ is the fraction of volume occupied by virialized objects with mass between M and $M + dM$. Since each one occupies a volume $V = M/\bar{\rho}$, where $\bar{\rho}$ is the average density of the universe, the number of virialized object $n(M)$ with mass between M and $M + dM$ is given by

$$\frac{dn}{dM}dM = \frac{\bar{\rho}}{M}\left|\frac{dF}{dM}\right|dM, \tag{17}$$

so

$$\frac{dn}{dM} = \frac{\bar{\rho}}{M}\frac{dF}{dS}\left|\frac{dS}{dM}\right| = \frac{\bar{\rho}}{M^2}\mathcal{F}(S)2\sigma^2\frac{d\ln\sigma^{-1}}{d\ln M}, \tag{18}$$

where we have defined $S = \sigma^2$.

We now express $\Pi(\delta_0; \delta_n; S)$ using a path integral formulation. We use the integral representation of the Dirac delta

$$\delta_{\rm D}(x) = \int_{-\infty}^{\infty}\frac{d\lambda}{2\pi}\,e^{-i\lambda x}, \tag{19}$$

and we write Eq. (13) as

$$W(\delta_0; \delta_1, \ldots, \delta_n; S_n) = \int_{-\infty}^{\infty}\frac{d\lambda_1}{2\pi}\cdots\frac{d\lambda_n}{2\pi}\,e^{i\sum_{i=1}^{n}\lambda_i\delta_i}\langle e^{-i\sum_{i=1}^{n}\lambda_i\delta(S_i)}\rangle. \tag{20}$$

Observe that the dependence on δ_0 here is hidden in the correlators of δ, *e.g.* $\langle\delta^2(S=0)\rangle = \delta_0^2$. It is convenient to set for simplicity $\delta_0 = 0$ from now on. For Gaussian fluctuations,

$$\langle e^{-i\sum_{i=1}^{n}\lambda_i\delta(S_i)}\rangle = e^{-\frac{1}{2}\sum_{i,j=1}^{n}\lambda_i\lambda_j\langle\delta(S_i)\delta(S_j)\rangle_c}, \tag{21}$$

as can be checked immediately by performing the Taylor expansion of the exponential on the left-hand side, and using the fact that, for Gaussian fluctuations, the generic correlator factorizes into sum of products of two-points correlators. This gives

$$W(\delta_0; \delta_1, \ldots, \delta_n; S_n) = \int\mathcal{D}\lambda\;e^{i\sum_{i=1}^{n}\lambda_i\delta_i - \frac{1}{2}\sum_{i,j=1}^{n}\lambda_i\lambda_j\langle\delta_i\delta_j\rangle_c}, \tag{22}$$

where

$$\int\mathcal{D}\lambda \equiv \int_{-\infty}^{\infty}\frac{d\lambda_1}{2\pi}\cdots\frac{d\lambda_n}{2\pi}, \tag{23}$$

and $\delta_i \equiv \delta(S_i)$. Then

$$\Pi(\delta_n; S_n) = \int_{-\infty}^{\delta_c} d\delta_1 \dots d\delta_{n-1} \int \mathcal{D}\lambda \exp\left\{ i\sum_{i=1}^{n} \lambda_i \delta_i - \frac{1}{2}\sum_{i,j=1}^{n} \lambda_i \lambda_j \langle \delta_i \delta_j \rangle_c \right\}. \tag{24}$$

Let us now compute the two-point correlator. Using as initial condition Eq. (11) integrates to

$$\delta(S) = \int_0^S dS'\, \eta(S'), \tag{25}$$

so the two-point correlator is given by

$$\langle \delta(S_i)\delta(S_j)\rangle_c = \int_0^{S_i} dS \int_0^{S_j} dS' \langle \eta(S)\eta(S')\rangle = \min(S_i, S_j). \tag{26}$$

Let us now take the derivative of Eq. (24) with respect to S_n

$$\begin{aligned} \frac{\partial}{\partial S_n}\Pi(\delta_n; S_n) &= \int_{-\infty}^{\delta_c} d\delta_1 \dots d\delta_{n-1} \int \mathcal{D}\lambda \left(-\frac{\lambda_n^2}{2}\right) \\ &\times \exp\left\{ i\sum_{i=1}^{n} \lambda_i \delta_i - \frac{1}{2}\sum_{i,j=1}^{n} \lambda_i \lambda_j \langle \delta_i \delta_j \rangle_c \right\} \\ &= \frac{1}{2}\frac{\partial^2}{\partial \delta_n^2} \int_{-\infty}^{\delta_c} d\delta_1 \dots d\delta_{n-1} \int \mathcal{D}\lambda \exp\left\{ i\sum_{i=1}^{n} \lambda_i \delta_i - \frac{1}{2}\sum_{i,j=1}^{n} \lambda_i \lambda_j \langle \delta_i \delta_j \rangle_c \right\}. \end{aligned} \tag{27}$$

We then discover that the probability $Pi(\delta_0; \delta_n; S_n)$ satisfies a Fokker-Planck equation (setting $S_n = S$ and $\delta_n = \delta$)

$$\frac{\partial}{\partial S}\Pi(\delta; S) = \frac{1}{2}\frac{\partial^2}{\partial \delta^2}\Pi(\delta; S), \tag{28}$$

which has to be solved with the following boundary conditions

$$\Pi(\delta; 0) = \delta_{\mathrm{D}}(\delta - \delta_0), \quad \text{and } \Pi(\delta_c; S) = 0, \tag{29}$$

where we have restored a non vanishing initial condition. The first condition says that the trajectory has to start from $\delta(0) = \delta_0$, while the second condition simply states that as the random walk reaches the barrier at δ_c for the first time the motion should stop, or in other words, the trajectory should be removed.

The solution of the Fokker-Planck equation with such boundary conditions is given by

$$\Pi(\delta;S)=\frac{1}{\sqrt{2\pi S}}\left(e^{-(\delta-\delta_0)^2/2S}-e^{(2\delta_c-\delta+\delta_0)^2/2S}\right), \tag{30}$$

a result already derived by Chandrasekhar in 1943 using the method of images [10]. Correspondingly

$$\mathcal{F}(S)=-\frac{1}{2}\left.\frac{\partial\Pi}{\partial\delta}\right|_{\delta=\delta_c}=\frac{1}{\sqrt{2\pi}S^{3/2}}e^{-(\delta_c-\delta_0)^2/2S}, \tag{31}$$

and (now setting δ_0 back to zero)

$$\left(\frac{dn}{dM}\right)_{\rm PS}=\left(\frac{2}{\pi}\right)^{1/2}\frac{\delta_c}{\sigma}e^{-\delta_c^2/(2\sigma^2)}\frac{\bar{\rho}}{M^2}\frac{d\ln\sigma^{-1}}{d\ln M}. \tag{32}$$

This result can be extended to arbitrary redshift z reabsorbing the evolution of the variance into δ_c, so that δ_c in the above result is replaced by $\delta_c(z)=\delta_c(0)/D_+(z)$, where $D_+(z)$ is the linear growth factor. Compared to detailed and computer-expensive N-body simulations, the PS theory predicts a little bit too many low-mass halos, roughly by a factor of two, and too few high-mass halos, by a factor $\mathcal{O}(10)$ or so. Nevertheless, the simple logic behind the PS theory work surprisingly well if we think of how complicated the collapse is. In particular, the exponential tail of the halo mass function is correctly obtained, even though not with quite the correct coefficient in the exponent. Nowadays analytic techniques generally go beyond the PS approach and model the collapse as, *e.g.* ellipsoidal. However, the PS theory is able to reproduce, at least qualitatively, several properties of dark matter halos such as their conditional and unconditional mass function, halo accretion histories, merger rates and halo bias [8].

3. Beyond the Press-Schecther theory: non-markovian effects in cosmology

As we have seen in the previous section, the cosmological problem of finding how many halos there are per given mass is reduced to a first-passage problem. One writes a Fokker-Planck equation describing the probability $\Pi(\delta,S)$ that the density perturbation acquires a given value δ at a given "time" S, supplemented by the absorbing barrier boundary condition that the probability vanishes when $\delta=\delta_c$. However, this procedure still raises some technical questions. In particular, the derivation in the previous section only works for a sharp k-space filter. However, there is no unambiguous way of associating a mass to a region of size R smoothed with a sharp k-space filter. The only unambiguous way of associating a mass M to a smoothing scale R is using a sharp filter in x-space, proportional to $\theta(R-r)$, in which case one has the obvious relation $M=(4/3)\pi R^3\rho$. This is also the relation used in numerical simulations and by observers.

The crucial point is how to associate a mass M to the filter scale R. For the sharp filter in x-space this is clear. The mass associated to a spherical region of radius R and density ρ is $M=(4/3)\pi R^3\rho$. For the other filters there is no unambiguous definition. A possibility often used is the following. One first normalizes W so that its maximum value is one. Calling W' this new dimensionless filter, one can define the volume V associated to the filter as $V=\int d^3x\, W'$, and $M=\rho V$. This procedure seems reasonable, but still it is somewhat arbitrary, since one might as well chose a different normalization for W'. For a gaussian filter, this gives $V=(2\pi)^{3/2}R^3$. For a sharp k-space filter, on top of this ambiguity, there is also the fact that such a volume is not even well defined. In fact, the k-space filter in coordinate space reads

$$W_{\text{sharp}-k}(r,R)=\frac{1}{2\pi^2R^3}\,\frac{\sin u-u\cos u}{u^3}\,, \tag{33}$$

where $u=r/R=k_fr$ and $r=|\mathbf{x}-\mathbf{x}'|$, which gives

$$W'=3\,\frac{\sin u-u\cos u}{u^3}\,, \tag{34}$$

and

$$V=4\pi\int_0^\infty dr\,r^2W'=12\pi R^3\int_0^\infty du\left[\frac{\sin u}{u}-\cos u\right]. \tag{35}$$

The integral of $\sin u/u$ gives $\pi/2$, but the limit for $\Lambda\to\infty$ of the integral of $\cos u$ from $u=0$ to $u=\Lambda$ does not exist. If one just sets it to zero, without much justification, one finds the result $V=6\pi^2R^3$. In any case, it is clear that it is difficult to give unambiguous numerical predictions for the halo mass function with a filter different from the sharp x-space filter.

The standard practice in the literature is to use the PS mass function, which can be derived from excursion set theory but only if one works with a sharp k-space filter, and at the same time to use $M=(4/3)\pi R^3\rho$, which is only valid for a sharp x-space filter. Of course this is not consistent and cannot be a good starting point for the inclusion of the non-Gaussianities, since one would attribute to primordial non-Gaussianities features in the mass function which are due, more trivially, to the filter function.

The problem, however, is that as soon as one uses a different filter (such as the tophat in real space), the Langevin equation with gaussian Dirac-delta noise, that describes a simple markovian process, is replaced by a very complicated non-markovian dynamics dictated by a colored noise. The system acquires memory properties and the probability $\Pi(\delta, S)$ no longer satisfy a simple diffusion equation such as the Fokker-Planck equation. The same is true if the density perturbation is non-Gaussian.

3.1. *The path-integral formulation and non-markovianity*

How to solve the problem described above? One possibility is to study the two-point correlator with tophat filter in coordinate space and see if it can be split into two parts (markovian and non-markovian). Luckily, in the cosmological setting this is possible, as we will now describe. One first studies the correlator $\langle\delta(R_1)\delta(R_2)\rangle$ with a tophat filter in coordinate space. It has been realized that one can split it as markovian plus a non-markovian piece

$$\langle\delta_i\delta_j\rangle = \min(S_i, S_j) + \Delta(S_i, S_j). \tag{36}$$

If one simply neglects $\Delta(S_i, S_j)$, the problems become formally identical to the one that we have solved previously.

By definition $\Delta(S_i, S_j)$ is symmetric, $\Delta(S_i, S_j) = \Delta(S_j, S_i)$, so it is sufficient to study it for $S_i \leq S_j$. We also use the notation $\Delta_{ij} = \Delta(S_i, S_j)$. Since, by definition, $\langle\delta_i^2\rangle = S_i$, we see that $\Delta(S_i, S_j)$ vanishes when $S_i = S_j$. Furthermore, at $S_i = 0$, $\delta_i = \delta_0$ is the same constant for all trajectories, so $\langle\delta_i\delta_j\rangle_c = \delta_0\langle\delta_j\rangle_c = 0$, and therefore $\Delta(S_i, S_j)$ vanishes when $S_i = 0$.

For the standard ΛCDM model (the model where the dark energy is made of a cosmological constant and the DM is cold) it turns out that a good approximation is represented by

$$\Delta(S_i, S_j) \simeq \kappa\,\frac{S_i(S_j - S_i)}{S_j}, \tag{37}$$

with $\kappa \simeq 0.45$. In the ΛCDM model that we are using, our numerical results display a very weak linear dependence of κ on R. Taking for instance the data in the range $R \in [1, 60]\,\mathrm{Mpc}/h$, the result of the numerical evaluation of is very well fitted by

$$\kappa(R) \simeq 0.4592 - 0.0031\,R, \tag{38}$$

where R is measured in Mpc/h.

Let us know see how the path-integral formulation can take care of the non-markovianity by dealing with it perturbatively. We can write the path integral in the form

$$\Pi(\delta_0;\delta_n;S_n) = \int_{-\infty}^{\delta_c} d\delta_1 \ldots d\delta_{n-1} \int \mathcal{D}\lambda \\ \times \exp\left\{ i\sum_{i=1}^{n} \lambda_i\delta_i - \frac{1}{2}\sum_{i,j=1}^{n} [\min(S_i,S_j) + \Delta(S_i,S_j)]\lambda_i\lambda_j \right\}. \tag{39}$$

One can now expand in powers of Δ_{ij}. The zeroth-order term is simply the Gaussian markovian piece recovered in the previous section $\Pi^{\rm gm}(\delta_0;\delta_n;S_n)$, whose corresponding first-crossing rate is

$$\mathcal{F}^{\rm gm} = -\int_{-\infty}^{\delta_c} d\delta\, \frac{\partial \Pi^{\rm gm}}{\partial S} \\ = \frac{1}{\sqrt{2\pi}}\, \frac{\delta_c}{S^{3/2}}\, e^{-\delta_c^2/(2S)}. \tag{40}$$

We now discuss the non-markovian corrections, to first order, using the analytical approximation (37) for Δ_{ij}. Expanding to first order in Δ_{ij} and using $\lambda_i e^{i\sum_k \lambda_k\delta_k} = -i\partial_i e^{i\sum_k \lambda_k\delta_k}$, where $\partial_i = \partial/\partial\delta_i$, the first-order correction to Π is

$$\Pi^{\Delta}(\delta_0;\delta_n;S_n) \equiv \int_{-\infty}^{\delta_c} d\delta_1 \ldots d\delta_{n-1}\, \frac{1}{2}\sum_{i,j=1}^{n} \Delta_{ij}\partial_i\partial_j \\ \times \int \mathcal{D}\lambda\, \exp\left\{ i\sum_{i=1}^{n} \lambda_i\delta_i - \frac{1}{2}\sum_{i,j=1}^{n} \min(S_i,S_j)\lambda_i\lambda_j \right\} \\ = \frac{1}{2}\sum_{i,j=1}^{n} \Delta_{ij} \int_{-\infty}^{\delta_c} d\delta_1 \ldots d\delta_{n-1}\, \partial_i\partial_j W^{\rm gm}(\delta_0;\delta_1,\ldots,\delta_n;S_n). \tag{41}$$

We may rewrite the term $\Delta_{ij}\partial_i\partial_j$ separating explicitly the derivative $\partial_n \equiv \partial/\partial\delta_n$ from the derivatives ∂_i with $i<n$, so (using $\Delta_{ij} = \Delta_{ji}$)

$$\frac{1}{2}\sum_{i,j=1}^{n} \Delta_{ij}\partial_i\partial_j = \frac{1}{2}\Delta_{nn}\partial_n^2 + \sum_{i=1}^{n-1} \Delta_{in}\partial_i\partial_n + \frac{1}{2}\sum_{i,j=1}^{n-1} \Delta_{ij}\partial_i\partial_j. \tag{42}$$

Since $\Delta_{ij} = 0$ when $i=j$, the above equation simplifies to

$$\frac{1}{2}\sum_{i,j=1}^{n} \Delta_{ij}\partial_i\partial_j = \sum_{i=1}^{n-1} \Delta_{in}\partial_i\partial_n + \sum_{i<j} \Delta_{ij}\partial_i\partial_j,$$

where

$$\sum_{i<j} \equiv \sum_{i=1}^{n-2} \sum_{j=i+1}^{n-1} . \tag{43}$$

When inserted into eq. (41) the term $\sum_{i=1}^{n-1} \Delta_{in}\partial_i\partial_n$ brings a factor $\sum_i$ that, in the continuum limit, produces an integral over an intermediate time S_i. Because of this dependence on the past history, we call this the "memory term". Similarly, the term $\sum_{i<j} \Delta_{ij}\partial_i\partial_j$ gives, in the continuum limit, a double integral over intermediate times S_i and S_j, and we call it the "memory-of-memory" term. Thus,

$$\Pi^{\Delta 1} = \Pi^{\rm mem} + \Pi^{\rm mem-mem}, \tag{44}$$

where

$$\begin{aligned} &\Pi^{\rm mem}(\delta_0; \delta_n; S_n) \\ &= \sum_{i=1}^{n-1} \Delta_{in}\partial_n \int_{-\infty}^{\delta_c} d\delta_1 \ldots d\delta_{n-1}\, \partial_i W^{\rm gm}(\delta_0; \delta_1, \ldots, \delta_n; S_n), \end{aligned} \tag{45}$$

and

$$\begin{aligned} &\Pi^{\rm mem-mem}(\delta_0; \delta_n; S_n) \\ &= \sum_{i<j} \Delta_{ij} \int_{-\infty}^{\delta_c} d\delta_1 \ldots d\delta_{n-1} \partial_i\partial_j\, W^{\rm gm}(\delta_0; \delta_1, \ldots, \delta_n; S_n). \end{aligned} \tag{46}$$

If we expand to quadratic and higher orders in Δ_{ij}, we get terms with a higher and higher number of summations (or, in the continuum limit, of integrations) over intermediate time variables. To compute the memory term we integrate ∂_i by parts,

$$\begin{aligned} &\int_{-\infty}^{\delta_c} d\delta_1 \ldots d\delta_{n-1}\, \partial_i W^{\rm gm}(\delta_0; \delta_1, \ldots, \delta_n; S_n) \\ &= \int_{-\infty}^{\delta_c} d\delta_1 \ldots \widehat{d\delta_i} \ldots d\delta_{n-1} W(\delta_0; \delta_1, \ldots, \delta_i = \delta_c, \ldots, \delta_{n-1}, \delta_n; S_n), \end{aligned} \tag{47}$$

where the notation $\widehat{d\delta_i}$ means that we must omit $d\delta_i$ from the list of integration variables. We next observe that, because of the markovianiy property, $W^{\rm gm}$ satisfies

$$\begin{aligned} &W^{\rm gm}(\delta_0; \delta_1, \ldots, \delta_{i-1}, \delta_c, \delta_{i+1}, \ldots, \delta_n; S_n) \\ &= W^{\rm gm}(\delta_0; \delta_1, \ldots, \delta_{i-1}, \delta_c; S_i) W^{\rm gm}(\delta_c; \delta_{i+1}, \ldots, \delta_n; S_n - S_i), \end{aligned} \tag{48}$$

so

$$\begin{aligned}&\int_{-\infty}^{\delta_c} d\delta_1 \ldots d\delta_{i-1} \int_{-\infty}^{\delta_c} d\delta_{i+1} \ldots d\delta_{n-1} \\ &\times W^{\rm gm}(\delta_0;\delta_1,\ldots,\delta_{i-1},\delta_c;S_i) W^{\rm gm}(\delta_c;\delta_{i+1},\ldots,\delta_n;S_n-S_i) \\ &= \Pi^{\rm gm}(\delta_0;\delta_c;S_i)\Pi^{\rm gm}(\delta_c;\delta_n;S_n-S_i),\end{aligned} \tag{49}$$

and we get

$$\Pi^{\rm mem}(\delta_0;\delta_n;S_n) = \sum_{i=1}^{n-1} \Delta_{in}\partial_n \left[\Pi^{\rm gm}(\delta_0;\delta_c;S_i)\Pi^{\rm gm}(\delta_c;\delta_n;S_n-S_i)\right]. \tag{50}$$

In the continuum limit we write

$$\sum_{i=1}^{n-1} \to \frac{1}{\epsilon}\int_0^{S_n} dS_i, \tag{51}$$

and we find

$$\begin{aligned}\Pi^{\rm mem}(\delta_0=0;\delta_n;S_n) &= \frac{1}{\pi}\partial_n \int_0^{S_n} dS_i\, \Delta(S_i,S_n)\frac{\delta_c(\delta_c-\delta_n)}{S_i^{3/2}(S_n-S_i)^{3/2}} \\ &\quad\times \exp\left\{-\frac{\delta_c^2}{2S_i}-\frac{(\delta_c-\delta_n)^2}{2(S_n-S_i)}\right\} \\ &= \kappa\partial_n\left[\frac{\delta_c(\delta_c-\delta_n)}{S_n}\,{\rm Erfc}\left(\frac{2\delta_c-\delta_n}{\sqrt{2S_n}}\right)\right].\end{aligned} \tag{52}$$

For the memory-of-memory term, proceeding as for the memory term, we get

$$\begin{aligned}&\Pi^{\rm mem-mem}(\delta_0;\delta_n;S_n) \qquad (53)\\ &= \sum_{i<j}\Delta_{ij}\Pi^{\rm gm}(\delta_0;\delta_c;S_i)\Pi^{\rm gm}(\delta_c;\delta_c;S_j-S_i)\Pi^{\rm gm}(\delta_c;\delta_n;S_n-S_j). \\ &= \frac{\kappa}{\pi\sqrt{2\pi}}\,\delta_c(\delta_c-\delta_n) \\ &\times\int_0^{S_n} dS_i\,\frac{1}{S_i^{1/2}}\,e^{-\delta_c^2/(2S_i)} \qquad (54)\\ &\times\int_{S_i}^{S_n} dS_j\frac{1}{S_j(S_j-S_i)^{1/2}(S_n-S_j)^{3/2}}\exp\left\{-\frac{(\delta_c-\delta_n)^2}{2(S_n-S_j)}\right\}.\end{aligned}$$

It is convenient to use the identity

$$\frac{(\delta_c-\delta_n)}{(S_n-S_j)}\exp\left\{-\frac{(\delta_c-\delta_n)^2}{2(S_n-S_j)}\right\} = \partial_n\exp\left\{-\frac{(\delta_c-\delta_n)^2}{2(S_n-S_j)}\right\}, \tag{55}$$

to write $\Pi^{\rm mem-mem}$ as a total derivative with respect to δ_n. The inner integral can now be computed rewriting it in terms of the variable $z = (\delta_c - \delta_n)^2/[2(S_n - S_j)]$, and gives

$$\Pi^{\rm mem-mem}(\delta_0 = 0; \delta_n; S_n) = \frac{\kappa\delta_c}{\sqrt{2\pi S_n}}\,\partial_n\Bigg[e^{-(\delta_c-\delta_n)^2/(2S_n)} \times \int_0^{S_n} \frac{dS_i}{S_i}\, e^{-\delta_c^2/(2S_i)}\,\mathrm{Erfc}\left((\delta_c - \delta_n)\sqrt{\frac{S_i}{2(S_n - S_i)S_n}}\right)\Bigg]. \tag{56}$$

We can now compute the first crossing rate using eq. (16). Since both $\Pi^{\rm mem}$ and $\Pi^{\rm mem-mem}$ have been expressed as a derivative with respect to δ_n, the integral over $d\delta_n$ is performed trivially, and we finally get

$$\mathcal{F}^{\rm mem}(S) = -\frac{\partial}{\partial S}\int_{-\infty}^{\delta_c} d\delta_n\,\Pi^{\rm mem}(\delta_0 = 0; \delta_n; S) = 0, \tag{57}$$

$$\begin{aligned}
\mathcal{F}^{\rm mem-mem}(S) &= -\frac{\partial}{\partial S}\int_{-\infty}^{\delta_c} d\delta_n\,\Pi^{\rm mem-mem}(\delta_0 = 0; \delta_n; S)\\
&= -\frac{\partial}{\partial S}\left[\frac{\kappa\delta_c}{\sqrt{2\pi S}}\int_0^S dS_i\,\frac{1}{S_i}\,e^{-\delta_c^2/(2S_i)}\right]\\
&= -\frac{\kappa\delta_c}{\sqrt{2\pi}}\frac{\partial}{\partial S}\left[\frac{1}{S^{1/2}}\Gamma\left(0, \frac{\delta_c^2}{2S}\right)\right]\\
&= -\frac{\kappa\delta_c}{\sqrt{2\pi}}\left[\frac{1}{S^{3/2}}e^{-\delta_c^2/(2S)} - \frac{1}{2S^{3/2}}\Gamma\left(0, \frac{\delta_c^2}{2S}\right)\right],
\end{aligned} \tag{58}$$

where $\Gamma(0, z)$ is the incomplete Gamma function. Putting everything together, we find the first-crossing rate to first order in the non-markovian corrections,

$$\mathcal{F}(S) = \frac{1-\kappa}{\sqrt{2\pi}}\frac{\delta_c}{S^{3/2}}e^{-\delta_c^2/(2S)} + \frac{\kappa}{2\sqrt{2\pi}}\frac{\delta_c}{S^{3/2}}\Gamma\left(0, \frac{\delta_c^2}{2S}\right). \tag{59}$$

The halo mass function in this approximation is therefore

$$f(\sigma) = (1-\kappa)\left(\frac{2}{\pi}\right)^{1/2}\frac{\delta_c}{\sigma}\,e^{-\delta_c^2/(2\sigma^2)} + \frac{\kappa}{\sqrt{2\pi}}\frac{\delta_c}{\sigma}\,\Gamma\left(0, \frac{\delta_c^2}{2\sigma^2}\right). \tag{60}$$

For large values of $\delta_c^2/2\sigma^2$

$$\Gamma\left(0, \frac{\delta_c^2}{2\sigma^2}\right) \simeq \frac{2\sigma^2}{\delta_c^2}\,e^{-\delta_c^2/(2\sigma^2)}. \tag{61}$$

Thus the incomplete Gamma function gives the same exponential factor as PS theory but with a smaller prefactor, so for large halo masses it is subleading, and the final result is a factor $(1-\kappa)$ times the PS prediction. A nice consistency check of this result is obtained by computing the fraction of volume occupied by virialized objects. In hierarchical power spectra, all the mass of the universe must finally end up in virialized objects, so we must have $F(S)=1$ when $\delta_c/\sigma\to 0$. Formally, the limit $\delta_c/\sigma\to 0$ can be obtained sending $\delta_c\to 0$ for fixed σ, so we require that

$$\lim_{\delta_c\to 0}\int_{-\infty}^{\delta_c} d\delta\,\Pi(\delta,S)=0. \tag{62}$$

In our case $\Pi=\Pi^{\rm gm}+\Pi^{\rm mem}+\Pi^{\rm mem-mem}$. Since $\Pi^{\rm gm}$ is the same as in the standard excursion set result, it already satisfies eq. (62), so we must find that, in the limit $\delta_c\to 0$, the integral of $\Pi^{\rm mem}+\Pi^{\rm mem-mem}$ from $-\infty$ to δ_c vanishes. Using eq. (52) we see that

$$\int_{-\infty}^{\delta_c} d\delta\,\Pi^{\rm mem}(\delta,S)=\kappa\left[\frac{\delta_c(\delta_c-\delta)}{S}\,{\rm Erfc}\left(\frac{2\delta_c-\delta}{\sqrt{2S}}\right)\right]_{\delta=\delta_c}=0, \tag{63}$$

for all values of δ_c. For the memory-of-memory term we find

$$\int_{-\infty}^{\delta_c} d\delta\,\Pi^{\rm mem-mem}(\delta,S)=\frac{\kappa}{\sqrt{2\pi}}\,\frac{\delta_c}{S^{1/2}}\Gamma\left(0,\frac{\delta_c^2}{2S}\right). \tag{64}$$

Since, for $z\to 0$, $\Gamma(0,z)\to -\ln z$, we have

$$\lim_{\delta_c\to 0}\delta_c\,\Gamma\left(0,\frac{\delta_c^2}{2S}\right)=0, \tag{65}$$

so Eq. (62) is indeed satisfied. In this section we have therefore described a powerful analytical formalism that allows us to compute consistently the halo mass function when non-markovian effects are present. We have applied it to the corrections generated by the tophat filter function in coordinate space. However, the same formalism allows one to compute perturbatively the effect of the non-Gaussianities on the halo mass function. This is a currently florishing line of research which unfortunately goes beyond the scope of this chapter. We prefer instead to describe another important cosmological application of the excursion set in cosmology, the galaxy bias.

4. The bias

In order to make full use of the cosmological information encoded in large-scale structure, it is essential to understand the relation between the number density of galaxies and the DM mass density field. It was first appreciated during the 1980s that these two fields need not be strictly proportional. So it is useful to introduce the linear bias parameter b

$$\left(\frac{\delta\rho}{\rho}\right)_{\rm g} = b\left(\frac{\delta\rho}{\rho}\right)_{\rm DM}. \tag{66}$$

This seems a reasonable assumption when $\delta\rho/\rho \ll 1$, although it leaves open the question of how the effective value of b would be expected to change on nonlinear scales. Galaxy clustering on large scales therefore allows us to determine mass fluctuations only if we know the value of the bias parameter b. We now consider the central mechanism of biased clustering, in which a rare high density fluctuation, corresponding to a massive object, collapses sooner if it lies in a region of large-scale overdensity. This helping hand from the long-wavelength modes means that overdense regions contain an enhanced abundance of massive objects with respect to the mean, so that these systems display enhanced clustering. The basic mechanism can be immediately understood via the Fig. 2 which explains the peak-background split model [9]. If we decompose a density field into a fluctuating component on galaxy scales, together with a long-wavelength as well, then those regions of density that lie above a threshold in density of ν times the variance σ will be strongly clustered. If proto-objects are presumed to form at the sites of these high peaks, then this is a population with Lagrangian bias, *i.e.* a non-uniform spatial distribution even prior to dynamical evolution of the density field. The key question is the physical origin of the threshold: for massive objects such as clusters, the requirement of collapse by the present imposes a threshold of $\nu > 2$. For galaxies, there will be no bias without additional mechanisms to cause star formation to favour those objects that collapse first. The excursion set formalism provides a neat framework to understand how the clustering of dark matter halos differs from the overall clustering of matter. Consider the solution to the excursion set problem in we derived in the previous section. This gives the probability distribution of δ given that on a smoothing scale S_0, the smoothed density fluctuation is δ_0. Notice that the important quantity is the relative height of the density threshold $(\delta_c - \delta_0)$ so that in regions with $\delta_0 > 0$ on large scales, trajectories are more likely to penetrate the barrier at δ_c and conversely for $\delta_0 < 0$.

The fraction of mass in collapsed halos of mass greater than M in a region that has a smoothed density fluctuation δ_0 on scale S_0 (corresponding

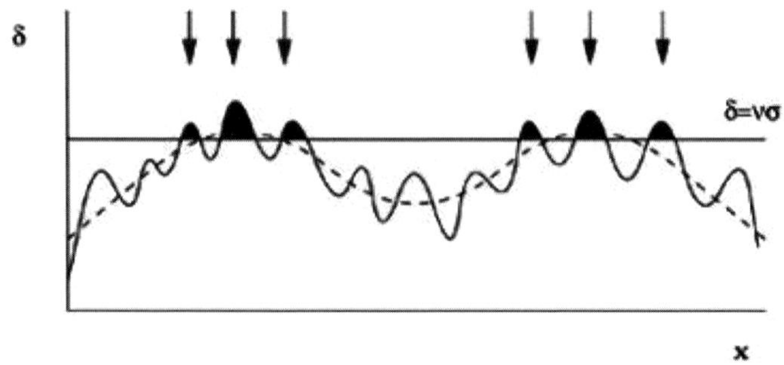

Fig. 2. The peak-background split model.

to mass M_0 and volume V_0) is given

$$F(M|\delta_0, S_0) = \text{Erfc}\left(\frac{\delta_c - \delta_0}{\sqrt{2(S - S_0)}}\right). \tag{67}$$

Notice that as the density of the region increases, F increases because smaller upward excursions are needed to cross the threshold. When $\delta_0 \to \delta_c$, $F \to 1$ because the entire region will then be interpreted as a collapsed halo of mass M_0. The fraction of mass in halos with mass in the range M to $M + \delta M$ is

$$\begin{aligned} f(M|\delta_0, S_0)\left|\frac{\delta S}{\delta M}\right|\delta M &\equiv \frac{\delta F(M|\delta_0, S_0)}{\delta M}\delta M \\ &= \frac{1}{\sqrt{2\pi}}\frac{\delta_c - \delta_0}{(S - S_0)^{3/2}}\left|\frac{\delta S}{\delta M}\right| \exp\left[-\frac{(\delta_c - \delta_0)^2}{2(S - S_0)}\right]\delta M, \end{aligned} \tag{68}$$

so that regions with smoothed density δ_0 on scale S_0 contain, on average,

$$\mathcal{N}(M|\delta_0, S_0)\delta M = \frac{M_0}{M}\, f(M|\delta_0, S_0)\left|\frac{\delta S}{\delta M}\right|\delta M \tag{69}$$

halos in this mass range. The quantity of interest is the relative over-abundance of halos in dense regions compared to the mean abundance of halos,

$$\delta_{\text{halo}}^{\text{L}} = \frac{\mathcal{N}(M|\delta_0, S_0)}{(dn(M)/dM)V_0} - 1, \tag{70}$$

where $dn(M)/dM$ is the mean number density of halos in a mass range of width dM about M. The superscript L indicates that this is the overdensity in the initial Lagrangian space determined by the mass distribution at some very early time, ignoring the dynamical evolution of the overdense patch.

The relative overdensity of halos in large overdense and underdense patches is easy to compute. In sufficiently large regions, $S_0 \ll S$, $\delta_0 \ll \delta_c$. Expanding Eq. (70) to first order in the variables S_0/S and δ_0/δ_c gives a simple relation between halo abundance and dark matter density

$$\delta^{\rm L}_{\rm halo} = \frac{\nu^2 - 1}{\delta_c}\delta_0, \tag{71}$$

where $\nu = \delta_c/S^{1/2} = \delta_c/\sigma(M)$ as before. The overdensity in the initial Lagrangian space is proportional to the dark matter overdensity and is a function of halo mass through ν. The final ingredient needed to relate the abundance of halos to the matter density is a model for the dynamics that can map the initial Lagrangian volume to the final Eulerian space. Let V and δ represent the Eulerian space variables corresponding to the Lagrangian space variables V_0 and δ_0. The final halo abundance is

$$\delta_{\rm h} = \frac{\mathcal{N}(M|\delta_0, S_0)}{(dn(M)/dM)V} - 1. \tag{72}$$

Mass conservation implies $V(1+\delta) = V_0(1+\delta_0)$, but In the limit of a small overdensity, $\delta_0 \ll 1$, $\delta \simeq \delta_0$, and $V \simeq V_0$, and therefore

$$\begin{aligned} \delta_{\rm h} &= \left(1 + \frac{\nu^2 - 1}{\delta_c}\right)\delta & (73)\\ &\equiv b_{\rm h}\delta. & (74) \end{aligned}$$

This expression states that the overdensity of halos is linearly proportional to the overdensity of the mass. The constant of proportionality $b_{\rm h}(M, z)$ depends on the masses of the halos, and the redshifts they virialized, but is independent of the size of the cells. Furthermore, its says that low-mass haloes are antibiased and high mass haloes are positively biased. We can now understand the observation that there are much more strongly clustered than galaxies in general: regions of large-scale overdensity contain systematically more high-mass haloes than expected if the haloes traced the mass. Indeed, by defining the correlation function

$$\begin{aligned} \xi(r) = \Big\langle \delta(\mathbf{x}+\mathbf{r})\delta(\mathbf{x}) \Big\rangle &= \int \frac{dk^3}{(2\pi)^3 V} \left|\delta_{\mathbf{k}}\right|^2 e^{-i\mathbf{k}\cdot\mathbf{r}} = \int_0^\infty d\ln k\, \frac{k^3 \left|\delta_{\mathbf{k}}\right|^2}{2\pi^2} \frac{\sin kr}{kr}, \\ &= \int_0^\infty d\ln k\, \Delta(k) \frac{\sin kr}{kr} & (75) \end{aligned}$$

for which, if $\Delta^2(k) \sim k^{n+3}$ we have $\xi(r) \sim r^{-n-3}$, it turns out observationally that

$$\xi_{\rm cc}(r) \simeq \left(\frac{r}{25\,h^{-1}\,{\rm Mpc}}\right)^{-1.8} \simeq 20\xi_{\rm gg}(r) \simeq 20\left(\frac{r}{5\,h^{-1}\,{\rm Mpc}}\right)^{-1.8}, \tag{76}$$

that is clusters are more correlated than galaxies. However, one should be careful that applying the idea to galaxies is not straightforward: we have shown that enhanced clustering is only expected for massive fluctuations with $\sigma < 1$, but galaxies at $z = 0$ fail this criterion. The high-peak idea applies will at high redshift, where massive galaxies are still assembling, but today there has been time for galaxy-scale haloes to collapse in all environments.

5. Conclusions

The computation of the probability of the first-passage time through a given threshold of a stochastic process is a classic problem that appears in many branches of physics. In this contribution we have seen that it even finds important applications to structure formation in cosmology. When the stochastic dynamics is markovian, the computation of the probability density admits elegant analytic solutions derived from the Fokker-Planck equation with an absorbing boundary condition. However, often the underlying dynamics is non-markovian. This is indeed the case in the cosmological application that we have discussed. We have seen that the dynamics is in fact markovian only if one smooths the density field with a window function sharp in momentum space, which does not have a well-defined volume in physical space, so it is not possible to consistently associate a mass to the DM halos defined in this way. Physically sensible definitions of the smoothed density field unavoidably lead to non-markovian evolution with the smoothing scale. A further source of non-markovianity can be given by non-gaussianities in the statistics of the primordial density fields. To cope with these issues, we have developed a reformulation of the first-passage time problem in terms of a path-integral with boundaries, and we have developed a technique for evaluating perturbatively this path integral, order by order in the non-Markovian terms. Beside the application to cosmology that we have presented, this technique can find a wider application in statistical physics.

The work of MM and AR is supported by the Fond National Suisse.

References

1. P. A. R. Ade *et al.* [Planck Collaboration], arXiv:1303.5076 [astro-ph.CO].
2. Press, W. H. & Schechter, P. 1974, ApJ 187, 425.
3. Bond, J. R., Cole, S., Efstathiou, G. & Kaiser, N. 1991, ApJ. 379, 440.
4. Redner, S. 2001, "A guide to first-passage processes", Cambridge University Press.
5. Maggiore M. & Riotto A. 2010a, ApJ, 711, 907.
6. Maggiore M. & Riotto A., 2010b, ApJ 717, 515.
7. Maggiore M. & Riotto A. 2010c, ApJ 717, 526.
8. Zentner, A. R. 2007, Int. J. Mod. Phys. D 16 763.
9. Bardeen J.M., Bond J.R., Kaiser N. & Szalay A.S. 1986, ApJ 304, 15.
10. Chandrasekhar, S. 1943, Rev. Mod. Phys. 15, 1.

Chapter 22

Self-Organized Escape Processes of Linear Chains in Nonlinear Potentials

Torsten Gross*, Dirk Hennig, and Lutz Schimansky-Geier

Department of Physics, Humboldt-University at Berlin, Newtronstr.15, D-12489 Berlin, Germany

Department of Mathematics, University of Portsmouth, Portsmouth, PO1 3HF, United Kingdom

**e-mail: tgross@physik.hu-berlin.de*

An enhancement of localized nonlinear modes in coupled systems gives rise to a novel type of escape process. We study a spatially one dimensional set-up consisting of a linearly coupled oscillator chain of N mass-points situated in a metastable nonlinear potential. The Hamilton-dynamics exhibits breather solutions as a result of modulational instability of the phonon states. These breathers localize energy by freezing other parts of the chain. Eventually this localised part of the chain grows in amplitude until it overcomes the critical elongation characterized by the transition state. Doing so, the breathers ignite an escape by pulling the remaining chain over the barrier. Even if the formation of singular breathers is insufficient for an escape, coalescence of moving breathers can result in the required concentration of energy. Compared to a chain system with linear damping and thermal fluctuations the breathers help the chain to overcome the barriers faster in the case of low damping. With larger damping, the decreasing life time of the breathers effectively inhibites the escape process.

1. Introduction

We study the first passage time problem of coupled oscillators in a Hamiltonian and also in a thermal setting, which includes fluctuations and linear damping. The typical setup is as follows: Initially, the coordinates of the considered oscillators will be near a potential minimum. Then, in its dynamical evolution the oscillator chain potentially surmounts an energetic barrier within the first passage time. This process will be called escape in the following. Our special interest lies in the distinctive contribution of nonlinear excitations in this passage over the barrier.

In particular, a chain of binary interacting units is a simple model for discussing the emergence of collective phenomena. Despite its simplicity, such a setup appears frequently in various physical contexts such as for the description of mechanical and electrical systems, polymers, networks of superconducting elements, chemical reactions in connected discrete boxes, to name but a few [1–7].

The escape processes of the chain out of a metastable state [8, 9] is also known as the nucleation of a kink-antikink pair [10–16] in biased sinusoidal potentials. In our study we place a chain with linear coupling springs in a nonlinear potential modelled by a polynomial of 3rd degree. As will be seen, the interplay between the effects arising from the nonlinear potential and the springs between the units will distribute energy inhomogeneously. Parts of the chain with large elongations will collect energy from their neighbouring regions. Such localized modes of energy are known as breather-solutions and have been studied intensively in the past in various contexts including micro-mechanical cantilever arrays [1, 17, 18], arrays of coupled Josephson junctions [19, 20], coupled optical wave guides [21, 22], Bose-Einstein condensates in optical lattices [23], in coupled torsion pendula [24], electrical transmission lines [25, 26], and granular crystals [27].

We concentrate here on the escape process and elaborate how these localized breathers modify this process [28, 29]. Certain aspects of the effect of breathers on first passage problems have not yet been considered in the literature. In this study we will determine the dependence of the mean first passage times or of escape rates on parameters of our problem. In this context, the magnitude of damping will be of particular interest.

For this purpose we consider first the pure deterministic set-up and study the properties of breathers arising on the chain whose units evolve in the nonlinear potential. In the second set-up we investigate thermally activated escape dynamics. The chain will be exposed to a thermal bath with temperature T. Consequently damping and noise is added to the deterministic dynamics accounting for coupled Langevin equations. Our main new findings concern the study of how a change of the friction coefficient modifies the escape process. While for stronger damping breather solutions do not play a significant role, in case of weak damping the escape times become even shorter compared to the deterministic case. Notably, the establishment of breathers along the chain helps the emergence of critical elongations from thermal fluctuations.

This work is structured as follows: In Sec. 2 we introduce the model of a linearly coupled chain situated in an external nonlinear potential describing a metastable situation. We study the critical transition state, i.e. the bottleneck configuration which the chain has to cross in order that a transition over the potential barrier takes place. In Sec. 3 we derive conditions for the modulational instability which determine the time scale for the growth of the breathers. We find two generic scenarios which govern the transition. With larger energy singular breathers achieve large elongations and can surpass the transition states alone. Differently, if the elongation of the breathers are too small, they undergo an erratic motion. On collision, breathers tend to merge. Thereby they cumulatively localize energy which can eventually cause the barrier crossing. In Sec. 4 we study the thermally activated escape of the interacting chain in the metastable potential landscape. Finally, we summarize our findings.

2. The one-dimensional chain model

We study a one-dimensional chain of N linearly coupled oscillators of mass m with elongations $q_n(t), n = 1, \ldots, N$. The chain is positioned in a cubic external potential. Every mass point experiences a nonlinear force caused by the potential

$$V(q_n) = \frac{m\,\omega_0^2}{2} q_n^2 - \frac{a}{3} q_n^3$$

and spring forces created from the neighbours with spring constant κ. Periodic boundary conditions are applied. Positions of particles perpendicular to the potential variation are kept constant [28], (for an alternative case see [30, 31]). First, we assume that there is no noise and no damping, hence yielding a canonic situation with a Hamiltonian dynamics and corresponding momenta $p_n(t), n = 1, \ldots, N$ canonically conjugate to the positions $q_n(t)$. Consequently the total energy of the chain is conserved.

In order to obtain dimensionless quantities we rescale units and parameters, $\widetilde{q_n} = a/(m\,\omega_0^2)\, q_n$, $\widetilde{p_n}^{\,2} = a^2/(m^4\,\omega_0^6) p_n$ and $\widetilde{t}^{\,2} = \omega_0^2\, t^2$. It results a dimensionless Hamiltonian with one remaining parameter only, which is the effective coupling strength $\widetilde{\kappa} = \kappa/(m\,\omega_0^2)$. In what follows we omit the tildes.

The Hamiltonian of the considered chain reads:

$$\mathcal{H} = \sum_{n=0}^{N-1} \left[\frac{p_n^2}{2} + \frac{\kappa}{2} \left(q_n - q_{n+1}\right)^2 + V\left(q_n\right) \right], \qquad V(q_n) = \frac{q_n^2}{2} - \frac{q_n^3}{3}.$$

The resulting equations of motion become

$$\ddot{q}_n + q_n - q_n^2 - \kappa\,(q_{n+1} + q_{n-1} - 2\,q_n) = 0, \qquad q_{N+1} = q_1. \tag{1}$$

For all numerical simulations, we will consider chains comprising $N = 100$ units. A study of the dependence of the escape process on the number of oscillators can be found in [29].

Initially, the units of the chain are placed close to the bottom of the external potential, that is nearby $q_{\min} = 0$ and are provided with energy E. As will be seen, the chain eventually generates critical elongations surpassing the potential's local maximum. This initiates a transition of the chain into the unbounded regime $q_n > q_{\max} = 1$, $n = 1, \ldots, N$, which we refer to as an escape. For a single particle to overcome the potential barrier it needs to be supplied with an energy $\Delta E > V(1) - V(0) = 1/6$.

In the following we want to illustrate that the generation of these critical states is efficient even in cases where the chain energy E is small compared to $E \ll N \cdot \Delta E$. This low-energy setting is obtained through the following initial preparation of the system

$$q_i^{\mathrm{IC}} = \Delta q_i + \Delta \qquad p_i^{\mathrm{IC}} = \Delta p_i,$$

where Δq_i and Δp_i are small random perturbations taken from a uniform distribution within the intervals

$$\Delta q_i \in \left[-\Delta q^{\mathrm{IC}}, \Delta q^{\mathrm{IC}}\right] \text{ and } \Delta p_i \in \left[-\Delta p^{\mathrm{IC}}, \Delta p^{\mathrm{IC}}\right],$$

and the coordinate shift, $0 < \Delta < 1$, is chosen to increase the system energy to a desired value. This procedure results in a variety of perturbed flat initial states comprising a specific energy which we view as a statistical ensemble.

2.1. *Transition states*

For Hamiltonian systems the local minima of the energy surface (in phase space) are Lyapunov stable. That is, orbits in the vicinity of a local minimum never leave it as their associated energy is conserved. Therefore, orbits with an energy exceeding the energy associated with a neighbouring saddle point of first order are no longer bound to the basin. Thus the saddle point is referred to as transition state, as it separates bounded from unbounded orbits. Concerning our problem, the system's energy has to exceed the transition state energy to make escape events possible. To determine this transition states, we have to solve $\nabla U(q_1, q_2, \ldots) = 0$, where

U denotes the potential energy (thus the transition state is a fixed-point) and the solution must render all eigenvalues of the Hessian matrix of U positive, except for a single negative one. In general, this is a non-trivial task requiring sophisticated numerical methods. Here, we used the dimer method. It is a minimum-mode following method that solely makes use of gradients of the potential surface. It was first introduced in [32] and its computational effort scales favorable with the system size.

In the one-dimensional chain model the transition state configurations solve the stationary equation[a]

$$q_n - q_n^2 - \kappa\,(q_{n+1} + q_{n-1} - 2\,q_n) = 0 \tag{2}$$

and fulfill the condition on the eigenvalues, λ^H, of the Hessian matrix H

$$H_{i,j} = \delta_{i,j}(2\,\kappa + 1 - 2\,q_i) - \kappa\;(\delta_{i,j+1} + \delta_{i,j-1})\,.$$

In the case of a vanishing coupling strength the oscillators and thus the equations of motion (1) are no longer coupled. Consequently, the fixed points of the system consist of all the configurations where each oscillators is placed either on the maximum of the potential barrier or the potential valley, $q_n^* = \{0, 1\}$. In this case also the Hessian matrix, H, becomes diagonal and we can directly read off its eigenvalues, $\lambda_n^H = 1 - 2\,q_n$. Demanding all eigenvalues to be negative except for one positive the transition states are found to be all the configurations where all oscillator are positioned in the potential valley except for one that is placed on the potential barrier. The according energy reads $E_{\text{act}}(\kappa = 0) = \Delta E = 1/6$.

In contrast, a very large coupling strength corresponds to a situation where the chain effectively becomes a single oscillator so that the transition state refers to a chain configuration where all oscillators are placed on the maximum of the potential. This can be shown by taking the limit $\kappa \to \infty$ in Eq. (2). If we want q_n^* to take on values within a bounded regime we must have $q_{n+1}^* + q_{n-1}^* - 2\,q_n^* = 0$ in order to satisfy Eq. (2) in this limit. In the case of periodic boundary conditions this becomes equivalent to $q_n^* = q^*$ so that Eq. (2) becomes $q^*(1 - q^*) = 0$. Which of its two roots corresponds to the transition state becomes clear from the linear stability analysis of Eq. (1) for this effective one oscillator problem.

$$\ddot{q} = -q + q^2 \approx -q^* + {q^*}^2 + (-1 + 2\,q^*)\,q = (-1 + 2\,q^*)\,q.$$

[a] An alternative approach for the one-dimensional chain model is presented in [28]. It casts the stationary equation into a two-dimensional map and links the localized lattice solutions to its homoclinic orbits.

Only the case $q_n = q^* = 1$ is associated with the inherent instability of a transition state. Accordingly, the transition state energy is found to be $E_{\text{act}}(\kappa \to \infty) = N\,\Delta E = N/6$.

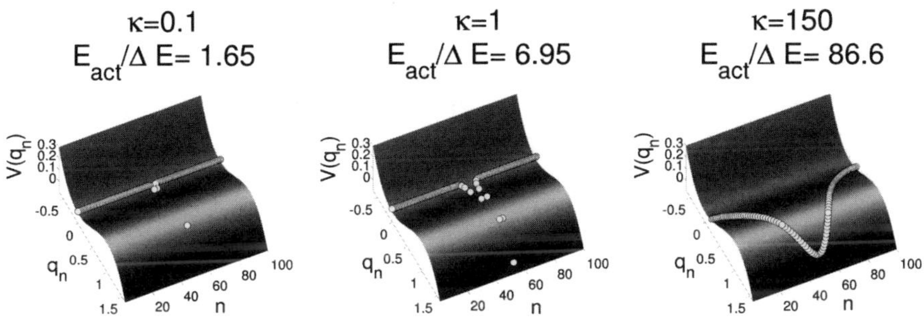

Fig. 1. Transition state chain configurations for different values of κ. $N = 100$.

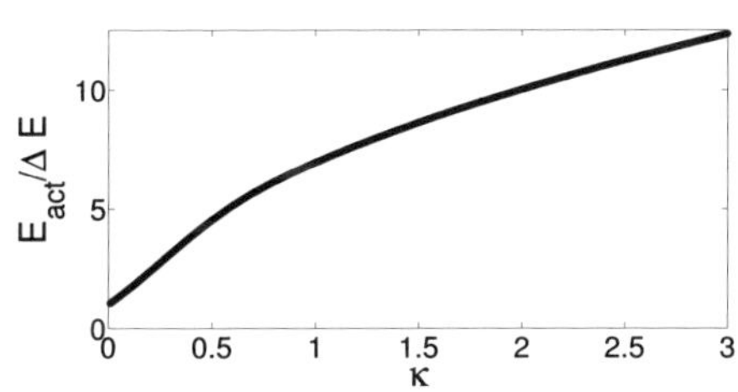

Fig. 2. Activation energy (energy of transition state configurations), $N = 100$.

The intermediate parameter regime has been evaluated using the dimer method and the results are represented in Fig. 1 and Fig. 2. The maximal amplitude of the hair pin-like transition state configuration grows with increasing κ until it reaches a critical elongation from which on it decreases until the entire chain approaches the maximum of the potential barrier as described above.

3. The formation of breathers

3.1. *Modulational instability of a chain in a nonlinear potential*

The energy that is initially homogeneously distributed along the entire chain quickly concentrates into local excitations of single oscillators. This process is governed by the formation of regularly shaped wave patterns, so-called breathers which are spatially localized and time-periodically varying solutions. Their emergence is due to a modulation instability. The way in which this mechanism applies to our situation is described below. We follow [33] and [34] in this paragraph.

As an approximation for small oscillation amplitudes we can neglect the nonlinear term in Eq. (1). The resulting equation in linear approximation exhibits phonon solutions with frequency ω and wave number $k = 2\pi\, k_0/N$ (with $k_0 \in \mathbb{Z}$ and $-N/2 \leq k_o \leq N/2$) related by the dispersion relation

$$\omega^2 = 1 + 4\,\kappa\,\sin^2\left(\frac{k}{2}\right).$$

We make an Ansatz that only takes into account the first harmonics (rotating wave approximation)

$$q_i = F_{1,i}(t)\,e^{-it} + F_{0,i}(t) + F_{2,i}(t)\,e^{-2it} + c.c.$$

The amplitudes of the harmonics are expected to be of a lower order of magnitude ($|F_{0,i}| \ll |F_{1,i}|$, $|F_{2,i}| \ll |F_{1,i}|$). Furthermore, we assume our envelope functions to vary slowly ($|\dot{F}_{m,i}| \ll |F_{m,i}|$) as well as the phonon band to be small ($1 > 4\kappa$). Within the limits of these assumptions we obtain a discrete nonlinear Schrödinger equation (DNLS) for the amplitudes of the first harmonic.

$$2i\dot{F}_{1,i} = \kappa\,((F_{1,i-1} + F_{1,i+1}) + 2F_{1,i}) - \frac{10}{3}\,|F_{1,i}|^2\,F_{1,i}. \tag{3}$$

We want to study the stability of this equation's plane wave solutions in the presence of small perturbations $|\delta B_i(t)| \ll 1$ and $|\delta\Psi_i(t)| \ll 1$, leading to a new Ansatz for the envelope function

$$F_{1,n}{}^{pert.} = (A + \delta B_n(t))\,e^{i((k\,n - \Delta\omega\,t) + \delta\Psi_n(t))}. \tag{4}$$

The perturbations are sufficiently small so that we can expand the envelope function up to the first order in δ and neglect all terms of higher order. Using the Ansatz (4) in Eq. (3) leads to a complex differential equation for the perturbation functions $B(t)$ and $\Psi(t)$. The real and imaginary part of this equation are independent. Hence, collecting all terms of first order in δ results in two linear relations.

$$\begin{aligned}
-A\,\dot{\Psi}_i &= -\frac{\kappa}{2}\,\{A\,\sin k\,(\Psi_{i-1} - \Psi_{i+1}) + \cos k\,(B_{i+1} + B_{i-1})\} - \frac{10}{3}\,A^2\,B_i \\
2\dot{B}_i &= -\kappa\,\{A\,\cos k\,(\Psi_{i+1} + \Psi_{i-1} - 2\Psi_i) + \sin k\,(B_{i+1} - B_{i-1})\}.
\end{aligned}$$

Again, the solution to those coupled equations are plane waves

$$\Psi_n = \Psi^0 e^{i(Q\,n - \Omega\,t)} \qquad B_n = B^0 e^{i(Q\,n - \Omega\,t)}$$

with the dispersion relation

$$\begin{aligned}(\Omega - \kappa\,\sin k\,\sin Q)^2 \\ = \kappa\,\cos k\,\sin^2\left(\frac{Q}{2}\right)\left(4\,\kappa\,\cos k\,\sin^2\left(\frac{Q}{2}\right) - \frac{20}{3}A^2\right)\end{aligned} \tag{5}$$

which describes the stability of the Q-mode perturbation on the k-mode carrier wave. Q and k have a $2\,\pi$ periodicity and can therefore be chosen to be in the first Brillouin zone. Furthermore, we can restrict the range of k and Q: $k, Q \in \{0, \pi\}$, because negative values correspond to waves with the opposite direction of propagation.

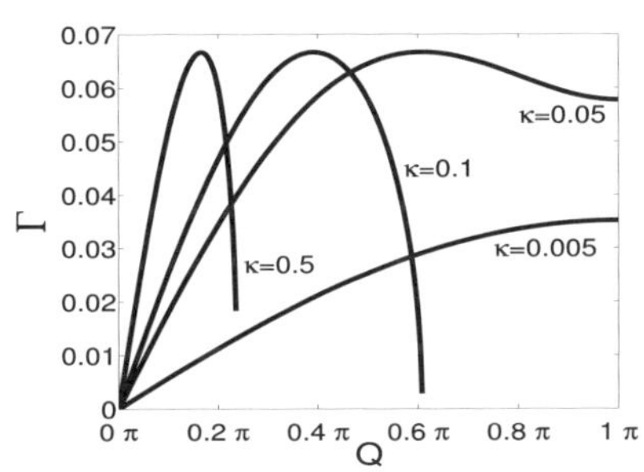

Fig. 3. Growth rates of unstable Q-modes according to Eq. (7) from a k=0 carrier mode with A=0.2

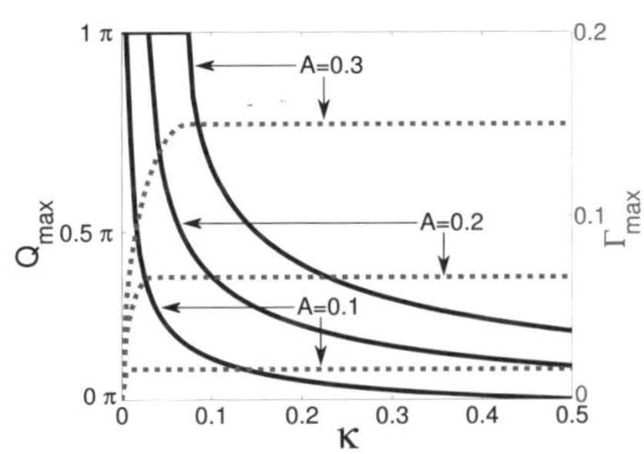

Fig. 4. Fastest growing modes (solid line) - Eq. (8) - and their growth rates (dashed line), k=0.

The perturbations are stable for $\Omega \in \mathbb{R}$ which is the case when the right hand side of Eq. (5) is positive. Therefore, all carrier waves with $k \in \{\pi/2, \pi\}$ are stable with respect to all perturbation modes. For $k \in \{0, \pi/2\}$ perturbations will grow, provided that

$$\cos k\,\sin^2\left(\frac{Q}{2}\right) \leq \frac{5\,A^2}{3\,\kappa}. \tag{6}$$

We can then find an according growth rate

$$\Gamma(Q) = |\mathrm{Im}(\Omega)| = \sin\left(\frac{Q}{2}\right)\sqrt{\frac{20}{3}\kappa\,\cos k\,\left(A^2 - \frac{3}{5}\kappa\sin^2\left(\frac{Q}{2}\right)\cos k\right)} \tag{7}$$

which, for the case that $A^2 \leq \frac{6}{5}\kappa \cos k$, has its maximum at

$$Q_{\max} = 2 \arcsin \sqrt{\frac{5\,A^2}{6\,\kappa\, \cos k}}. \tag{8}$$

Otherwise the maximum growth rate is found at $Q = \pi$. The corresponding growth rates become

$$\Gamma_{\max} = \begin{cases} \Gamma(Q_{\max}) = \frac{5}{3}A^2 & \text{if } A^2 \leq \frac{6}{5}\kappa \cos k \\ \Gamma(\pi) = \sqrt{\frac{20}{3}\kappa \cos k (A^2 - \frac{3}{5}\kappa \cos k)} < \Gamma(Q_{\max}) & \text{if } A^2 > \frac{6}{5}\kappa \cos k. \end{cases} \tag{9}$$

We recall from Sec. 2 that our system is initially prepared in a slightly perturbed $k = 0$ mode. This is thus the only possible carrier wave mode as the amplitudes, A, of all other modes (which scale with the amplitude of the perturbation) are likely to be too small to generate growing modes – see inequality (6) – or the arising maximal growth rates are suppressed. Evaluating the growth rate of instabilities on the $k = 0$ carrier mode for different values of κ (Fig. 3), we find that the modulational instability becomes more mode selective with increasing κ. Hence, for large values of κ the only relevant unstable modes are near the fastest growing mode depicted in Fig. 4. In such a situation we expect the emergence of a regular wave pattern (an array of breathers) that efficiently localizes energy and thereby enhances the escape of the chain.

3.2. *Optimal coupling*

According to our findings in the previous section the appearing breather array becomes more regular with increasing κ . In particular its prominent mode number, and therefore the number of breathers, gets smaller, all of which results in an efficient energy localization. However, an increase in the coupling strength comes along with an increase in the activation energy – see Fig. 2 – which hinders a swift escape for a given system energy. Therefore, we can expect to find an intermediate κ that optimizes the escape rate.

This optimal κ can be determined by an analytical approximation which assumes the entire system energy to be distributed evenly among N_B non-interacting oscillators, where N_B is the number of breathers related to the prominent wave length, $Q_{\max}$, of the modulational instability. The ratio of such an oscillator's energy, E_B, to the activation energy can be regarded as a measure of escape efficiency. It is dependent on $Q_{\max}$ which in turn depends on the $k = 0$ phonon amplitude A which we relate to the system energy via $E(A) = N\,V(A)$. Thus we can write

$$\frac{E_B(\kappa)}{E_{\rm act}} \propto \frac{1}{Q_{\max}(E,\kappa)\,E_{\rm act}(\kappa)}.$$

This escape efficiency takes on its maximum for the optimal coupling strength, κ^*, formally

$$\kappa^*(E) = \arg\max_{\kappa\in\mathbb{R}+} \frac{1}{Q_{\max}(E,\kappa)\,E_{\rm act}(\kappa)}. \tag{10}$$

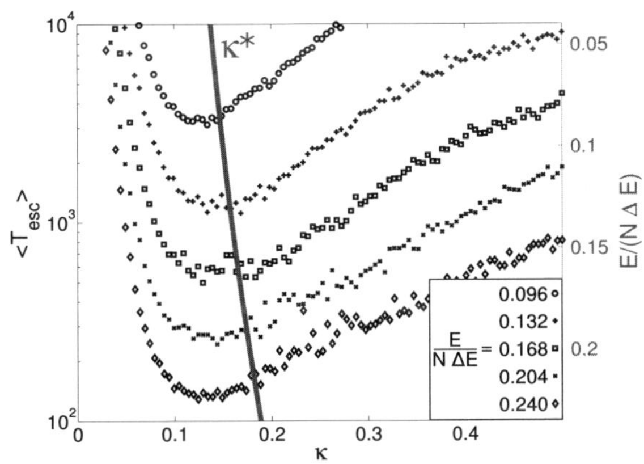

Fig. 5. Average escape times (marker symbols) for 500 realizations with randomized initial conditions as described in Sec. 2. Parameters: $\Delta q^{\rm IC} = 0.05, \Delta p^{\rm IC} = 0.05, N = 100$. The solid grey line represents the analytical approximation for the optimal coupling strength – Eq. (10) – for energy values given by the right-hand axis.

This approximation can now be compared to the numerical evaluation of average escape times in dependence of κ, see Fig. 5. Equation (1) has been integrated using a fourth order Runge-Kutta scheme. Numerical accuracy was obtained by ensuring the energy deviation to remain smaller than the order of 10^{-12}. The average escape times were determined from 500 realization of randomized initial conditions at a given energy according to Sec. 2 for each marker symbol. The escape time measures the time it takes from the initialization to the moment when all oscillators have surpassed the potential barrier. For all depicted values of κ at least 95 % of the chain realizations escaped in the maximal integration time of $5 \cdot 10^5$ time units.

Figure 5 clearly shows the predicted resonance behaviour and also reveals a fairly good accordance of the analytical approximation of the op-

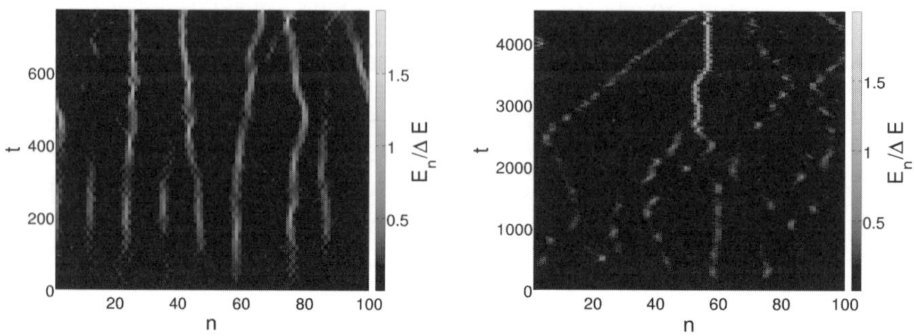

(a) $\frac{E}{N\,\Delta E} = 0.12$ (b) $\frac{E}{N\,\Delta E} = 0.07$

Fig. 6. Temporal evolution of the energy distribution $E_n(t)$. The localization of energy from an initially homogeneous state causes in both cases an escape at the end of the depicted time frame. While the system energy in Fig. 6(a) is sufficiently high to let an individual breather of the early regular breather array surpass the potential barrier, the lower energy in Fig. 6(b) necessitates a merging of breathers to cause the critical chain elongation. Parameters: $\kappa = 0.15, \Delta q^{\mathrm{IC}} = 0.05, \Delta p^{\mathrm{IC}} = 0.05, N = 100$.

timal κ with the simulation results. This seems to verify our initial assumption of a regular breather array that fully concentrates the energy into single oscillators. But this reasoning fails to explain the pronounced variation of the average escape times (ranging over several orders of magnitude) for different energies. Additionally, the equipartition of a low system energy will not allocate single breathers with an energy sufficient to trigger an escape event. E.g. for $\kappa = 0.15$ and $E/(N\,\Delta E) = 0.1$ we expect an array with ten or more breathers so that each one could only hold an energy $E/N_B < \Delta E$. Nevertheless, an escape takes place, eventually. This implies a further concentration of energy beyond the initial creation of the breather array. In order to study this process, we look at the snapshots of the energy distribution

$$E_n = \frac{p_n^2}{2} + V(q_n) + \frac{\kappa}{4}\left\{(q_n - q_{n+1})^2 + (q_{n-1} - q_n)^2\right\}.$$

In Fig. 6, E_n has been tracked in time (upwards) for two exemplary cases. Energy is localized in both cases starting from an initially homogeneous state. In Fig. 6(a) we see the appearance of a regular breather pattern. Every breather concentrates enough energy to certain oscillators in order to trigger an escape. In contrast, the lower system energy in Fig. 6(b) does not allow for a direct escape of the initial breathers. Instead, breathers start an erratic movement. After an inelastic interaction they merge and can thereby eventually result in a configuration exceeding the critical chain elongation, see also [35]. However, this secondary process is slow compared to the (direct) breather formation which explains the different orders of magnitude of the escape times scale in Fig. 5.

4. Thermally activated escapes supported by breathers

In the previous sections we have been concerned with the deterministic chain dynamics leading to an escape event. In this section we study how a thermal bath with temperature T will modify the transition over the barrier. For this purpose we consider the associated Langevin equation,

$$\ddot{q}_n + q_n - q_n^2 - \kappa\,(q_{n+1} + q_{n-1} - 2\,q_n) + \gamma\dot{q}_n + \xi_n(t) = 0, \qquad (11)$$

with the friction parameter γ and a Gaussian white noise term $\xi_n(t)$. In order to be able to compare the deterministic situation to the thermally activated setting, the associated conserved energy E in the Hamiltonian case and the average energy $\overline{E}$ transferred from the bath need to be equal. The latter is governed by the correlation function of the noisy force $\xi(t)$. Its permanent variation acts as a source of energy for the chain, which is counterbalanced by the dissipative friction forces.

The transferred energy is defined by the autocorrelation function of the noise sources

$$\langle \xi_n(t)\xi_{n'}(t')\rangle = 2\,\gamma\,\overline{E}/N\,\delta_{n,n'}\,\delta(t-t')\,. \qquad (12)$$

This relation, known as fluctuation dissipation theorem, implies that the mean energy of all particles is given by $\overline{E}$. If expressed by the bath temperature, every particle gets in average $k_B\,T$, *i.e.* $\overline{E} = N\,k_B\,T$ with k_B being the Boltzmann constant.

In numerical simulations of the Langevin equation, taking into account relation (12), we have assured that the full average energy converges to $\overline{E}$. After the chain has relaxed to a stationary situation around the metastable minimum of the potential $V(q_{min})$, the oscillators obey the canonic distribution in phase space near to this minimum.

We measure average escape times for the system described by Eq. (11). The latter is numerically integrated using an Euler scheme, again with a maximal integration time of $5 \cdot 10^5$ time units. The system is initialized with all oscillators set to the minimum of the potential and zero momenta. We then let the system thermalize until its energy reaches $\overline{E}$ for the first

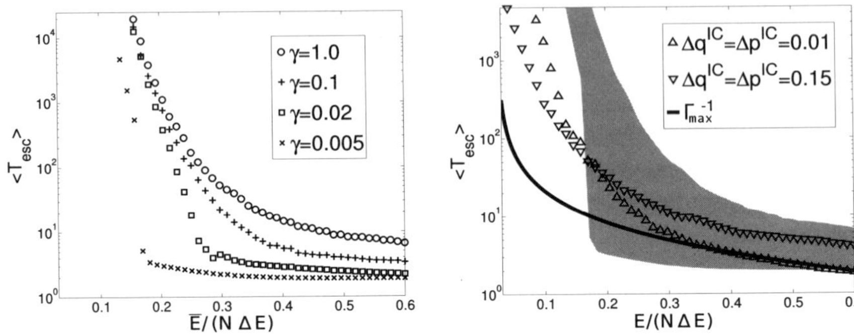

(a) Thermally activated setting – eq. (11) (b) Deterministic setting – Eq. (1).

Fig. 7. Average escape times in the thermally activated case for different values of the friction constant and their comparison to the deterministic setting for 500 realizations each. The grey area in Fig. 7(b) sketches the average escape times for the thermally activated case for $0.005 < \gamma < 1.0$ as shown in Fig. 7(a) where as the symbols shows results from the deterministic set-up with initial conditions given in the inset. Parameters: $\kappa = 0.15, N = 100$.

time. The time from this moment until all oscillators have surpassed the potential's maximum counts as the escape time.

We study the system for the optimal coupling constant, $\kappa = 0.15$, as described in Sec. 3.2. Figure 7(a) shows the average escape times of 500 realizations for different values of the friction constant in dependence of $\overline{E}$. Figure 7(b) compares these times (depicted as the grey surface) to the according average escape times of the deterministic system. It additionally shows the characteristic time constant for the formation of breathers, $\Gamma^{-1}_{\max}$, taken from Eq. (9), where again we relate the $k = 0$ phonon amplitude, A, to the flat state energy, that is $E(A) = N\,V(A)$, with external potential V.

Especially for smaller energies the deterministic escape is considerably faster than the thermally activated one. Notably for $E/(N\,\Delta E) < 0.1$ and quite contrary to the deterministic setting, escape events are practically absent during our simulations time in the thermal case. For larger energy values this picture can change to a higher efficiency of the thermal escape process when damping is weak. Also two deterministic settings with different magnitudes of the random initial perturbations have a converse behaviour for low and high energies.

How can this result be explained by observations of the chain dynamics? For small values of γ the system approaches the deterministic setting. This entails an observable tendency towards a more localized energy distribution

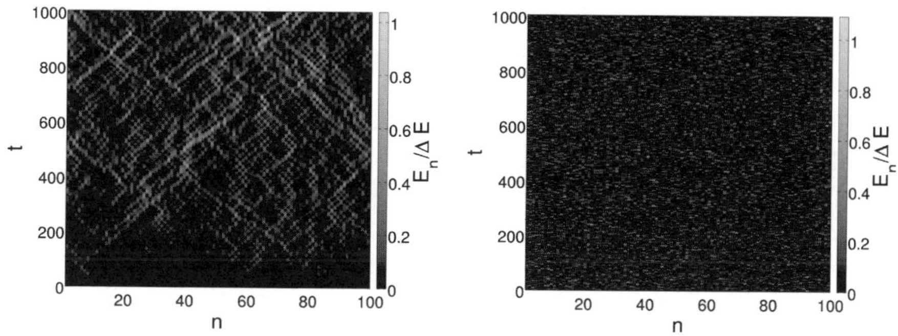

(a) $\gamma = 0.005$ (b) $\gamma = 1.0$

Fig. 8. Temporal evolution of the energy distribution $E_n(t)$ in the thermally activated case. A smaller friction constant leads to a higher degree of energy localization. Parameters: $\overline{E}/(N\,\Delta E) = 0.15$, $\kappa = 0.15$.

as seen in comparing Fig. 8(a) and Fig. 8(b). The relaxation time of the chain scales with the inverse of the damping constant. Correspondingly, the life times of local excitations grow with decreasing γ. The outcome is a more heterogeneous energetic structure where thermal fluctuations are more likely to cause critical chain elongations. This explains the faster escape comparing small with large damping constants. But even in the case of very small γ the relaxation time is still much shorter than the time needed for the coalescence of multiple breathers (which is of the order of several hundred time units, see Fig. 6(b)). Therefore, the long term cumulative concentration of energy, as described in Sec. 3.2, is generally inhibited in the thermal case. This explains the virtual impossibility of a thermal escape for small energies.

In the opposite case of higher energies the deterministic escape is mostly proceeded by initial breathers. The formation time of the initial breather array can be estimated by the inverse of the maximal growth rate, $\Gamma_{\max}$ from Eq. (9). Figure 7(b) shows the expected convergence of $\Gamma_{\max}^{-1}$ to the deterministic escape times for large energies. We recall from Sec. 3 that $\Gamma_{\max}$ was determined starting with a linear expansion in the perturbations. We believe that this explains the better match in the case of smaller initial perturbations. The divergence of $\Gamma_{\max}^{-1}$ with the average escape times for small energies again gives evidence to the fact that the initial breather array does not induce an escape and other mechanisms are needed.

In the thermally activated case the large energy setting holds two characteristic scenarios. The mobility of the breathers becomes amplified if noise acts. If the breathers possess longer life times, *i.e.* for smaller damping, a few (usually not more the two) breathers can temporarily merge and thus approach a critical chain elongation. Starting with a heterogeneous energy structure after thermalization, this then leads to smaller escape times compared to the deterministic setting that first has to re-allocate energy from an initially homogeneous state. In the opposite case of strong damping, the life times of breather is too short and the energy distribution is mostly homogeneous (see Fig. 8(b)). The escapes then relies entirely on rare, large enough spontaneous fluctuation of the noise term and the average escape times become comparatively large.

Finally, we want to examine the converse behaviour for different magnitudes of the initial perturbations in the deterministic case. As the initial conditions for smaller perturbations are closer to the $k = 0$ phonon mode the initial breather array emerges more quickly so that the escape times are smaller when the energy is high enough for initial breathers to ignite the escape. Contrarily, stronger perturbations lead to a higher breather mobility which accelerates the breather coalescence so that the escape times become smaller when low energy necessitates coalescence.

5. Summary

We have studied various examples of collective escape processes in many-particle systems. First we started with a noise-free escape in a chain of coupled oscillators evolving in a metastable potential. The second part of this work examined a thermally activated escape. To this end the original system has been augmented by a linear friction term and Gaussian thermal, white noise of vanishing mean satisfying the fluctuation dissipation theorem.

While a thermally activated escape becomes virtually impossible at low system energy, the deterministic system remains capable of efficiently crossing the potential barrier. In more detail, the deterministic dynamics of interacting chain units leads to the formation of breather solutions localizing energy such that the chain passes through a transition state and crosses the potential barrier. In particular at low system energy, the interaction between several formed breathers resulting in their coalescence eventually enables the chain to accumulate sufficient energy to overcome the potential barrier. Interestingly, in the thermally activated setting with a sufficiently weak damping, the breather formation and thermal fluctuation cooperate to localize energy which accomplishes effective barrier crossings.

We underline that the dynamics of interacting particles exhibiting collective behaviour such as breather formation and their interaction has a huge impact on the escape and activation dynamics in such many-body systems. Hence, our study intends to offer new perspectives on the understanding of such collective escape processes.

Acknowledgments

L. Schimansky-Geier thanks for support from IRTG 1640 of the Deutsche Forschungsgemeinschaft. The authors acknowledge previous co-authors for a successful cooperation.

References

1. M. Sato, B. E. Hubbard, and A. J. Sievers, *Rev. Mod. Phys.* **78**, 137–157 (2006).
2. S. Flach and C. R. Willis, *Phys. Rep.* **295** (5), 181–264 (1998).
3. T. Dauxois, M. Peyrard, and C. Willis, *Physica D* **57**, 267–282 (1992).
4. P. Reineker and R. G. Winkler, *Phys. Lett. A* **141**, 264–268 (1989).
5. M. Doi and H. See, *Introduction to Polymer Physics*, Oxford University Press (Oxford, 1995).
6. T. Erneux and G. Nicolis, *Physica D* **67**, 237 (1993).
7. V.A. Makarov, V.I. Nekorkin, V.B. Kazantsev, and M.G. Velarde, Physica D **100**, 330 (1997).
8. J.S. Langer, *Ann. Phys. (N.Y.)* **54**, 258 (1969).
9. P. Hänggi, P. Talkner and M. Borkovec, *Rev. Mod. Phys.* **62**, 251 (1990).
10. F. Marchesoni, P. Hänggi, and P. Sodano, In *Universalities in condensed matter*, eds. L. R. Jullien, L. Peliti and N. Boccara, *Springer Proc. in Physics* vol. 32, p. 88, Springer Verlag (1988).
11. P. Hänggi, F. Marchesoni and P. Riseborough, *Europhys. Lett.* **13**, 217–222 (1990).
12. M. Büttikker and T. Christen, *Phys. Rev. Lett.* **75**, 1895–1898 (1995).
13. T. Christen, *Europhys. Lett.* **31**, 181–186 (1995).
14. T. Christen, *Phys. Rev. E* **51**, 604–612 (2009).
15. K. Fedorov and A. Pankratov, *Phys. Rev. Lett.* **103**, 260601 (2009).
16. D.R. Gulevich, M.B.Gaifullin and F.V. Kusmartsev, *Eur. Phys. Journ. B* **85**, 24 (2012).
17. M. Sato, B.E. Hubbard, A.J. Sievers, B. Ilic, D.A. Czaplewski, and H.G. Craighead, *Phys. Rev. Lett.* **90** 044102 (2003).
18. M. Sato, B.E. Hubbard, L.Q. English, A.J. Sievers, B. Ilic, D.A. Czaplewski, and H.G. Craighead, *Chaos* **13** 702 (2003).
19. E. Trias, J. Mazo, and T. Orlando, *Phys. Rev. Lett.* **84** 741 (2000).

20. P. Binder, D. Abraimov, A.V. Ustinov, S. Flach, and Y. Zolotaryuk, *Phys. Rev. Lett.* **84** 745 (2000).
21. J.W. Fleischer, M. Segev, N.K. Efremidis, and D.N. Christodoulides, *Nature* **422** 147 (2003)
22. H.S. Eisenberg, Y. Silberberg, R. Morantotti, A.R. Boyd, and J.S. Atchinson, *Phys. Rev. Lett.* **81** 3383 (1998).
23. B. Eiermann, T. Anker, M. Albiez, M. Taglieber, P. Treutlein, K.P. Marzlin, and M.K. Oberthaler, *Phys. Rev. Lett.* **92** 23041 (2004).
24. J. Cuevas, L.Q. English, P.G. Kevredikis, and M. Anderson, *Phys. Rev. Lett.* **102**, 224101 (2009).
25. L.Q. English, M. Sato, and A.J. Sievers, *Phys. Rev. E* **77**, 066601 (2008).
26. P. Marquié, J.M. Bilbaut and M. Remoissenet, *Phys. Rev. E***51**, 6127 (1995).
27. N. Boechler, G. Theocharis, S. Job, and P.G. Kevredikis, *Phys. Rev. Lett.* **104**, 244302 (2010).
28. D. Hennig, L. Schimansky-Geier and P. änggi, *Europhys. Lett.* **78** (2007) 20002
29. D. Hennig, S. Fugmann, L. Schimansky-Geier, and P. Hänggi, *Phys. Rev. E* **76**, 041110 (2007).
30. S. Martens, D. Hennig, S. Fugmann, and L. Schimansky-Geier, *Phys. Rev. E* **78**, 041121 (2008).
31. S. Fugmann, D. Hennig, L. Schimansky-Geier, and P. Hänggi, *Phys. Rev. E* **77**, 061135 (2007).
32. G. Henkelman and H. Jonsson, *Journ. of Chem. Phys.* **111**, 7010–7022 (1999).
33. Y. S. Kivshar and M. Peyrard, *Phys. Rev. A* **46**, 3198–3205 (1992).
34. I. Daumont, T. Dauxois, and M. Peyrard, *Nonlinearity* **10**, 617 (1997).
35. M. Peyrard, *Physica D***119**, 184 – 199 (1998).

Chapter 23

Efficient Monte Carlo Methods for Simulating Diffusion-Reaction Processes in Complex Systems

Denis Grebenkov

Laboratoire de Physique de la Matière Condensée, CNRS – Ecole Polytechnique, 91128 Palaiseau, France

denis.grebenkov@polytechnique.edu

We present the principles, mathematical bases, numerical shortcuts and applications of fast random walk (FRW) algorithms. This Monte Carlo technique allows one to simulate individual trajectories of diffusing particles in order to study various probabilistic characteristics (harmonic measure, first passage/exit time distribution, reaction rates, search times and strategies, etc.) and to solve the related partial differential equations. The adaptive character and flexibility of FRWs make them particularly efficient for simulating diffusive processes in porous, multiscale, heterogeneous, disordered or irregularly-shaped media.

1. Introduction

Diffusion in complex systems often invokes numerical simulations. Except few simple shapes (such as a disk or a sphere) for which diffusion equations possess explicit solutions [1, 2], one needs to resort to numerical methods that can be roughly divided into two groups:

(1) **Finite differences, finite elements, boundary elements, etc.** A domain and/or its boundary are discretized with a regular or adaptive mesh. The original continuous problem is then replaced by a set of linear equations to be solved numerically. The solution is obtained at all mesh nodes at successive time moments. Since the accuracy and efficiency of these deterministic numerical schemes significantly rely on the discretization, mesh construction turns out to be the key issue and often a limiting factor, especially in three dimensions.

(2) **Monte Carlo simulations.** A probabilistic interpretation of diffusion equations is employed [3–9] to represent the original continuous problem

as the expectation of a functional of an appropriate stochastic process. Many random trajectories of this process are then generated and used to approximate the expectation and thus the solution. Since there is no discretization, neither of the domain, nor of boundary conditions, Monte Carlo techniques are flexible and easy to implement, especially for studying diffusion in complex geometries [10–12].

In basic Monte Carlo simulations, one fixes a time step δ and approximates each trajectory $\mathbf{r}(t)$ by a sequence of t/δ independent normally distributed random jumps along each coordinate, with mean zero and variance $2D\delta$. Note that other jump distributions may be used, e.g., discrete displacements to neighboring sites of a lattice, ballistic displacements in random direction, etc. At each jump, one has to check whether the trajectory remains inside the confining domain. If not, the jump has to be modified according to the imposed boundary condition. For instance, Neumann boundary condition is implemented by reflecting the trajectory into the domain, while the simulation is stopped for Dirichlet boundary condition. Partial absorption/reflections, trapping, splitting and other local mechanisms can also be implemented. These fixed-time step simulations are easy to implement but are inefficient in hierarchical or *multiscale* porous media. In fact, the time step δ must be chosen so small to ensure that the average one-step displacement $\sqrt{2D\delta}$ is much smaller than the smallest geometrical feature of the medium. Since most particles are released inside large pores, a very large number of steps, t/δ, may be required for simulating each trajectory. This is the major drawback of basic Monte Carlo schemes.

In order to overcome this limitation, Müller proposed the concept of variable time-step, geometry-adapted or fast random walks (FRWs) [13]. This technique was broadly employed by many authors, for instance, to simulate diffusion-limited growth phenomena [14–16] and to study diffusion-reaction processes and related first-passage problems in random packs of spheres [17–19] or near prefractal boundaries [20–23]. The idea is to replace Brownian motion by an equivalent "spherical process" that explores the medium as fast as possible. A particle starts to diffuse from an initial position $\mathbf{r}_0$ which may be prescribed (fixed) or chosen randomly. One draws the largest disk (or ball in three dimensions) which is centered at $\mathbf{r}_0$ and inscribed in the confining medium (Fig. 1). Its radius ℓ_0 is the distance between $\mathbf{r}_0$ and the boundary. After wandering inside the disk during a random time τ_1, the particle exits the disk at random point $\mathbf{r}_1$. Since there was no "obstacles" inside the disk, all the exit points of the disk are

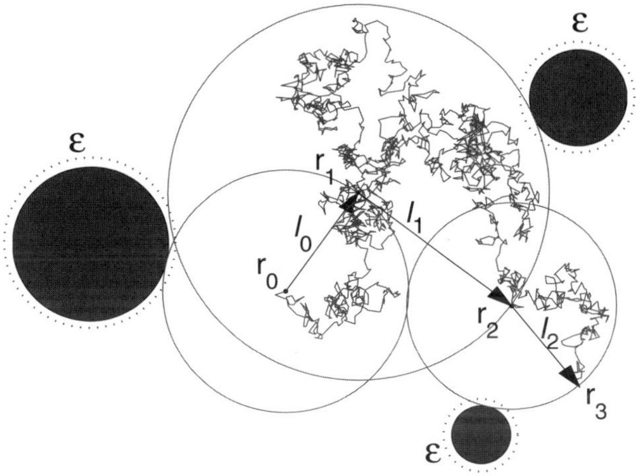

Fig. 1. A fast random walk in a medium with obstacles (dark disks). From an initial position $\mathbf{r}_0$, one determines the distance ℓ_0 to the obstacles (or to the boundary of the medium) and draws a circle of radius ℓ_0 centered at $\mathbf{r}_0$. A "jump" to a randomly (uniformly) chosen point on the circle is then executed. This single large displacement (shown by an arrow) replaces a detailed simulation of Brownian trajectory inside the disk of radius ℓ_0. From the new point $\mathbf{r}_1$, one determines the distance again and executes the next jump, and so on (only three jumps are shown) [24].

equally accessible for isotropic Brownian motion so that the exit point $\mathbf{r}_1$ has a uniform distribution on the circle of radius ℓ_0. From $\mathbf{r}_1$, the new largest disk of radius ℓ_1 is inscribed in the medium. After wandering inside the disk during a random time τ_2, the particle exits at random point $\mathbf{r}_2$, and so on. Following the Brownian trajectory of the particle, one can construct the sequence of inscribed disks (i.e., their centers $\mathbf{r}_n$ and radii ℓ_n) and the associated exit times τ_n.

The fundamental idea behind FRWs is that the sequence $\{\mathbf{r}_n, \ell_n, \tau_n\}$ can be constructed directly, without simulating the underlying Brownian trajectory at all. At each step, one determines the distance ℓ_n between the current position $\mathbf{r}_n$ and the boundary of the medium and chooses the next position $\mathbf{r}_{n+1}$ randomly and uniformly on the circle of radius ℓ_n. The time τ_{n+1} needed to exit from the disk (i.e., to jump from $\mathbf{r}_n$ to $\mathbf{r}_{n+1}$) is a random variable which can be generated from the well-known probability distribution (Section 2). A detailed time-consuming simulation of a Brownian trajectory with high spatial resolution is therefore replaced by generation of random jumps which are adapted to the local geometrical structure of the medium. In other words, the spatial resolution of the spherical process is constantly adapted to the distance to the boundary: closer the particle to the boundary, finer the simulation scale. Performing each jump at largest possible distance yields a tremendous gain in computational time.

In this chapter, we review some practical aspects for an efficient implementation of FRWs and discuss several extensions and applications to first-passage phenomena. Section 2 introduces the formulas for generating first exit times and positions. In Section 3, several strategies for estimating the distance to the boundary are discussed. Finally, Section 4 presents extensions, applications and conclusions.

2. Exit time and position distributions

We start by recalling the computation of the first exit time and position distributions for a general bounded domain $\Omega \subset \mathbb{R}^d$. Since the derived distributions will be used to generate jumps in a FRW algorithm, the underlying "jump" domain Ω has to be simple (e.g., to be a disk or a sphere, as described in Section 1). We emphasize that the simple shape of Ω and the specific condition on its boundary $\partial\Omega$ have nothing to do with the shape of the medium and the boundary mechanism that are going to be modeled with the FRW algorithm.

We introduce the diffusion propagator $G_t(\mathbf{r}_0, \mathbf{r})$ (also known as heat kernel, or Green function of diffusion equation) as the probability density for Brownian motion to move from a point $\mathbf{r}_0$ to a vicinity of a point $\mathbf{r}$ in time t, *without hitting the boundary* $\partial\Omega$ *of the domain* Ω *during this time.* The diffusion propagator satisfies the diffusion equation

$$\frac{\partial}{\partial t} G_t(\mathbf{r}_0, \mathbf{r}) = D\Delta G_t(\mathbf{r}_0, \mathbf{r}), \tag{1}$$

where $\Delta = \partial^2/\partial x_1^2 + ... + \partial^2/\partial x_d^2$ is the Laplace operator, and D the diffusion coefficient. This equation is completed by the initial condition with Dirac δ-distribution, $G_{t=0}(\mathbf{r}_0, \mathbf{r}) = \delta(\mathbf{r}_0 - \mathbf{r})$, stating that $\mathbf{r}_0$ is the starting point, and by Dirichlet boundary condition, $G_t(\mathbf{r}_0, \mathbf{r}) = 0$ at $\mathbf{r} \in \partial\Omega$, that "excludes" Brownian trajectories that hit or crossed the boundary $\partial\Omega$ prior to time t. The diffusion propagator can be expressed in terms of the (L_2-normalized) Laplace operator eigenfunctions $u_m(\mathbf{r})$ and eigenvalues λ_m [1, 2, 26]

$$G_t(\mathbf{r}_0, \mathbf{r}) = \sum_{m=1}^{\infty} e^{-D\lambda_m t} u_m(\mathbf{r}_0) u_m^*(\mathbf{r}), \tag{2}$$

where asterisk denotes the complex conjugate, and the eigenvalue equation $\Delta u_m(\mathbf{r}) + \lambda_m u_m(\mathbf{r}) = 0$ in Ω is completed by Dirichlet boundary condition $u_m(\mathbf{r}) = 0$ at $\mathbf{r} \in \partial\Omega$.

The diffusion propagator allows one to compute the first exit time and position distributions. For Brownian motion started from a point $\mathbf{r}_0$, we denote $\tau = \inf\{t > 0 \ : \ \mathbf{r}(t) \notin \Omega\}$ the first exit time τ from the domain Ω. Since τ is the *first* exit time, Brownian motion should not hit the boundary $\partial\Omega$ at earlier times $t < \tau$. The probability of this event (so-called survival probability) is obtained by integrating the probability density $G_t(\mathbf{r}_0, \mathbf{r})$ over all the arrival points $\mathbf{r}$:

$$\mathbb{P}_{\mathbf{r}_0}\{\tau > t\} \equiv S_{\mathbf{r}_0}(t) = \int\limits_{\Omega} d\mathbf{r} \; G_t(\mathbf{r}_0, \mathbf{r}). \tag{3}$$

The probability density for the first exit time is simply

$$\rho_{\mathbf{r}_0}(t) \equiv -\frac{\partial}{\partial t} S_{\mathbf{r}_0}(t) = -\int\limits_{\Omega} d\mathbf{r} \; D\Delta G_t(\mathbf{r}_0, \mathbf{r}) = \int\limits_{\partial\Omega} d\mathbf{s} \left(-D\frac{\partial}{\partial n} G_t(\mathbf{r}_0, \mathbf{r}) \right)_{\mathbf{r}=\mathbf{s}}, \tag{4}$$

where we first used the diffusion equation and then Green formula to integrate by parts. Here $\partial/\partial n$ is the normal derivative directed outwards the domain Ω. The probability density $\rho_{\mathbf{r}_0}(t)$ characterizes the time of the exit, regardless its position. Alternatively, one can consider the exit position $\mathbf{s} \in \partial\Omega$, regardless its time, whose distribution is known as the harmonic measure [25]. The harmonic measure density, $\rho_{\mathbf{r}_0}(\mathbf{s})$, can be expressed through the diffusion propagator as

$$\rho_{\mathbf{r}_0}(\mathbf{s}) = \int\limits_{0}^{\infty} dt \left(-D\frac{\partial}{\partial n} G_t(\mathbf{r}_0, \mathbf{r}) \right)_{\mathbf{r}=\mathbf{s}}. \tag{5}$$

Looking at Eqs. (4, 5), one can see that the diffusive flux $-D\frac{\partial}{\partial n} G_t(\mathbf{r}_0, \mathbf{r})$ plays the role of the joint probability density for the exit time and position, while $\rho_{\mathbf{r}_0}(t)$ and $\rho_{\mathbf{r}_0}(\mathbf{s})$ are marginal densities for time and position obtained after integration over the other variable. When both the exit time and position are needed, one can use the joint distribution. In practice, one can first generate the exit time and then generate the exit position conditioned to be at the prescribed exit time t:

$$\rho_{\mathbf{r}_0,t}(\mathbf{s}) = \frac{1}{\rho_{\mathbf{r}_0}(t)} \left(-D\frac{\partial}{\partial n} G_t(\mathbf{r}_0, \mathbf{r}) \right)_{\mathbf{r}=\mathbf{s}}, \tag{6}$$

where the prefactor $1/\rho_{\mathbf{r}_0}(t)$ reflects the conditional character of this distribution. Finally, one may also need the probability density of the arrival positions $\mathbf{r}$ inside the domain conditioned to survive up to time t, namely, $G_t(\mathbf{r}_0, \mathbf{r})/S_{\mathbf{r}_0}(t)$.

The above probability densities can be analytically computed for simple domains for which the Laplace operator eigenfunctions are known explicitly. In the following three subsections, we summarize the formulas for interval, disk and sphere. For convenience, we will use the rescaled (dimensionless) time and space variables, $t \to Dt/L^2$ and $x \to xL$, where L is the length of the interval, or the radius of the disk/sphere.

2.1. *Interval*

For the unit interval $(0, 1)$, the eigenfunctions and eigenvalues of the Laplace operator with Dirichlet boundary condition are

$$u_m(x) = \sqrt{2}\,\sin(\pi m x), \qquad \lambda_m = \pi^2 m^2 \quad (m = 1, 2, 3, \ldots), \tag{7}$$

so that the one-dimensional propagator is simply

$$G_t(x_0, x) = 2\sum_{m=1}^{\infty} \sin(\pi m x_0)\ \sin(\pi m x)\ e^{-\pi^2 m^2 t}. \tag{8}$$

It is also useful to write an alternative representation by image method:

$$G_t(x_0, x) = \frac{1}{\sqrt{4\pi t}} \sum_{k=-\infty}^{\infty} \left[e^{-(x-x_0+2k)^2/(4t)} - e^{-(x+x_0+2k)^2/(4t)}\right]. \tag{9}$$

The first expansion (8) rapidly converges for large t, while Eq. (9) rapidly converges for small t.

Using Eq. (3), one gets two alternative representations for the cumulative distribution function for the exit time:

$$\begin{aligned} S_{x_0}(t) &= 2\sum_{m=1}^{\infty} \sin(\pi m x_0)\frac{1-(-1)^m}{\pi m}\ e^{-\pi^2 m^2 t} \\ &= 1 - \operatorname{erfc}\left(\frac{x_0}{\sqrt{4t}}\right) + \sum_{k=1}^{\infty}(-1)^k \left[\operatorname{erfc}\left(\frac{k-x_0}{\sqrt{4t}}\right) - \operatorname{erfc}\left(\frac{k+x_0}{\sqrt{4t}}\right)\right], \end{aligned} \tag{10}$$

where erfc(x) is the complementary error function. The density is

$$\rho_{x_0}(t) = 2\pi \sum_{m=1}^{\infty} \sin(\pi m x_0)(1-(-1)^m)m\, e^{-\pi^2 m^2 t} = \frac{1}{\sqrt{4\pi t^3}} \sum_{k=-\infty}^{\infty} (-1)^k (x_0+k) \exp\left(-\frac{(x_0+k)^2}{4t}\right). \tag{11}$$

As the boundary of an interval consists of two endpoints, the harmonic measure density is reduced to two functions, namely, the probability $\rho_{x_0}(\mathbf{s}=0)$ to reach the left endpoint before the right one, and that of the complementary event:

$$\rho_{x_0}(\mathbf{s}=0) = 2\sum_{m=1}^{\infty} \frac{\sin(\pi m x_0)}{\pi m} = 1 - x_0, \qquad \rho_{x_0}(\mathbf{s}=1) = x_0. \tag{12}$$

This is the classical result for the gambler's ruin [6]. Other probability densities can also be explicitly written.

2.2. *Disk*

For the unit disk with Dirichlet boundary condition, the Laplace operator eigenvalues and eigenfunctions are [1, 2]

$$\lambda_{nk} = \alpha_{nk}^2, \qquad u_{nk}(r,\theta) = \frac{\epsilon_n}{\sqrt{\pi}} \frac{1}{-J_n'(\alpha_{nk})} J_n(\alpha_{nk} r)\cos n\theta, \tag{13}$$

where $\epsilon_n = \sqrt{2}$ for $n > 0$ and $\epsilon_0 = 1$, $J_n'(z)$ is the derivative of the Bessel function $J_n(z)$ of the first kind, and $\{\alpha_{nk}\}_{k=0,1,2,\ldots}$ is the set of all positive zeros of the function $J_n(z)$ (with $n = 0,1,2,\ldots$). For convenience, the double index nk is used instead of the single index m to enumerate the eigenfunctions and eigenvalues. The asymptotic behavior of zeros α_{nk} is well known while their numerical computation is straightforward by bisection method or Newton's method. After this preliminary step, the spectral decomposition (2) yields the diffusion propagator $G_t(\mathbf{r}_0,\mathbf{r})$ and the consequent probability densities: $\rho_{\mathbf{r}_0}(t)$, $\rho_{\mathbf{r}_0}(\mathbf{s})$ and $\rho_{\mathbf{r}_0,t}(\mathbf{s})$. When the starting point $\mathbf{r}_0$ is located in the center of the disk, only the terms with $n = 0$ (invariant under rotation) do contribute. In particular, one gets

$$S_c(t) = 2\sum_{k=0}^{\infty} \frac{e^{-\alpha_{0k}^2 t}}{\alpha_{0k} J_1(\alpha_{0k})}, \tag{14}$$

while the harmonic measure density is uniform: $\rho_c(\mathbf{s}) = \frac{1}{2\pi}$, as all the boundary points are equivalent from the center (here and throughout the text, the subscript 'c' refers to the starting point at the center of the jump domain).

2.3. *Sphere*

The Laplace operator eigenvalues and eigenfunctions for the unit sphere with Dirichlet boundary condition are [1, 2]

$$\lambda_{nk} = \alpha_{nk}^2, \qquad u_{nkl}(r,\theta,\varphi) = \frac{\sqrt{2n+1}}{\sqrt{2\pi}} \, \frac{1}{-j_n'(\alpha_{nk})} \, j_n(\alpha_{nk} r) P_n^l(\cos\theta) e^{il\varphi}, \tag{15}$$

where $j_n'(z)$ is the derivative of the spherical Bessel function $j_n(z)$ of the first kind, $P_n^l(z)$ the associated Legendre polynomial, and $\{\alpha_{nk}\}_{k=0,1,2,\ldots}$ the set of all positive zeros of the function $j_n(z)$ (with $n = 0, 1, 2, \ldots$). The last index l ranges from $-n$ to n. The spectral decomposition (2) allows one to compute all the necessary distributions. When the starting point $\mathbf{r}_0$ is located at the center, the formulas are simplified, e.g.,

$$S_c(t) = 2 \sum_{k=1}^{\infty} (-1)^{k+1} e^{-\pi^2 k^2 t} \tag{16}$$

(as $\alpha_{0k} = \pi(k+1)$, $k = 0, 1, 2, \ldots$), while the harmonic measure density is uniform: $\rho_c(\mathbf{s}) = \frac{1}{4\pi}$. Using the techniques of Laplace transforms and series summation, one can get an alternative representation for the survival probability as [27]

$$S_c(t) = 1 - \frac{2}{\sqrt{\pi t}} \sum_{k=1}^{\infty} e^{-(2k-1)^2/(4t)}, \tag{17}$$

which rapidly converges for small t.

2.4. *Behavior of the exit time distribution*

By definition, the survival probability monotonously decreases from 1 at $t = 0$ to 0 as t goes to infinity, as illustrated on Fig. 2a. In the short-time limit, the exit probability $1 - S_c(t)$ is extremely small,

$$1 - S_c(t) \simeq \begin{cases} 2e^{-1/(16t)} \frac{4\sqrt{t}}{\sqrt{\pi}} \left(1 - 8t + 192t^2 + O(t^3)\right) & (d = 1), \\ 2e^{-1/(4t)} \left(1 - t + 4t^2 + O(t^3)\right) & (d = 2), \\ 2e^{-1/(4t)} \frac{1}{\sqrt{\pi t}} & (d = 3), \end{cases} \tag{18}$$

because very few particles can travel the distance from the origin to the boundary during a short time[a]. In turn, for large t, the survival probability decays exponentially,

$$S_c(t) \simeq \begin{cases} \frac{4}{\pi} e^{-\pi^2 t} & (d=1), \\ \frac{2}{\alpha_{00} J_1(\alpha_{00})} e^{-\alpha_{00}^2 t} & (d=2), \\ 2e^{-\pi^2 t} & (d=3), \end{cases} \tag{19}$$

since it is unlikely for diffusing particles to avoid the encounter with the boundary during a long time (note that $\alpha_{00} \approx 2.4048$ and $J_1(\alpha_{00}) \approx 0.5191$).

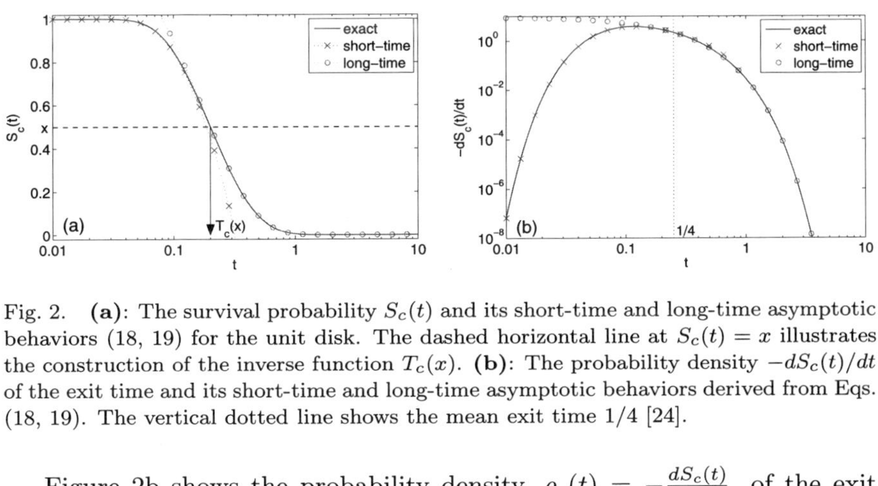

Fig. 2. **(a)**: The survival probability $S_c(t)$ and its short-time and long-time asymptotic behaviors (18, 19) for the unit disk. The dashed horizontal line at $S_c(t) = x$ illustrates the construction of the inverse function $T_c(x)$. **(b)**: The probability density $-dS_c(t)/dt$ of the exit time and its short-time and long-time asymptotic behaviors derived from Eqs. (18, 19). The vertical dotted line shows the mean exit time 1/4 [24].

Figure 2b shows the probability density, $\rho_c(t) = -\frac{dS_c(t)}{dt}$, of the exit time. This figure and the above asymptotic behaviors clearly indicate that the exit time is localized around its mean value which is equal to $1/(2d)$ (and 1/8 in 1D). In particular, the probability that τ does not belong to an interval $(t_{\min}, t_{\max})$, can be made negligible by choosing $t_{\min}$ and $t_{\max}$ appropriately. For instance, if $t_{\min} = 0.001$ and $t_{\max} = 10$, one has

$$\mathbb{P}\{\tau \notin (0.001, 10)\} = 1 - S_c(0.001) + S_c(10) < 10^{-12d} \qquad (d = 2, 3). \tag{20}$$

As a consequence, exit times beyond this interval can be ignored.

[a]The prefactor is different for $d = 1$ as the diameter of the unit interval is twice smaller than that of the unit disk and sphere. Note also that, according to Eq. (17), the first correction term for the case $d = 3$ vanishes as $\exp(-9/(4t))$, i.e., much faster than polynomial corrections in other dimensions.

2.5. *Generation of the exit times*

The explicit form of Eqs. (14, 16) allows one to generate exit times by inversion method. Inverting numerically the function $S_c(t)$ (i.e., finding a function $T_c(x)$ such that $S_c(T_c(x)) = x$ for any x between 0 and 1), one obtains a mapping from random variables χ_n with a uniform distribution on the unit interval, to the exit times

$$\tau_n = \frac{\ell_n^2}{D} T_c(\chi_n) \tag{21}$$

for the interval of length ℓ_n, or for the disk or sphere of radius ℓ_n, with a given diffusion coefficient D. The uniform random variables χ_n are generated by a standard routine for pseudo-random numbers.

In summary, two preliminary numerical procedures are required for generating the exit times:

(1) In 2D, one needs to find a finite number of positive zeros $\{\alpha_{0k}\}$ of the Bessel function $J_0(z)$. The inequalities $\pi k < \alpha_{0k} < \pi(k+1)$ allow one to search for a single zero on each interval $(\pi k, \pi k + \pi)$ by bisection or Newton's method. Since smaller times require larger truncation sizes, the asymptotic formula (18) can be used instead of the truncated series in Eqs. (14, 16) to improve the accuracy at short times. In one and three dimensions, this step is skipped because the eigenvalues are known explicitly.

(2) One constructs the function $T_c(x)$ as a numerical solution $T_c(x) = t$ of the equation $S_c(t) = x$ for a mesh of points x. The monotonous decrease of $S_c(t)$ ensures, for any x, the unique solution which can be computed by bisection method or Newton's method.

Both procedures rely on classical numerical methods. Moreover, these procedures have to be performed once and forever while the stored values of the function $T_c(x)$ can be loaded before starting Monte Carlo simulations. As a consequence, the generation of the exit times during simulations is reduced, through Eq. (21), to a routine generation of uniformly distributed pseudo-random numbers.

2.6. *Boundary condition*

By construction, a generated sequence of disks or spheres may become arbitrarily close to the boundary of the medium but it never hits the boundary. It is therefore necessary to set a threshold distance ε below which the trajectory is considered to hit the boundary (Fig. 1). This threshold is typically set to be much smaller than any relevant length scale. Since the average number of jumps needed to reach the boundary scales as $\ln(1/\varepsilon)$, setting ε to very small values would not significantly slow down simulations.

After the hit, the next step depends on the boundary condition or mechanism to be modeled. The two simplest conditions, Dirichlet and Neumann, represent purely absorbing and purely reflecting boundaries, respectively. In the former case, once a particle arrives onto the boundary, it is immediately "killed" and the simulation of the trajectory is stopped. This absorbing condition may model various physical, chemical or biological mechanisms, e.g., relaxation of local magnetization on paramagnetic impurities dispersed on the interface in NMR experiments; one-way permeation through cellular membranes; chemical transformation on the catalytic surface, etc. [28–32]. Whatever the microscopic mechanism is, the interaction with the boundary changes the state of the particle and thus removes it from the transport process. In the opposite case of Neumann boundary condition, the particle is reflected back into the medium and continues its diffusive motion, without changing its state. One can also simulate the so-called Robin boundary condition for partially absorbing/reflecting boundary when absorption and reflection events are chosen randomly [22, 33, 34]. Note that the boundary may have variable properties in space and time, e.g., some parts of the boundary may be reflecting while the other absorbing (this is the typical situation for search problems when a target is located on the boundary [35, 36]); moreover, these properties may evolve with time, e.g., during passivation/deactivation of catalysts [37]. One can also consider multi-compartment media with permeable boundary (e.g., living cells and the extracellular space). Once the trajectory hits this boundary, it may be reflected on both sides (with equal or different probabilities). Apart from these standard boundary mechanisms, one can model trapping of a particle for a fixed or randomly distributed waiting time [38], multistage surface kinetics [39], and surface transfer mechanisms [40, 41]. Finally, the FRW algorithm for bulk diffusion can be combined with other numerical techniques for modeling transport on the boundary. Such a flexibility of FRW algorithms makes them attractive for simulating various diffusive phenomena.

The implementation of Dirichlet boundary condition is trivial: simulation of the trajectory is simply stopped after hitting the boundary. One can record the hitting position (for computing the harmonic measure [20, 21]), the hitting time (for computing the exit time distribution [23]), the traveled

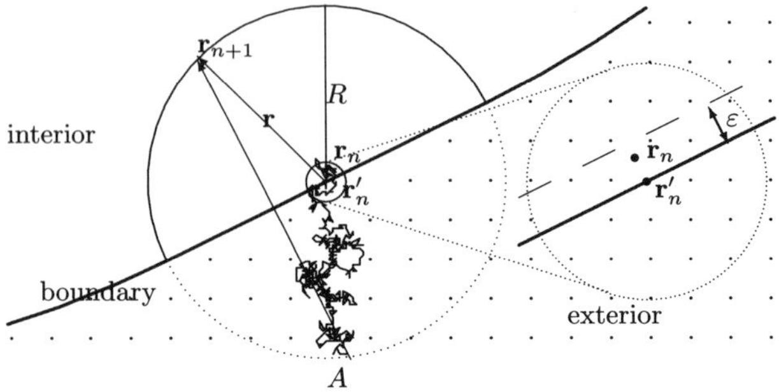

Fig. 3. When the particle has approached the reflecting boundary of the medium closer than a prescribed threshold ε, it is "released" on a circle of radius R centered at the encounter boundary point $\mathbf{r}'_n$. If the released point A does not belong to the interior of the medium, one uses its mirror reflected point $\mathbf{r}_{n+1}$ inside the medium. The radius R should be chosen as large as possible, but small enough in comparison to the characteristic length of the boundary (so that the boundary is almost flat at scale R) [24].

distance (for computing the time-dependent diffusion coefficient [42–46]), etc.

Neumann boundary condition is much more delicate, as the trajectory has to be reflected back into the medium. In the simplest implementation, one just moves the current position to the interior point of the medium at a fixed (small) distance from the boundary. This reflection distance has to be larger than the hitting threshold ε but much smaller than other relevant length scales. The choice of the reflection distance is a compromise between the accuracy and rapidity of simulations: a large distance would bias the results, while a small distance would lead to multiple small jumps after reflection. Another, much more efficient way to reflect back the trajectory is to make a relatively large jump from the boundary of the domain to a half-circle or half-sphere (Fig. 3). Indeed, if the boundary is flat (or may be approximated as flat up to a certain length scale), diffusion in the related half-disk (half-ball) with the reflecting base is equivalent to diffusion in the whole disk (ball). For instance, the distribution of the exit times is the same, while the distribution of the exit position is still uniform. An implementation of this reflection jump allows one to significantly speed up simulations.

2.7. *Incomplete jump*

If the reflecting condition is set over the whole boundary, the trajectory would never stop. One adds therefore a condition to stop the simulation when the time counter $t = \tau_1 + \tau_2 + \ldots + \tau_n$ exceeds a prescribed time T. More generally, whatever the boundary condition is, one may need to stop the trajectory after or at time T. If the trajectory has to be stopped *precisely at* T (e.g., to compute the traveled distance at time T), the last jump has to be simulated differently. In fact, we are interested in the *intermediate* position $\mathbf{r}(T)$ of the trajectory which is started at the center of the jump domain at time $t_0 = t - \tau_n < T$ and conditioned to exit from this domain at time $t > T$. The trajectory can be split into two independent parts: from t_0 to T, and from T to t. The conditional probability density for the intermediate position $\mathbf{r}$ is

$$\frac{G_{T-t_0}(\mathbf{r}_0, \mathbf{r})\ \rho_{\mathbf{r}}(t-T)}{\rho_{\mathbf{r}_0}(t-t_0)}, \tag{22}$$

where the first factor is the probability density for moving from $\mathbf{r}_0$ to $\mathbf{r}$ during the time $T - t_0$ (first part), the second factor is the probability density for exiting the domain at time $t - T$ after starting at $\mathbf{r}$ (second part), while the denominator is the probability density for exiting at time $t - t_0$ after starting at $\mathbf{r}_0$. The intermediate position $\mathbf{r}$ can be generated from this density.

2.8. *Extension: rectangular domains*

The boundary of many model or image-reconstructed structures is formed by horizontal and vertical segments (in 2D) or by rectangular plates oriented along three coordinate axes (in 3D). For such structures, it may be more convenient to split the trajectory into "elementary blocks" inside rectangular jump domains rather than spherical ones, as illustrated on Fig. 4 [47, 48]. The use of rectangles/parallelepipeds as jump domains allows one to significantly reduce the number of small jumps near the boundary of the medium. On one hand, the generation of jumps becomes more complicated because the arrival points over the edges are not uniformly distributed any more. One needs therefore to generate not only the exit times but also the conditional exit positions according to Eq. (6). On the other hand, Brownian motion inside rectangles/parallelepipeds is formed by independent one-dimensional Brownian motions along each coordinate, and the problem is essentially reduced to one-dimensional setting on the interval. In particular, the diffusion propagator $G_t^{(d)}(\mathbf{r}_0, \mathbf{r})$ in the jump domain $\Omega = [0, L_1] \times [0, L_2] \times ... \times [0, L_d] \subset \mathbb{R}^d$ is factored out as

$$G_t^{(d)}(\mathbf{r}_0, \mathbf{r}) = \prod_{j=1}^{d} \frac{1}{L_j} \, G_{Dt/L_j^2}(\mathbf{r}_{0j}/L_j, \mathbf{r}_j/L_j), \tag{23}$$

where $G_t(x_0, x)$ is the one-dimensional propagator in the unit interval $[0, 1]$ given by Eqs. (8, 9), with dimensionless time Dt/L_j^2 and dimensionless space coordinates x_0 and x, representing the starting and arrival points.

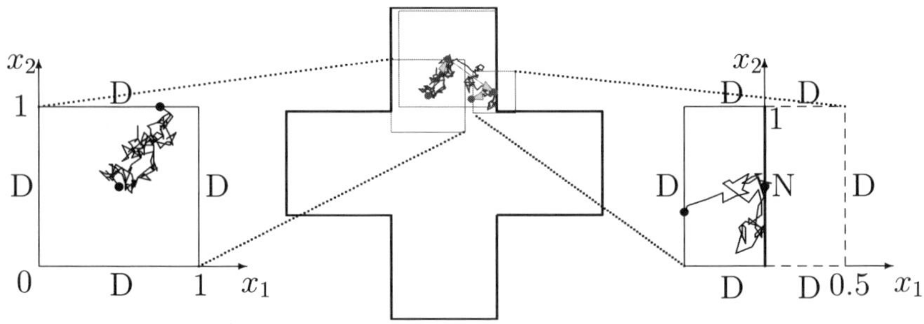

Fig. 4. (Color online) A FRW simulation of reflected Brownian motion inside a cross-shaped medium by using square jump domains (shown in red) whose centers (full circles) belong to the trajectory (shown by solid blue line and generated by basic Monte Carlo simulations). On the left, we show a zoom of the first square jump domain for which the distributions of the exit time and position can be explicitly found by solving the appropriate PDE equation with Dirichlet boundary condition (denoted by 'D'). Each jump starts from the center. On the right, we show the third jump domain for reflection from the boundary. The right vertical edge contains the starting point and coincides with the reflecting boundary of the medium. The mirror reflection across this vertical edge reduces the analysis to that on the square with Dirichlet boundary condition.

The cumulative distribution of the exit time is the survival probability in Ω:

$$S_{\mathbf{r}_0}^{(d)}(t) = \int\limits_{\Omega} d\mathbf{r} \; G_t^{(d)}(\mathbf{r}_0, \mathbf{r}) = \prod_{j=1}^{d} S_{\mathbf{r}_{0j}/L_j}(Dt/L_j^2), \tag{24}$$

where $S_{x_0}(t)$ is given by Eq. (10). As usual, the probability density is $\rho_{\mathbf{r}_0}^{(d)}(t) = -\frac{\partial}{\partial t} S_{\mathbf{r}_0}^{(d)}(t)$. In the case of the (hyper)cube (with $L_j = L$) with the starting point in the center (i.e., $\mathbf{r}_0 = [L, L, ..., L]/2$), one gets

$$S_c(t) = \left[S_{\frac{1}{2}}(Dt/L^2)\right]^d, \qquad \rho_c(t) = \frac{Dd}{L^2} \left[S_{\frac{1}{2}}(Dt/L^2)\right]^{d-1} \rho_{\frac{1}{2}}(Dt/L^2). \tag{25}$$

Once the exit time is generated, one needs to generate the exit position at this time. For the sake of clarity, we only consider the starting point from the center of the domain. Starting from Eq. (6) and skipping technical derivations, we obtain the conditional probability density of the exit point $\mathbf{s}$ to lie on the face which is orthogonal to the coordinate axis j:

$$\omega_{c,t}^{j}(\mathbf{s}) = \frac{q_j(t)}{2} \prod_{k \neq j}^{d} \frac{G_{Dt/L_k^2}(1/2, s_k/L_k)}{L_k \; S_{\frac{1}{2}}(Dt/L_k^2)}, \tag{26}$$

where the last factor is the product of conditional probability densities for (independent) displacements in the orthogonal directions $k \neq j$, and

$$q_j(t) \equiv \frac{L_j^{-2} \; \frac{\rho_{\frac{1}{2}}(Dt/L_j^2)}{S_{\frac{1}{2}}(Dt/L_j^2)}}{\sum_{k=1}^{d} L_k^{-2} \; \frac{\rho_{\frac{1}{2}}(Dt/L_k^2)}{S_{\frac{1}{2}}(Dt/L_k^2)}} \tag{27}$$

is the conditional probability to exit through one of two faces at direction j. When all lengths are equal to each other, $L_1 = ... = L_d = L$, one gets $q_j(t) = 1/d$ (all faces are equivalent from the center).

The displacements along each coordinate can be generated by numerically inverting the cumulative distribution

$$\begin{aligned} F_t(x) &\equiv \frac{1}{S_{\frac{1}{2}}(t)} \int_0^x dx' G_t(1/2, x') \\ &= \frac{2/\pi}{S_{\frac{1}{2}}(t)} \sum_{m=1}^{\infty} (-1)^{m+1} \frac{1 - \cos(\pi(2m-1)x)}{2m-1} \; e^{-\pi^2(2m-1)^2 t} \\ &= \frac{1/2}{S_{\frac{1}{2}}(t)} \sum_{k=-\infty}^{\infty} \left[\operatorname{erfc}\left(\frac{x + 2k + 1/2}{\sqrt{4t}}\right) - \operatorname{erfc}\left(\frac{2k + 1/2}{\sqrt{4t}}\right) \right. \\ &\qquad \left. - \operatorname{erfc}\left(\frac{x + 2k - 1/2}{\sqrt{4t}}\right) + \operatorname{erfc}\left(\frac{2k - 1/2}{\sqrt{4t}}\right) \right]. \end{aligned} \tag{28}$$

A technical difficulty is that this function depends on two variables x and t, i.e., one needs to invert this function versus x for various values of t as a parameter. Other methods such as rejection sampling can also be applied to generate these random variables.

As discussed in Section 2.6, the boundary condition has to be implemented. When the particle is close to the reflecting boundary, one needs to perform a reflection jump, as discussed in Section 2.6. In this case, one or several edges (faces) of the jump domain may be reflecting (Fig. 4). In order to generate the exit time and position, one can consider diffusion in the extended jump domain which is composed of the original jump domain and its mirror image with respect to the reflecting boundary. The previous formulas can be directly applied.

3. Distance computation

The adaptive splitting of the trajectory into independent "elementary blocks" relies on estimating the distance from any interior point to the boundary. We emphasize that the jump distance is allowed to be smaller than the distance to the boundary, but it must not exceed this distance. In other words, the problem of computing the exact distance can be relaxed to a simpler problem of finding a lower bound of the distance. Depending on the studied geometry, this purely geometrical problem can be solved in different ways. In this section, we briefly describe two strategies for efficient distance estimation.

3.1. *Self-similar (fractal) domains*

Fractals are often used as a paradigm of complex domains [49]. On one hand, fractals are very irregular shapes which exhibit geometrical details at various length scales, "exploding" the classical notions of length, surface area or volume. On the other hand, self-similar or self-affine hierarchical structures help to perform accurate theoretical and numerical analysis on fractals. In particular, the self-similarity of von Koch curves and surfaces allows one for a rapid estimation of the distance from any interior point to these boundaries [20, 21].

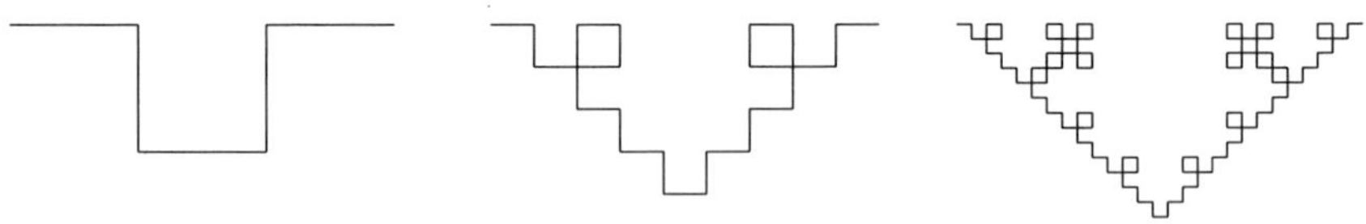

Fig. 5. Three generations of the quadratic von Koch curve of fractal dimension $\ln 5/\ln 3 \approx 1.465$. At each iteration, one replaces all linear segments by the rescaled generator (first generation).

To illustrate this idea, we consider the quadratic von Koch curve (Fig. 5). When a random walker is far from the boundary, it does not "distinguish" its geometrical details. One can thus estimate the distance by considering the coarsest generation (Fig. 6). Getting closer and closer to the boundary, the random walker starts to "recognize" smaller and smaller geometrical details. But at the same time, when small details appear in view, the rest of the boundary becomes "invisible". Consequently, one can explicitly determine the distance by examining only the local geometrical environment (see [21] for details). The advantage and eventual drawback of this geometry-adapted algorithm is the need for a specific implementation for each studied geometry. Relying on self-similarity of von Koch fractals, we were able to generate up to 10^{10} random trajectories for highly irregular boundaries with up to 10-12 iterations (in 2D). Note that these boundaries present geometrical features at length scales from 1 to $(1/3)^{10}$ (i.e., over five orders of magnitude).

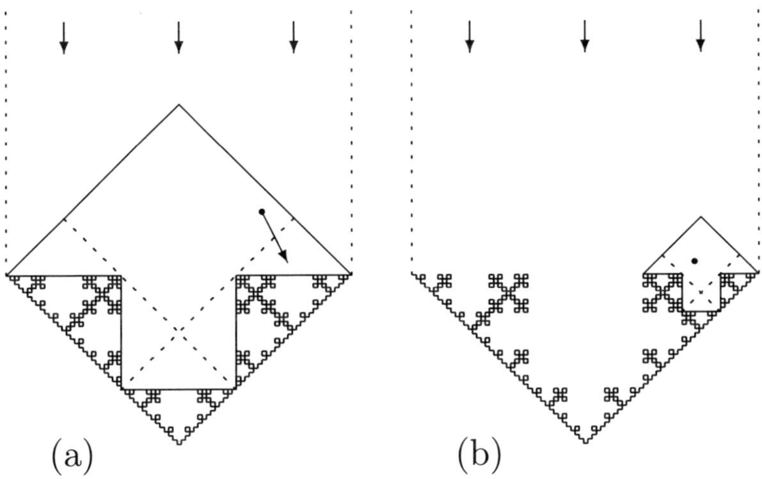

Fig. 6. **(a)** Basic arrow-like cell which is divided into the rotated square and five small triangles. Once a Brownian particle arrived into the rotated square, the distance between its current position (full circle) and the boundary of the arrow-like cell (the generator) can be computed explicitly. Random jumps inside the rotated square can therefore be executed, until the particle either exits from the arrow-like cell, or enters into a small triangle. **(b)** In the latter case, the particle starts to "see" the geometrical details of the next generation. The rescaled arrow-like cell is then used to compute the distance to the boundary.

3.2. *Sphere packs*

Sphere packs are often used to model porous and granular media, colloidal solutions, gels and polymer networks. This is indeed a quite generic model as spheres (or disks in two dimensions) can be mono- or polydisperse, overlapping or not, impermeable or not, while their locations can be regular or random. From the numerical point of view, sphere packs are particularly attractive as the distance from any point to the surface of a sphere can be easily calculated by knowing only the location and radius of the sphere. When the number of spheres is large (say, above few thousands), the computation of the distance to the boundary of the pack as the minimum over all distances to individual spheres becomes too time-consuming. In that case, one needs additional "tools" in order to rapidly locate a limited number of spheres that are the closest to a given point (the current position of Brownian trajectory). Among such tools, one can mention "coarse-grained" distance maps [15], Whitney decomposition and Voronoi cells. We illustrate a Whitney decomposition which is a partition of the medium into squares/cubes such that the size of each square/cube is proportional to the distance from that element to the boundary of the medium (Fig. 7). Once such a decomposition of the domain is constructed (prior to launching Monte Carlo simulations), the distance estimation is very rapid: one determines the square/cube to which a given point belongs, and takes the size of this element as an estimate for the distance. In practice, one uses dyadic division into squares/cubes, while the decomposition is stopped when a desired finest resolution is reached. The adaptive character of Whitney decompositions or other "coarse-grained" distance maps allows one to rapidly compute distances for sphere packs with millions of spheres. The related FRW algorithm can be used for simulating growth phenomena such as diffusion-limited aggregation [14–16], diffusion-reaction processes, search problems, etc.

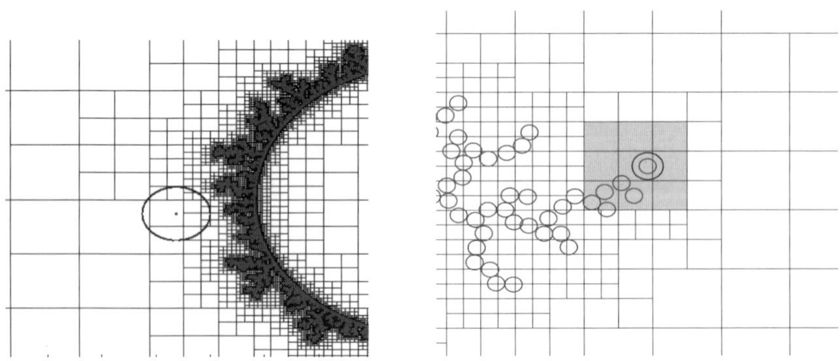

Fig. 7. (Color online) Whitney decomposition into dyadic squares near a DLA aggregate (shown in blue), with a zoom on the right. The finest resolution is set to the diameter of small disks forming the aggregate (courtesy by D. D. Nguyen).

4. Applications, extensions and perspectives

4.1. *Diffusion-weighted magnetic resonance imaging*

Diffusion-weighted MRI is a widespread experimental technique in which the random trajectories of diffusing nuclei (e.g., water molecules) are encoded by applying an inhomogeneous magnetic field [50–52]. When the nuclei diffuse inside a heterogeneous medium, the statistics of random trajectories is affected by the presence of walls or obstacles. Although these microscopic restrictions are not visible at the spatial resolution of DMRI, these geometrical features are statistically aggregated into the macroscopic signal. Measuring the signal at different diffusion times and magnetic field gradients, one aims to infer the morphological structure of a sample and to characterize the dynamics of a system. The non-invasive character of DMRI made this technique the gold standard in material sciences, neurosciences and medicine, with numerous applications to mineral samples (e.g., sedimentary rocks in oil industry; soils in agriculture; concrete, cements and gypsum in building industry) and biological samples (e.g., brain, lungs, bones).

Until nowadays, modeling DMRI experiments was mainly restricted to simple structures [53–56], while truly multiscale three-dimensional porous media such as concretes or sedimentary rocks remained out of reach due to the lack of efficient numerical techniques. The efficiency and flexibility of FRW algorithms make them promising tools for modeling DMRI. An approximate implementation of the gradient encoding into FRW algorithms was recently proposed and opened new opportunities in the study of mineral and biological samples [24, 57].

4.2. *Continuous-time random walks*

Many physical, chemical and biological transport processes exhibit anomalous features, e.g., the subdiffusive scaling of the mean square displacement [58–60]. Examples are the motion of organelles, vesicles or tracers in living cells [61–66], animals searching for food [67], contaminant or pollution spreading [68, 69], etc. For instance, overcrowding in living cells and colloidal gels, or deep wells in the interaction energy landscape [58, 70], may result in a heavy-tailed distribution of waiting times between jumps. Such long stalling periods would yield subdiffusive dynamics that can be described by Continuous-Time Random Walk (CTRW) with diverging mean waiting time but finite-variance spatial displacements. In that frame, the diffusion equation (1) has to be replaced by fractional diffusion equation, in which the history of trapping or caging is incorporated through the fractional Riemann-Liouville derivative in time. Since the spatial dynamics is still governed by the Laplace operator, only the temporal dependence in the spectral representation (2) has to be modified as

$$G_t(\mathbf{r}_0, \mathbf{r}) = \sum_{m=1}^{\infty} E_\alpha(-D\lambda_m t) u_m(\mathbf{r}_0) u_m^*(\mathbf{r}), \tag{29}$$

where $E_\alpha(z)$ is the Mittag-Leffler function [71]. Using this extension, one can recompute the exit time and position distributions in order to simulate restricted subdiffusion in multiscale or complex structures [72–74].

4.3. *Intermittent processes*

Another important extension of FRW is related to diffusive processes with two successively alternating phases (e.g., active and passive transport of vesicles in living cells [75–78]). These so-called intermittent processes have been intensively studied during the last decade. In particular, the alternation of phases was shown to facilitate search processes and speed up biochemical kinetics (see the review [41] and references therein). For instance, in the case of surface-mediated diffusion in spherical domains, the mean time for finding a target was shown to be minimal when the phases of bulk and surface diffusion alternate at certain rate [79–82]. At the same time, the question of optimality for the mean search time remains open for porous media or irregularly-shaped domains such as catalysts or biological structures. The FRW algorithm can be adapted to simulate such intermittent diffusive processes.

4.4. *Conclusions and perspectives*

In this chapter, we described the basic principles and several applications of fast random walk algorithms. The algorithm relies on adaptive splitting of the random trajectory inside a complex medium into independent "elementary blocks" in simple jump domains for which simulations can be performed much more efficiently. In other words, the algorithm eliminates the geometrical complexity of the system and reduces the original problem to the study of diffusion inside a disk, a sphere or a rectangle, for which many mathematical tools (PDE, spectral methods, etc.) are particularly efficient [1, 2]. This "trick" drastically speeds up Monte Carlo simulations since at each step the largest possible exploration is performed. In practice, the computation is reduced to estimating the distance from any interior point to the boundary that can be solved in different ways depending on the studied geometry. The adaptive character of FRW techniques makes them well suited for simulating diffusion-reaction processes in complex, multiscale, disordered, heterogeneous or irregularly-shaped media. The FRW algorithms can be applied for calculating the harmonic measures and their extensions, the first passage and exit time distributions, search times, reaction rates [83, 84], residence times [85], etc. Many growth processes such as diffusion-limited aggregation and related models can be efficiently realized by using these techniques. We also illustrated one application of these algorithms for computing DMRI signals and discussed extensions for modeling restricted anomalous diffusion and intermittent processes.

References

1. J. Crank, *The Mathematics of Diffusion*, 2nd Ed. Clarendon, Oxford (1975).
2. H. S. Carslaw and J. C. Jaeger, *Conduction of Heat in Solids*, 2nd Ed. Clarendon, Oxford (1959).
3. K. Itô and H. P. McKean, *Diffusion Processes and Their Sample Paths.* 2nd Ed., Springer-Verlag, Berlin, Heidelberg (1996).
4. M. Freidlin, *Functional Integration and Partial Differential Equations.* Annals of Mathematics Studies, Princeton University Press, Princeton, New Jersey (1985).
5. R. F. Bass, *Diffusions and Elliptic Operators.* Springer, New York (1998).
6. W. Feller, *An Introduction to Probability Theory and Its Applications.* Third ed. New York: John Wiley (1968).
7. S. Redner, *A Guide to First-Passage Processes.* Cambridge University Press, Cambridge, England (2001).
8. B. D. Hughes, *Random Walks and Random Environments.* Clarendon Press, Oxford (1995).
9. G. H. Weiss, *Aspects and Applications of the Random Walk.* North-Holland, Amsterdam (1994).
10. K. K. Sabelfeld, *Monte Carlo Methods in Boundary Value Problems.* Springer-Verlag: New York - Heidelberg, Berlin (1991).
11. K. K. Sabelfeld and N. A. Simonov, *Random Walks on Boundary for Solving PDEs.* Utrecht, The Netherlands (1994).
12. G. N. Milshtein, *Numerical Integration of Stochastic Differential Equations.* Kluwer, Dordrecht, the Netherlands (1995).
13. M. E. Müller, Some Continuous Monte Carlo Methods for the Dirichlet Problem, *Ann. Math. Statist.* **27**, 569-589 (1956).
14. T. A. Witten and L. M. Sander, Diffusion-Limited Aggregation, a Kinetic Critical Phenomenon, *Phys. Rev. Lett.* **47**, 1400 (1981).

15. P. Ossadnik, Multiscaling Analysis of Large-Scale Off-Lattice DLA, *Physica A* **176**, 454 (1991).
16. T. C. Halsey, Diffusion-Limited Aggregation: A Model for Pattern Formation, *Phys. Today* (2000).
17. S. Torquato and I. C. Kim, Efficient simulation technique to compute effective properties of heterogeneous media, *Appl. Phys. Lett.* **55**, 1847 (1989).
18. L. H. Zheng and Y. C. Chiew, Computer simulation of diffusion-controlled reactions in dispersions of spherical sinks, *J. Chem. Phys.* **90**, 322-327 (1989).
19. S. B. Lee, I. C. Kim, C. A. Miller, and S. Torquato, Random-walk simulation of diffusion-controlled processes among static traps, *Phys. Rev. B* **39**, 11833-11839 (1989).
20. D. S. Grebenkov, What Makes a Boundary Less Accessible, *Phys. Rev. Lett.* **95**, 200602 (2005).
21. D. S. Grebenkov, A. A. Lebedev, M. Filoche and B. Sapoval, Multifractal Properties of the Harmonic Measure on Koch Boundaries in Two and Three Dimensions, *Phys. Rev. E* **71**, 056121 (2005).
22. D. S. Grebenkov, Scaling Properties of the Spread Harmonic Measures, *Fractals* **14**, 231-243 (2006).
23. P. Levitz, D. S. Grebenkov, M. Zinsmeister, K. Kolwankar and B. Sapoval, Brownian flights over a fractal nest and first passage statistics on irregular surfaces, *Phys. Rev. Lett.* **96**, 180601 (2006).
24. D. S. Grebenkov, A fast random walk algorithm for computing the pulsed-gradient spin-echo signal in multiscale porous media, *J. Magn. Reson.* **208**, 243-255 (2011).
25. J. B. Garnett and D. E. Marshall, *Harmonic Measure.* Cambridge University Press (2005).
26. D. S. Grebenkov and B.-T. Nguyen, Geometrical structure of Laplacian eigenfunctions, *SIAM Rev.* **55**, 601-667 (2013).
27. D. S. Grebenkov, Multiple Correlation Function Approach: Rigorous Results for Simple Geometries, *Diff. Fundam.* **5**, 1-34 (2007).
28. K. R. Brownstein and C. E. Tarr, Importance of Classical Diffusion in NMR Studies of Water in Biological Cells, Phys. Rev. A **19**, 2446-2453 (1979).
29. G. C. Bond, *Heterogeneous Catalysis: Principles and Applications.* Clarendon, Oxford (1987).
30. O. G. Berg, R. B. Winter, and P. H. von Hippel, Diffusion-driven mechanisms of protein translocation on nucleic acids. 1. Models and theory. *Biochemistry* **20**, 6929-6948 (1981).
31. G. Adam and M. Delbrück, *Reduction of Dimensionality in Biological Diffusion Processes.* W. H. Freeman Co., Publishers, San Francisco (1968).
32. H. Sano and M. Tachiya, Theory of diffusion-controlled reactions on spherical surfaces and its application to reactions on micellar surfaces, *J. Chem. Phys.* **75**, 2870-2878 (1981).
33. B. Sapoval, General Formulation of Laplacian Transfer Across Irregular Surfaces, *Phys. Rev. Lett.* **73**, 3314 (1994).
34. D. S. Grebenkov, "Partially Reflected Brownian Motion: A Stochastic Approach to Transport Phenomena", in "Focus on Probability Theory", Ed.

L. R. Velle, pp. 135-169, Nova Science Publishers (2006).

35. A. Singer, Z. Schuss, D. Holcman, and R. S. Eisenberg, Narrow Escape, Part I, *J. Stat. Phys.* **122**, 437 (2006).
36. Z. Schuss, A. Singer, and D. Holcman, The narrow escape problem for diffusion in cellular microdomains, *Proc. Nat. Acad. Sci. USA* **104**, 16098 (2007).
37. M. Filoche, D. S. Grebenkov, J. S. Andrade, and B. Sapoval, Passivation of irregular surfaces accessed by diffusion, *Proc. Nat. Acad. Sci. USA* **105**, 7636 (2008).
38. D. Holcman, A. Marchewka, and Z. Schuss, Survival probability of diffusion with trapping in cellular neurobiology, *Phys. Rev. E* **72**, 031910 (2005).
39. T. Chou and M. R. D'Orsogna, Multistage adsorption of diffusing macromolecules and viruses, *J. Chem. Phys.* **127**, 105101 (2007).
40. O. Bénichou, C. Loverdo, M. Moreau, and R. Voituriez, Optimizing intermittent reaction paths, *Phys. Chem. Chem. Phys.* **10**, 7059-7072 (2008).
41. O. Bénichou, C. Loverdo, M. Moreau, and R. Voituriez, Intermittent search strategies, *Rev. Mod. Phys.* **83**, 81-130 (2011).
42. L. L. Latour, P. P. Mitra, R. L. Kleinberg, and C. H. Sotak, Time-dependent diffusion coefficient of fluids in porous media as a probe of surface-to-volume ratio, *J. Magn. Reson. A* **101**, 342-346 (1993).
43. L. L. Latour, K. Svoboda, P. P. Mitra, and C. H. Sotak, Time-dependent diffusion of water in a biological model system, *Proc. Nat. Acad. Sci. USA* **91**, 1229-1233 (1994).
44. T. M. de Swiet and P. N. Sen, Time dependent diffusion coefficient in a disordered medium, *J. Chem. Phys.* **104**, 206-209 (1996).
45. P. N. Sen, Time-dependent diffusion coefficient as a probe of geometry, *Conc. Magn. Reson.* **23** A, 1-21 (2004).
46. D. S. Novikov, E. Fieremans, J. H. Jensen and J. A. Helpern, Random walks with barriers, *Nat. Phys.* **7**, 508-514 (2011).
47. M. Deaconu and A. Lejay, A Random Walk on Rectangles Algorithm, *Method Comput. Appl. Probab.* **8**, 135-151 (2006).
48. S. Zein, A. Lejay, and M. Deaconu, An efficient algorithm to simulate a Brownian motion over irregular domains, *Commun. Comput. Phys.* **8**, 901-916 (2010).
49. B. Mandelbrot, *The Fractal Geometry of Nature.* San Francisco, Freeman (1982).
50. P. T. Callaghan, *Principles of Nuclear Magnetic Resonance Microscopy.* Clarendon Press, Oxford (1991).
51. W. S. Price, *NMR Studies of translational Motion: Principles and Applications.* Cambridge University Press (2009).
52. D. S. Grebenkov, NMR survey of reflected Brownian motion, *Rev. Mod. Phys.* **79**, 1077-1137 (2007).
53. B. Balinov, B. Jönsson, P. Linse, and O. Söderman, The NMR Self-Diffusion Method Applied to Restricted Diffusion. Simulation of Echo Attenuation from Molecules in Spheres and Between Planes, *J. Magn. Reson. A* **104**, 17-25 (1993) [see Erratum: J. Magn. Reson. A **108**, 130 (1994)].
54. A. Coy and P. T. Callaghan, Pulsed Gradient Spin Echo nuclear magnetic

resonance for molecules diffusing between partially reflecting rectangular barriers, *J. Chem. Phys.* **101**, 4599-4609 (1994).

55. A. Duh, A. Mohorič, and J. Stepišnik, Computer Simulation of the Spin-Echo Signal Distribution in the Case of Restricted Self-Diffusion, *J. Magn. Reson.* **148**, 257-266 (2001).
56. R. M. E. Valckenborg, H. P. Huinink, J. J. v. d. Sande, and K. Kopinga, Random-walk simulations of NMR dephasing effects due to uniform magnetic-field gradients in a pore, *Phys. Rev. E* **65**, 021306 (2002).
57. D. S. Grebenkov, H. T. Nguyen and J.-R. Li, A fast random walk algorithm for computing diffusion-weighted NMR signals in multi-scale porous media: a feasibility study for a Menger sponge, *Micropor. Mesopor. Mater.* **178**, 56-59 (2013).
58. J.-P. Bouchaud and A. Georges, Anomalous diffusion in disordered media: Statistical mechanisms, models and physical applications, *Phys. Rep.* **195**, 127-293 (1990).
59. R. Metzler and J. Klafter, The random walk's guide to anomalous diffusion: a fractional dynamics approach, *Phys. Rep.* **339**, 1-77 (2000).
60. S. Havlin and D. ben Avraham, Diffusion in disordered media, *Adv. Phys.* **51**, 187-292 (2002).
61. M. J. Saxton and K. Jacobson, Single-particle tracking: Applications to Membrane Dynamics, *Annu. Rev. Biophys. Biomol. Str.* **26**, 373-399 (1997).
62. I. Golding and E. C. Cox, Physical Nature of Bacterial Cytoplasm, *Phys. Rev. Lett.* **96**, 098102 (2006).
63. R. Metzler, V. Tejedor, J.-H. Jeon, Y. He, W. H. Deng, S. Burov, and E. Barkai, Analysis of Single Particle Trajectories: From Normal to Anomalous Diffusion, *Acta Phys. Pol. B* **40**, 1315-1331 (2009).
64. J. Szymanski and M. Weiss, Elucidating the Origin of Anomalous Diffusion in Crowded Fluids, *Phys. Rev. Lett.* **103**, 038102 (2009).
65. J.-H. Jeon, V. Tejedor, S. Burov, E. Barkai, C. Selhuber-Unkel, K. Berg-Sørensen, L. Oddershede, and R. Metzler, In Vivo Anomalous Diffusion and Weak Ergodicity Breaking of Lipid Granules, *Phys. Rev. Lett.* **106**, 048103 (2011).
66. E. Bertseva, D. S. Grebenkov, P. Schmidhauser, S. Gribkova, S. Jeney, and L. Forró, Optical Trapping Microrheology in Cultured Human Cells, *Eur. Phys. J. E* **35**, 63 (2012).
67. G. M. Viswanathan, V. Afanasyev, S. V. Buldyrev, E. J. Murphy, P. A. Prince, and H. E. Stanley, Lévy flight search patterns of wandering albatrosses, *Nature* **381**, 413-415 (1996).
68. J. W. Kirchner, X. Feng, and C. Neal, Fractal stream chemistry and its implications for contaminant transportin catchments, *Nature* **403**, 524 (2000).
69. H. Scher, G. Margolin, R. Metzler, J. Klafter, and B. Berkowitz, The dynamical foundation of fractal stream chemistry: The origin of extremely long retention times, *Geophys. Res. Lett.* **29**, 1061 (2002).
70. J.-P. Korb, Y. Goddard, J. Pajski, G. Diakova, and R. G. Bryant, Extreme-Values Statistics and Dynamics of Water at Protein Interfaces, *J. Phys. Chem. B* **115**, 12845-12858 (2011).

71. H. J. Haubold, A. M. Mathai, and R. K. Saxena, Mittag-Leffler Functions and Their Applications, *J. Appl. Math.* **2011**, 298628 (2011).
72. S. B. Yuste and K. Lindenberg, Subdiffusive target problem: Survival probability, *Phys. Rev. E* **76**, 051114 (2007).
73. D. G. Grebenkov, Subdiffusion in a bounded domain with a partially absorbing-reflecting boundary, *Phys. Rev. E* **81**, 021128 (2010).
74. D. G. Grebenkov, Searching for partially reactive sites: Analytical results for spherical targets, *J. Chem. Phys.* **132**, 034104 (2010).
75. A. Caspi, R. Granek, and M. Elbaum, Enhanced Diffusion in Active Intracellular Transport, *Phys. Rev. Lett.* **85**, 5655 (2000).
76. D. Arcizet, B. Meier, E. Sackmann, J. O. Rädler, and D. Heinrich, Temporal Analysis of Active and Passive Transport in Living Cells, *Phys. Rev. Lett.* **101**, 248103 (2008).
77. C. P. Brangwynne, G. H. Koenderink, F. C. MacKintosh, and D. A. Weitz, Intracellular transport by active diffusion, *Trends Cell. Biol.* **19**, 423 (2009).
78. D. A. Kenwright, A. W. Harrison, T. A. Waigh, P. G. Woodman, and V. J. Allan, First-passage-probability analysis of active transport in live cells, *Phys. Rev. E* **86**, 031910 (2012).
79. O. Bénichou, D. S. Grebenkov, P. Levitz, C. Loverdo and R. Voituriez, Optimal reaction time for surface-mediated diffusion, *Phys. Rev. Lett.* **105**, 150606 (2010).
80. O. Bénichou, D. S. Grebenkov, P. Levitz, C. Loverdo and R. Voituriez, Mean first-passage time of surface-mediated diffusion in spherical domains, *J. Stat. Phys.* **142**, 657-685 (2011).
81. T. Calandre, O. Bénichou, D. S. Grebenkov, and R. Voituriez, The interfacial territory covered by surface-mediated diffusion, *Phys. Rev. E* **85**, 051111 (2012).
82. J.-F. Rupprecht, O. Bénichou, D. S. Grebenkov, and R. Voituriez, Kinetics of active surface-mediated diffusion in spherically symmetric domains, *J. Stat. Phys.* **147**, 891-918 (2012).
83. N. Agmon, Residence times in diffusion processes, *J. Chem. Phys.* **81**, 3644 (1984).
84. G. H. Weiss, Overview of theoretical models for reaction rates, *J. Stat. Phys.* **42**, 3-36 (1986).
85. D. S. Grebenkov, Residence times and other functionals of reflected Brownian motion, *Phys. Rev. E* **76**, 041139 (2007).